Organic Reactions

Organic Reactions

VOLUME 45

JOHN WILEY & SONS, INC.

New York • Chichester • Brisbane • Toronto • Singapore

This text is printed on acid-free paper.

Published by John Wiley & Sons, Inc.

Library of Congress Catalog Card Number 42-20265

ISBN 0-471-03161-5

Printed in the United States of America

10 9 8 7 6 5 4 3 2 1

PREFACE TO THE SERIES

In the course of nearly every program of research in organic chemistry the investigator finds it necessary to use several of the better-known synthetic reactions. To discover the optimum conditions for the application of even the most familiar one to a compound not previously subjected to the reaction often requires an extensive search of the literature; even then a series of experiments may be necessary. When the results of the investigation are published, the synthesis, which may have required months of work, is usually described without comment. The background of knowledge and experience gained in the literature search and experimentation is thus lost to those who subsequently have occasion to apply the general method. The student of preparative organic chemistry faces similar difficulties. The textbooks and laboratory manuals furnish numerous examples of the application of various syntheses, but only rarely do they convey an accurate conception of the scope and usefulness of the processes.

For many years American organic chemists have discussed these problems. The plan of compiling critical discussions of the more important reactions thus was evolved. The volumes of *Organic Reactions* are collections of chapters each devoted to a single reaction, or a definite phase of a reaction, of wide applicability. The authors have had experience with the processes surveyed. The subjects are presented from the preparative viewpoint, and particular attention is given to limitations, interfering influences, effects of structure, and the selection of experimental techniques. Each chapter includes several detailed procedures illustrating the significant modifications of the method. Most of these procedures have been found satisfactory by the author or one of the editors, but unlike those in *Organic Syntheses* they have not been subjected to careful testing in two or more laboratories.

Each chapter contains tables that include all the examples of the reaction under consideration that the author has been able to find. It is inevitable, however, that in the search of the literature some examples will be missed, especially when the reaction is used as one step in an extended synthesis. Nevertheless, the investigator will be able to use the tables and their accompanying bibliographies in place of most or all of the literature search so often required.

Because of the systematic arrangement of the material in the chapters and the entries in the tables, users of the books will be able to find information desired by reference to the table of contents of the appropriate chapter. In the interest of economy the entries in the indices have been kept to a minimum, and, in particular, the compounds listed in the tables are not repeated in the indices.

The success of this publication, which will appear periodically, depends upon the cooperation of organic chemists and their willingness to devote time and effort to the preparation of the chapters. They have manifested their interest already by the almost unanimous acceptance of invitations to contribute to the work. The editors will welcome their continued interest and their suggestions for improvements in *Organic Reactions.*

Chemists who are considering the preparation of a manuscript for submission to *Organic Reactions* are urged to write either secretary before they begin work.

CONTENTS

Organic Reactions

CHAPTER 1

THE NAZAROV CYCLIZATION

KARL L. HABERMAS AND SCOTT E. DENMARK

Department of Chemistry, University of Illinois, Urbana, Illinois

TODD K. JONES

Ligand Pharmaceuticals Inc., San Diego, California

CONTENTS

Organic Reactions, Vol. 45, Edited by Leo A. Paquette et al.
ISBN 0-471-03161-5

INTRODUCTION

The Nazarov cyclization is named after the eminent Russian chemist I. N. Nazarov (1900–1957). In the course of an extensive study on the formation of allyl vinyl ketones by the mercuric ion and acid-catalyzed hydration of dienynes,

Nazarov and his co-workers discovered a secondary reaction to form 2-cyclopentenones (Eq. 1).[1–38] Nazarov initially formulated a direct acid-catalyzed closure of the allyl vinyl ketones and demonstrated the preparation of 2-cyclopentenones from these precursors in dozens of cases. However in 1952, Braude and Coles[39] suggested the intermediacy of carbocations and demonstrated that the formation of 2-cyclopentenones actually proceeds via the α,α'-divinyl ketones. This fact, together with further mechanistic clarification, has led to the current formulation of the Nazarov cyclization as the acid-catalyzed closure of divinyl ketones to 2-cyclopentenones.

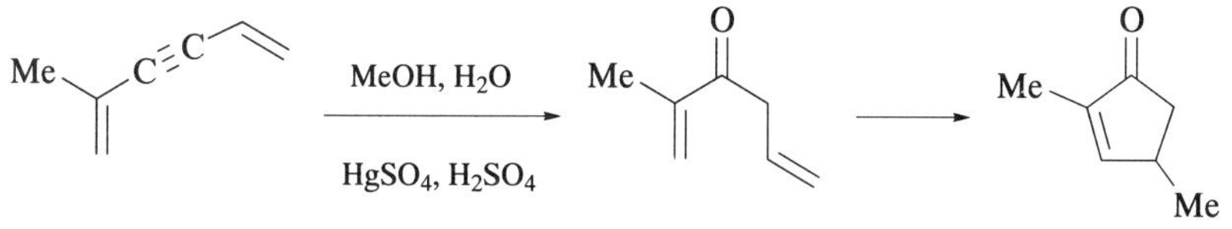

MeOH, H_2O; $HgSO_4$, H_2SO_4

(Eq. 1)

H_3PO_4, HCO_2H; 90°

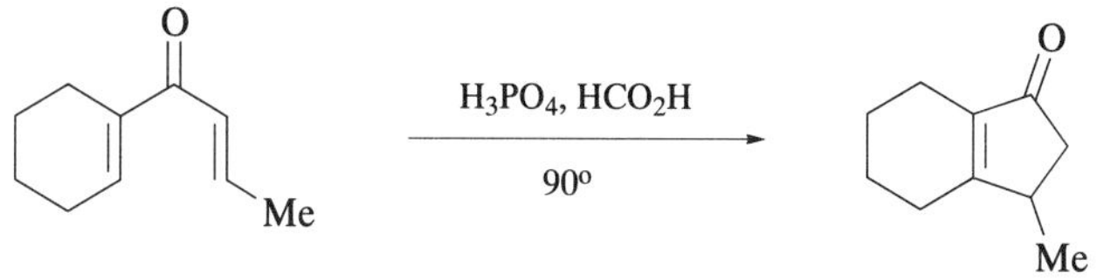

(Eq. 2)

This process was already documented in 1903.[40] Treatment of dibenzylideneacetone (**1**) with concentrated sulfuric acid and acetic anhydride followed by mild alkaline hydrolysis afforded a ketol **2**, the correct structure of which was finally proposed in 1955 (Eq. 3).[41] Other examples of acid-catalyzed cyclizations of divinyl ketones exist in the early literature, as do reactions which must have proceeded via divinyl ketones.[42,43]

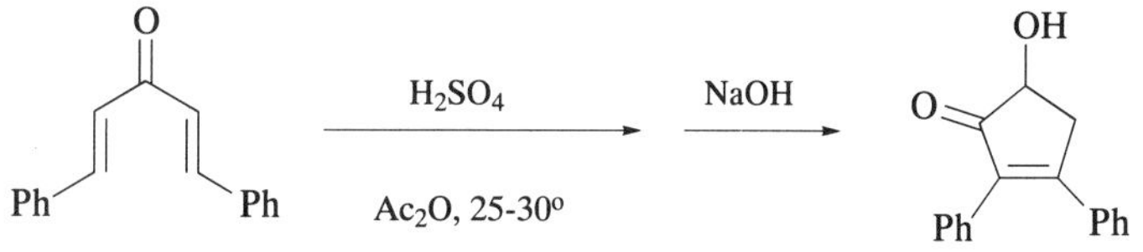

H_2SO_4; Ac_2O, 25-30°; NaOH

1 **2**

(Eq. 3)

A broader definition of the Nazarov cyclization includes a wide variety of precursors that under specific reaction conditions also produce 2-cyclopentenones via divinyl ketones or their functional equivalents. A case in point is the formation of cyclopentenone **3** by treatment of either divinyl ketone **4** or tetrahydropyrone **5** with ethanolic hydrochloric acid (Eq. 4).[44] It is the structural variety of precursors that lends versatility to the Nazarov cyclization and which also serves as the organizational framework for this chapter.

(Eq. 4)

To facilitate presentation, the reaction is divided into six categories defined by the structure of the precursor: (1) cyclization of divinyl and allyl vinyl ketones, (2) cyclization of silylated (stannylated) divinyl ketones, (3) in situ generation/cyclization of divinyl ketones or equivalents, (4) generation/cyclization of divinyl ketone equivalents by solvolysis, (5) alkyne-based precursors of divinyl ketones, and (6) coupling reactions to form and cyclize divinyl ketones. The logic of this sequence follows from the order of decreasing structural similarity of the precursors to divinyl ketones. This review is intended to be comprehensive in the coverage of cyclizations which produce 2-cyclopentenones. The related electrocyclic closure of less oxidized pentadienylic cations[45] is not covered. The reaction was most recently reviewed in 1991.[46] Prior to that, the Nazarov cyclization had been reviewed in 1983[47] and also in the context of pentannulation.[48,49]

MECHANISM AND STEREOCHEMISTRY

The first reasonable proposals for the mechanism of the Nazarov cyclization suggested the intermediacy of a β ketocarbocation from *C*-protonation of either divinyl or allyl vinyl ketones.[39] The ring is formed by intramolecular attack on the enone with concomitant generation of an α-ketocarbocation. Loss of a β proton from this intermediate affords the product. Nazarov provided support for this proposal by demonstrating the incorporation of deuterium from D_3PO_4 in different positions from divinyl or allyl vinyl ketones.[32,35,36]

From both stereochemical and spectroscopic studies it is now well established that the Nazarov cyclization is a pericyclic reaction belonging to the class of electrocyclizations, specifically a 4π electrocyclic closure of a 3-hydroxypentadienylic cation.[50] Simple pentadienylic cations had been observed early on to undergo facile cyclization to cyclopentenylic cations.[51,52] More recently, the *O*-protonated divinyl ketones and the cyclopentenone rearrangement products have been spectroscopically observed.[53] Moreover, the intermediacy of carbocations is consistent with the observation of Wagner–Meerwein rearrangement products and anomalous cyclization pathways.

As with all pericyclic reactions, mechanism and stereochemistry are intimately coupled. Thus discussion of one feature must invoke the other. Therefore, for a clear description of the mechanism, the stereospecific[54] aspects of the Nazarov cyclization are discussed below.

The basic tenets of the theory of electrocyclic reactions[50] makes very clear predictions about the relative configuration of the substituents on the newly formed bond of the five-membered ring. Unfortunately, secondary rearrangements thwarted early attempts to verify the stereochemical predictions of orbital symmetry control in the parent system.[52] Subsequent studies with the pentamethyl derivatives were successful.[55,56]

The most convincing evidence for a pericyclic mechanism was the demonstration of complementary rotatory pathways for the thermal (conrotatory) and photochemical (disrotatory) cyclizations of bis(1-cyclohexenyl)ketones (Eq. 5) precisely as predicted by the conservation of orbital symmetry.[57]

R = H, Me

254 nm, benzene, *disrotatory* → *syn*

AcOH, H_3PO_4, 50°, *conrotatory* → *anti*

(Eq. 5)

Additional confirmation came from reinvestigation of the early work on the thermal cyclization of dibenzylideneacetone (**1**)[58] and dibenzylidene-3-pentanone (**6**)[44] in the presence of acid. Careful examination revealed the conrotatory electrocyclic pathway for both of these substrates.

1 → H_2SO_4, Ac_2O, 25-30° → HO_3SO, OAc, Ph, Ph

6 → HI → (28%) + (72%)

(Eq. 6)

The rules for conservation of orbital symmetry also predict a change in the configuration of the products from either a given cyclization mode or a permutation of starting olefin geometry. This prediction is difficult to test under the normal reaction conditions (acid, light) that would isomerize dienone double bonds. However, fluoride treatment of the silyl epoxide **7** leads to the formation of the *cis*-disubstituted cyclopentenone **9** from a *Z,E* precursor, the putative allene oxide **8**, which cyclizes via the 2-oxido pentadienylic cation.[59]

Me_3Si, O, C_6H_{13}-*n*, CF_3CO_2, H, C_5H_{11}-*n*, H, H **7** —CsF, MeCN, 23°, (20-35%)→ [O, C_6H_{13}-*n*, H, C_5H_{11}-*n*, H, H **8**] → [O^-, C_6H_{13}-*n*, +, C_5H_{11}-*n*] → O, C_6H_{13}-*n*, C_5H_{11}-*n* **9**

(Eq. 7)

The predicted photochemical disrotatory closure of protonated divinyl ketones has been documented in several laboratories.[60–62] The examples in Eq. 8 show

O, H, H —*t*-BuOH, H_2SO_4, *hν*→ *t*-BuO, O, H, H (30%) + *t*-BuO, O, H, H (30%)

O, D, D —H_2SO_4, *hν*→ D, H, H, O, D (71%) + O, D, D, H (29%)

Me, Me, O, Me, Me —*i*-PrOH, *hν*→ Me, Me, O, OPr-*i*, Me, Me

(Eq. 8)

how secondary processes allowed for determination of the sense of electrocyclization, for example, cation quenching and hydride or alkyl migration.

A theoretical study of the Nazarov cyclization has evaluated the relative stability of the 3-oxopentadienylic cation and the product cyclopentenylic cations. The lowest energy conformation was the *syn–anti* isomer and the overall reaction was found to be exothermic by 2.3 kcal/mol. A transition state was also located corresponding to an activation enthalpy of 20.6 kcal/mol. The C(1)–C(5) distance is 2.09 Å in the transition structure.[63,64]

The cationic electrocyclization mechanism allows prediction of substituent effects. In the rate-limiting step, the distribution of positive charge changes from C(1), C(3), and C(5) in the pentadienylic cation to C(2) and C(4) in the cyclopentenylic cation (Eq. 9). Thus substituents that stabilize positive charge should ac-

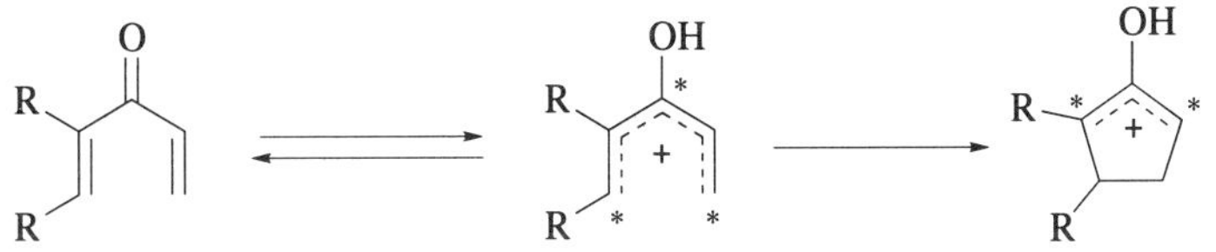

(Eq. 9)

celerate (α position) or decelerate (β position) the electrocyclization depending upon location. Moreover, the effects for α substitution should be greater in magnitude for groups of similar stabilizing capacity since the charge is less delocalized. This empirical analysis has been verified experimentally as well as theoretically by Mulliken population analysis.[64]

Beyond the mechanistically mandated stereochemical imperative of conrotatory or disrotatory closure, there exists a secondary stereochemical feature that arises because of the duality of allowed electrocyclization pathways. This feature, called torquoselectivity,[65] is manifest when the divinyl ketone is chiral by virtue of a remote substituent, and the two pathways lead to diastereomers (Eq. 10). The

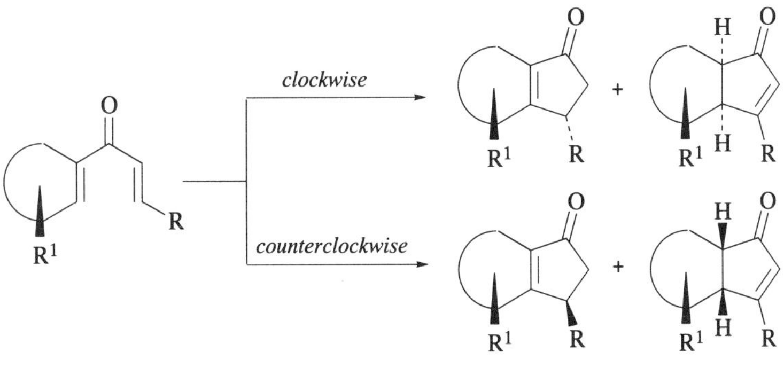

(Eq. 10)

nature of the relationship between the newly created centers and preexisting centers depends upon the location of the cyclopentenone double bond. In the classical cyclization, the double bond normally occupies the most substituted position, corresponding to a Saytzeff elimination process. The sense of rotation is defined by clockwise (*R*) rotation or counterclockwise (*S*) rotation viewing down the C–O bond. Thus, depending on the final placement of the double bond, the newly created center may be proximal or distal to the preexisting center.

The factors that control the sense of torquoselection in the Nazarov cyclization are primarily steric in origin. Most significant are the torsional and nonbonded interactions between the substituents in the vicinity of the newly forming bond during the electrocyclization event. The phosphoric acid induced closure of triquinane **10** in Eq. 11 illustrates the stereochemical consequences of opposite conrotatory closures.[66] Interestingly the major product **11** arises from bond formation on the concave face of the diquinane unit.

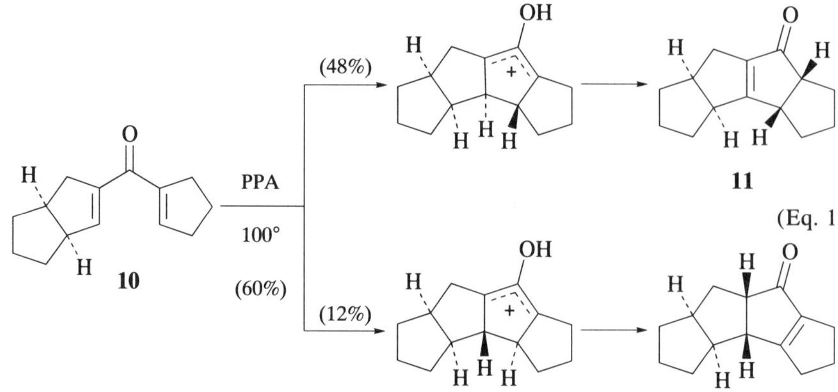

10 → PPA, 100° (60%) → (48%) **11** ; (12%) **12** (Eq. 11)

SCOPE AND LIMITATIONS

Divinyl and Allyl Vinyl Ketones

The tautomeric divinyl and allyl vinyl ketones are equivalent precursors for acid-promoted cyclization. Deuterium-labeling studies established the isomerization of the allyl to a vinyl group prior to cyclization.[32,35,36] The preparation of these substrates differs, however, since most allyl vinyl ketones are produced by the mercuric ion catalyzed hydration of dienynes, which in turn are prepared by dehydration of vinylacetylide adducts of ketones. This approach allows for the incorporation of many substituents. Since divinyl ketones are implied as intermediates in other variants of the Nazarov cyclization, only those reactions where a divinyl ketone is used directly are discussed below.

Acyclic Precursors. Simple cyclopentenones have been prepared in modest to good yield using the Nazarov cyclization. Many substitution patterns are available by this protocol, most commonly 2,3,4-trisubstituted and 2,3,4,4-tetrasubstituted 2-cyclopentenones. The usually vigorous reaction conditions (phosphoric acid, heat) lead to the thermodynamically most stable cyclopentenone double bond isomer. Aromatic substituents are compatible with the reaction conditions and have a beneficial effect on the yield and, in the α position, on the rate as well.

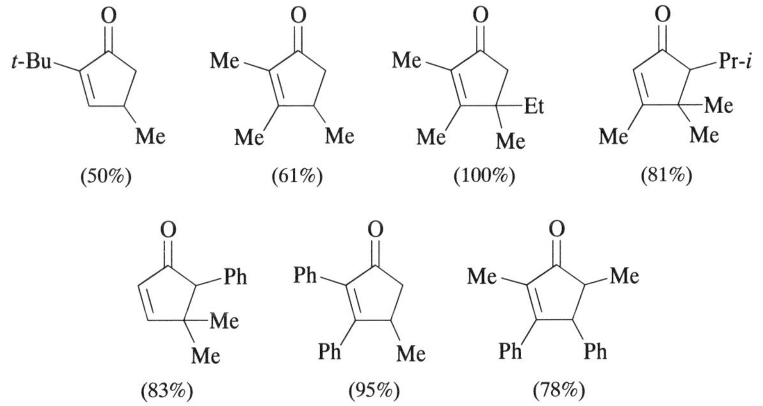

(50%) (61%) (100%) (81%)

(83%) (95%) (78%)

Figure 1

Monocyclic Precursors. The Nazarov cyclization has been extensively used as a cyclopentenone annulation method. For the construction of fused cyclic systems, one of the groups attached to the ketone must be cyclic. Classically, this was accomplished by vinylacetylide addition to cycloalkanones followed by dehydration and alkyne hydrolysis (Eq. 12). However, with unsymmetrically

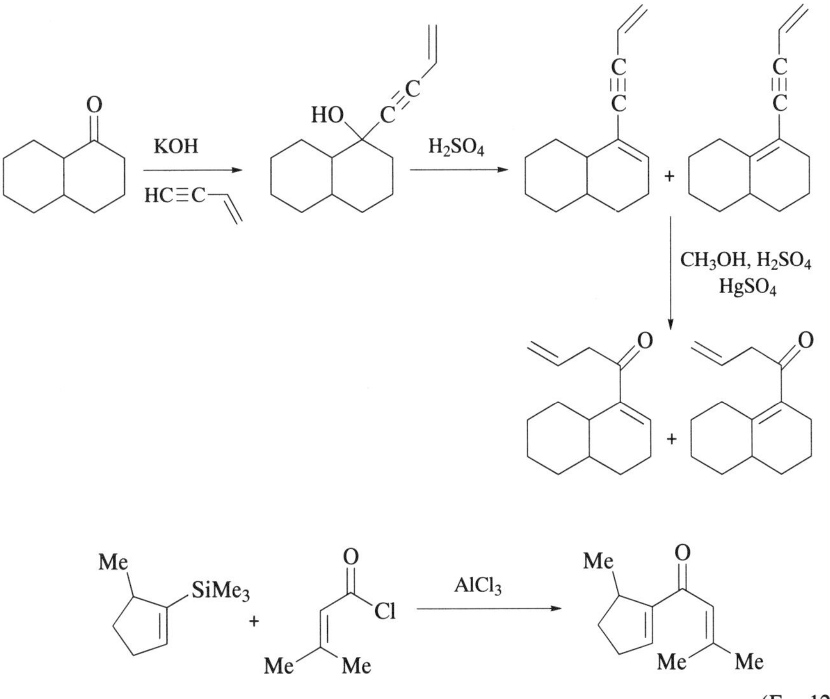

KOH, HC≡C– ; H_2SO_4 ; CH_3OH, H_2SO_4, $HgSO_4$

$AlCl_3$

(Eq. 12)

substituted cycloalkanones, regioisomers are formed.[11,22] An alternative construction, based on the acylation of cycloalkenylsilanes, provides divinyl ketones with predictable structure.[67,68]

Annulations onto preexisting rings of 5, 6, 7, and 12 members have been reported. The yields are generally lowest for 5-membered rings. The vinyl (allyl) appendage can accommodate substituents on both α and β positions as well as β,β' disubstitution. The compatibility of ester and imide functional groups is also noteworthy.[69,69a,70]

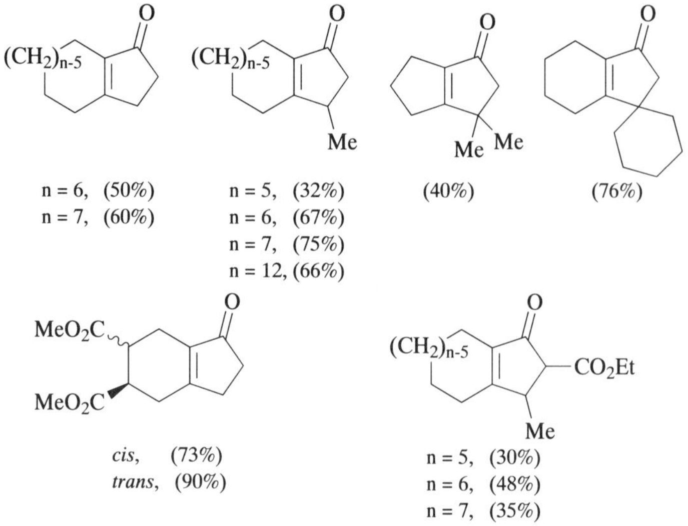

$(CH_2)_{n-5}$ n = 6, (50%) n = 7, (60%)

$(CH_2)_{n-5}$ Me n = 5, (32%) n = 6, (67%) n = 7, (75%) n = 12, (66%)

Me Me (40%)

(76%)

MeO_2C MeO_2C *cis*, (73%) *trans*, (90%)

$(CH_2)_{n-5}$ CO_2Et Me n = 5, (30%) n = 6, (48%) n = 7, (35%)

Figure 2

The issue of torquoselectivity is illustrated in the cyclization of the *cis* and *trans* diesters **13** and **14**. In each case a single diastereomer of the ring fusion unsaturated isomers **15** and **16** is produced (Eq. 13).[70]

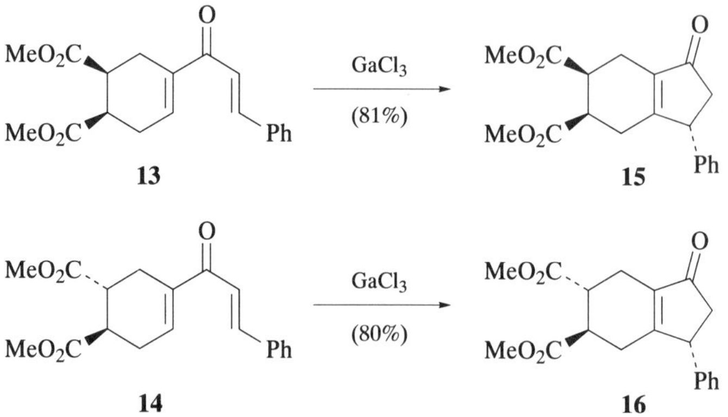

13 $\xrightarrow[(81\%)]{GaCl_3}$ **15**

14 $\xrightarrow[(80\%)]{GaCl_3}$ **16**

(Eq. 13)

Bis Cyclic Precursors. Formation of a cyclopentenone imbedded in a linear polycyclic array has been accomplished by Nazarov cyclization of bis(cycloalkenyl) ketones. Various combinations of five-, six-, and seven-membered ring systems have been employed.[71,72] Since the reaction operates under rather harsh conditions, the location of the double bond and the ring fusion stereochemistry are difficult to control.

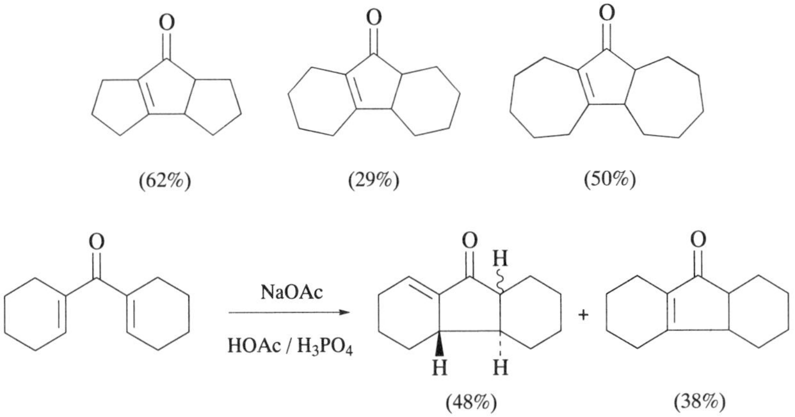

(62%) (29%) (50%)

NaOAc

HOAc / H_3PO_4

(48%) (38%)

Figure 3

This mode of cyclization has been employed in the establishment of the conrotatory electrocyclization mechanism[57,73,74] and in synthetic efforts directed toward trichodiene[75] and ophiobolins.[66]

Aromatic Precursors. Aryl vinyl ketones can also undergo Nazarov cyclization under sufficiently vigorous conditions. Both monocyclic and bis-cyclic types have been documented.[76,77] In the monocyclic series, the presence of aromatic substituents at the α position of the vinyl group facilitates cyclization and leads to high yields.[78]

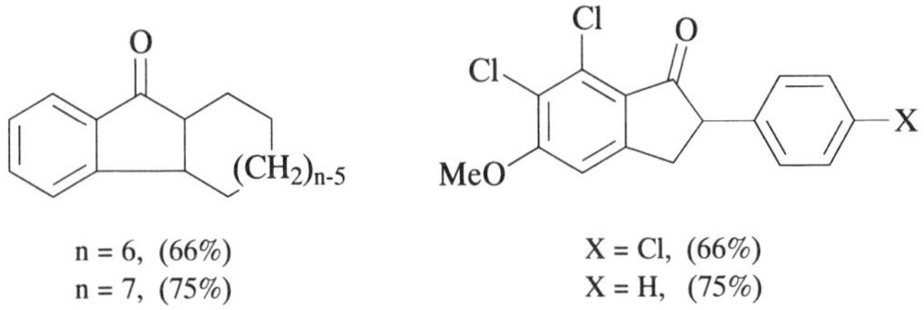

$(CH_2)_{n-5}$

n = 6, (66%)
n = 7, (75%)

X = Cl, (66%)
X = H, (75%)

Figure 4

Anomalous Cyclizations. The intermediacy of stable carbocations in the Nazarov cyclization has led to the observation of a number of rearrangement products[58] and alternative reaction pathways. Three categories have been identified. The first is simple Wagner–Meerwein rearrangement of the first-formed cyclopentenylic cation. This pathway is characteristic of β,β,β'-trisubstituted precursors **17** and **18** (Eq. 14).[53,62]

17 — p-TsOH, 180° → (40%) + (60%)

18 — H_2SO_4 →
R = Et, (39%)
R = *i*-Pr, (55%)
R = *n*-Bu, (47%)

(Eq. 14)

The second category is the "abnormal" Nazarov cyclization, which affords transposed 2-cyclopentenones arising from nucleophilic capture of the intermediate cyclopentenylic cations with carboxylic acids (Eq. 15).[79,80] Both ketones and

H_3PO_4 / HCO_2H →
X = O, (77%)
X = (OCH_2CH_2O), (55%)

H_3PO_4 / HCO_2H →
X = O (58%) (6%)
X = (OCH_2CH_2O) (47%) (37%)

(Eq. 15)

their derived dioxolane acetals have been employed in this transformation. An interesting consequence of this pathway is the production of fused bicyclic ketones from macrocyclic divinyl ketones.

The third category of anomalous cyclizations involves the intervention of an electrocyclic closure of linearly conjugated dienyl ketones. The monocyclic dienyl vinyl ketones **19** produce exclusively the angularly substituted hydrindenones **20** (Eq. 16).[81] These products must arise from preferential electrocyclic closure of the dienyl ketone followed by rapid Wagner–Meerwein shift of the vinyl group. The substituent effects on rate of rearrangement are consistent with this pathway.

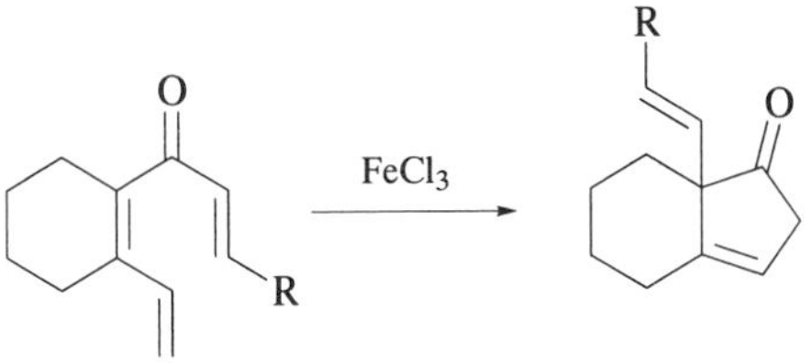

R = H, (77%)
R = $SiMe_3$, (55%)
R = Me, (61%)
R = Ph, (65%)

(Eq. 16)

19 $\xrightarrow{FeCl_3}$ **20**

Photochemical Cyclizations. Although a number of photo-induced electrocyclic closures of divinyl ketones have been reported, they are of little preparative significance.[60–62,82,82a,82b] For the most part these reactions have provided convincing evidence for the electrocyclization mechanism.[73] The products of disrotatory closure are isolated in these cases (see Eqs. 5 and 8). The photoisomerization of 4-pyrones in alcoholic medium affords 5-alkoxy-4-hydroxy-2-cyclopentenones. These products most probably arise from a photo-Nazarov reaction followed by capture of the zwitterionic intermediate (Eq. 17).[83] Intramolecular capture of the zwitterionic intermediate has also been reported.[83a]

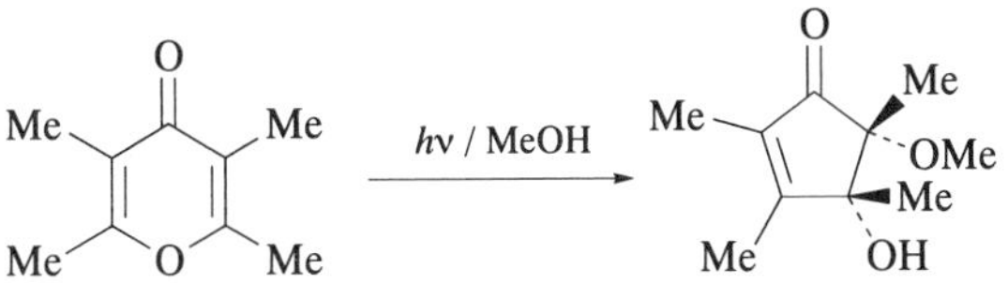

$h\nu$ / MeOH

(Eq. 17)

Silylated (Stannylated) Divinyl Ketones

The preparative utility of the Nazarov cyclization is greatly enhanced by employing β-silyl- or α'-silyldivinyl ketones as precursors.[81,84–90] The trialkylsilyl groups control the collapse of the intermediate cyclopentenylic cations,[91] thus providing two important benefits: (1) secondary cationic rearrangements are suppressed and (2) the final position of the cyclopentenone double bond is controlled. The latter feature is particularly significant in cyclopentenone annulations with monocyclic precursors because the double bond can be placed in the thermodynamically less stable position, therefore preserving the stereocenters created at the ring fusion.

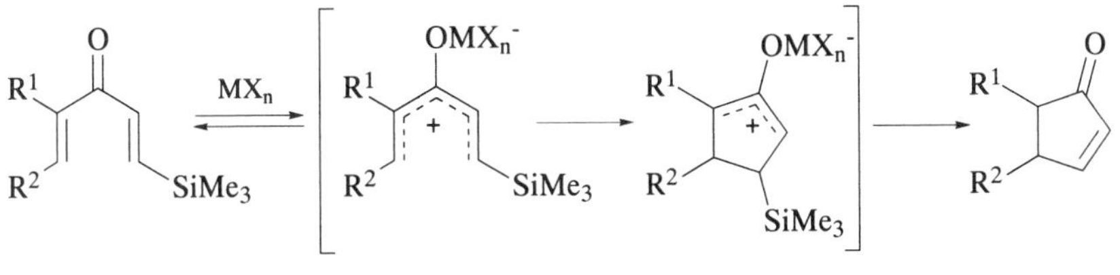

MX_n

(Eq. 18)

Vinylsilanes. A number of general methods for the preparation of β-silyldivinyl ketones have been developed.[85,88,92–96] Trimethylsilyl groups have been most often employed, though larger alkyl and aryl groups have also been used successfully.[86,87] The loss of the silicon electrofuge in the cyclization is dependent on the geometry of the β-silylvinyl unit. The triethylsilyl group in *Z*-**21** is partially retained in the product of cyclization,[97] whereas *E*-**21** undergoes highly selective silicon-directed closure (Eq. 19).[98]

FeCl$_3$

(*Z*)-**21**: R^1 = SiEt$_3$; R^2 = H (18%) (24%)

(*E*)-**21**: R^1 = H; R^2 = SiEt$_3$ (85%) (0%)

(Eq. 19)

The silicon-directed Nazarov cyclization is effectively promoted by Lewis acids, most commonly anhydrous iron(III) chloride, at temperatures below ambient.[85] For slow reacting substrates, boron trifluoride etherate and zirconium tetrachloride can also be employed where the oxidizing properties of iron(III) chloride are problematic.[88] Even with the extremely reactive α'-silyl substrates such as **22**, oxidized and chlorinated products can be isolated from reactions with iron(III) chloride at low temperature (Eq. 20).[89]

22 + dimers

FeCl$_3$, -20°, 0.08 M (58%) (10%) (30%)

BF$_3$•OEt$_2$, 0°, 0.08 M (71%) (0%) (21%)

(Eq. 20)

The utility of the silicon-directed Nazarov cyclization is illustrated by the preparation of simple cyclopentenones in which the double bond resides in the least substituted position. Examples of α and β monosubstitution and α,β disubstitution have been described (Eq. 21).[85]

FeCl$_3$

R^1 = H; R^2 = Me, (54%)

R^1 = *t*-Bu; R^2 = H, (95%)

R^1 = Me; R^2 = Me, (95%)

(Eq. 21)

In an analogous fashion, β-tributylstannyldivinyl ketones **23** are employed for tin-directed Nazarov cyclizations, exclusively in the acyclic mode (Eq. 22).[99]

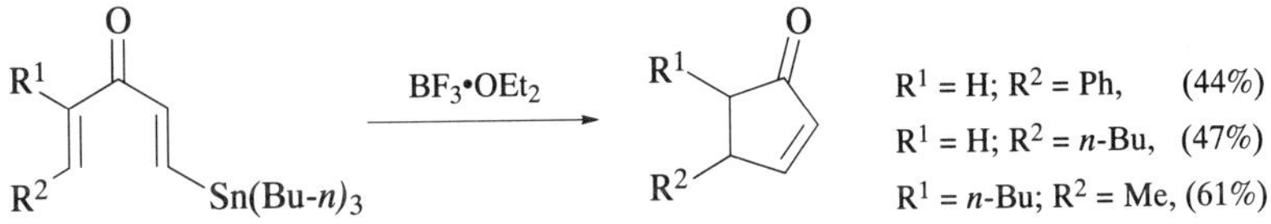

R^1 = H; R^2 = Ph, (44%)
R^1 = H; R^2 = *n*-Bu, (47%)
R^1 = *n*-Bu; R^2 = Me, (61%)

(Eq. 22)

These reactions are best promoted by boron trifluoride etherate. The construction of the substrates by aldol condensation of the β-stannylvinyl ketones is noteworthy.

The silicon-directed Nazarov cyclization is ideally suited for cyclopentenone annulation. Precursors containing 5-, 6-, 7-, and 12-membered rings afford bicyclic products in good yields wherein the double bond is exclusively in the less substituted position.[85] The ring fusion is predominantly in the *cis* configuration. Both heterocycles and unsaturated carbocycles can be employed as substrates for the annulation (Eq. 23).[88]

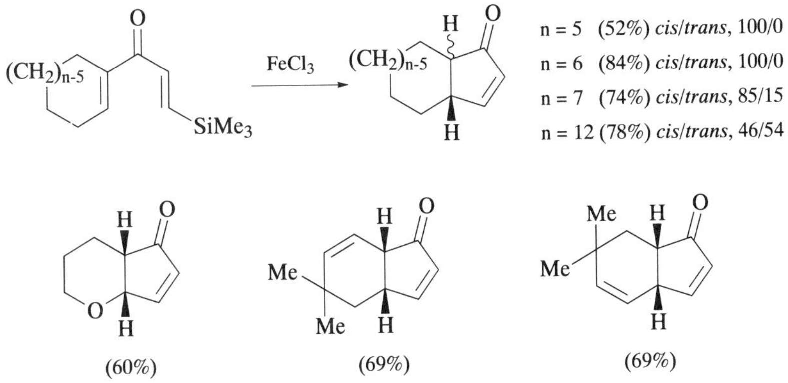

n = 5 (52%) *cis/trans*, 100/0
n = 6 (84%) *cis/trans*, 100/0
n = 7 (74%) *cis/trans*, 85/15
n = 12 (78%) *cis/trans*, 46/54

(60%) (69%) (69%)

(Eq. 23)

Cyclopentenone annulation with chiral substrates has been extensively examined. The stereochemical course of conrotatory closure is dependent upon ring size, substituent location, substituent size, and silyl group size.[86–88] In the cyclopentenyl series the selectivity is modest, though construction of linear (**24**)[95] and angular (**25**)[100,100a] triquinanes has been reported to proceed with excellent stereoselectivity (Eq. 24).

In the cyclohexenyl series, the directing effect is greatest when substituents are located at the γ position of the endocyclic enone **26** (Eq. 25).[86] The major isomer produced bears a *trans* relationship between the vicinal stereocenters, and the

BF$_3$•OEt$_2$

24

(Eq. 24)

BF$_3$•OEt$_2$

25

FeCl$_3$

26

R = Me; (99%) α/β, 78/22
R = vinyl; (66%) α/β, 70/30
R = OBn; (76%) α/β, 90/10
R = Ph; (76%) α/β, 94/6

(Eq. 25)

level of selection is highly dependent upon the substituent size. The bulk of silicon substituents was also shown to have an influence on stereoselectivity.[87]

Both cycloheptenyl and cyclooctenyl systems bearing a γ-benzyloxymethyl substituent have been examined and show remarkable levels of stereoselectivity. In these cases, the major product bears a *cis* relationship between the vicinal stereocenters at the ring fusion and the benzyloxymethyl group. The *cis/trans* ratio is 92/8 in **27** and 93/7 in **28**.[98]

27 **28**

Figure 5

Allylsilanes. The use of silicon direction has also been adapted to construction of linearly fused tricyclic systems.[89] In this variant the placement of the double bond is directed away from both ring fusions by placement of the trialkylsilyl group at the α' position. These reactions proceed much more rapidly than those of the β-silyldivinyl ketones and are compatible with various combinations of ring sizes (Eq. 26).

$FeCl_3$ or $BF_3{\cdot}OEt_2$

n = m = 6, (70%) *trans*
n = 6; m = 5, (88%) *cis*
n = m = 5, (77%) *cis*

(Eq. 26)

The stereodirecting effect of the remote silicon electrofuge has been demonstrated in the highly stereoselective electrocyclization of an optically active sample of **29** (Eq. 27).[90] The enantiomeric purity and absolute configuration of **30** (as shown) establish the exclusive conrotatory pathway, which corresponds to an *anti* S_E' substitution.

$FeCl_3$ (Eq. 27)

29 88% ee **30** 88% ee

In Situ Generation of Divinyl Ketones or Equivalents

From α-Alkoxy Enones. Construction of cyclopentenones by the Nazarov cyclization can be effected by generation of divinyl ketones under conditions that effect closure. Of the myriad of precursors of divinyl ketones, the simplest are α'- and β'-heterosubstituted, α,β-unsaturated ketones. The α'-oxygenated enones are readily prepared by the addition of acyl anion equivalents (propenal d^1 reagents) to ketones (Eq. 28). The elimination–cyclization is effected by treatment of the derived tertiary acetates **31** with acid at elevated temperatures.[101] The parent alcohols or silyl ethers are also converted, albeit in lower yield.

p-TsOH, toluene, 112°

31

R = H (48%)
R = Me_3Si, (48%)
R = OAc, (53%)

(Eq. 28)

From β-Substituted Precursors. β′-Substituted enones suffer facile β elimination of chlorine,[102,103] nitrogen,[104] or oxygen groups under acidic conditions required to cyclize the resulting divinyl ketones. The double β elimination of tetrahydro-4-pyranones **32** to form 2-cyclopentenone-4-carboxylates is induced by trimethylsilyl iodide[105] or trimethylsilyl triflate[69a] (Eq. 29). Interestingly, the putative intermediate α-carboalkoxy divinyl ketones have been independently cyclized with trimethylsilyl iodide[69] and triflate.[69a]

Me_3SiCl, NaI, DMF, 120° (77%) (Eq. 29)

32

From α-Vinylcyclobutanones. Divinylketones are also generated by an unusual acid catalyzed rearrangement of α-vinylcyclobutanones. Treatment of β,β-disubstituted α-vinylcyclobutanones **33** with a catalytic amount of boron trifluoride etherate or methanesulfonic acid results in the formation of ring opened divinyl ketones. By the use of a full equivalent of acid, the cyclopentenone is the major product (Eq. 30). The reaction is of limited preparative value as four different classes of products have been identified arising from different substitution patterns.[106–108]

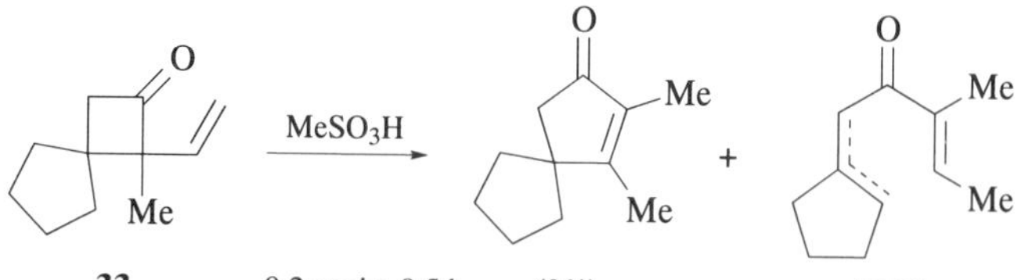

$MeSO_3H$ (Eq. 30)

33 0.2 equiv, 0.5 h (0%) (67%)
1.0 equiv, 26 h (71%) (8%)

Solvolytic Generation of Divinyl Ketones or Equivalents

From *gem*-Dichlorohomoallyl Alcohols. The structural diversity of divinyl ketone equivalents as precursors to cyclopentenones found in this class is remarkable. While this diversity lends breadth to the scope of the ring forming process, the reaction conditions required to unveil the precursor are often harshly acidic. Such is the case in the transformation of dichloro homoallylic alcohols **34**. These compounds (available from addition of 1,1-dichloroallyllithium to ketones) undergo solvolysis in neat trifluoroacetic acid to afford cyclopentenones in good yield (Eq. 31).[109,110] The reactions likely proceed by initial dehydration to the divinyl dichloride followed by ionization to a 3-chloropentadienylic cation.

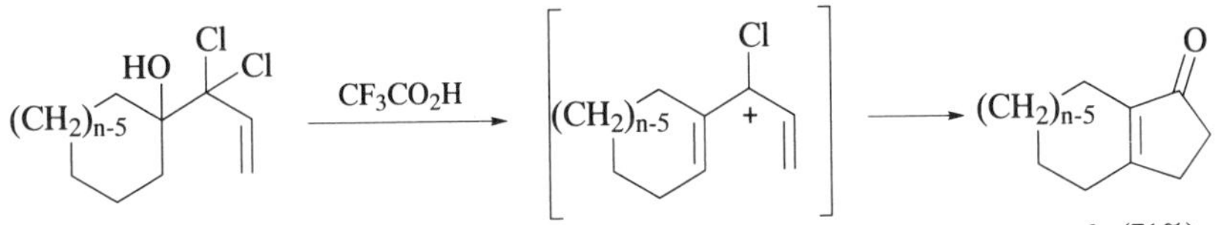

34

CF_3CO_2H

n = 6 (71%)
n = 7 (87%)
n = 8 (90%)

(Eq. 31)

From *gem*-Dichlorocyclopropylmethanols. The same manifold of intermediates can be accessed from dichlorocyclopropanes in two different ways (Eq. 32).

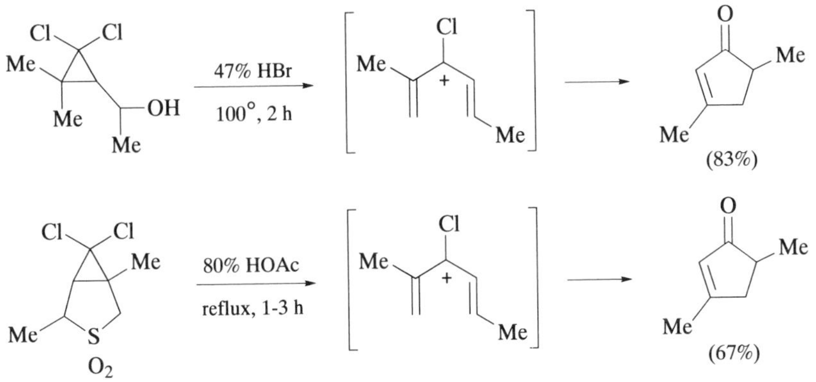

47% HBr, 100°, 2 h — (83%)

80% HOAc, reflux, 1-3 h — (67%)

(Eq. 32)

The first involves solvolysis of the dichlorocyclopropyl carbinols. Heating the carbinols in 47% hydrobromic acid affords cyclopentenones in good yields. The cyclopropylcarbinyl cation rearranges with loss of a proton to a divinyl dichloride as in the previous reaction. Because of the ready availability of the precursors from dichlorocyclopropanation of allylic alcohols, this procedure is best suited for the preparation of simple cyclopentenones.[110,111] In the second procedure, the dichlorocarbene/3-sulfolene adduct suffers extrusion of sulfur dioxide to form the divinyl dichloride which (under the solvolysis conditions) proceeds analogously to cyclopentenones.[112]

Rearrangement of Vinylallene Oxides. Peracid epoxidation of vinylallenes bearing allenic substituents results in the formation of cyclopentenones (Eq. 33). If no allenic substituents are present, epoxidation occurs on the vinyl group, leading to a stable oxirane. The details of this intriguing process are still unknown, but the intermediacy of a vinylallene oxide has gained support recently.[59] Rearrangement of the allene oxide may proceed via a 2-oxidopentadienyl cation or a cyclobutanone. The procedure is applicable to both simple and fused cyclopentenones with various substitution patterns using either peracids or singlet oxygen.[113–122]

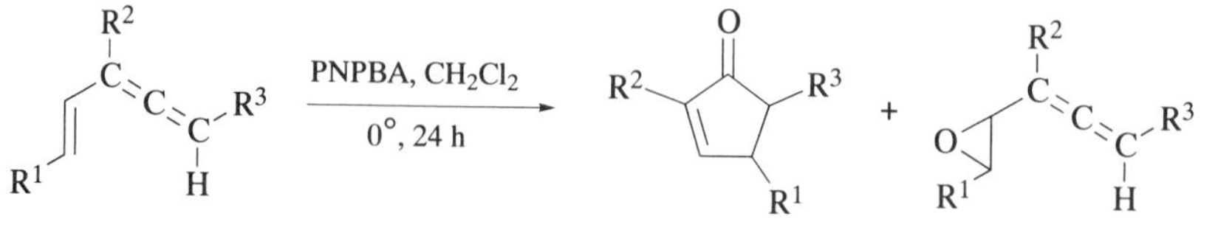

PNPBA, CH_2Cl_2, 0°, 24 h

R^1 = *n*-Pr; R^2 = R^3 = H	(55%)	(0%)
R^1 = R^3 = H; R^2 = *i*-Pr	(0%)	(60%)
R^1 = R^2 = H; R^3 = *n*-Pr	(3%)	(52%)

(Eq. 33)

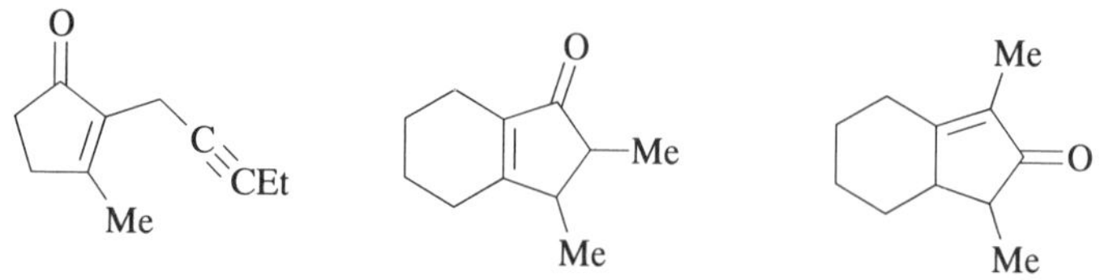

(70%) (50%) (50%)

Figure 6

The unproductive oxidation of the vinyl double bond can be suppressed by taking advantage of hydroxy-directed epoxidation of allylic and homoallylic alcohols. Attachment of allylic or homoallylic type hydroxy groups at either position of the allene unit is effective.[123–126] At the 1 position, for example, **35** (Eq. 34),

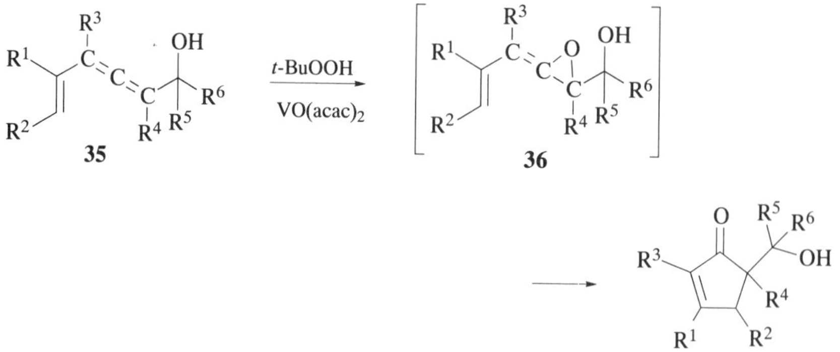

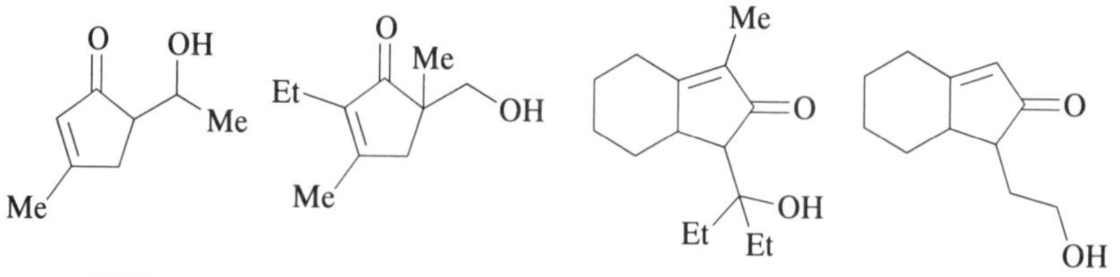

35 —*t*-BuOOH, VO(acac)$_2$→ [**36**] →

(55%) (63%) (45%) (40-70%)

(Eq. 34)

primary, secondary, and tertiary allylic alcohols and primary homoallylic alcohols have been examined. Presumably, the cyclopentenones are formed via the intermediacy of vinylallene oxide **36**. Finally, an important stereocontrol feature has been noted in the homoallylic alcohols (Eq. 35) that is consistent with either a concerted or zwitterionic mechanism.

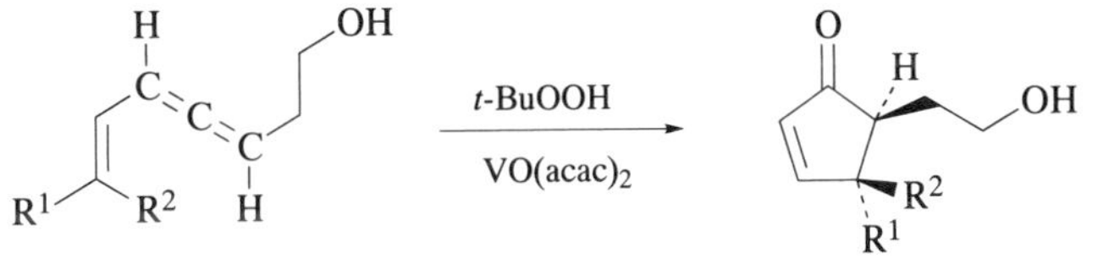

t-BuOOH, VO(acac)$_2$ (Eq. 35)

Placement of the epoxidation-directing substituent at the 3 position also leads to cyclopentenones, but the regioselectivity of epoxidation is dependent on the location of and substitution at the hydroxy-bearing group.[124,126] With allylic alcohols, secondary hydroxy groups lead primarily to cyclopentenones, while tertiary hydroxy groups lead exclusively to allenyl epoxides. With homoallylic alcohols, cyclopentenones are the major products independent of the substitution pattern (Eq. 36).

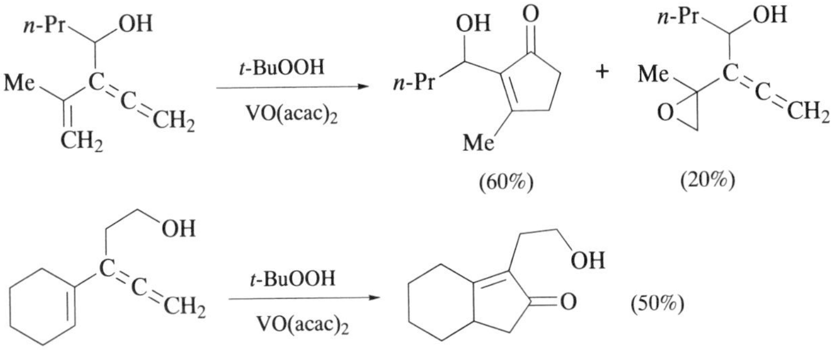

t-BuOOH, VO(acac)$_2$ (60%) + (20%)

t-BuOOH, VO(acac)$_2$ (50%)

(Eq. 36)

Vinylallenes can also be converted to cyclopentenones by solvometallation in the presence of mercury(II) acetate or thallium(III) acetate (Eq. 37). Electrophilic

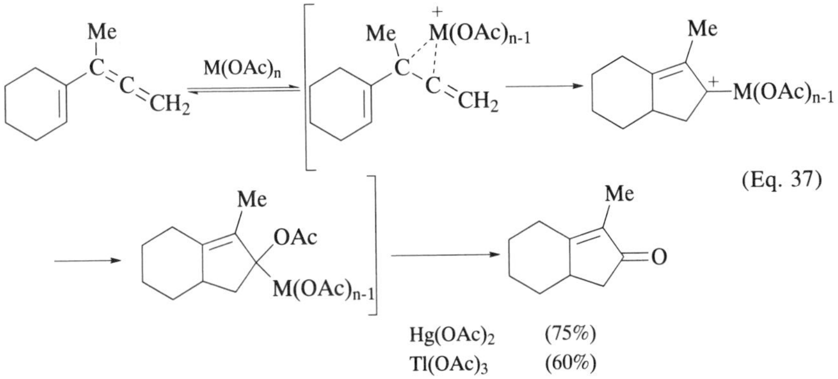

M(OAc)$_n$ (Eq. 37)

Hg(OAc)$_2$	(75%)
Tl(OAc)$_3$	(60%)

activation of allenes by these metal salts is a well-precedented Markownikoff process. In this case a spontaneous demetallation takes place to afford the cyclopentenone and the nascent metal. The yields are improved in general by the use of acetoxymercuration compared to acetoxythallation.[127,128]

Although mechanistically unrelated to the foregoing vinylallene processes, the solvolysis of methyoxyallenyl alcohol **37** in the presence of boron trifluoride etherate constitutes an efficient construction of α-methylenecyclopentenones (Eq. 38).[129–131] This transformation has been generalized to incorporate different

OMe HO R1 C C CH2 R2 CHO **37** $BF_3 \cdot OEt_2$ R1 OMe C C CH2 R2 OBF_3 → R1 OMe + R2 O F_3B^- → O R1 R2 OH

Me O Me OH (68%) O OH (80%) O SPh (82%)

(Eq. 38)

functional groups in the product and is ideally suited for the synthesis of α-methylenecyclopentanoid natural products. The electrocyclic process is formulated as a closure of a 1-oxypentadienylic cation closely related to the solvolysis of 2-furyl carbinols described in the next section.

From 2-Furyl Carbinols. The acid-catalyzed rearrangement of 2-furyl carbinols constitutes a versatile, albeit modest synthesis of 4-hydroxy-2-cyclopentenones (Eq. 39).[132–139] An important advantage of this synthesis is the ready availability of the precursors from a Grignard reaction of furfural. The choice of acid catalyst for the reaction is guided by the substitution on the furan ring. For bromo- or unsubstituted furans the reaction is sluggish, and sulfuric acid is used. For alkyl-substituted furans, zinc chloride is recommended.

Mechanistically, the reaction is intriguing as it corresponds to the in situ formation of a conjugated dienone rather than a cross-conjugated divinyl ketone. Thus, cyclization in a Nazarov sense (as a 1,4-dihydroxypentadienylic cation)

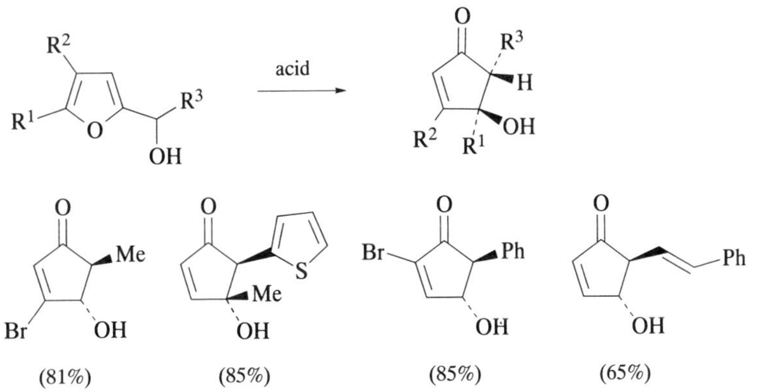

(81%) (85%) (85%) (65%)

(Eq. 39)

leads to a 4-hydroxy-2-cyclopentenone. An important consequence of the electrocyclization is the *trans* relationship of the 4-hydroxy group and the 5 substituent.

Acetylene-Containing Precursors

The acid-catalyzed hydration of acetylenes to ketones provides general access to carbonyl compounds from hydrocarbon precursors. The functionality needed to provide the additional double bonds to produce the equivalent of divinyl ketones can be derived from alkenes, alcohols, amines, or acetals.

Dienynes. The simplest progenitors of divinyl ketones are dienynes.[13–15,22–25,31] These compounds are easily prepared by addition of vinylacetylenes to ketones followed by dehydration (usually in a Saytzeff sense, Eq. 40). In certain cases,

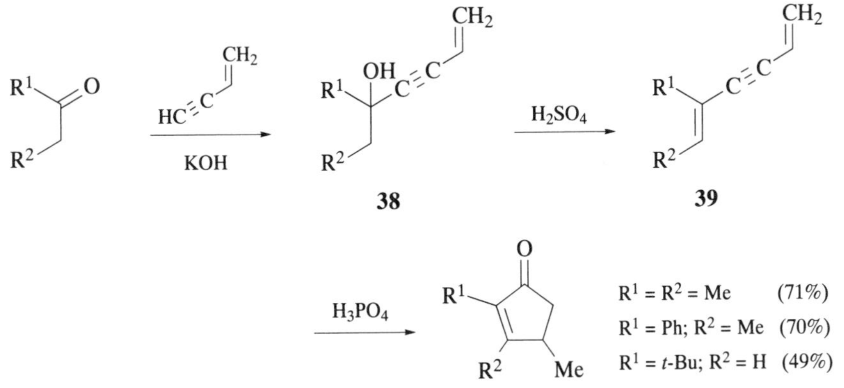

38 **39**

$R^1 = R^2 = Me$ (71%)
$R^1 = Ph; R^2 = Me$ (70%)
$R^1 = t\text{-Bu}; R^2 = H$ (49%)

(Eq. 40)

the vinylacetylenic alcohol **38** is directly employed, since the dehydration, hydration, and cyclization are all acid-catalyzed steps.[11,31,33,34,140] The preparation allows for general access to both cyclic and acyclic dienynes **39** at various levels of substitution. The conditions required for transformation to cyclopentenones are rather harsh (concentrated phosphoric or hydrochloric acid, 50-80°) since both hydration of the acetylene and cyclization must take place.

The majority of examples of this process employ vinylacetylene itself, which affords 1,2-disubstituted 4-methyl-2-cyclopentenones. α-Branched ketones afford 3-methyl-4,4-disubstituted 2-cyclopentenones.[24,25] If cyclic ketones are employed as precursors, the overall process constitutes a cyclopentenone annulation.[22,26,31,141]

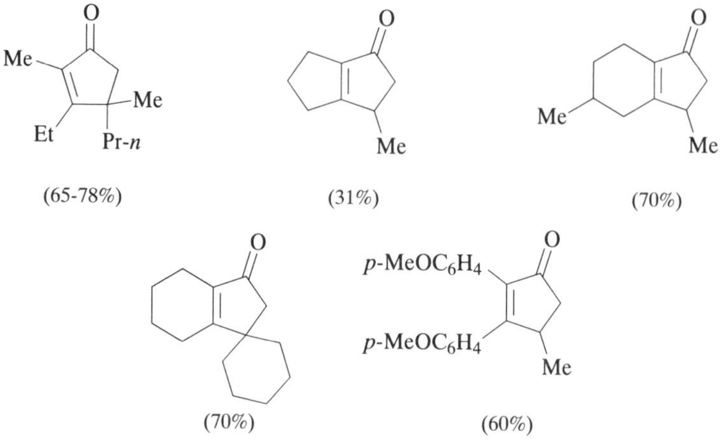

(65-78%) (31%) (70%)

(70%) (60%)

Figure 7

Although the enynols **38** are readily accessible and can be transformed into cyclopentenones, their molecular connectivity is not optimal since an array of six atoms is required to produce the requisite divinyl ketone equivalent that minimally needs only five atoms. This concept is illustrated by the rearrangement of the isomeric enynol acetate **40**. The cyclization is promoted by a palladium(II) catalyst in warm acetonitrile by the mechanistically intriguing process outlined in Eq. 41.[142] The reaction bears a resemblance to the silicon-directed Nazarov cy-

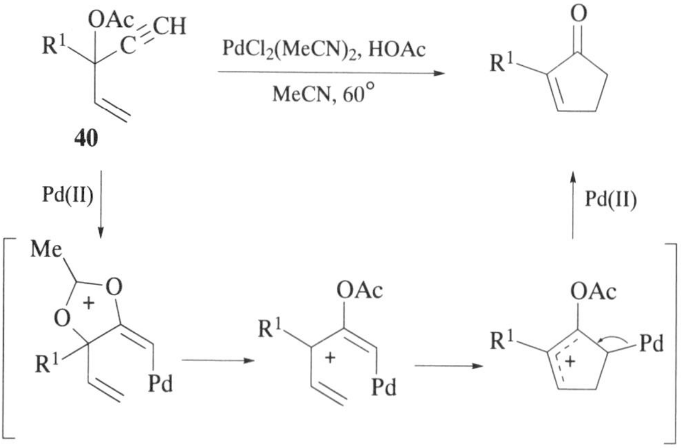

$PdCl_2(MeCN)_2$, HOAc; MeCN, 60°

40

Pd(II)

(Eq. 41)

clization in the ejection of the palladium(II) electrofuge. Both secondary and tertiary acetates can be employed, but substitution has been examined only at the α-vinyl position.

Propargylic Alcohols and Derivatives. A variety of other structural modifications that involve replacement of the terminal double bond with an oxidized carbon group (alcohol, amine, acetal) are also amenable to cyclization. The most common variation on this theme involves the acid-catalyzed transformation of ynediols available from the addition of propargylic alcohols and ethers to ketones (Eq. 42).[143–154] The advantages of this method are that the conditions are milder

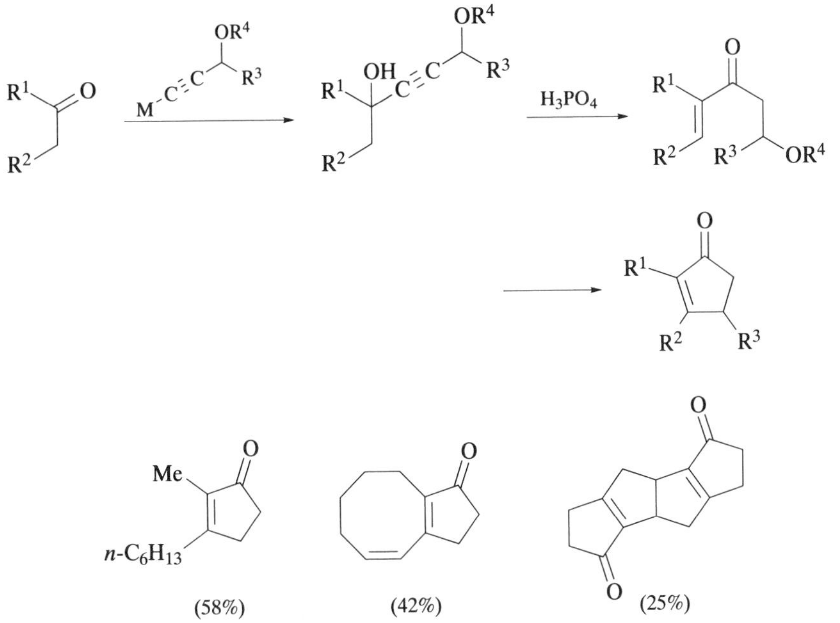

(Eq. 42)

than those for dienynes, rarely requiring heating (most likely proceeding via a Rupe rearrangement), and the substitution at C-4 is easily varied and can be hydrogen. For many years this reaction constituted the cyclopentanone annulation method of choice. Acyclic and monocyclic substrates containing 5-, 6-, 7-, 8-, and 12-membered rings are compatible. Double annulations are also possible for the rapid construction of polyquinanes. The analogous transformation of acetylenic amino alcohols for cyclopentenone annulation proceeds albeit in lower yields.[155]

If the higher oxidation state propargyl acetals are employed, an additional double bond is incorporated in the product.[156] Unfortunately, the yields from this variant are generally low and the position of the double bond is not controllable.

Respectable levels of relative asymmetric induction (torquoselectivity) can be obtained in this variant.[147] The major diastereomer obtained from conrotatory closure depends on the position and nature of the substituent. If the substituent is vicinal to the newly forming stereocenter as in **41** the extent of 1,2-induction is very high. However, if the substituent is more remote, as in **42** and **43**, the extent

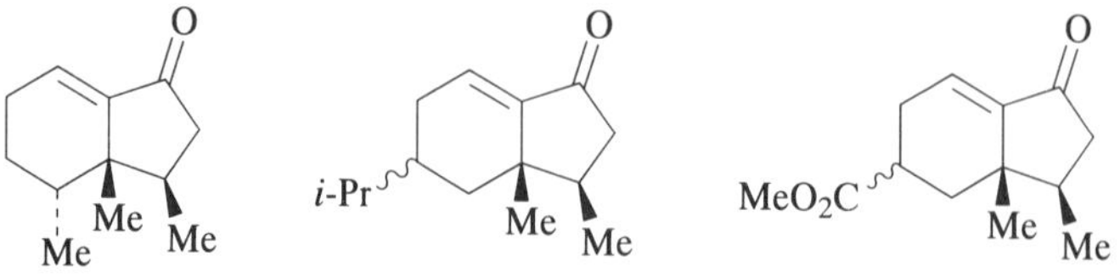

41 (60%)

42 (60%)
cis/trans 87/13

43 (60%)
cis/trans 60/40

Figure 8

of stereoselection is dependent on the size of the substituent, with the selectivity decreasing from isopropyl to methoxycarbonyl.

Divinyl Ketones from Coupling

In this category the precursors of the cyclopentenone are the most structurally remote because the carbon skeleton of the divinyl ketone is constructed during the operation. In all other variants, the carbon skeleton is assembled first and then, through functional group manipulations, the divinyl ketone equivalent is revealed under conditions that induce cyclization. With one exception, the carbon–carbon bond forming reactions are acylations of alkenes or acetylenes.

From α,β-Unsaturated Acids. α,β-Unsaturated acids or anhydrides undergo aliphatic Friedel-Crafts reaction with cycloalkenes to ultimately afford fused cyclopentenones (Eq. 43). The conditions for generation of the unsaturated acylium

polyphosphoric acid

R = H (16%)
R = Me (60%)
R = Ph (26%)

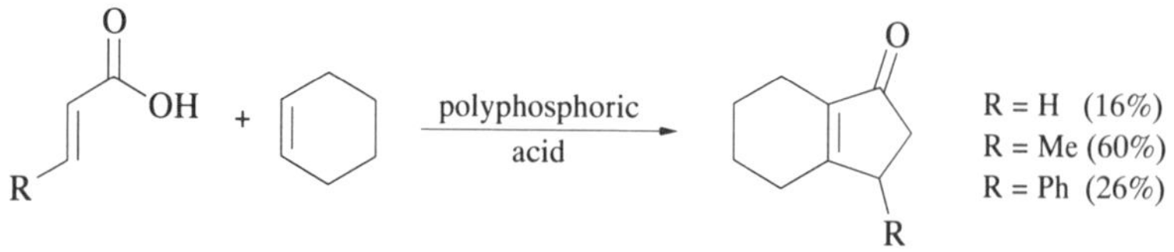

(Eq. 43)

ions are sufficiently acidic (polyphosphoric acid, 40–60°) to effect the Nazarov cyclization of the intermediate divinyl ketones. Yields are in general modest.

From α,β-Unsaturated Esters. Treatment of α,β-unsaturated esters with polyphosphoric acid also produces cyclopentenones. This variant is an alternative entry to the unsaturated acylium ion/alkene mixture that produces divinyl ketones by Friedel-Crafts acylation. The action of hot polyphosphoric acid on olefinic esters causes ionization of the alkyl-oxygen bond, leading to a mixture of unsaturated acid and alkene (Eq. 44). Although the yields are better by this in situ

polyphosphoric acid

(58-67%) (60%) (60%) (59%)

(Eq. 44)

generation protocol, a disadvantage of this approach is the potential for generation of different alkenes in the cleavage reaction. Another problem (intrinsic to all acylations) is the regiochemical ambiguity in the reaction of unsymmetrically substituted alkenes ($R^1 \neq R^2$). The ease of preparation of the starting esters allows for ready alteration of the substitution pattern in the cyclopentenone.

From *α,β*-Unsaturated Acid Chlorides. The most commonly employed precursors of unsaturated acylium ions are acid chlorides and bromides. The acylation of alkenes with these reagents constitutes a general synthesis of divinyl ketones which, under the conditions of acylation, suffer Nazarov cyclization to cyclopentenones.[157] As with the other carboxy precursors, the regioselectivity of electrophilic attack is ambiguous unless the alkene is either symmetrical or strongly biased toward Markownikoff addition. Both acid chlorides and bromides have been employed together with both cyclic and acyclic alkenes. Aluminum trichloride is the preferred Lewis acid promoter. Occasionally, the intermediate divinyl ketone or a *β*-chloroenone can be isolated.

Isoalkanes have also been employed as precursors of the alkenes in combination with alkenoyl chlorides (Eq. 45).[157a] The isoalkanes are oxidized to alkenes by hydride transfer to acetyl chloride/aluminum chloride or copper(II) sulfate in nitromethane. The isoalkanes (methylcyclopentane, methylcyclohexane, and 2-methylbutane) afford trisubstituted alkenes which undergo in situ

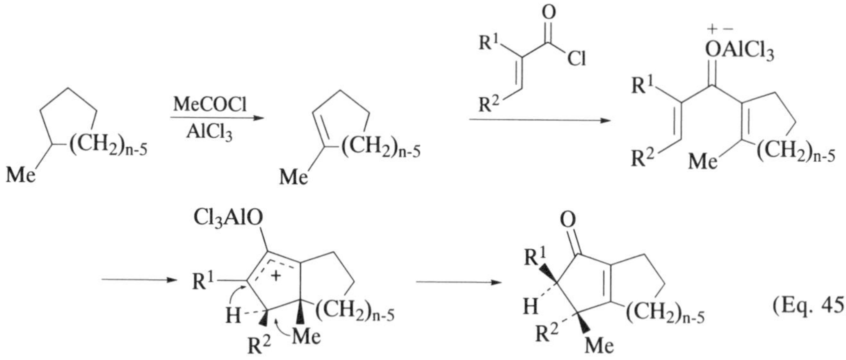

(Eq. 45)

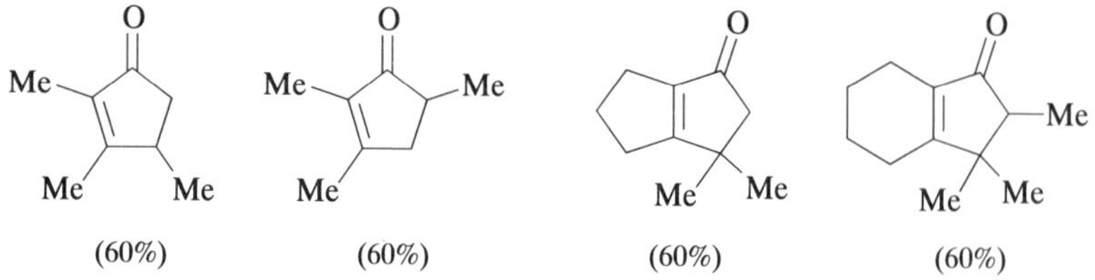

(60%) (60%) (60%) (60%)

acylation/Nazarov cyclization as described above. Labeling studies have revealed the involvement of hydride and alkyl migrations in the formation of the cyclopentenones.

A significant improvement in the utility of this approach is the employment of vinylsilanes for the preparation of the intermediate divinyl ketones. Owing to the ability of the silicon moiety to direct the site of electrophilic substitution, the problem of regiochemical ambiguity is resolved. Vinylsilane reagents have been used in two different modes. The first employs vinyltrimethylsilane itself as an ethylene equivalent in combination with olefinic acid chlorides.[158,159] This is used primarily for the annulation of the cyclopentenone ring. Tin tetrachloride is the reagent of choice to promote both electrophilic substitution and the Nazarov cyclization. The double-bond position is thermodynamically controlled. Substituted acryloyl chlorides also react with vinyltrimethylsilane to afford simple cyclopentenones.[160] For this variant, aluminum chloride is the preferred reagent (Eq. 46).

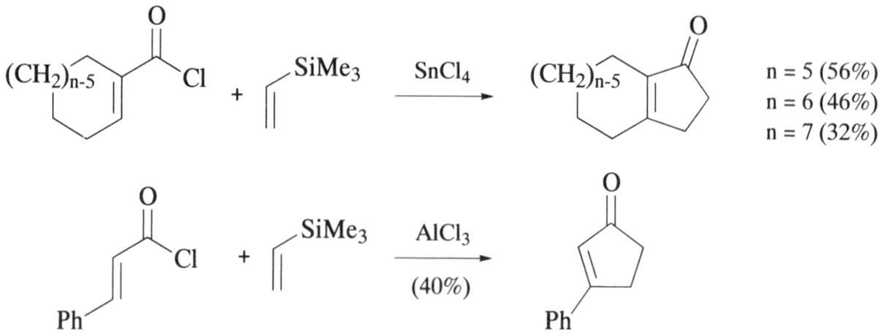

n = 5 (56%)
n = 6 (46%)
n = 7 (32%)

(40%)

(Eq. 46)

1- and 2-Phenylthio-substituted vinylsilanes are also useful reagents for cyclopentenone annulation.[161–163] The products of reaction with these reagents are very different (Eq. 47). In annulation reactions with a cyclopentenoyl chloride,

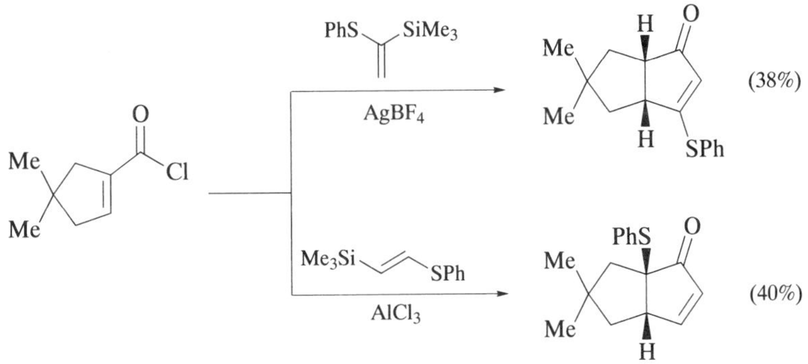

(Eq. 47)

1-phenylthiovinyltrimethylsilane affords 3-phenylthio-2-cyclopentenones while the 2-phenylthiovinyltrimethylsilane affords 5-phenylthio-2-cyclopentenones. In both cases the yields are modest.

The second application of vinylsilanes in the Nazarov cyclization uses substituted acryloyl chlorides in combination with cyclic vinylsilanes.[67,68] In general the vinylsilanes are obtained by silylation of the vinyllithium reagents generated from the corresponding hydrazones. In this modification the two steps are promoted by different Lewis acids: aluminum trichloride for the acylation and tin tetrachloride or boron trifluoride etherate for the Nazarov cyclization. The yield of the reaction is dependent on the degree of substitution of the alkenoyl chloride in the following order: β,β-disubstituted > β-monosubstituted > unsubstituted. With β,β-disubstituted acryloyl chlorides, the position of the cyclopentenone double bond is not well controlled (Eq. 48).

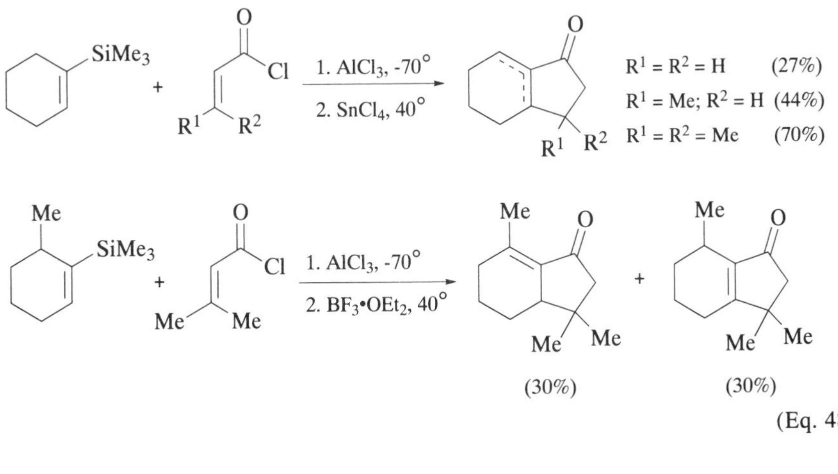

(Eq. 48)

The acylation of acetylenes with alkenoyl chlorides constitutes an alternative construction of cyclopentenones in a higher oxidation state. This reaction is most likely not mechanistically related to the Nazarov cyclization, rather involving electrophilic attack of the vinyl cation on the enone double bond. The major products are 5-chloro-2-cyclopentenones and 3-alkylidene-2-cyclopentenones along with chlorinated divinyl ketones. The ratio of chloro- to alkylidenecyclopentenones is dependent on the type of alkyne employed as illustrated for hexyne isomers: terminal alkynes afford mostly alkylidenecyclopentenones while internal alkynes (and acetylene itself) afford mostly chlorocyclopentenones (Eq. 49).[164–167]

R¹ R² O Cl —($R^3C{\equiv}CCH_2R^4$, $AlCl_3$)→ Cl R¹ R² O R³ R⁴ + R¹ R² O R³ R⁴

Me O Cl —(n-BuC≡CH, $AlCl_3$)→ Me O Pr-n (55%) + Cl Me O Bu-n (9%)

Me O Cl —(EtC≡CEt, $AlCl_3$)→ Cl Me O Et Et (55%)

(Eq. 49)

Miscellaneous Couplings

The construction of cyclopentenones from metal carbonyl compounds is a powerful technology, albeit unrelated to the Nazarov cyclization. There are however, two organometallic carbonylation reactions that most likely involve the formation of divinyl ketones which cyclize under the reaction conditions to afford cyclopentenones (Eq. 50).[168,169]

SYNTHETIC UTILITY

Several variants of the Nazarov cyclization have been employed in synthesis endeavors. For example, syntheses of simple cyclopentanoids such as *cis*-jasmone,[111] prostaglandin analogs[135,136,170] and (±)-valleranal,[122] (±)-methyleneomycin B,[130] and (±)-xanthocin[131] feature the Nazarov cyclization as a key step. More recently

$$n\text{-PrC}\equiv\text{CPr-}n \xrightarrow[\text{HNO}_3,\ \text{EtOH}]{\text{Ni(CO)}_4}$$ (25%)

$$\text{cyclohexene oxide} + \text{PhC}\equiv\text{CH} \xrightarrow[\text{2. HCO}_2\text{H, H}_3\text{PO}_4]{\text{1. Me}_3\text{SiMn(CO)}_5\text{, 1 bar}}$$ (15%) + (6%)

(Eq. 50)

the Nazarov cyclization has been used in the synthesis of polyquinane natural products such as (±)-hirsutene,[163] (±)-modhephene,[171,172] (±)-silphinene,[100,148] (±)-pentalenene,[100a] (±)-$\Delta^{9(12)}$-capnellene,[95] and (±)-cedrene.[173] The synthesis of (±)-$D^{9(12)}$-capnellene is noteworthy for the use of the silicon-directed Nazarov cyclization in an iterative fashion. The synthesis of yuehchukene alkaloid analogs has also employed a Nazarov cyclization as a key step.[173a]

The Nazarov cyclization employing propargyl alcohols has been successfully applied to the synthesis of (±)-strigol,[145] (±)-nor-sterepolide,[150] (±)-nookatone,[147] (±)-muscone,[152,174] and (±)-muscopyridine.[147] Industrially, an aromatic Nazarov cyclization is used in the synthesis of (+)-indacrinone.[78]

A Nazarov-type cyclization has been postulated in the biosynthetic pathways for *cis*-jasmonic acid[175,176] and marine-derived prostanoids such as preclavulone A.[177,178]

EXPERIMENTAL CONDITIONS

The diversity of substrates that are employed in the Nazarov cyclization and their differing propensity to be transformed into the divinyl ketone equivalents preclude generalizations about the reaction conditions. Since the cyclization of divinyl ketones requires the formation of a 3-oxypentadienylic cation, protic acids or Lewis acids are usually involved. The classical reagent for Nazarov cyclizations is polyphosphoric acid, usually in formic acid solution, or sulfuric acid in methanol. Modern variants employ Lewis acids such as tin tetrachloride, boron trifluoride etherate, aluminum trichloride, or ferric chloride in chlorocarbon solvents.

EXPERIMENTAL PROCEDURES

The procedures described below are chosen to be representative of each of the important structural classes of precursors for cyclopentenones. The generic structural class of precursor is provided parenthetically after the compound name.

3,4,4-Trimethyl-5-phenyl-2-cyclopentenone (Cyclization of an Allyl Vinyl Ketone).[24] 2-Methyl-3-phenyl-2,6-heptadien-4-one (12 g, 0.60 mol) was slowly added with stirring to 15 g of conc. phosphoric acid (sp. gr. 1.82) . The mixture evolved enough heat to reach a temperature of 85° and quickly became homogeneous. After being stirred at 70° for 40 minutes, the reaction mixture was diluted with water and then extracted with diethyl ether. The organic extracts were washed with sodium bicarbonate solution, dried over magnesium sulfate, and then concentrated. The residue was vacuum distilled to afford 10 g of the product (83%); bp 129–130° (2 torr); n_D^{20} 1.5520; d_4^{20} 1.044. Anal. Calcd. for $C_{14}H_{16}O$: C, 83.92; H, 7.99. Found: C, 83.69; H, 8.15. The ketone solidified after distillation and was recrystallized from 80% ethanol; mp 49°. The semicarbazone of this ketone was also prepared; mp 221.5°.

***cis*-Tricyclo[6.3.0.0^{3,7}]undec-1(8)-en-2-one (Cyclization of a Divinyl Ketone).**[71] 1,1′-Dicyclopentenyl ketone (19 g, 0.12 mol) was added with good stirring to hot (100°) polyphosphoric acid (100 g) under nitrogen. The colorless solution immediately turned dark brown. The reaction mixture was stirred for 30 minutes at 100°. After this time, the reaction solution was cooled in an ice bath, and ice (100 g) was added immediately to the hot acid. The mixture was stirred for 5 minutes. A dark precipitate formed during the addition of ice, but dissolved on the addition of diethyl ether. Standard extractive workup with ether gave a brown oil (19 g) which was distilled carefully to give 11.9 g (62% yield) of a the tricyclic product of greater than 95% isomeric purity by gas chromatography (OV-225) as a colorless oil; bp 60-63° (0.05 torr); UV nm max (ε) 244 (3700), 308 (72); IR (Nujol) 1690, 1630 cm^{-1}; ^{1}H NMR ($CDCl_3$, 270 MHz) δ multiplets centered at ca. 3.2 (1H), 3.1 (1H), 2.5 (2H), 2.4 (4H), 1.9 (1H), 1.6 (4H), and 1.3 (1H). A 2,4-dinitrophenylhydrazone of the product was also prepared, mp 201–202° (chloroform/ethanol).

***cis,trans*-1,3,4,5,6,7,8,8a-Octahydroazulen-1-one (Silicon-Directed Cyclization).**[85] Anhydrous iron trichloride (345 mg, 2.13 mmol) was added in one portion to a cold (−5°) solution of (*E*)-1-(1-cycloheptenyl)-3-trimethylsilyl-2-propen-1-one (450 mg, 2.02 mmol) in 25 mL of dichloromethane. The mixture was stirred at −5° for 50 minutes by which time the starting material had been consumed. Water (20 mL) was added, the mixture was diluted with dichloromethane (10 mL), and the organic layer was removed. The aqueous phase was extracted with dichloromethane (2 × 30 mL) and the individual organic extracts were washed with saturated aqueous ammonium chloride solution and brine. The combined organic extracts were dried (K_2CO_3) and concentrated. The residue was purified by flash column chromatography on silica gel (eluting with hexane-ethyl acetate 4:1) followed by distillation, bp 110° (0.01 torr) to afford 225 mg (74%) of the azulenone. GC analysis revealed the product to be an 85/15 mixture of *cis* and *trans* isomers; IR ($CHCl_3$) 3010, 2940, 1705 (C=O), 1580 (C=C) cm^{-1}; ^{1}H NMR ($CDCl_3$, 90 MHz) δ 7.85 (dd, J = 5.0, 2.0 Hz, 1H), 6.40 (dd, J = 5.0, 1.5 Hz, 1H), 3.55–3.25 (m, 1H), 2.92–2.62 (m, 1H), 2.48–1.38

(br m, 10H); mass spectrum m/z (rel. intensity) 150 (68), 135 (28), 108 (65), 107 (78), 95 (100), 94 (30), 93 (32), 83 (35), 79 (71), 77 (36), 68 (32), 67 (40), 66 (32), 53 (42). Anal. Calcd. for $C_{10}H_{14}O$: C, 79.96; H, 9.39. Found: C, 79.85, H, 9.50.

(4ab,4ba,9aa)-1,2,3,4,4a,4b,5,6,7,9a-Decahydro-1*H*-fluoren-9-one (Silicon-Directed Cyclization of an Allylsilane).[89] To a cold, (−50°) stirred mixture of anhydrous iron trichloride (170 mg, 1.05 mmol) in dry dichloromethane (40 mL, 0.02 M) was added dropwise a solution of (1-cyclohexenyl) (6-trimethylsilyl-1-cyclohexenyl) ketone (262 mg, 1.00 mmol) in 10 mL of dichloromethane. The reddish-brown mixture was allowed to stir 1 minute and then quenched by the addition of brine (50 mL) and diluted with diethyl ether (50 mL). The water layer was separated, extracted with diethyl ether (2 × 50 mL) and the combined diethyl ether extracts were washed with water (75 mL) and brine (75 mL), and then dried ($MgSO_4$) and evaporated to afford 157 mg (79%) of the product as a clear and colorless oil; bp 85° (0.3 torr); mp 78.5–79.5° (pentane); R_f: 0.23 (hexane/EtOAc 19/1); IR (neat): 2921s, 2854s, 2813w, 1713s, 1653s, 1449s, 1415m, 1363w, 1349w, 1293m, 1244m, 1232w, 1222w, 1208m, 1196w, 1173w, 1144w cm^{-1}; ^1H NMR ($CDCl_3$, 300 MHz) δ 6.64 (d, $J = 2.6$ Hz, 1H), 2.35–2.19 (m, 1H), 2.19–1.61 (m, 6H), 1.70–1.91 (m, 3H), 1.55–1.37 (m, 1H), 1.30–0.84 (m, 6H); ^{13}C NMR ($CDCl_3$, 75.5 MHz) δ 204.7, 141.0, 131.5, 54.9, 48.1, 42.4, 30.3, 26.5, 25.9, 25.5, 25.1, 21.4. MS (70 eV) m/z (rel. intensity) 190 (M^+, 82), 163 (11), 162 (84), 161 (32), 149 (10), 148 (14), 147 (18), 134 (11), 133 (19), 109 (15), 108 (100), 98 (20), 94 (50), 81 (24), 80 (55), 79 (39). Anal. Calcd. for $C_{13}H_{18}O$: C, 82.06; H, 9.53. Found: C, 82.11; H, 9.54.

Bicyclo[10.3.0]pentadec-1(12)-en-13-one (Solvolysis of a Dichloro Homoallyl Alcohol).[109] 1-(1,1-Dichloro-2-propenyl)cyclodecanol (105 g, 0.36 mmol) was added in one portion to trifluoroacetic acid (1.3 mL) at room temperature. The dark red solution was stirred vigorously for 1.5 hours and then was diluted with diethyl ether (10 mL) and neutralized with aqueous sodium bicarbonate solution. The mixture was extracted with diethyl ether (4 × 20 mL) and the combined organic extracts were dried (Na_2SO_4) and concentrated in vacuo to give an oil, which was purified by preparative TLC (silica gel, dichloromethane-diethyl ether, 10/1) to afford 71 mg (90%) of the product; ^1H NMR ($CDCl_3$ 60 MHz) δ 2.6–1.9 (m, 8H), 1.9–0.9 (m, 16H); IR (neat) 1690, 1634 cm^{-1}; mass spectrum m/z (rel. intensity) 220 (M^+, 65), 177 (93), 149 (100), 110 (65).

2-Pentyl-2-cyclopentenone (Epoxidation of a Vinylallene).[119] 4-Nitroperoxybenzoic acid (3.66 g, 0.02 mol) was added in small portions to a cold (0°), stirred solution of 3-pentyl-1,2,4-pentatriene (2.73 g, 0.02 mol) in dichloromethane (50 mL). After being stirred at 0° for 24 hours, the suspension was filtered and the filtrate was washed with 5% aqueous sodium hydroxide (3 × 20 mL) and water and was then dried ($MgSO_4$). The dichloromethane was evaporated and the residue was purified by silica gel chromatography (petroleum ether–ether, 20/1) to afford 2.42 g (80%) of the product; IR (neat) 3040, 1705, 1445,

1050, 1000 cm^{-1}; ^1H NMR (CCl_4) δ 7.15 (m, 1H), 2.75–1.90 (m, 6H), 1.80–1.10 (m, 6H), 0.89 (t, 3H).

Methylenomycin B (Solvolysis of a Methoxyallenyl Vinyl Carbinol).[130] Trifluoroacetic anhydride (2.3 mL, 16.3 mmol) was added dropwise over 15–30 minutes to a cold (−20°) solution of 2,3-dimethyl-3-hydroxy-4-[(methoxymethyl)oxy]-1,4,5-hexatriene (1.0 g, 5.4 mmol) containing 2,6-lutidine (3.1 mL, 27 mmol). After 5–10 minutes at −20°, the reaction was quenched by the addition of water (3 mL) and the product was extracted into diethyl ether. The ether layer was washed with water and brine and dried over magnesium sulfate. Filtration followed by evaporation of the solvent furnished a residue which was purified by column chromatography on silica gel to afford 490 mg (74%) of methylenomycin B as a pale yellow oil which crystallized in the freezer; mp 4°; IR (neat) 1690, 1660, 1625 cm^{-1}; ^1H NMR ($CDCl_3$, 300 MHz) δ 6.02 (m, J = 1Hz, 1H), 5.32 (m, J = 1.5 Hz, 1H), 3.05 (br s, 2H), 2.04 (m, J = 1.0 Hz, 3H), 1.78 (m, J = 1.0 Hz, 3H); ^{13}C NMR (C_6D_6, 75 MHz) δ 194.79, 162.30, 142.35, 138.18, 114.11, 36.59, 15.90, 8.28; mass spectrum m/z (rel. intensity) 122, 107, 93, 86, 84, 79.

3a-Methyl-2,3,3a,4,5,6-hexahydro-(1*H*)-inden-1-one (Cyclization of a Propargylic Diol).[147] n-Butyllithium solution in hexane (2.0 M, 15.8 mL, 32 mmol) was added to a cold (−78°) solution of propargyl alcohol (0.85 g, 15.1 mmol) in THF (40 mL). After being stirred at −78° for 3 hours a solution of 2-methylcyclohexanone (1.12 g, 10.0 mmol) in THF (10 mL) was added and the solution was allowed to stir at −78° for 1 hour and then at room temperature for 0.5 hour. After aqueous workup, the crude oil was purified by silica gel column chromatography to afford the adduct (1.48 g, 88% yield) as a 3:2 mixture of diastereomers. The mixture was used for the subsequent cyclization.

Concentrated sulfuric acid (1.5 mL, 28 mmol) was added dropwise at 0° over 15 minutes to a solution of the adducts (162 mg, 0.96 mmol) in methanol (1.5 mL). After being stirred at 0° for 1.5 hours the reaction was diluted with diethyl ether (15 mL) and neutralized with aqueous sodium hydrogen carbonate solution. Extractive workup afforded a crude oil that was purified by preparative TLC (dichloromethane) to afford 101 mg (70%) of the product: bp 78–80° (0.04 torr); IR (neat) 1716, 1646 cm^{-1}; ^1H NMR (CCl_4, 90 MHz) δ 6.37 (t, J = 3.6 Hz, 1H), 2.5–1.2 (m, 10H), 1.08 (s, 3H); mass spectrum m/z (rel. intensity) 151 (M^+ +1, 8), 150 (M^+, 42), 135 (32), 122 (33), 108 (75), 93 (88), 79 (100). Anal. Calcd. for $C_{10}H_{14}O$: C, 79.95; H, 9.39. Found: C, 79.69; H, 9.33.

1,2,3,4,5,6-Hexahydropentalen-1-one (Coupling of an α,β-Unsaturated Acid Chloride with Vinyltrimethylsilane).[68] Tin tetrachloride (26.35 g, 101 mmol) was added dropwise to a cold (−30°) solution of cyclopentene-1-carbonyl chloride (12.00 g, 92 mmol) and vinyltrimethylsilane (10.13 g, 101 mmol) in dichloromethane (100 mL). The reaction mixture was allowed to stir at −30° for 1 hour and then warmed to 25° and stirred for 6 hours. The solution was poured onto water (100 mL) and extracted with dichloromethane (3 × 100 mL).

The combined organic extracts were washed with saturated aqueous sodium bicarbonate solution (100 mL), dried (Na_2SO_4), and evaporated in vacuo to afford 5.90 g (52.5%) of the product; bp 114–115° (13 torr); IR (neat) 2920, 2820, 1695, 1640, 1385, 1025 cm^{-1}; ^1H NMR (CCl_4, 60 MHz) δ 2.82–2.15 (m, 8H), 2.15–1.6 (m, 2H); mass spectrum *m/z* calcd. for $C_8H_{10}O$: 122.073, found 122.073.

TABULAR SURVEY

The following tables contain examples of the Nazarov cyclization reaction in its various manifestations as defined in the preceding sections. The tables are arranged following the organizational format described in the Introduction. The table headings are self explanatory following the order of decreasing structural similarity to divinyl ketones. Within each structural subclass the listing of examples follows the order of increasing complexity from acyclic to monocyclic to polycyclic. Where established in the original articles, the configuration of the products is indicated. The literature survey includes articles appearing up to December 1991.

Since all of the carbon atoms for the final products appear in the starting material (with the exception of Tables V.A., V.C., and V.D.) the ordering of increasing carbon count for the starting materials and the products coincide. To further maintain this order, the carbon atoms of the silyl and tin substituents in the silicon- and tin-directed Nazarov cyclizations (Table II) are not counted. For the in situ construction of divinyl ketones (Tables V.A., V.C., and V.D.) the order follows increasing carbon count in the carboxylic acid derivative. For similar derivatives, the order of increasing complexity (acyclic, cyclic) followed by increasing carbon count (not including silyl substituents) of the olefinic or acetylenic component is observed.

The following abbreviations are used in the tables:

acac	acetylacetonate
Bn	benzyl
BOM	benzyloxymethyl
DME	1,2-dimethoxyethane
MCPBA	*m*-chloroperoxybenzoic acid
MOM	methoxymethyl
MPPA	monoperoxyphthalic acid
PNPBA	*p*-nitroperoxybenzoic acid
PPA	polyphosphoric acid
TFA	trifluoroacetic acid
TFAA	trifluoroacetic anhydride
TfO	trifluoromethanesulfonate
thexyl	*i*-$PrCMe_2$
TMS	trimethylsilyl
Ts	*p*-toluenesulfonyl

TABLE I. CYCLIZATION OF ALLYL VINYL AND DIVINYL KETONES: A. ACYCLIC PRECURSORS

Reactant	Conditions	Product(s) and Yield(s) (%)	Refs.
C_7			
O, Me	HCO_2H, H_3PO_4, 80°, 7 h	O, Me, Me (75)	5, 20
	D_3PO_4, 20°, 4-7 h	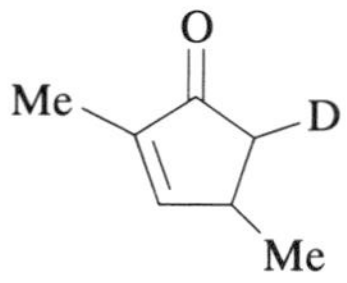 O, Me, D, Me (30)	32, 35
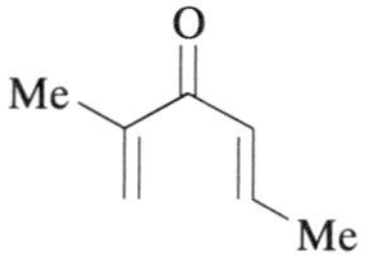 O, Me, Me	H_3PO_4, 20°, 10 h	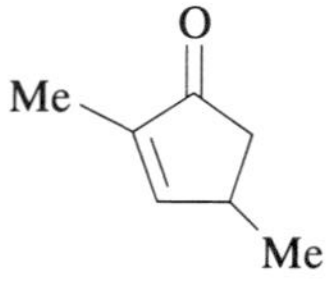 O, Me, Me (21) 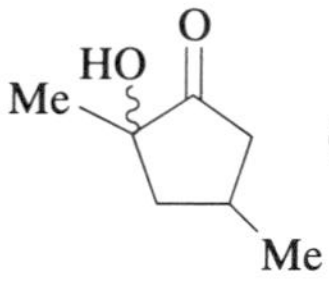 + HO, O, Me, Me (11)	38
C_8			
O, Me, Me	Anhyd. HCO_2H, H_3PO_4, 40-50°, 2 h, 70-75°, 5 h	O, Me, Me, Me (25)	9
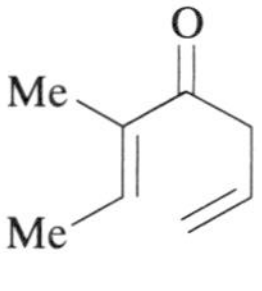 O, Me, Me	H_3PO_4 or HCl, 20°	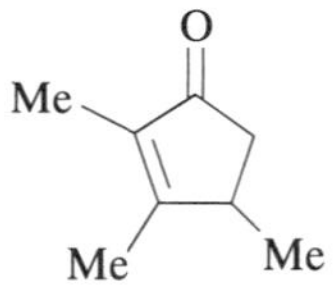 O, Me, Me, Me (—)	6

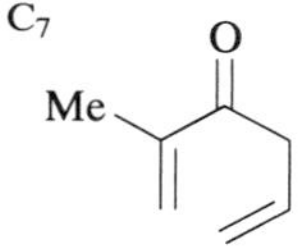

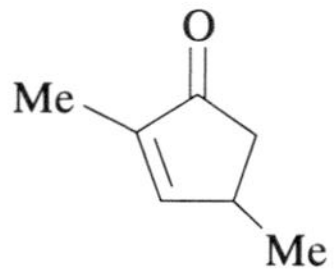

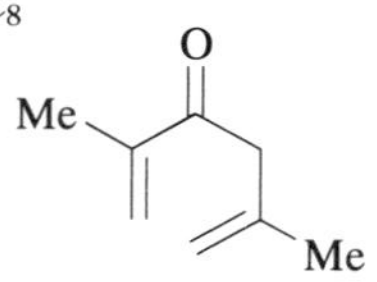

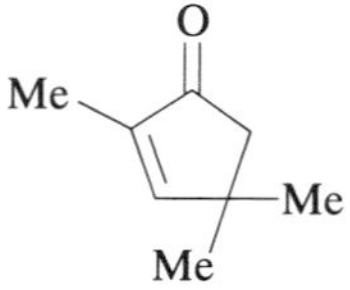

TABLE I. CYCLIZATION OF ALLYL VINYL AND DIVINYL KETONES: A. ACYCLIC PRECURSORS (*Continued*)

Reactant	Conditions	Product(s) and Yield(s) (%)		Refs.
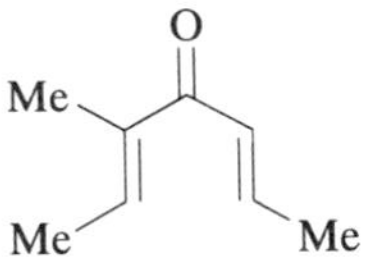 O, Me, Me, Me	H_2SO_4, 60°, 6 h	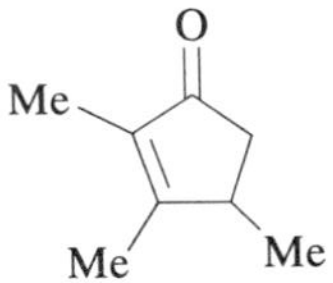 O, Me, Me, Me	(61)	53
C_9 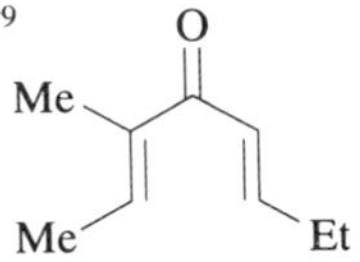 O, Me, Me, Et	H_2SO_4, 60°, 6 h	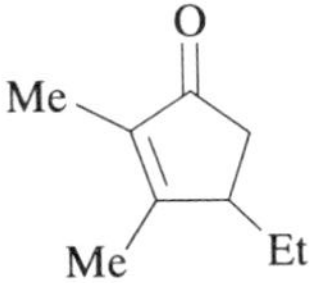 O, Me, Me, Et	(37)	53
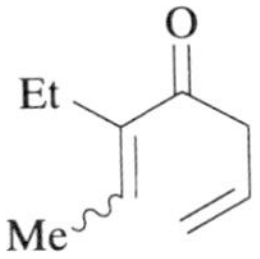 O, Et, Me	Conc. HCl, 60°, 3 h	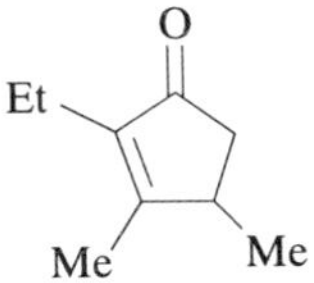 O, Et, Me, Me	(75)	12
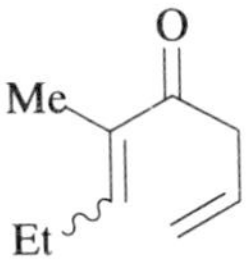 O, Me, Et	H_3PO_4, 60°, 2 h	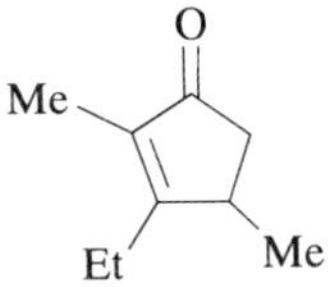 O, Me, Et, Me	(60)	12
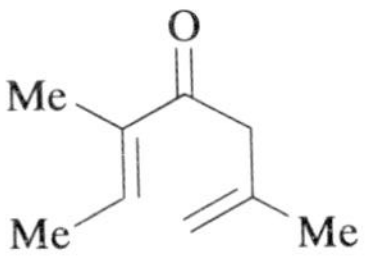 O, Me, Me, Me	H_3PO_4, 60°, 4 h	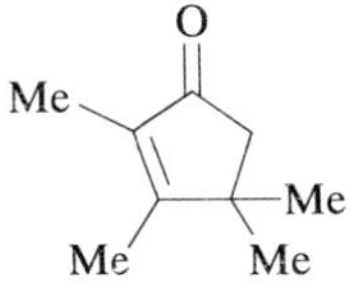 O, Me, Me, Me, Me	(60)	18

TABLE I. CYCLIZATION OF ALLYL VINYL AND DIVINYL KETONES: A. ACYCLIC PRECURSORS (*Continued*)

Reactant	Conditions	Product(s) and Yield(s) (%)	Refs.
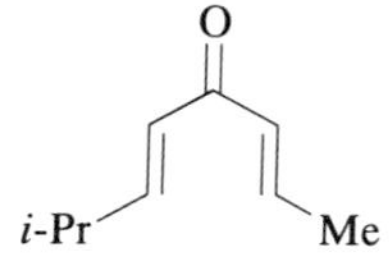 O; i-Pr; Me	96% H_2SO_4, 25°	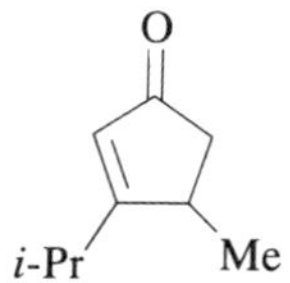 O; i-Pr; Me (17) 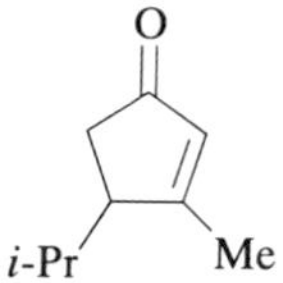 + O; i-Pr; Me (11)	52
C_{10}			
O; Me; Me; Me; Me	H_3PO_4 or conc. HCl, 20°, 30 min, 60°, 4 h	O; Me; Me; Me; Et (100)	10
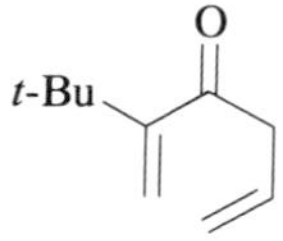 O; t-Bu	$HgSO_4$, H_2SO_4, MeOH 65°, 10 h	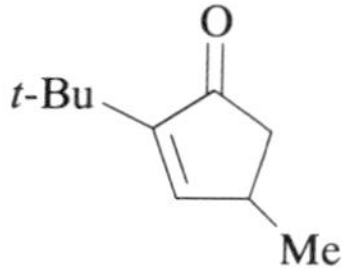 O; t-Bu; Me (50)	15
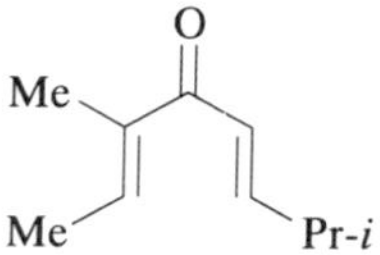 O; Me; Me; Pr-*i*	H_2SO_4 , 60°, 6 h	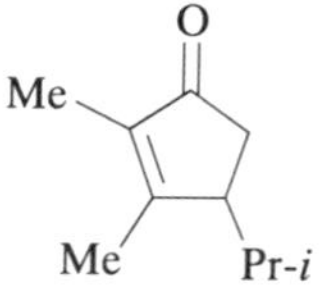 O; Me; Me; Pr-*i* (58)	53
C_{11}			
O; Me; Me; Bu-*n*	H_2SO_4, 60°, 6 h	O; Me; Me; Bu-*n* (63)	53

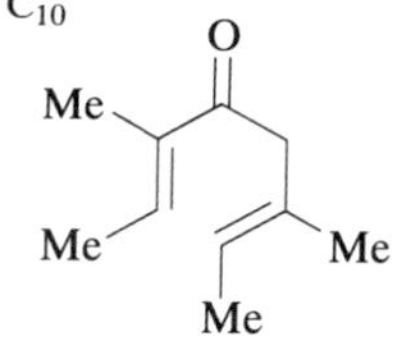

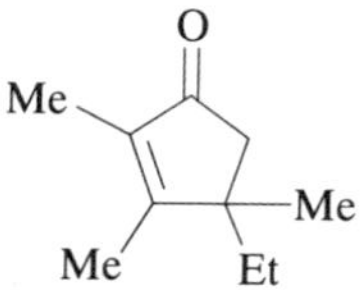

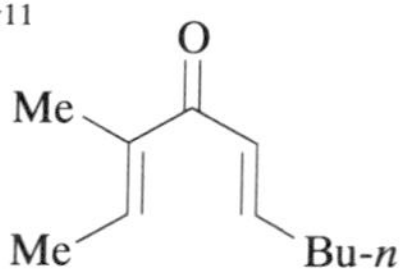

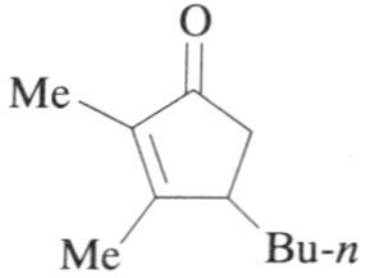

TABLE I. CYCLIZATION OF ALLYL VINYL AND DIVINYL KETONES: A. ACYCLIC PRECURSORS (*Continued*)

Reactant	Conditions	Product(s) and Yield(s) (%)	Refs.
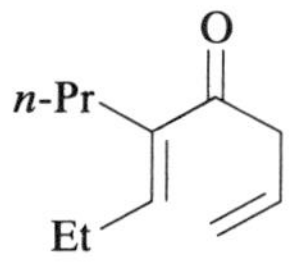 n-Pr, Et, O	H_3PO_4, 60-65°, 2.5 h 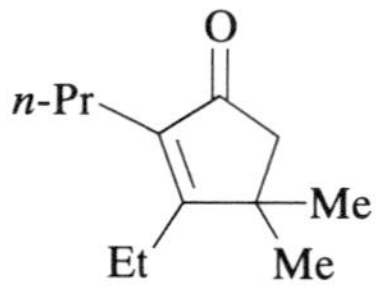	n-Pr, O, Et, Me, Me (80)	28
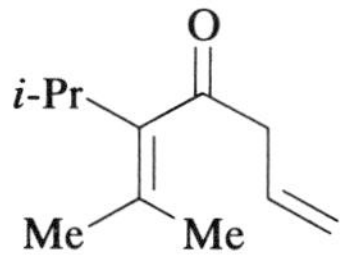 i-Pr, O, Me, Me	H_3PO_4, 60-65°, 6 h	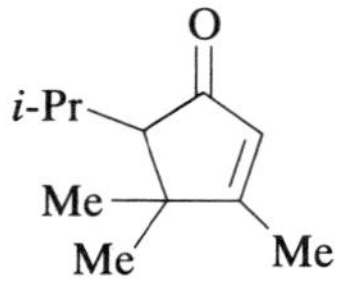 i-Pr, O, Me, Me, Me (81)	27, 38
	D_2O, P_2O_5, 20°, 8 h	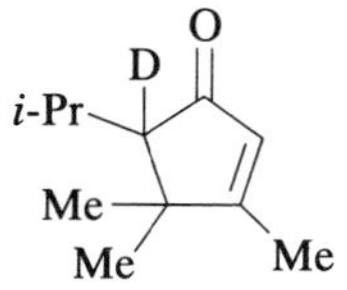 D, i-Pr, O, Me, Me, Me (83)	36
C_{12}			
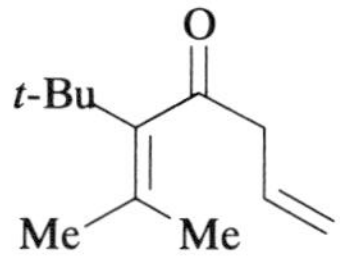 t-Bu, O, Me, Me	H_3PO_4, 65°, 45 min 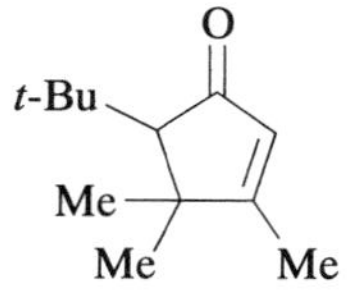	t-Bu, O, Me, Me, Me (68)	25
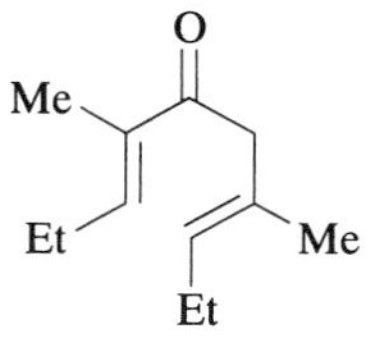 Me, O, Et, Me, Et	HCO_2H, H_3PO_4, 90°, 5 h 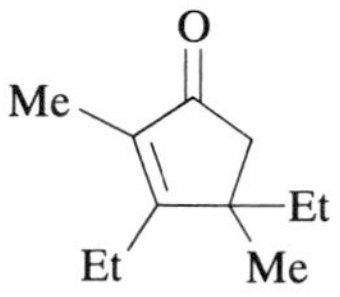	Me, O, Et, Et, Me (64)	19

TABLE I. CYCLIZATION OF ALLYL VINYL AND DIVINYL KETONES: A. ACYCLIC PRECURSORS (*Continued*)

Reactant	Conditions	Product(s) and Yield(s) (%)		Refs.
C_{13}				
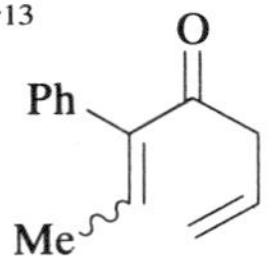 Ph, Me (structure)	H_3PO_4, 60°, 1.5 h	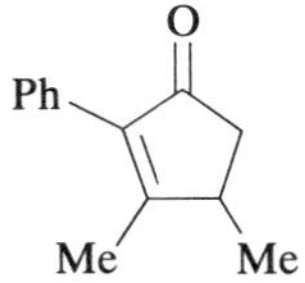 Ph, Me, Me (structure)	(95)	17
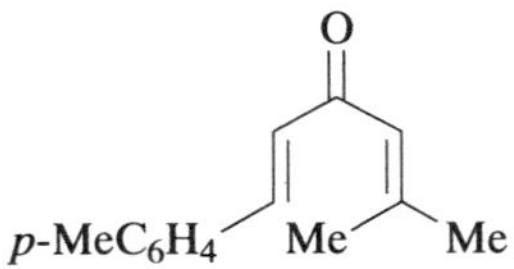 p-MeC_6H_4, Me, Me (structure)	BF_3•Et_2O, C_6H_6, reflux, 72 h	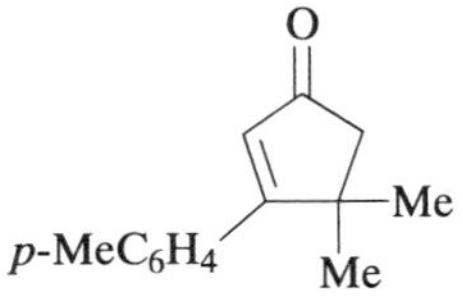 p-MeC_6H_4, Me, Me (structure)	(10)	68
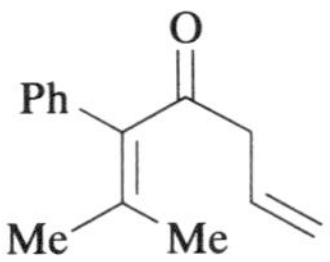 Ph, Me, Me (structure)	H_3PO_4, 70°, 40 min	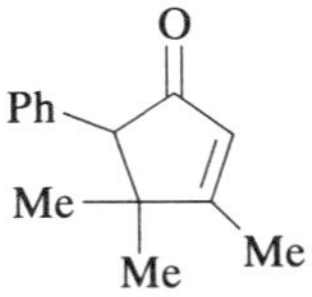 Ph, Me, Me, Me (structure)	(83)	24
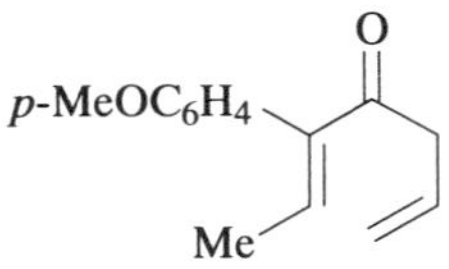 p-$MeOC_6H_4$, Me (structure)	TsOH, 155°, 10 min	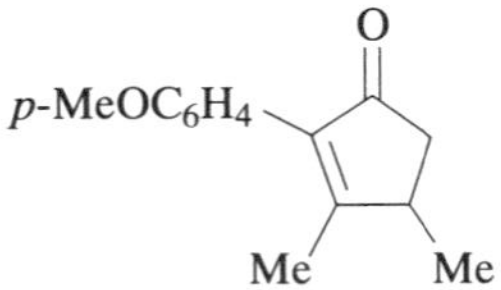 p-$MeOC_6H_4$, Me, Me (structure)	(—)	23
C_{15}				
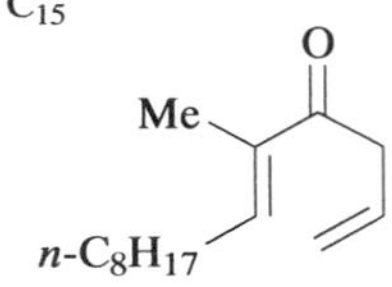 Me, n-C_8H_{17} (structure)	H_3PO_4, 60°, 3 h	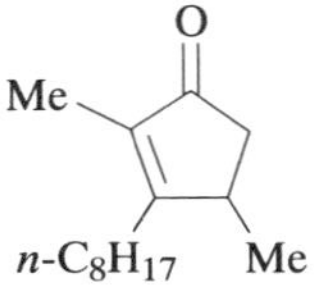 Me, n-C_8H_{17}, Me (structure)	(94)	29

TABLE I. CYCLIZATION OF ALLYL VINYL AND DIVINYL KETONES: A. ACYCLIC PRECURSORS (*Continued*)

Reactant	Conditions	Product(s) and Yield(s) (%)		Refs.
C_{17}				
O, Ph, Ph	1. H_2SO_4, Ac_2O 2. K_2CO_3	OAc, OSO_3K, Ph, Ph	(88)	58
C_{18}				
O, Ph, Ph	H_3PO_4, 70-80°, 15 min	O, Ph, Ph, Me	(95)	30
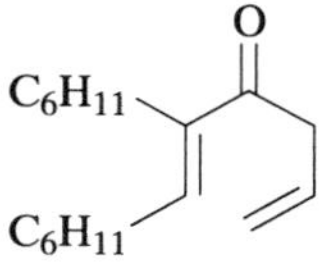 O, C_6H_{11}, C_6H_{11}	H_3PO_4, 70-75°, 5 h	O, C_6H_{11}, C_6H_{11}, Me	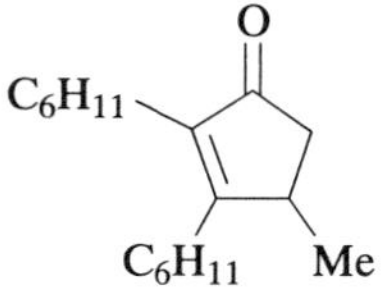 (50)	26
C_{19}				
O, Me, Me, Ph, Ph	Conc. HCl, EtOH, reflux, 2 h	O, Me, Me, Ph, Ph	(78)	25

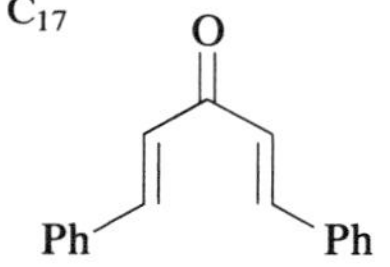

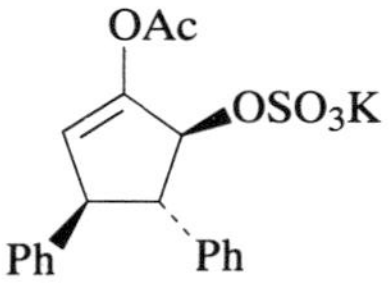

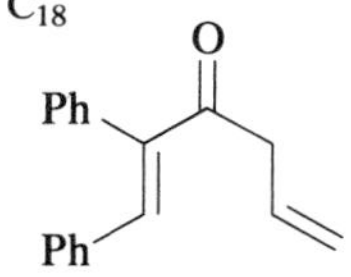

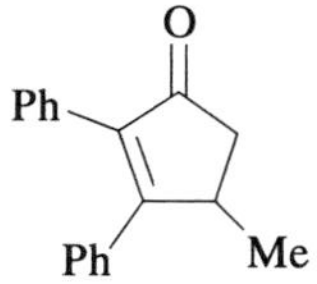

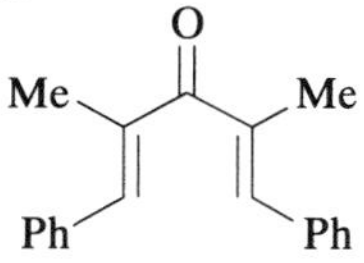

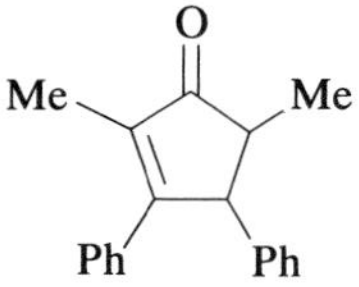

TABLE I. CYCLIZATION OF ALLYL VINYL AND DIVINYL KETONES: A. ACYCLIC PRECURSORS (*Continued*)

Reactant	Conditions	Product(s) and Yield(s) (%)	Refs.
	Conc. HI, red P, reflux, 24 h	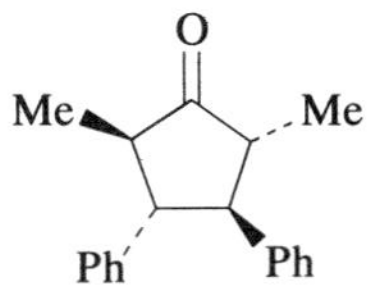 Me, Me, Ph, Ph (60) 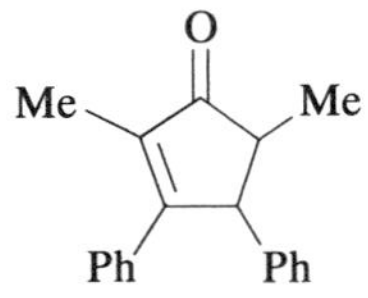 + Me, Me, Ph, Ph (4)	44
C_{20}			
p-$MeOC_6H_4$, p-$MeOC_6H_4$	TsOH, 180°, 20 min	p-$MeOC_6H_4$, p-$MeOC_6H_4$, Me (59)	23

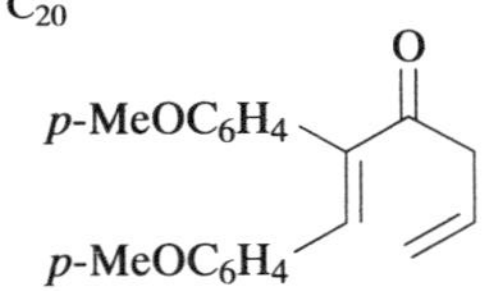

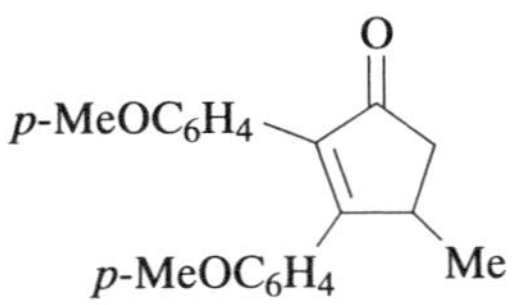

TABLE I. CYCLIZATION OF ALLYL VINYL AND DIVINYL KETONES: B. CYCLIC PRECURSORS

Reactant	Conditions	Product(s) and Yield(s) (%)	Refs.
C_9			
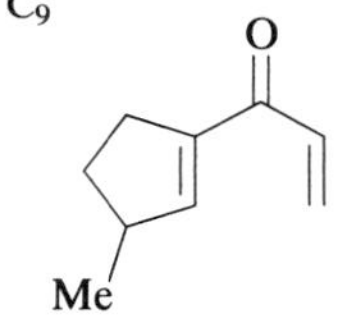 Me, O	Polyphosphoric Acid (PPA), 55-60°, 10 min	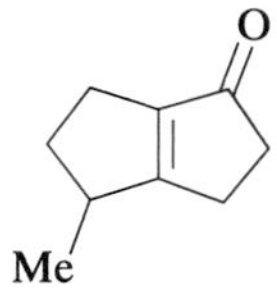 O, Me (32)	173
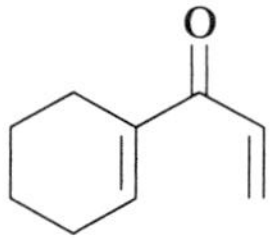 O	H_3PO_4, HCO_2H, 80-90°, 4 h	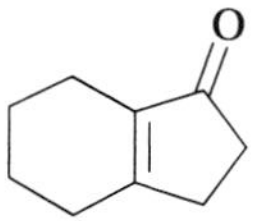 O (50)	39
	$FeCl_3$, CH_2Cl_2, 0°, 1.25 h	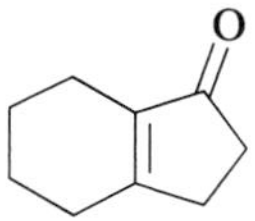 O (61) 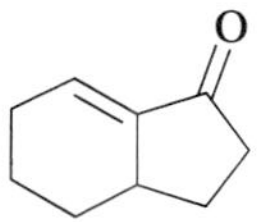 + O (17)	85
C_{10}			
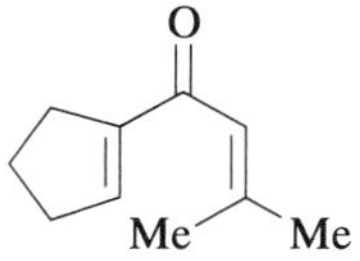 O, Me, Me	1. $SnCl_4$, CH_2Cl_2, reflux, 24 h 2. $RhCl_3$, EtOH, reflux, 2 h	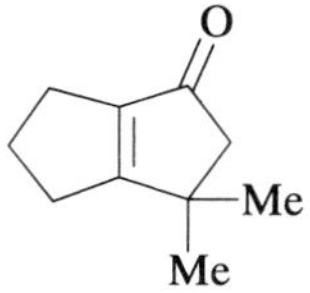 O, Me, Me (40)	171, 179, 172
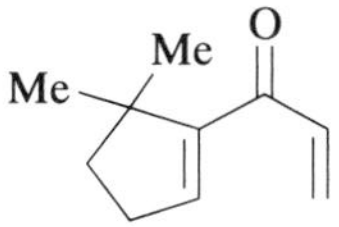 Me, Me, O	P_2O_5, $MeSO_3H$, rt, 2 min	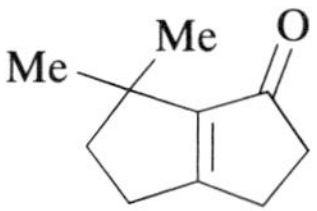 Me, Me, O (65)	179, 180

TABLE I. CYCLIZATION OF ALLYL VINYL AND DIVINYL KETONES: B. CYCLIC PRECURSORS (*Continued*)

Reactant	Conditions	Product(s) and Yield(s) (%)	Refs.
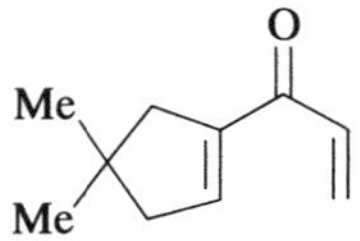 Me, Me, O	P_2O_5, $MeSO_3H$, 0-20°, 5 min	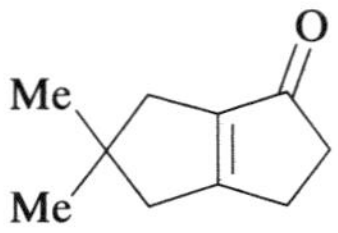 Me, Me, O (40)	148
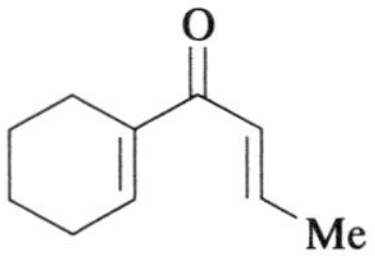 O, Me	H_3PO_4, HCO_2H, 90°, 7 h	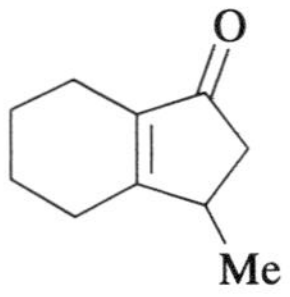 O, Me (67)	39
	H_3PO_4, 15-20°, 4.5 h	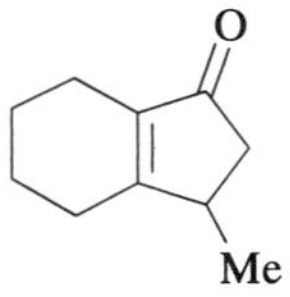 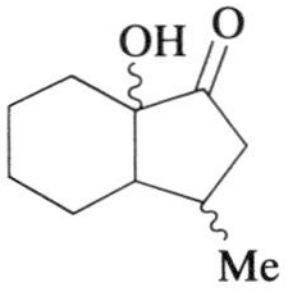 O, Me (—) + OH, O, Me (—)	38
	H_3PO_4, 60-65°, 6 h	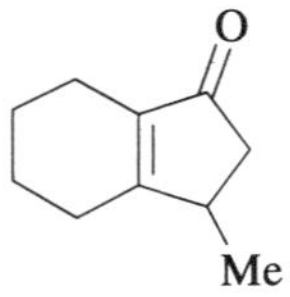 O, Me (41) 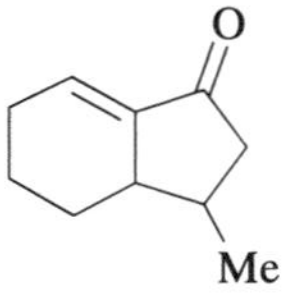 + O, Me (17)	38
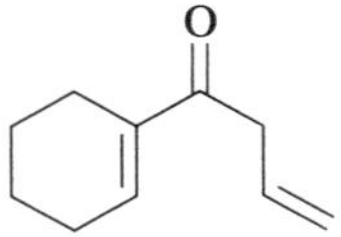 O	H_3PO_4, HCO_2H, 90°, 7 h	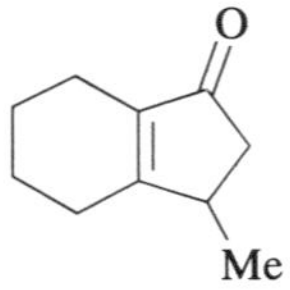 O, Me (70)	7

TABLE I. CYCLIZATION OF ALLYL VINYL AND DIVINYL KETONES: B. CYCLIC PRECURSORS (*Continued*)

Reactant	Conditions	Product(s) and Yield(s) (%)	Refs.
O	H_3PO_4, HCO_2H, 90°, 3 h	O (60)	77
C_{11} Me O Me Me	$SnCl_4$, CH_2Cl_2, reflux, 3 d	Me O Me Me (33) + Me O Me Me (19)	68, 67
O	PPA, 100°, 30 min	O H H (62)	71
Me Me O O	H_3PO_4, 65°, 4 h	Me Me O O Me (24)	21
Me O	H_3PO_4, 65°, 5.5 h	Me O Me (73)	104

TABLE I. CYCLIZATION OF ALLYL VINYL AND DIVINYL KETONES: B. CYCLIC PRECURSORS (*Continued*)

Reactant	Conditions	Product(s) and Yield(s) (%)	Refs.
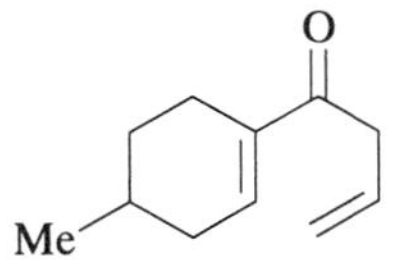 Me, O	H_3PO_4, 60-65°, 6 h 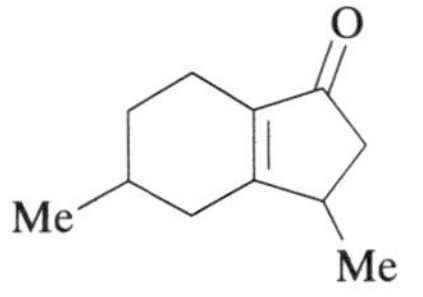	O, Me, Me (73)	13
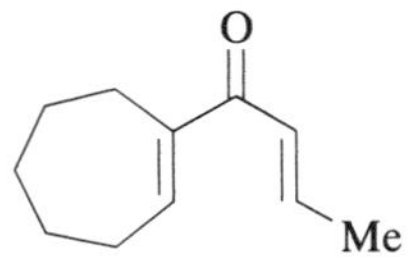 O, Me	H_3PO_4, HCO_2H, 90°, 6 h 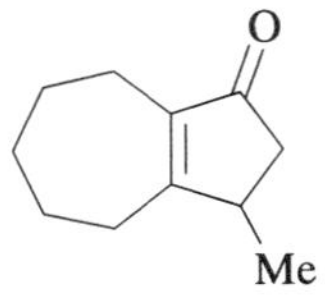	O, Me (75)	77
O, Me, Me, CO_2R, Et		O, Me, Me, CO_2R, Et	
R			
C_{12} Et	3 equiv $SnCl_4$, CH_2Cl_2, rt, 24 h	(30)	69
Et	3 equiv TMSOTf, CH_2Cl_2, rt, 2 h	(31)	69a
C_{20} (–)-menthyl	3 equiv TMSOTf, CH_2Cl_2, rt, 2 h	(39)	69a
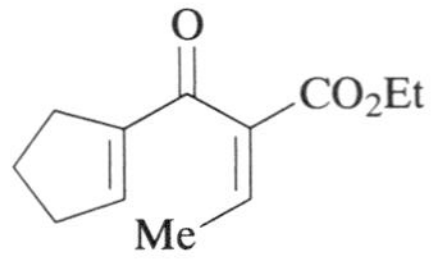 O, CO_2Et, Me	3 equiv $SnCl_4$, CH_2Cl_2, rt, 24 h 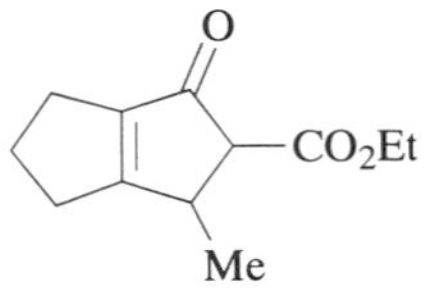	O, CO_2Et, Me (30)	69

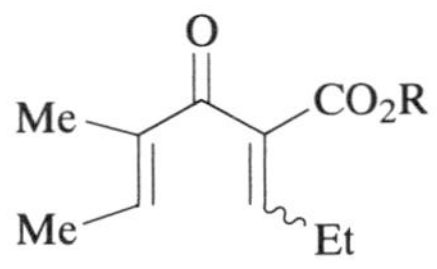

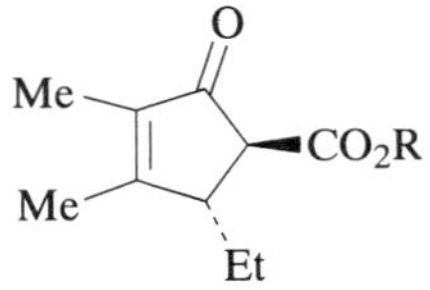

TABLE I. CYCLIZATION OF ALLYL VINYL AND DIVINYL KETONES: B. CYCLIC PRECURSORS (*Continued*)

Reactant	Conditions	Product(s) and Yield(s) (%)	Refs.
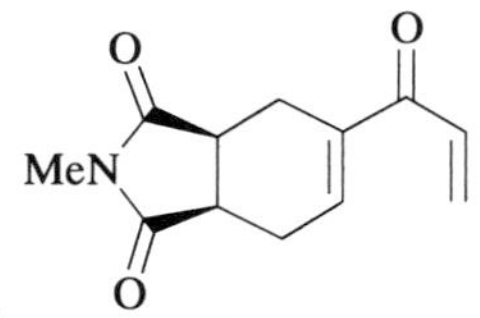 MeN, O, O, O	$GaCl_3$, CH_2Cl_2, rt, 6 h	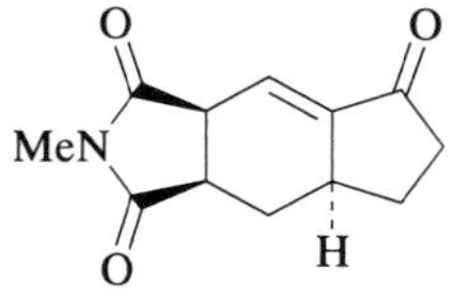 MeN, O, O, O, H (17)	70
C_{13} 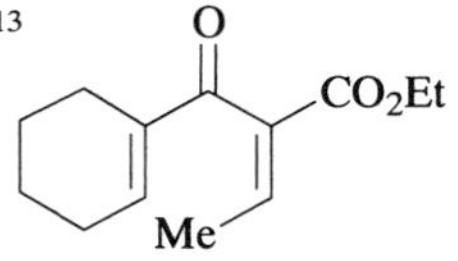 O, CO_2Et, Me	3 equiv TMSI, CCl_4, rt, 24 h	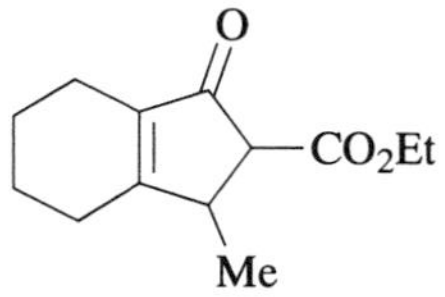 O, CO_2Et, Me (48)	69
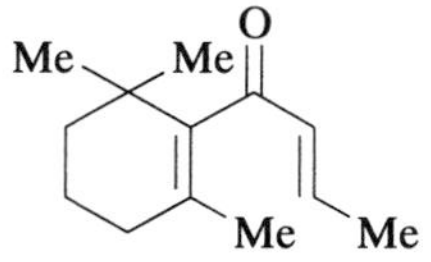 Me, Me, O, Me, Me	H_3PO_4, rt	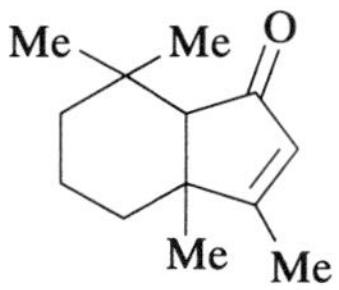 Me, Me, O, Me, Me (80)	181
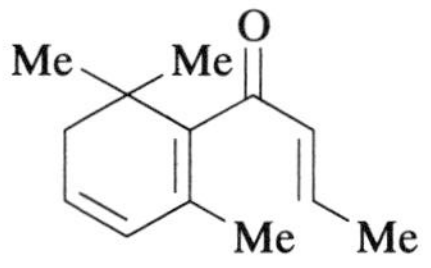 Me, Me, O, Me, Me	H_3PO_4, rt	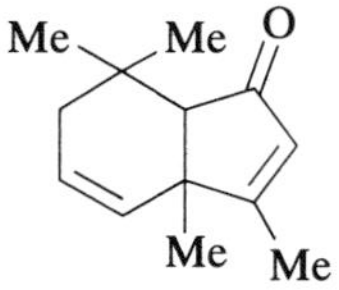 Me, Me, O, Me, Me (80)	181
	1% TsOH, 180°	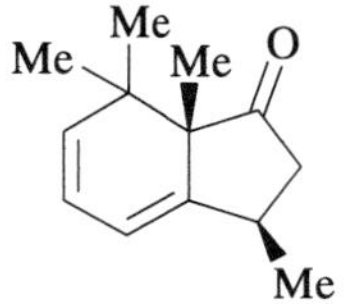 Me, Me, Me, O, Me (40) 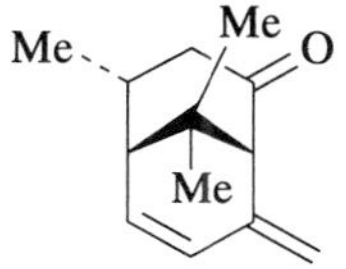 + Me, Me, O, Me (60)	181

TABLE I. CYCLIZATION OF ALLYL VINYL AND DIVINYL KETONES: B. CYCLIC PRECURSORS (*Continued*)

Reactant	Conditions	Product(s) and Yield(s) (%)	Refs.
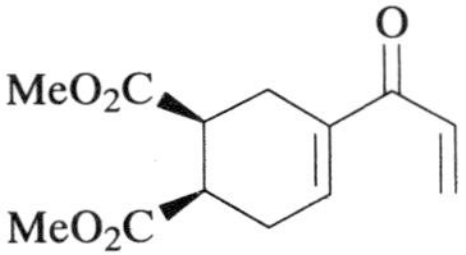 MeO_2C, MeO_2C	$GaCl_3$, CH_2Cl_2, -20 to 0°, 30 min	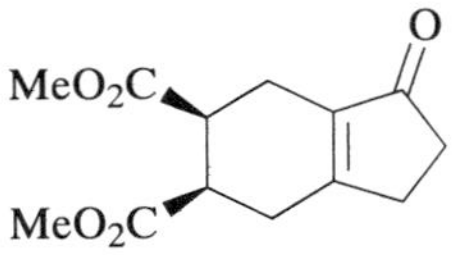 MeO_2C, MeO_2C (73)	70
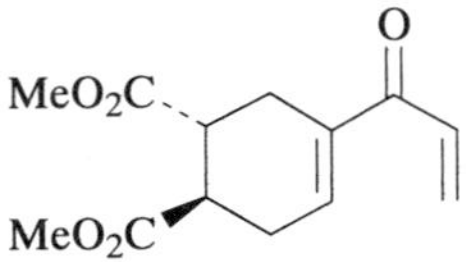 MeO_2C, MeO_2C	$GaCl_3$, CH_2Cl_2, rt, 3 h	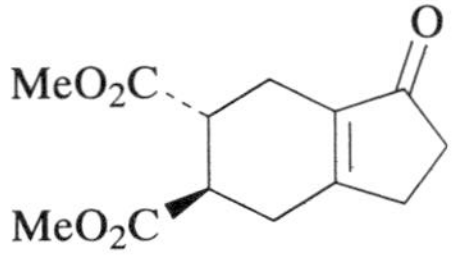 MeO_2C, MeO_2C (90)	70
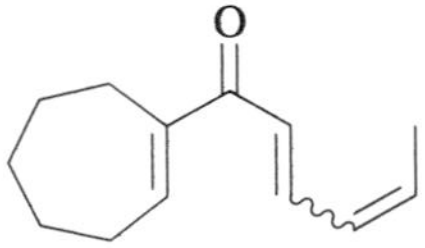	H_3PO_4, HCO_2H, 90°	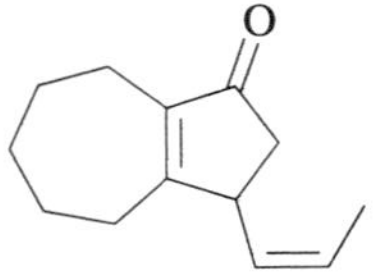 (24)	77
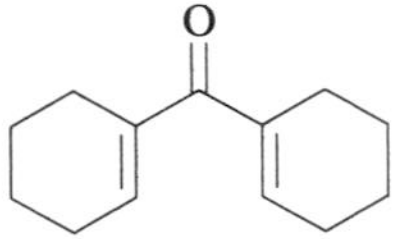	H_3PO_4, HCO_2H, 90°, 10 h	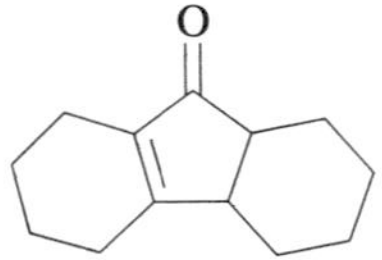 (29)	39
	HOAc, NaOAc, H_3PO_4, 68°, 3.5 h	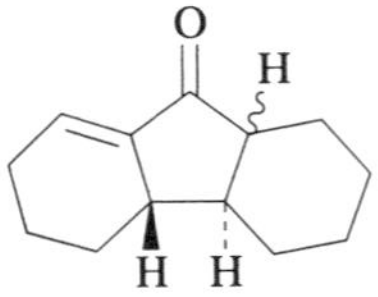 **I** α:β 11:89 (48) + 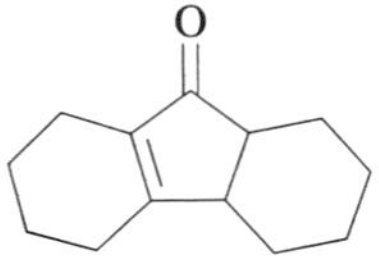 **II** (38)	73

TABLE I. CYCLIZATION OF ALLYL VINYL AND DIVINYL KETONES: B. CYCLIC PRECURSORS (*Continued*)

Reactant	Conditions	Product(s) and Yield(s) (%)	Refs.
	C_6H_6, 210°, flow system	**I** α:β 41:59 (33) + **II** (3)	73
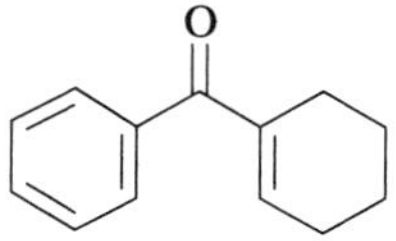 O	PPA, 100°, 15 min	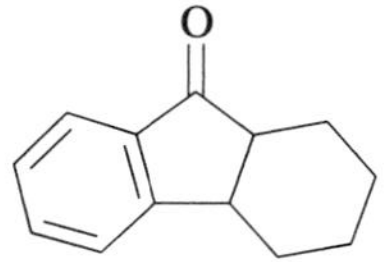 O (65)	76
C_{14}			
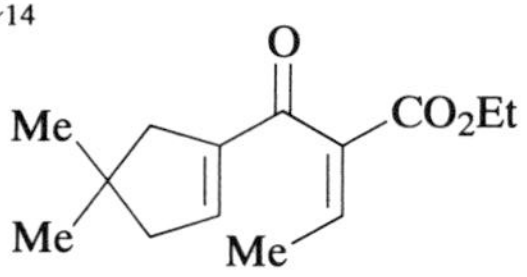 O, Me, Me, Me, CO_2Et	3 equiv $SnCl_4$, CH_2Cl_2, rt, 20 h	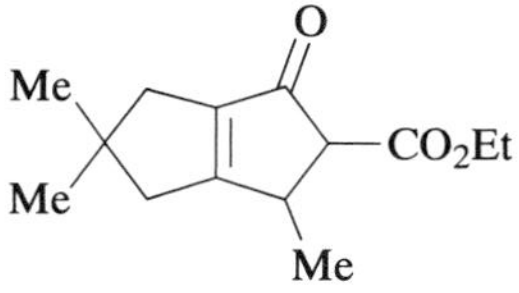 O, Me, Me, Me, CO_2Et (30)	41
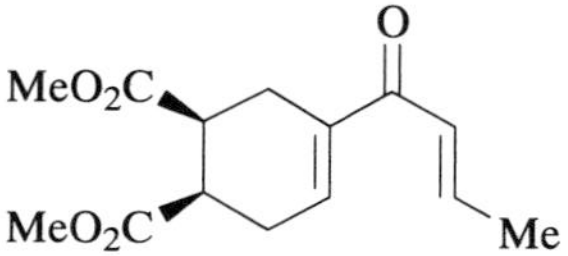 O, MeO_2C, MeO_2C, Me	$GaCl_3$, CH_2Cl_2, rt, 24 h	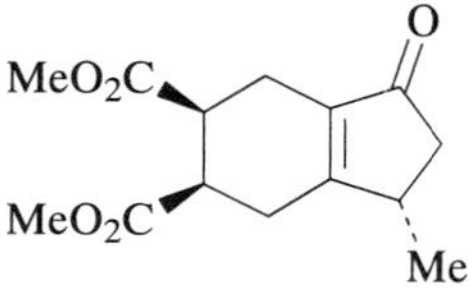 O, MeO_2C, MeO_2C, Me (40) 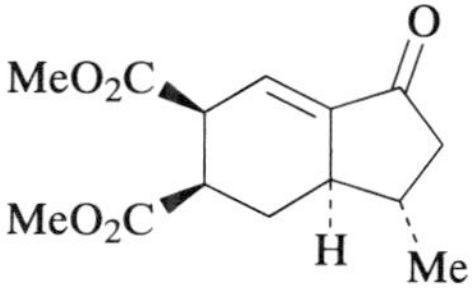 + O, MeO_2C, MeO_2C, H, Me (49)	70
	TsOH, C_6H_6, reflux, 6.5 h	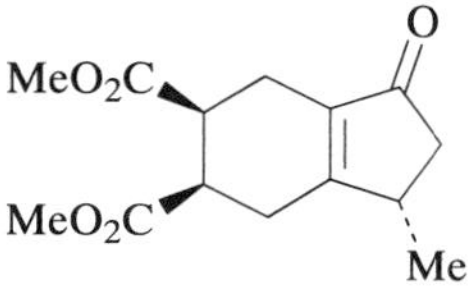 O, MeO_2C, MeO_2C, Me (29) 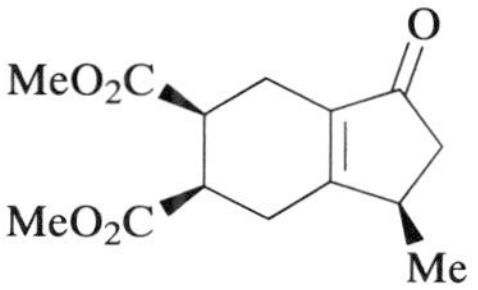 + O, MeO_2C, MeO_2C, Me (29)	70

TABLE I. CYCLIZATION OF ALLYL VINYL AND DIVINYL KETONES: B. CYCLIC PRECURSORS (*Continued*)

Reactant	Conditions	Product(s) and Yield(s) (%)	Refs.
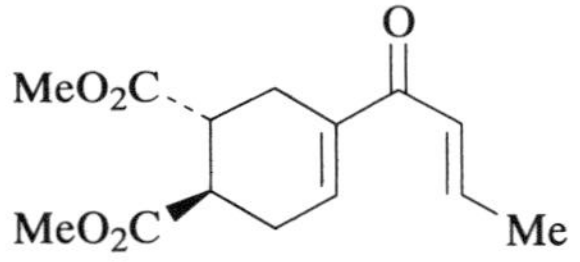 MeO_2C, MeO_2C, O, Me	$GaCl_3$, CH_2Cl_2, rt, 24 h	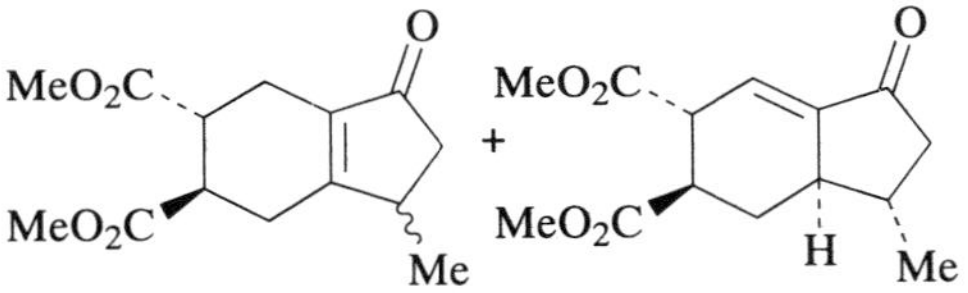 MeO_2C, MeO_2C, O, Me; α:β 50:50 (62) + MeO_2C, MeO_2C, O, H, Me (26)	70
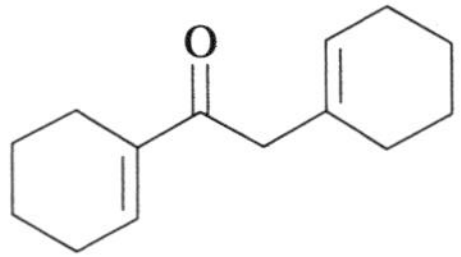 O	H_3PO_4, 60°, 6 h	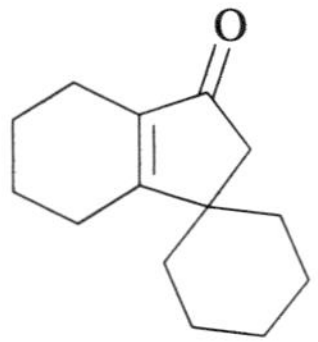 O (76)	26
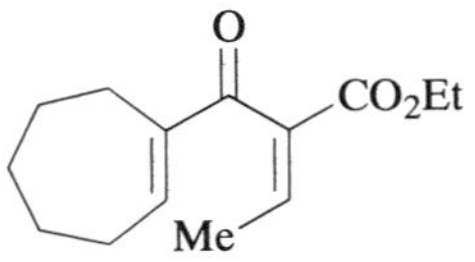 O, CO_2Et, Me	2 equiv TMSI, CCl_4, rt, 24 h	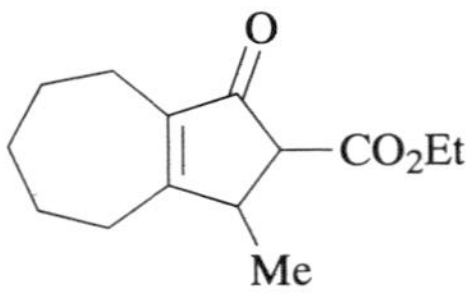 O, CO_2Et, Me (35)	69
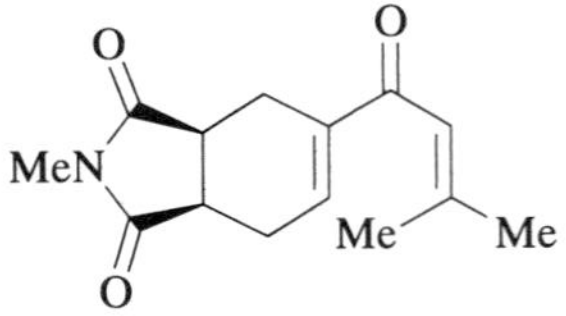 O, MeN, O, O, Me, Me	$GaCl_3$, CH_2Cl_2, rt, 6 days	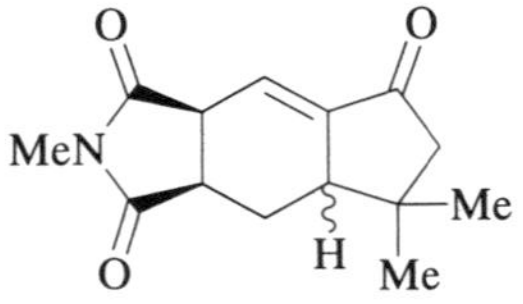 O, MeN, O, O, H, Me, Me (90) α:β 50:50	70

TABLE I. CYCLIZATION OF ALLYL VINYL AND DIVINYL KETONES: B. CYCLIC PRECURSORS (*Continued*)

Reactant	Conditions	Product(s) and Yield(s) (%)		Refs.
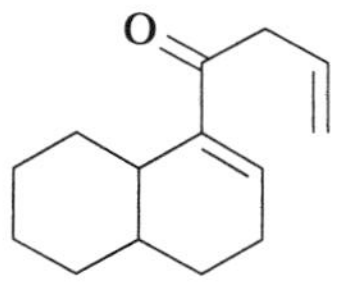	H_3PO_4, 65°, 6.5 h	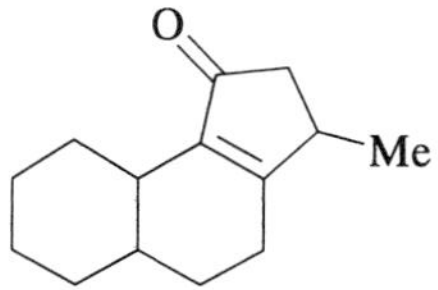 Me	(30)	22
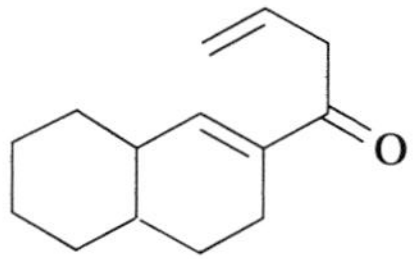	H_3PO_4, HCO_2H, 60-65°, 6.5 h	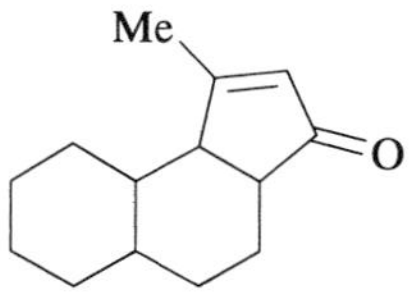 Me	(73)	11
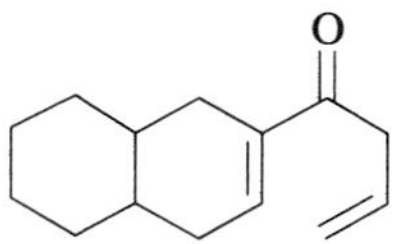	H_3PO_4, HCO_2H, 60-65°, 6h	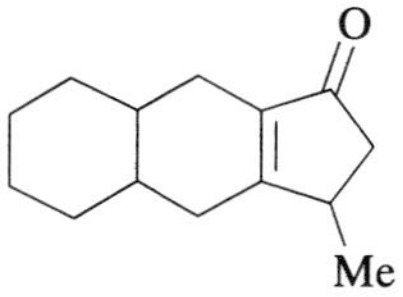 Me	(—)	11
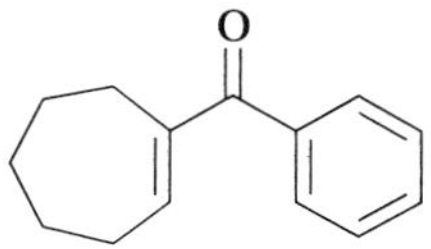	H_3PO_4, HCO_2H, 90°, 6 h	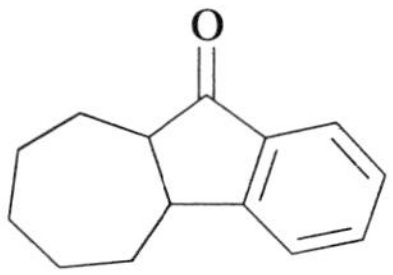	(75)	77

TABLE I. CYCLIZATION OF ALLYL VINYL AND DIVINYL KETONES: B. CYCLIC PRECURSORS (*Continued*)

Reactant	Conditions	Product(s) and Yield(s) (%)	Refs.
	PPA, 100°	(48) + (12)	66
C_{15}			
	$GaCl_3$, CH_2Cl_2, rt, 5 days	(68) + (29) R = CO_2Me	70
	TsOH, toluene, reflux, 24 h	(57)	70
	$BF_3{\bullet}OEt_2$, $CHCl_3$, reflux 5 days	(80) α:β 70:30	75

TABLE I. CYCLIZATION OF ALLYL VINYL AND DIVINYL KETONES: B. CYCLIC PRECURSORS (*Continued*)

Reactant	Conditions	Product(s) and Yield(s) (%)	Refs.
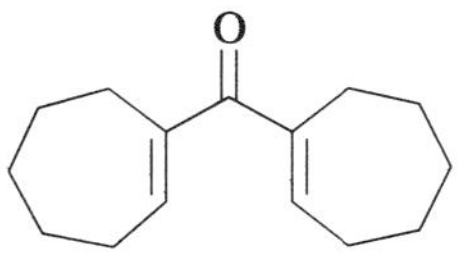 O	H_3PO_4, HCO_2H, 0-90°, 6 h	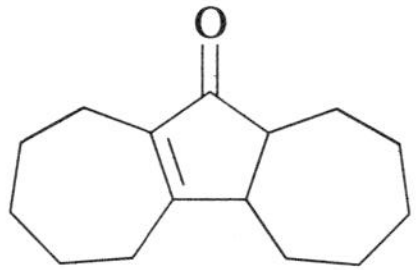 O (50)	72
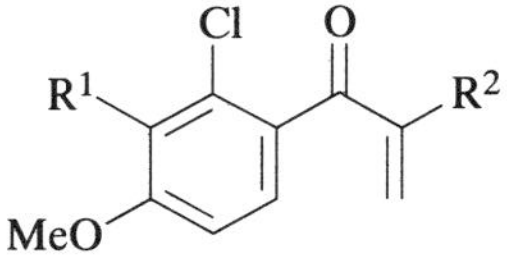 Cl O R[1] R[2] MeO	H_2SO_4, CH_2Cl_2, 5°, 3.5 h	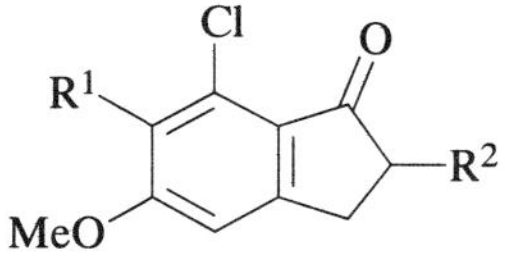 Cl O R[1] R[2] MeO	78
R^1 / R^2			
C_{16} Cl / Ph		(75)	
Cl / p-FC_6H_4		(100)	
Cl / p-ClC_6H_4		(89)	
Cl / p-BrC_6H_4		(94)	
C_{17} Me / Ph		(100)	
C_{16} 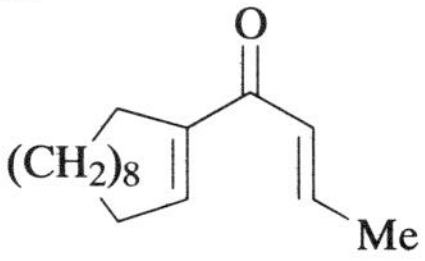 O $(CH_2)_8$ Me	BF_3•OEt_2, C_6H_6, heat	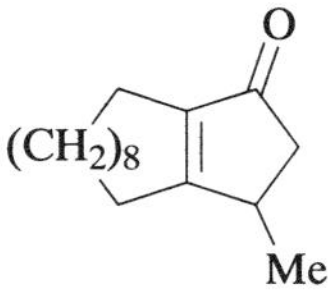 O $(CH_2)_8$ Me (48) 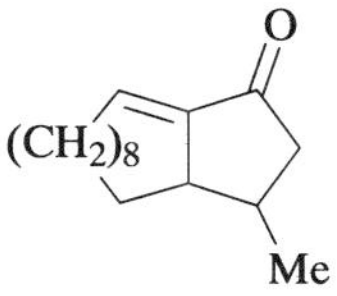 + O $(CH_2)_8$ Me (18)	67

TABLE I. CYCLIZATION OF ALLYL VINYL AND DIVINYL KETONES: B. CYCLIC PRECURSORS (*Continued*)

Reactant	Conditions	Product(s) and Yield(s) (%)	Refs.
C_{18}			
MeN, O, O, O, Ph	$GaCl_3$, CH_2Cl_2, rt, 2 days	MeN, O, O, O, H, Ph (15) + MeN, O, O, O, H, Ph (31)	70
H, H, Me, O, Pr-*i*	TsOH, toluene, 110°	H, H, Me, O, H, Pr-*i* (20)	66
C_{19}			
MeO_2C, MeO_2C, O, Ph	$GaCl_3$, CH_2Cl_2, rt, 5 h	MeO_2C, MeO_2C, O, Ph (81) + MeO_2C, MeO_2C, O, H, Ph (5)	70
	TsOH, toluene, reflux, 12 h	R, R, O, Ph α:β 50:50 (63) + R, R, O, H, Ph (33) R = CO_2Me	70

TABLE I. CYCLIZATION OF ALLYL VINYL AND DIVINYL KETONES: B. CYCLIC PRECURSORS (*Continued*)

Reactant	Conditions	Product(s) and Yield(s) (%)	Refs.
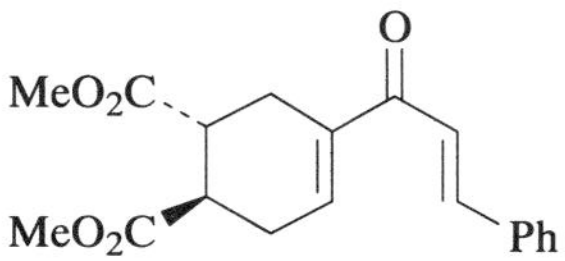 MeO_2C, MeO_2C (structure), Ph	$GaCl_3$, CH_2Cl_2, rt, 18 h	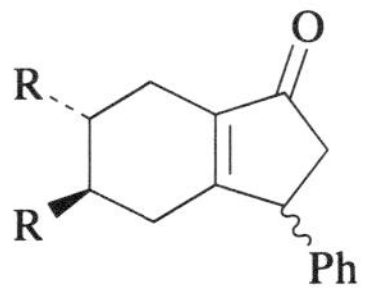 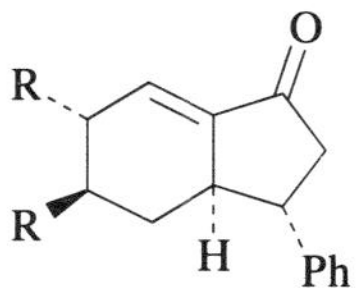 R, R, O, Ph (80) + R, R, O, H, Ph (4); R = CO_2Me	70
C_{22}			
N, SO_2Ph, O, Me (structure)	$AlCl_3$, benzene, rt, 20 h	H, Me, H, O, N, SO_2Ph (49) + H, Me, H, O, N, SO_2Ph (24)	173a

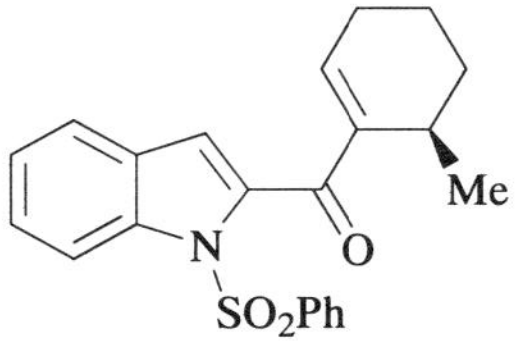

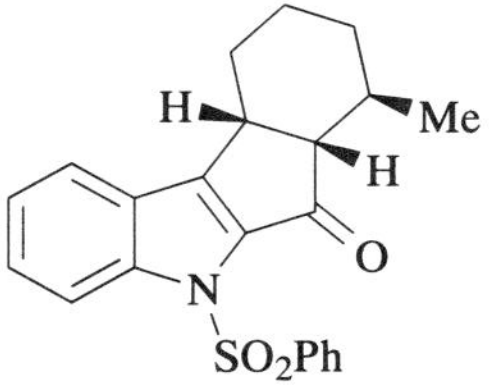

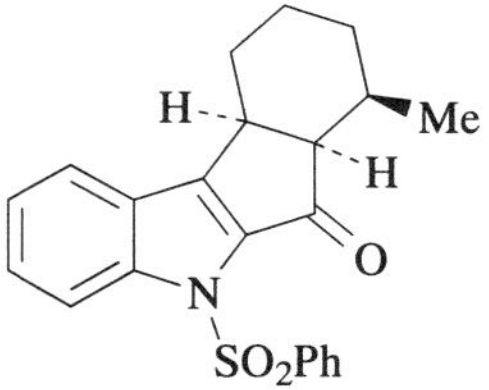

TABLE I. CYCLIZATION OF ALLYL VINYL AND DIVINYL KETONES: C. ANOMALOUS CYCLIZATIONS

Reactant	Conditions	Product(s) and Yield(s) (%)	Refs.
C_8			
O; Me, Me, Me	H_2SO_4, 60°, 6 h	O; Me, Me, Me (46)	53
O; CD_3, CD_3, Me	H_2SO_4, 60°, 6 h	O; Me, CD_3, CD_3 (14) + O; CD_3, CD_3, Me (6)	53
O; Me, Me, CD_3	H_2SO_4, 60°, 6 h	O; D, Me, Me, CD_3 (50) + O; D, Me, Me, CD_3 (5)	53
C_9			
O; Me, Me, Et	H_2SO_4, 60°, 6 h	O; Et, Me, Me (39)	53
O; Et, Et	H_3PO_4, HCO_2H, 90°, 2-3 h	O; Et, Et (77)	80, 79

TABLE I. CYCLIZATION OF ALLYL VINYL AND DIVINYL KETONES: C. ANOMALOUS CYCLIZATIONS (*Continued*)

Reactant	Conditions	Product(s) and Yield(s) (%)	Refs.
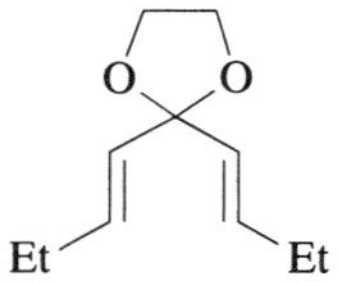 Et, Et (dioxolane)	H_3PO_4, HCO_2H, 90°, 2-3 h	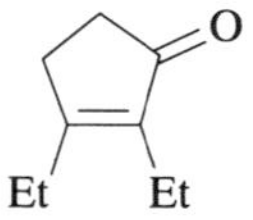 Et, Et (67)	79
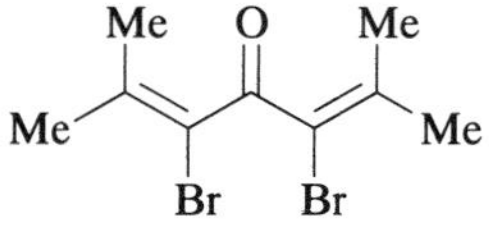 Me, Me, O, Me, Me, Br, Br	H_2SO_4, 20°, 10 min	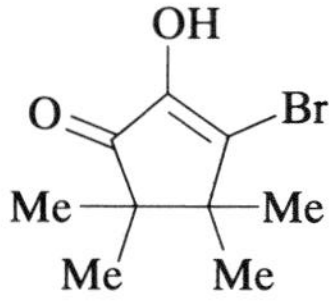 OH, O, Br, Me, Me, Me, Me (75)	43
C$_{10}$			
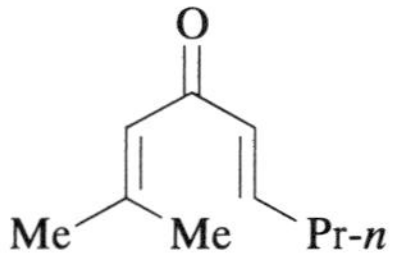 O, Me, Me, Pr-*n*	H_2SO_4, 60°, 6 h	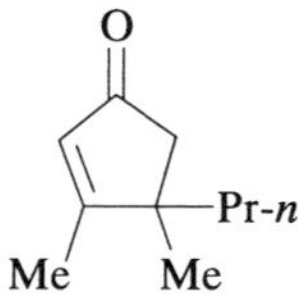 O, Pr-*n*, Me, Me (55)	53
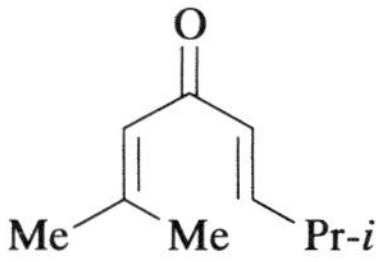 O, Me, Me, Pr-*i*	H_2SO_4, 60°, 6 h	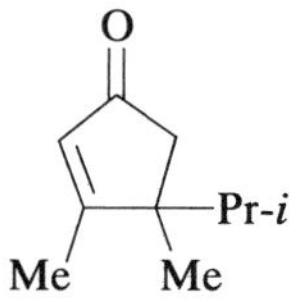 O, Pr-*i*, Me, Me (31)	53
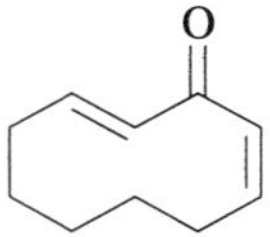 O	H_3PO_4, HCO_2H, 90°, 2-3 h	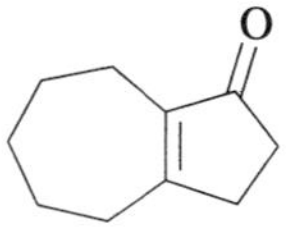 O (58) 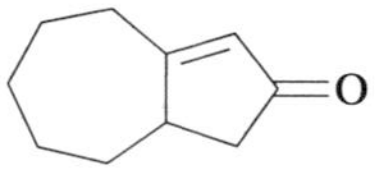 + O (6)	80, 79

TABLE I. CYCLIZATION OF ALLYL VINYL AND DIVINYL KETONES: C. ANOMALOUS CYCLIZATIONS (*Continued*)

Reactant	Conditions	Product(s) and Yield(s) (%)	Refs.
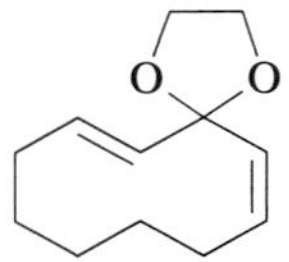	H_3PO_4, HCO_2H, 90°, 2-3 h	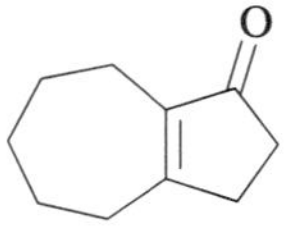 (47) + 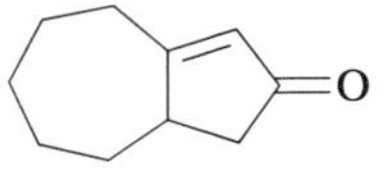 (37)	80, 79
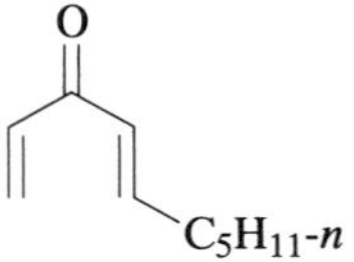 C_5H_{11}-*n*	H_3PO_4, HCO_2H, 90°, 2-3 h	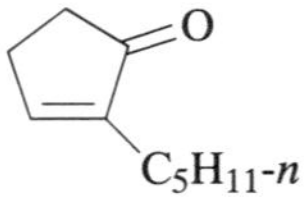 C_5H_{11}-*n* (18) 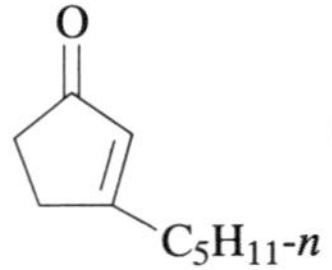 + C_5H_{11}-*n* (2)	80, 79
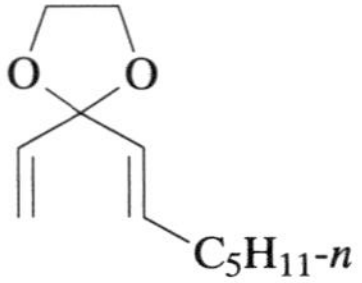 C_5H_{11}-*n*	H_3PO_4, HCO_2H, 90°, 2-3 h	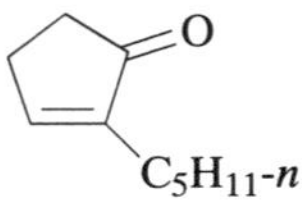 C_5H_{11}-*n* (14) 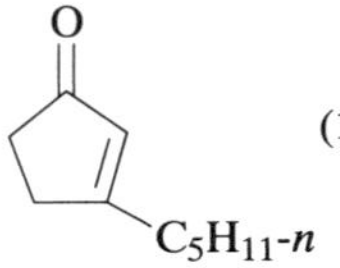 + C_5H_{11}-*n* (1)	80, 79
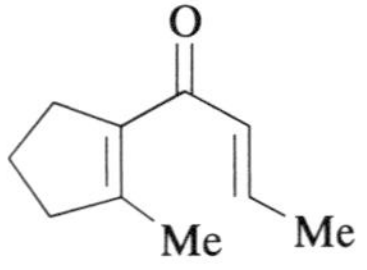 Me Me	$AlCl_3$, CH_2Cl_2, reflux, 4 h	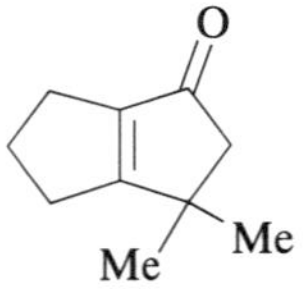 Me Me (40)	157a
C_{11}			
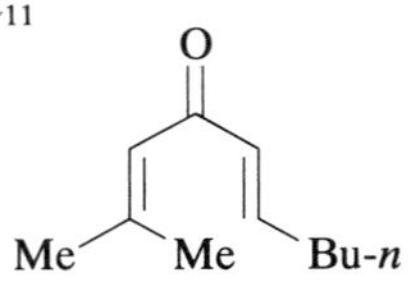 Me Me Bu-*n*	H_2SO_4, 60°, 6 h	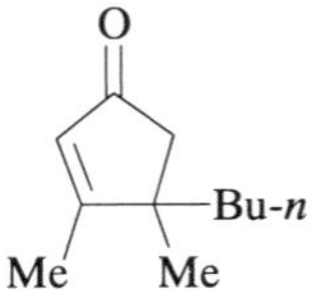 Bu-*n* Me Me (47)	53

TABLE I. CYCLIZATION OF ALLYL VINYL AND DIVINYL KETONES: C. ANOMALOUS CYCLIZATIONS (*Continued*)

Reactant	Conditions	Product(s) and Yield(s) (%)		Refs.
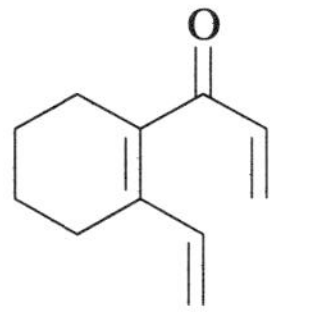	$FeCl_3$, CH_2Cl_2, -15°, 30 min		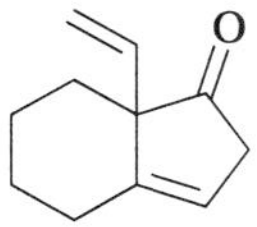 (19)	81
C_{12} 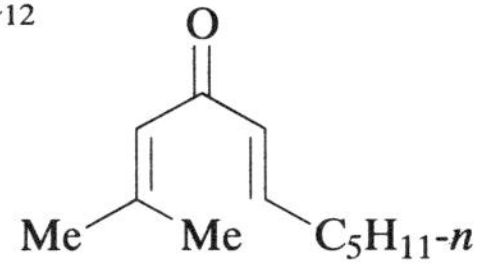 Me, Me, C_5H_{11}-*n*	H_2SO_4, 60°, 6 h	Me, Me, C_5H_{11}-*n*	(47)	53
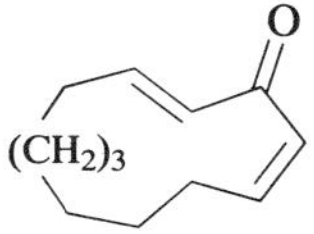 $(CH_2)_3$	H_3PO_4, HCO_2H, 90°, 2-3 h	$(CH_2)_3$	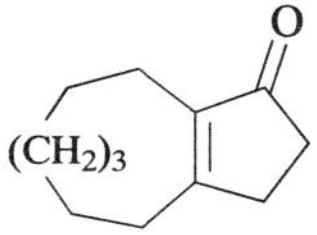 (67)	79, 80
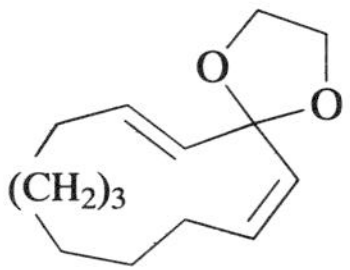 $(CH_2)_3$	H_3PO_4, HCO_2H, 90°, 2-3 h	$(CH_2)_3$	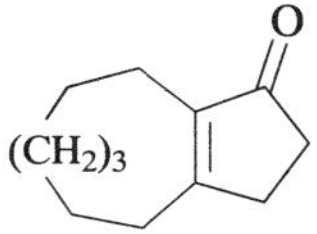 (63)	80, 79
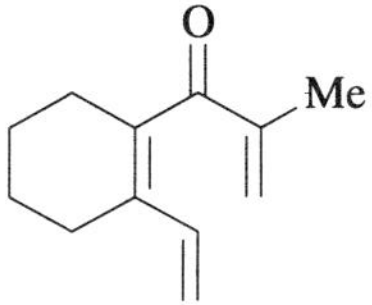 Me	$FeCl_3$, CH_2Cl_2, -15°, 30 min	Me	(16)	81

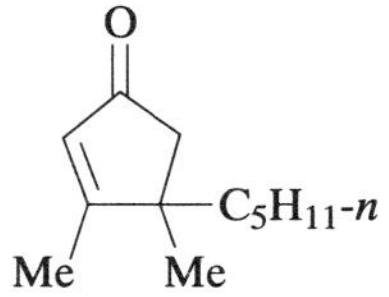

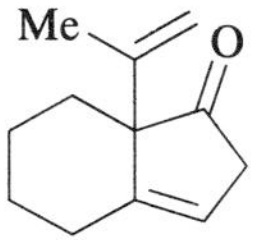

TABLE I. CYCLIZATION OF ALLYL VINYL AND DIVINYL KETONES: C. ANOMALOUS CYCLIZATIONS (*Continued*)

Reactant	Conditions	Product(s) and Yi eld(s) (%)	Refs.
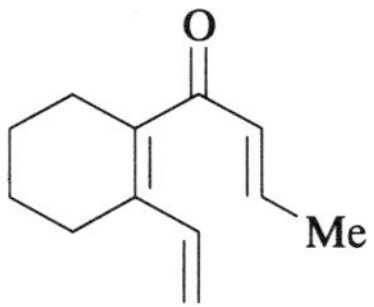 O, Me	$FeCl_3$, CH_2Cl_2, 0°, 30 min	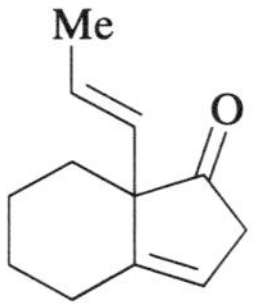 Me, O (61)	81
C_{14} 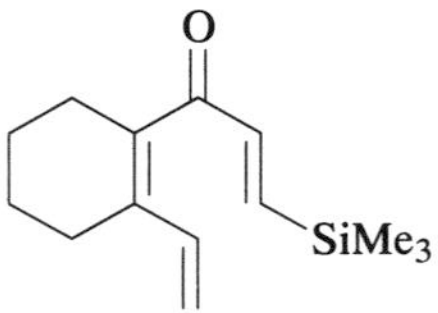 O, $SiMe_3$	$FeCl_3$, CH_2Cl_2, -10°, 30 min	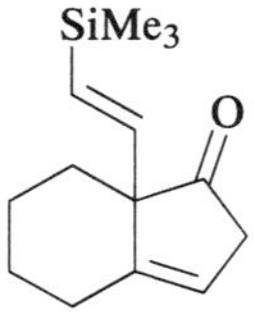 $SiMe_3$, O (59)	81
C_{17} 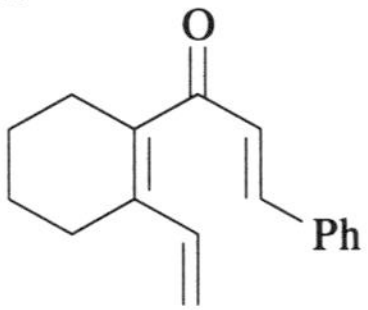 O, Ph	$FeCl_3$, CH_2Cl_2, 0°, 30 min	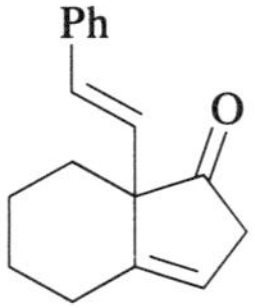 Ph, O (65)	81
C_{20} 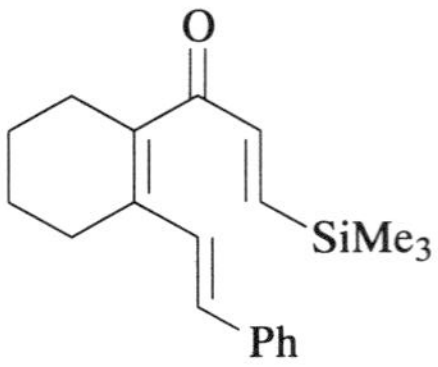 O, $SiMe_3$, Ph	$FeCl_3$, CH_2Cl_2, 0°, 5 min	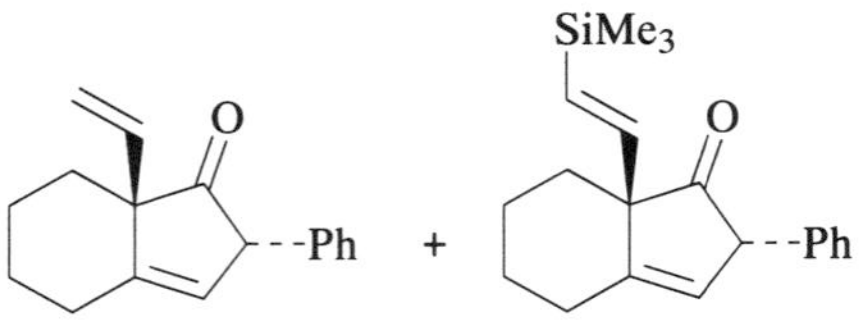 O, - - -Ph (44) + $SiMe_3$, O, - - -Ph (34)	81

TABLE I. CYCLIZATION OF ALLYL VINYL AND DIVINYL KETONES: D. PHOTOCHEMICAL CYCLIZATIONS

Reactant	Conditions	Product(s) and Yield(s) (%)	Refs.
C_7 3,5-dimethyl-4H-pyran-4-one (Me, Me)	$h\nu$	2,4-dimethylcyclopent-4-ene-1,3-dione (Me, Me), minor + 3,6-dimethyl-2H-pyran-2-one (Me, Me), major (—)	182
	$h\nu$, ROH	5-OR-4-OH-2,5-dimethylcyclopent-2-enone (Me, Me, OR, OH)	182
		R = Me (—)	
		R = CH_2CF_3 (—)	
4H-pyran-4-one (R^1, R^2, R^3, R^4)	$h\nu$, MeOH	cyclopentenone (R^3, MeO, HO, R^1, R^2, R^4)	83
R^1 = H, R^2 = Me, R^3 = OMe, R^4 = H		(—)	
R^1 = Me, R^2 = H, R^3 = OMe, R^4 = H		(—)	
C_8 R^1 = Me, R^2 = Me, R^3 = Me, R^4 = H		(—)	
C_9 R^1 = Me, R^2 = Me, R^3 = Me, R^4 = Me		(—)	

TABLE I. CYCLIZATION OF ALLYL VINYL AND DIVINYL KETONES: D. PHOTOCHEMICAL CYCLIZATIONS (*Continued*)

Reactant	Conditions	Product(s) and Yield(s) (%)	Refs.
C_7 O		RO, H, H, O + RO, H, H, O	60
	$h\nu$, HOAc, 20 h	R = OAc (30) (30)	
	$h\nu$, t-BuOH, H_2SO_4	R = OBu-t (30) (30)	
O, R	1. $h\nu$, FSO_3H, CCl_4 2. $NaHCO_3$	O, Me, R	183
		R = H (—)	
		R = Me (—)	
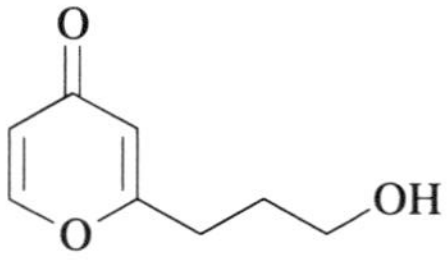 C_8 O, O, OH	$h\nu$, CF_3CH_2OH	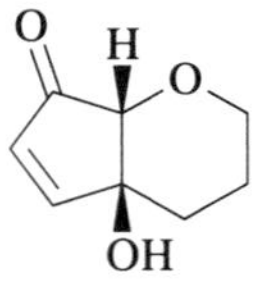 O, H, O, OH (43)	83a
C_9 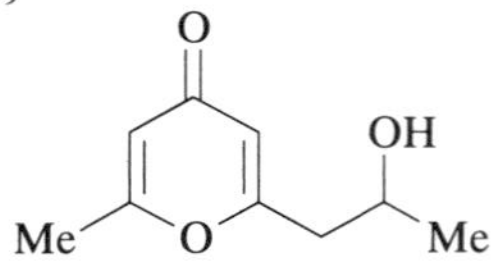 O, OH, Me, O, Me	$h\nu$, CF_3CH_2OH	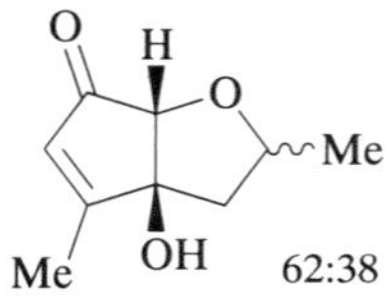 O, H, O, Me, Me, OH 62:38 (75)	83a

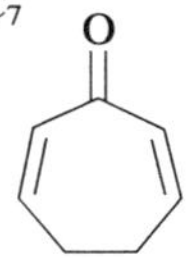

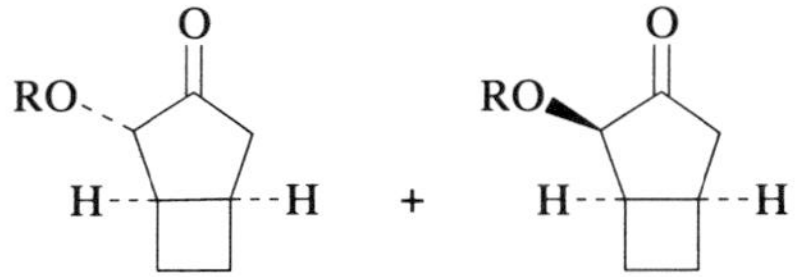

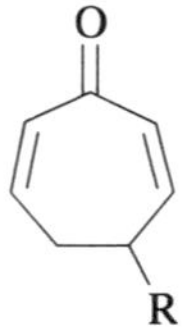

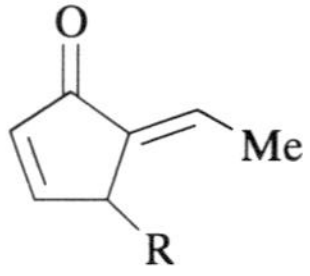

TABLE I. CYCLIZATION OF ALLYL VINYL AND DIVINYL KETONES: D. PHOTOCHEMICAL CYCLIZATIONS (*Continued*)

Reactant	Conditions	Product(s) and Yield(s) (%)	Refs.
R, R (cyclononadienone)	$h\nu$, H_2SO_4	+ R R = H (71) (29) R = D (71) (29)	61
C_{10} Me, Me, Me, O, OH		Me, O, Me, O, Me, OH + O, Me, Me, OMe, Me, OH, OH	
	$h\nu$, CH_3OH	(60) (14-16)	83a
	$h\nu$, CF_3CH_2OH	(75) (0)	83a
C_{11} 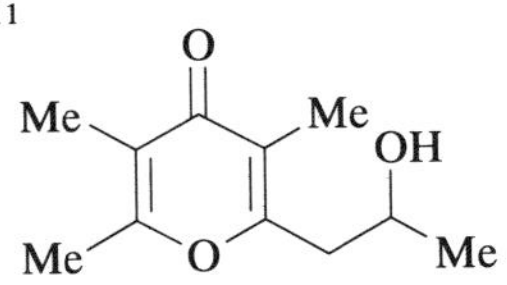 Me, Me, OH, Me, O, Me 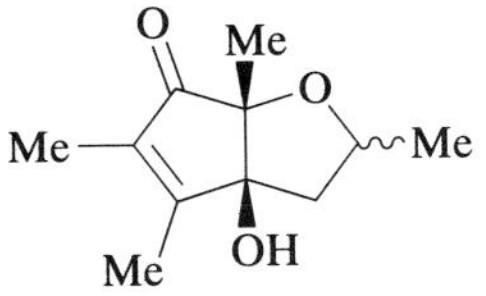 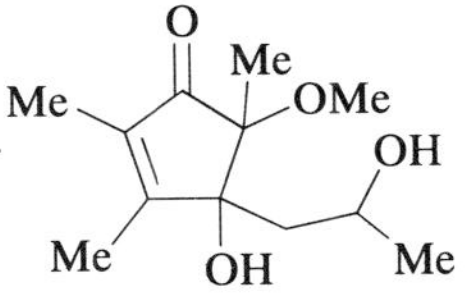		Me, O, Me, O, Me, OH, Me + O, Me, Me, OMe, OH, Me, OH, Me	
	$h\nu$, CH_3OH	(64) 66:34 (14-16)	83a
	$h\nu$, CF_3CH_2OH	(84) 66:34 (0)	83a
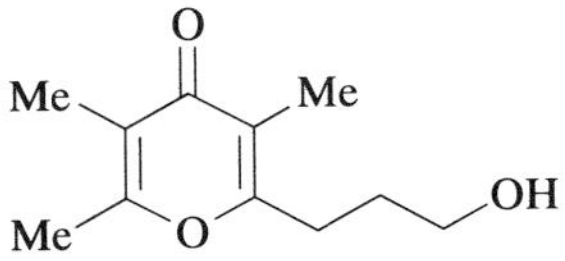 Me, Me, Me, O, OH	$h\nu$, CF_3CH_2OH 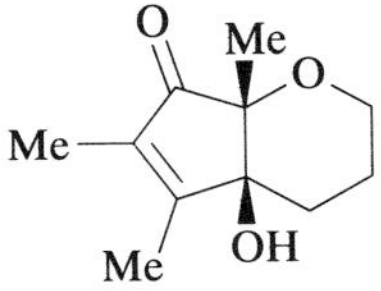	O, Me, O, Me, Me, OH (99)	83a

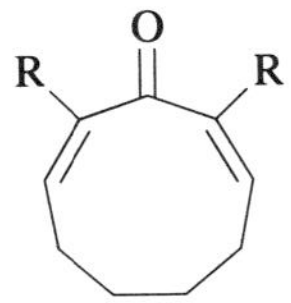

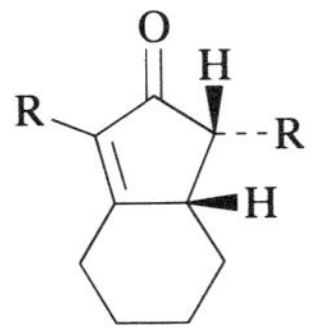

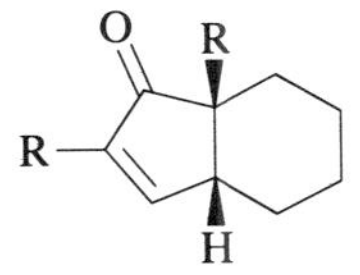

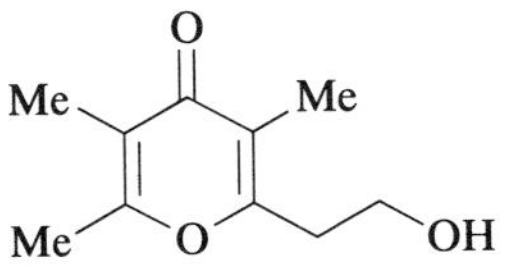

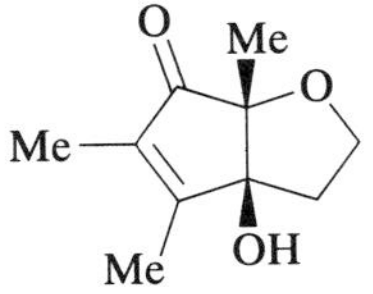

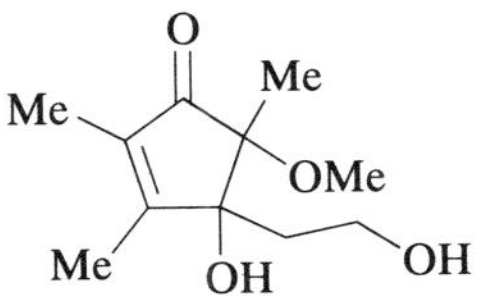

TABLE I. CYCLIZATION OF ALLYL VINYL AND DIVINYL KETONES: D. PHOTOCHEMICAL CYCLIZATIONS (*Continued*)

Reactant	Conditions	Product(s) and Yield(s) (%)	Refs.
C_{12}			
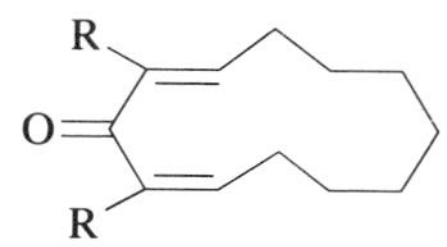 R = H	$h\nu$, H_2SO_4	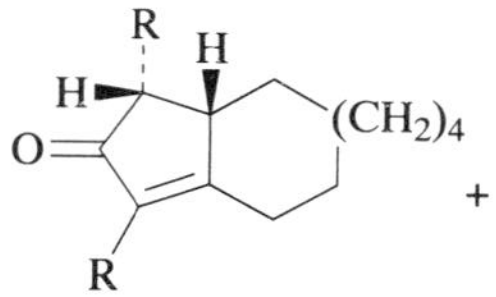 (70) + 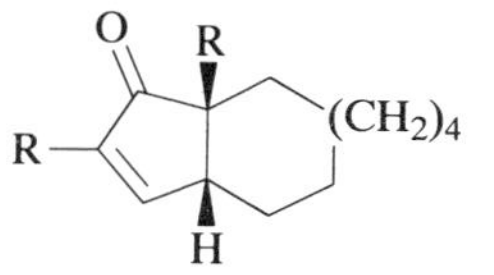 (30)	61
R = D		(70) + (30)	
C_{13}			
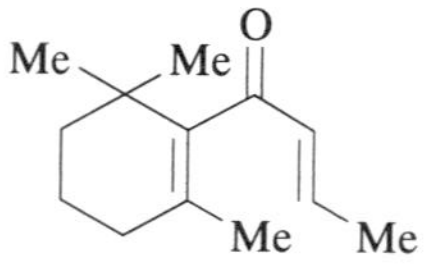	$h\nu$, *i*-PrOH	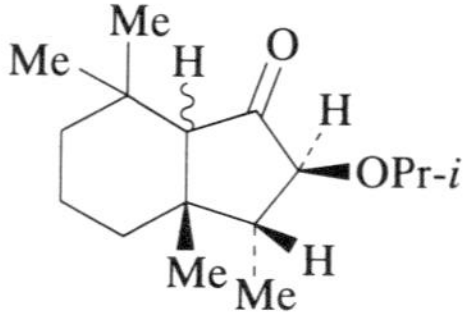 OPr-*i* (—)	62
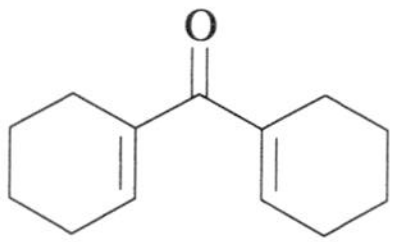	$h\nu$, C_6H_6	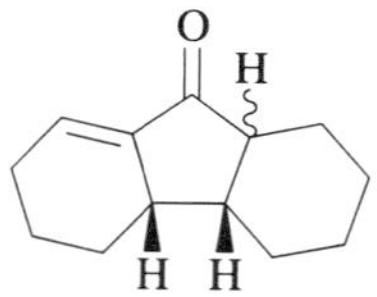 (56) + 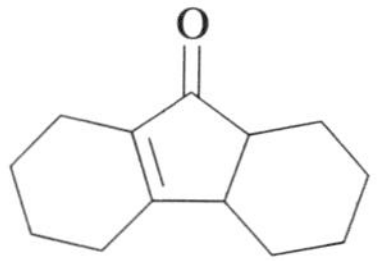 (7)	73
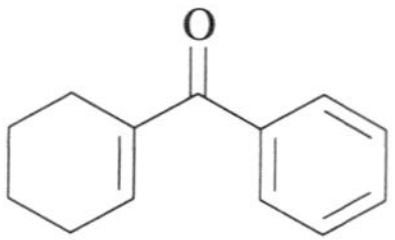	$h\nu$, C_6H_6	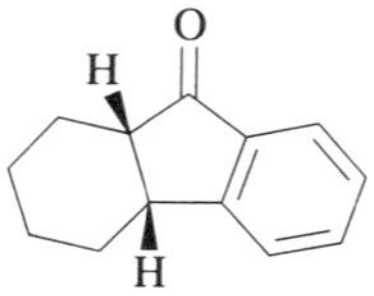 (65)	82a

TABLE I. CYCLIZATION OF ALLYL VINYL AND DIVINYL KETONES: D. PHOTOCHEMICAL CYCLIZATIONS (*Continued*)

Reactant	Conditions	Product(s) and Yield(s) (%)	Refs.
C_{14}			
Me, Me, O, Me, O, Pr-*i*, OH	$h\nu$, CF_3CH_2OH	O, Me, O, Pr-*i*, Me, Me, OH (61) 60:40	83a
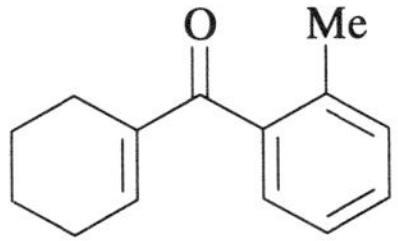 O, Me	$h\nu$, C_6H_6	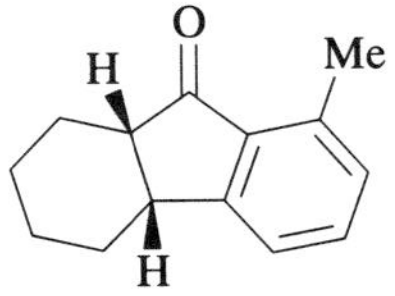 H, O, Me, H (57)	82a
C_{15}			
HO_2C, Me, O, OMe	$h\nu$, C_6H_6	HO_2C, Me, H, O, H, OMe (50) + Me, HO_2C, H, O, H, OMe (50)	82b
C_{16} 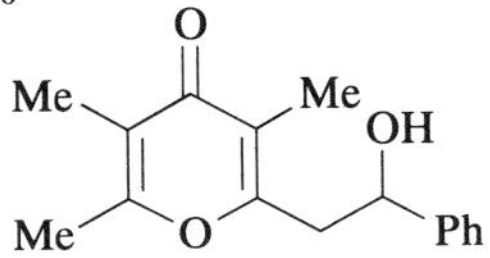		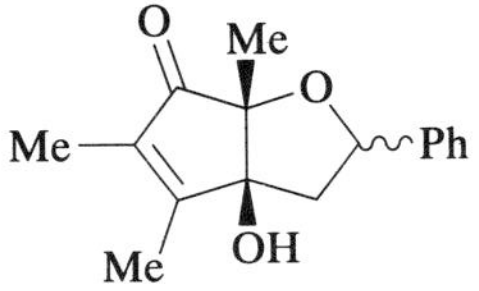 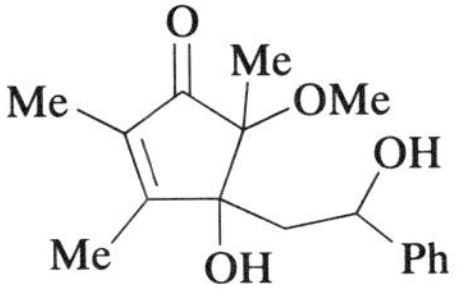	
Me, Me, O, Me, OH, O, Ph		O, Me, O, Me, Ph, Me, OH + Me, O, Me, OMe, OH, Me, OH, Ph	
	$h\nu$, CH_3OH	(76) 50:50 + (14-16)	83a
	$h\nu$, CF_3CH_2OH	(92) 50:50 + (0)	83a

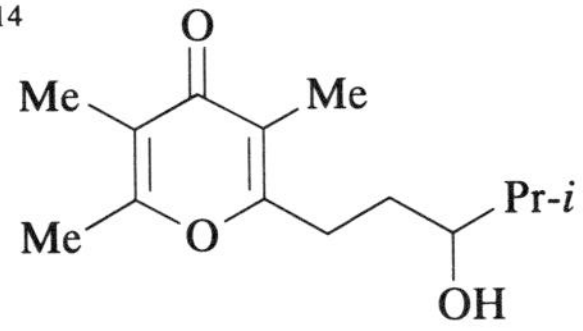

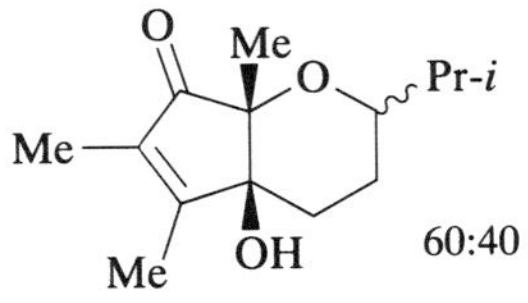

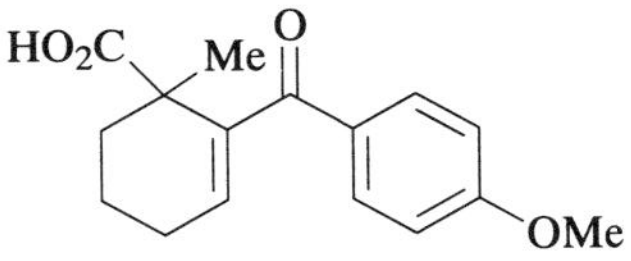

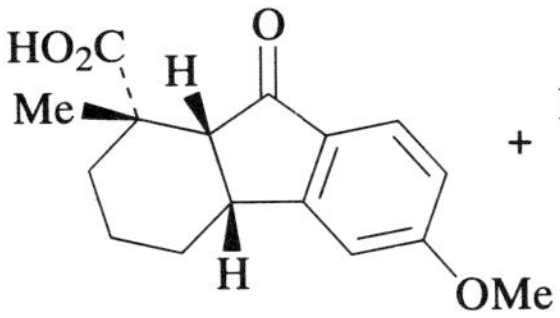

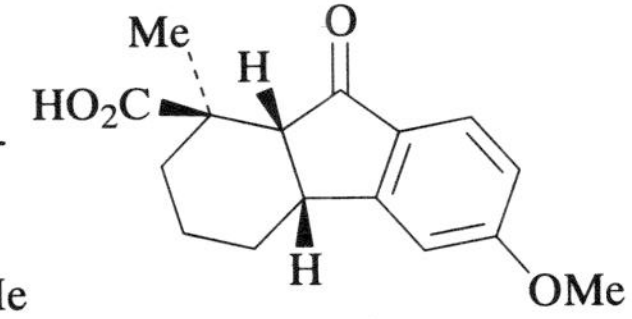

TABLE II. SILICON(TIN)-DIRECTED NAZAROV CYCLIZATIONS: A. ACYCLIC PRECURSORS

Reactant	Conditions	Product(s) and Yield(s) (%)	Refs.
C_6			
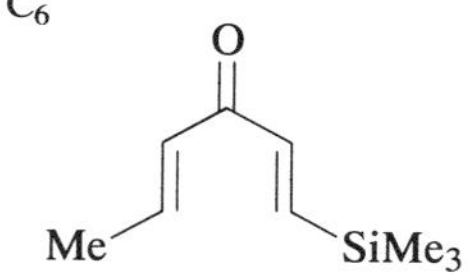 Me, SiMe3	$FeCl_3$, CH_2Cl_2, 20°, 12 h	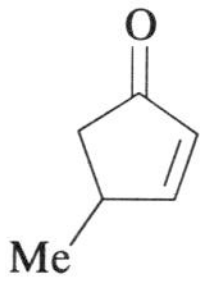 Me (54)	85
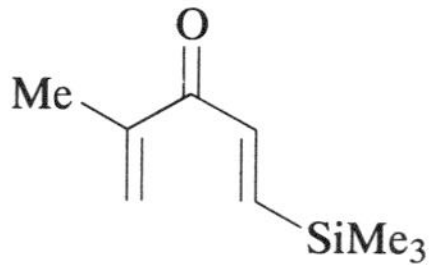 Me, SiMe3	$FeCl_3$, CH_2Cl_2, -20°, 6 h	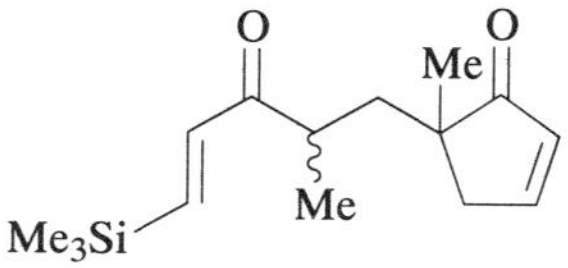 Me3Si, O, Me, Me, O (42)	85
C_7			
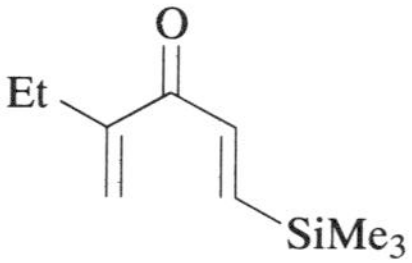 Et, SiMe3	$FeCl_3$, CH_2Cl_2, 0°	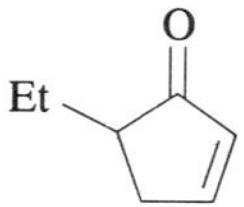 Et, O	85
	Concentration of divinyl ketone (M)		
	0.08 2 h	(—)	
	0.02 8 h	(33)	
	0.004 48 h	(51)	
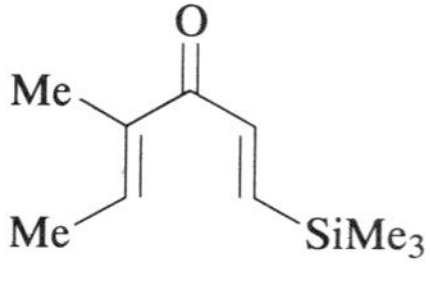 Me, Me, SiMe3	$FeCl_3$, CH_2Cl_2, -10°, 0.5 h	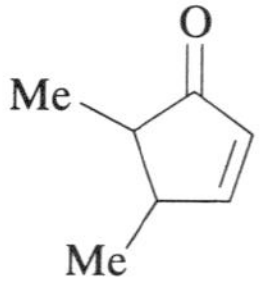 Me, Me, O *cis:trans* 59:41 (95)	85

TABLE II. SILICON(TIN)-DIRECTED NAZAROV CYCLIZATIONS: A. ACYCLIC PRECURSORS (*Continued*)

Reactant	Conditions	Product(s) and Yield(s) (%)	Refs.
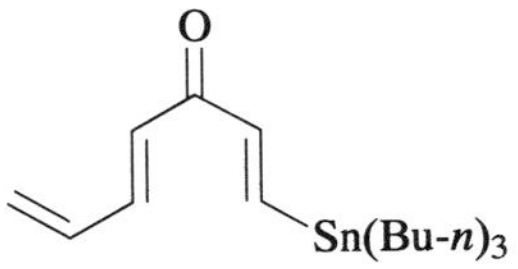 O; Sn(Bu-*n*)$_3$	BF_3•OEt_2, CH_2Cl_2, 20°	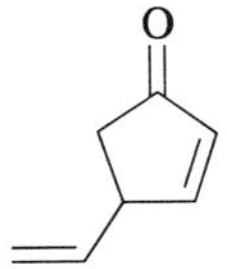 O (21)	99
C_8			
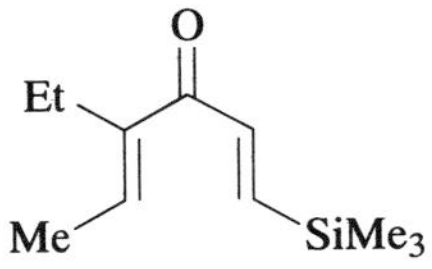 O; Et; Me; $SiMe_3$	$FeCl_3$, CH_2Cl_2, 0°, 1 h	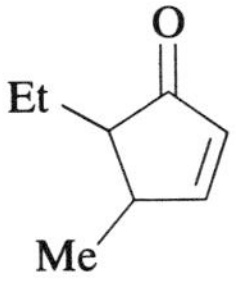 O; Et; Me *cis*:*trans* 41:59 (70)	85
C_9			
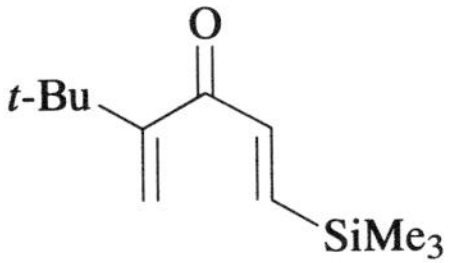 O; *t*-Bu; $SiMe_3$	$FeCl_3$, CH_2Cl_2, 0°, 1 h	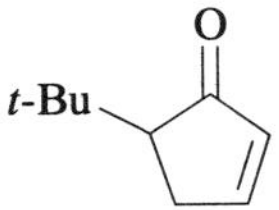 O; *t*-Bu (97)	85
C_{10}			
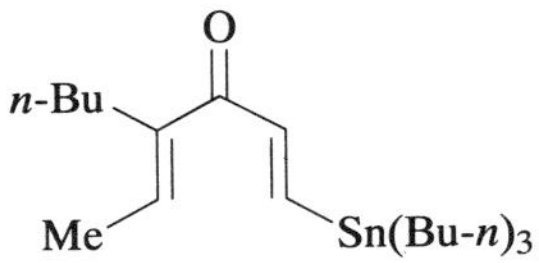 O; *n*-Bu; Me; Sn(Bu-*n*)$_3$	BF_3•OEt_2, CH_2Cl_2, 20°	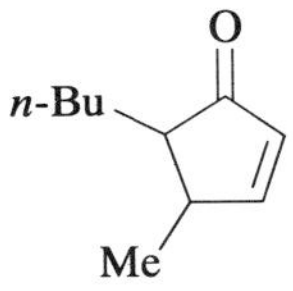 O; *n*-Bu; Me *cis*:*trans* 37:63 (93)	99

TABLE II. SILICON(TIN)-DIRECTED NAZAROV CYCLIZATIONS: A. ACYCLIC PRECURSORS (*Continued*)

Reactant	Conditions	Product(s) and Yield(s) (%)	Refs.
C_{11}			
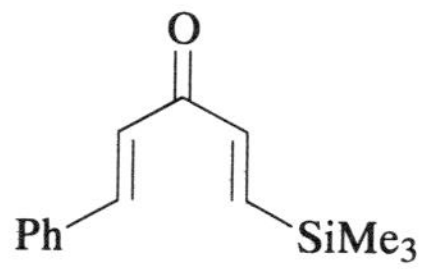 Ph, $SiMe_3$	$FeCl_3$, toluene, 20°, 12 h	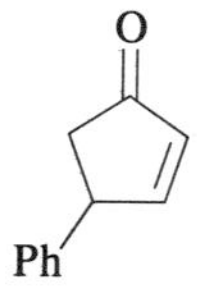 Ph (27)	85
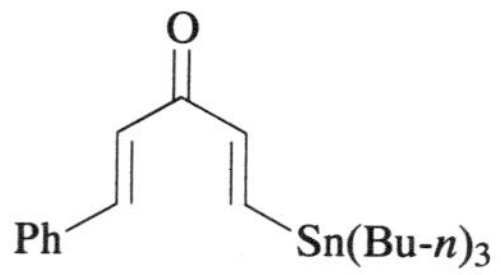 Ph, $Sn(Bu\text{-}n)_3$	$BF_3 \cdot OEt_2$, CH_2Cl_2, 20°	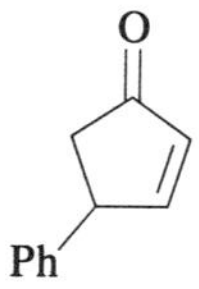 Ph (44)	99
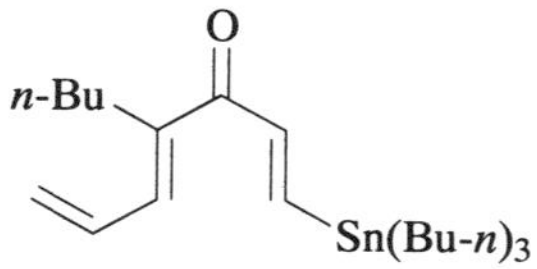 *n*-Bu, $Sn(Bu\text{-}n)_3$	$BF_3 \cdot OEt_2$, CH_2Cl_2, 20°	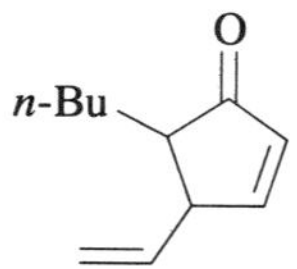 *n*-Bu *cis*:*trans* 43:57 (97)	99
C_{12}			
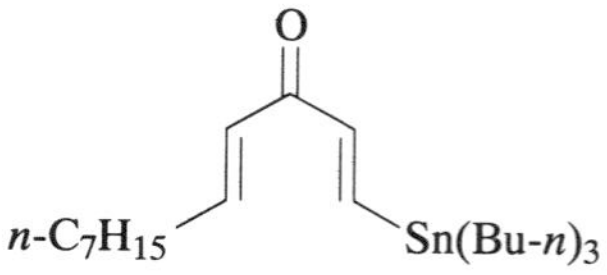 $n\text{-}C_7H_{15}$, $Sn(Bu\text{-}n)_3$	$BF_3 \cdot OEt_2$, CH_2Cl_2, 20°	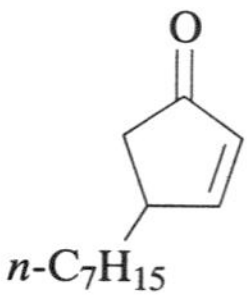 $n\text{-}C_7H_{15}$ (47)	99

TABLE II. SILICON(TIN)-DIRECTED NAZAROV CYCLIZATIONS: A. ACYCLIC PRECURSORS (*Continued*)

Reactant	Conditions	Product(s) and Yield(s) (%)	Refs.
C_{15} n-Bu, O, Ph, Sn(Bu-n)$_3$	BF_3•OEt_2, CH_2Cl_2, 20°	n-Bu, O, Ph *cis:trans* 42:58 (92)	99
C_{16} n-Bu, O, n-C_7H_{15}, Sn(Bu-n)$_3$	BF_3•OEt_2, CH_2Cl_2, 20°	n-Bu, O, n-C_7H_{15} *cis:trans* 42:58 (87)	99
C_{17} n-Bu, O, Ph, Sn(Bu-n)$_3$	BF_3•OEt_2, CH_2Cl_2, 20°	n-Bu, O, Ph *cis:trans* 40:60 (83)	99
C_{27} n-C_7H_{15}, O, BnO, C_5H_{15}-n, Sn(Bu-n)$_3$	1. BF_3•OEt_2, CH_2Cl_2, 20° 2. Basic Al_2O_3, CH_2Cl_2, rt 24 h	n-C_7H_{15}, O, BnO, C_5H_{11}-n (56)	99

TABLE II. SILICON(TIN)-DIRECTED NAZAROV CYCLIZATIONS: B. CYCLIC PRECURSORS

Reactant	Conditions	Product(s) and Yield(s) (%)		Refs.
C_8				
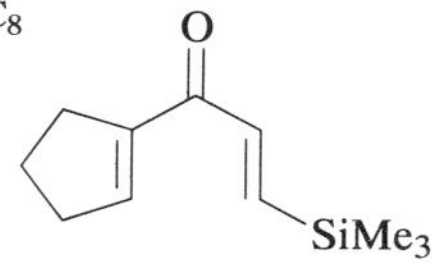 O, SiMe$_3$	$FeCl_3$, $(CH_2Cl)_2$, 20°, 2.5 h	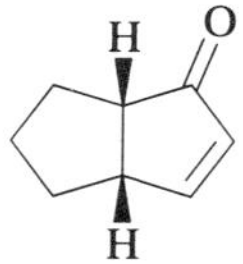 H, O, H	(52)	85
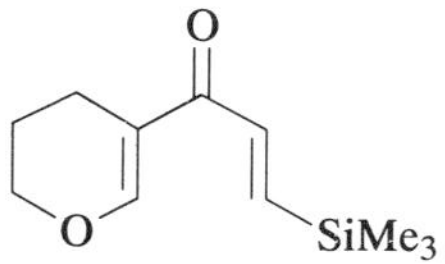 O, O, SiMe$_3$	$FeCl_3$, CH_2Cl_2, 20°, 8 h	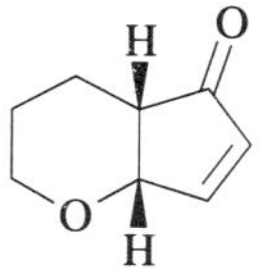 H, O, O, H	(60)	88
C_9				
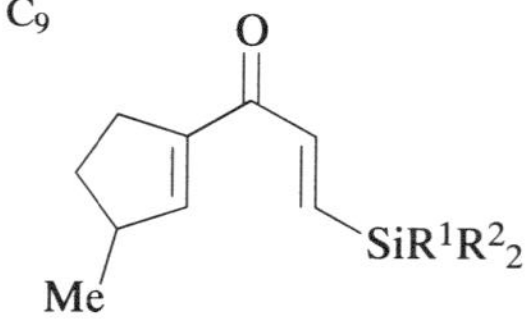 O, SiR^{1}R$^2_2$, Me	$FeCl_3$, CH_2Cl_2, 20°	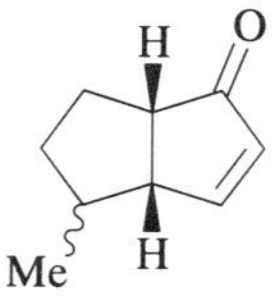 H, O, Me, H		87
R^1 R^2		β:α		
Me Me	2.5 h	54:46	(50)	
Ph Me	3 h	59:41	(46)	
Me Ph	2 h	62:38	(41)	
Ph Ph	4 h	76:24	(13)	
i-Pr *i*-Pr	4 h	79:21	(13)	
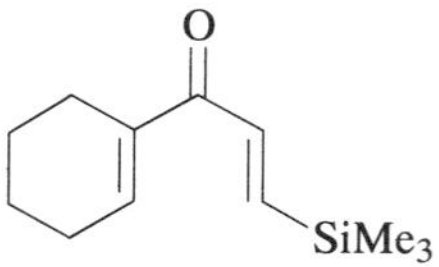 O, SiMe$_3$	$FeCl_3$, CH_2Cl_2, 0°, 4 h	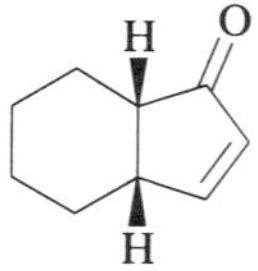 H, O, H	(84)	85

TABLE II. SILICON(TIN)-DIRECTED NAZAROV CYCLIZATIONS: B. CYCLIC PRECURSORS (*Continued*)

Reactant	Conditions	Product(s) and Yield(s) (%)		Refs.
C_{10}				
Me, Me, O, $SiMe_3$	$FeCl_3$, CH_2Cl_2, 20°, 2.5 h	Me, Me, H, O, H	(83)	184
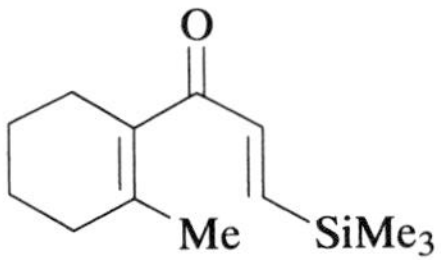 O, Me, $SiMe_3$	$FeCl_3$, CH_2Cl_2, 20°, 13 h	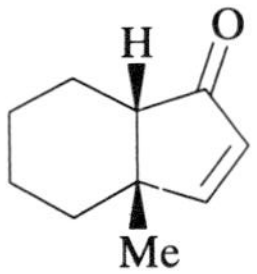 H, O, Me	(70)	88
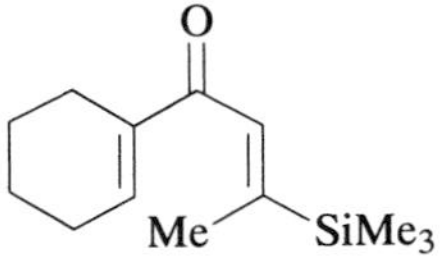 O, Me, $SiMe_3$	$FeCl_3$, CH_2Cl_2, 20°, 12 h	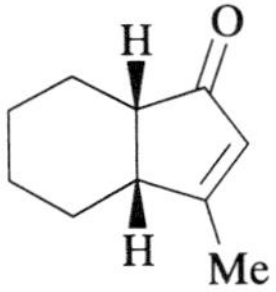 H, O, H, Me	(76)	88
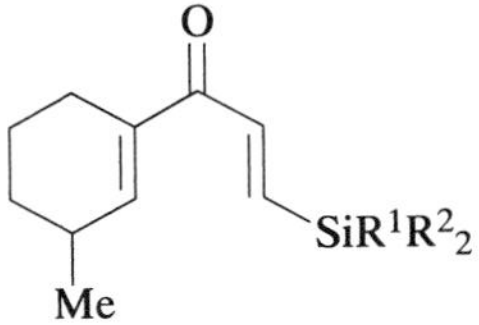 O, $SiR^1R^2_2$, Me	$FeCl_3$, CH_2Cl_2	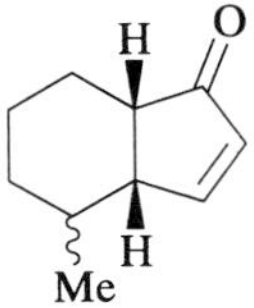 H, O, H, Me		87
R^1 R^2		β:α		
Me Me	0°, 4 h	78:22	(99)	
Ph Me	20°, 2 h	84:16	(63)	
Me Ph	20°, 2 h	86:14	(83)	
Ph Ph	20°, 2 h	87:13	(15)	
i-Pr *i*-Pr	20°, 4 h	90:10	(70)	

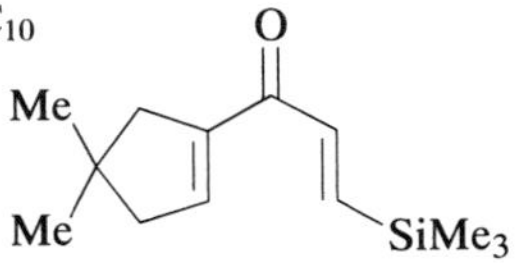

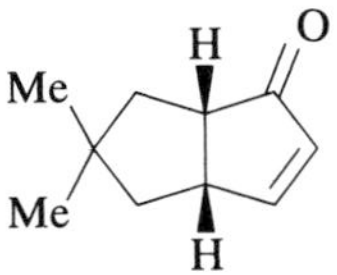

TABLE II. SILICON(TIN)-DIRECTED NAZAROV CYCLIZATIONS: B. CYCLIC PRECURSORS (*Continued*)

Reactant	Conditions	Product(s) and Yield(s) (%)	Refs.
N-CO_2Me, $SiMe_3$	$ZrCl_4$, $(CH_2Cl)_2$, 60°, 36 h	H, O, N, H, CO_2Me (76)	88
Me, O, SiR_3	$FeCl_3$, CH_2Cl_2	Me, H, O, H	86
R		β:α	
Me	0°, 4 h	72:28 (85)	
i-Pr	20°, 2 h	78:22 (78)	
O, $SiMe_3$	$FeCl_3$, CH_2Cl_2, 0°, 1 h	O, *cis*:*trans* 85:15 (74)	85
C_{11}			
Me, Me, O, Me, $SiMe_3$	BF_3•OEt_2, toluene, 100°, 36 h	Me, Me, H, O, Me (70)	95

TABLE II. SILICON(TIN)-DIRECTED NAZAROV CYCLIZATIONS: B. CYCLIC PRECURSORS (*Continued*)

Reactant	Conditions	Product(s) and Yield(s) (%)	Refs.
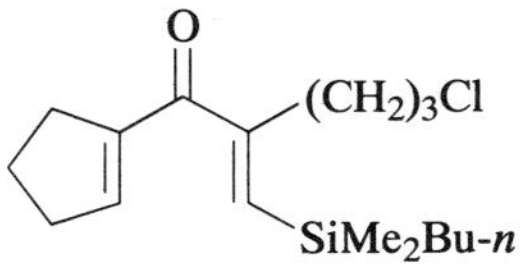 $(CH_2)_3Cl$, $SiMe_2Bu$-*n*	$SnCl_4$, -78°, 1 h	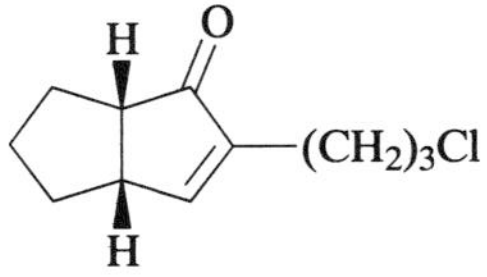 $(CH_2)_3Cl$ (100)	97
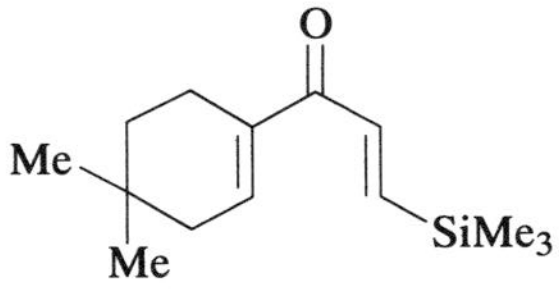 Me, Me, $SiMe_3$	$FeCl_3$, CH_2Cl_2, 0°, 4 h	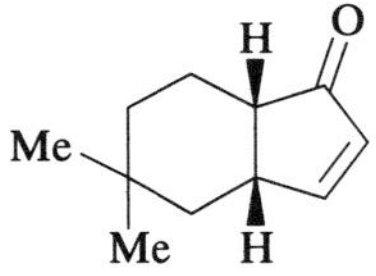 Me, Me (78)	88
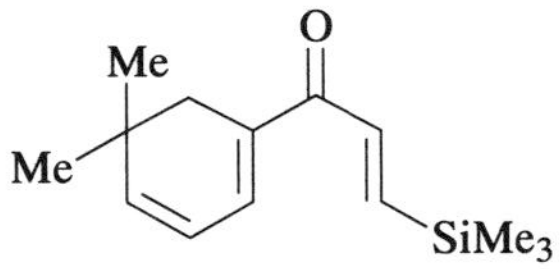 Me, Me, $SiMe_3$	$FeCl_3$, CH_2Cl_2, 20°, 2 h	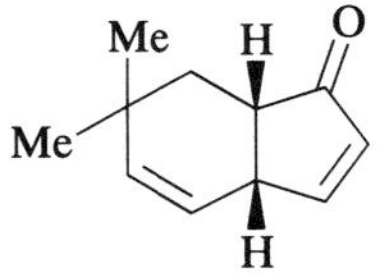 Me, Me (69)	88
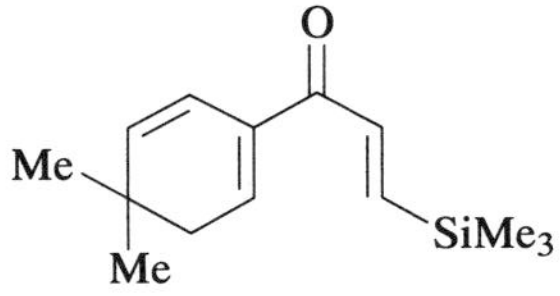 Me, Me, $SiMe_3$	$FeCl_3$, CH_2Cl_2, -15°, 1 h	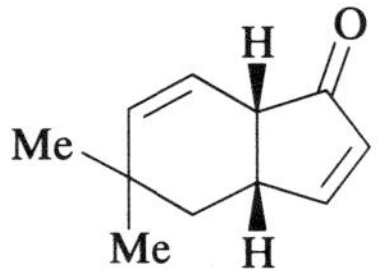 Me, Me (69)	88
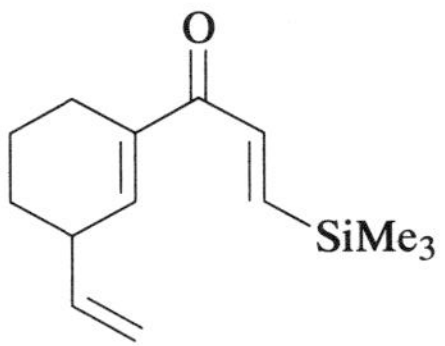 $SiMe_3$	$FeCl_3$, CH_2Cl_2, -10°, 2 h	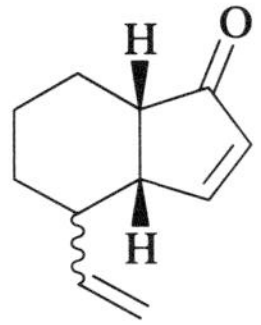 β:α 78:22 (66)	87

TABLE II. SILICON(TIN)-DIRECTED NAZAROV CYCLIZATIONS: B. CYCLIC PRECURSORS (*Continued*)

Reactant	Conditions	Product(s) and Yield(s) (%)	Refs.
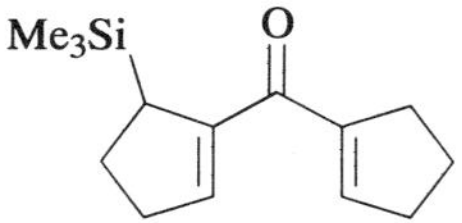 Me_3Si, O	BF_3•OEt_2, CH_2Cl_2, 10°, 10 min	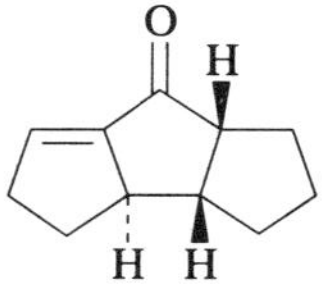 O, H, H H (77)	89
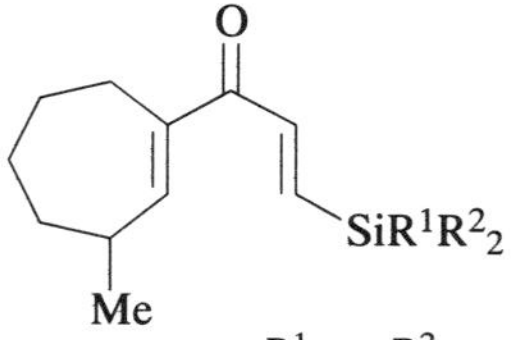 O, $SiR^1R^2_2$, Me	$FeCl_3$, CH_2Cl_2, 0°	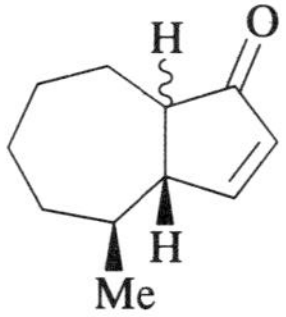 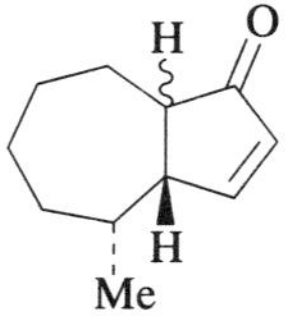 H, O, H, Me + H, O, H, Me	98
R^1 R^2		β:α β:α	
Me Me	1 h	(47) 71:29 (36) 100:0	
Ph Me	1 h	(44) 71:29 (33) 98:2	
Me Ph	1 h	(29) 73:27 (29) 98:2	
C_{12} Ph Ph	2 h	(27) 70:30 (17) 100:0	
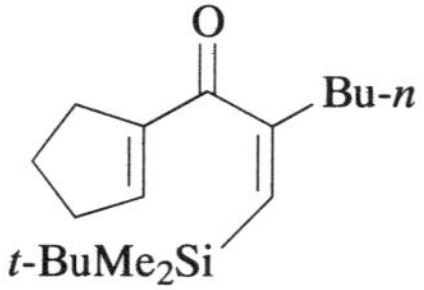 O, Bu-*n*, *t*-$BuMe_2Si$	BF_3•OEt_2, CH_2Cl_2,-78°; -20°, 4 h; 0°, 24 h	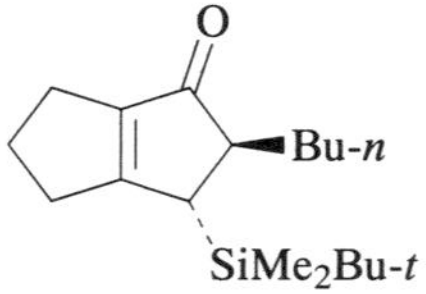 O, Bu-*n*, $SiMe_2Bu$-*t* (69)	97
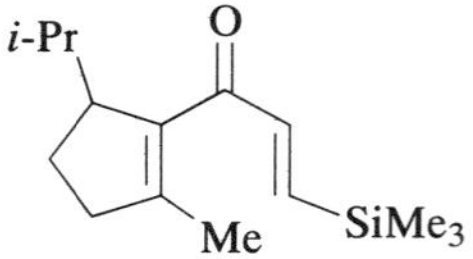 *i*-Pr, O, Me, $SiMe_3$	BF_3•OEt_2, C_6H_6, 90°, 20 h	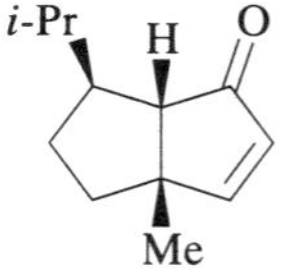 *i*-Pr, H, O, Me (60) 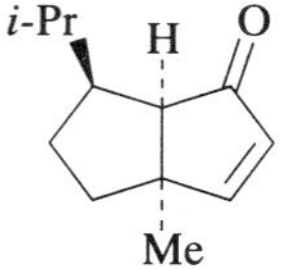 + *i*-Pr, H, O, Me (12)	185

TABLE II. SILICON(TIN)-DIRECTED NAZAROV CYCLIZATIONS: B. CYCLIC PRECURSORS (*Continued*)

Reactant	Conditions	Product(s) and Yield(s) (%)	Refs.
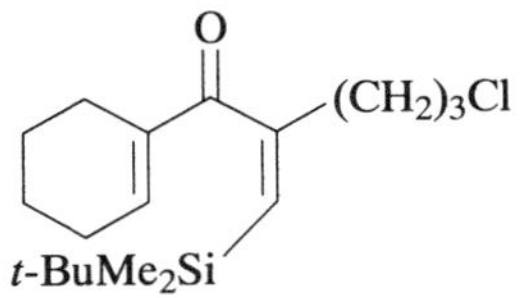 (CH$_2$)$_3$Cl; t-BuMe$_2$Si	BF_3•OEt_2, CH_2Cl_2,-78°; -20°, 4 h; 0°, 24 h	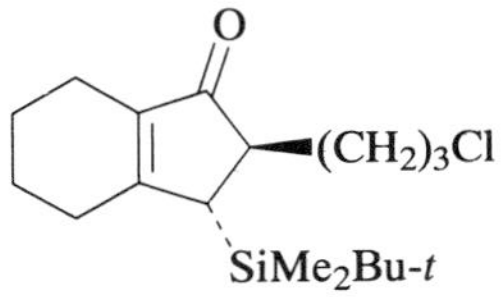 (CH$_2$)$_3$Cl; SiMe$_2$Bu-t (80)	97
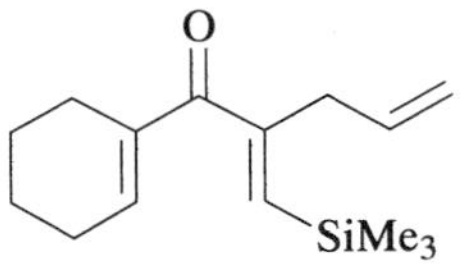 SiMe$_3$	$FeCl_3$, CH_2Cl_2, -25°, 1 h	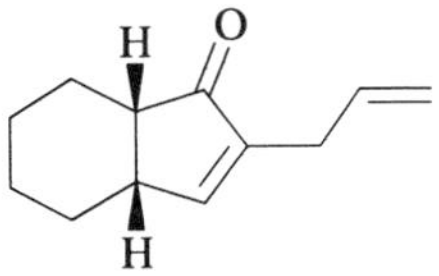 H; H (76)	88
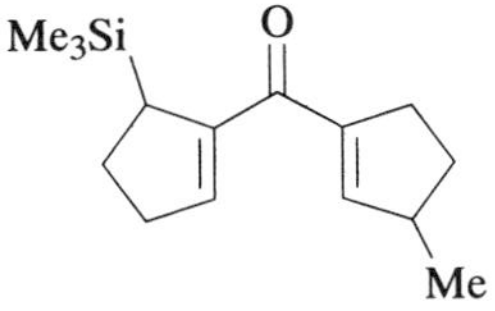 Me$_3$Si; Me	BF_3•OEt_2, CH_2Cl_2, 10°, 1 h	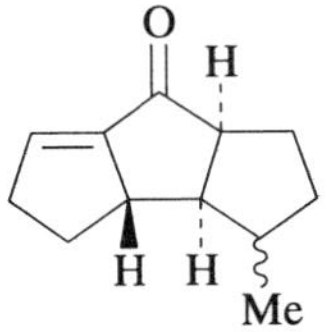 H; H H; Me 53:47 (83)	89
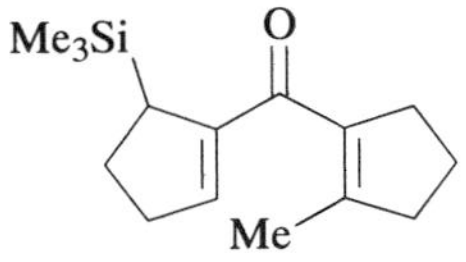 Me$_3$Si; Me	BF_3•OEt_2, CH_2Cl_2, 20°, 3 h	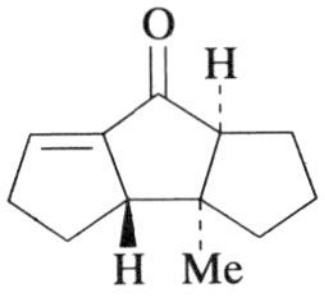 H; H Me (79)	89
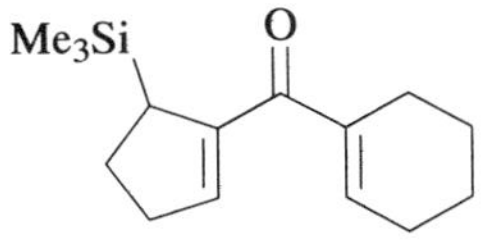 Me$_3$Si	$FeCl_3$, CH_2Cl_2, -20°, 45 min	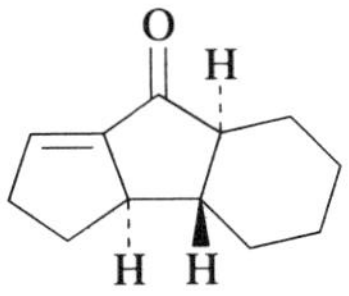 H; H H (36)	89

TABLE II. SILICON(TIN)-DIRECTED NAZAROV CYCLIZATIONS: B. CYCLIC PRECURSORS (*Continued*)

Reactant	Conditions	Product(s) and Yield(s) (%)	Refs.
Me$_3$Si, O	$FeCl_3$, CH_2Cl_2, -50°, 15 min	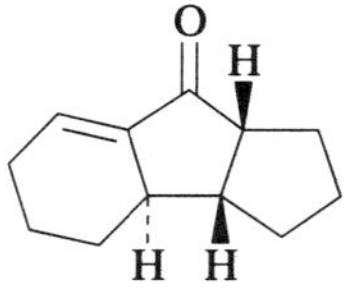 O, H, H, H (88)	89
C_{13}			
O, *t*-Bu, $SiMe_3$	$FeCl_3$, CH_2Cl_2, 0°, 4 h	H, O, *t*-Bu, H β:α 76:24 (82)	86
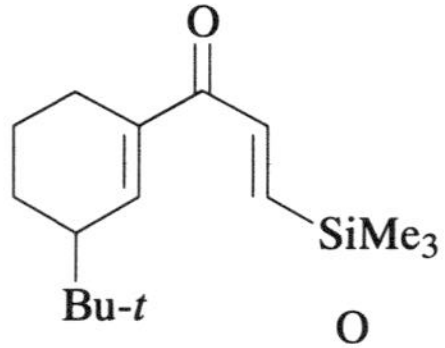 O, $SiMe_3$, Bu-*t*	$FeCl_3$, CH_2Cl_2, 0°, 8 h	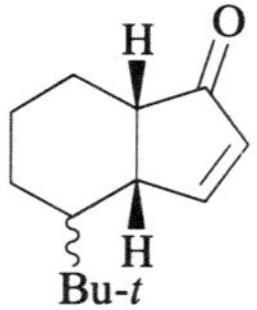 H, O, H, Bu-*t* β:α 94:6 (63)	87
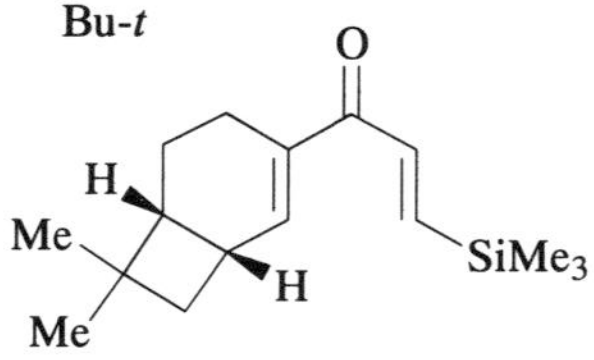 O, H, Me, Me, H, $SiMe_3$	$FeCl_3$, CH_2Cl_2, -25°, 2 h	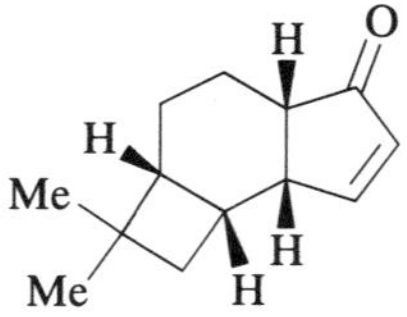 H, O, H, Me, Me, H, H (60) + H, O, H, Me, Me, H, H (9)	88
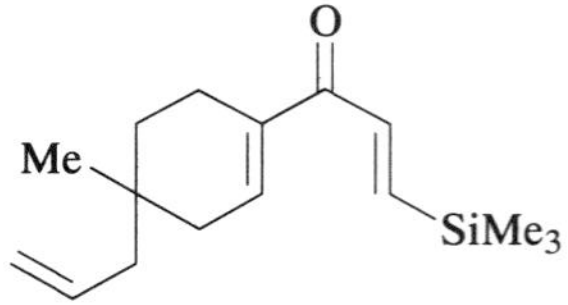 O, Me, $SiMe_3$	$FeCl_3$, CH_2Cl_2, 0-20°, 1 h	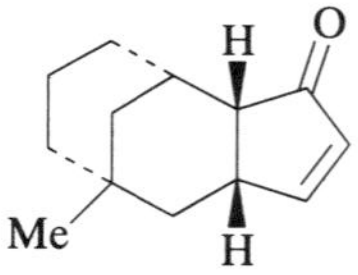 H, O, Me, H (44)	88

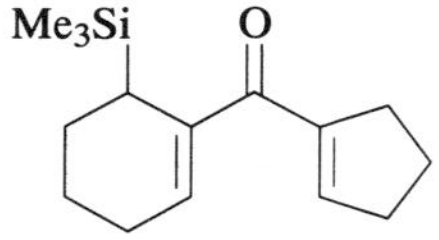

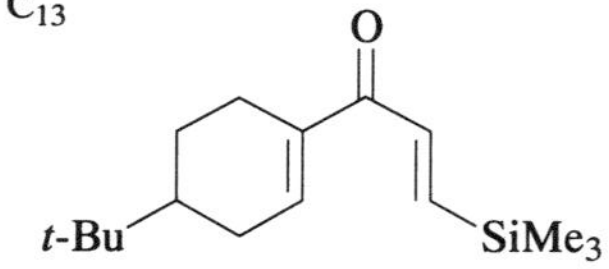

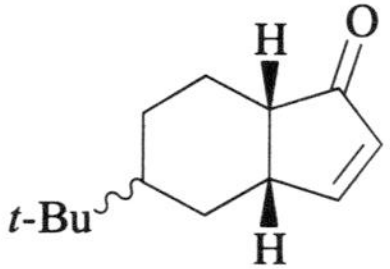

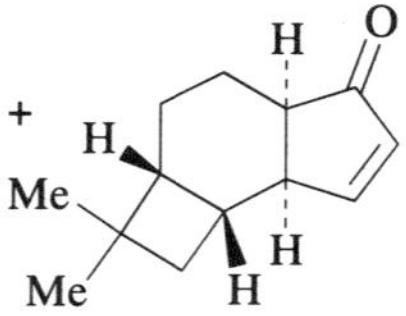

TABLE II. SILICON(TIN)-DIRECTED NAZAROV CYCLIZATIONS: B. CYCLIC PRECURSORS (*Continued*)

Reactant	Conditions	Product(s) and Yield(s) (%)	Refs.
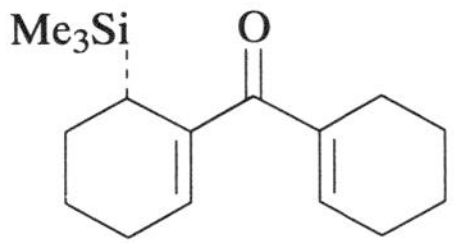 Me_3Si, O; (*S*)-(–)	$FeCl_3$, CH_2Cl_2, -50°, 1 min	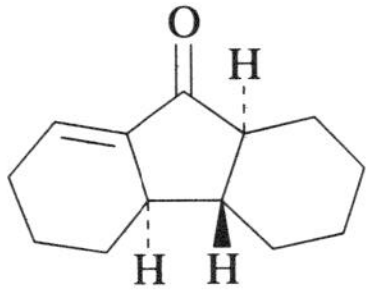 O, H, H H (72)	90, 89
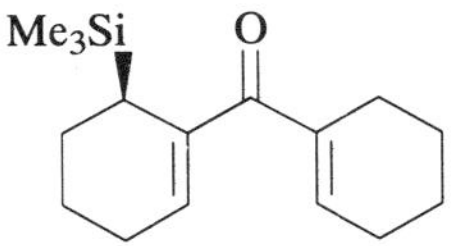 Me_3Si, O; (*R*)-(+)	$FeCl_3$, CH_2Cl_2, -50°, 1 min	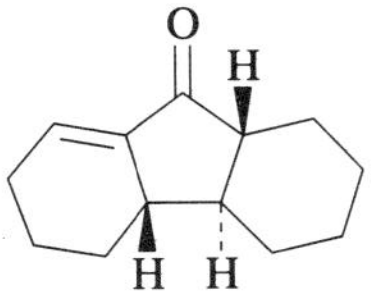 O, H, H H (58)	90, 89
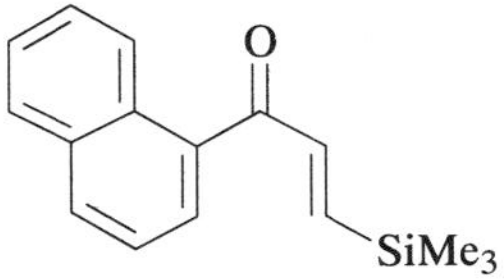 O, $SiMe_3$		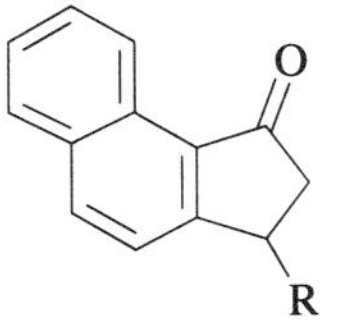 O, R	85
	$FeCl_3$, $(CH_2Cl)_2$, 20°, 12 h	R = $SiMe_3$ (70)	
	$FeCl_3$, toluene, 20°, 48 h	R = H (60)	
C_{14}			
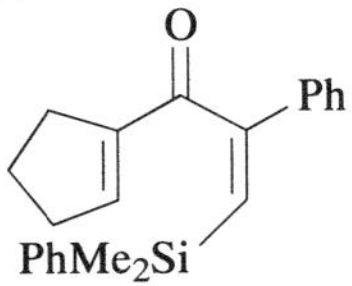 O, Ph, $PhMe_2Si$	$FeCl_3$, CH_2Cl_2, -25°, 3 h	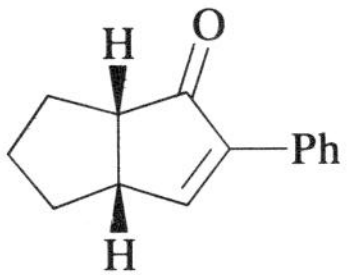 H, O, Ph, H (86)	88

TABLE II. SILICON(TIN)-DIRECTED NAZAROV CYCLIZATIONS: B. CYCLIC PRECURSORS (*Continued*)

Reactant	Conditions	Product(s) and Yield(s) (%)	Refs.
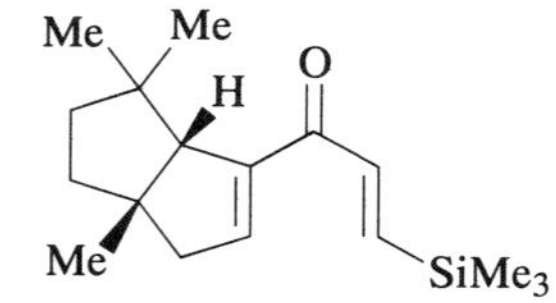 Me, Me, H, O, Me, $SiMe_3$	BF_3•OEt_2, toluene, 20°, 6 h	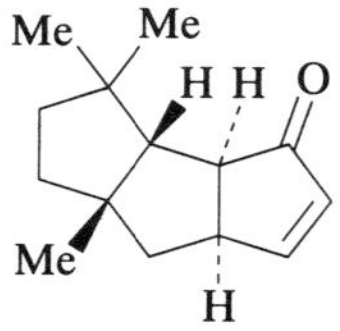 Me, Me, H H O, Me, H (80)	95
C_{15}			
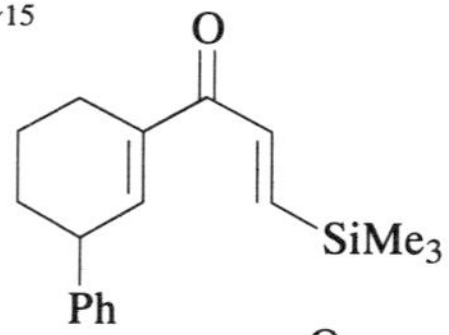 O, $SiMe_3$, Ph	$FeCl_3$, CH_2Cl_2, 0°, 4 h	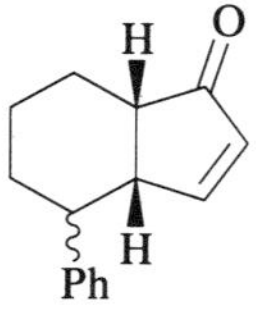 H O, H, Ph β:α 94:6 (76)	87
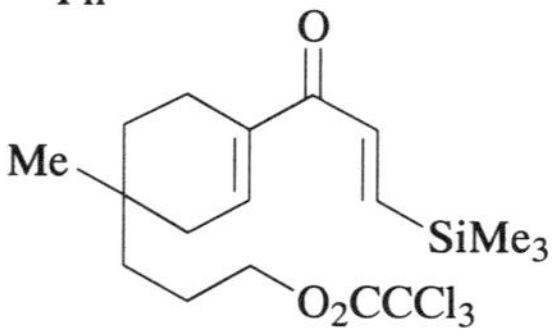 O, Me, $SiMe_3$, O_2CCCl_3	$FeCl_3$, CH_2Cl_2, 0°, 4 h	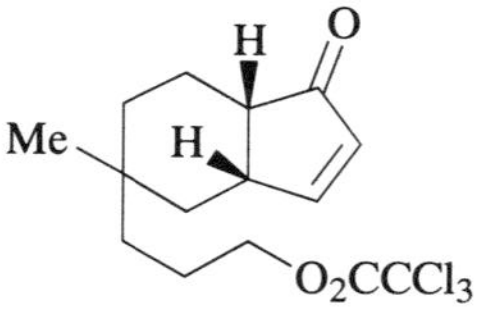 H O, Me, H, O_2CCCl_3 (78)	88
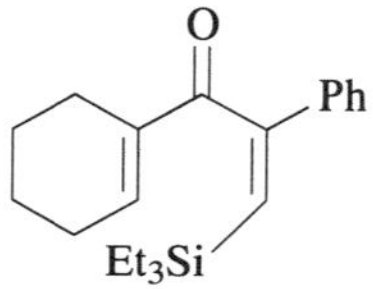 O, Ph, Et_3Si		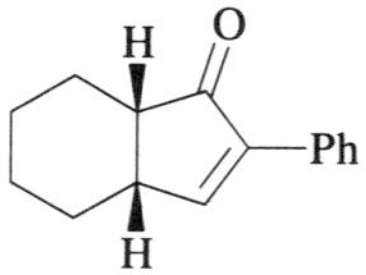 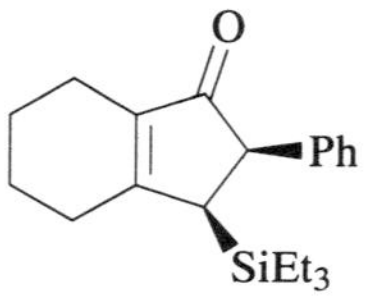 H O, Ph, H + O, Ph, $SiEt_3$	97
	BF_3•OEt_2, CH_2Cl_2, -78°; -20°, 4 h; 0°, 24 h	(30) + (37)	
	$SnCl_4$, CH_2Cl_2, -78°, 2 h	(18) + (18)	
	$FeCl_3$, CH_2Cl_2, 0°, 4 h	(18) + (24)	

TABLE II. SILICON(TIN)-DIRECTED NAZAROV CYCLIZATIONS: B. CYCLIC PRECURSORS (*Continued*)

Reactant	Conditions	Product(s) and Yield(s) (%)	Refs.
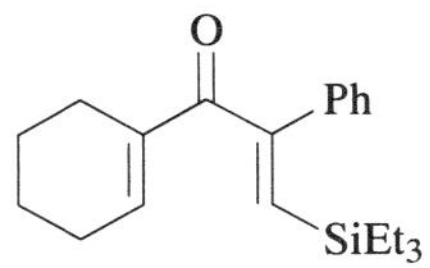 Ph, $SiEt_3$	$FeCl_3$, CH_2Cl_2, -25°, 3 h	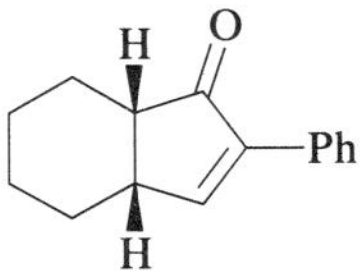 H, H, Ph (86)	98
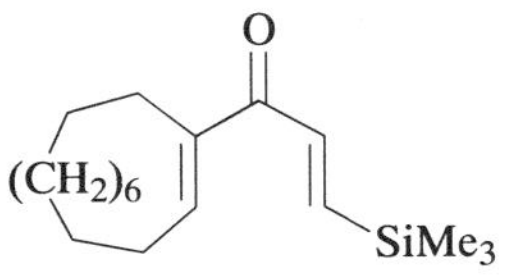 $(CH_2)_6$, $SiMe_3$	$FeCl_3$, CH_2Cl_2, 0°, 2 h	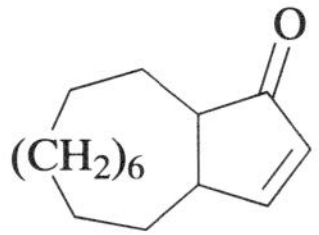 $(CH_2)_6$ *cis:trans* 46:54 (78)	85
C_{16}			
OBn, $SiMe_3$	$FeCl_3$, CH_2Cl_2, 0°, 2 h	H, H, OBn β:α 94:6 (76)	87
C_{17}			
OBOM, $SiMe_3$	$FeCl_3$, CH_2Cl_2, 0°, 2 h	H, H, OBOM β:α 93:7 (40)	87

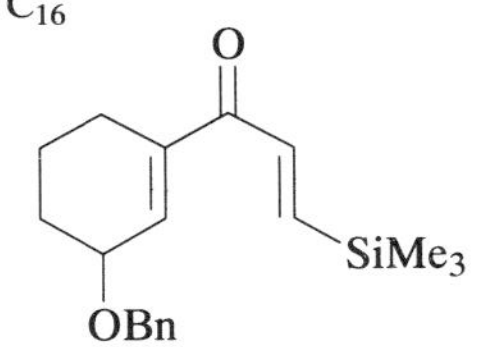

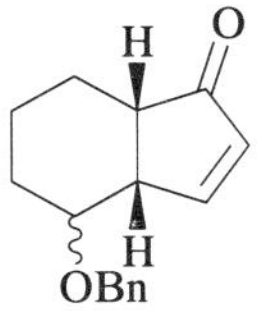

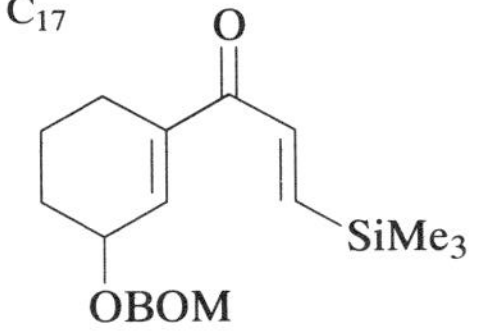

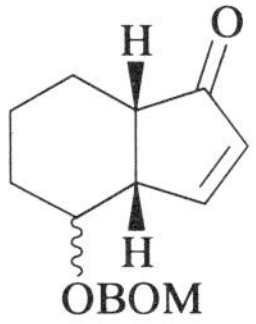

TABLE II. SILICON(TIN)-DIRECTED NAZAROV CYCLIZATIONS: B. CYCLIC PRECURSORS (*Continued*)

Reactant	Conditions	Product(s) and Yield(s) (%)	Refs.
C_{18} 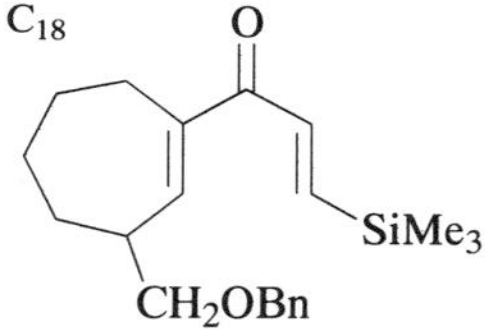 (structure: SiMe$_3$, CH$_2$OBn)	$FeCl_3$, CH_2Cl_2, 0°, 1 h	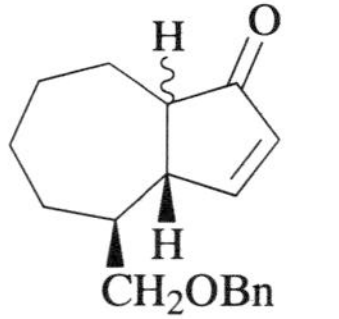 (structure: H, O, H, CH$_2$OBn) β:α 62:38 (8) 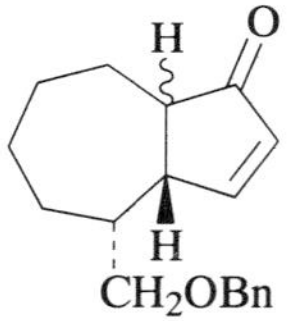 + (structure: H, O, H, CH$_2$OBn) β:α 85:15 (72)	98
C_{19} 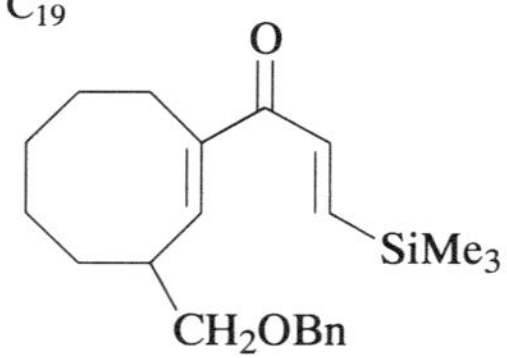 (structure: SiMe$_3$, CH$_2$OBn)	$FeCl_3$, CH_2Cl_2, 0°, 1 h	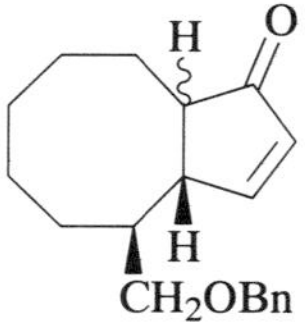 (structure: H, O, H, CH$_2$OBn) β:α 57:43 (15) 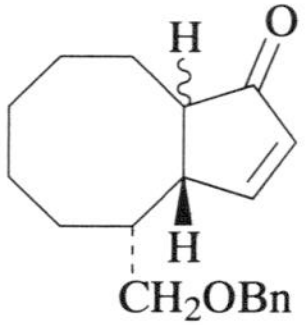 + (structure: H, O, H, CH$_2$OBn) β:α 97:3 (73)	98
C_{22} 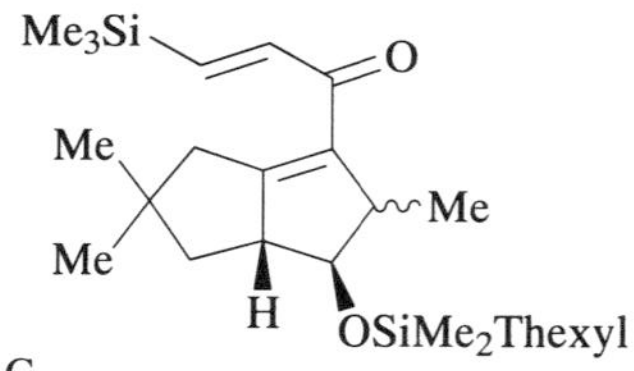 (structure: Me$_3$Si, O, Me, Me, Me, H, OSiMe$_2$Thexyl)	BF_3•OEt_2, C_6H_5Et, reflux	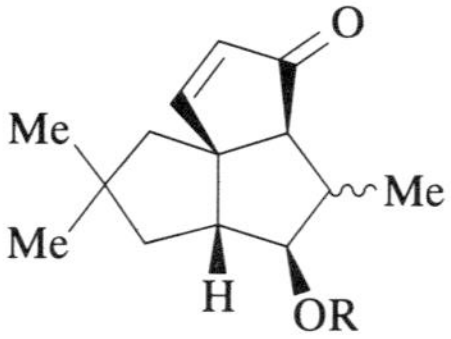 (structure: O, Me, Me, Me, H, OR) R = SiMe$_2$Thexyl (38); R = H (22)	100a
C_{28} 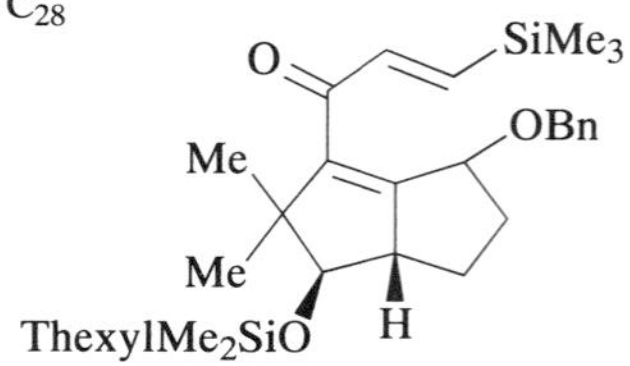 (structure: O, SiMe$_3$, OBn, Me, Me, ThexylMe$_2$SiO, H)	BF_3•OEt_2, C_6H_5Et, reflux	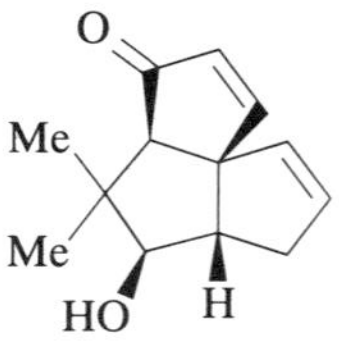 (structure: O, Me, Me, HO, H) (53)	100

TABLE III. CYCLIZATIONS OF IN SITU GENERATED DIVINYL KETONES: A. NON-ACETYLENES; α-ELIMINATION

Reactant	Conditions	Product(s) and Yield(s) (%)	Refs.
C_9			
AcO, O	TsOH, toluene, 112°	O (41)	101
C_{10}			
RO, O, Me	TsOH, toluene, 112°	O, Me	101
R = H		(48)	
R = Me_3Si		(48)	
R = Ac		(53)	
RO, O, Me	TsOH, toluene, 112°	O, Me	101
R = Me_3Si		(19)	
R = Ac		(65)	
C_{11}			
Me_3SiO, O, Me, Me	TsOH, toluene, 112°	Me, O, Me (22)	186

TABLE III. CYCLIZATIONS OF IN SITU GENERATED DIVINYL KETONES: A. α-ELIMINATION *(Continued)*

Reactant	Conditions	Product(s) and Yield(s) (%)	Refs.
RO, O, Me; R = H; R = Ac	TsOH, toluene, 112°	O, Me (41) (60)	101
C_{12}			
Me_3SiO, O, Ph, Me, Me	TsOH, toluene, 112°	O, Ph, Me (37) + O, Ph, Me (41)	101
C_{13}			
Me, Me_3SiO, O, Me, Me, Me	TsOH, toluene, 112°	Me, O, Me, Me, Me (56) + Me, O, Me, Me (24)	101
C_{14}			
AcO, O, t-Bu, Me	TsOH, toluene, 112°	O, t-Bu, Me (66)	101

TABLE III. CYCLIZATIONS OF IN SITU GENERATED DIVINYL KETONES: A. α-ELIMINATION *(Continued)*

Reactant	Conditions	Product(s) and Yield(s) (%)	Refs.
R, $OSiMe_3$, O, Me	TsOH, toluene, 112°	R, O, Me + R, O, Me	101
C_{14} R = H		(50) (50)	
C_{15} R = OMe		(23) (77)	
C_{16} RO, O, $(CH_2)_6$, Me	TsOH, toluene, 112°	$(CH_2)_6$, O, Me	101
R = H		(47)	
R = OAc		(91)	

TABLE III. CYCLIZATIONS OF IN SITU GENERATED DIVINYL KETONES: B. NON-ACETYLENES; β-ELIMINATION

Reactant	Conditions	Product(s) and Yield(s) (%)	Refs.
C_8 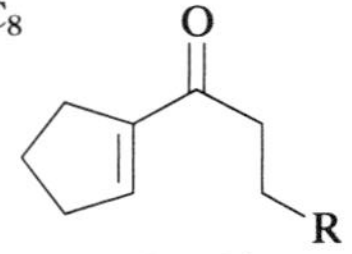		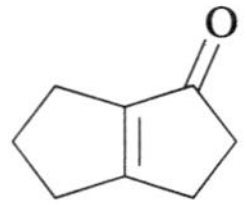	
R = Cl	H_3PO_4, HCO_2H, 90°, 12 h	(97)	103
R = NEt_2	H_3PO_4, HCO_2H, 80-90°, 6 h	(15)	104
C_9 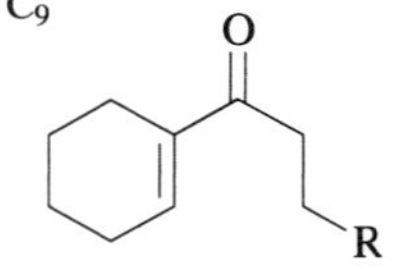		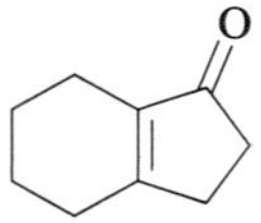	
R = Cl	H_3PO_4, HCO_2H, 90°, 3 h	(72)	102
R = NEt_2	H_3PO_4, HCO_2H, 80-90°, 6 h	(45)	104
C_{11} 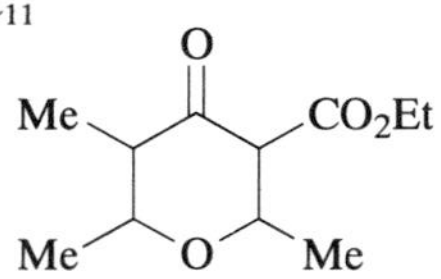		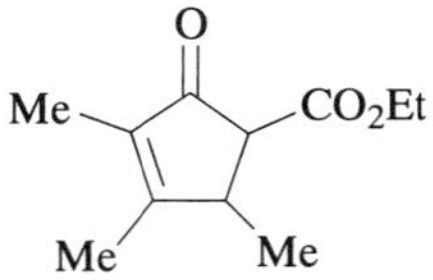	
Me, Me, O, Me, CO_2Et		Me, Me, Me, CO_2Et, O	
	Me_3SiCl, NaI, DMF, 120°, 6 h	(59)	105
	Me_3SiOTf, CH_2Cl_2, rt, 2 h	(55)	69a
C_{12}			
O, Me, Ph, NEt_2	H_3PO_4, HCO_2H, 80-90°, 6 h	O, Me, Ph (50)	104

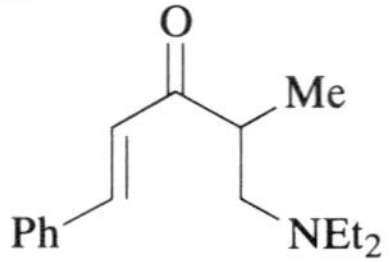

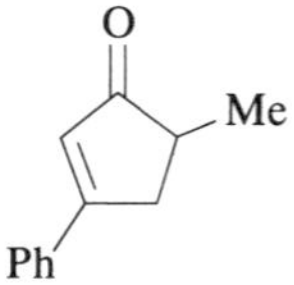

TABLE III. CYCLIZATIONS OF IN SITU GENERATED DIVINYL KETONES: B. β-ELIMINATION *(Continued)*

Reactant	Conditions	Product(s) and Yield(s) (%)	Refs.
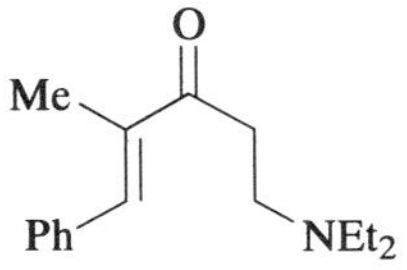 Me, Ph, O, NEt_2	H_3PO_4, HCO_2H, 80-90°, 6 h	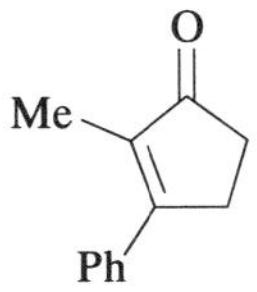 Me, O, Ph (45)	104
C_{13}			
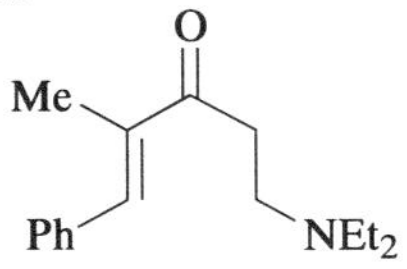 Me, Ph, O, NEt_2	H_3PO_4, HCO_2H, 80-90°, 6 h 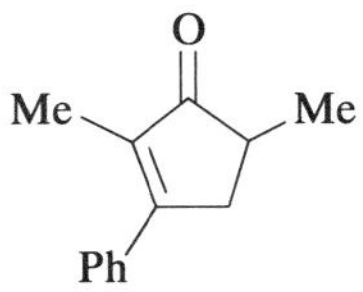	Me, O, Me, Ph (30)	104
O, CO_2Et, O, Me	Me_3SiCl, NaI, DMF, 120°, 5 h Me_3SiOTf, CH_2Cl_2, rt, 2 h	O, Me (77) (77)	105 69a
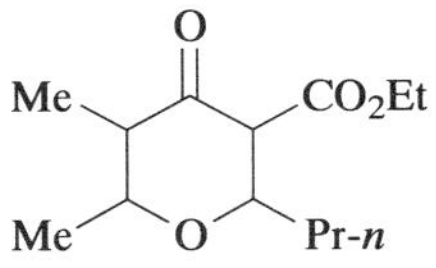 Me, O, CO_2Et, Me, O, Pr-*n*	Me_3SiCl, NaI, DMF, 120°, 7 h 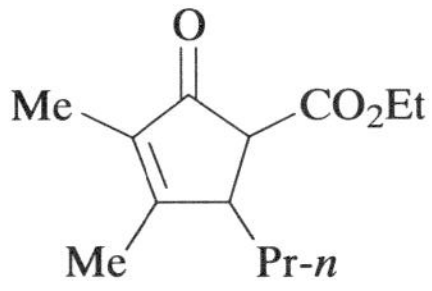	Me, O, CO_2Et, Me, Pr-*n* (68)	105

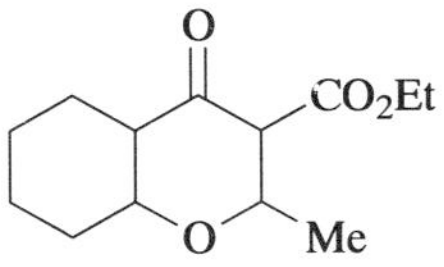

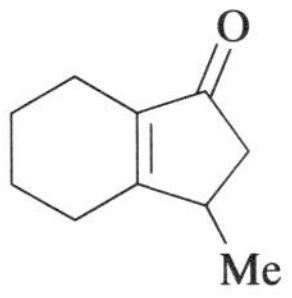

TABLE III. CYCLIZATIONS OF IN SITU GENERATED DIVINYL KETONES: B. β-ELIMINATION *(Continued)*

Reactant	Conditions	Product(s) and Yield(s) (%)	Refs.
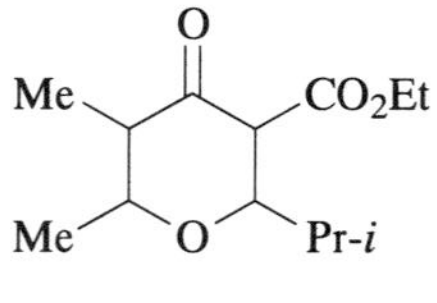 O, Me, CO_2Et, Me, O, Pr-*i*	Me_3SiCl, NaI, DMF, 120°, 10 h	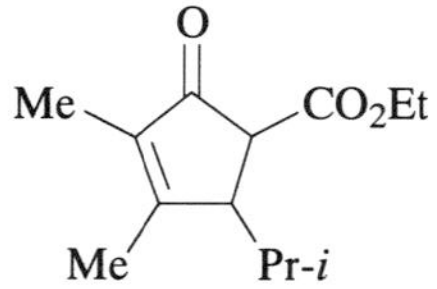 O, Me, CO_2Et, Me, Pr-*i* (32) 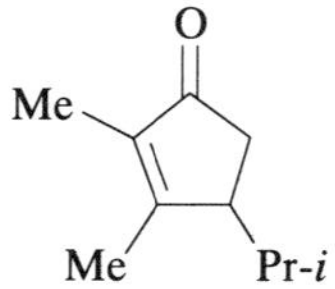 + O, Me, Me, Pr-*i* (44)	105
C_{16}			
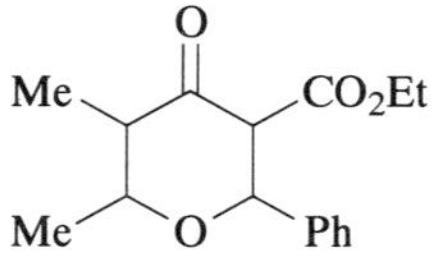 O, Me, CO_2Et, Me, O, Ph	Me_3SiCl, NaI, DMF, 120°, 10 h	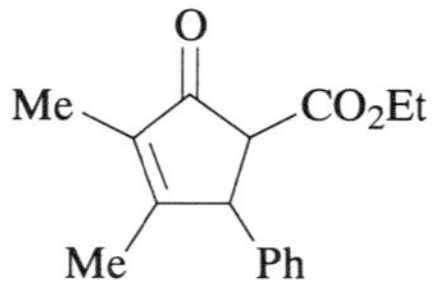 O, Me, CO_2Et, Me, Ph (64)	105

TABLE III. CYCLIZATIONS OF IN SITU GENERATED DIVINYL KETONES: C. ACYCLIC ACETYLENES

Reactant	Conditions	Product(s) and Yield(s) (%)		Refs.
C_6 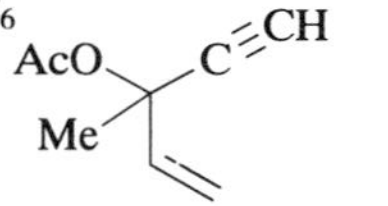 AcO, Me, C≡CH	$PdCl_2(MeCN)_2$, HOAc, MeCN, 60-80°	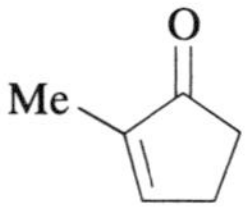 O, Me	(50-61)	142
C_8 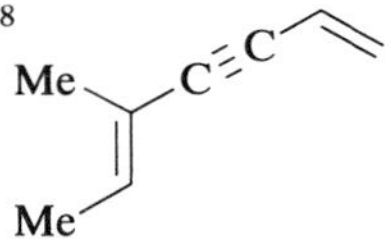 Me, Me, C≡C	conc. HCl, 60-70°, 4 h	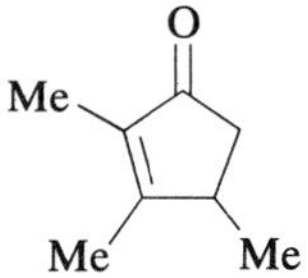 O, Me, Me, Me	(71)	4, 6
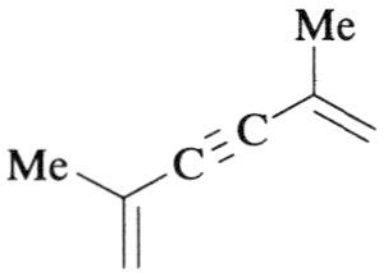 Me, Me, C≡C	H_3PO_4, 40-50°, 2 h; 60-65°, 5 h	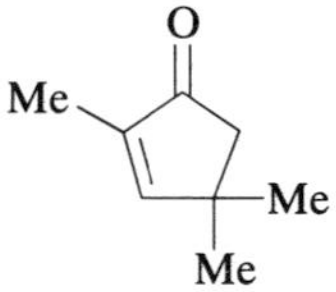 O, Me, Me, Me	(52)	9
C_9 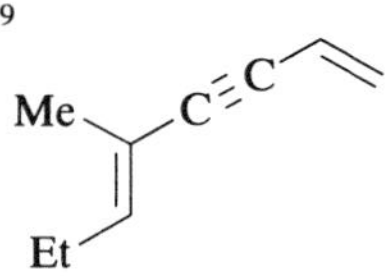 Me, Et, C≡C	conc. HCl, 60-70°, 4 h	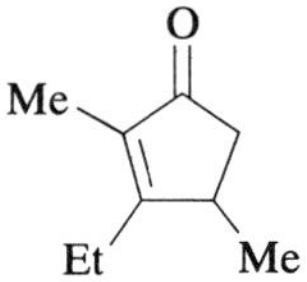 O, Me, Et, Me	(70)	4
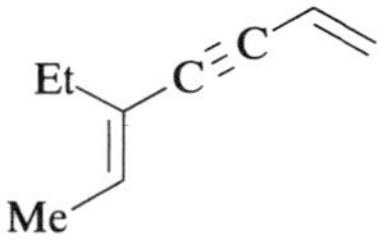 Et, Me, C≡C	conc. HCl, 60-70°, 4 h	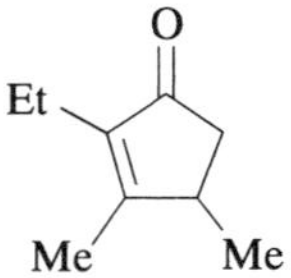 O, Et, Me, Me	(70)	4

TABLE III. CYCLIZATIONS OF IN SITU GENERATED DIVINYL KETONES: C. ACYCLIC ACETYLENES (*Continued*)

Reactant	Conditions	Product(s) and Yield(s) (%)	Refs.
C_{10}			
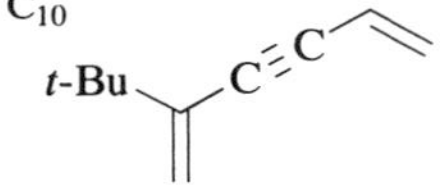 t-Bu, C≡C	90% MeOH, H_2SO_4, $HgSO_4$, 60-70°, 4 h	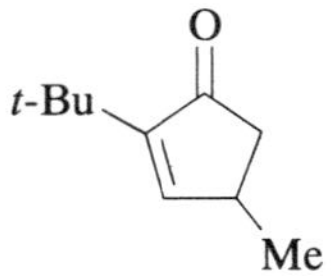 t-Bu, O, Me (49) 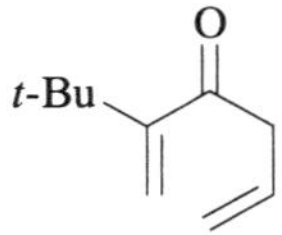 + t-Bu, O (14)	15
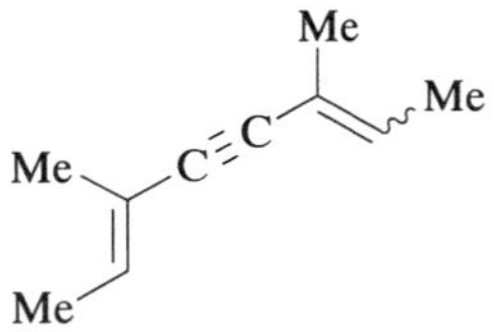 Me, Me, C≡C, Me, Me	H_3PO_4, 70°, 5 h	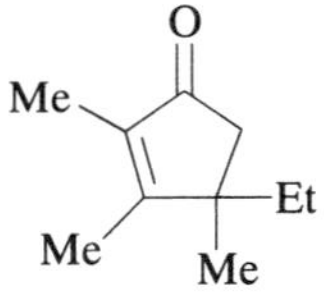 O, Me, Et, Me, Me (100)	10
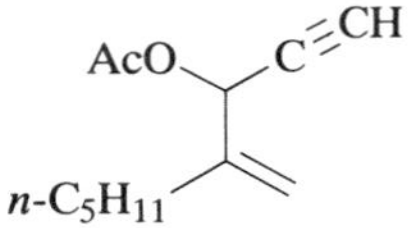 AcO, C≡CH, n-C_5H_{11}	$PdCl_2(MeCN)_2$, HOAc, MeCN, 60-80°	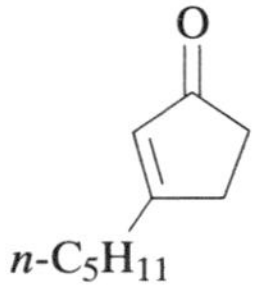 O, n-C_5H_{11} (65-73)	31
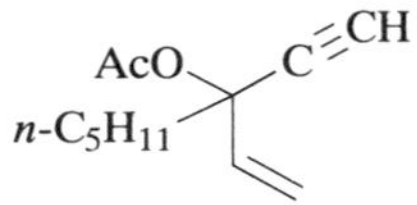 AcO, C≡CH, n-C_5H_{11}	$PdCl_2(MeCN)_2$, HOAc, MeCN, 60-80°	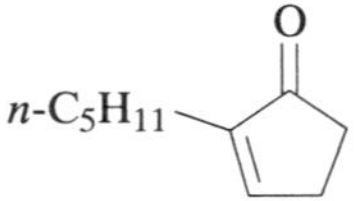 O, n-C_5H_{11} (48-66)	31
C_{11}			
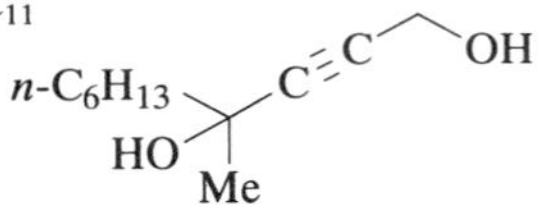 n-C_6H_{13}, HO, Me, C≡C, OH	MeOH, H_2SO_4, 0°, 1.7 h	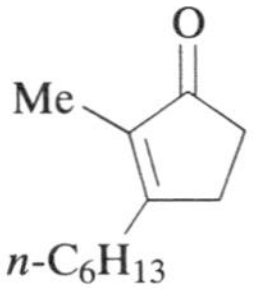 O, Me, n-C_6H_{13} (58)	147

TABLE III. CYCLIZATIONS OF IN SITU GENERATED DIVINYL KETONES: C. ACYCLIC ACETYLENES (*Continued*)

Reactant	Conditions	Product(s) and Yield(s) (%)	Refs.
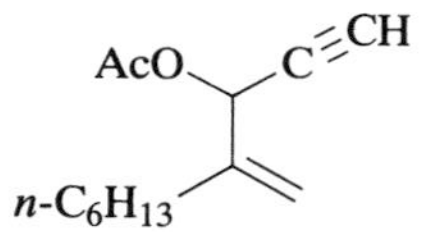 AcO, C≡CH, n-C_6H_{13}	$PdCl_2(MeCN)_2$, HOAc, MeCN, 60-80°	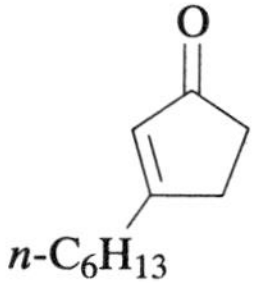 n-C_6H_{13} (63)	142
C_{12}			
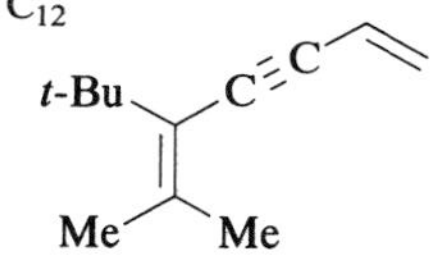 t-Bu, C≡C, Me, Me	H_3PO_4, 60-65°, 30 min 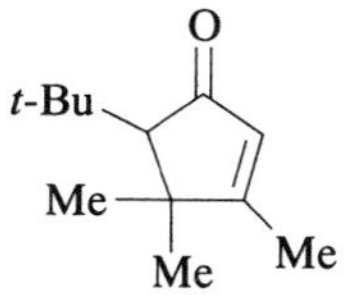	O, t-Bu, Me, Me, Me (9)	25
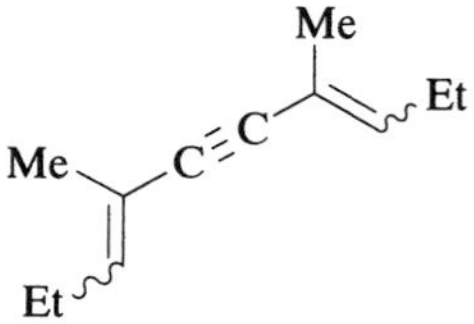 Me, C≡C, Me, Et, Et	HCO_2H, H_3PO_4, 90°	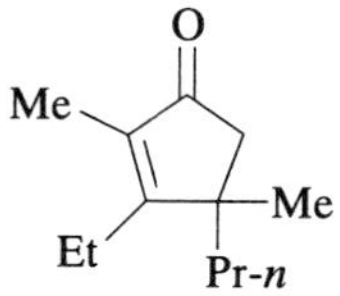 O, Me, Me, Et, Pr-n (65-70)	19
C_{14}			
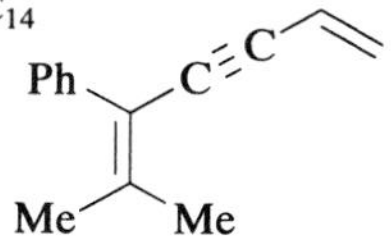 Ph, C≡C, Me, Me	H_3PO_4, 70-80°, 3 h 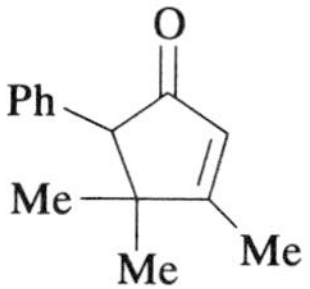	O, Ph, Me, Me, Me (9)	24
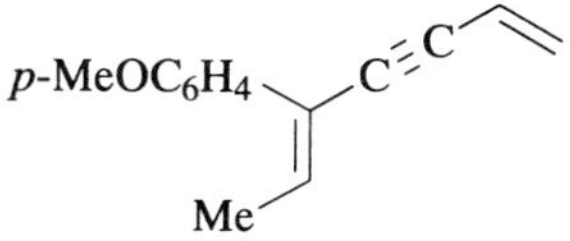 p-$MeOC_6H_4$, C≡C, Me	1. $HgSO_4$, H_2SO_4, MeOH, 65°, 12 h 2. p-TsOH, 155°, 10 min 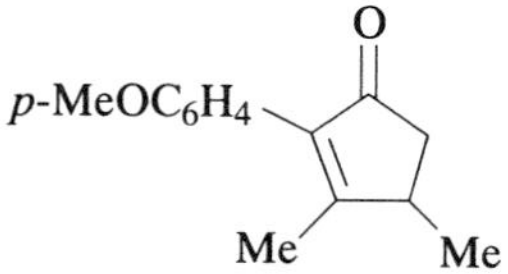	O, p-$MeOC_6H_4$, Me, Me (57)	23

TABLE III. CYCLIZATIONS OF IN SITU GENERATED DIVINYL KETONES: D. CYCLIC ACETYLENES

Reactant	Conditions	Product(s) and Yield(s) (%)	Refs.
C_8 AcO, C≡C, NEt_2	H_3PO_4, HCO_2H, reflux, 5 h	O (7)	155
C_9 C≡C	H_3PO_4, 55-60°	O, Me (31)	8
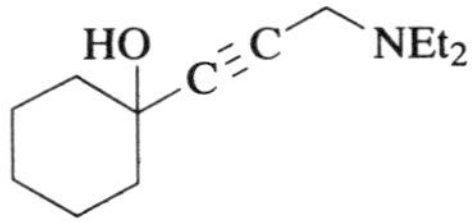 HO, C≡C, NEt_2	H_3PO_4, HCO_2H, reflux; $Hg(OAc)_2$, reflux, 4 h	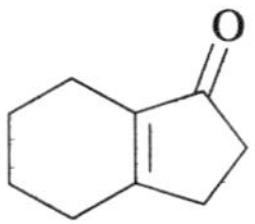 O (52)	155
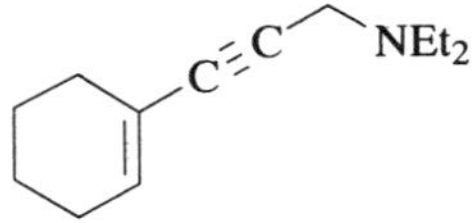 C≡C, NEt_2	H_3PO_4, HCO_2H, reflux; $Hg(OAc)_2$, reflux, 4 h	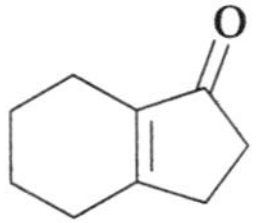 O (48)	155
C_{10} HO, Me, Me, C≡C, OTHP	P_2O_5, $MeSO_3H$, 0°, 5 min	O, Me, Me (—) + O, Me, Me (—) 1:1	148

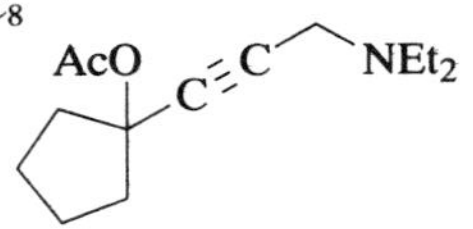

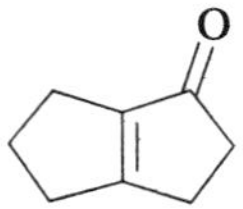

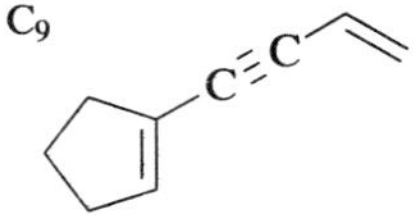

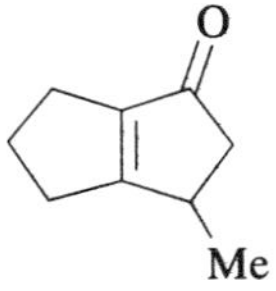

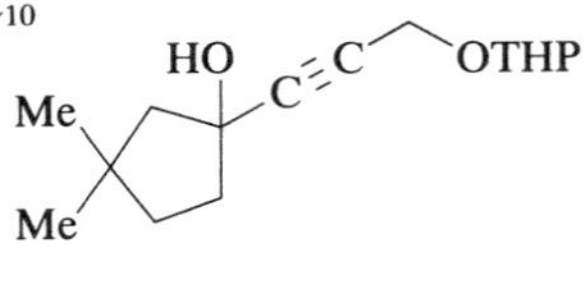

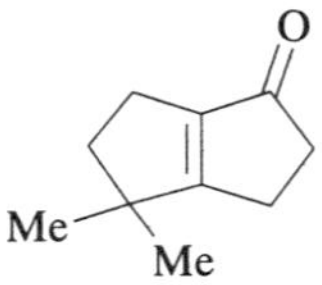

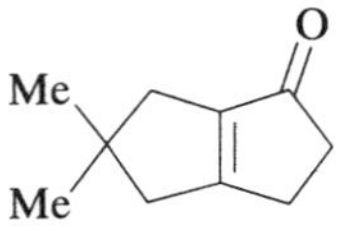

TABLE III. CYCLIZATIONS OF IN SITU GENERATED DIVINYL KETONES: D. CYCLIC ACETYLENES (*Continued*)

Reactant	Conditions	Product(s) and Yield(s) (%)	Refs.
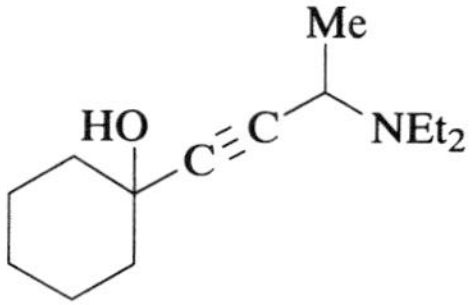 HO, C≡C, Me, NEt_2	H_3PO_4, HCO_2H, reflux, 6 h 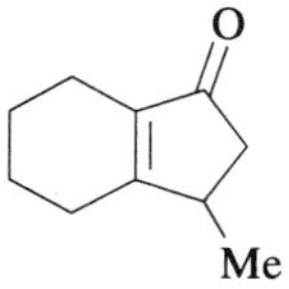	O, Me (51)	155
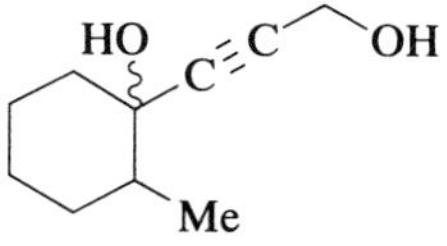 HO, C≡C, OH, Me	H_2SO_4, MeOH, 0°, 30 min	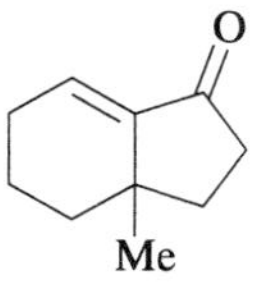 O, Me (70)	147, 149
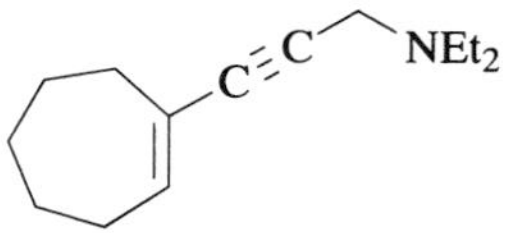 C≡C, NEt_2	H_3PO_4, HCO_2H, reflux, 6 h 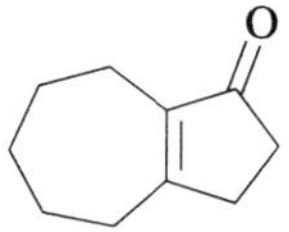	O (8)	155
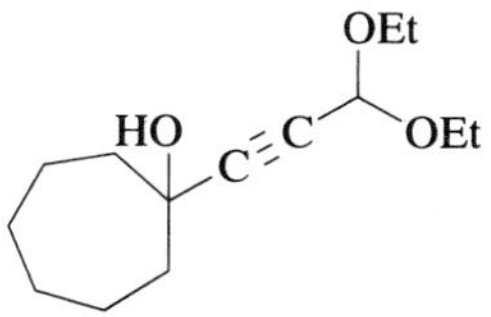 HO, C≡C, OEt, OEt	1. $MeSO_2Cl$, Et_3N, CH_2Cl_2, 0°, 0.5 h 2. H_2SO_4, EtOH, 0°, 0.5-1 h 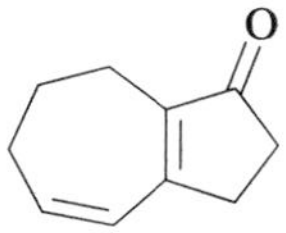	O (25)	156
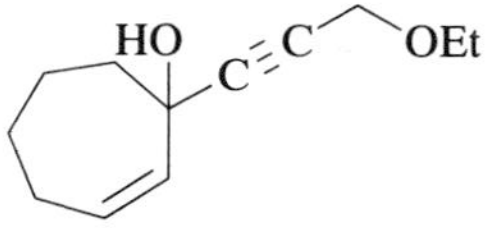 HO, C≡C, OEt	1. Ac_2O, pyridine, rt, 1.5 h 2. H_2SO_4, MeOH, -15°, 2.5 h 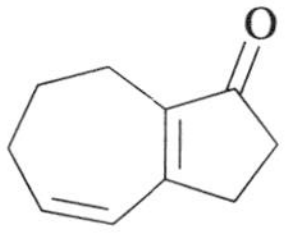	O (49)	147

TABLE III. CYCLIZATIONS OF IN SITU GENERATED DIVINYL KETONES: D. CYCLIC ACETYLENES (*Continued*)

Reactant	Conditions	Product(s) and Yield(s) (%)	Refs.
C_{11} Me, C≡C	HgO, BF_3•OEt_2, Cl_3CCO_2H, MeOH, 20-50°	O, Et (28)	187
Me, C≡C	H_3PO_4, 65°, 6 h	O, Me, Me (70)	13
Me, C≡C	H_3PO_4, 65°, 6 h	O, Me, Me (70)	14
C≡C, Me	H_3PO_4, 60-65°, 6 h	O, Me, Me (65)	31
HO, C≡C, Me	H_3PO_4	O, Me, Me (~60)	31

TABLE III. CYCLIZATIONS OF IN SITU GENERATED DIVINYL KETONES: D. CYCLIC ACETYLENES (*Continued*)

Reactant	Conditions	Product(s) and Yield(s) (%)	Refs.
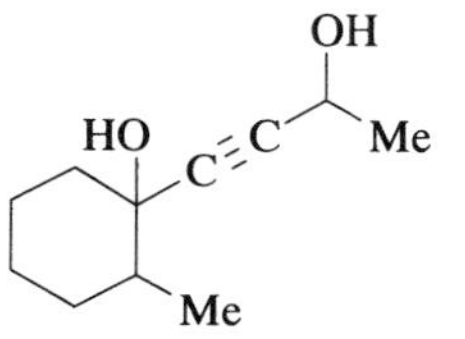 [structure: HO, C≡C, OH, Me, Me]	H_2SO_4, MeOH, 0°, 30 min	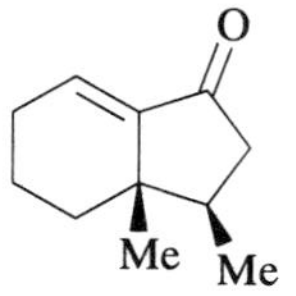 [structure: O, Me, Me] (67) 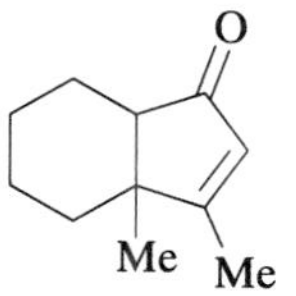 + [structure: O, Me, Me] (10)	147, 149
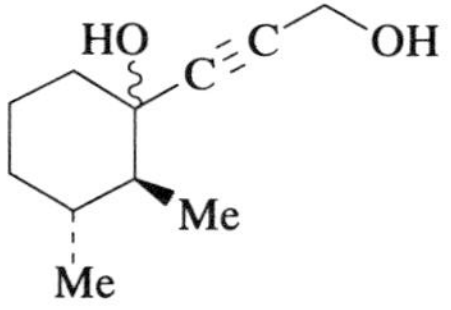 [structure: HO, C≡C, OH, Me, Me]	1. Ac_2O, pyridine, rt, 50, min 2. H_2SO_4, CF_3CH_2OH, 0-20°, 12 h	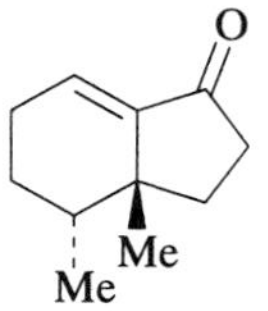 [structure: O, Me, Me] (60)	147, 149
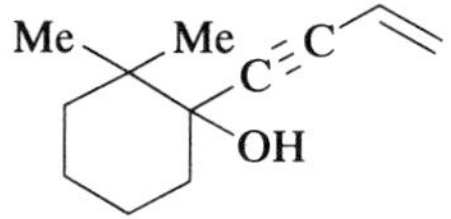 [structure: Me, Me, C≡C, OH]	P_2O_5, $MeSO_3H$, -15°, 10 min	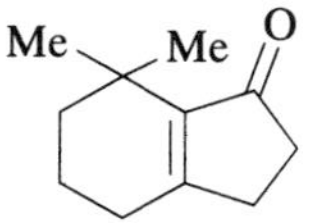 [structure: Me, Me, O] (53)	145
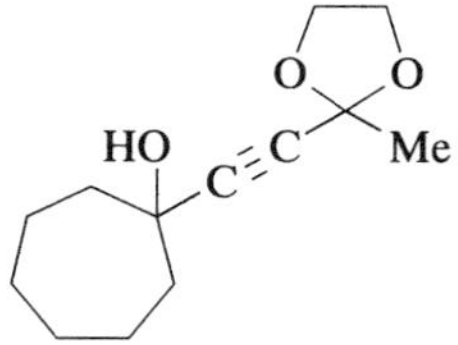 [structure: HO, C≡C, O, O, Me]	1. $MeSO_2Cl$, Et_3N, CH_2Cl_2, 0°, 0.5 h 2. H_2SO_4, EtOH, 0°, 0.5-1 h	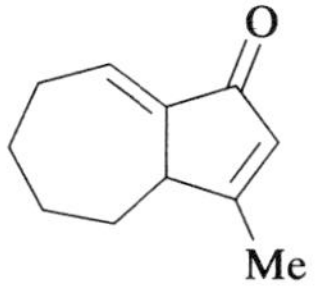 [structure: O, Me] (27) 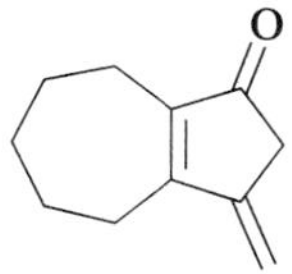 + [structure: O] (23)	156

TABLE III. CYCLIZATIONS OF IN SITU GENERATED DIVINYL KETONES: D. CYCLIC ACETYLENES (*Continued*)

Reactant	Conditions	Product(s) and Yield(s) (%)	Refs.
HO, C≡C, OEt, OEt, Me	1. $MeSO_2Cl$, Et_3N, CH_2Cl_2, 0°, 0.5 h 2. H_2SO_4, EtOH, 0°, 0.5-1 h	O, Me (51)	156
HO, C≡C, OH, Me	H_2SO_4, MeOH, 0°, 4 h	O, Me (67) + Me, O (14)	147
Me, C≡C, NEt_2	H_3PO_4, HCO_2H, reflux, 4 h	O, Me (35)	155
C≡C, OH, OH	1. Ac_2O, pyridine, rt, 1 h 2. H_2SO_4, MeOH, -15°, 25 min	O (42)	147
C_{12}			
OH, HO, C≡C, Me, MeO_2C, Me	H_2SO_4, MeOH, 50°, 30 min	O, MeO_2C, Me, Me (49-60) β:α 60:40	188, 147

TABLE III. CYCLIZATIONS OF IN SITU GENERATED DIVINYL KETONES: D. CYCLIC ACETYLENES (*Continued*)

Reactant	Conditions	Product(s) and Yield(s) (%)	Refs.
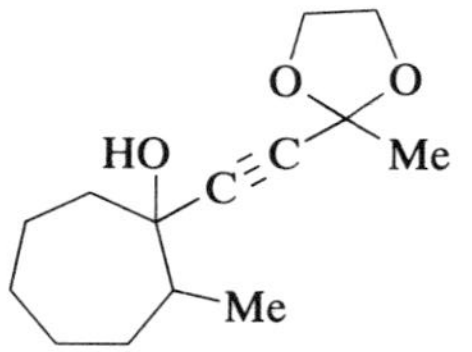	1. $MeSO_2Cl$, Et_3N, CH_2Cl_2, 0°, 0.5 h 2. H_2SO_4, EtOH, 0°, 0.5-1.0 h	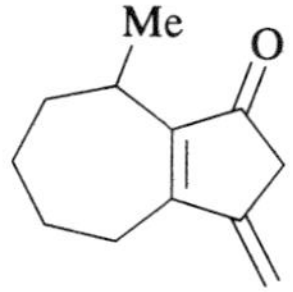 (23)	156
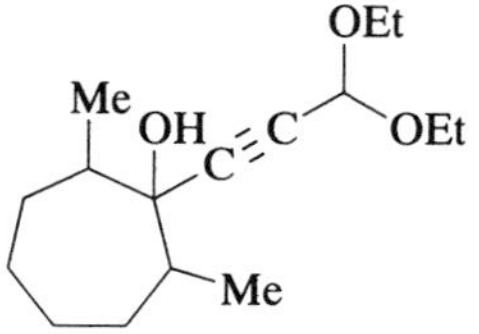	H_2SO_4, EtOH, 0°, 0.5-1.0 h	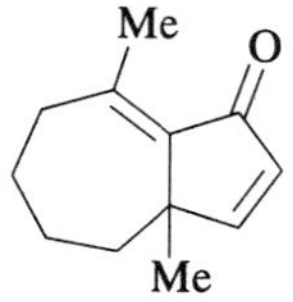 (73)	156
C_{13} 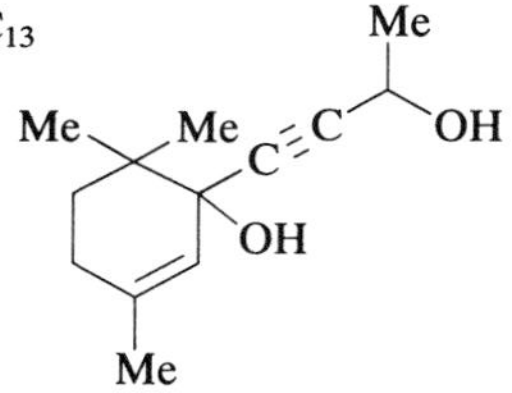	HCO_2H, H_2O, 90°, 20 min	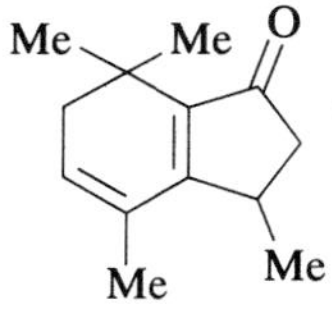 (10) + 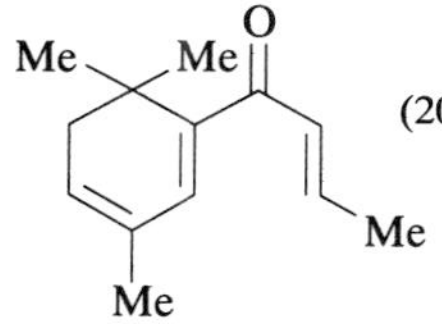 (20)	151
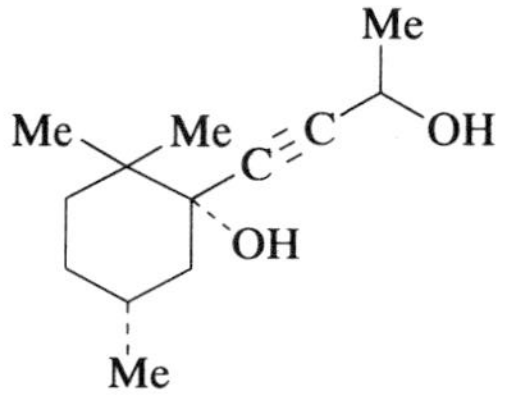	HCO_2H, H_2O, 90°, 20 min	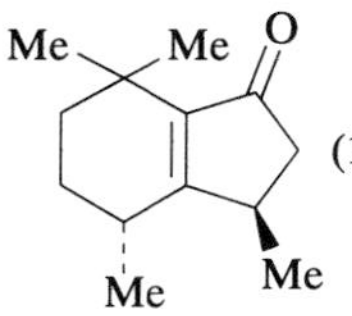 (19) + 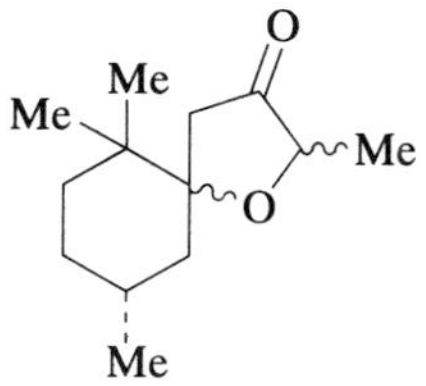 (6)	151

TABLE III. CYCLIZATIONS OF IN SITU GENERATED DIVINYL KETONES: D. CYCLIC ACETYLENES (*Continued*)

Reactant	Conditions	Product(s) and Yield(s) (%)	Refs.
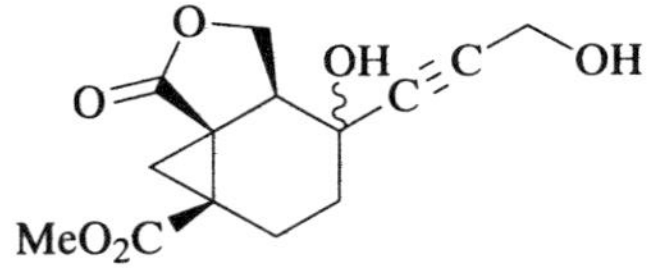 MeO_2C (structure)	P_2O_5, $MeSO_3H$, rt, 2.5 h	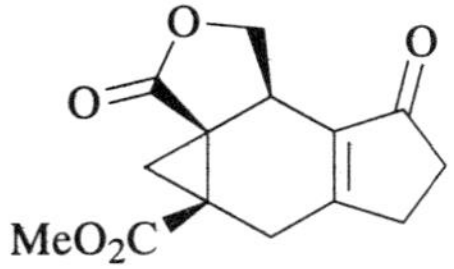 MeO_2C (structure) (34)	150
C_{14}			
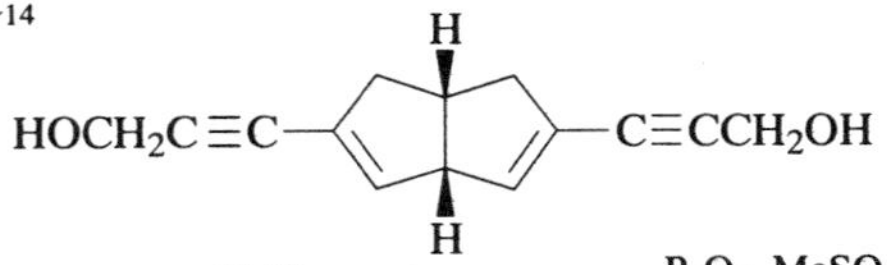 $HOCH_2C{\equiv}C$–(bicycle)–$C{\equiv}CCH_2OH$ + $HOCH_2C{\equiv}C$–(bicycle)–$C{\equiv}CCH_2OH$ (47:53) 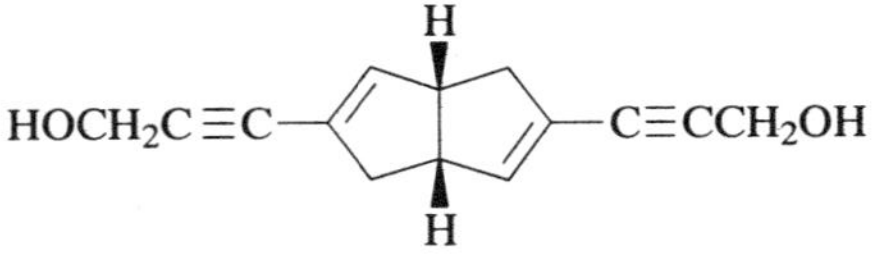	P_2O_5, $MeSO_3H$, rt	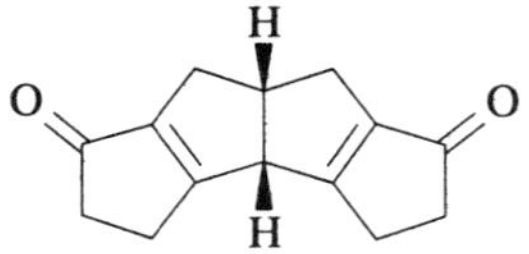 (31) 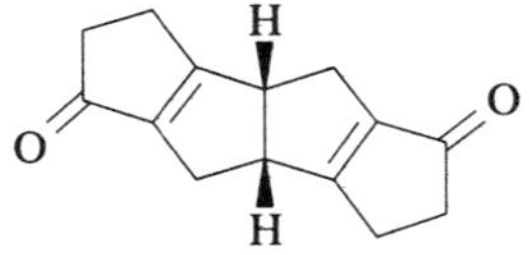 + (46)	143
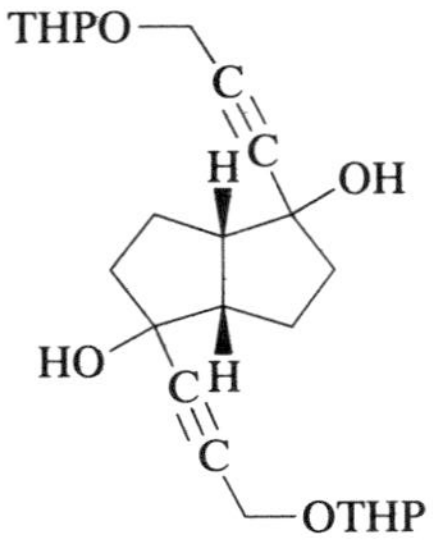 THPO– … –OTHP (diol structure)	1. Amberlite, MeOH 2. P_2O_5, $MeSO_3H$, rt, 5 h	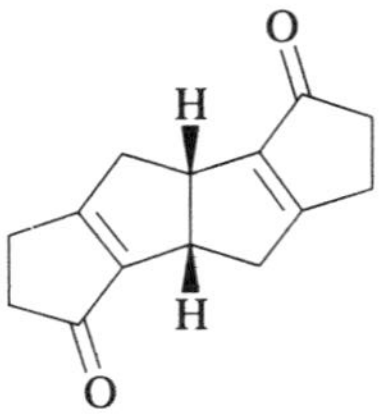 (25)	146

TABLE III. CYCLIZATIONS OF IN SITU GENERATED DIVINYL KETONES: D. CYCLIC ACETYLENES (*Continued*)

Reactant	Conditions	Product(s) and Yield(s) (%)	Refs.
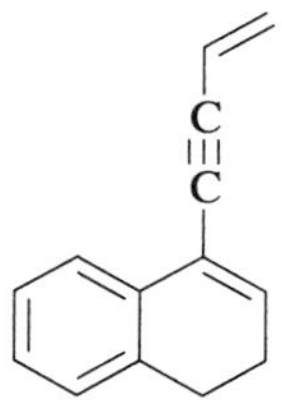 C≡C	H_2SO_4, $HgSO_4$, MeOH, 60°, 10 h	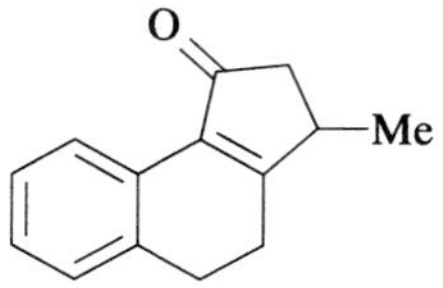 O, Me (48)	22
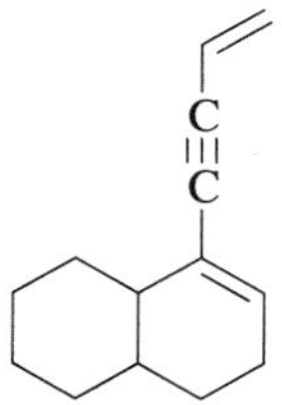 C≡C	H_3PO_4, 65°, 7 h	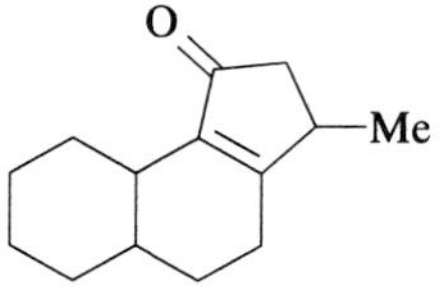 O, Me (27)	22
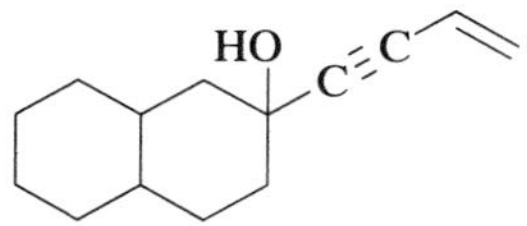 HO, C≡C	H_3PO_4, 70°, 12 h	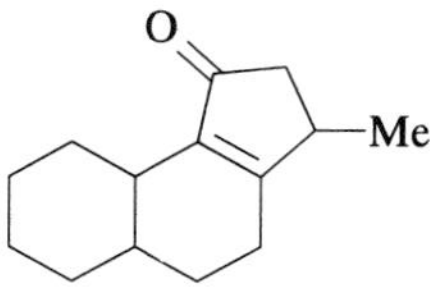 O, Me (40) 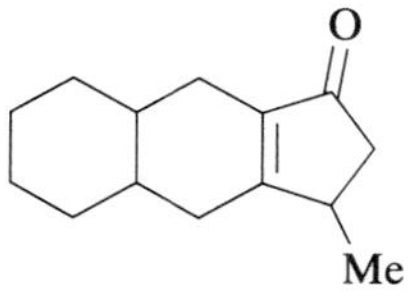 + O, Me (50)	34, 11
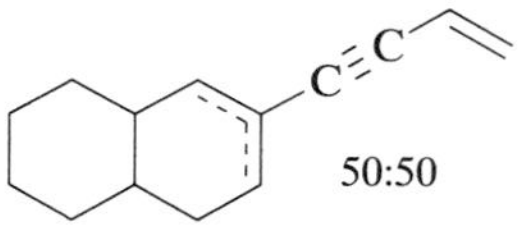 C≡C 50:50	H_3PO_4, 65-70°, 6 h	" (—) + " (—)	11

TABLE III. CYCLIZATIONS OF IN SITU GENERATED DIVINYL KETONES: D. CYCLIC ACETYLENES (*Continued*)

Reactant	Conditions	Product(s) and Yield(s) (%)	Refs.
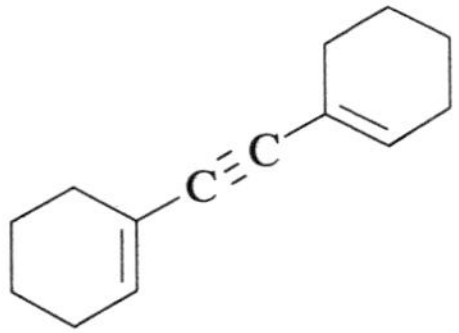	HCO_2H, 55-60°, 10 h	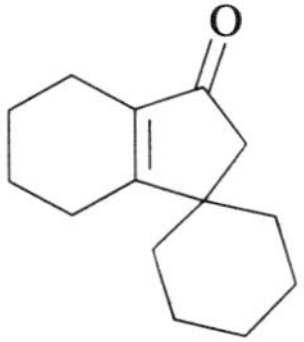 (70)	26
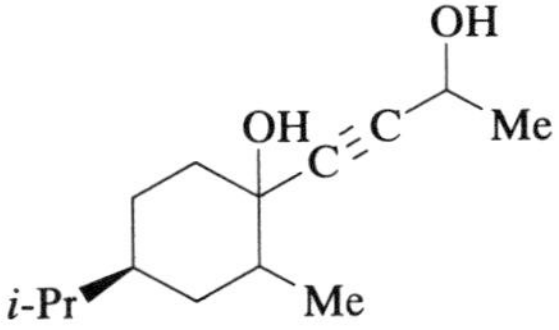	H_2SO_4, MeOH, 0°, 30 min	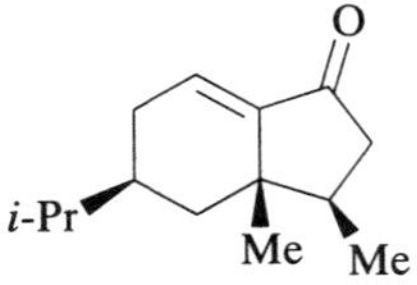 (54) + 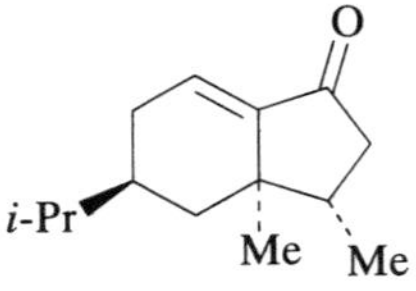 (10)	147, 149
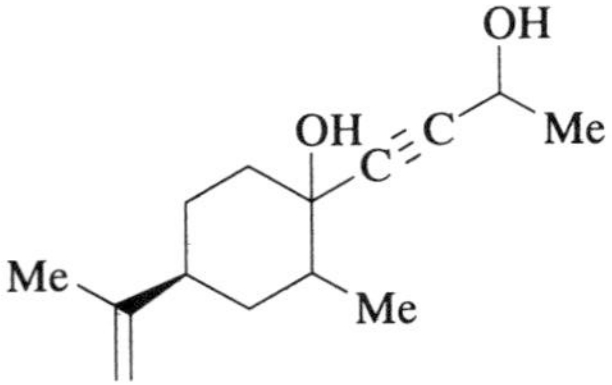	H_2SO_4, MeOH, 0°, 30 min	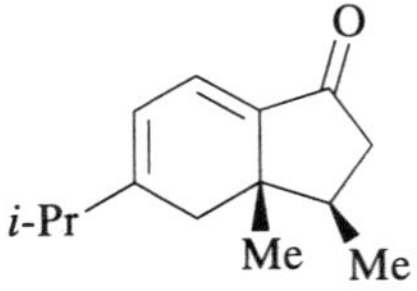 (55)	147, 188
C_{15}			
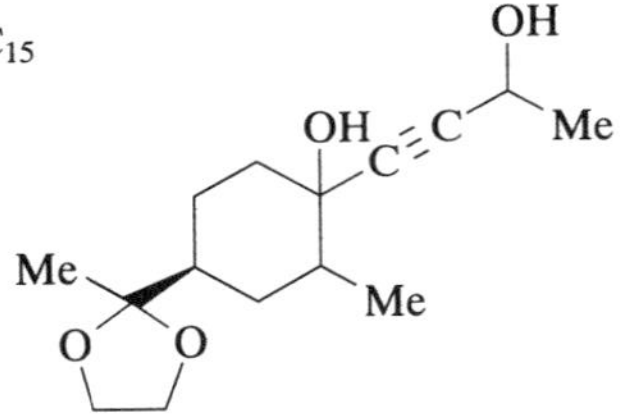	H_2SO_4, MeOH, 0°, 30 min	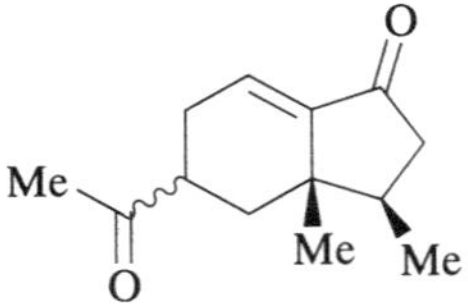 (34)	147

TABLE III. CYCLIZATIONS OF IN SITU GENERATED DIVINYL KETONES: D. CYCLIC ACETYLENES (*Continued*)

Reactant	Conditions	Product(s) and Yield(s) (%)	Refs.
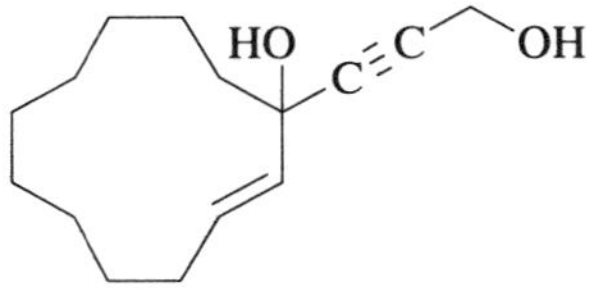 HO, C≡C, OH	H_2SO_4, MeOH, 0°, 30 min	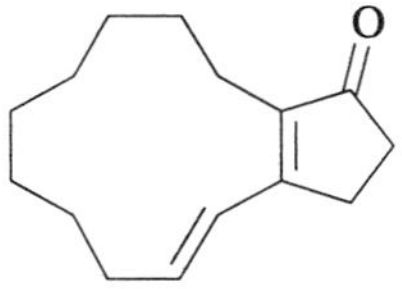 O (65)	147, 153
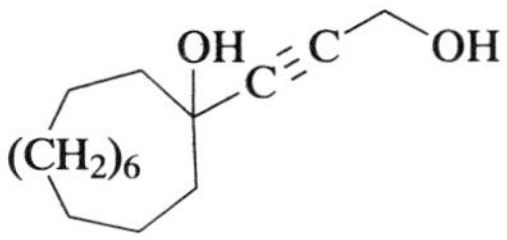 $(CH_2)_6$, OH, C≡C, OH	Cation resin KU-2, HOAc, reflux, 15 h	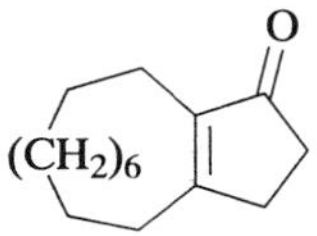 O, $(CH_2)_6$ (73)	154
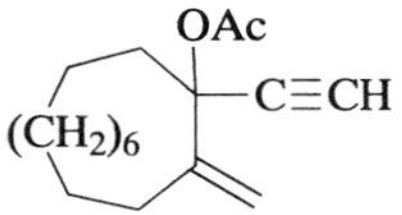 OAc, $(CH_2)_6$, C≡CH	$PdCl_2(MeCN)_2$, HOAc, MeCN, 60-80°	" (78-89)	142
C_{16}			
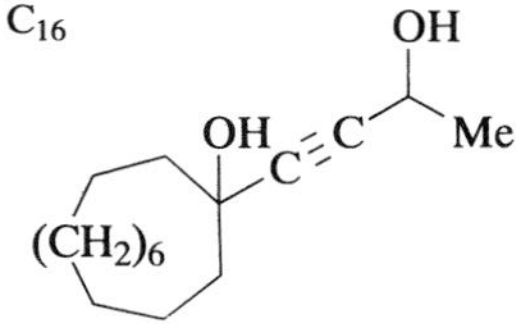 OH, OH, C≡C, Me, $(CH_2)_6$	HOAc, H_2SO_4	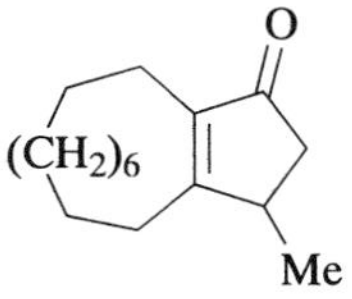 O, $(CH_2)_6$, Me (—)	152
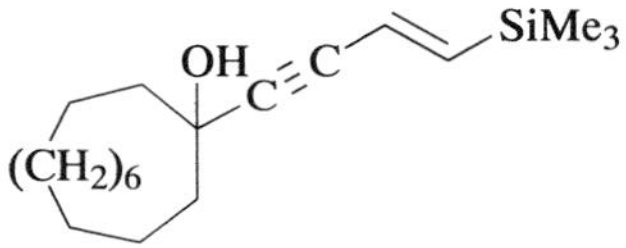 OH, C≡C, SiMe$_3$, $(CH_2)_6$	HOAc, H_2SO_4, 50-60°, 10 min	" (64)	141

259640

TABLE III. CYCLIZATIONS OF IN SITU GENERATED DIVINYL KETONES: D. CYCLIC ACETYLENES (*Continued*)

Reactant	Conditions	Product(s) and Yield(s) (%)	Refs.
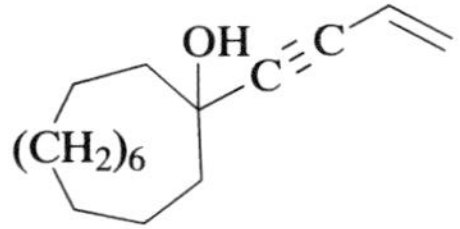 (CH2)6 ring, OH, C≡C–CH=CH2	H_2SO_4, HOAc, (1:10), reflux, 2 h H_2SO_4, HOAc, (1:20), reflux, 2 h Cation resin KU-2, HOAc, reflux, 15 h	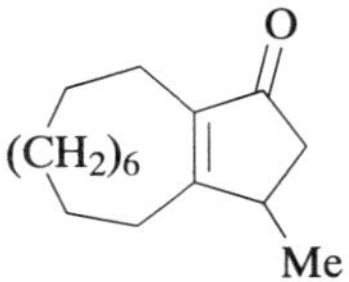 (CH2)6, O, Me (73) (75) (80)	140
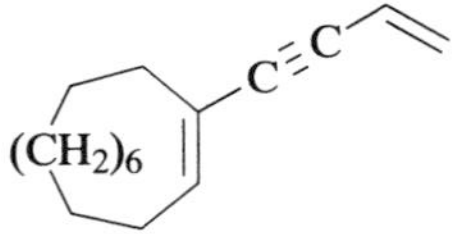 (CH2)6 ring, C≡C–CH=CH2	P_2O_5, H_3PO_4, heptane, 90°, 8 h H_2SO_4, HOAc (1:9), 100°, 13 h	" (79) " (45)	140
C_{20}			
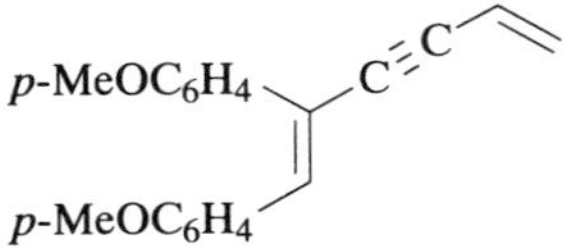 p-MeOC6H4, p-MeOC6H4, C≡C–CH=CH2	H_2SO_4, $HgSO_4$, MeOH, 65°, 6 h	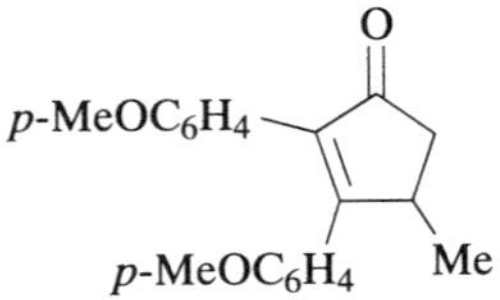 p-MeOC6H4, p-MeOC6H4, O, Me (60)	23
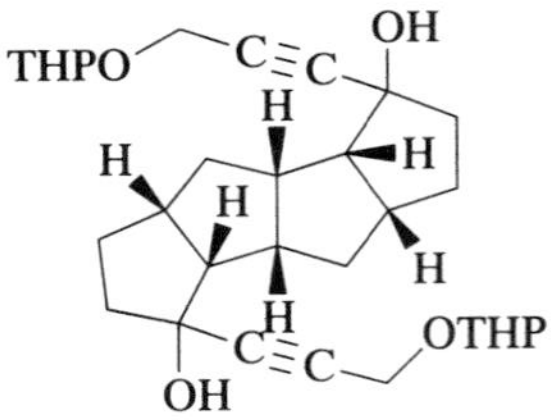 THPO, C≡C, OH, H, H, H, H, H, H, C≡C, OTHP, OH	1. Amberlite, MeOH 2. P_2O_5, $MeSO_3H$, rt, 5 h	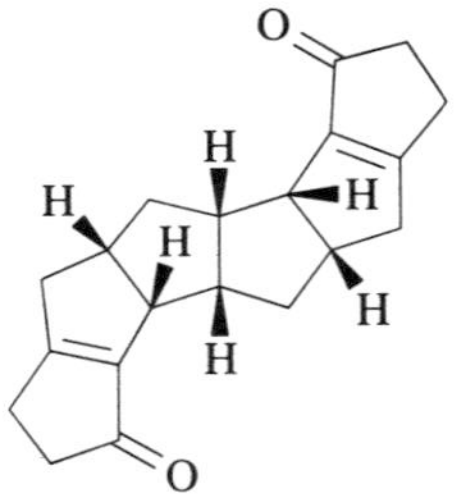 O, H, H, H, H, H, H, O (45)	146

TABLE III. CYCLIZATIONS OF IN SITU GENERATED DIVINYL KETONES: E. α-VINYLCYCLOBUTANONES

Reactant	Conditions	Product(s) and Yield(s) (%)	Refs.
C_{10}			
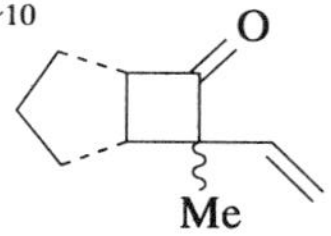 Me α:β 70:30	$MeSO_3H$, rt, 30 min $MeSO_3H$, $CDCl_3$, 60°, 30 min	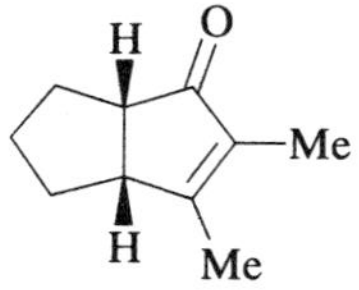 (46) (76)	107, 108
C_{11}	$MeSO_3H$, CH_2Cl_2, rt, 26 h	(71) + (8)	107, 108
C_{12}	$MeSO_3H$, CH_2Cl_2, rt, 2 h	(51) + (20)	107, 108
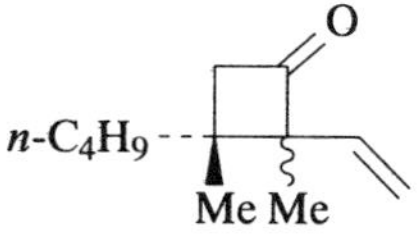 n-C_4H_9, Me Me Me α:β 50:50	$MeSO_3H$, CH_2Cl_2, rt, 16 h	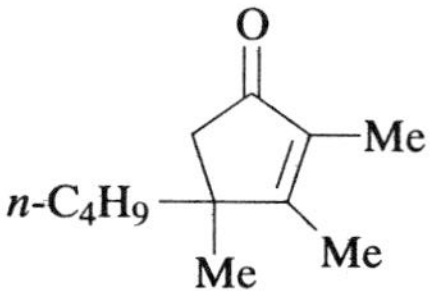 (65) 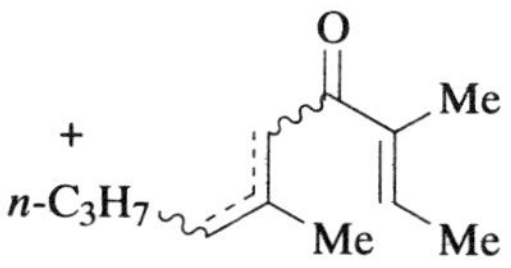 + (11)	107, 108

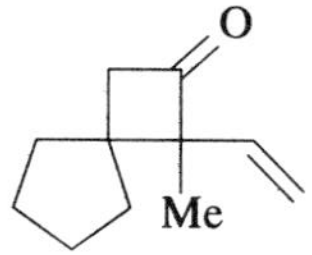

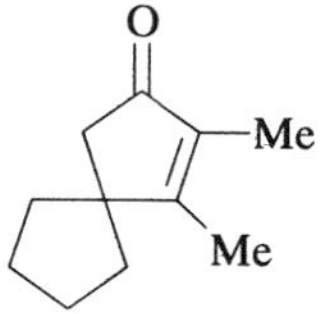

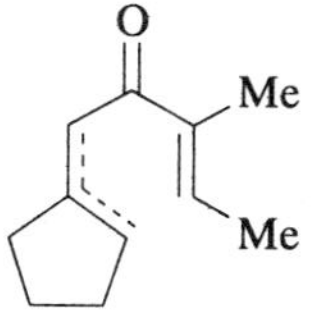

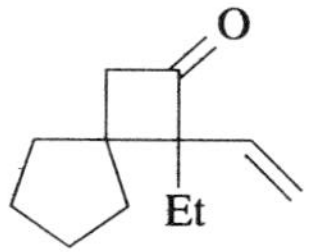

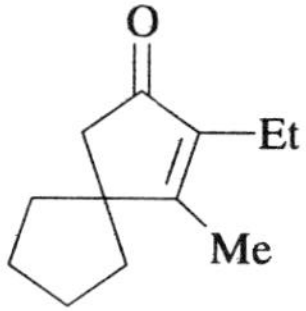

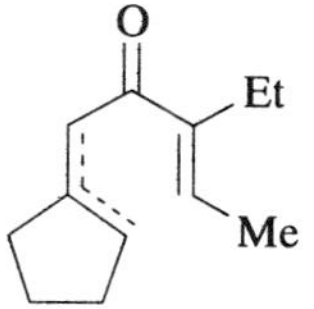

TABLE III. CYCLIZATIONS OF IN SITU GENERATED DIVINYL KETONES: E. α-VINYLCYCLOBUTANONES (*Continued*)

Reactant	Conditions	Product(s) and Yield(s) (%)	Refs.
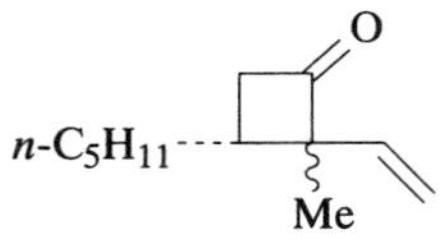 Me α:β 70:30	$MeSO_3H$, CH_2Cl_2, 0°, 65 min	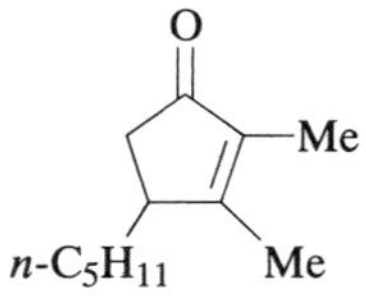 (51) + 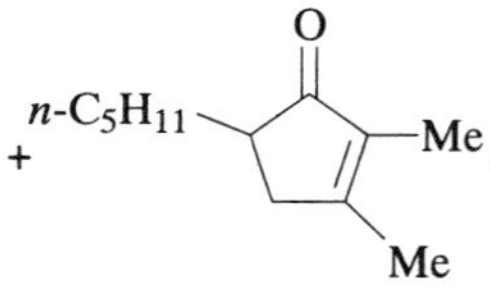 (4)	107, 108
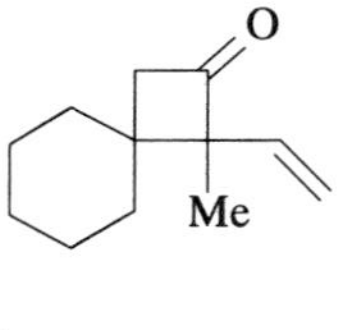	$MeSO_3H$, CH_2Cl_2, rt, 24 h	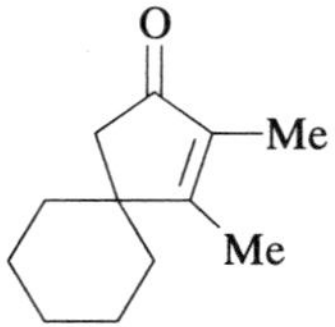 (66) + 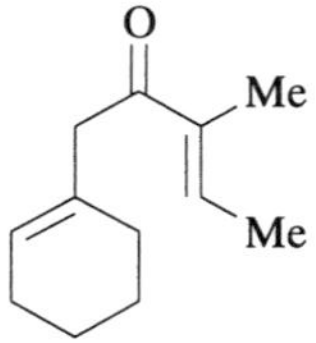 (4)	107, 108
C_{13} 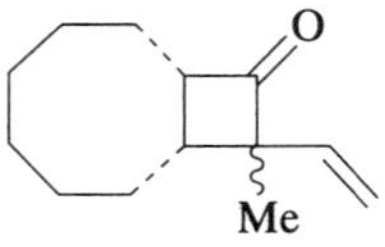 Me α:β 75:25	$MeSO_3H$, rt, 30 min	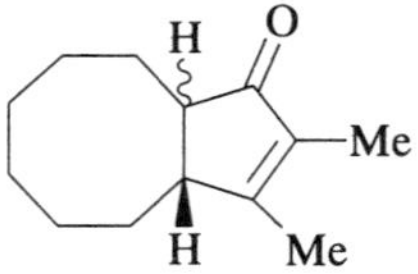 (43) 76:24 + 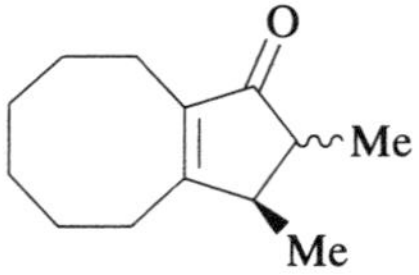 (15) 58:42	107, 108
C_{14} 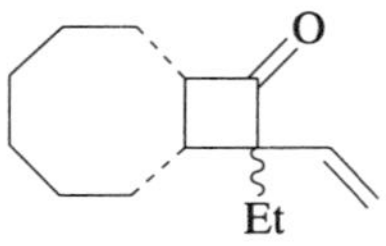 Et α:β 85:15	$MeSO_3H$, rt, 4 h	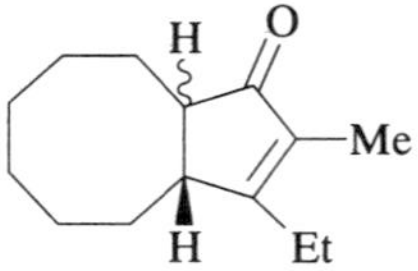 (39) 69:31 + 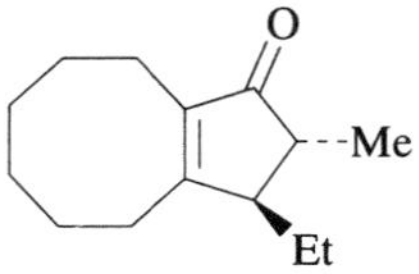 (15)	107, 108

TABLE IV. CYCLIZATION OF DIVINYL KETONE EQUIVALENTS FROM SOLVOLYSIS: A. GEMINAL DICHLORIDES

Reactant	Conditions	Product(s) and Yield(s) (%)		Refs.
C_6				
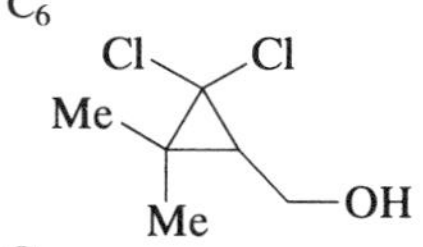 Cl, Cl, Me, Me, OH	47% HBr, 100°, 2 h	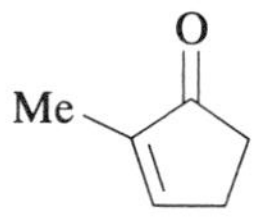 O, Me	(44)	141, 109
C_7				
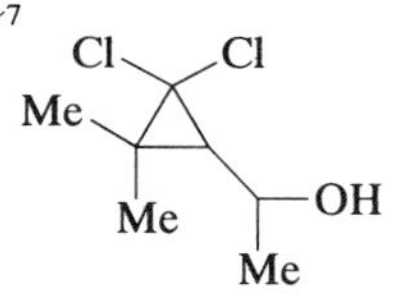 Cl, Cl, Me, Me, OH, Me	47% HBr, 100°, 2 h	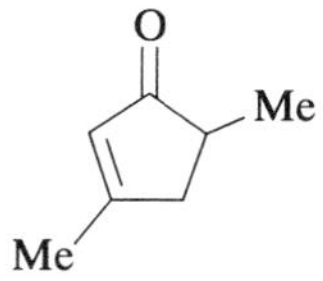 O, Me, Me	(83)	141, 109
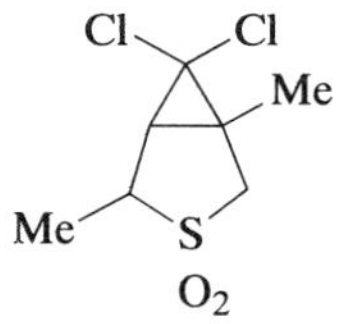 Cl, Cl, Me, Me, S, O_2	80% HOAc, reflux, 1-3 h	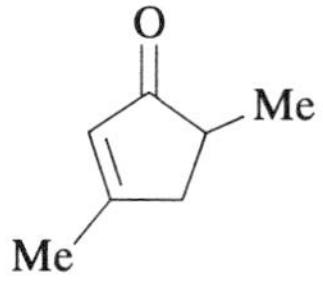 O, Me, Me	(67)	112
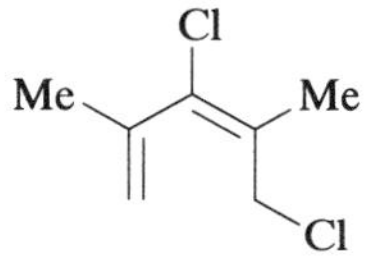 Cl, Me, Me, Cl	80% HOAc, reflux	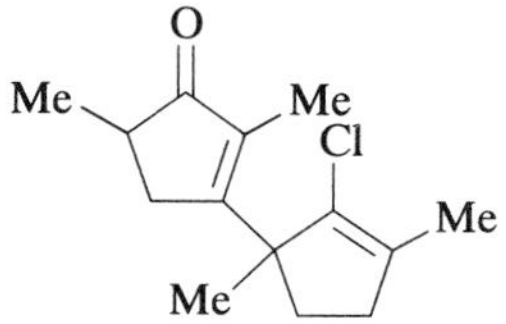 O, Me, Me, Cl, Me, Me	(55-60)	189
C_9				
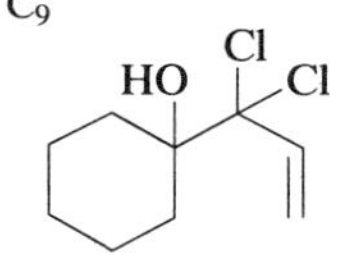 HO, Cl, Cl	TFA, rt, 1.5 h	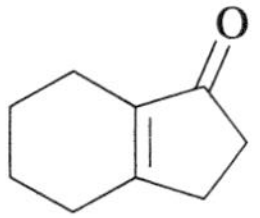 O	(71)	110, 109

TABLE IV. CYCLIZATION OF DIVINYL KETONE EQUIVALENTS FROM SOLVOLYSIS: A. DICHLORIDES (*Continued*)

Reactant	Conditions	Product(s) and Yield(s) (%)	Refs.
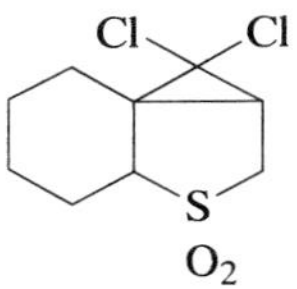 Cl, Cl, S, O_2	80% HOAc, reflux 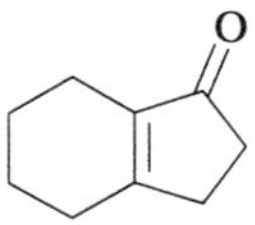	O (48)	112
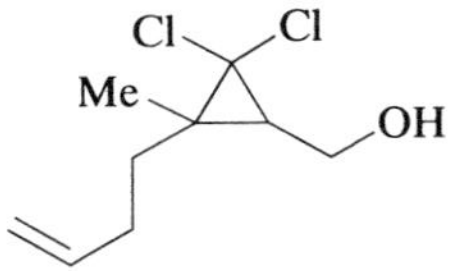 Cl, Cl, Me, OH	47% HBr, 100°, 2 h	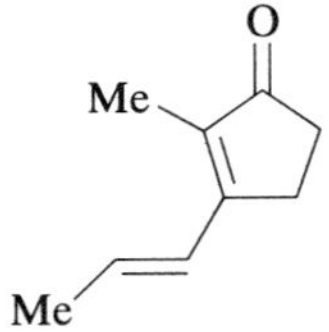 O, Me, Me (70)	111, 109
C_{10}			
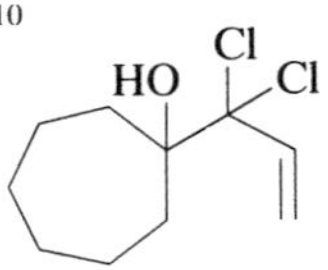 HO, Cl, Cl	TFA, rt, 1.5 h	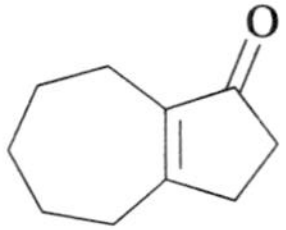 O (87)	110, 109
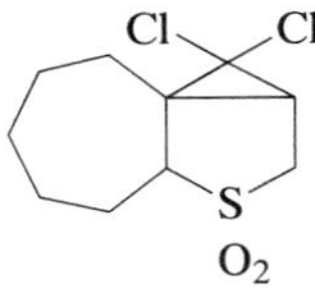 Cl, Cl, S, O_2	80% HOAc, reflux 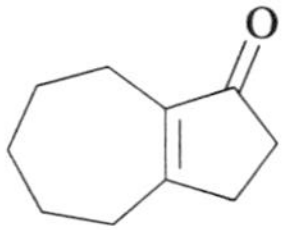	O (80)	112
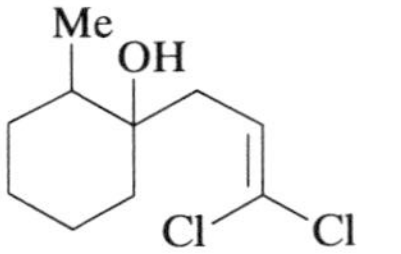 Me, OH, Cl, Cl	TFA, rt, 2.5 h 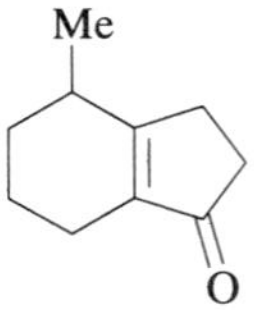	Me, O (47)	109

TABLE IV. CYCLIZATION OF DIVINYL KETONE EQUIVALENTS FROM SOLVOLYSIS: A. DICHLORIDES (*Continued*)

Reactant	Conditions	Product(s) and Yield(s) (%)	Refs.
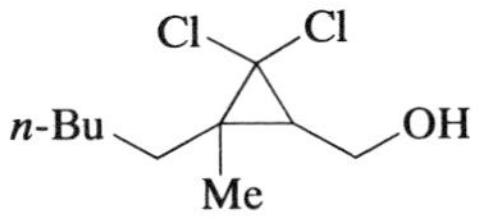 Cl, Cl, *n*-Bu, OH, Me	47% HBr, 100°, 2 h	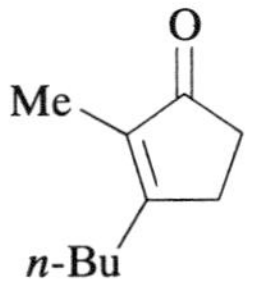 O, Me, *n*-Bu (56)	111, 109
C_{11}			
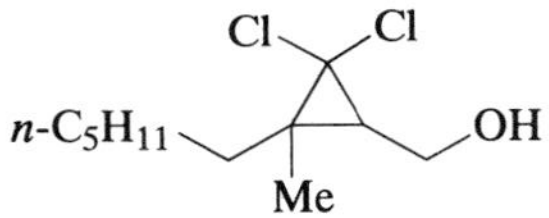 Cl, Cl, *n*-C_5H_{11}, OH, Me	47% HBr, 100°, 2 h	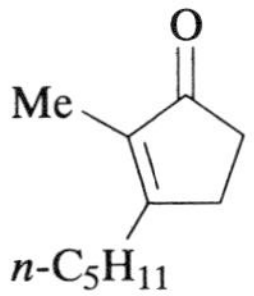 O, Me, *n*-C_5H_{11} (59)	111, 109
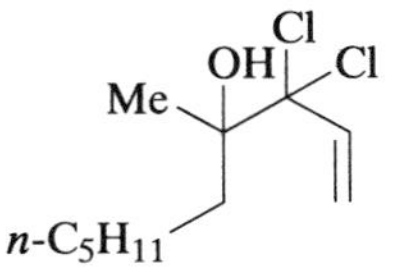 Cl, OH, Cl, Me, *n*-C_5H_{11}	TFA, rt, 1.5 h	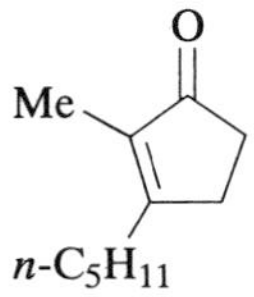 O, Me, *n*-C_5H_{11} (87)	109
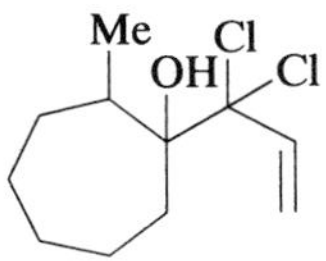 Me, Cl, OH, Cl	TFA, rt, 1.5 h	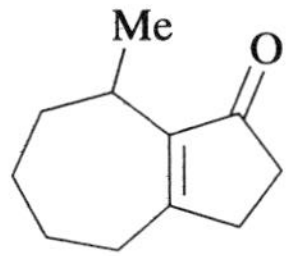 Me, O (58)	110, 109
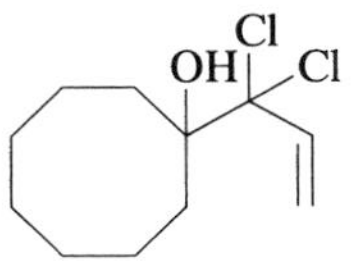 Cl, OH, Cl	TFA, rt, 1.5 h	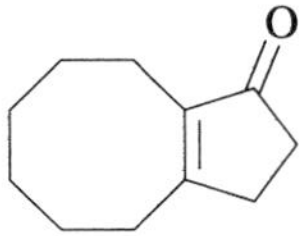 O (80)	110, 109

TABLE IV. CYCLIZATION OF DIVINYL KETONE EQUIVALENTS FROM SOLVOLYSIS: A. DICHLORIDES (*Continued*)

Reactant	Conditions	Product(s) and Yield(s) (%)	Refs.
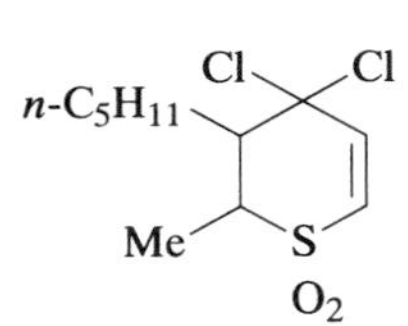 n-C_5H_{11}, Cl, Cl, Me, S, O_2	80% HOAc, reflux	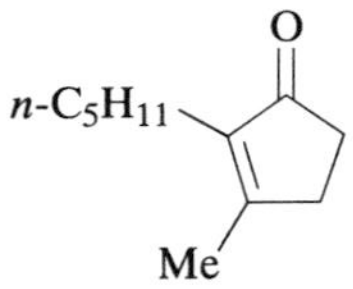 n-C_5H_{11}, O, Me (70)	112
C_{13} 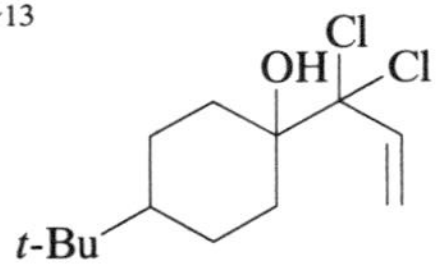 OH, Cl, Cl, t-Bu	TFA, rt, 1.5 h	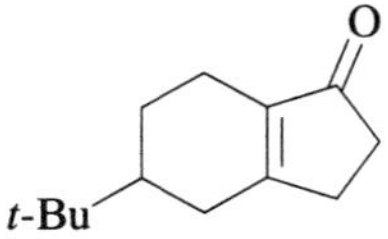 O, t-Bu (72)	110, 109
C_{15} 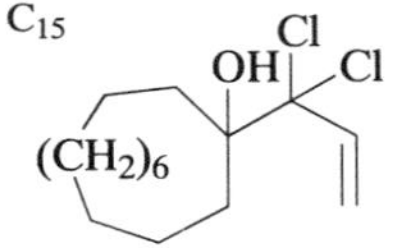 OH, Cl, Cl, $(CH_2)_6$	TFA, rt, 2.5 h	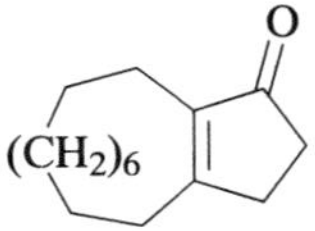 O, $(CH_2)_6$ (90)	110, 109
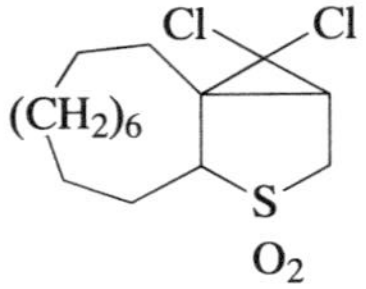 Cl, Cl, $(CH_2)_6$, S, O_2	80% HOAc, reflux	" (60)	112
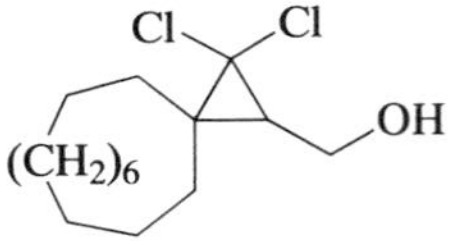 Cl, Cl, OH, $(CH_2)_6$	47% HBr, 100°, 2 h	" (37)	111, 109

TABLE IV. CYCLIZATION OF DIVINYL KETONE EQUIVALENTS FROM SOLVOLYSIS: B. 2-FURYLCARBINOLS

Reactant	Conditions	Product(s) and Yield(s) (%)		Refs.
C_6				
OH, Me, Br	H_2SO_4, H_2O, DME, 85-90°, 6 h	O, Me, Br, OH	(81)	137
OH, Me	PPA, 50°, 24 h	O, Me, OH	(30)	132
C_7				
OH, Me, Me, Br	H_2SO_4, H_2O, DME, 85-90°, 12 h	O, Me, Me, Br, OH	(28)	137
C_9				
Cl, $P(OEt)_2$, O, R^1, R^2	DME, H_2O, rt	O, O, R^2, $P(OEt)_2$, R^1, OH		139
R^1 = H, R^2 = H		(54)		
R^1 = Br, R^2 = H		(60)		
R^1 = H, R^2 = Br		(78)		

TABLE IV. CYCLIZATION OF DIVINYL KETONE EQUIVALENTS FROM SOLVOLYSIS: B. 2-FURYLCARBINOLS (*Continued*)

Reactant	Conditions	Product(s) and Yie ld(s) (%)		Refs.
OH, Me, O	$ZnCl_2$, H_2O, acetone, 60°, 72 h	O, OH, Me	(35)	135
C_{10}				
Cl, Me, O, P(OEt)$_2$, O, Br	DME, H_2O, rt	O, O, P(OEt)$_2$, OH, Br, Me	(33)	139
OH, Me, O, Bu-*n*	$ZnCl_2$, H_2O, acetone, 60°, 72 h	O, Bu-*n*, OH, Me	(18)	135
OH, Me, O, S	$ZnCl_2$, H_2O, acetone, 60°, 4 h	O, S, OH, Me	(85)	135
C_{11}				
OH, O, C_6H_{13}-*n*	PPA, 50°, 24 h	O, C_6H_{13}-*n*, OH	(70)	132

TABLE IV. CYCLIZATION OF DIVINYL KETONE EQUIVALENTS FROM SOLVOLYSIS: B. 2-FURYLCARBINOLS (*Continued*)

Reactant	Conditions	Product(s) and Yield(s) (%)		Refs.
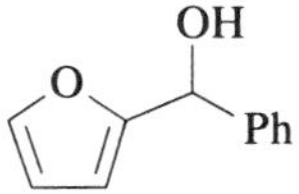 OH, Ph, Br	PPA, 50°, 24 h	O, Ph, OH	(65)	132
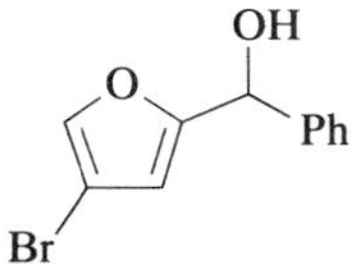 OH, Ph, Br	DME, H_2O, H_2SO_4, 85-90°, 1 h	O, Ph, OH, Br	(85)	137
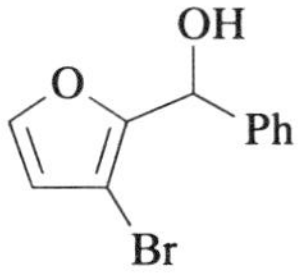 OH, Ph, Br	DME, H_2O, H_2SO_4, 85-90°, 0.3 h	O, Br, Ph, OH	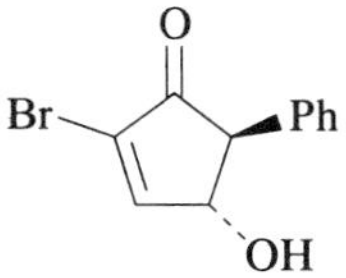 (85)	137
C_{12}				
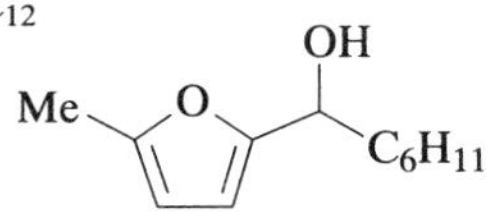 Me, OH, C_6H_{11}	$ZnCl_2$, H_2O, acetone, 60°, 72 h	O, C_6H_{11}, OH, Me	(16)	135
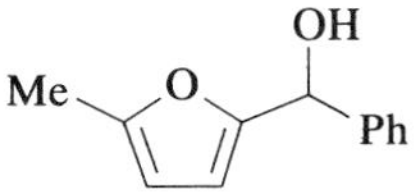 Me, OH, Ph	$ZnCl_2$, H_2O, acetone, 60°, 24 h	O, Ph, OH, Me	(70)	135

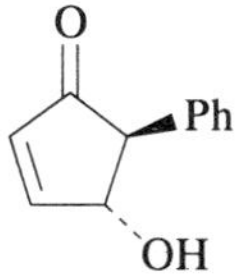

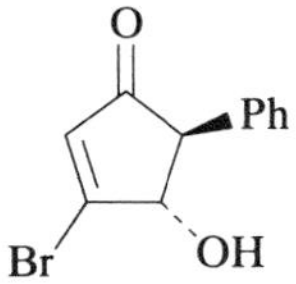

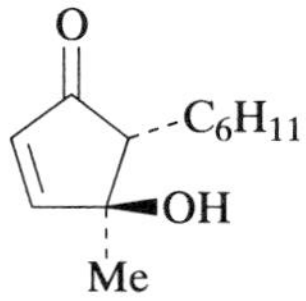

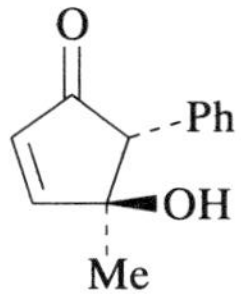

TABLE IV. CYCLIZATION OF DIVINYL KETONE EQUIVALENTS FROM SOLVOLYSIS: B. 2-FURYLCARBINOLS (*Continued*)

Reactant	Conditions	Product(s) and Yield(s) (%)	Refs.
Ph, Br, O, OH, Me	DME, H_2O, H_2SO_4, 85-90°, 48 h	O, Me, OH, Br, Ph (30)	137
C_{13}			
Me, O, OH, C_6H_4Me-*p*	$ZnCl_2$, H_2O, acetone, 60°, 4 h	O, C_6H_4Me-*p*, OH, Me (65)	135
O, OH, Ph	Acetone, H_2O, 80°, 1.5 h	O, Ph, OH (65)	138
C_{14}			
Me, O, OH, Ph	Acetone, H_2O, 70°, 48 h	O, Ph, OH, Me (30)	138
	MeCN, H_2O, 55-60°	O, Ph, OH, Me (28) + O, Me, O, Ph (2 0)	138

TABLE IV. CYCLIZATION OF DIVINYL KETONE EQUIVALENTS FROM SOLVOLYSIS: B. 2-FURYLCARBINOLS (*Continued*)

Reactant	Conditions	Product(s) and Yield(s) (%)	Refs.
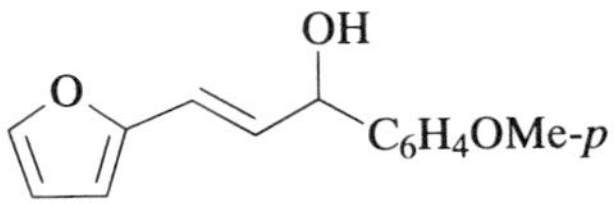 OH, C_6H_4OMe-*p*	Acetone, H_2O, 50°, 17 h	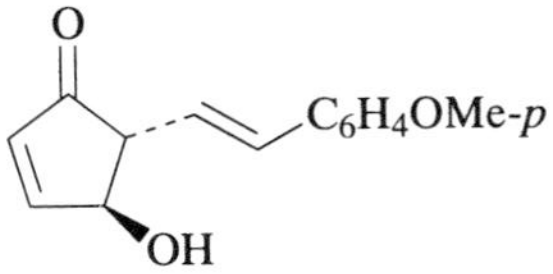 O, C_6H_4OMe-*p*, OH (40)	138
C_{15}			
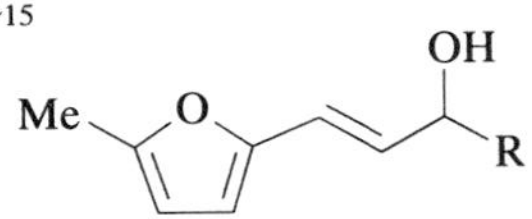 Me, O, OH, R; R = C_6H_4Me-*p*	MeCN, H_2O, 55-60°	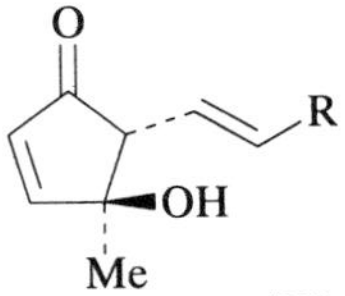 O, R, OH, Me (27) 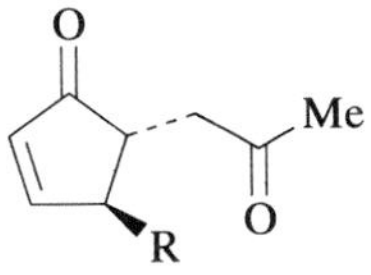 + O, Me, R, O (20)	138
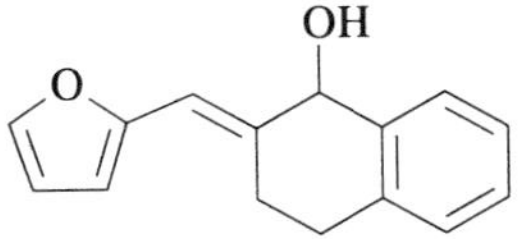 O, OH	Acetone, H_2O, 50°, 48 h	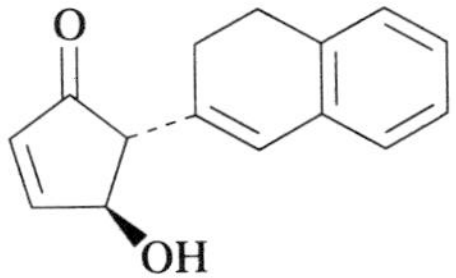 O, OH (43)	138
	MeCN, H_2O, 55-60°	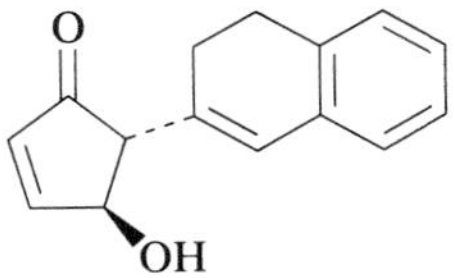 O, OH (37) 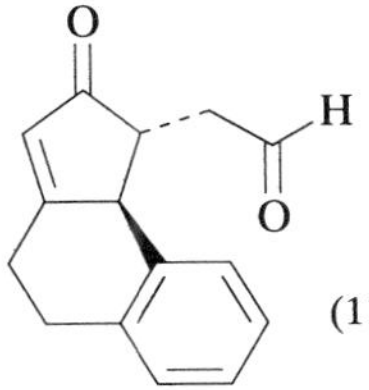 + O, H, O (11)	138

TABLE IV. CYCLIZATION OF DIVINYL KETONE EQUIVALENTS FROM SOLVOLYSIS: B. 2-FURYLCARBINOLS (*Continued*)

Reactant	Conditions	Product(s) and Yield(s) (%)	Refs.
C_{16}			
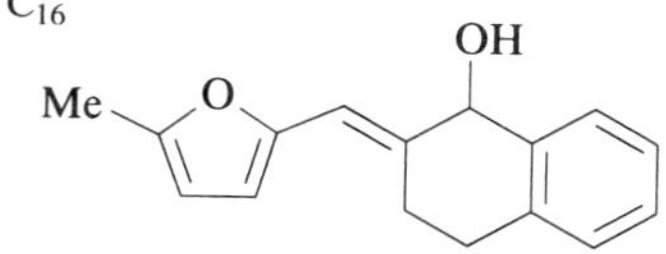 OH, Me, O	Acetone, H_2O, 70°, 48 h	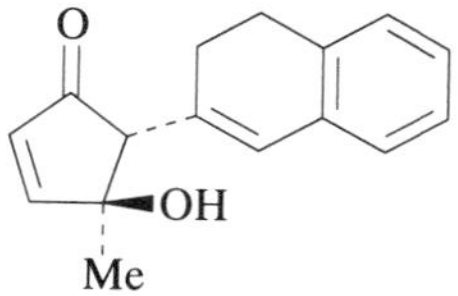 O, OH, Me (32)	138
	MeCN, H_2O, 55-60°	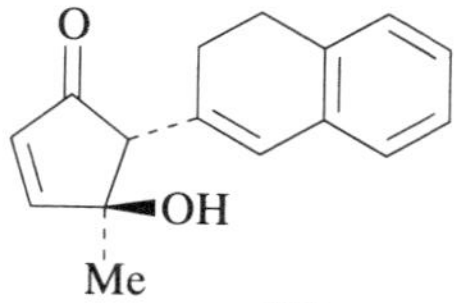 O, OH, Me (25) 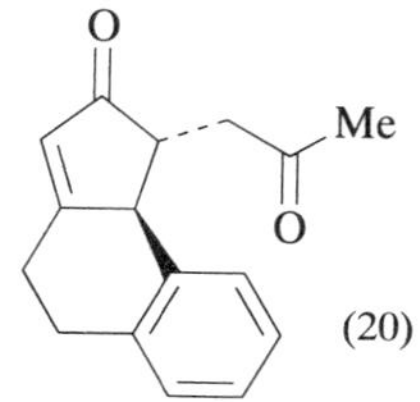 + O, Me, O (20)	138
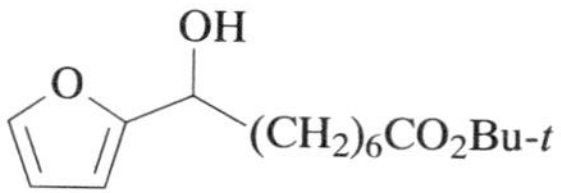 OH, O, $(CH_2)_6CO_2Bu$-*t*	PPA, 50°, 24 h	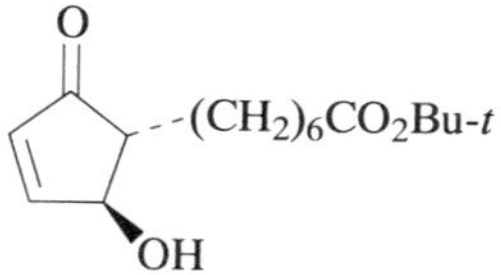 O, $(CH_2)_6CO_2Bu$-*t*, OH (51)	133
C_{17}			
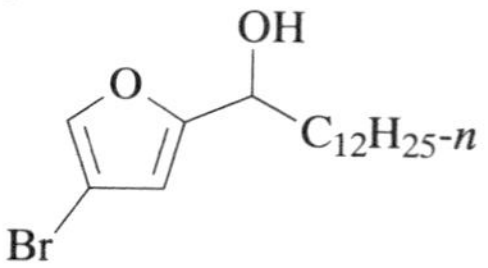 OH, O, $C_{12}H_{25}$-*n*, Br	DME, H_2O, H_2SO_4, 85-90°, 12 h	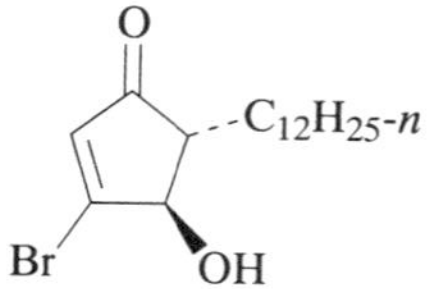 O, $C_{12}H_{25}$-*n*, Br, OH (75)	137

TABLE IV. CYCLIZATION OF DIVINYL KETONE EQUIVALENTS FROM SOLVOLYSIS: B. 2-FURYLCARBINOLS (*Continued*)

Reactant	Conditions	Product(s) and Yield(s) (%)	Refs.
OH, $C_{12}H_{25}$-*n*, Br	DME, H_2O, H_2SO_4, 85-90°, 6 h	O, Br, $C_{12}H_{25}$-*n*, OH (82)	137
C_{26} OH, R	H_2SO_4, H_2O, acetone, 50°, 30 h	O, R, OH (90)	134
R = Me, Me, Me, H, H, H, H, HO			

TABLE IV. CYCLIZATION OF DIVINYL KETONE EQUIVALENTS FROM SOLVOLYSIS: C. VINYLALLENES

Reactant	Conditions	Product(s) and Yield(s) (%)	Refs.
C_6			
Me, C=C=CH2 (structure)	PNPBA, CH_2Cl_2, 0°, 24 h	Me, O (structure) (60)	114, 113
	$Hg(OAc)_2$, $HClO_4$, HOAc, 25°, 1 h	(31)	127
H, C=C=C, H, OH (structure)	*t*-BuOOH, $VO(acac)_2$, CH_2Cl_2, rt	O, OH (structure) (40-70)	125
C_7			
Me, C=C=CH2, Me (structure)	PNPBA, CH_2Cl_2, 0°, 24 h	Me, O, Me (structure) (39) + O, Me, Me, C=C=CH2 (structure) (21)	114, 113
Me, C=C=C, Me, H (structure) racemic	$Hg(OAc)_2$, $HClO_4$, HOAc, 25°, 1 h	Me, O, Me (structure) racemic (20)	127
(*R*)	MCPBA, CH_2Cl_2, 0°, 24 h	(*S*) (—)	116

TABLE IV. CYCLIZATION OF DIVINYL KETONE EQUIVALENTS FROM SOLVOLYSIS: C. VINYLALLENES (*Continued*)

Reactant	Conditions	Product(s) and Yield(s) (%)	Refs.
C_8			
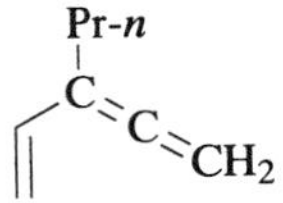 [structure]	PNPBA, CH_2Cl_2, 0°, 24 h	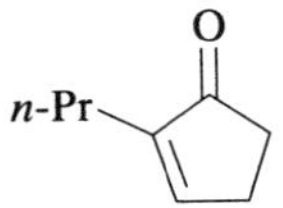 [structure] (60)	114, 113
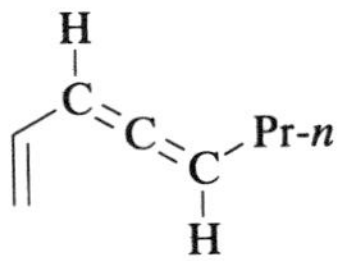 [structure]	PNPBA, CH_2Cl_2, 0°, 24 h	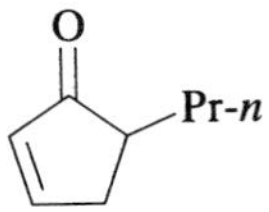 [structure] (57) 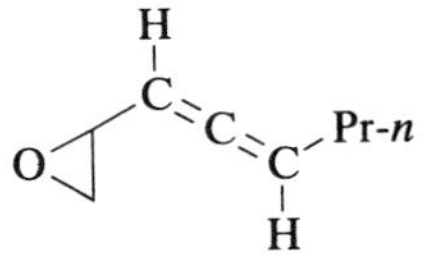 + [structure] (3)	114, 113
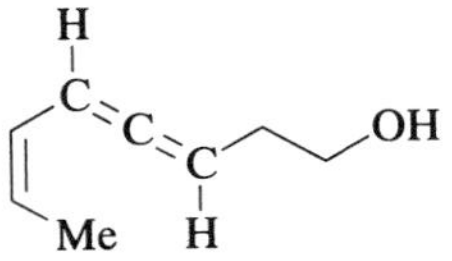 [structure]	*t*-BuOOH, $VO(acac)_2$, CH_2Cl_2, rt	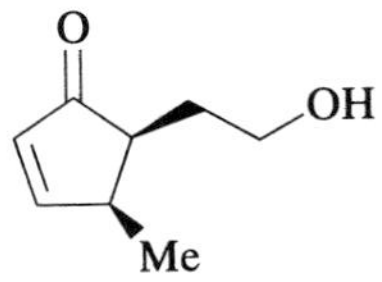 [structure] (40-70)	125
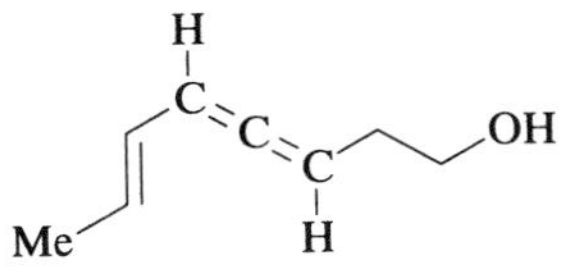 [structure]	*t*-BuOOH, $VO(acac)_2$, CH_2Cl_2, rt	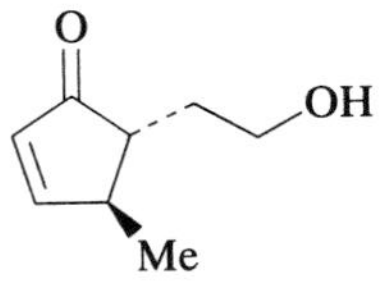 [structure] (40-70)	125
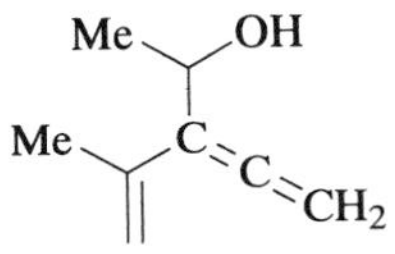 [structure]	*t*-BuOOH, $VO(acac)_2$, CH_2Cl_2, rt	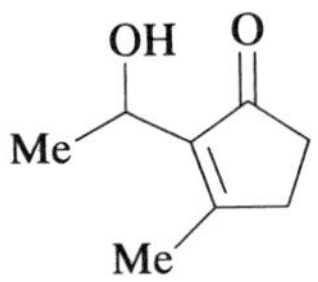 [structure] (65) 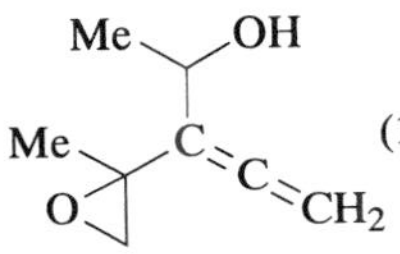 + [structure] (15)	124

TABLE IV. CYCLIZATION OF DIVINYL KETONE EQUIVALENTS FROM SOLVOLYSIS: C. VINYLALLENES (*Continued*)

Reactant	Conditions	Product(s) and Yield(s) (%)		Refs.
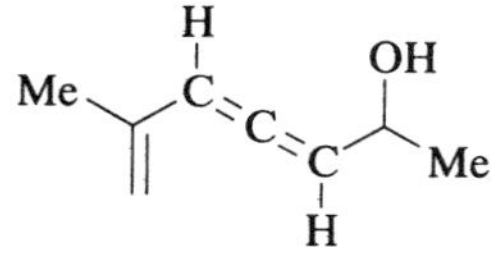 H, Me, OH, Me, H	t-BuOOH, VO(acac)$_2$, C_6H_6, 80°, 1.5 h	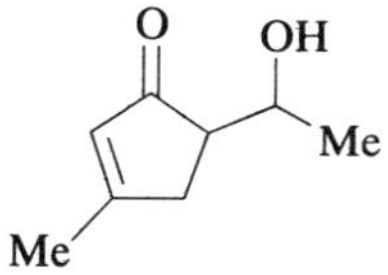 O, OH, Me, Me	(55)	123
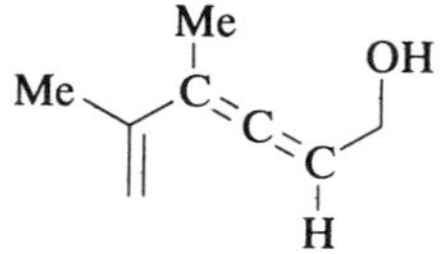 Me, Me, OH, H	t-BuOOH, VO(acac)$_2$, C_6H_6, 20°, 1 h; 50°, 3 h	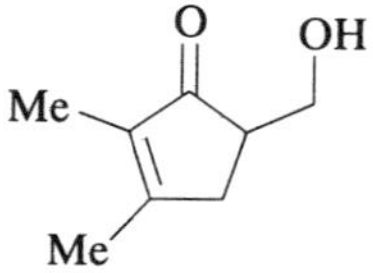 O, OH, Me, Me	(50)	123
C_9				
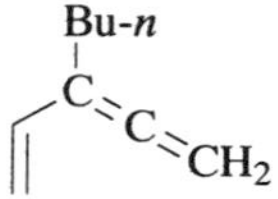 Bu-n, C=C=CH_2	$h\nu$, O_2, $(MeCO)_2$, CH_2Cl_2	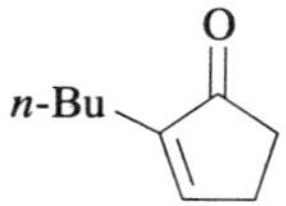 O, n-Bu	(60)	117
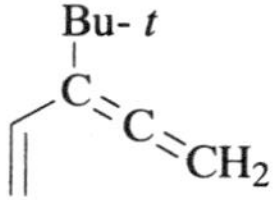 Bu-t, C=C=CH_2	$h\nu$, O_2, $(MeCO)_2$, CH_2Cl_2	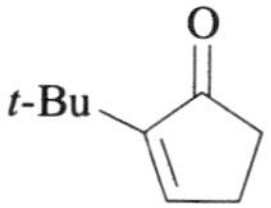 O, t-Bu	(40)	117
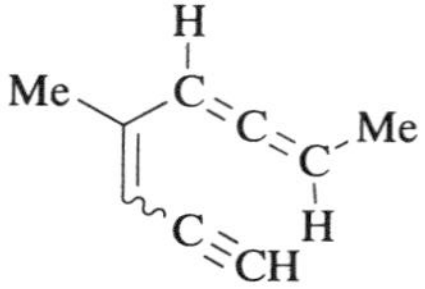 H, Me, Me, H, C≡CH	MCPBA, $NaHCO_3$, H_2O, 0°, 24 h	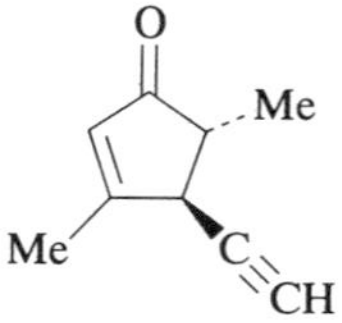 O, Me, Me, C≡CH	(45)	126

TABLE IV. CYCLIZATION OF DIVINYL KETONE EQUIVALENTS FROM SOLVOLYSIS: C. VINYLALLENES (*Continued*)

Reactant	Conditions	Product(s) and Yield(s) (%)	Refs.
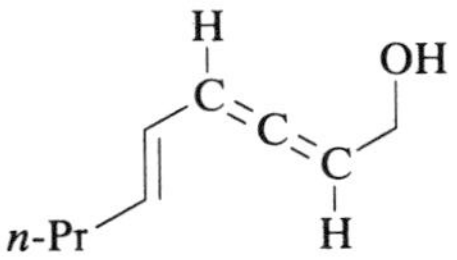 [structure]	*t*-BuOOH, VO(acac)$_2$, CH$_2$Cl$_2$, rt	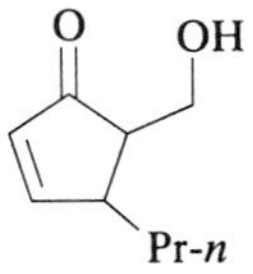 [structure] (40-70)	125
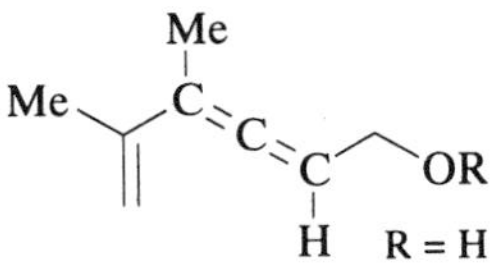 [structure] R = H; R = Ac	*t*-BuOOH, VO(acac)$_2$, benzene 20°, 4 h 80°, 4 h	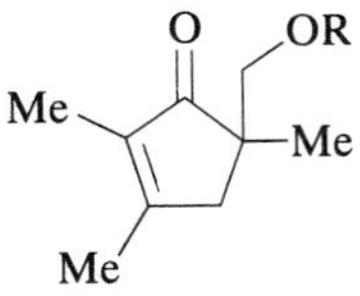 [structure] R = H (45); R = Ac (20)	123
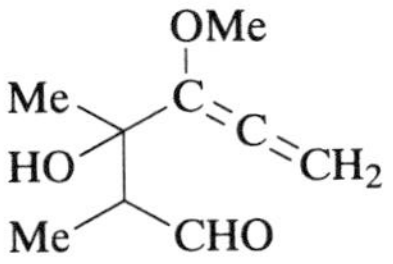 [structure]	BF$_3$•OEt$_2$, CH$_2$Cl$_2$, -78°, 10 min	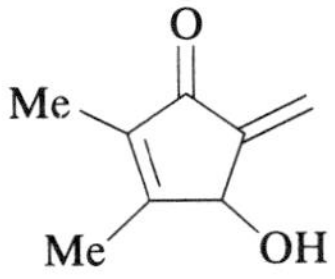 [structure] (68)	129
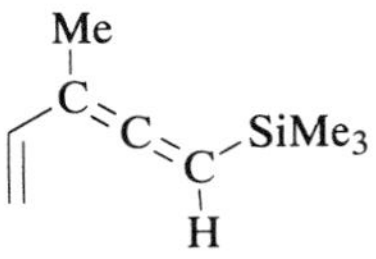 [structure]	H$_2$O$_2$, PhCN	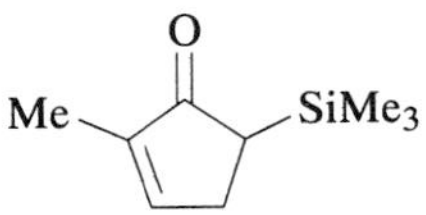 [structure] (60)	190
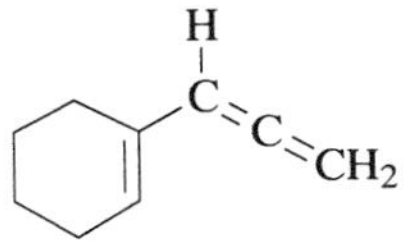 [structure]	1. Tl(OAc)$_3$, HOAc, rt 2. HCl, rt, 10 min	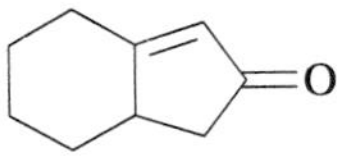 [structure] (36)	128

TABLE IV. CYCLIZATION OF DIVINYL KETONE EQUIVALENTS FROM SOLVOLYSIS: C. VINYLALLENES (*Continued*)

Reactant	Conditions	Product(s) and Yield(s) (%)	Refs.
C_{10}			
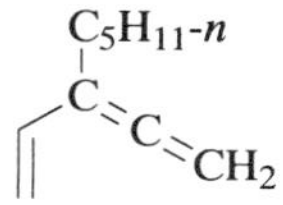 C_5H_{11}-*n*, C=C=CH_2		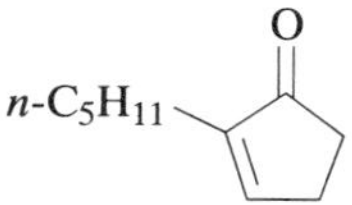 *n*-C_5H_{11}, O	
	PNPBA, CH_2Cl_2, 0°, 24 h	(80)	119
	$Hg(OAc)_2$, $HClO_4$, HOAc, rt, 1 h	(59)	127
	$HClO_4$, 80°, 1 h	(70)	127
	*h*ν, O_2, $(MeCO)_2$, CH_2Cl_2	(55)	117
	1. $Tl(OAc)_3$, HOAc, rt 2. HCl, rt, 10 min	(60)	128
	1. $Hg(OAc)_2$, HOAc, rt, 30 min 2. $HClO_4$, 70°, 1 h	(70)	128
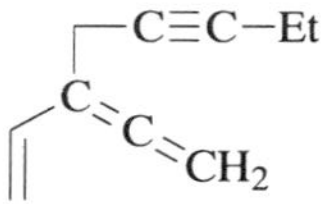 C≡C−Et, C=C=CH_2	PNPBA, $NaHCO_3$, CH_2Cl_2, 0°, 24 h	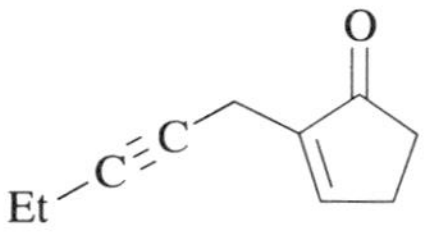 Et, C≡C, O (45)	120
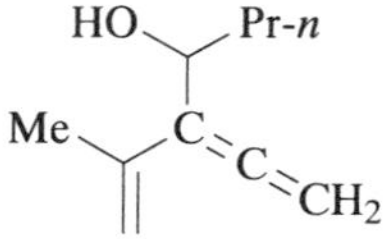 HO, Pr-*n*, Me, C=C=CH_2	*t*-BuOOH, $VO(acac)_2$, CH_2Cl_2, rt, 90 min	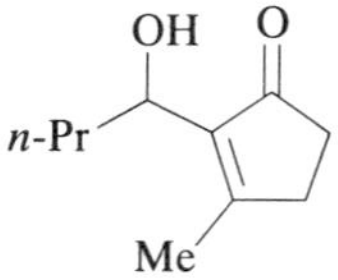 OH, O, *n*-Pr, Me (60) 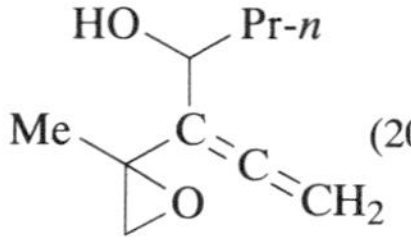 + HO, Pr-*n*, Me, C=C=CH_2, O (20)	124
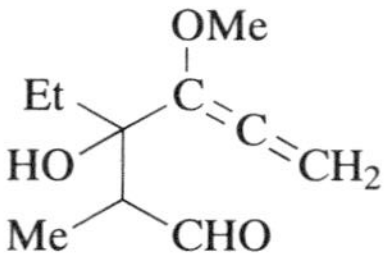 OMe, Et, HO, Me, CHO, C=C=CH_2	BF_3•OEt_2, CH_2Cl_2, -78°, 10 min	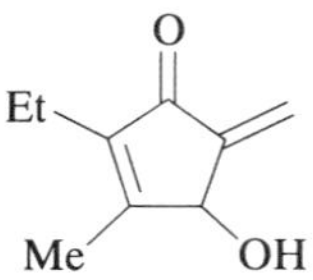 O, Et, Me, OH (72)	129

TABLE IV. CYCLIZATION OF DIVINYL KETONE EQUIVALENTS FROM SOLVOLYSIS: C. VINYLALLENES (*Continued*)

Reactant	Conditions	Product(s) and Yield(s) (%)		Refs.
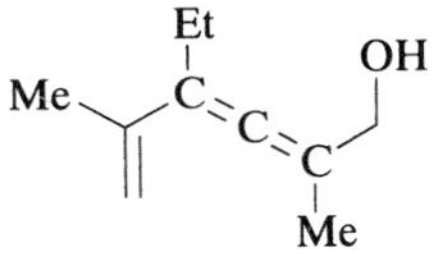 Me, Et, OH, Me (vinylallene alcohol)	*t*-BuOOH, $VO(acac)_2$, C_6H_6, 20°, 4 h	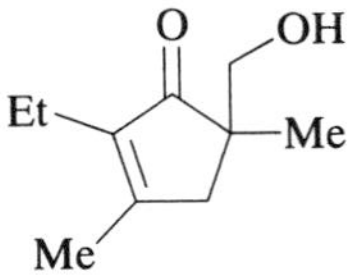 Et, Me, O, OH, Me (cyclopentenone)	(63)	123
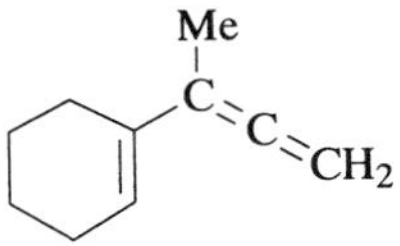 Me, CH_2 (cyclohexenyl allene)	1. $Tl(OAc)_3$, HOAc, rt 2. HCl, rt, 10 min	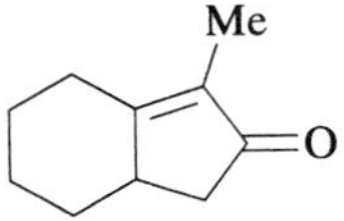 Me, O (bicyclic enone)	(60)	128
	1. $Hg(OAc)_2$, HOAc, rt, 30 min 2. $HClO_4$, 70°, 1 h		(75)	128
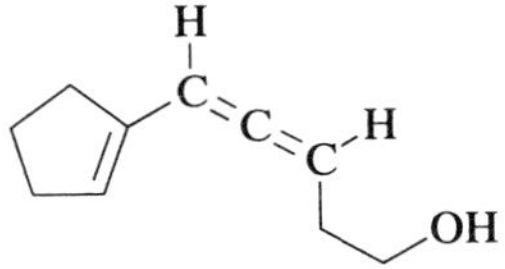 H, H, OH (cyclopentenyl allene)	*t*-BuOOH, $VO(acac)_2$, CH_2Cl_2, rt	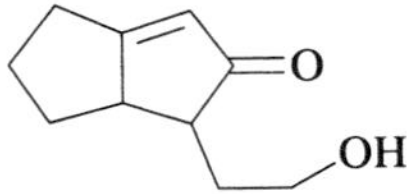 O, OH (bicyclic enone)	(40-70)	125
C_{11}				
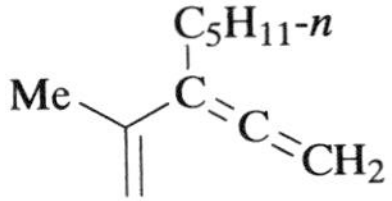 C_5H_{11}-*n*, Me, CH_2	MCPBA, CH_2Cl_2, 0°, 24 h	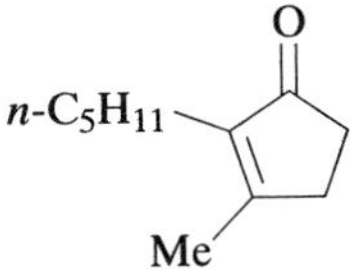 *n*-C_5H_{11}, O, Me (cyclopentenone)	(65)	119
	hν, O_2, $(MeCO)_2$, CH_2Cl_2		(35)	117
	1. $Tl(OAc)_3$, HOAc, 40° 2. HCl, rt, 10 min		(45)	128
	1. $Hg(OAc)_2$, HOAc, rt, 30 min 2. $HClO_4$, 70°, 1 h		(54)	128

TABLE IV. CYCLIZATION OF DIVINYL KETONE EQUIVALENTS FROM SOLVOLYSIS: C. VINYLALLENES (*Continued*)

Reactant	Conditions	Product(s) and Yield(s) (%)	Refs.
C_5H_{11}-*n*, Me (wavy bond), C=C=CH_2		*n*-C_5H_{11}, O, Me	
E, Z	1. $Hg(OAc)_2$, HOAc, rt, 30 min 2. $HClO_4$, 70°, 1 h	(51)	128
E	"	(78)	128
Z	"	(50)	128
E,Z	1. $Tl(OAc)_3$, HOAc, 40° 2. HCl, rt, 10 min	(25)	128
C_5H_{11}-*n*, C=C=C, Me, H	$Hg(OAc)_2$, HOAc, $HClO_4$, rt, 1 h	*n*-C_5H_{11}, O, Me (42)	127
	$HClO_4$, 80°, 1 h	(50)	127
	1. $Hg(OAc)_2$, HOAc, rt, 30 min 2. $HClO_4$, 70°, 1 h	(50)	128
	1. $Tl(OAc)_3$, HOAc, 40° 2. HCl, rt, 10 min	(68)	128
C≡CEt, Me, C=C=CH_2	PNPBA, $NaHCO_3$, CH_2Cl_2, 0°, 24 h	EtC≡C, O, Me (70)	120
H, Me, C=C=C, Et, Me, C≡CH	MCPBA, $NaHCO_3$, H_2O, 0°, 24 h	O, Et, Me, Me, C≡CH (58)	126

TABLE IV. CYCLIZATION OF DIVINYL KETONE EQUIVALENTS FROM SOLVOLYSIS: C. VINYLALLENES (*Continued*)

Reactant	Conditions	Product(s) and Yield(s) (%)	Refs.
H, C, C, C, Me, Me, H	PNPBA, CH_2Cl_2, 0°, 24 h	O, Me, Me (50)	118
Me, C, C, C, Me, H	PNPBA, CH_2Cl_2, 0°, 24 h 1. $Hg(OAc)_2$, HOAc, rt, 30 min 2. $HClO_4$, 70°, 1 h 1. $Tl(OAc)_3$, HOAc, 40° 2. HCl, rt, 10 min	Me, O, Me (50) (49) (44)	118 128 128
H, C, C, C, OH, H	*t*-BuOOH, $VO(acac)_2$, CH_2Cl_2, rt	O, OH (40-70)	125
OH, C, C, CH_2	*t*-BuOOH, $VO(acac)_2$, CH_2Cl_2, rt	OH, O (50)	124
C_{12}			
C_5H_{11}-*n*, Me, C, C, C, Me, H	1. $Hg(OAc)_2$, HOAc, rt, 30 min 2. $HClO_4$, 70°, 1 h 1. $Tl(OAc)_3$, HOAc, 40° 2. HCl, rt, 10 min	O, *n*-C_5H_{11}, Me, Me (79) (61)	128 128

TABLE IV. CYCLIZATION OF DIVINYL KETONE EQUIVALENTS FROM SOLVOLYSIS: C. VINYLALLENES (*Continued*)

Reactant	Conditions	Product(s) and Yield(s) (%)	Refs.
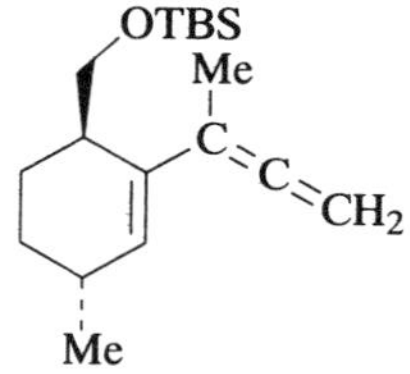 OTBS, Me, C=C=CH$_2$, Me	Tl(OAc)$_3$, HOAc, rt, 30 min	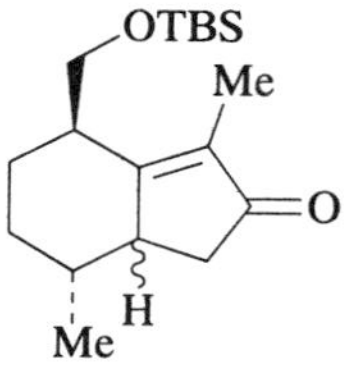 OTBS, Me, O, H, Me 55:45 (70)	122
C$_{13}$ 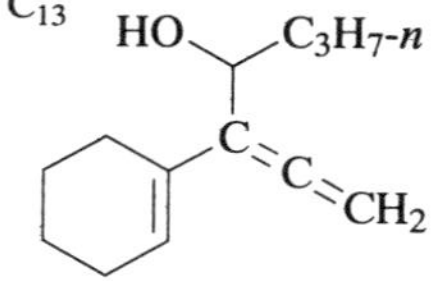 HO, C$_3$H$_7$-*n*, C=C=CH$_2$	*t*-BuOOH, VO(acac)$_2$, CH$_2$Cl$_2$, rt	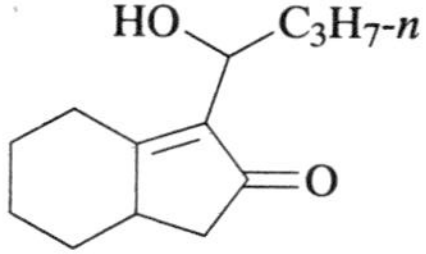 HO, C$_3$H$_7$-*n*, O (32) 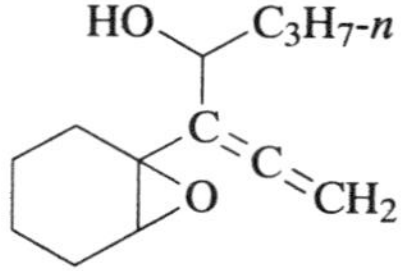 + HO, C$_3$H$_7$-*n*, C=C=CH$_2$, O (33)	124
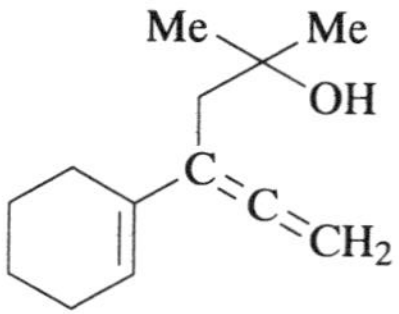 Me, Me, OH, C=C=CH$_2$	*t*-BuOOH, VO(acac)$_2$, CH$_2$Cl$_2$, rt	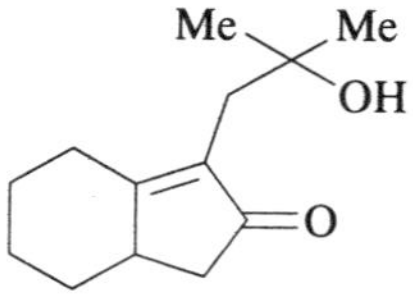 Me, Me, OH, O (40)	124
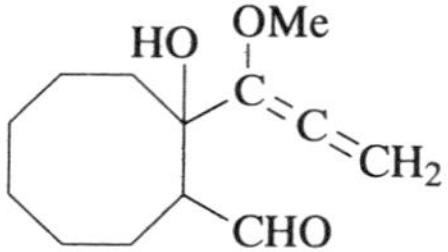 HO, OMe, C=C=CH$_2$, CHO	BF$_3$•OEt$_2$, CH$_2$Cl$_2$, 0°, 10 min	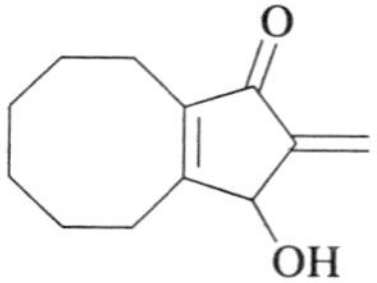 O, OH (80)	129

TABLE IV. CYCLIZATION OF DIVINYL KETONE EQUIVALENTS FROM SOLVOLYSIS: C. VINYLALLENES (*Continued*)

Reactant	Conditions	Product(s) and Yield(s) (%)	Refs.
C_{14}			
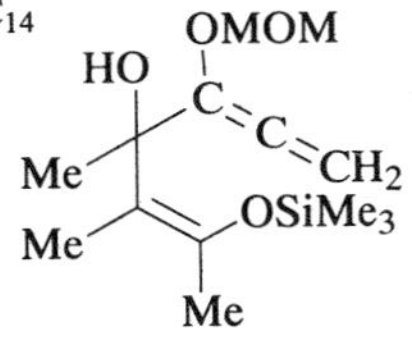 HO, OMOM, Me, Me, Me, $C=C=CH_2$, $OSiMe_3$	TFAA, 2,6-lutidine, CH_2Cl_2, -30°	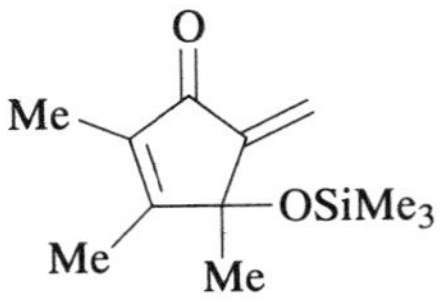 Me, Me, Me, O, $OSiMe_3$ (83)	191
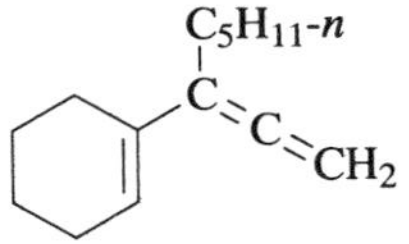 C_5H_{11}-*n*, $C=C=CH_2$	*h*ν, O_2, $(MeCO)_2$, CH_2Cl_2	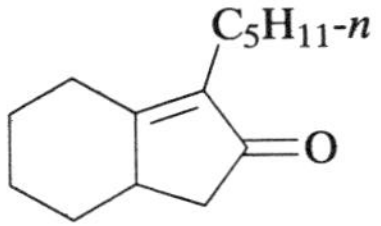 C_5H_{11}-*n*, O (40) 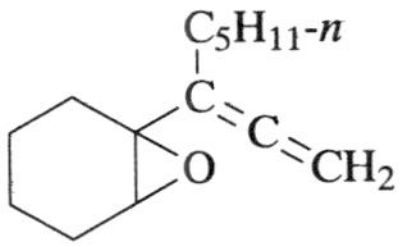 + C_5H_{11}-*n*, $C=C=CH_2$, O (25)	117
C_{15}			
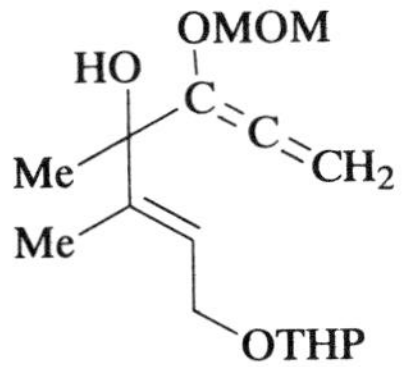 HO, OMOM, Me, Me, $C=C=CH_2$, OTHP	$MeSO_2Cl$, Et_3N, THF, -20 - 0°, 12 h	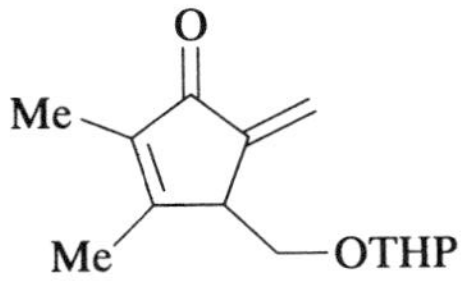 Me, Me, O, OTHP (50)	130
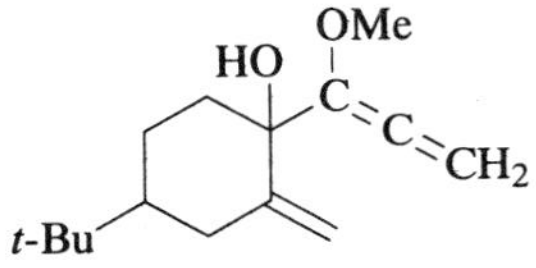 HO, OMe, $C=C=CH_2$, *t*-Bu	BF_3•OEt_2, CH_2Cl_2, 0°, 10 min	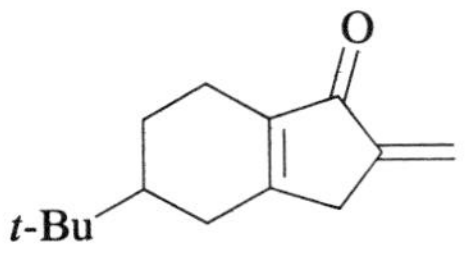 O, *t*-Bu (56)	129

TABLE IV. CYCLIZATION OF DIVINYL KETONE EQUIVALENTS FROM SOLVOLYSIS: C. VINYLALLENES (*Continued*)

Reactant	Conditions	Product(s) and Yield(s) (%)	Refs.
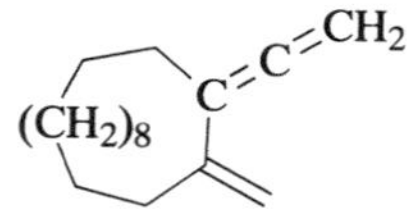 $(CH_2)_8$ ring; $C{=}C{=}CH_2$	MPPA, CH_2Cl_2, Et_2O	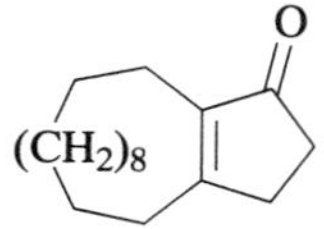 $(CH_2)_8$ (50)	121
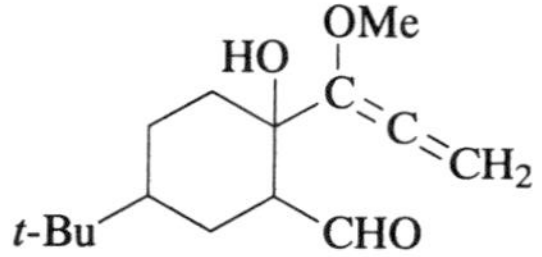 HO, OMe, $C{=}C{=}CH_2$, CHO, *t*-Bu	$BF_3{\cdot}OEt_2$, CH_2Cl_2, 0°, 10 min	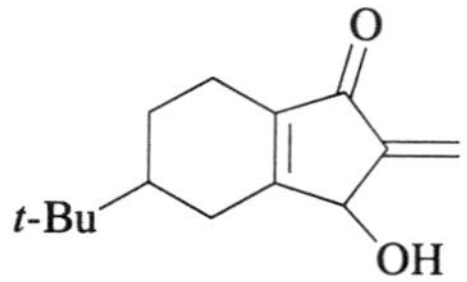 O, *t*-Bu, OH (76)	129
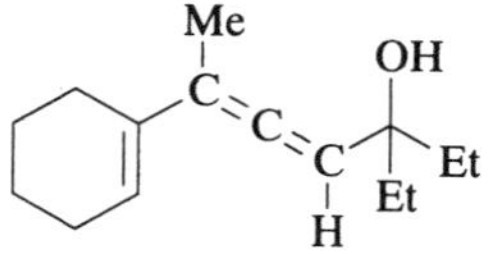 Me, $C{=}C{=}C$, OH, Et, Et, H	*t*-BuOOH, $VO(acac)_2$, C_6H_6, 20°, 1 h	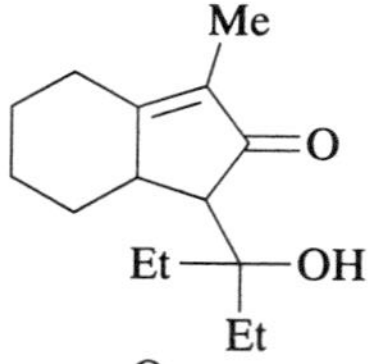 Me, O, Et, OH, Et (45)	123
C_{16}			
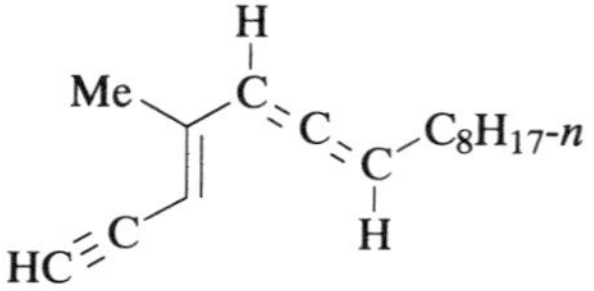 H, Me, $C{=}C{=}C$, C_8H_{17}-*n*, H, $HC{\equiv}C$	MCPBA, CH_2Cl_2, $NaHCO_3$, 0°, 24 h	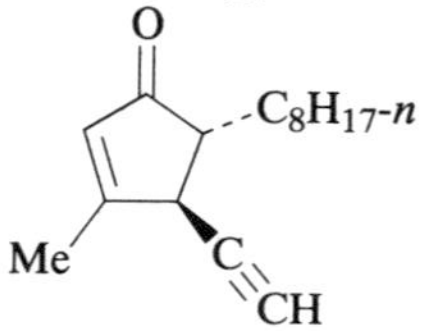 O, C_8H_{17}-*n*, Me, $C{\equiv}CH$ (58)	126
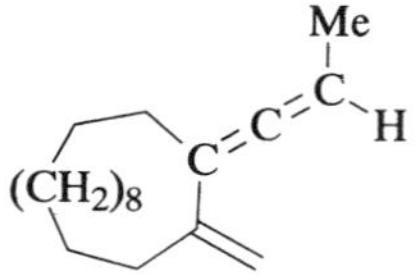 Me, $(CH_2)_8$ ring; $C{=}C{=}C$, H	MPPA, CH_2Cl_2, Et_2O	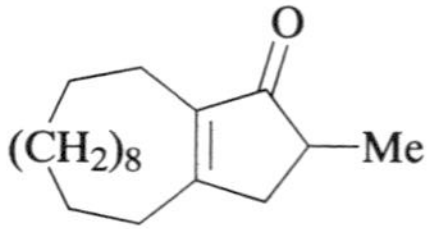 O, $(CH_2)_8$, Me (52)	121

TABLE IV. CYCLIZATION OF DIVINYL KETONE EQUIVALENTS FROM SOLVOLYSIS: C. VINYLALLENES (*Continued*)

Reactant	Conditions	Product(s) and Yield(s) (%)	Refs.
C_{17}			
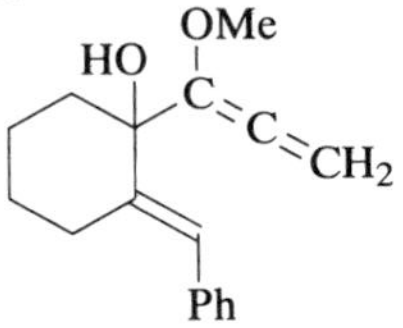 HO, OMe, CH_2, Ph	$BF_3 \cdot OEt_2$, CH_2Cl_2, 0°, 10 min 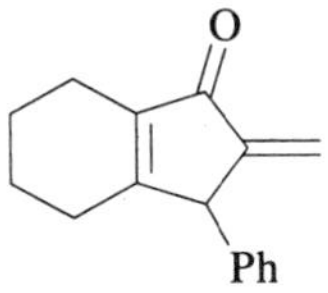	O, Ph (78)	129
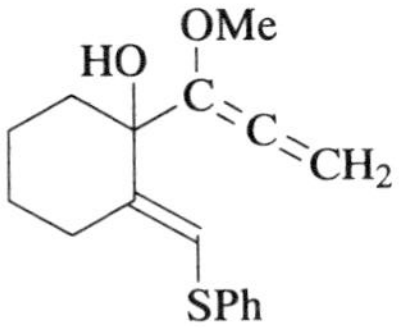 HO, OMe, CH_2, SPh	$BF_3 \cdot OEt_2$, CH_2Cl_2, 0°, 10 min	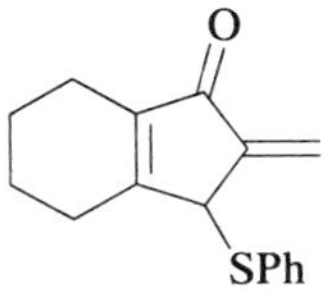 O, SPh (82)	129
C_{18}			
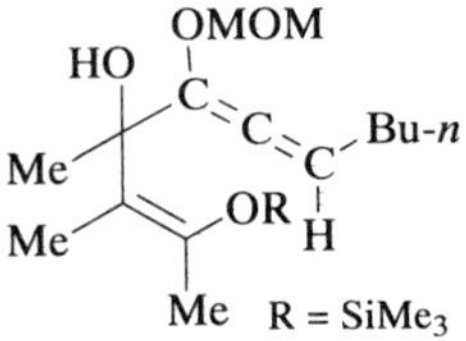 HO, OMOM, Bu-*n*, Me, Me, OR, H, Me, R = $SiMe_3$	TFAA, 2,6-lutidine, CH_2Cl_2, -30°	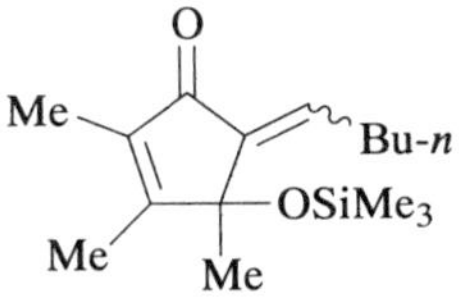 O, Me, Bu-*n*, $OSiMe_3$, Me, Me (70)	191
C_{20}			
O, $(CH_2)_3CO_2H$, H, C_5H_{11}-*n*, H	rt 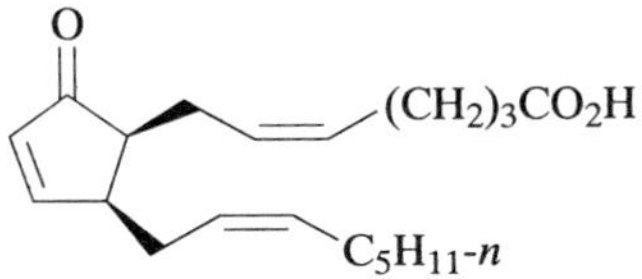	O, $(CH_2)_3CO_2H$, C_5H_{11}-*n* (-)	192

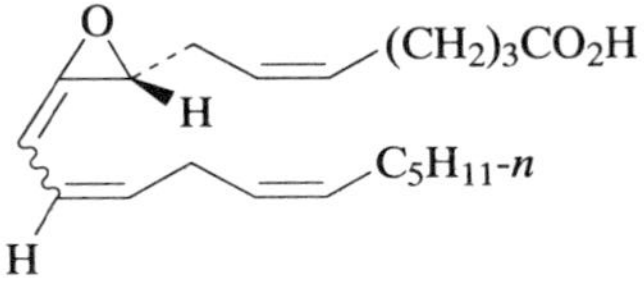

TABLE IV. CYCLIZATION OF DIVINYL KETONE EQUIVALENTS FROM SOLVOLYSIS: C. VINYLALLENES (*Continued*)

Reactant	Conditions	Product(s) and Yield(s) (%)	Refs.
C_{21}			
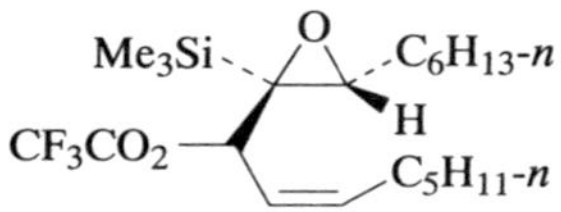 Me_3Si, O, C_6H_{13}-*n*, H, CF_3CO_2, C_5H_{11}-*n*	CsF, MeCN, rt, 10-14 h	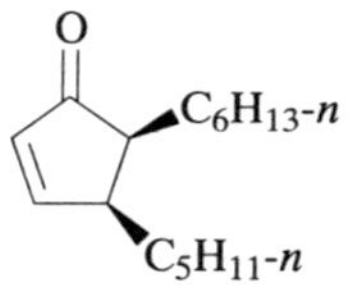 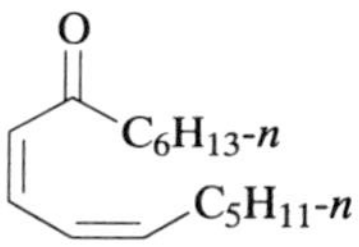 O, C_6H_{13}-*n*, C_5H_{11}-*n* (20-35) + O, C_6H_{13}-*n*, C_5H_{11}-*n* (20-25)	59
C_{25}			
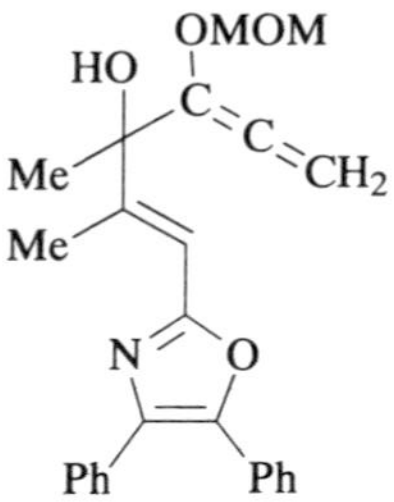 OMOM, HO, C=C=CH_2, Me, Me, N, O, Ph, Ph	TFAA, 2,6-lutidine, CH_2Cl_2, -30°	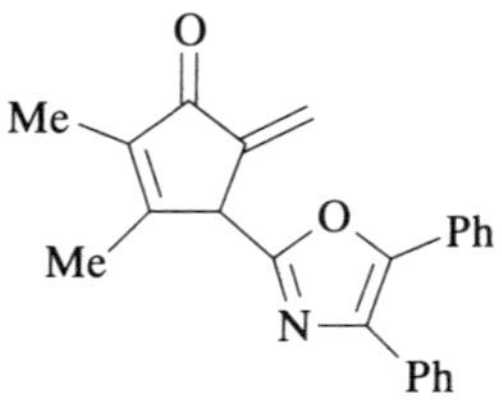 O, Me, Me, O, N, Ph, Ph (65)	130
C_{27}			
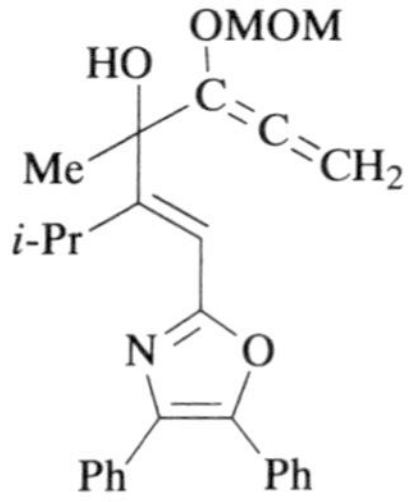 OMOM, HO, C=C=CH_2, Me, *i*-Pr, N, O, Ph, Ph	TFAA, 2,6-lutidine, CH_2Cl_2, -10 to 0°	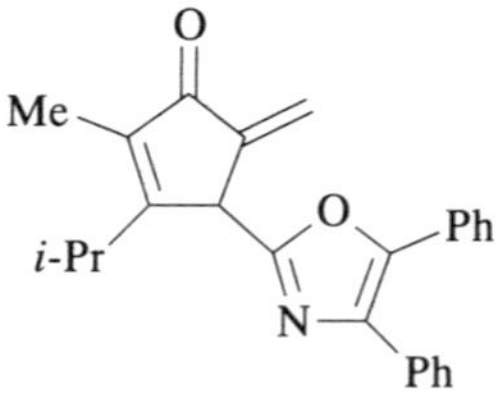 O, Me, *i*-Pr, O, N, Ph, Ph (72)	131, 130

TABLE V. IN SITU CONSTRUCTION OF DIVINYL KETONES: A. OLEFINIC ACIDS AND ANHYDRIDES

Reactant	Conditions	Product(s) and Yield(s) (%)	Refs.
C_3			
O, OH	Cyclohexene, PPA, 57°, 30 min	O (16)	76
C_4			
O, OH, Me	Cyclopentene, PPA, 40°, 1 h	O, Me (22)	76
	Cyclohexene, PPA, 40°, 1 h	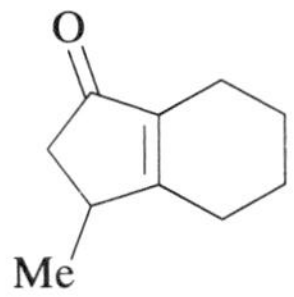 O, Me (60)	76
	Cycloheptene, PPA, 50°	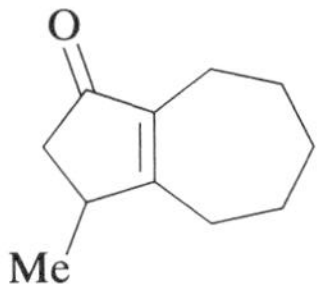 O, Me (6) + O, Me, Me (18)	174

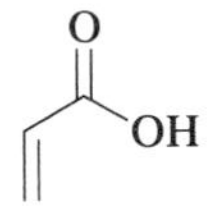

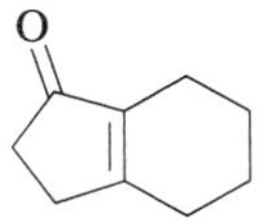

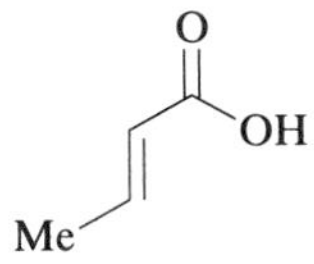

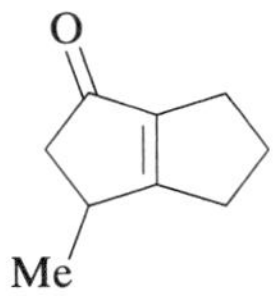

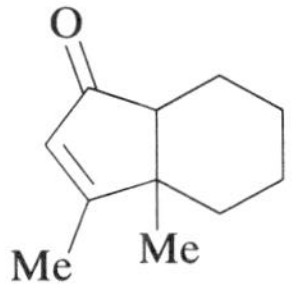

TABLE V. IN SITU CONSTRUCTION OF DIVINYL KETONES: A. OLEFINIC ACIDS AND ANHYDRIDES (*Continued*)

Reactant	Conditions	Product(s) and Yield(s) (%)	Refs.
	Cycloheptene, PPA, 70°	(4) + (29) β:α 62:38	193
	1-Methylcyclohexene, PPA, 70°	(4) + (29) β:α 62:38	193
	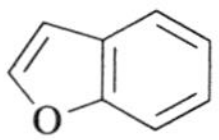 [benzofuran], PPA, 130°	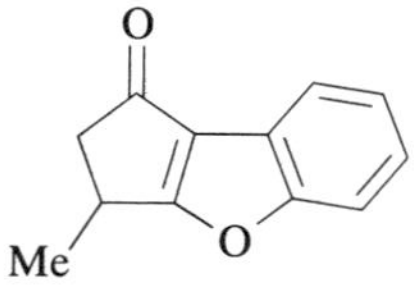 Me (34)	194
C_7			
Ph, Ph (anhydride)	Cyclohexene, PPA, 57°, 30 min	(42)	76
C_9			
Ph, OH (acid)	Cyclohexene, PPA, 57°, 30 min	Ph (26)	76

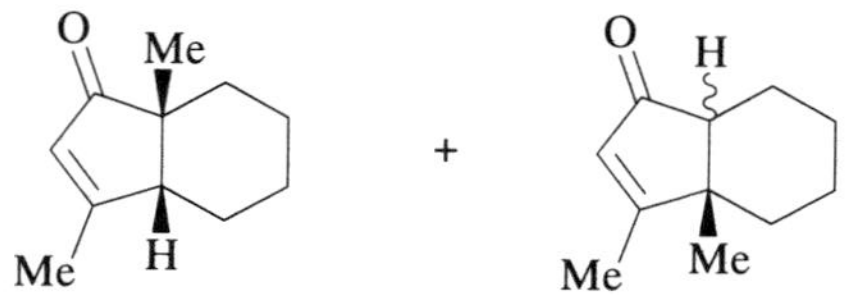

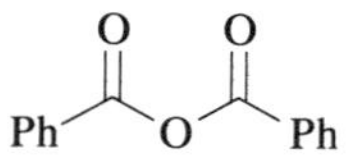

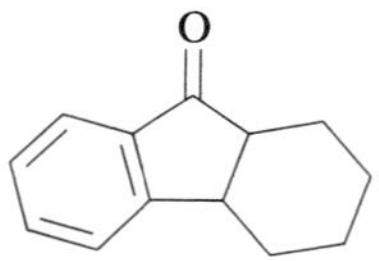

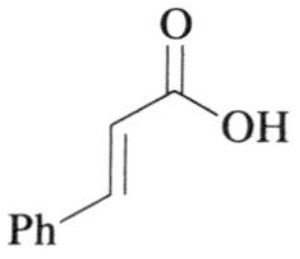

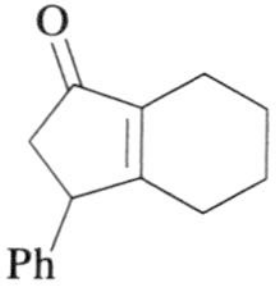

TABLE V. IN SITU CONSTRUCTION OF DIVINYL KETONES: B. OLEFINIC ESTERS

Reactant	Conditions	Product(s) and Yield(s) (%)		Refs.
C_6 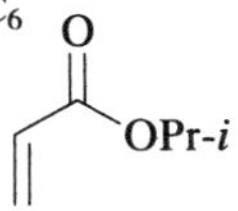 O, OPr-*i*	PPA, 100°, 1 h	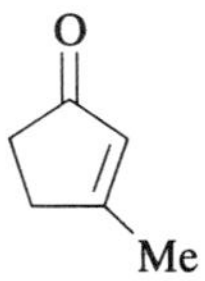 O, Me	(15-20)	195
C_7 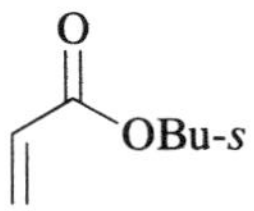 O, OBu-*s*	PPA, 100°, 1h	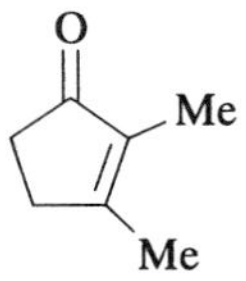 O, Me, Me	(20)	195
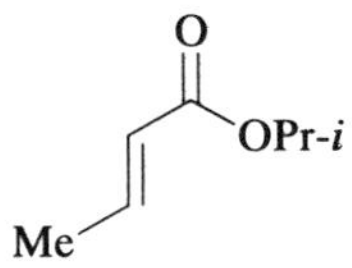 O, OPr-*i*, Me	PPA, 100°, 1 h	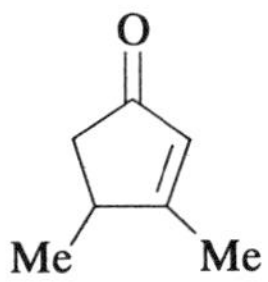 O, Me, Me	(58-67)	195
C_8 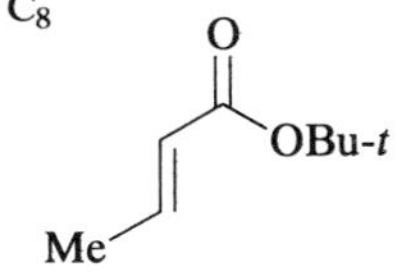 O, OBu-*t*, Me	PPA, 100°, 1 h	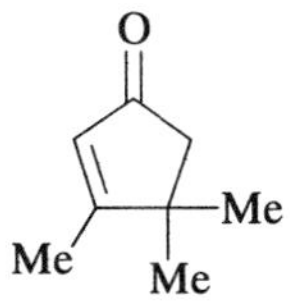 O, Me, Me, Me	(45)	195
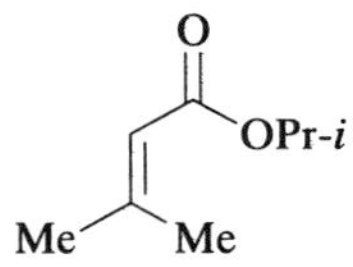 O, OPr-*i*, Me, Me	PPA, 100°, 1 h	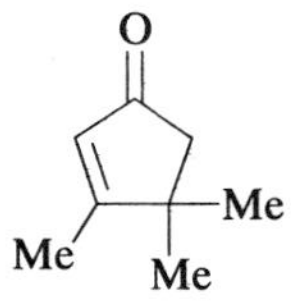 O, Me, Me, Me	(60)	195

TABLE V. IN SITU CONSTRUCTION OF DIVINYL KETONES: B. OLEFINIC ESTERS (*Continued*)

Reactant	Conditions	Product(s) and Yield(s) (%)	Refs.
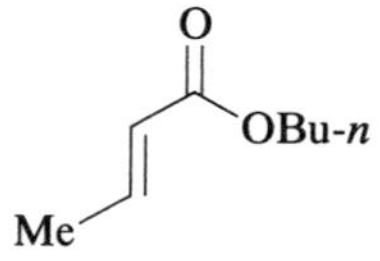	PPA, 100°, 12 h	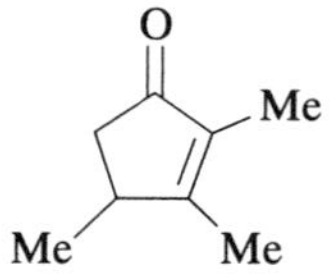 (6) + 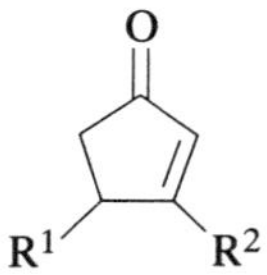 (4) R^1 = Me or Et, R^2 = Et or Me	195
C_9			
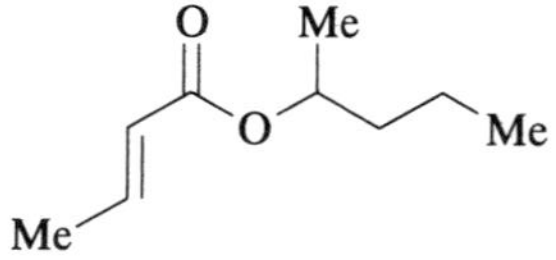	PPA, 100°, 7 h	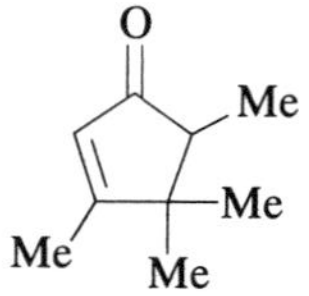 (36) + 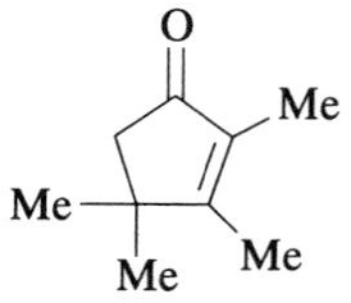 (9)	195
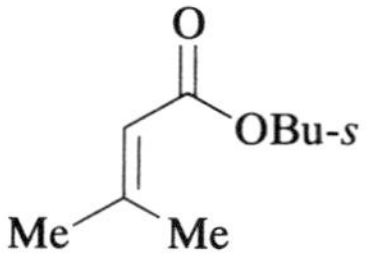	PPA, 100°, 1 h	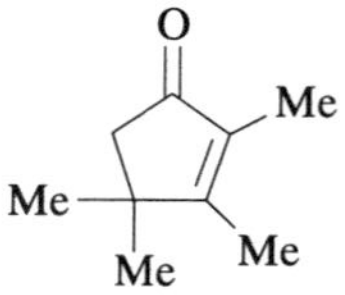 (60)	195
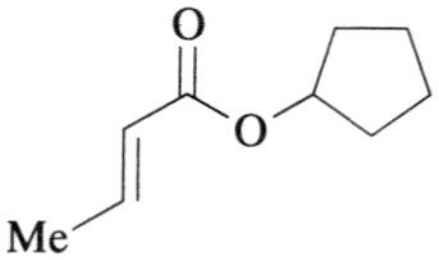	PPA, 45-55°, 90 min	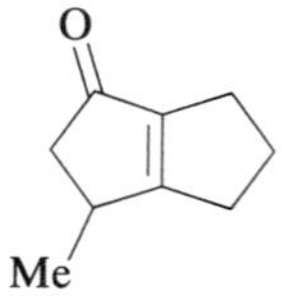 (20)	195

TABLE V. IN SITU CONSTRUCTION OF DIVINYL KETONES: B. OLEFINIC ESTERS (*Continued*)

Reactant	Conditions	Product(s) and Yield(s) (%)	Refs.
C_{10} 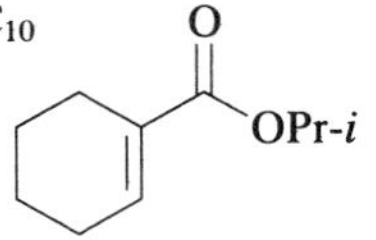 OPr-*i*	PPA, 100°, 1 h 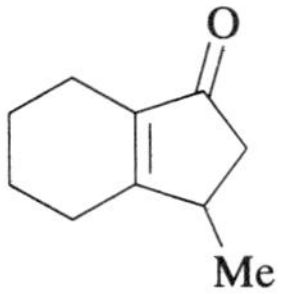	Me (60)	195
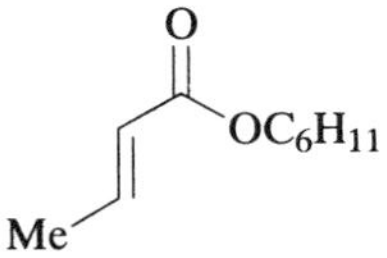 Me OC$_6$H$_{11}$	PPA, 60°, 90 min 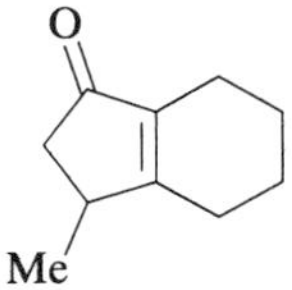	Me (60)	195
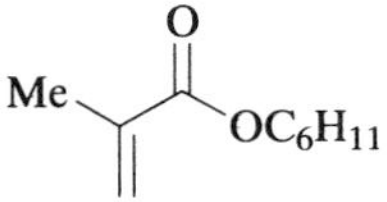 Me OC$_6$H$_{11}$	PPA, 100°, 1 h	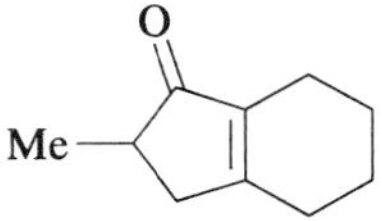 Me (17)	195
C_{11} 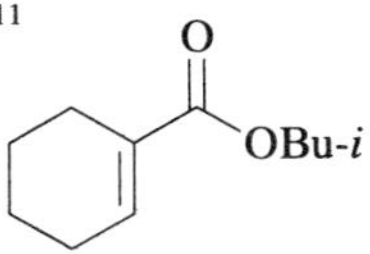 OBu-*i*	PPA, 100°, 2 h	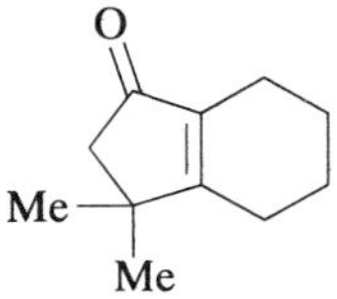 Me Me (59)	195
C_{13} 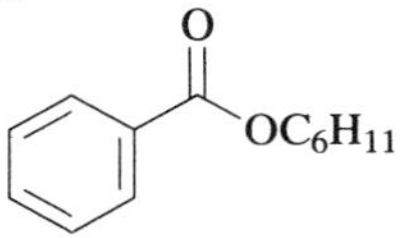 OC$_6$H$_{11}$	PPA, 100°, 20 min	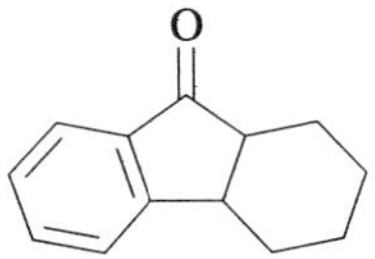 (40)	195

TABLE V. IN SITU CONSTRUCTION OF DIVINYL KETONES: C. OLEFINIC ACID HALIDES AND OLEFINS

Reactant	Conditions	Product(s) and Yield(s) (%)	Refs.
C_3			
O, Cl	Me_3Si, Pr-*n*, Me , $AlCl_3$, CH_2Cl_2, rt, 2 h	O, Et, Me (19) + O, Et, Me (5)	68, 67
	1-Trimethylsilylcyclohexene, $AlCl_3$, CH_2Cl_2, rt, 2 h	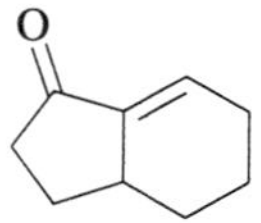 O (15) 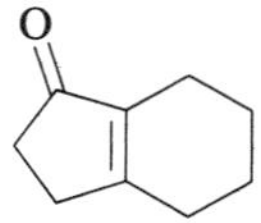 + O (12)	68, 67
	1. 1-Trimethylsilylcycloheptene, $AlCl_3$, NaOAc, CH_2Cl_2, -45°, 2 h 2. TFA, rt, 3 h	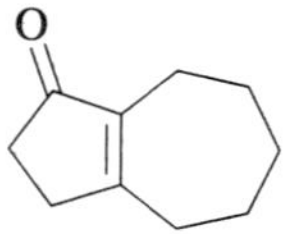 O (10)	68, 67
C_4			
Me, O, Br	Cyclohexene, $AlCl_3$, 0°	Me, O (36)	157
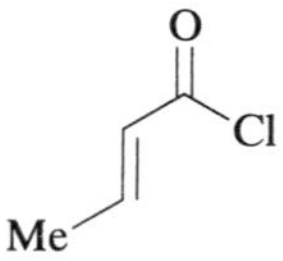 O, Cl, Me	$Me_3SiCH{=}CH_2$, CCl_4, 77°, 30 min	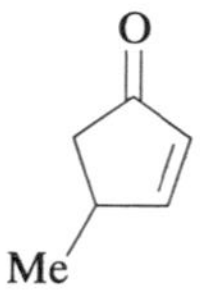 O, Me (63)	160

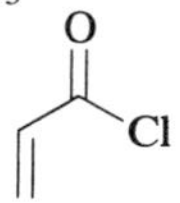

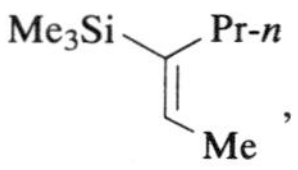

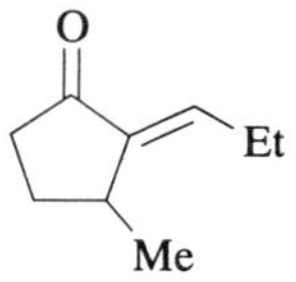

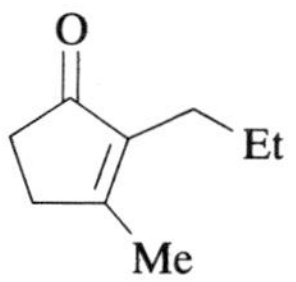

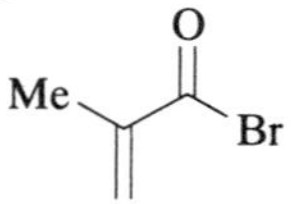

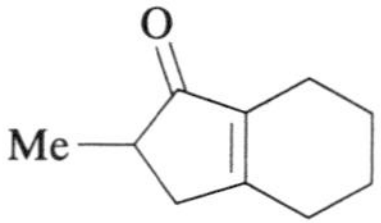

TABLE V. IN SITU CONSTRUCTION OF DIVINYL KETONES: C. OLEFINIC ACID HALIDES AND OLEFINS (*Continued*)

Reactant	Conditions	Product(s) and Yield(s) (%)	Refs.
	1. 1-Trimethylsilylcyclopentene, $AlCl_3$, CH_2Cl_2, rt, 1 h 2. BF_3•OEt_2, C_6H_6, 80°	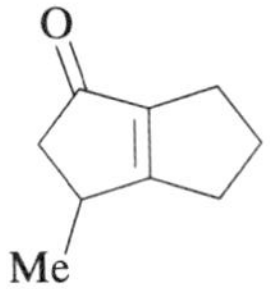 O, Me (58)	68, 67
	1. 1-Trimethylsilylcyclohexene, $AlCl_3$, CH_2Cl_2, -78°, 15 min 2. BF_3•OEt_2, C_6H_6, 80°, 3 d	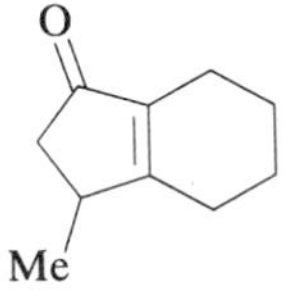 O, Me (44)	68, 67
	1. Cyclohexene, $AlCl_3$, CH_2Cl_2, -10° 2. H_3PO_4, HCO_2H, 80-90°, 4 h	(30)	102
	1. 1-Trimethylsilylcyclododecene, $AlCl_3$, CH_2Cl_2, rt 2. BF_3•OEt_2, C_6H_6, 80°	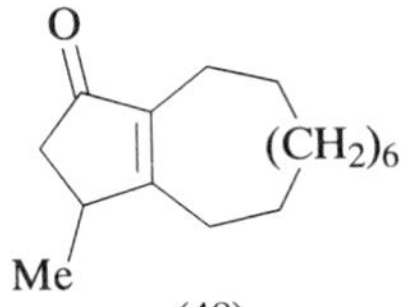 O, Me, $(CH_2)_6$ (48) 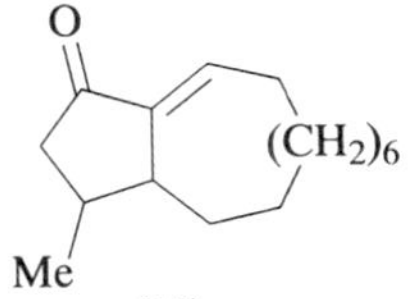 + O, Me, $(CH_2)_6$ (18)	68, 67
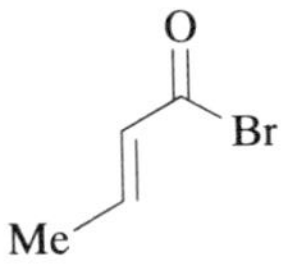 O, Br, Me	MeCH=CMe$_2$, $AlCl_3$, 0°	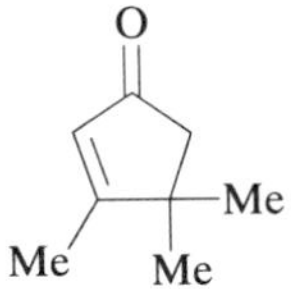 O, Me, Me, Me (15) 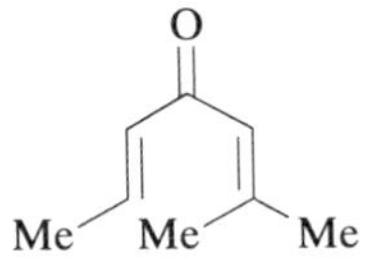 + O, Me, Me, Me (50)	157
	Cyclohexene, $AlCl_3$, -78°	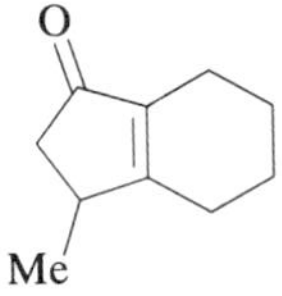 O, Me (72) 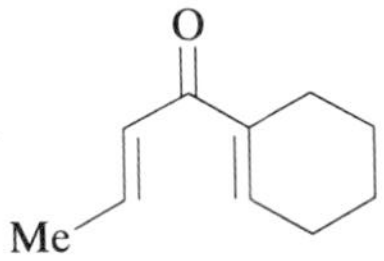 + O, Me (8)	157

TABLE V. IN SITU CONSTRUCTION OF DIVINYL KETONES: C. OLEFINIC ACID HALIDES AND OLEFINS (*Continued*)

Reactant	Conditions	Product(s) and Yield(s) (%)	Refs.
	cis,trans-Cyclododecene, $AlCl_3$, -78°	O, Me, $(CH_2)_6$ (16) + O, Me, Cl, $(CH_2)_6$ (64)	157
C_5 O, Cl, Me, Me	1. Me$_3$Si, Pr-*n*, Et, $AlCl_3$, CH_2Cl_2, -78° 2. BF_3•OEt_2, C_6H_6, 80°, 1-3 d	O, Et, Me, Me, Et (41) + O, Et, Me, Me, Et (21)	68, 67
	1. 1-Trimethylsilylcyclopentene, $AlCl_3$, -78°, 15 min 2. $SnCl_4$, CH_2Cl_2, reflux, 8-12 h	O, Me, Me (25) + O, Me, Me (37)	68, 67
	1. 1-Trimethylsilylcyclohexene, $AlCl_3$, CH_2Cl_2, -78°, 15 min 2. BF_3•OEt_2, CH_2Cl_2, reflux, 8-12 h	O, Me, Me (42) + O, Me, Me (28)	68, 67

TABLE V. IN SITU CONSTRUCTION OF DIVINYL KETONES: C. OLEFINIC ACID HALIDES AND OLEFINS (*Continued*)

Reactant	Conditions	Product(s) and Yield(s) (%)	Refs.
	1. 6-Methyl-1-trimethylsilylcyclohexene, $AlCl_3$, CH_2Cl_2, -78°, 15 min 2. $BF_3{\bullet}OEt_2$, CH_2Cl_2, reflux, 8-12 h	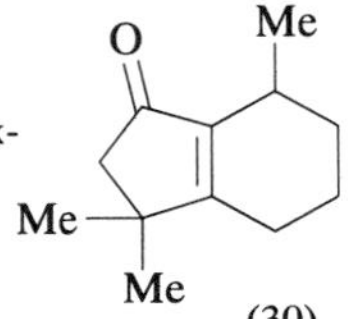 O, Me, Me, Me (30) 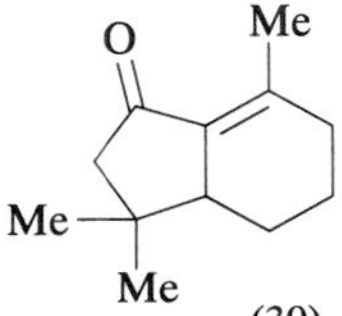 + O, Me, Me, Me (30)	68, 67
	1. 1-Trimethylsilylcycloheptene, $AlCl_3$, CH_2Cl_2, -78°, 15 min 2. $BF_3{\bullet}OEt_2$, CH_2Cl_2, reflux, 8-12 h	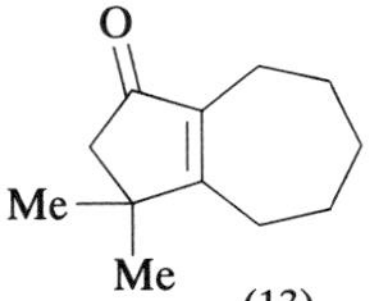 O, Me, Me (13) 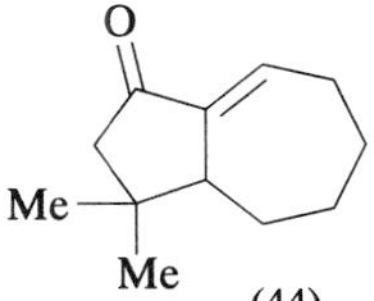 + O, Me, Me (44)	68, 67
	1. 1-Trimethylsilylcyclododecene, $AlCl_3$, CH_2Cl_2, -78°, 15 min 2. $BF_3{\bullet}OEt_2$, C_6H_6, 80°, 1-3 d	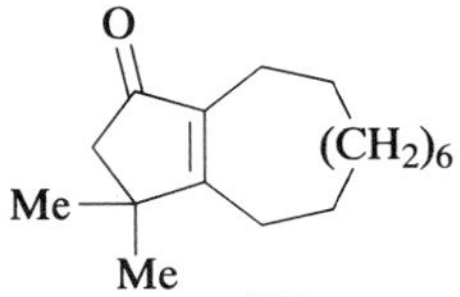 O, $(CH_2)_6$, Me, Me (17) 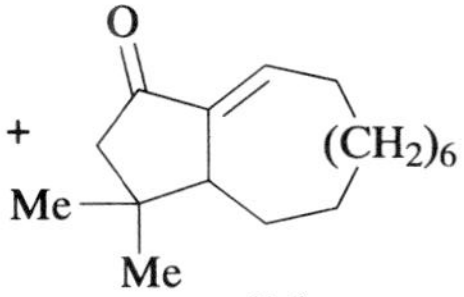 + O, $(CH_2)_6$, Me, Me (26)	68, 67
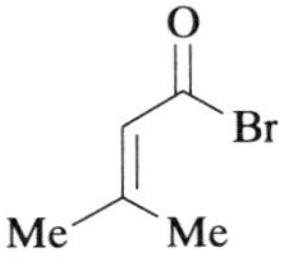 O, Br, Me, Me	Cyclohexene, $AlCl_3$, -78°	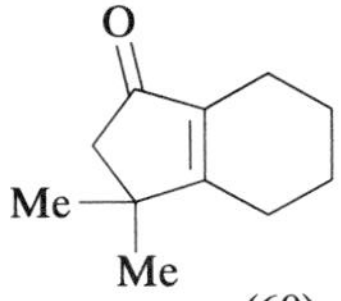 O, Me, Me (60) 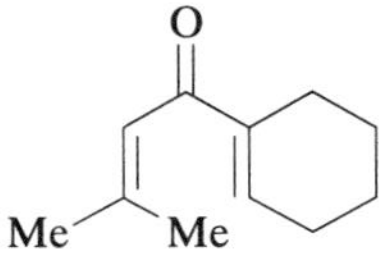 + O, Me, Me (20)	157

TABLE V. IN SITU CONSTRUCTION OF DIVINYL KETONES: C. OLEFINIC ACID HALIDES AND OLEFINS (*Continued*)

Reactant	Conditions	Product(s) and Yield(s) (%)	Refs.
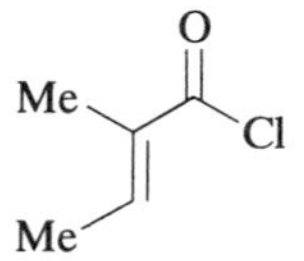 Me, Me, O, Cl	Cyclohexene, $AlCl_3$, -78°	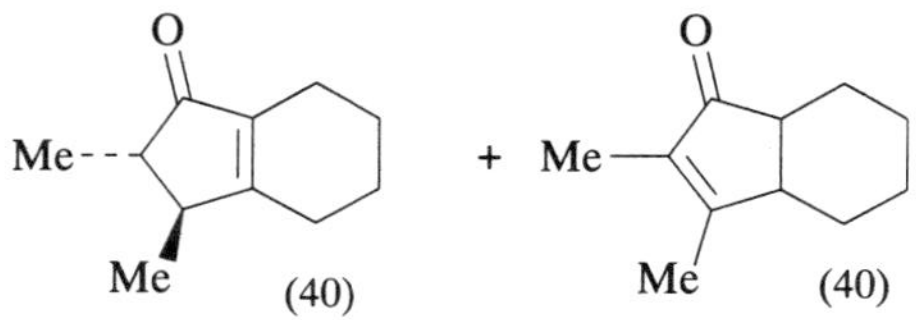 Me, O, Me (40) + Me, O, Me (40)	157
C_6			
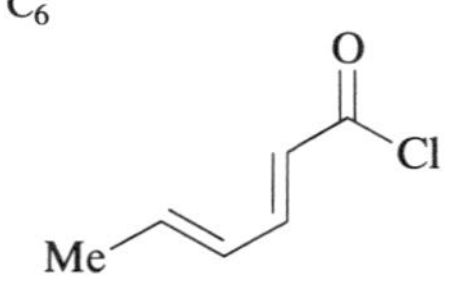 Me, O, Cl	$Me_3SiCH{=}CH_2$, $AlCl_3$, $(CH_2Cl)_2$, 84°, 1 h	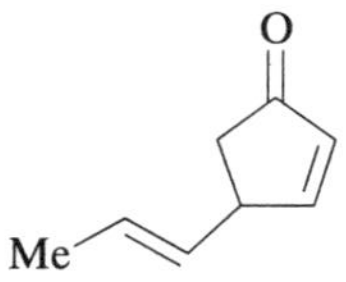 Me, O (18-80)	160
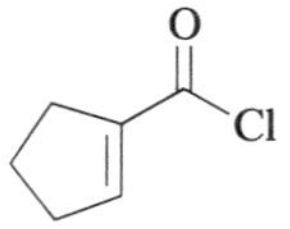 O, Cl	$Me_3SiCH{=}CH_2$, $SnCl_4$, CH_2Cl_2, -30 to 25°, 6 h	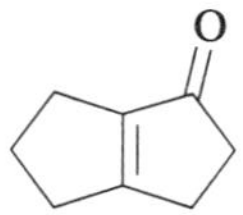 O (53)	159, 158
	$Me_3SiCH{=}CHSPh$, $AlCl_3$, $(CH_2Cl)_2$, 80°, 18 h	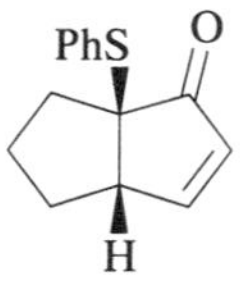 PhS, O, H (55)	163, 161
	$Me_3Si(ArS)C{=}CH_2$, $AgBF_4$, CH_2Cl_2, $(CH_2Cl)_2$, -50 to 25°, 14-20 h	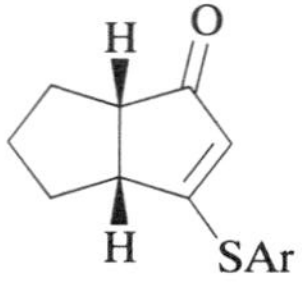 H, O, H, SAr Ar = Ph (35-45) Ar = 2,4-$(O_2N)_2C_6H_3$ (58) Ar = 4-ClC_6H_4 (15)	163, 161

TABLE V. IN SITU CONSTRUCTION OF DIVINYL KETONES: C. OLEFINIC ACID HALIDES AND OLEFINS (*Continued*)

Reactant	Conditions	Product(s) and Yield(s) (%)	Refs.
C_7			
O, Cl	$Me_3SiCH{=}CH_2$, $SnCl_4$, CH_2Cl_2, -30 to 25°, 6 h	O (46)	159, 158
	1. Cyclohexene, $AlCl_3$, CH_2Cl_2, -5° 2. H_3PO_4, HCO_2H, 90°, 18 h	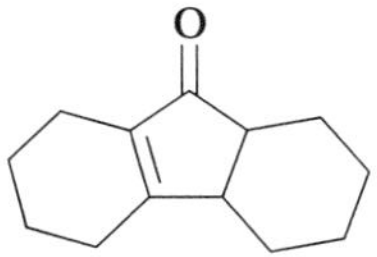 O (42)	102
C_8			
Me, Me, O, Cl	$Me_3Si(PhS)C{=}CH_2$, $AgBF_4$, $(CH_2Cl)_2$, -50 to 20°, 20 h	Me, Me, H, H, O, SPh (38)	163, 161, 162
	$Me_3SiCH{=}CHSPh$, $AlCl_3$, $(CH_2Cl)_2$, 80°, 18 h	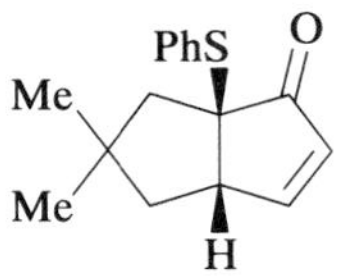 Me, Me, PhS, H, O (55)	163, 161
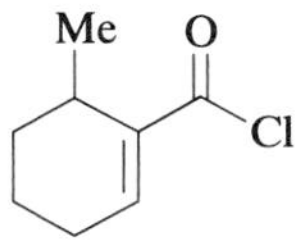 Me, O, Cl	$Me_3SiCH{=}CH_2$, $SnCl_4$, CH_2Cl_2, -30 to 25°, 6 h	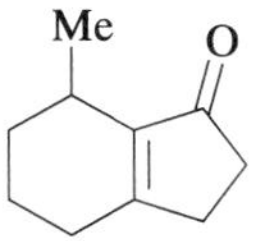 Me, O (64)	159

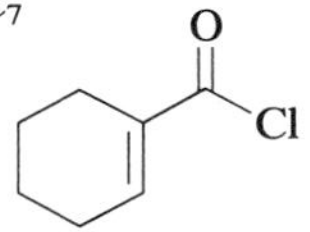

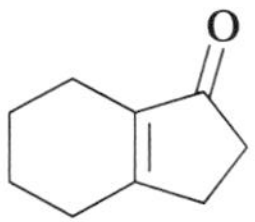

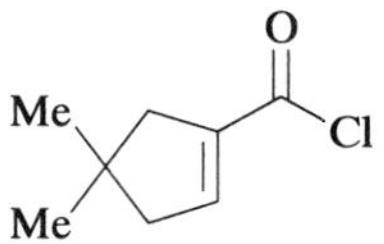

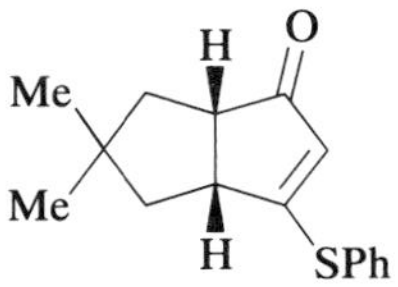

TABLE V. IN SITU CONSTRUCTION OF DIVINYL KETONES: C. OLEFINIC ACID HALIDES AND OLEFINS (*Continued*)

Reactant	Conditions	Product(s) and Yield(s) (%)	Refs.
	$Me_3Si(PhS)C{=}CH_2$, $AgBF_4$, CH_2Cl_2, $(CH_2Cl)_2$, -50 to 20° 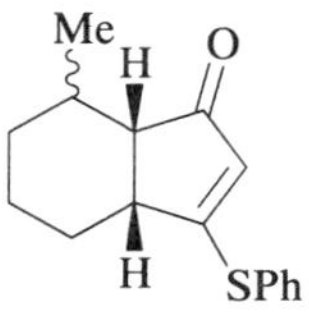	Me, H, O, H, SPh (42)	163
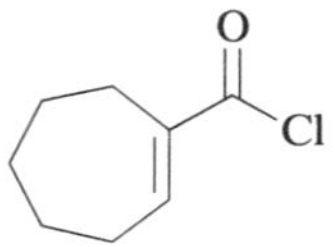 O, Cl	$Me_3SiCH{=}CH_2$, $SnCl_4$, CH_2Cl_2, -30 to 20°, 6 h	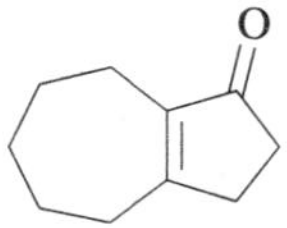 O (32)	159, 158
C_9			
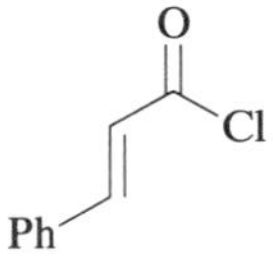 O, Cl, Ph	$Me_3SiCH{=}CH_2$, $AlCl_3$, CCl_4, 77°, 30 min	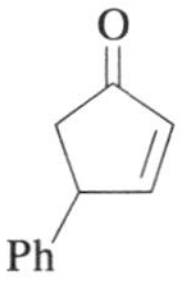 O, Ph (46)	160
C_{10}			
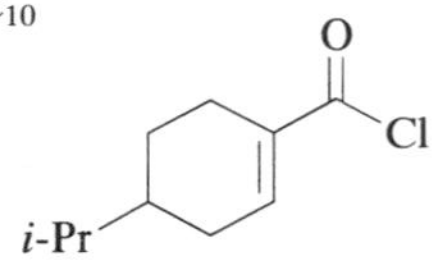 O, Cl, *i*-Pr	$Me_3SiCH{=}CH_2$, $SnCl_4$, CH_2Cl_2, -30 to 25°, 6 h	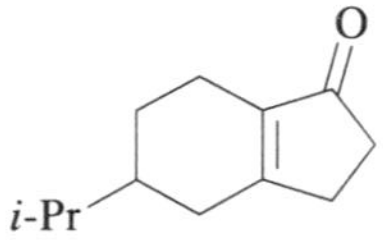 O, *i*-Pr (56)	159, 158

TABLE V. IN SITU CONSTRUCTION OF DIVINYL KETONES: D. ACID HALIDES AND PARAFFINS

Reactant	Conditions	Product(s) and Yield(s) (%)	Refs.
C_4 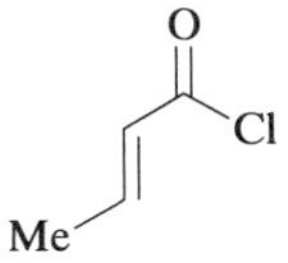 O, Cl, Me	1. $AlCl_3$, CH_2Cl_2, 0°, 0.5 h 2. 2-Methylbutane, reflux, 4 h	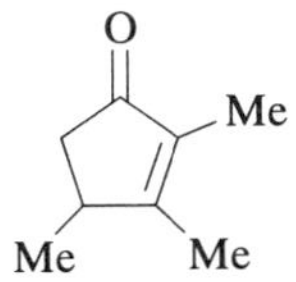 O, Me, Me, Me (60)	157a
	1. $AlCl_3$, CH_2Cl_2, 0°, 0.5 h 2. Methylcyclopentane, AcCl, CH_2Cl_2, reflux, 2 h	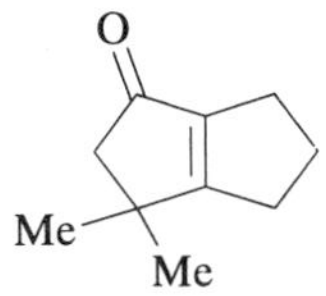 O, Me, Me (60) 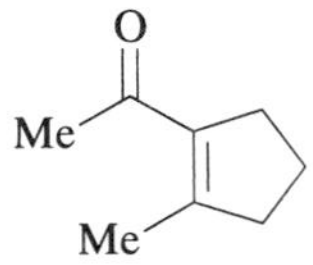 + O, Me, Me (15)	157a
	1. $AlCl_3$, CH_2Cl_2, 0°, 0.5 h 2. Cyclohexane-d_{12}, AcCl, CH_2Cl_2, reflux, 8 h	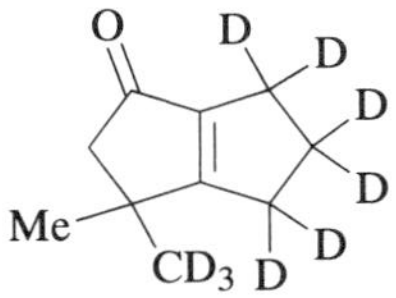 O, D, D, D, D, D, D, Me, CD_3 (45) 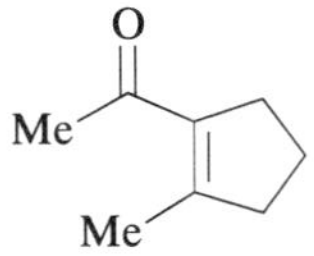 + O, Me, Me (15)	157a
	1. $AlCl_3$, CH_2Cl_2, 0°, 0.5 h 2. Methylcyclohexane, reflux, 6 h	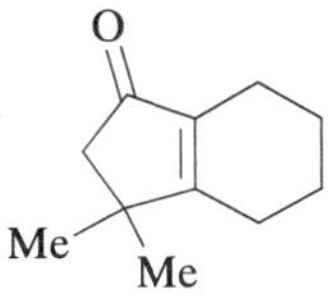 O, Me, Me (60)	157a

TABLE V. IN SITU CONSTRUCTION OF DIVINYL KETONES: D. ACID HALIDES AND PARAFFINS (*Continued*)

Reactant	Conditions	Product(s) and Yield(s) (%)	Refs.
C_5			
Me, Me, O, Cl	1. $AlCl_3$, CH_2Cl_2, 0°, 0.5 h 2. 2-Methylbutane, reflux, 4 h	Me, Me*, Me, Me*, Me*, O; * = Me (60)	157a
	1. $AlCl_3$, CH_2Cl_2, 0°, 0.5 h 2. $(CD_3)_2CHCH_3$, reflux, 4 h	* = CD_3, (d_6) (60)	157a
	1. $AlCl_3$, CH_2Cl_2, 0°, 0.5 h 2. Methylcyclopentane, AcCl reflux, 10 h	Me, Me, Me, O (60) + Me, Me, O (15)	157a
	1. $AlCl_3$, CH_2Cl_2, 0°, 0.5 h 2. 2-Methylcyclohexane, reflux, 6 h	Me, Me, Me, O (60)	157a
C_6			
Me, Et, O, Cl	1. $AlCl_3$, CH_2Cl_2, 0°, 0.5 h 2. 2-Methylbutane, reflux, 4 h	Me, Me, Et, Me, Me, O (47) + Me, Et, Me, Me, Me, O (14)	157a

TABLE V. IN SITU CONSTRUCTION OF DIVINYL KETONES: D. ACID HALIDES AND PARAFFINS (*Continued*)

Reactant	Conditions	Product(s) and Yield(s) (%)	Refs.
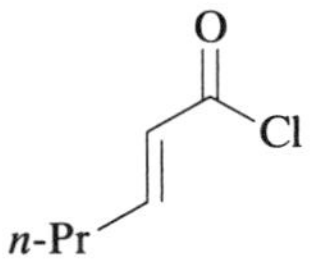 n-Pr–CH=CH–COCl	1. $AlCl_3$, CH_2Cl_2, 0°, 0.5 h 2. Methylcyclopentane, AcCl CH_2Cl_2, reflux, 4 h	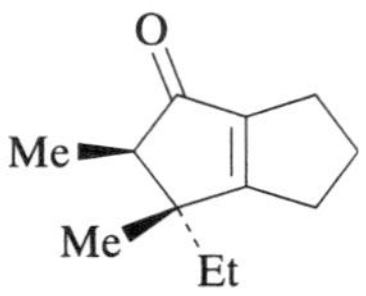 (60) 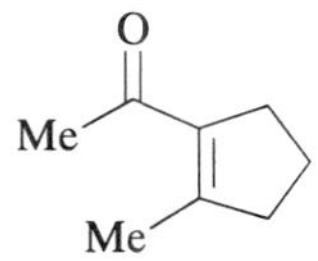 + (15)	157a
	1. $AlCl_3$, CH_2Cl_2, 0°, 0.5 h 2. Methylcyclohexane, CH_2Cl_2, reflux, 4 h	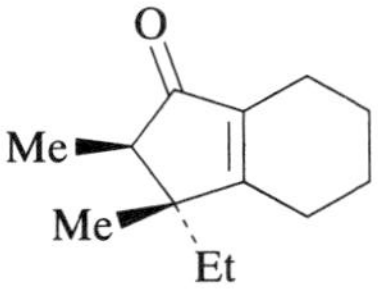 (60)	157a
	1. $AlCl_3$, CH_2Cl_2, 0°, 0.5 h 2. 2-Methylbutane, reflux, 4 h	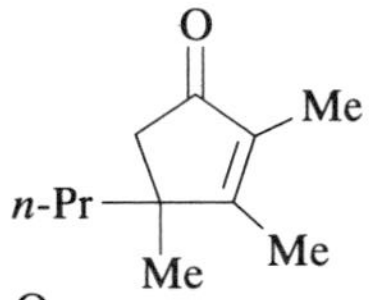 (60)	157a
	1. $AlCl_3$, CH_2Cl_2, 0°, 0.5 h 2. Methylcyclopentane, AcCl, CH_2Cl_2, reflux, 5 h	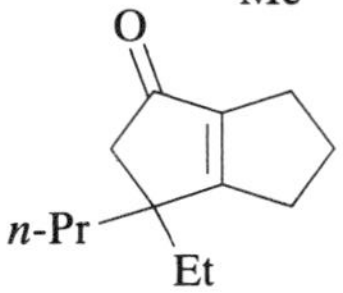 (60) 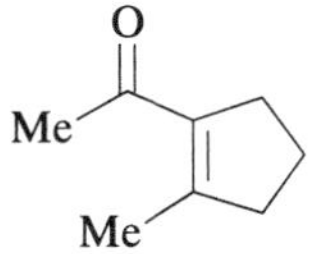 + (15)	157a
	1. $AlCl_3$, CH_2Cl_2, 0°, 0.5 h 2. Cyclohexane-d_{12}, AcCl, CH_2Cl_2, reflux, 5 h	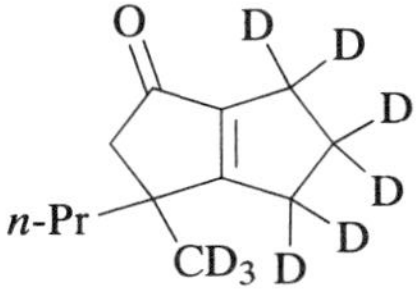 (40) 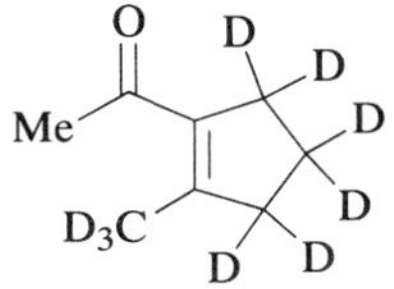 + (15)	157a

TABLE V. IN SITU CONSTRUCTION OF DIVINYL KETONES: D. ACID HALIDES AND PARAFFINS (*Continued*)

Reactant	Conditions	Product(s) and Yield(s) (%)	Refs.
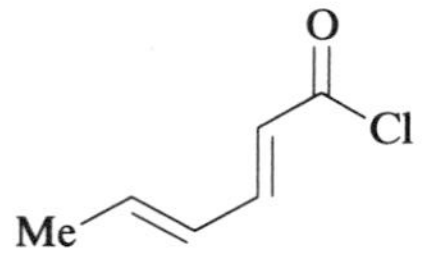 Me, O, Cl	1. $AlCl_3$, EtOH, CH_2Cl_2, 0°, 0.5 h 2. 2-Methylbutane, rt, 28 h	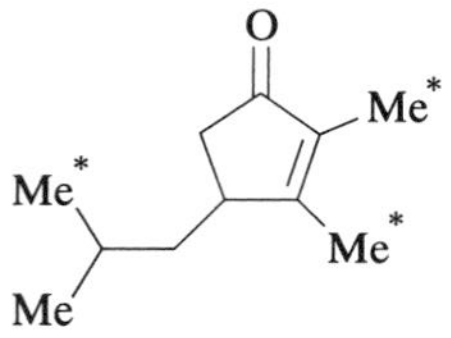 O, Me*, Me*, Me*, Me; * = Me (42)	157a
	1. $AlCl_3$, EtOH, CH_2Cl_2, 0°, 0.5 h 2. $(CD_3)_2CHCH_3$, rt, 28 h	* = CD_3, (d_6) (-)	157a
	1. $AlCl_3$, CH_2Cl_2, $MeNO_2$, H_2O, 0°, 0.25 h 2. 2-Methylbutane, reflux, 3 h	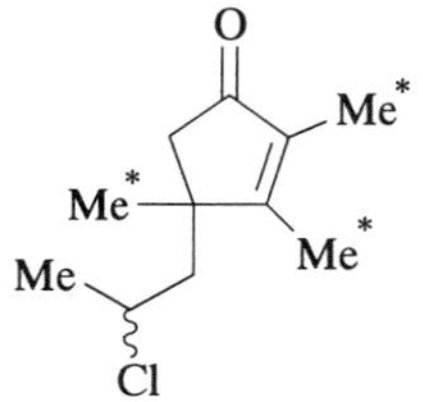 O, Me*, Me*, Me*, Me, Cl; * = Me (46)	157a
	1. $AlCl_3$, CH_2Cl_2, $MeNO_2$, H_2O, 0°, 0.25 h 2. $(CD_3)_2CHCH_3$, reflux, 3 h	* = CD_3, (d_6) (-)	157a
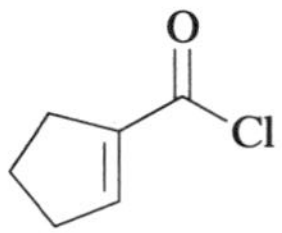 O, Cl	1. $AlCl_3$, CH_2Cl_2, 0°, 0.5 h 2. 2-Methylbutane, reflux, 12 h	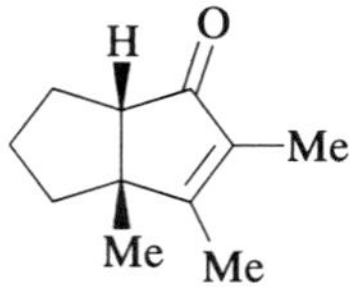 H, O, Me, Me, Me (25)	157a
C_7 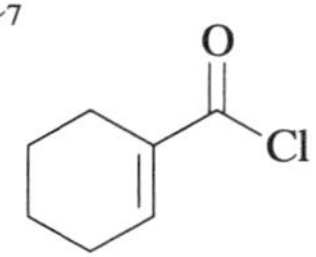 O, Cl	1. $AlCl_3$, CH_2Cl_2, 0°, 0.5 h 2. 2-Methylcyclopentane, AcCl, reflux, 3.5 h	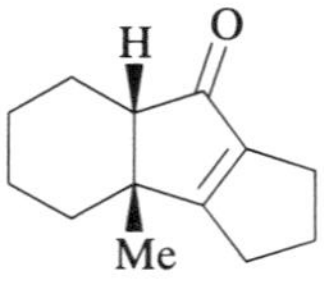 H, O, Me (60) 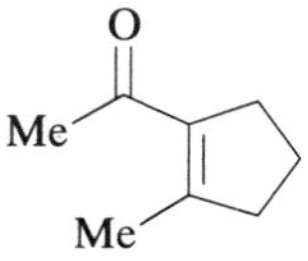 + O, Me, Me (15)	157a

TABLE V. IN SITU CONSTRUCTION OF DIVINYL KETONES: E. ACID HALIDES AND ACETYLENES

Reactant	Conditions	Product(s) and Yield(s) (%)	Refs.
C_3			
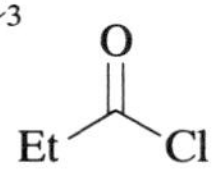 EtC(O)Cl	$MeC\equiv C(CH_2)_7CO_2Me$, $AgBF_4$, CH_2Cl_2, -20 to 0°, 10 min	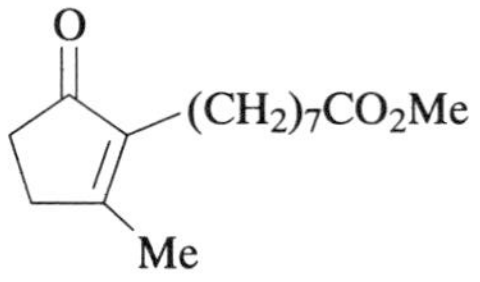 2-$(CH_2)_7CO_2Me$-3-Me-cyclopent-2-enone (35)	170
	$Me(CH_2)_7C\equiv C(CH_2)_7CO_2Me$, $AgBF_4$, CH_2Cl_2, $(CH_2Cl)_2$, -20 to 0°, 10 min	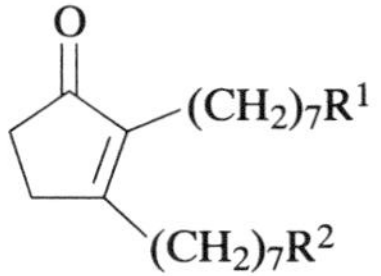 2-$(CH_2)_7R^1$-3-$(CH_2)_7R^2$-cyclopent-2-enone, R^1 = Me or CO_2Me, R^2 = CO_2Me or Me (15)	170
	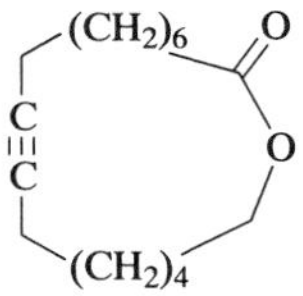 macrocyclic alkynyl lactone [$(CH_2)_6$, $C\equiv C$, $(CH_2)_4$], $AgBF_4$, CH_2Cl_2, $(CH_2Cl)_2$, -20 to 0°, 10 min	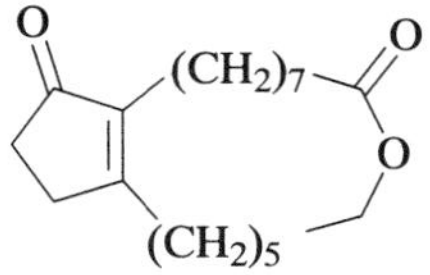 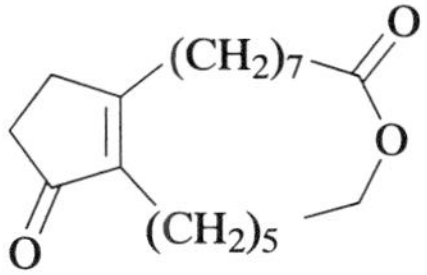 fused cyclopentenone macrolactone [$(CH_2)_7$, $(CH_2)_5$] or regioisomer (44)	170
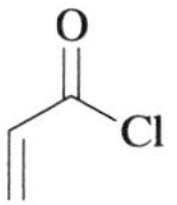 CH_2=CHC(O)Cl	$HC\equiv CH$, $AlCl_3$, $(CH_2Cl)_2$, rt, 6-7 h	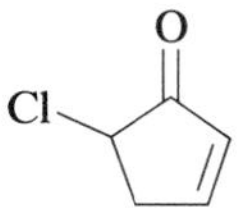 5-chlorocyclopent-2-enone (38) 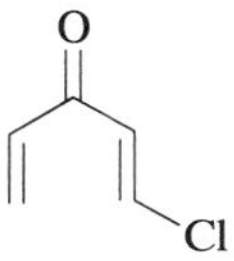 + CH_2=CHC(O)CH=CHCl (2)	164
C_4			
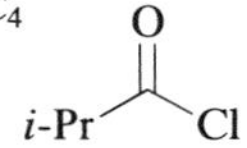 i-PrC(O)Cl	$MeC\equiv CMe$, $AgBF_4$, CH_2Cl_2, $(CH_2Cl)_2$, -60°, 5 min	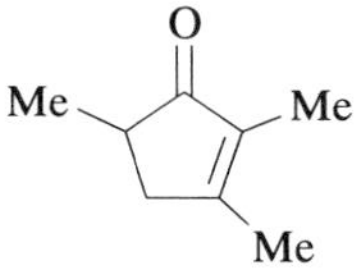 2,3,5-trimethylcyclopent-2-enone (55)	196

TABLE V. IN SITU CONSTRUCTION OF DIVINYL KETONES: E. ACID HALIDES AND ACETYLENES (*Continued*)

Reactant	Conditions	Product(s) and Yield(s) (%)	Refs.
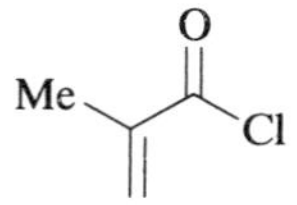 Me, Cl (O)	$HC\equiv CH$, $AlCl_3$, $(CH_2Cl)_2$, rt, 6-7 h	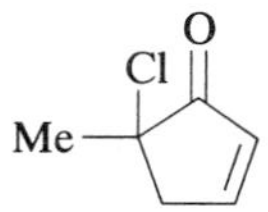 Me, Cl, O (71) 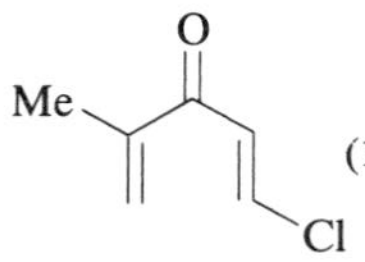 + Me, O, Cl (1)	164
	$HC\equiv CMe$, $AlCl_3$, $(CH_2Cl)_2$, 0-40°, 2 h	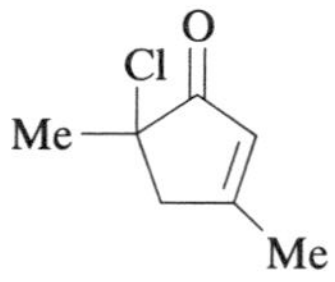 Me, Cl, O, Me (62) 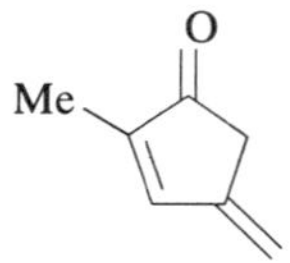 + Me, O (3)	166
	$HC\equiv CEt$, $AlCl_3$, $(CH_2Cl)_2$, 0-40°, 2 h	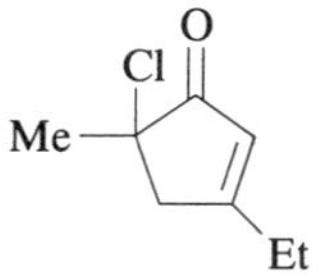 Me, Cl, O, Et (8) 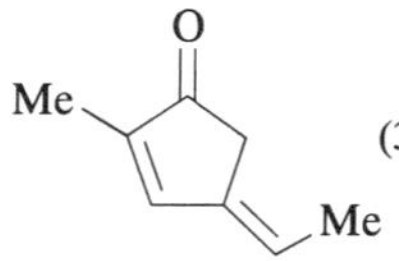 + Me, O, Me (33)	166
	$MeC\equiv CMe$, $AlCl_3$, $(CH_2Cl)_2$, 0-40°, 2 h	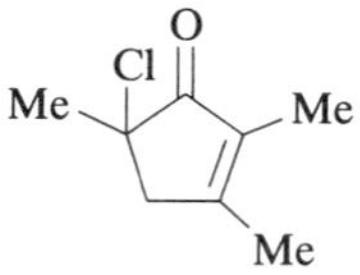 Me, Cl, O, Me, Me (40)	166
	$HC\equiv CPr$-*n*, $AlCl_3$, $(CH_2Cl)_2$, 0-40°, 2 h	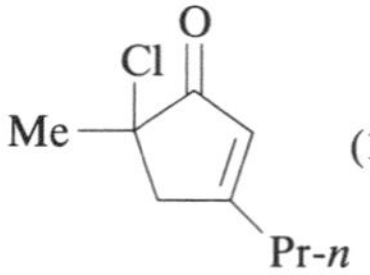 Me, Cl, O, Pr-*n* (10) 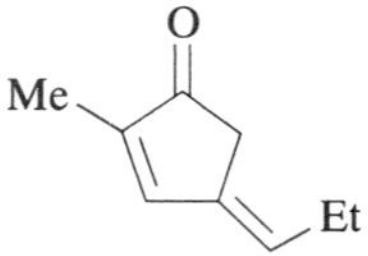 + Me, O, Et (42)	166

TABLE V. IN SITU CONSTRUCTION OF DIVINYL KETONES: E. ACID HALIDES AND ACETYLENES (*Continued*)

Reactant	Conditions	Product(s) and Yield(s) (%)	Refs.
	MeC≡CEt, $AlCl_3$, $(CH_2Cl)_2$, 0-40°, 2 h	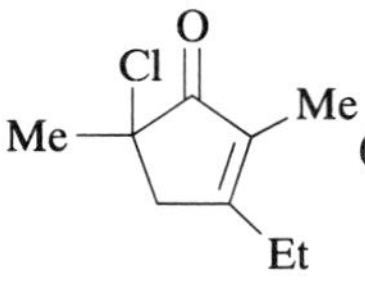 (17) + 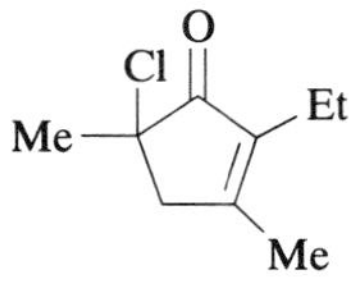 (17)	166
	HC≡CBu-*n*, $AlCl_3$, $(CH_2Cl)_2$, 0-40°, 2 h	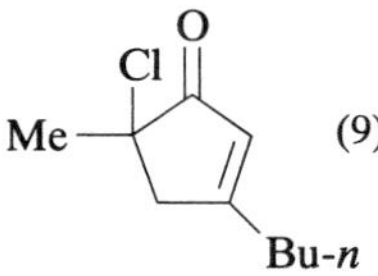 (9) + 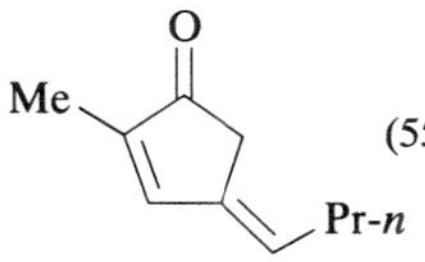 (55)	166
	MeC≡CPr-*n* , $AlCl_3$, $(CH_2Cl)_2$, 0-40°, 2 h	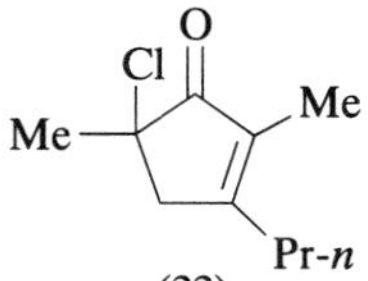 (22) + 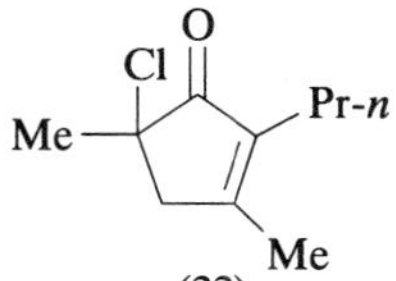 (22)	166
	EtC≡CPr-*n* , $AlCl_3$, $(CH_2Cl)_2$, 0-40°, 2 h	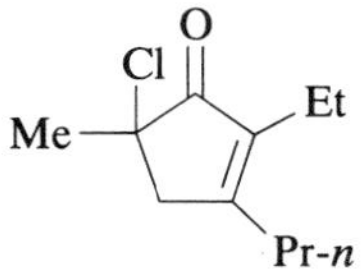 (55)	166
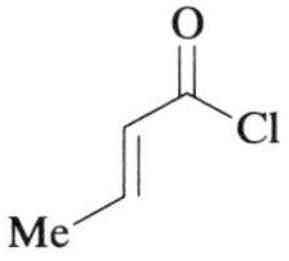	HC≡CH, $AlCl_3$, $(CH_2Cl)_2$, rt, 6-7 h	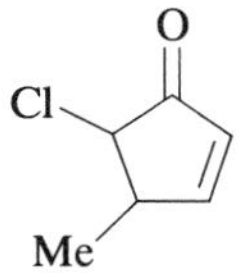 (53) + 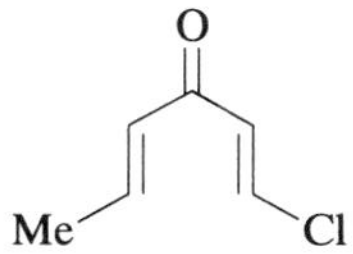 (7)	166

TABLE V. IN SITU CONSTRUCTION OF DIVINYL KETONES: E. ACID HALIDES AND ACETYLENES (*Continued*)

Reactant	Conditions	Product(s) and Yield(s) (%)	Refs.
	$HC\equiv CEt$, $AlCl_3$, $(CH_2Cl)_2$, 0-40°, 2 h	(9) + (55)	166
	$MeC\equiv CMe$, $AlCl_3$, $(CH_2Cl)_2$, 0-40°, 2 h	(30)	166
	$HC\equiv CPr$-*n*, $AlCl_3$, $(CH_2Cl)_2$, 0-40°, 2 h	(14) + (31)	166
	$HC\equiv CBu$-*n*, $AlCl_3$, $(CH_2Cl)_2$, 0-40°, 2 h	(13) + (29)	166
	$MeC\equiv CPr$-*n*, $AlCl_3$, $(CH_2Cl)_2$, 0-40°, 2 h	R^1 = Me, R^2 = Pr-*n*; R^1 = Pr-*n*, R^2 = Me (23) + (7)	166

TABLE V. IN SITU CONSTRUCTION OF DIVINYL KETONES: E. ACID HALIDES AND ACETYLENES (*Continued*)

Reactant	Conditions	Product(s) and Yield(s) (%)	Refs.
	MeC≡CBu-*n*, $AlCl_3$, $(CH_2Cl)_2$, 0-40°, 2 h	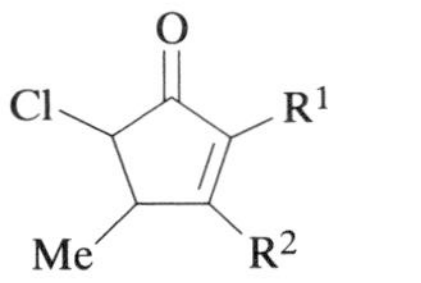 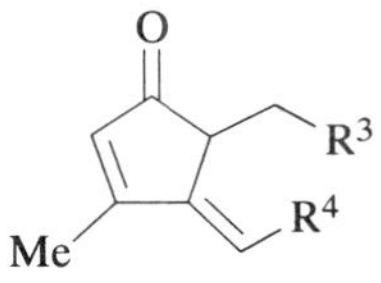 Cl, Me, R^1, R^2 cyclopentenone + Me, R^3, R^4 cyclopentenone; R^1 = Me, R^2 = Bu-*n*; R^1 = Bu-*n*, R^2 = Me (25); R^3 = H, R^4 = Pr-*n*; R^3 = Pr-*n*, R^4 = H (5)	166
	EtC≡CPr-*n*, $AlCl_3$, $(CH_2Cl)_2$, 0-40°, 2 h	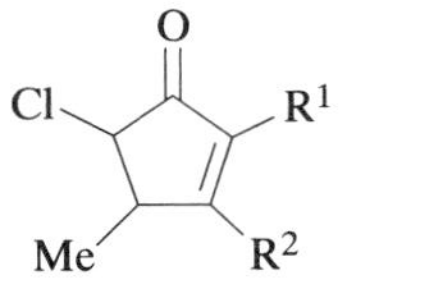 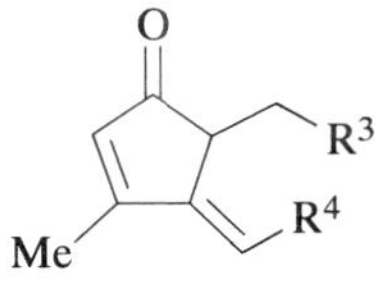 Cl, Me, R^1, R^2 cyclopentenone + Me, R^3, R^4 cyclopentenone; R^1 = Et, R^2 = Pr-*n*; R^1 = Pr-*n*, R^2 = Et (43); R^3 = Me, R^4 = Et; R^3 = Et, R^4 = Me (8)	166
C_5			
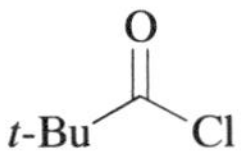 *t*-BuCOCl	HC≡CMe, $AgBF_4$, $(CH_2Cl)_2$, -60°, 5 min	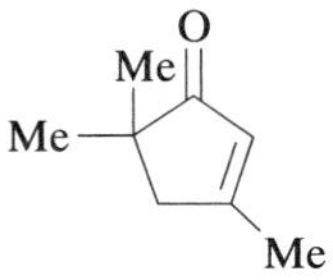 5,5-Me,Me-3-Me-cyclopentenone (66)	196
	MeC≡CMe, $AgBF_4$, $(CH_2Cl)_2$, -60°, 5 min	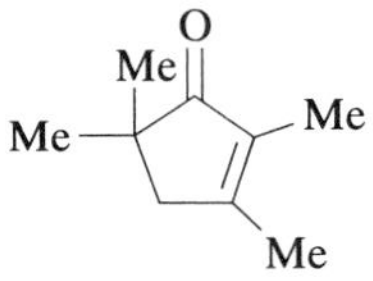 5,5-Me,Me-2,3-Me,Me-cyclopentenone (73)	196

TABLE V. IN SITU CONSTRUCTION OF DIVINYL KETONES: E. ACID HALIDES AND ACETYLENES (*Continued*)

Reactant	Conditions	Product(s) and Yield(s) (%)	Refs.
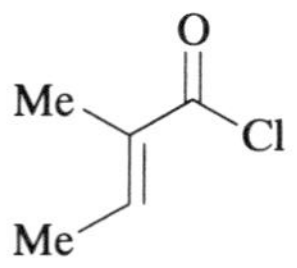 Me, Me, O, Cl (structure)	$HC{\equiv}CH$, $AlCl_3$, $(CH_2Cl)_2$, rt, 6-7 h	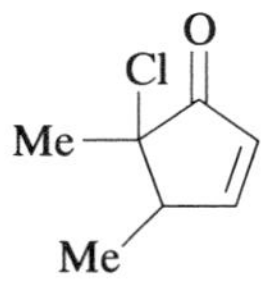 Cl, O, Me, Me (structure) (70)	164
	$HC{\equiv}CMe$, $AlCl_3$, $(CH_2Cl)_2$, 0-40°, 2 h	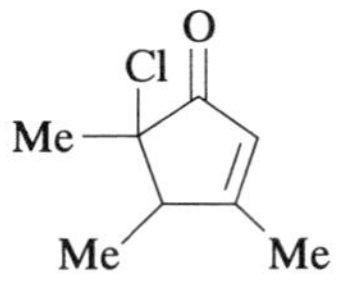 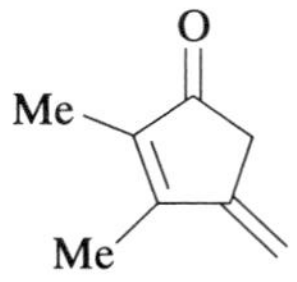 Cl, O, Me, Me, Me (structure) (26) + Me, Me, O (structure) (40)	166
	$HC{\equiv}CEt$, $Al\ Cl_3$, $(CH_2Cl)_2$, 0-40°, 2 h	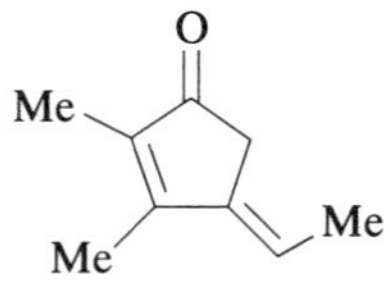 O, Me, Me, Me (structure) (68)	166
	$MeC{\equiv}CMe$, $AlCl_3$, $(CH_2Cl)_2$, 0-40°, 2 h 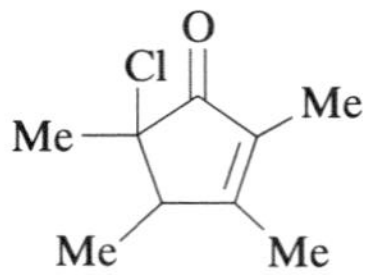	Cl, O, Me, Me, Me, Me (structure) (40)	166

TABLE V. IN SITU CONSTRUCTION OF DIVINYL KETONES: E. ACID HALIDES AND ACETYLENES (*Continued*)

Reactant	Conditions	Product(s) and Yield(s) (%)	Refs.
	$HC{\equiv}CPr$-*n*, $AlCl_3$, $(CH_2Cl)_2$, rt, 6-7 h	Me, Me, O, Et (62) + Me, O, Me, Cl, Et (5)	166
	$MeC{\equiv}CEt$, $AlCl_3$, $(CH_2Cl)_2$, 0-40°, 2 h	Cl, O, Me, R^1, Me, R^2; R^1 = Et, R^2 = Me; R^1 = Me, R^2 = Et (54) + Me, O, R^3, Me, R^4; R^3 = H, R^4 = Me; R^3 = Me, R^4 = H (6)	166
	$HC{\equiv}CBu$-*n*, $AlCl_3$, $(CH_2Cl)_2$, 0-40°, 2 h	Me, O, Me, Pr-*n* (74)	166
Et, O, Cl	$HC{\equiv}CH$, $AlCl_3$, $(CH_2Cl)_2$, rt, 6-7 h	Cl, O, Et (42) + O, Et, Cl (8)	164

TABLE V. IN SITU CONSTRUCTION OF DIVINYL KETONES: E. ACID HALIDES AND ACETYLENES (*Continued*)

Reactant	Conditions	Product(s) and Yield(s) (%)	Refs.
C_6			
n-Pr, Cl, O	$HC{\equiv}CH$, $AlCl_3$, $(CH_2Cl)_2$, rt, 6-7 h	Cl, O, n-Pr (30) + O, n-Pr, Cl (10)	164
i-Pr, Cl, O	$HC{\equiv}CH$, $AlCl_3$, $(CH_2Cl)_2$, rt, 6-7 h	Cl, O, i-Pr (21) + O, i-Pr, Cl (9)	164
C_7			
n-Bu, Cl, O	$HC{\equiv}CH$, $AlCl_3$, $(CH_2Cl)_2$, rt, 6-7 h	Cl, O, n-Bu (44) + O, n-Bu, Cl (2)	164
C_8			
n-Pr, Et, Cl, O	$HC{\equiv}CMe$, $AlCl_3$, $(CH_2Cl)_2$, 0-40°, 2 h	Cl, O, n-Pr, Et, Me (16) + O, n-Pr, Et (30)	166

TABLE V. IN SITU CONSTRUCTION OF DIVINYL KETONES: E. ACID HALIDES AND ACETYLENES (*Continued*)

Reactant	Conditions	Product(s) and Yield(s) (%)	Refs.
	HC≡CEt, $AlCl_3$, $(CH_2Cl)_2$, 0-40°, 2 h	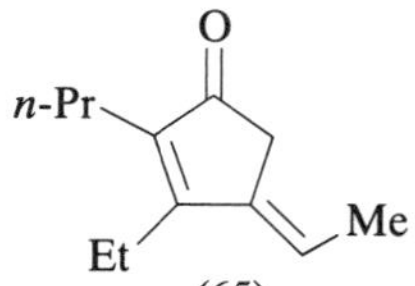 (65) + 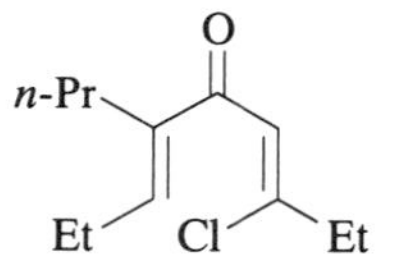 (7)	166
	HC≡CPr-*n*, $AlCl_3$, $(CH_2Cl)_2$, 0-40°, 2 h	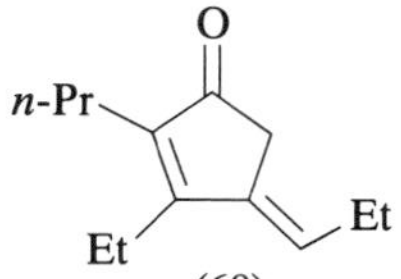 (60) + 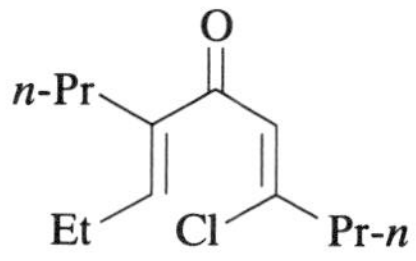 (23)	166
	HC≡CBu-*n*, $AlCl_3$, $(CH_2Cl)_2$, 0-40°, 2 h	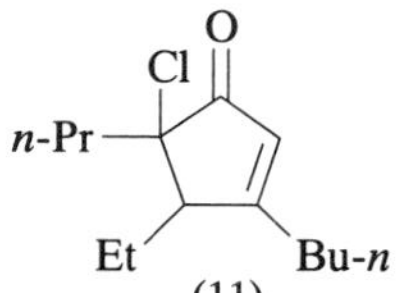 (11) + 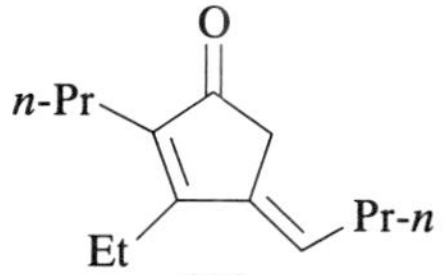 (60)	166
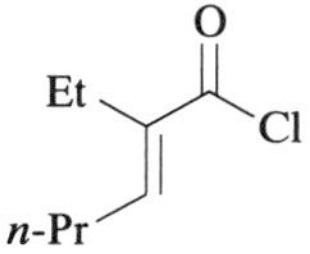	HC≡CH, $AlCl_3$, $(CH_2Cl)_2$, rt, 6-7 h	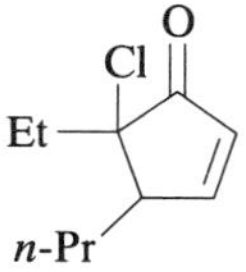 (65)	164

TABLE V. IN SITU CONSTRUCTION OF DIVINYL KETONES: F. ORGANOMETALLICS

Reactant	Conditions	Product(s) and Yield(s) (%)	Refs.
C_2			
HC≡CH	$Ni(CO)_4$, EtOH, HNO_3, HCl, 40-60°, 2 h	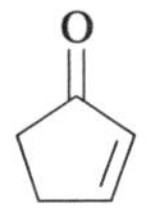 O (15)	168
C_4			
MeC≡CMe	$Ni(CO)_4$, EtOH, HNO_3, HCl, 40-60°, 2 h	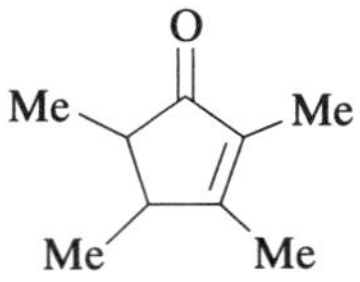 O, Me, Me, Me, Me (61)	168
C_6			
EtC≡CEt	$Ni(CO)_4$, EtOH, HNO_3, HCl, 40-60°, 2 h	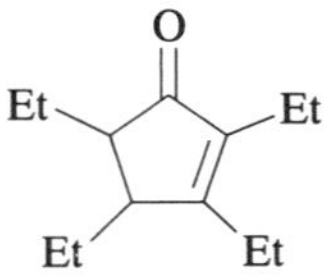 O, Et, Et, Et, Et (70)	168
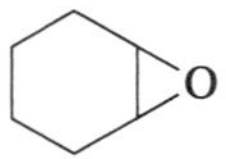 O	1. $Me_3SiMn(CO)_5$,PhC≡CH, Et_2O, 1 bar, rt 2. HCO_2H, H_3PO_4, 90°, 2 h	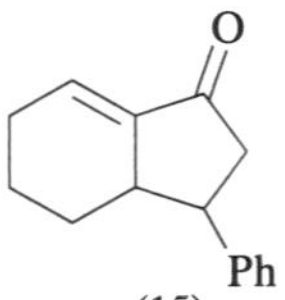 O, Ph (15) 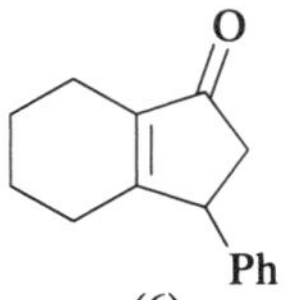 + O, Ph (6)	169

TABLE V. IN SITU CONSTRUCTION OF DIVINYL KETONES: F. ORGANOMETALLICS (*Continued*)

Reactant	Conditions	Product(s) and Yield(s) (%)		Refs.
C_8				
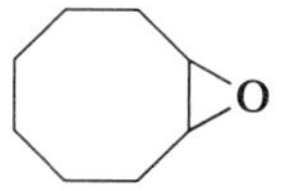 O	1. $Me_3SiMn(CO)_5$,PhC≡CH, Et_2O, 1 bar, rt 2. HCO_2H, H_3PO_4, 90°, 2 h	O Ph	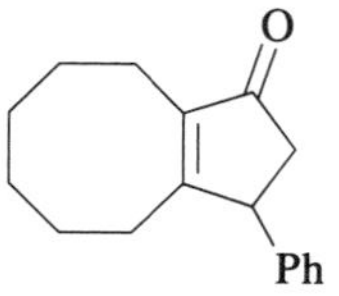 (35)	169
n-PrC≡CPr-*n*	$Ni(CO)_4$, EtOH, HNO_3, HCl, 40-60°, 2 h	O *n*-Pr Pr-*n* *n*-Pr Pr-*n*	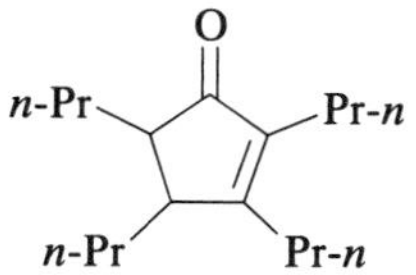 (25)	168
C_{10}				
n-BuC≡CBu-*n*	$Ni(CO)_4$, EtOH, HNO_3, HCl, 40-60°, 2 h	O *n*-Bu Bu-*n* *n*-Bu Bu-*n*	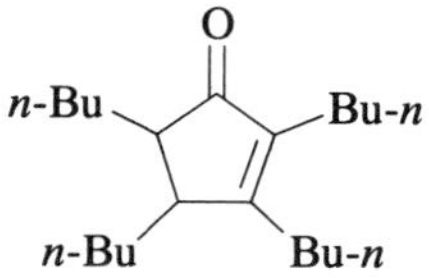 (25)	168
C_{14}				
PhC≡CPh	$Ni(CO)_4$, EtOH, HNO_3, HCl, 40-60 3, 2 h	O Ph Ph Ph Ph	(14)	168

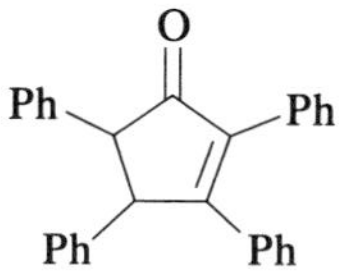

REFERENCES

[1] I. N. Nazarov and I. I. Zaretskaya, *Izv. Akad. Nauk SSSR, Ser. Khim.*, **1941**, 211 [*C.A.*, **37**, 6243[7] (1943)].

[2] I. N. Nazarov and I. I. Zaretskaya, *Izv. Akad. Nauk SSSR, Ser. Khim.*, **1942**, 200 [*C.A.*, **39**, 1619[4] (1945)].

[3] I. N. Nazarov and A. I. Kuznetsova, *Izv. Akad. Nauk SSSR, Ser. Khim.*, **1942**, 392 [*C.A.*, **39**, 1620[3] (1945)].

[4] I. N. Nazarov and Ya. M. Yanbikov, *Izv. Akad. Nauk SSSR, Ser. Khim.*, **1943**, 389 [*C.A.*, **39**, 503a (1945)].

[5] I. N. Nazarov and I. I. Zaretskaya, *Izv. Akad. Nauk SSSR, Ser. Khim.*, **1944**, 65 [*C.A.*, **39**, 1620[9] (1945)].

[6] I. N. Nazarov and I. I. Zaretskaya, *Izv. Akad. Nauk SSSR, Ser. Khim.*, **1946**, 529 [*C.A.*, **42**, 7731a (1948)].

[7] I. N. Nazarov and L. N. Pinkina, *Izv. Akad. Nauk SSSR, Ser. Khim.*, **1946**, 633 [*C.A.*, **42**, 7731h (1948)].

[8] I. N. Nazarov and M. S. Burmistrova, *Izv. Akad. Nauk SSSR, Ser. Khim.*, **1947**, 51 [*C.A.*, **42**, 7732g (1948)].

[9] I. N. Nazarov and S. S. Bukhmutskaya, *Izv. Akad. Nauk SSSR, Ser. Khim.*, **1947**, 205 [*C.A.*, **42**, 7733e (1948)].

[10] I. N. Nazarov and G. P. Verkholetova, *Izv. Akad. Nauk SSSR, Ser. Khim.*, **1947**, 277 [*C.A.*, **42**, 7734a (1948).

[11] I. N. Nazarov and M. S. Burmistrova, *Izv. Akad. Nauk SSSR, Ser. Khim.*, **1947**, 353 [*C.A.*, **42**, 7734i (1948)].

[12] I. N. Nazarov and I. I. Zaretskaya, *Zh. Obshch. Khim.*, **18**, 665 (1948) [*C.A.*, **43**, 115f (1949)].

[13] I. N. Nazarov and L. N. Pinkina, *Zh. Obshch. Khim.*, **18**, 675 (1948) [*C.A.*, **43**, 116h (1949)].

[14] I. N. Nazarov and L. N. Pinkina, *Zh. Obshch. Khim.*, **18**, 681 (1948) [*C.A.*, **43**, 117c (1949)].

[15] I. N. Nazarov and I. L. Kotlyarevskii, *Zh. Obshch. Khim.*, **18**, 896 (1948) [*C.A.*, **43**, 117h (1949)].

[16] I. N. Nazarov and I. L. Kotlyarevskii, *Zh. Obshch. Khim.*, **18**, 903 (1948) [*C.A.*, **43**, 118d (1949)].

[17] I. N. Nazarov and I. L. Kotlyarevskii, *Zh. Obshch. Khim.*, **18**, 911 (1948) [*C.A.*, **43**, 119a (1949)].

[18] I. N. Nazarov and S. S. Bakhmutskaya, *Zh. Obshch. Khim.*, **18**, 1077 (1948) [*C.A.*, **43**, 1332g (1949)].

[19] I. N. Nazarov and G. P. Verkholetova, *Zh. Obshch. Khim.*, **18**, 1083 (1948) [*C.A.*, **43**, 1333b (1949)].

[20] I. N. Nazarov and T. D. Nagibina, *Zh. Obshch. Khim.*, **18**, 1090 (1948) [*C.A.*, **43**, 1333h (1949)].

[21] I. N. Nazarov and I. V. Torgov, *Zh. Obshch. Khim.*, **18**, 1338 (1948) [*C.A.*, **43**, 2161h (1949)].

[22] I. N. Nazarov and M. S. Burmistrova, *J. Gen. Chem. USSR*, **20**, 1335 (1950).

[23] I. N. Nazarov and I. L. Kotlyarevsky, *J. Gen. Chem. USSR*, **20**, 1491 (1950).

[24] I. N. Nazarov and I. L. Kotlyarevsky, *J. Gen. Chem. USSR*, **20**, 1501 (1950).

[25] I. N. Nazarov and I. L. Kotlyarevsky, *J. Gen. Chem. USSR*, **20**, 1509 (1950).

[26] I. N. Nazarov and S. S. Bakhmutskaya, *J. Gen. Chem. USSR*, **19**, a223 (1949).

[27] I. N. Nazarov and L. N. Pinkina, *J. Gen. Chem. USSR*, **19**, a331 (1949).

[28] I. N. Nazarov and I. I. Zaretskaya, *Izv. Akad. Nauk SSSR, Ser. Khim.*, **1949**, 178 [*C.A.*, **43**, 6623c (1949)].

[29] I. N. Nazarov and I. I. Zaretskaya, *Izv. Akad. Nauk SSSR, Ser. Khim.*, **1949**, 184 [*C.A.*, **43**, 6623h (1949)].

[30] I. N. Nazarov and I. L. Kotlyarevskii, *Izv. Akad. Nauk SSSR, Ser. Khim.*, **1949**, 293 [*C.A.*, **43**, 6624f (1949)].

[31] I. N. Nazarov and L. N. Pinkina, *J. Gen. Chem. USSR*, **20**, 2079 (1950).

[32] D. N. Kursanov, Z. N. Parnes, I. I. Zaretskaya, and I. N. Nazarov, *Izv. Akad. Nauk SSSR, Ser. Khim.*, **1953**, 114 [*C.A.*, **48**, 3271 (1954)].

[33] I. N. Nazarov and M. S. Burmistrova, *J. Gen. Chem. USSR*, **20**, 2091 (1950).

[34] I. N. Nazarov and T. D. Nagibina, *J. Gen. Chem. USSR*, **23**, 839 (1953).

[35] I. N. Nazarov, I. I. Zaretskaya, Z. N. Parnes, and D. N. Kursanov, *Izv. Akad. Nauk SSSR, Ser. Khim.*, **1953**, 519 [*C.A.*, **48**, 9930b (1954)].
[36] D. N. Kursanov, Z. N. Parness, I. I. Zaretskaya, and I. N. Nazarov, *Izv. Akad. Nauk SSSR, Ser. Khim.*, **1954**, 859 [*C.A.*, **49**, 13910a (1955)].
[37] I. N. Nazarov and I. I. Zaretskaya, *J. Gen. Chem. USSR*, **27**, 693 (1957).
[38] I. N. Nazarov, I. I. Zaretskaya, and T. I. Sorkina, *J. Gen. Chem. USSR*, **30**, 765 (1960).
[39] E. A. Braude and J. A. Coles, *J. Chem. Soc.*, **1952**, 1430.
[40] D. Vorländer and G. Schroeter, *Chem. Ber.*, **36**, 1490 (1903).
[41] C. F. H. Allen, J. A. Van Allen, and J. F. Tinker, *J. Org. Chem.*, **20**, 1387 (1955).
[42] F. Francis and F. G. Willson, *J. Chem. Soc.*, **1913**, 2238.
[43] C. W. Shoppee and R. E. Lack, *J. Chem. Soc.*, **1969**, 1346.
[44] C. W. Shoppee and B. J. A. Cooke, *J. Chem. Soc., Perkin Trans. 1*, **1973**, 1026.
[45] T. S. Sorensen and A. Rauk, in *Pericyclic Reactions*, Vol. II, A. P. Marchand and R. E. Lehr, Eds., Academic Press, New York, 1977, pp. 1–67.
[46] S. E. Denmark, in *Comprehensive Organic Synthesis*, Vol. 5, L. A. Paquette, Ed., Pergamon Press, Oxford, 1991, pp. 751–784.
[47] C. Santelli-Rouvier and M. Santelli, *Synthesis*, **1983**, 429.
[48] M. Ramaiah, *Synthesis*, **1984**, 529.
[49] M. Ono, J. Syn. *Org. Chem. Jpn.*, **39**, 872 (1981).
[50] R. B. Woodward and R. Hoffman, in *The Conservation of Orbital Symmetry*, Verlag Chemie, Weinheim, 1971, pp. 38–64.
[51] N. C. Deno, C. U. Pittman, Jr., and J. O. Turner, *J. Am. Chem. Soc.*, **87**, 2153 (1965).
[52] T. S. Sorensen, *J. Am. Chem. Soc.*, **89**, 3782 (1967).
[53] J. Motoyoshiya, T. Yazaki, and S. Hayashi, *J. Org. Chem.*, **56**, 735 (1991).
[54] E. L. Eliel, in *The Stereochemistry of Carbon Compounds*, McGraw Hill, New York, 1960, p. 436.
[55] P. H. Campbell, N. W. K. Chiu, K. Deugau, I. J. Miller, and T. S. Sorensen, *J. Am. Chem. Soc.*, **91**, 6404 (1969).
[56] N. W. K. Chiu and T. S. Sorensen, *Can. J. Chem.*, **51**, 2776 (1973).
[57] R. B. Woodward, *Chem. Soc. Special Publication No. 21*, **1967**, pp. 237–239.
[58] C. W. Shoppee and B. J. A. Cooke, *J. Chem. Soc., Perkin Trans. 1*, **1972**, 2271.
[59] E. J. Corey, K. Ritter, M. Yus, and C. Nájera, *Tetrahedron Lett.*, **28**, 3547 (1987).
[60] H. Nozaki and M. Kurita, *Tetrahedron Lett.*, **1968**, 3635.
[61] R. Noyori, Y. Ohnishi, and M. Katô, *J. Am. Chem. Soc.*, **97**, 928 (1975).
[62] C. P. Visser and H. Cerfontain, *Recl. Trav. Chim. Pays-Bas*, **102**, 307 (1983).
[63] D. A. Smith and C. W. Ullmer, III, *Tetrahedron Lett.*, **32**, 725 (1991).
[64] D. A. Smith and C. W. Ullmer, III, *J. Org. Chem.*, **56**, 4444 (1991).
[65] K. N. Houk, in *Strain and Its Implications In Organic Chemistry*, A. deMeijere and S. Blechert, Eds., Kluwer Academic Publishers, Dodrecht, 1989, pp. 25–37.
[66] G. Mehta and N. Krishnamurthy, *J. Chem. Soc., Chem. Commun.*, **1986**, 1319.
[67] W. E. Fristad, D. S. Dime, T. R. Bailey, and L. A. Paquette, *Tetrahedron Lett.*, **1979**, 1999.
[68] L. A. Paquette, W. E. Fristad, D. S. Dime, and T. R. Bailey, *J. Org. Chem.*, **45**, 3017 (1980).
[69] J. P. Marino and R. J. Linderman, *J. Org. Chem.*, **46**, 3696 (1981).
[69a] J. P. F. Andrews and A. C. Regan, *Tetrahedron Lett.*, **32**, 7731 (1991).
[70] E. Wada, I. Funiuara, S. Kanemasa, and O. Tsuge, *Bull. Chem. Soc. Jpn.*, **60**, 325 (1987).
[71] P. E. Eaton, C. Giordano, G. Schloemer, and U. Vogel, *J. Org. Chem.*, **41**, 2238 (1976).
[72] E. A. Braude, W. F. Forbes, and E. A. Evans, *J. Chem. Soc.*, **1953**, 2202.
[73] R. E. Lehr, Ph.D. Thesis, Harvard University, 1968.
[74] D. B. Kurland, Ph.D. Thesis, Harvard University, 1967.
[75] K. E. Harding and K. S. Clement, *J. Org. Chem.*, **49**, 3870 (1984).
[76] S. Dev, *J. Indian Chem. Soc.*, **34**, 169 (1957).
[77] E. A. Braude and W. F. Forbes, *J. Chem. Soc.*, **1953**, 2208.
[78] S. J. de Solms, O. W. Woltersdorf, Jr., E. J. Cragoe, Jr., L. S. Watson, and G. M. Fanelli, Jr., *J. Med. Chem.*, **21**, 437 (1978).

[79] S. Hirano, T. Hiyama, and H. Nozaki, *Tetrahedron Lett.*, **1974**, 1429.
[80] S. Hirano, S. Takagi, T. Hiyama, and H. Nozaki, *Bull. Chem. Soc. Jpn.*, **53**, 169 (1980).
[81] S. E. Denmark and G. A. Hite, *Helv. Chim. Acta,* **71**, 195 (1988).
[82] K. Schaffner and M. Demuth, in *Rearrangements in Ground and Excited States,* Vol. III, P. de Mayo, Ed., Academic Press, New York, 1980, p. 281
[82a] A. B. Smith, III and W. C. Agosta, *J. Am. Chem. Soc.*, **95**, 1961 (1973).
[82b] L. M. Jackman, E. F. M. Stephenson, and H. C. Yick, *Tetrahedron Lett.*, **1970**, 3325.
[83] J. W. Pavlik and L. T. Kavlinkonis, *Tetrahedron Lett.*, **1976**, 1939.
[83a] F. G. West, P. V. Fisher, and C. A. Willoughby, *J. Org. Chem.* **55**, 5936 (1990).
[84] S. E. Denmark and T. K. Jones, *J. Am. Chem. Soc.*, **104**, 2642 (1982).
[85] T. K. Jones and S. E. Denmark, *Helv. Chim. Acta,* **66**, 2377 (1983).
[86] T. K. Jones and S. E. Denmark, *Helv. Chim. Acta,* **66**, 2397 (1983).
[87] S. E. Denmark, K. L. Habermas, G. A. Hite, and T. K. Jones, *Tetrahedron,* **42**, 2821 (1986).
[88] S. E. Denmark, K. L. Habermas, and G. A. Hite, *Helv. Chim. Acta,* **71**, 168 (1988).
[89] S. E. Denmark and R. C. Klix, *Tetrahedron,* **44**, 4043 (1988).
[90] S. E. Denmark, M. A. Wallace, and C. B. Walker, Jr., *J. Org. Chem.*, **55**, 5543 (1990).
[91] J. B. Lambert, G. Wang, R. B. Finzel, and D. H. Teramura, *J. Am. Chem. Soc.*, **109**, 7838 (1987).
[92] T. K. Jones and S. E. Denmark, *Org. Syn.*, **64**, 182 (1985).
[93] T. K. Jones and S. E. Denmark, *J. Org. Chem.*, **47**, 4595 (1982)
[94] R. F. Cunico and F. J. Clayton, *J. Org. Chem.*, **41**, 1480 (1976).
[95] G. T. Crisp, W. J. Scott, and J. K. Stille, *J. Am. Chem. Soc.*, **106**, 7500 (1984).
[96] I. Fleming, T. W. Newton, and F. Roessler, *J. Chem. Soc., Perkin Trans. 1,* **1981**, 2527.
[97] B. L. Chenard, C. M. Van Zyl, and D. R. Sanderson, *Tetrahedron Lett.*, **27**, 2801 (1986).
[98] T. K. Jones, K. L. Habermas, and S. E. Denmark, unpublished results.
[99] M. R. Peel and C. R. Johnson, *Tetrahedron Lett.*, **27**, 5947 (1986).
[100] M. Franck-Neumann, M. Miesch, and E. Lacroix, *Tetrahedron Lett.*, **30**, 3533 (1989).
[100a] M. Franck-Neumann, M. Miesch, and L. Gross, *Tetrahedron Lett.*, **33**, 3879 (1992).
[101] R. M. Jacobson, G. P. Lahm, and J. W. Clader, *J. Org. Chem.*, **45**, 395 (1980).
[102] G. Baddeley, H. T. Taylor, and W. Pickles, *J. Chem. Soc.*, **1953**, 124.
[103] N. Jones and H. T. Taylor, *J. Chem. Soc.*, **1959**, 4017.
[104] R. Jacquier, M. Mousseron, and S. Boyer, *Bull. Soc. Chim. Fr.*, **1956**, 1653.
[105] T. Sakai, K. Miyata, and A. Takeda, *Chem. Lett.*, **1985**, 1137.
[106] D. A. Jackson, M. Rey, and A. S. Dreiding, *Helv. Chim. Acta,* **66**, 2330 (1983).
[107] D. A. Jackson, M. Rey, and A. S. Dreiding, *Helv. Chim. Acta,* **68**, 439 (1985).
[108] D. A. Jackson, M. Rey, and A. S. Dreiding, *Tetrahedron Lett.*, **24**, 4817 (1983).
[109] T. Hiyama, M. Shinoda, M. Tsukanaka, and H. Nozaki, *Bull. Chem. Soc. Jpn.*, **53**, 1010 (1980).
[110] T. Hiyama, M. Shinoda, and H. Nozaki, *Tetrahedron Lett.*, **1978**, 771.
[111] T. Hiyama, M. Tsukanaka, and H. Nozaki, *J. Am. Chem. Soc.*, **96**, 3713 (1974).
[112] Y. Gaoni, *Tetrahedron Lett.*, **1978**, 3277.
[113] J. Grimaldi and M. Bertrand, *Bull. Soc. Chim. Fr.*, **1971**, 957.
[114] J. Grimaldi and M. Bertrand, *Tetrahedron Lett.*, **1969**, 3269.
[115] W. Smadja, *Chem. Rev.*, **83**, 263 (1983).
[116] M. Bertrand, J. P. Dulcere, and G. Gil, *Tetrahedron Lett.*, **1977**, 4403.
[117] M. Malacria and J. Goré, *J. Org. Chem.*, **44**, 885 (1979).
[118] M.Bertrand, J.-P. Dulcère, J. Grimaldi, and M. Malacria, *C. R. Seances Acad. Sci., Ser. C,* **279**, 805 (1974).
[119] M. L. Roumestant, M. Malacria, and J. Goré, *Synthesis,* **1976**, 755.
[120] M. Malacria and M. L. Roumestant, *Tetrahedron,* **33**, 2813 (1977).
[121] M. Bertrand, J. P. Dulcere, G. Gil, and M. L. Roumestant, *Tetrahedron Lett.*, **1979**, 1845.
[122] R. Baudouy, J. Sartoretti, and F. Choplin, *Tetrahedron,* **39**, 3293 (1983).
[123] A. Doutheau, J. Sartoretti, and J. Goré, *Tetrahedron,* **39**, 3059 (1983).
[124] A. Doutheau, J. Goré, and J. Diab, *Tetrahedron,* **41**, 329 (1985).
[125] S. J. Kim and J. K. Cha, *Tetrahedron Lett.*, **29**, 5613 (1988).

[126] A. Doutheau, J. Goré, and M. Malacria, *Tetrahedron,* **33**, 2393 (1977).
[127] F. Delbecq and J. Goré, *Tetrahedron Lett.*, **1976**, 3459.
[128] R. Baudouy, F. Delbecq, and J. Goré, *Tetrahedron,* **36**, 189 (1980).
[129] M. A. Tius and D. P. Astrab, *Tetrahedron Lett.*, **25**, 1539 (1984).
[130] M. A. Tius, D. P. Astrab, A. H. Fauq, J. B. Ousset, and S. Trehan, *J. Am. Chem. Soc.*, **108**, 3438 (1986).
[131] M. A. Tius and D. P. Astrab, *Tetrahedron Lett.*, **30**, 2333 (1989).
[132] G. Piancatelli, A. Scettri, and S. Barbadoro, *Tetrahedron Lett.*, **1976**, 3555.
[133] G. Piancatelli and A. Scettri, *Tetrahedron Lett.*, **1977**, 1131.
[134] G. Piancatelli and A. Scettri, *Tetrahedron,* **33**, 69 (1977).
[135] G. Piancatelli, A. Scettri, G. David, and M. D'Auria, *Tetrahedron,* **34**, 2775 (1978).
[136] G. Piancatelli, *Heterocycles,* **19**, 1735 (1982).
[137] M. D'Auria, F. D'Onofrio, G. Piancatelli, and A. Scettri, *Gazz. Chim. Ital.*, **116**, 173 (1986).
[138] R. Antonioletti, A. De Mico, G. Piancatelli, A. Scettri, and O. Ursini, *Gazz. Chim. Ital.*, **116**, 745 (1986).
[139] E. Castagnino, M. D'Auria, A. De Mico, F. D'Onofrio, and G. Piancatelli, *J. Chem. Soc., Chem. Commun.*, **1987**, 907.
[140] L. I. Zakharkin, A. P. Pryanishnikov, and V. V. Guseva, *J. Org. Chem. USSR,* **18**, 80 (1981).
[141] T. Mandai, T. Yanagi, K. Araki, Y. Morisaki, M. Kawada, and J. Otera, *J. Am. Chem. Soc.*, **106**, 3670 (1984).
[142] V. Rautenstrauch, *J. Org. Chem.*, **49**, 950 (1984).
[143] P. E. Eaton, A. Srikrishna, and F. Uggeri, *J. Org. Chem.*, **49**, 1728 (1984).
[144] P. E. Eaton, R. H. Müller, G. R. Carlson, D. A. Cullison, G. F. Cooper, T.-C. Chou, and E. P. Krebs, *J. Am. Chem. Soc.*, **99**, 2751 (1977).
[145] G. A. MacAlpine, R. A. Raphael, A. Shaw, A. W. Taylor, and H.-J. Wild, *J. Chem. Soc., Perkin Trans. 1,* **1976**, 410.
[146] M. A. MacKervey, P. Vibuljan, G. Ferguson, and P. Y. Siew, *J. Chem. Soc., Chem. Commun.*, **1981**, 912.
[147] T. Hiyama, M. Shinoda, H. Saimoto, and H. Nozaki, *Bull. Chem. Soc. Jpn.*, **54**, 2747 (1981).
[148] L. A. Paquette and A. Leone-Bay, *J. Am. Chem. Soc.*, **105**, 7352 (1983).
[149] T. Hiyama, M. Shinoda, and H. Nozaki, *J. Am. Chem. Soc.*, **101**, 1599 (1979).
[150] Y. Arai, K. Takeda, K. Masuda, and T. Koizumi, *Chem. Lett.*, **1985**, 1531.
[151] J. G. Witteveen and A. J. A. van der Weerdt, *Recl. Trav. Chim. Pays-Bas,* **100**, 383 (1981).
[152] M. Baumann, W. Hoffmann, and N. Müller, *Tetrahedron Lett.*, **1976**, 3585.
[153] H. Saimoto, T. Hiyama, and H. Nozaki, *Tetrahedron Lett.*, **21**, 3897 (1980).
[154] L. I. Zakharkin, A. P. Pryanishikov, and V. V. Guseva, *J. Org. Chem. USSR,* **18**, 83 (1981).
[155] A. M. Islam and R. A. Raphael, *J. Chem. Soc.*, **1953**, 2247.
[156] D. R. Williams, A. Abbaspour, and R. M. Jacobson, *Tetrahedron Lett.*, **22**, 3565 (1981).
[157] S. Hacini, R. Pardo, and M. Santelli, *Tetrahedron Lett.*, **1979**, 4553.
[157a] C. Morel-Fourrier, J.-P. Dulcère, and M. Santelli, *J. Am. Chem. Soc.*, **113**, 8062 (1991).
[158] F. Cooke, J. Schwindeman, and P. Magnus, *Tetrahedron Lett.*, **1979**, 1995.
[159] F. Cooke, R. Moerck, J. Schwindeman, and P. Magnus, *J. Org. Chem.*, **45**, 1046 (1980).
[160] G. Kjeldsen, J. S. Knudsen, L. S. Ravn-Petersen, and K. B. G. Torssell, *Tetrahedron,* **39**, 2237 (1983).
[161] P. Magnus, D. A. Quagliato, and J. C. Huffman, *Organometallics,* **1**, 1240 (1982).
[162] P. Magnus and D. A. Quagliato, *Organometallics,* **1**, 1243 (1982).
[163] P. Magnus and D. Quagliato, *J. Org. Chem.*, **50**, 1621 (1985).
[164] G. J. Martin and G. Daviaud, *Bull. Soc. Chim. Fr.*, **1970**, 3098.
[165] G. J. Martin, C. Rabiller, and G. Mabon, *Tetrahedron Lett.*, **1970**, 3131.
[166] G. J. Martin, C. Rabiller, and G. Mabon, *Tetrahedron,* **28**, 4027 (1972).
[167] C. Rabiller, G. Mabon, and G. J. Martin, *Bull. Chim. Soc. Fr.*, **1973**, 3462.
[168] W. Best, B. Fell, and G. Schmidt, *Chem. Ber.*, **109**, 2914 (1976).
[169] P. R. DeShong and D. R. Sidler, *J. Org. Chem.*, **53**, 4892 (1988).
[170] K. S. Jadhav, S. B. Thakur, and S. C. Bhattacharyya, *Indian J. Chem.*, **16B**, 280 (1978).

[171] H. Schostarez and L. A. Paquette, *Tetrahedron,* **37**, 4431 (1981).
[172] W. Oppolzer and K. Bättig, *Helv. Chim. Acta,* **64**, 2489 (1981).
[173] M. Horton and G. Pattenden, *J. Chem. Soc., Perkin Trans. 1,* **1984**, 811.
[173a] K. F. Cheng, K.-P. Chan, and T.-F. Lai, *J. Chem. Soc., Perkin Trans. 1,* **1991**, 2461.
[174] L. Rand and R. J. Dolinski, *J. Org. Chem.,* **31**, 3063 (1966).
[175] B. A. Vick and D. C. Zimmerman, *Biochem. Biophys. Res. Commun.,* **111**, 470 (1983).
[176] M. Hamberg, *Biochim. Biophys. Acta,* **920**, 76 (1987).
[177] E. J. Corey, M. d'Alarcao, S. P. T. Matsuda, P. T. Lansbury, Jr., and Y. Yamada, *J. Am. Chem. Soc.,* **109**, 289 (1987).
[178] T. M. Harris, S. W. Baertschi, and A. R. Brash, *J. Am. Chem. Soc.,* **111**, 5003 (1989).
[179] L. A. Paquette and K. E. Stevens, *Can. J. Chem.,* **62**, 2415 (1984).
[180] K. E. Stevens and L. A. Paquette, *Tetrahedron Lett.,* **22**, 4393 (1981).
[181] G. Ohloff, K. H. Schulte-Elte, and E. Demole, *Helv. Chim. Acta,* **54**, 2913 (1971).
[182] J. A. Barltrop, A. C. Day, and C. J. Samuel, *J. Chem. Soc., Chem. Commun.,* **1977**, 598.
[183] R. Noyori, Y. Ohnishi, and M. Kato, *J. Am. Chem. Soc.,* **94**, 5105 (1972).
[184] S. E. Denmark and E. M. Carreira, unpublished results.
[185] L. A. Paquette and H.-J. Kang, *J. Am. Chem. Soc.,* **113**, 2610 (1991).
[186] R. M. Jacobson and G. P. Lahm, *J. Org. Chem.,* **44**, 462 (1979).
[187] J. C. Hamlet, H. B. Henbest, and E. R. H. Jones, *J. Chem. Soc.,* **1951**, 2652.
[188] T. Hiyama, M. Shinoda, and H. Nozaki, *Tetrahedron Lett.,* **1979**, 3529.
[189] Y. Gaoni, *Tetrahedron Lett.,* **1977**, 371.
[190] J.-P. Dulcere, J. Grimaldi, and M. Santelli, *Tetrahedron Lett.,* **22**, 3179 (1981).
[191] M. A. Tius and X.-M. Zhou, *Tetrahedron Lett.,* **30**, 4629 (1989).
[192] A. R. Brash, *J. Am. Chem. Soc.,* **111**, 1891 (1989).
[193] S. B. Kulkarni and S. Dev, *Tetrahedron,* **24**, 545 (1968).
[194] H. G. Grant, *J. Heterocyclic Chem.,* **15**, 1235 (1978).
[195] J.-M. Conia and M.-L. Leriverend, *Bull. Soc. Chim. Fr.,* **1970**, 2981.
[196] A. A. Schegolev, W. A. Smit, G. V. Roitburd, and V. F. Kucherov, *Tetrahedron Lett.,* **1974**, 3373.

CHAPTER 2

KETENE CYCLOADDITIONS

JOHN A. HYATT AND PETER W. RAYNOLDS

Research Laboratories, Eastman Chemical Co., Kingsport, Tennessee

CONTENTS

Organic Reactions, Vol. 45, Edited by Leo A. Paquette et al.
ISBN 0-471-03161-5

ACKNOWLEDGMENTS

The authors wish to thank the management of the Research Laboratories, Eastman Chemical Company, for their support in the preparation of this review. JAH in particular thanks Lois Epley, Debra Bledsoe, Delta Childress, Betty Hicks, Martha Yonce, and Paula Phillips for their excellent efforts in preparation of the manuscript. Both authors are grateful to Dr. R. H. Hasek, whose compilations of ketene literature in these laboratories provided a sound basis for starting the present work.

INTRODUCTION

The reaction of ketene with itself was described almost simultaneously in 1908 by Chick and Wilsmore in England,[1] and by Staudinger and Klever in Germany;[2] priority for the discovery was attributed to Wilsmore by the German group.[3]

Ketene cycloaddition was an early example (along with the Claisen, Diels-Alder, and ene reactions) of a peculiar process, one that formed carbon–carbon bonds with ease, often without the need for solvent, catalyst, or high heat. Subsequent work by others was done in the shadow of Staudinger's exhaustive and rigorous study of all phases of ketene reactivity.[4] Three factors led to a resurgence of interest in this reaction beginning in the 1960s. Haloketenes, which had previously eluded study, were found to have high reactivity, and the halogens could easily be removed after the reaction.[5–7] The increasing sophistication of powerful analytical methods, particularly nuclear magnetic resonance spectroscopy, led to discovery of the interesting stereochemical aspects of this reaction. Finally, the new theory of orbital symmetry conservation provided a conceptual framework to rationalize these puzzling "no mechanism" reactions.[8]

This is a review of cycloaddition reactions of ketenes. Here, a cycloaddition is defined as a reaction of a ketene with an unsaturated organic compound to give a cyclic product by a mechanism that, in principle, involves the almost simultaneous formation of two bonds between two reactants. While we are indifferent to whether bond formation is concerted or stepwise, no other chemical process can take place between the formation of the first and second bond. Our definition involves both structural and mechanistic factors, and it was difficult to avoid a certain amount of arbitrariness. We exclude products, such as dehydroacetic acid, which seem to us to arise by concatenation of ketene molecules followed by cyclization. These reactions, in our view, are more properly considered to arise from a series of ionic reactions, and the prediction of the eventual product does not take advantage of the special mechanistic features commonly associated with true cycloadditions. Additions to imines to give β-lactams are numerous and will be covered in a separate review. The literature has been searched to the end of 1988.

Many reviews devoted partially or exclusively to ketene cycloadditions have been published.[4,9–18] Specific topics that have been reviewed include haloketenes,[19,20] fluoroketenes,[21,22] cyanoketenes,[23] intramolecular cycloadditions,[24,25] conjugated ketenes,[26] and β-lactam antibiotics.[27,28] Ancillary topics pertinent to ketene cycloadditions that have been reviewed include cycloreversion reactions,[29] ketene equivalents, which provide ketene functionality with olefin-like reactivity especially in [4+2] reactions,[30] application of frontier molecular orbital theory to cycloadditions,[31] and a critical discussion of cycloadditions with polar intermediates.[32] Applications of cyclobutanones in synthesis have also been reviewed.[33–35]

MECHANISM AND STEREOCHEMISTRY

Ketene cycloadditions resemble certain other processes, such as the Cope rearrangement and the Diels-Alder, Claisen, and ene reactions, in that carbon–carbon bond formation occurs readily, even in the absence of catalyst, solvent, light, or high heat. Features of ketene cycloadditions that were difficult to rationalize with the mechanistic knowledge available at the middle of the twentieth century included the formation of a four-membered ring by a nonphotochemical process, even when, in the case of 1,3-dienes, a cyclohexene was possible;[36] the retention of stereochemistry about the olefin in the cyclobutanone product; the fact that *cis*-alkenes reacted much faster than *trans*-alkenes;[37] and that monosubstituted ketenes reacted with cyclopentadiene to put the substituent in the *endo* position of the bicyclo[3.2.0]hept-2-en-7-one.[38] When the concept of orbital symmetry conservation proved so useful in explaining many processes of this type,[39] it became natural to view ketene cycloadditions as one example of this general phenomenon. Much of the mechanistic research that has been conducted on ketene cycloadditions since then has focused on the timing of bond formation and other features of the reaction that would tend to support or disprove the orbital symmetry conservation theory. While such work was of great importance as the practical limits of symmetry conservation theory were explored, it had the potential to disguise the fact that most of the special features of ketene cycloadditions arise from the geometry of approach of the two reactants, and that the "orthogonal" mode of attack does not require symmetry theory.[40,41]

Geometry of Approach. The hallmark of ketene cycloaddition is the peculiar way that the two reactants approach each other. Rather than lining up in parallel fashion before forming bonds, the ketene and olefin come together at right angles, in what is termed an "orthogonal" approach.[37,38,42] This arrangement produces an unusual result: the bulkiest substituent on the ketene ends up on the most sterically hindered face of the cyclobutanone (Eq. 1). The larger the difference in bulk of the two substituents on the ketene, the larger the effect. In the nomenclature devised by Woodward and Hoffmann,[39] the ketene forms bonds antarafacially (on different sides of the π bond), while the olefin forms bonds suprafacially (on the same side of the π bond). It should be noted that while this unusual geometry is well explained by symmetry conservation theory, the same

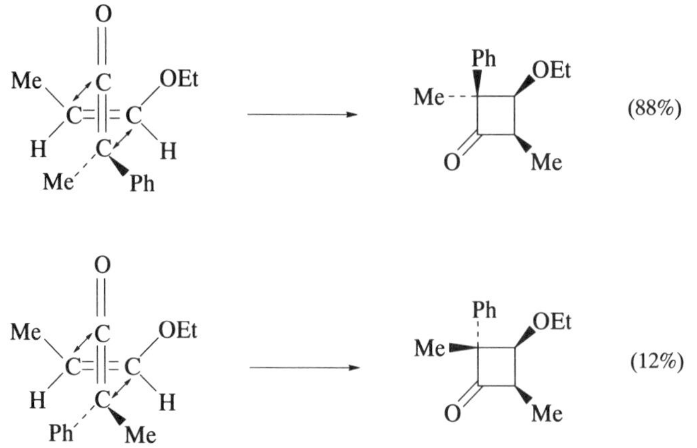

(88%)

(12%)

(Eq. 1)

geometry can occur with a dipolar intermediate as well.[43,44] This geometry immediately suggested explanations for the higher reactivity of *cis*- over *trans*-alkenes, and the fact that one substituent on the ketene or alkene often accelerates cycloaddition, possibly because of electronic effects, while two substituents on one of the reactants results in a rate reduction, presumably because of steric effects.[37,45]

The antarafacial involvement of the ketene was first noted in the addition of ketenes to cyclopentadiene. The larger substituent on the ketene tends to end up in the *endo* position, on the concave face of the bicyclo[3.2.0]hept-2-en-6-one.[38] This so-called "*endo* effect" is common with cyclic olefins, but the source of the

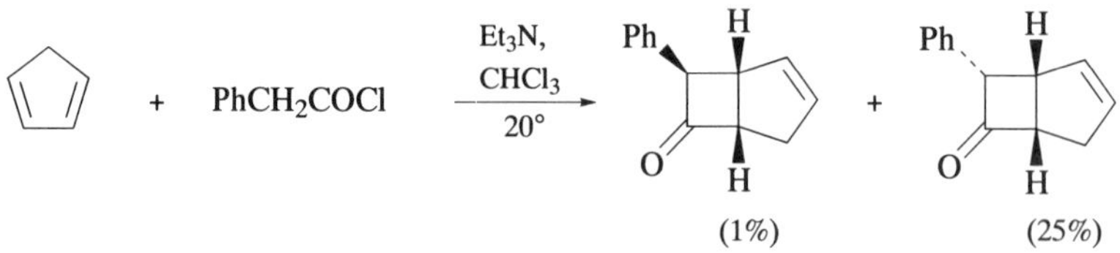

+ $PhCH_2COCl$ $\xrightarrow[20°]{Et_3N,\ CHCl_3}$ (1%) + (25%)

(Eq. 2)

selectivity may be due to a combination of factors. Equilibration experiments showed that most substituents are slightly more stable in the 7-*endo* position than in the 7-*exo* position of a bicyclo[3.2.1]heptane.[46] For example, at equilibrium, the *endo*/*exo* ratio for a 7-phenyl group in the product of Eq. 2 is 66/34. Thus the *endo* position is the more stable, but the thermodynamics of the product do not explain all of the selectivity.

Retention of Stereochemistry. Stereochemistry about the olefin is retained in the cyclobutanone product. Slight (<5%) stereochemical leakage has been noted in the reaction of diphenylketene with propenyl propyl ether (Eq. 3).[47] This result lends credence to, rather than weakens, the general conclusion that significant loss of stereochemistry is not to be expected in any except the most polar systems.

OPr (cis) + $Ph_2C{=}C{=}O$ → cis:trans = 96.2:3.8

(99%)

OPr (trans) + $Ph_2C{=}C{=}O$ → cis:trans = 2.5:97.5

(Eq. 3)

Solvent Effects. The rate of cycloaddition increases as the polarity of the solvent is increased.[48] Dimethylketene dimerizes 19 times faster in acetonitrile than in hexane.[49] This result is all the more noteworthy because if bond formation were completely synchronous, the reaction should have shown an inverse solvent effect, or a rate retardation with increasing solvent polarity, since the reactant has a larger dipole moment than the product. Solvent effects, while general, are not always of practical importance, since ketene cycloadditions are generally run without solvent if possible. The effect of a polar solvent on rate is often negated by dilution of the reactants. If two products can form, the ratio often depends on the polarity of the solvent. This effect is most important with more highly reactive polar systems.

Substituent Effects. As mentioned above, one substituent on either the ketene or the olefin generally accelerates the reaction, while two generally slow the reaction because of steric effects. This result can be rationalized by the geometry of approach: one bulky substituent can position itself away from the reaction center, causing little change in rate from steric effects, while a second substituent is placed in the immediate vicinity of the other reactant. The rate of cycloaddition is accelerated by electron-donating substituents on the olefin and by electron-withdrawing groups on the ketene.

Timing of Bond Formation. Both solvent and isotope effects on the rate of dimethylketene dimerization[50–54] indicate that cycloaddition of even relatively symmetrical reactants occurs with at least some nonsynchronous character. In more polar systems, such as cycloadditions of *tert*-butylcyanoketene, it is difficult to avoid explanations involving stepwise mechanisms with a dipolar intermediate.[55] Recent theoretical *ab initio* calculations do not support a concerted, supra-antarafacial mechanism for the addition of ketene to ethylene.[56] Nevertheless, alkenes without electron-donating groups cycloadd in a way that seems to allow no alternative reaction pathway between the formation of the first and second bonds, if indeed the two bonds are not formed at almost the same time. Bond rotation resulting in loss of stereochemistry does not occur, as illustrated in the impressive addition of diphenylketene to *trans*-cyclooctene with no loss of stereochemistry.[57] As exemplified in Table IV, addition to anti-Bredt olefins takes place in high yield; side reactions resulting from a charged intermediate are exceptional.

Reversibility. The formation of simple cyclobutanones is reversible under relatively mild conditions. For instance, 2,2,3-triphenylcyclobutanone gives diphenylketene and styrene upon pyrolysis at 200°, while the same compound cleaves to triphenylethene and presumably ketene upon photolysis.[58]

SCOPE AND LIMITATIONS

[2+2] Cycloadditions

Dimerization. The dimerization of ketenes gives 1,3-cyclobutanediones and β-propiolactones. Higher oligomers which likely arise from secondary reactions are also obtained. If the cyclobutanedione has an acidic hydrogen, the enol form generally predominates, and acylation of the enol may occur. Ketene, which can be handled as the pure, colorless liquid at $-78°$, dimerizes exothermically below room temperature to afford mostly the β-propiolactone (Eq. 4). The highly reactive

$CH_2{=}C{=}O$ —(< 0°)→ [cyclobutenone with OR] (4-5%), R = H, Ac + [diketene, CH2, O] (88-90%) (Eq. 4)

unsubstituted diketene, which is prepared on a multimillion pound per year scale by the pyrolysis of acetic acid, is an industrially important raw material used in the preparation of acetoacetates and acetoacetamides.[59]

Disubstituted ketenes, except those with electron-withdrawing groups, give 1,3-cyclobutanediones in very high yield, as exemplified by the dimerization of dimethylketene (Eq. 5).[49] The reaction rate is sensitive to the polarity of the solvent,

$(CH_3)_2C{=}C{=}O$ —(CCl_4, 35°)→ [tetramethyl-1,3-cyclobutanedione] (99%) (Eq. 5)

vent,[49] as well as to the electronic and steric nature of the substituents. For instance, dimethylketene is highly reactive at room temperature, whereas diphenyl- and ethylbutylketene are stable enough to be stored at room temperature for months without appreciable degradation.

A double bond in conjugation with the ketene can participate in the dimerization, either initially, as in the case of vinylketene (Eq. 6),[60] or in what could be a

[vinylketene, C=O] —(25°)→ [pyranone with allylidene substituent, O, O] (70%) (Eq. 6)

rearrangement of the primary dimerization product of diphenylketene (Eq. 7). The cyclobutanedione is the only product observed at lower temperatures.[61]

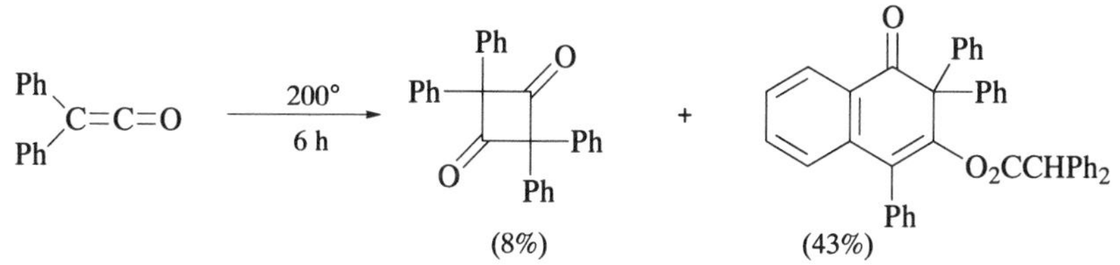

(Eq. 7)

Lewis acids such as aluminum chloride,[62] and bases such as triethyl phosphite,[48] alter the course of the uncatalyzed dimerization, and other products, probably arising from ionic processes, are formed (Eq. 8).

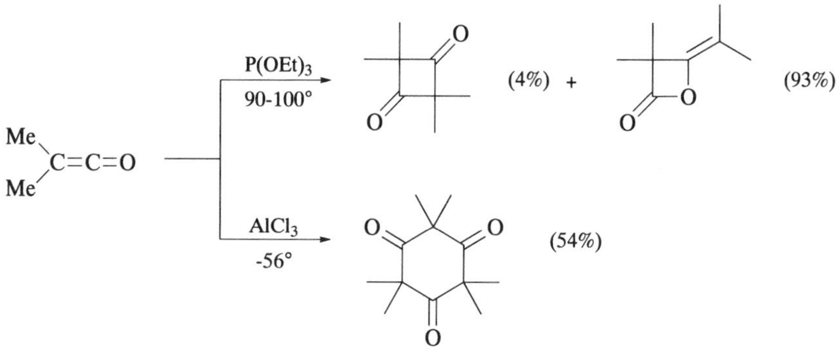

(Eq. 8)

Disubstituted ketenes with at least one electron-withdrawing group behave differently. A dimer of dichloroketene has not been described, although in one instance, a secondary product, possibly derived from a β-propiolactone dimer, was isolated (Eq. 9).[63]

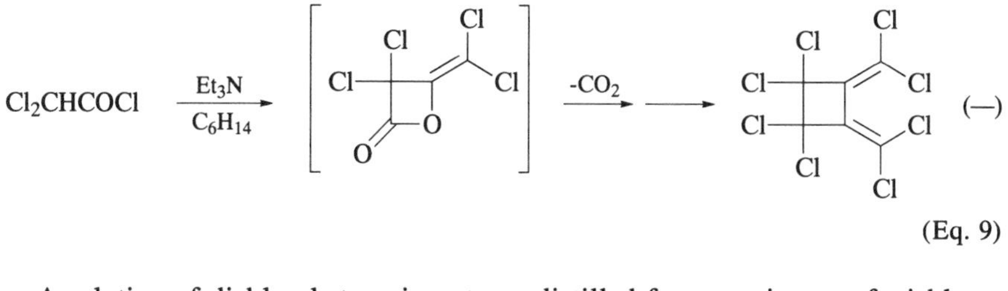

(Eq. 9)

A solution of dichloroketene in octane, distilled from a mixture of trichloroacetyl bromide and zinc, was stable for a week at room temperature or for 24 hours at 85°; no evidence for a dimer was found in the IR spectrum of the mixture.[7]

Bis(trifluoromethyl)ketene likewise does not dimerize thermally[64,65] but affords the β-propiolactone dimer with diethyl nitroxide.[66] A γ-lactone is obtained in the presence of triethylamine (Eq. 10).[67]

$$(CF_3)_2C{=}C{=}O \xrightarrow{Et_3N} \text{(}\gamma\text{-lactone: } CF_3,\ F,\ CF_3,\ CF_3,\ F,\ F\text{)} \quad (\text{—})$$ (Eq. 10)

Spontaneous dimerization is not observed with *tert*-butylcyanoketene, while the β-propiolactone dimer is obtained in the presence of triethylamine.[68] No dimer was reported for *tert*-butylcarboethoxyketene after 2 months at room temperature.[69] Various products are obtained from ketenes bearing one electron-withdrawing group and a second substituent less bulky than *tert*-butyl, as indicated in Table II.

Monosubstituted ketenes, like ketene itself, give mixtures of hydroxycyclobutenone (the enol form of the 1,3-cyclobutanedione product, which is often acylated), and the β-propiolactone. Although exceptions exist, it appears that the hydroxycyclobutenone is most likely to be formed under salt-free conditions [pyrolysis of an anhydride, or zinc dechlorination followed by distillation, (Eqs. 11 and 12)], while dehydrochlorination of an acid chloride with triethylamine almost always yields the lactone (Eq. 13).

$$(EtCO)_2O \xrightarrow[\substack{500^\circ \\ 40\text{-}100\ \text{torr}}]{-\,EtCO_2H} \mathbf{I}\ (76\%) + \mathbf{II}$$

R = H, COEt

(Eq. 11)

$$CH_3CH(Br)COBr \xrightarrow[\substack{THF \\ -30^\circ}]{Zn} \mathbf{I}\ (88\%)$$ (Eq. 12)

$$CH_3CH_2COCl \xrightarrow[Et_2O]{Et_3N} \mathbf{II}\ (74\%)$$ (Eq. 13)

Monoalkylketenes are much more reactive than either the unsubstituted or disubstituted species and, except in sterically hindered examples, dimerization occurs below 0°. They are generally not dealt with as pure compounds but rather are treated as transient species in solution below 0°. Trimethylsilylketene, which does not dimerize, is a noteworthy exception.[70]

Mixtures of unlike ketenes can produce unsymmetrical products in synthetically useful yields, especially when the electronic characteristics of the two reactants are different. Ketene and monosubstituted ketenes react with electron-poor ketenes to give β-propiolactones, with the oxygen of the electron-poor ketene becoming the lactone oxygen (Eqs. 14 and 15).[64,71,72]

Me(H)C=C=O + $(CF_3)_2C=C=O$ —(-80°)→ CF$_3$ CF$_3$ O O (61%)

(Eq. 14)

Me(H)C=C=O + NC(t-Bu)C=C=O —(PhMe)→ CN Bu-*t* O O (49%)

(Eq. 15)

Disubstituted ketenes react with other disubstituted ketenes to give predominately 1,3-cyclobutanediones, but, in contrast to homodimerizations, the β-propiolactone product is sometimes observed (Eq. 16).[73]

COCl + *t*-Bu(Cl)CHCOCl —(Et_3N, C_6H_6)→ O Cl O Bu-*t* (56%) + Cl Bu-*t* O O (13%)

(Eq. 16)

Cyclobutanedione dimers of unsubstituted and disubstituted ketenes are efficiently converted into the ketenes by pyrolysis in a hot tube.[74] If the dimer is available, thermal cycloreversion is often the best preparative method for the ketene. However, this method has not been extended to the less well studied monosubstituted ketenes.

Isolated Olefins. Ketenes react with simple alkenes to afford, in almost all cases, cyclobutanones. In the absence of electron-donating substituents, the order of reactivity is *trans* olefin < *cis* olefin < cyclic olefin < linear diene < cyclic diene.[37] The stereochemistry about the double bond is retained, even in the extreme case of cycloadditions to *trans*-cyclooctene (Eq. 17).[57]

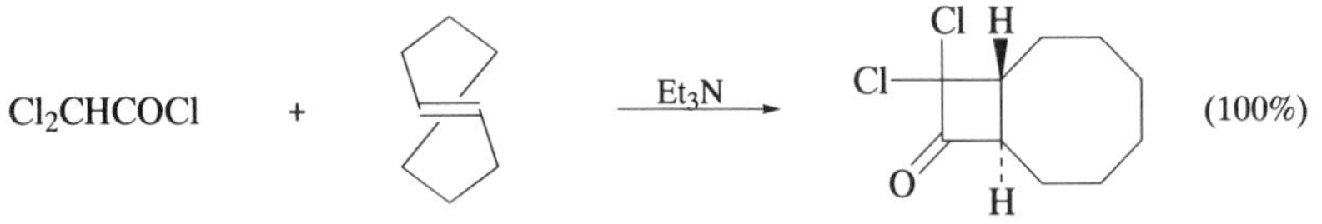

$Cl_2CHCOCl$ + [cyclooctyne-type alkene] $\xrightarrow{Et_3N}$ (100%)

(Eq. 17)

Regiochemistry is determined by the polarization of the double bond, even in the face of steric crowding by substituents on the 2 and 3 positions of the cyclobutanone. Except when the ketene is activated by electron-donating or withdrawing substituents, useful yields are obtained only when the ketene is disubstituted. Should the cycloaddition product of ketene itself be desired, it is common to use the highly reactive dichloroketene followed by dechlorination with zinc-copper, as in the synthesis of isocaryophyllene (Eq. 18).[75]

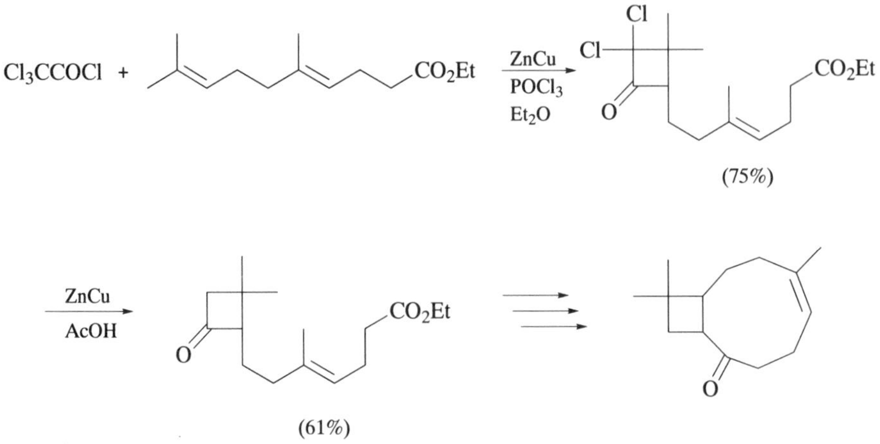

Cl_3CCOCl + ... CO_2Et $\xrightarrow[POCl_3, Et_2O]{ZnCu}$ ... CO_2Et (75%)

$\xrightarrow[AcOH]{ZnCu}$ ... CO_2Et (61%) $\Rightarrow$

(Eq. 18)

If the substrate contains two unsaturated centers, reaction at both sites is often observed (Eq. 19).[76]

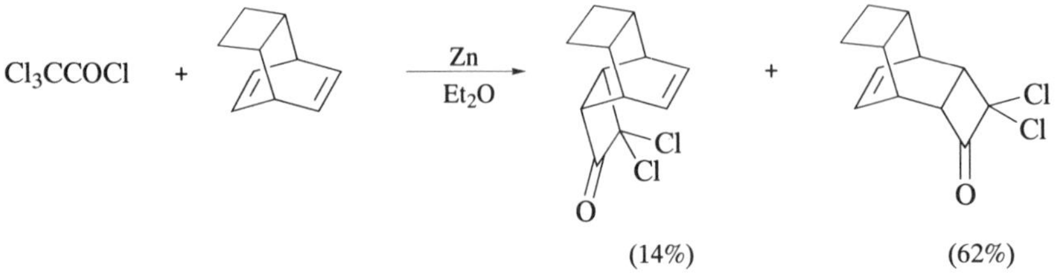

Cl_3CCOCl + $\xrightarrow[Et_2O]{Zn}$ (14%) + (62%)

(Eq. 19)

Side reactions are not common if the substrate does not contain an activating group. No rearrangement was seen with an "anti-Bredt" olefin,[77] or with a reactant prone to cyclopropylcarbinyl carbenium ion rearrangement (Eqs. 20 and 21).[78]

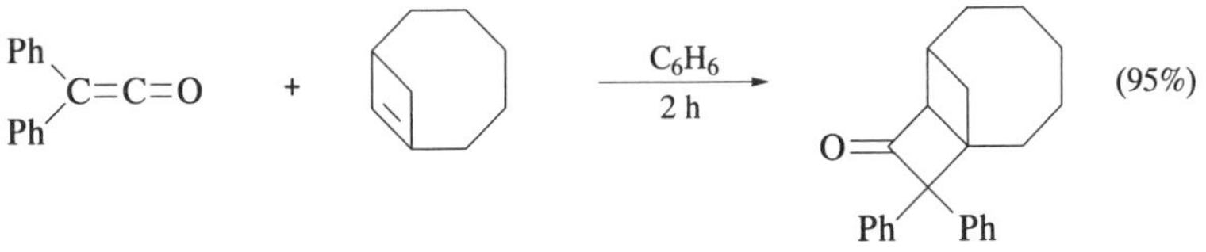

(95%) (Eq. 20)

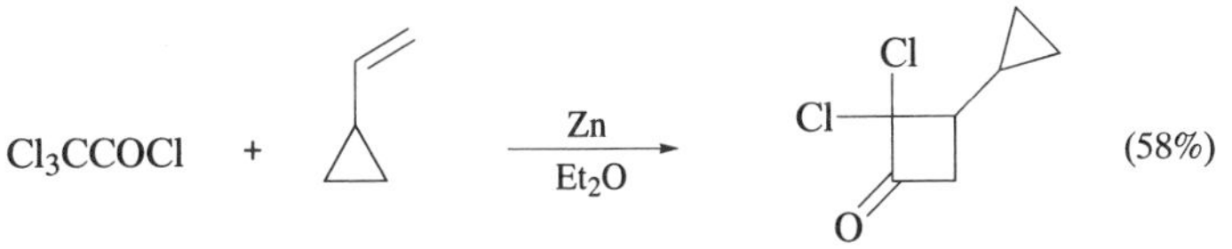

(58%) (Eq. 21)

On the other hand, methylcyclopropene gives ene-type and rearranged products, possibly by way of dipolar intermediates (Eq. 22).[79]

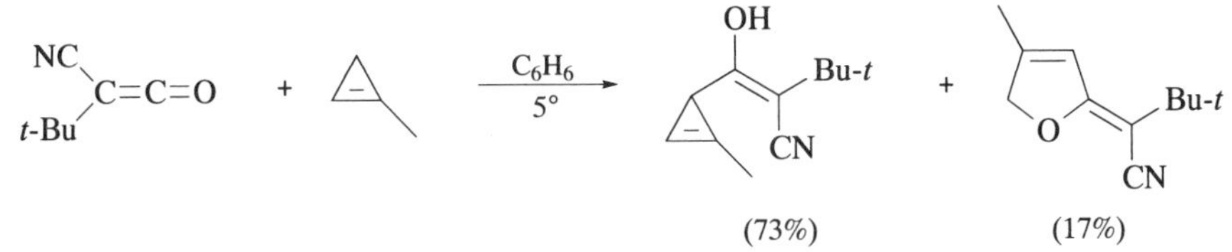

(73%) (17%) (Eq. 22)

Diphenylketene gives a 2:1 adduct with 1,1-diphenylethene (Eq. 23).[80,81] The cyclobutanone is not observed.

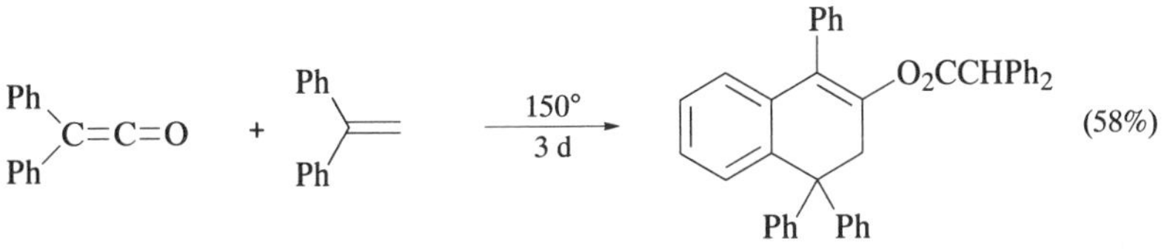

(58%) (Eq. 23)

Dichloroketene adducts can be converted to useful derivatives. Sequential treatment with *n*-butyllithium and acetic anhydride followed by oxidation with ruthenium dioxide and sodium metaperiodate affords vicinal dicarboxylic acids stereospecifically (Eq. 24).[82]

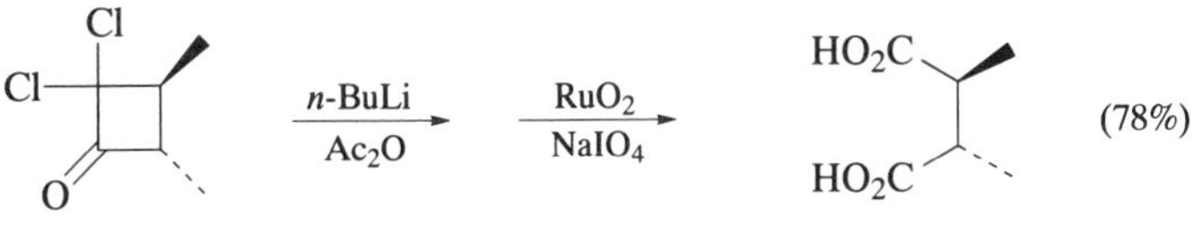

(78%) (Eq. 24)

A three-carbon annelation sequence, involving methylene insertion with diazomethane followed by removal of the chlorines, affords cyclopentanones (Eq. 25).[83]

CH_2N_2, Et_2O; Zn, AcOH (82%)

(Eq. 25)

Linear Dienes. The reaction of ketenes with unpolarized acyclic dienes gives cyclobutanones. Although formation of cyclobutanones instead of Diels–Alder adducts is one of the characteristics of ketene cycloadditions (Eq. 26), cyclic

4 d (95%)

(Eq. 26)

products with larger rings, often arising from addition across the ketene carbonyl, are common with perfluoroketenes[84,85] (Eqs. 27 and 29) or with alkoxybutadienes (Eq. 28).[86] The reaction of ketene with 1,3-butadiene has been reported,[87]

100°, 60 h (78%)

(Eq. 27)

C_6H_6, 48 h (—)

(Eq. 28)

100° (—)

(Eq. 29)

but useful yields of cyclobutanones are generally obtained only when the ketene has two substituents.[36]

Cyclic Dienes. As discussed in the Mechanism section, cyclic dienes react with ketenes to yield cyclobutanones. The larger of the two substituents on the ketene goes in the *endo* position.

Dehalogenation of the dichloroketene adduct to various cyclic dienes, followed by oxidation, affords a simple route to key intermediates in the synthesis of prostaglandins (Eq. 30).[88–91]

Cl, Cl, O; Zn; H_2O_2, AcOH; O, O (90%) (Eq. 30)

Most fulvenes react on the ring (Eq. 31.)[92]

OEt + $Me_2CHCOCl$ $\xrightarrow[C_6H_{14}]{Et_3N}$ O, OEt (15%) + O, EtO (45%)

(Eq. 31)

Allenes. Addition of ketenes to allenes yields methylenecyclobutanones. If the allene is unsymmetrical, all possible products are often formed, and there are no simple rules to predict which will predominate. The 1,3-disubstituted allenes can react to put the substituent on the methylene in a position either *Z* or *E* to the carbonyl, and both products are often observed (Eq. 32).[93] Acyclic 1,3-dialkylal-

$(CH_2)_8$, H, C=C=C, H + Ph, Ph, C=C=O $\xrightarrow[4\ h]{120^\circ}$ Ph, Ph, O, H, $(CH_2)_8$ (32%) + Ph, Ph, O, H, $(CH_2)_8$ (40%)

(Eq. 32)

lenes usually give more *Z* than *E* isomer,[93] and, since the optical activity of chiral allenes is mostly lost in the *Z* isomer, it has been argued by some that this reaction occurs by a stepwise mechanism.[94] On the other hand, optically enriched 1,3-diphenylallene affords only the two *E* isomers, which are optically active (Eq. 33).[95]

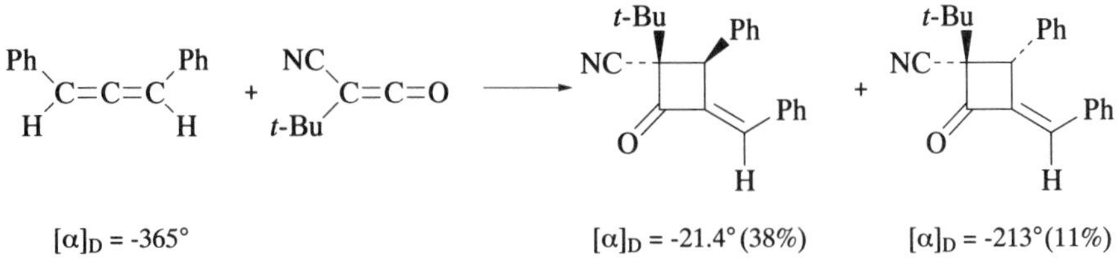

$[\alpha]_D$ = -365° $[\alpha]_D$ = -21.4° (38%) $[\alpha]_D$ = -213° (11%)

(Eq. 33)

Enamines. Enamines react readily with ketenes. When the enamine nitrogen atom bears two alkyl substituents, the cyclobutanone product is, in many cases, not isolated in a synthetically useful yield. Part of the reason for the relatively few examples of isolated [2+2] adducts may be that rearrangement of the intially formed cyclobutanone is often the desired outcome of the sequence, as in the preparation of 1,3-cyclotetradecanedione (Eq. 34).[96]

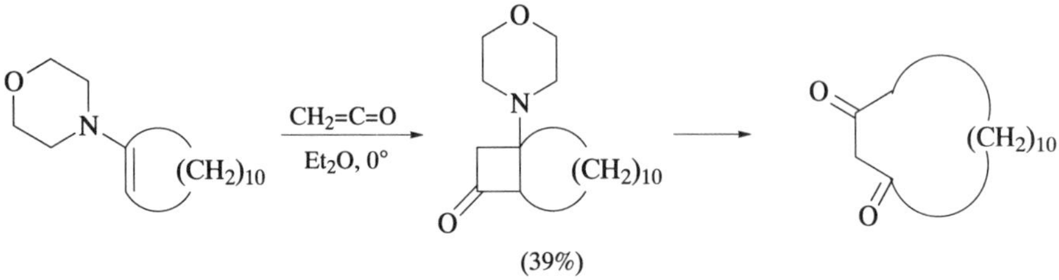

$CH_2{=}C{=}O$, Et_2O, 0°   (39%)

(Eq. 34)

Products other than the normal [2+2] adduct are common in reactions of enamines. Methylvinylketene gives mostly [4+2] products in methylene chloride at room temperature, with increasing amounts of cyclobutanone when the reaction is run in hexane at 70° (Eq. 35).[97]

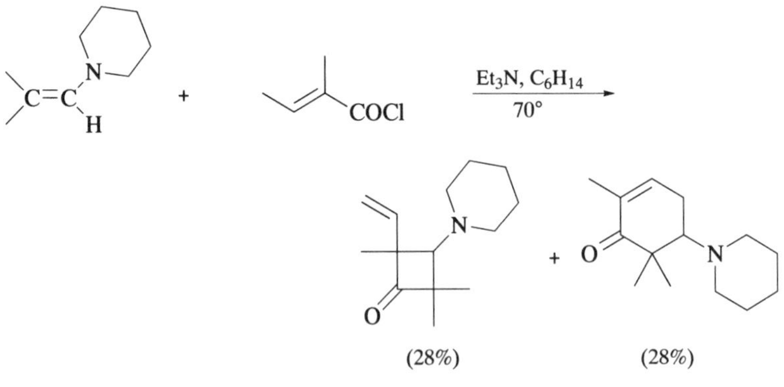

Et_3N, C_6H_{14}, 70°

(28%)   (28%)

(Eq. 35)

Two molecules of the ketene often react with one of an enamine with elimination of amine to give a pyrone (Eq. 36).[98]

Me_2N– (cyclopentenyl) + $CH_2{=}C{=}O$ ⟶ (pyrone) (20%) (Eq. 36)

Acylation of the enamine competes with cycloaddition, and linear products are often obtained (Eq. 37).[99]

(pyrrolidinyl)N(H)C=C(H)CO_2Me + $Ph_2C{=}C{=}O$ ⟶ (pyrrolidinyl)N(H)C=C(CO_2Me)$COCHPh_2$ (88%) (Eq. 37)

Enamides require higher temperatures to react than enamines, and give stable isolable cyclobutanones in good yield (Eq. 38).[100] Byproducts are not common.

MeCO(Me)N(H)C=CH_2 + (n-Bu)(Et)C=C=O ⟶(MeCN) 2-n-Bu-2-Et-3-[N(COMe)Me]cyclobutanone (62%) (Eq. 38)

Enol Ethers. Olefins substituted with an ether oxygen react much more rapidly than simple alkenes. Even ketene itself, which usually dimerizes faster than it cycloadds to simple alkenes, reacts readily with 1-methoxycyclohexene to give the cyclobutanone in 60% yield.[101] As with simple alkenes, stereochemistry is retained in the cycloadduct, and regiochemistry is consistent with electronic effects. As was noted in the Mechanism section, a small amount of stereochemical "leakage" occurs, even with a relatively simple reactant, propenyl propyl ether.[47] This small loss of stereospecificity does not detract from the general synthetic utility of the reaction. Linear products are also obtained. Ring-opened structures are occasionally seen with 1,1-disubstituted enol ethers, where steric bulk may inhibit ring closure with disubstituted ketenes. In one case, it has been argued that the linear species is not a secondary product arising from rearrangement of the initially formed cyclobutanone, but instead results from rearrangement of a zwitterionic intermediate (Eq. 39).[102]

Although examples of stable 2,2,3,3-tetraalkylcyclobutanones derived from alkenes are common, 2,2,3-trialkyl-3-alkoxycyclobutanones are not often reported,

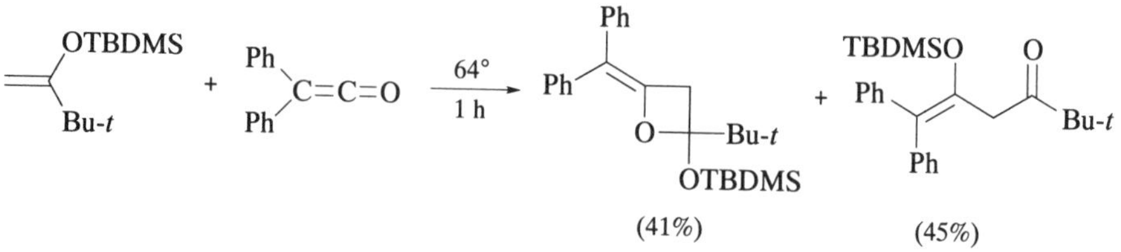

(Eq. 39)

perhaps because of the formation of linear products. In the case of 2-ethoxypropene, the cyclobutanone is observed at room temperature, but isomerizes to the linear product during isolation (Eq. 40).[103] Contrary to this generalization,

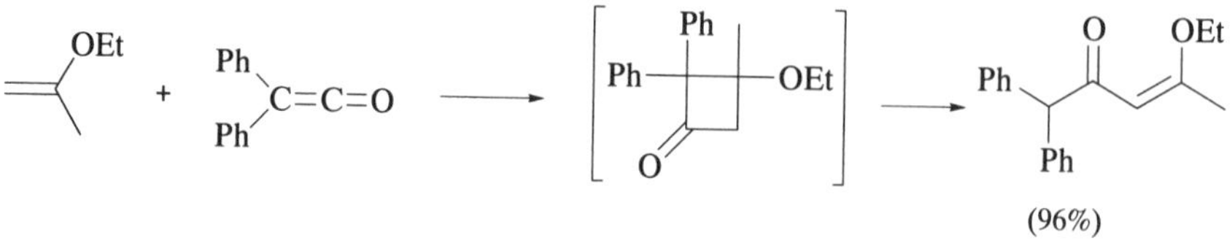

(Eq. 40)

2,2-dialkyl-3-aryl-3-alkoxycyclobutanones are formed with ease (Eq. 41),[104] as are cyclobutanones with three instead of four substituents in the 2 and 3 positions.

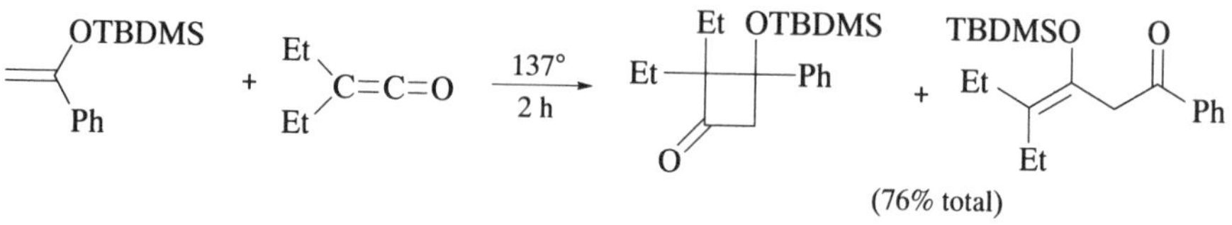

(Eq. 41)

In a few cases, addition across the carbonyl of the ketene is observed. The oxetane in Eq. 42, which can be isolated, isomerizes to the open-chain isomer.[105]

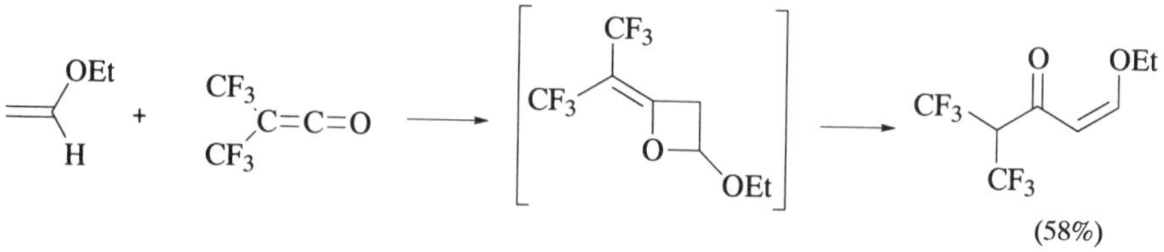

(Eq. 42)

3-Silyloxycyclobutanones revert to ketenes and enol ethers under relatively mild conditions (Eq. 43).[102] Similar behavior is observed for the diphenylmethyleneoxetane in Eq. 39.[102]

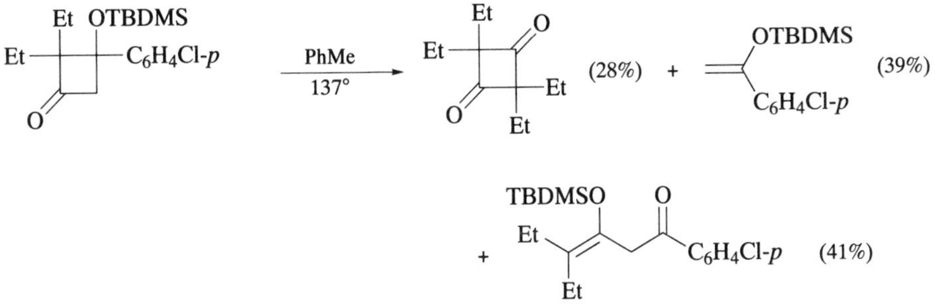

(Eq. 43)

Polyoxygenated Olefins. Olefins with two to four ether groups react in the same way as the monoethers, affording oxygenated cyclobutanones (Eq. 44).[105]

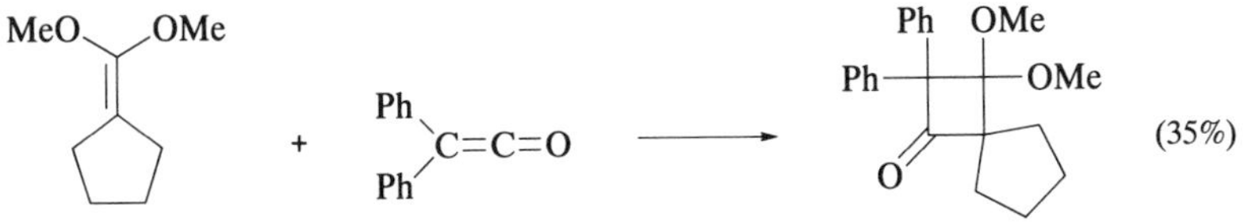

(Eq. 44)

The chief side reactions are addition across the carbonyl of the ketene[106] and formation of ring-opened products. Tetramethoxyethene reacts with diphenylketene to afford the methylenoxetane, which can be isolated if care is taken to avoid hydrolysis (Eq. 45).[106] Ring-opened ketones are common byproducts with polyalkoxy acetals. Although opening of an initially formed cycloadduct is always a possibility, no cyclobutanone is detected by IR when diphenylketene reacts with 1,1-diethoxyethene at −40° (Eq. 46).[103] Trialkylsilyloxy acetals afford

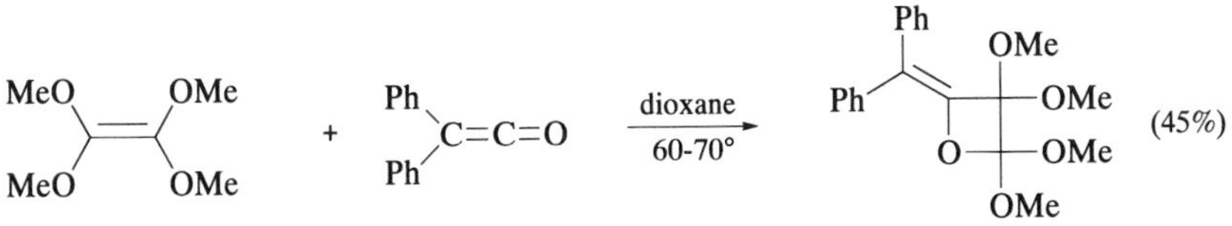

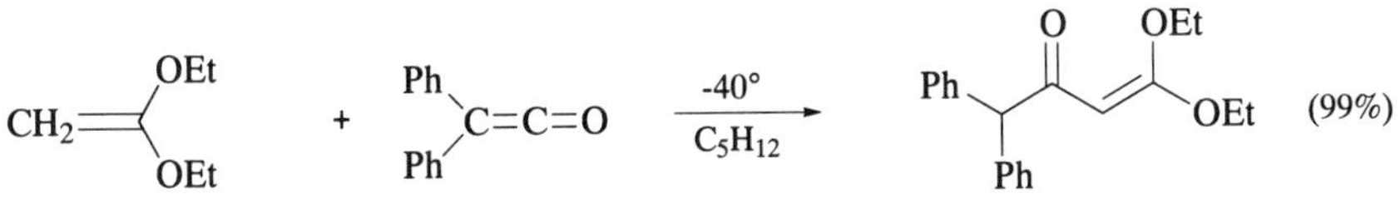

(Eq. 45)

(Eq. 46)

β,γ-unsaturated esters (Eq. 47),[105] even when the α,β isomer is structurally possible. The ring-opened product is easily rationalized by a mechanism involving a zwitterionic intermediate.

MeO OTMS + O=C= (cyclopentylidene ketene) → 200°, 4 h → OTMS, CO_2Me (75%)

(Eq. 47)

Intramolecular Cycloadditions. Internal reaction of a ketene with a suitable site of unsaturation on the same molecule occurs readily, and products can be predicted by the same rules used for bimolecular reactions. This reaction has been reviewed.[25] Two modes of internal addition are usually possible, and both products are seen in many instances (Eq. 48).[107]

R COCl → Et_3N, R = Ph → Ph, O (74%)

Et_3N, R = H → H, H, O ≡ H, H, O (9%)

(Eq. 48)

The chain connecting the ketene and the olefin can be as short as two atoms, but useful results are most often obtained when the connecting link consists of three atoms, yielding 5-membered rings, or four atoms, yielding 6-membered rings. α,β-Unsaturated ketenes form cyclobutenones reversibly when the system contains substitutents on both the ketene and the α carbon, and the β carbon is unsubstituted (Eq. 49).[108]

COCl → Et_3N, PhMe → O, C_5H_9-n (56%) → 130° → O (76%)

(Eq. 49)

In general, when the ketene is generated by dehydrohalogenation of an α,β-unsaturated acid chloride, the proton is removed from the less-substituted γ carbon, regardless of the stereochemistry of the double bond.[109,110] Exceptions to this rule, however, are not difficult to find (Eq. 50).[110]

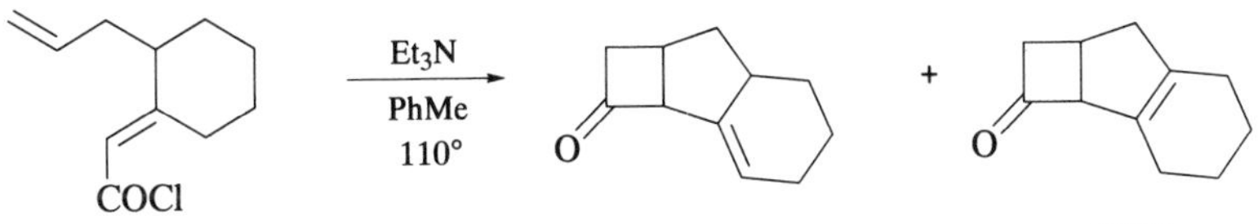

(Eq. 50)

Reactivity patterns follow those of bimolecular reactions of ketenes. Disubstituted ketenes react poorly with unactivated olefins. Good results are obtained with aldoketenes and with ketenes substituted with activating groups such as double bonds, halides, or alkoxy groups.

Acetylenes. The [2+2] cycloaddition of ketenes to acetylenes provides an efficient synthetic route to cyclobutenones. Additions of ketenes to acetylene have not been reported, but reactions of substituted ketenes with both mono- and disubstituted acetylenes proceed in moderate to good yield. Thus 1-hexyne and dichloroketene give a 65–77% yield of 1-butyl-4,4-dichlorocyclobuten-3-one (Eq. 51);[111–113] the regiochemistry demonstrated here is always observed with ter-

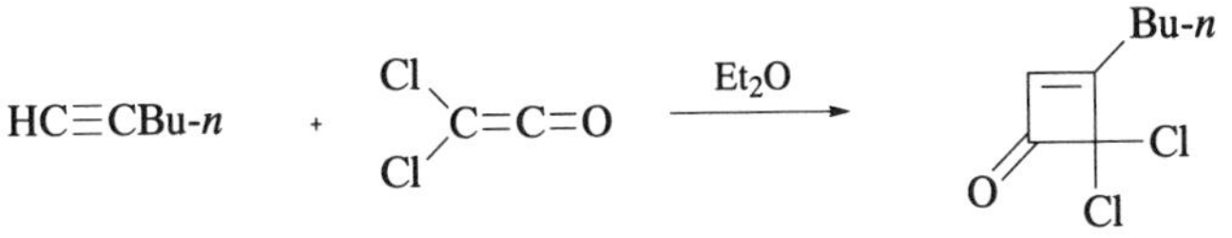

(65-77%) (Eq. 51)

minal alkynes. Internal alkynes generally give both possible regioisomers, but reaction of 1-phenyl-1-propyne with chlorocyanoketene in toluene affords the cyclobutenone in 84% yield (Eq. 52).[114] A thiepin alkyne adds to dichloroketene (Eq. 53) to provide a bicyclic product (80%).[60]

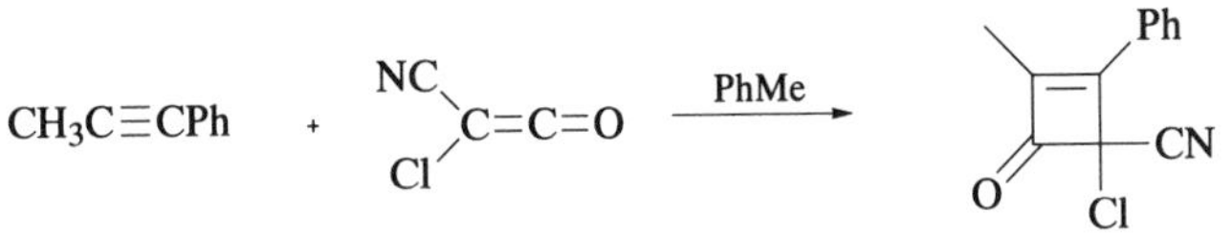

(84%) (Eq. 52)

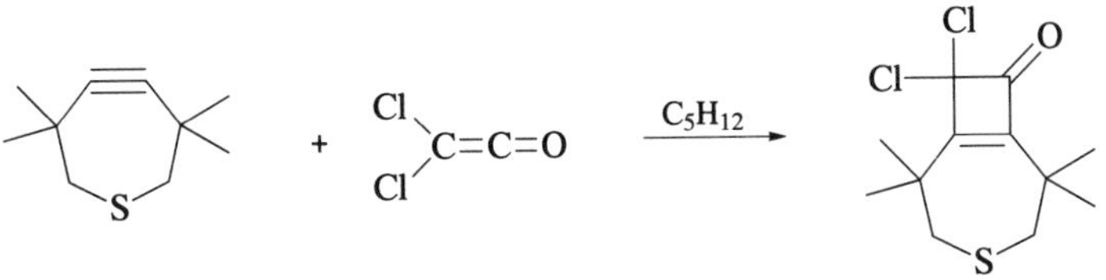

(80%)

(Eq. 53)

Alkynes are more reactive toward ketenes than olefins or even enol ethers, and considerable selectivity is seen in the reactions of enynes with ketenes. A cyclobutenone is the only product reported from dimethylketene and 4-methoxybutenyne (Eq. 54).[115] Similarly, 3-methylbutenyne and dichloroketene give only a cyclobutenone (Eq. 55).[111,113]

HC≡C(H)C=C(H)OMe + $Me_2C{=}C{=}O$ $\xrightarrow{\text{MeCN}}$ (38%)

(Eq. 54)

HC≡C(Me)C=CH_2 + $Cl_2C{=}C{=}O$ $\xrightarrow{\text{Et}_2\text{O}}$ (45%) (Eq. 55)

It should be noted that rearrangement of the primary dichloroketene-alkyne products to isomeric cyclobutenones can occur as illustrated by the formation of two products in the cycloaddition of dichloroketene to 3-hexyne (Eq. 56).[116] This

EtC≡CEt + $Cl_2C{=}C{=}O$ $\xrightarrow{\text{Et}_2\text{O}}$ (30%) + (56%)

(Eq. 56)

problem occurs when dichloroketene is generated by the trichloroacetyl chloride/Zn-Cu/$POCl_3$ system; the rearrangement is attributed to the catalytic effect of the byproduct zinc chloride.[116]

Another common rearrangement of an initial ketene–acetylene cycloadduct occurs when the ketene is phenyl-substituted and the reaction is run at elevated temperature. The primary cyclobutenone product undergoes electrocyclic ring opening to provide an intermediate styryl ketene which closes to yield a 1-naphthol derivative.[117,118] Reaction of the naphthol with excess ketene can lead to 1-naphthyl esters as the ultimate products; the preparation of naphthyl diphenylacetate in 82% yield illustrates the entire process (Eq. 57).[119]

Acetylenes bearing silyl, germyl, arsenyl, phosphoryl, and even transition metal substituents can serve as substrates for [2+2] cycloadditions with ketenes. Equations 58[120,121] and 59[122] illustrate such processes. The regiochemistry of cycloadditions involving organometallic acetylenes is generally that seen with simpler alkynes, but silylalkylacetylenes give both possible regioisomers. Thus trimethyl-

PhC≡CPh + $Ph_2C{=}C{=}O$ —(80°, 3 d)→ [cyclobutenone] → [ ] → [ ] → [ ] → O_2CCHPh_2 (82%)

(Eq. 57)

$Et_3GeC{\equiv}COEt$ + $Cl_2C{=}C{=}O$ → (—)

(Eq. 58)

$C{\equiv}CNi(Ph_3P)(\eta^5\text{-}C_5H_5)$ + $Ph_2C{=}C{=}O$ —(C_6H_6, 2 d)→ $Ni(Ph_3P)(\eta^5\text{-}C_5H_5)$ (70%)

(Eq. 59)

silylpropyne and dichloroketene give isomeric cyclobutenones in 27% and 47% yields, respectively (Eq. 60).[120]

MeC≡CTMS + $Cl_2C{=}C{=}O$ —(Et_2O)→ (27%) + (47%)

(Eq. 60)

Alkoxyacetylenes. The ketene cycloaddition chemistry of acetylenic ethers departs from that of alkynes through greater reactivity, both in the cycloaddition process and in subsequent rearrangements of the initial cycloadducts, as well as in the occurrence of occasional anomalous reactions.

The regiochemistry of [2+2] cycloaddition of ketenes to alkoxyacetylenes is dictated by the electronics of the system and invariably leads to the formation of 1-alkoxycyclobuten-3-ones. The preparation of 1-ethoxy-4,4-dimethylcyclobuten-3-one from dimethylketene and ethoxyacetylene serves to illustrate the process (Eq. 61).[115,123–128]

$$HC{\equiv}COEt + Me_2C{=}C{=}O \xrightarrow{MeCN} \text{1-ethoxy-4,4-dimethylcyclobuten-3-one} \quad (57\text{-}80\%) \qquad \text{(Eq. 61)}$$

The 1-alkoxycyclobuten-3-ones have found considerable synthetic utility as precursors to the otherwise inaccessible cyclobutane-1,3-diones (Eq. 62).[126,127]

$$HC{\equiv}COEt + CH_2{=}C{=}O \xrightarrow[0^\circ]{CH_2Cl_2} \text{1-ethoxycyclobuten-3-one } (31\%) \xrightarrow[H_2O]{H^+} \text{cyclobutane-1,3-dione } (64\%) \qquad \text{(Eq. 62)}$$

The ketene–alkoxyacetylene system appears insensitive to the method of ketene generation; yields are generally moderate to good and reactions are rapid near room temperature. An early but interesting and synthetically useful method of carrying out these reactions involves simply heating an ethyl 1-alkynyl ether to about 100°. Elimination of ethylene affords an alkylketene which immediately adds to remaining acetylene. Thus 1-ethoxy-1-heptyne gives an 83% yield of cyclobutenone when heated to 120–130° (Eq. 63).[129] In an interesting extension of

$$n\text{-}C_5H_{11}C{\equiv}COEt \xrightarrow[-\,C_2H_4]{\text{heat}} [n\text{-}C_5H_{11}CH{=}C{=}O] \xrightarrow{n\text{-}C_5H_{11}C{\equiv}COEt} \text{cyclobutenone } (n\text{-}C_5H_{11}, OEt, C_5H_{11}\text{-}n) \quad (83\%) \qquad \text{(Eq. 63)}$$

this method, di-*tert*-butoxyacetylene eliminates isobutylene in benzene at reflux; the resulting *tert*-butoxyketene adds to the acetylene to give a quantitative yield of 1,2,4-tri-*tert*-butoxycyclobuten-3-one (Eq. 64).[130]

t-BuOC≡COBu-t —heat, $-C_4H_8$→ (100%) (Eq. 64)

A useful, high-yielding synthesis of substituted phenols has resulted from the study of the cycloaddition of vinylketenes (generated by thermal rearrangement of cyclobutenones) to acetylenic ethers. The possible [4+2] cycloaddition process does not occur, and the initially formed [2+2] cycloadduct undergoes a cascade of electrocyclic processes at 80–160° to provide the phenolic product (Eq. 65).[131]

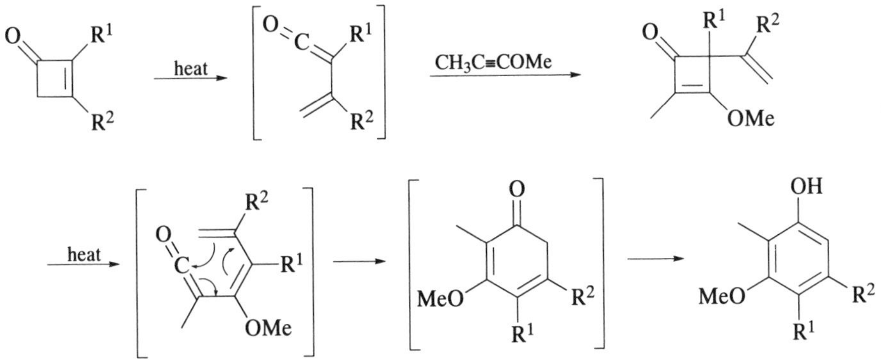

(Eq. 65)

The formation of isoprenoid phenol from methoxyacetylene and 1-methylcyclobutenone illustrates the use of this sequence in synthesis (Eq. 66).[131]

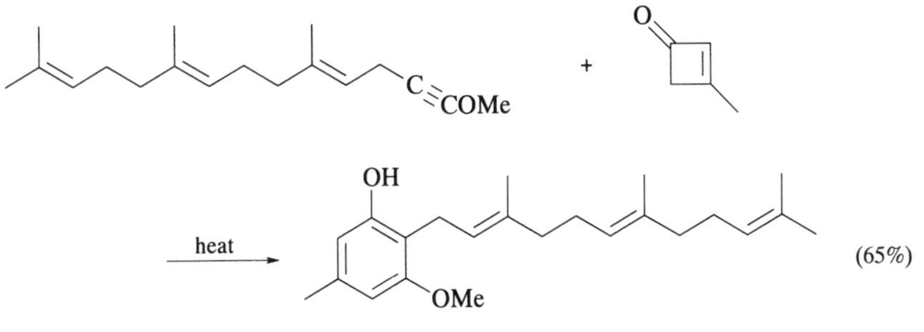

(65%)

(Eq. 66)

As was seen with simple acetylenes, reactions of alkoxyalkynes with phenylketenes can lead to substituted 1-naphthols via ring opening and rearrangement of the initially formed cyclobutenone. This process may occur at low temperatures, as in the case of 1-methoxypropyne and diphenylketene in benzene at 60° (Eq. 67):[132]

$MeC{\equiv}COMe + Ph_2C{=}C{=}O \xrightarrow[60°]{C_6H_6}$ OMe, Ph, Ph, O (24%) + OH, OMe, Ph (8%)

(Eq. 67)

Two types of anomalous ketene–alkoxyalkyne ether reactions are known. The first is [2+2] cycloaddition to the carbonyl group of perfluoroalkylketenes; the resulting oxete can be observed at low temperatures but rearranges to an allene ester upon isolation (Eq. 68).[133]

$HC{\equiv}COEt + (CF_3)_2C{=}C{=}O \longrightarrow$ [OEt, CF_3, O, CF_3] (Eq. 68)

$\longrightarrow (CF_3)_2C{=}C{=}C(CO_2Et)H$ (39%)

The second type of anomalous ketene–alkoxyalkyne ether reaction is that of diarylketenes and ethoxyacetylene in nitromethane solution at subzero temperatures (the expected [2+2] adducts and their naphthol rearrangement products are obtained in ether or benzene solution).[126,132] Thus ethoxyacetylene and phenyl-(*p*-bromophenyl)ketene give a mixture (2.4:1) of isomeric azulenes (Eq. 69).[134]

$HC{\equiv}COEt + Ph(p\text{-}BrC_6H_4)C{=}C{=}O \xrightarrow[-20°]{CH_3NO_2}$

OEt, *p*-BrC_6H_4, O + Br, OEt, Ph, O (Eq. 69)

Yields are typically low (ca. 35%); the mechanism of this process has been the subject of considerable research.[132,134–138]

Ynamines. Ketene–ynamine cycloadditions differ from those involving acetylenic ethers and simple acetylenes by their propensity toward [2+2] addition to both the olefin and carbonyl moieties of the ketene. The oxete products of reaction with ketene carbonyl groups often undergo rearrangement to allenes in situ (Eq. 70).

$$R^1C{\equiv}CN(R^2)_2 + (R^3)_2C{=}C{=}O \longrightarrow \text{cyclobutenone } [R^1, N(R^2)_2, R^3, R^3] + \text{oxete } [(R^2)_2N, R^1, {=}C(R^3)_2] \longrightarrow (R^3)_2C{=}C{=}C(R^1)CON(R^2)_2$$

(Eq. 70)

Although a sizable number of ketene cycloadditions with ynamines has been recorded, it is still difficult to predict the outcome of any proposed reaction. Neither steric nor electronic factors consistently appear to control the mode of cycloaddition. The problem is compounded by the fact that yields are often modest; when a low yield of either a cyclobutenone or an oxete (or allene) product is reported, one can only wonder whether the product arising from the alternative regiochemistry was present but not characterized. Furthermore, solvent and temperature effects are not well understood in these reactions. Thus diethylaminophenylacetylene and diphenylketene give a 95% yield of cyclobutenone product when mixed at $-50°$ in diethyl ether; but in benzene at 15°, an allene is obtained in 14% yield along with the cyclobutenone (54%) (Eq. 71).[139,140]

$$PhC{\equiv}CNEt_2 + Ph_2C{=}C{=}O \xrightarrow[-50°]{Et_2O} \text{cyclobutenone } [Ph, NEt_2, Ph, Ph] \; \mathbf{I}\ (95\%)$$

$$PhC{\equiv}CNEt_2 + Ph_2C{=}C{=}O \xrightarrow[15°]{C_6H_6} \mathbf{I}\ (54\%) + Ph_2C{=}C{=}C(Ph)CONEt_2\ (14\%)$$

(Eq. 71)

A number of selective cycloadditions involving complex ynamines has nevertheless been successfully carried out. Reaction of the enynamine of Eq. 72 with

TMSO(C=CMe2)C≡CN(Me)Ph + $Me_2C{=}C{=}O$ $\xrightarrow{Et_2O}$ cyclobutenone product (TMSO, N(Me)Ph) (38%)

(Eq. 72)

dimethylketene leads to a cyclobutenone in 38% yield; the product of the reaction of ketene with the enol ether moiety was not reported.[141]

That the ketene–ynamine system can be a preparatively useful route to allenic amides is shown by the formation of such a product (53%) by reaction of *tert*-butylcyanoketene with 1-(diethylamino)propyne (Eq. 73).[142]

$$MeC{\equiv}CNEt_2 + (NC)(t\text{-Bu})C{=}C{=}O \xrightarrow[80^\circ]{C_6H_6} (NC)(t\text{-Bu})C{=}C{=}C(CONEt_2)(Me)$$

(53%)

(Eq. 73)

The electrocyclic cascade of reactions leading from acetylenic ethers and vinylketenes to phenols has been successfully extended to ynamines. Thermolysis of 3-methylcyclobutenone in the presence of 1-(dimethylamino)octyne affords an 83% yield of aminophenol (Eq. 74).[131] This synthetically valuable reaction il-

n-$C_6H_{13}C{\equiv}CNMe_2$ + 3-methylcyclobutenone $\xrightarrow{150^\circ}$ aminophenol (OH, n-C_6H_{13}, Me_2N, Me) (83%)

(Eq. 74)

lustrates the potential of ynamine–ketene cycloadditions; however, until more systematic development of the reaction leads to the means to predict, if not control, the regiochemistry of addition, this potential will seldom be realized.

In the reaction of 1-(diethylamino)propyne with silylvinylketenes the initially formed cyclobutenone product rearranges in hexane at 25° to give a bicyclo[3.1.0] product (Eq. 75).[143] This is an alternative outcome of the electrocyclic cascade sequence which has been seen only in heavily silylated systems.[143,144]

Aldehydes and Ketones. The [2+2] cycloaddition of ketenes to aldehydes and ketones produces 2-oxetanones (β-lactones) in preparatively useful yields (Eq. 76).

$MeC{\equiv}CNEt_2$ + (MeO)(p-$CH_3C_6H_4$)C=C(TMS)C(TMS)=C=O $\xrightarrow[25^\circ]{C_6H_{14}}$ (—)

(Eq. 75)

$R^1R^2C{=}O$ + $R^3R^4C{=}C{=}O$ ⟶ (—) (Eq. 76)

Catalysis is an important factor in most cycloadditions of ketenes to carbonyl compounds. Only systems bearing strongly electronegative substituents, such as carbonyl cyanide or perfluoroketones, react readily with simple ketenes in the absence of catalysts (Eq. 77).[145,146]

$(CF_3)_2C{=}O$ + $CH_2{=}C{=}O$ $\xrightarrow[-78^\circ]{Et_2O}$ (96%)

(Eq. 77)

Most other ketene–carbonyl cycloadditions employ Lewis acids, amines, or amine salts as catalysts. Boron trifluoride etherate, aluminum chloride, and zinc chloride are generally useful; the last is generated in situ when ketenes are formed by dehalogenation of α-halo acid chlorides. Similarly, the trialkylamine and trialkylammonium chloride present when ketenes are produced by dehydrohalogenation of acid chlorides can catalyze the cycloaddition process. Amine catalysis of ketene–aldehyde cycloadditions forms the basis of an important asymmetric induction technique (see discussion below).

Solvents useful in these additions include ethers, halocarbons, and hydrocarbons. Most cycloadditions are carried out at subambient temperatures, but unreactive systems such as trimethylsilyketene and chloral (Eq. 78)[147] or perfluorodimethylketene with formaldehyde (Eq. 79)[148] require heating. In general, low

CCl_3CHO + H(TMS)C=C=O $\xrightarrow{90^\circ}$ (64%) (Eq. 78)

cis/trans = 0.25

$$\mathrm{HCHO} + (CF_3)_2C{=}C{=}O \xrightarrow[\mathrm{ZnCl_2}]{150^\circ} \text{2-oxetanone, 3,3-}(CF_3)_2 \quad (\text{—}) \qquad \text{(Eq. 79)}$$

reaction temperatures are preferred because excess heat can bring about decarboxylation of 2-oxetanones to give olefinic products (see below).

A wide variety of ketene structural types is amenable to this type of cycloaddition. Ketene itself, alkylketenes, haloketenes, and silyl- or germylketenes have been successfully employed.[147,149–151] In general, aryl- and diarylketenes are rather unreactive with most aldehydes and ketones, but give good yields of cycloadducts with quinones.[152–154] The cycloaddition process tolerates a wide range of carbonyl substituents, but β-lactone formation from very hindered ketones has not been reported. The preparation of the oxetanones in Eqs. 80,[155] 81,[156] 82,[157] 83,[158] 84,[159,160] 85,[161] and 86[162] illustrates the range of reactions possible. The pref-

$$\text{7-oxabicyclo[2.2.1]hept-5-en-2-one} + Cl_2C{=}C{=}O \xrightarrow{\mathrm{Et_2O}} \text{spiro-3,3-dichloro-2-oxetanone} \quad (87\%) \qquad \text{(Eq. 80)}$$

$$t\text{-Bu}C(O)CH_3 + CH_2{=}C{=}O \xrightarrow[\mathrm{CCl_4}]{\mathrm{BF_3{\bullet}Et_2O}} \text{4-methyl-4-}t\text{-Bu-2-oxetanone} \quad (67\%) \qquad \text{(Eq. 81)}$$

$$\text{tetramethylcyclopropanone} + Me_2C{=}C{=}O \xrightarrow[-78^\circ]{\mathrm{CH_2Cl_2}} \text{spiro-2-oxetanone} \quad (\text{—}) \qquad \text{(Eq. 82)}$$

$$HC{\equiv}CCHO + CH_2{=}C{=}O \xrightarrow[-20^\circ]{\mathrm{BF_3{\bullet}Et_2O}} \text{4-ethynyl-2-oxetanone} \quad (40\%) \qquad \text{(Eq. 83)}$$

$$\text{oxiranecarboxaldehyde (CHO)} + CH_2{=}C{=}O \xrightarrow[\mathrm{Et_2O}]{\mathrm{ZnCl_2}} \text{4-oxiranyl-2-oxetanone} \quad (42\%) \qquad \text{(Eq. 84)}$$

p-NCC$_6$H$_4$CHO + Cl$_2$C=C=O $\xrightarrow[CH_2Cl_2]{Et_2O}$ β-lactone (C$_6$H$_4$CN-p, Cl, Cl) (82%)

(Eq. 85)

(82%)

(Eq. 86)

erence for the aldehyde over ketone, alkyne, and olefin functionality (Eqs. 86, 83, and 80) is particularly noteworthy.

An industrially important application of the ketene–aldehyde cycloaddition is the preparation of sorbic acid from ketene and crotonaldehyde (Eq. 87). This pro-

CHO + CH$_2$=C=O $\xrightarrow{BF_3}$

heat

(Eq. 87)

cess is usually carried out in the presence of boron trifluoride or other Lewis acid catalysts under conditions in which the presumed β-lactone cycloadduct polymerizes to a low molecular weight polyester; formation of the latter is promoted by the presence of transition metal salts of fatty acids. The polyester is thermally degraded to sorbic acid.[163,164]

Cycloadditions of ketenes to quinones are generally carried out in hydrocarbon solvents at subambient temperature in the absence of catalysts. These reactions display considerable selectivity. Formation of β-lactones generally prevails over cycloaddition to the olefinic unsaturation; one rare example of poor selectivity is the reaction of durenequinone and diphenylketene, from which both lactone and cyclobutanone are obtained (Eq. 88).[153]

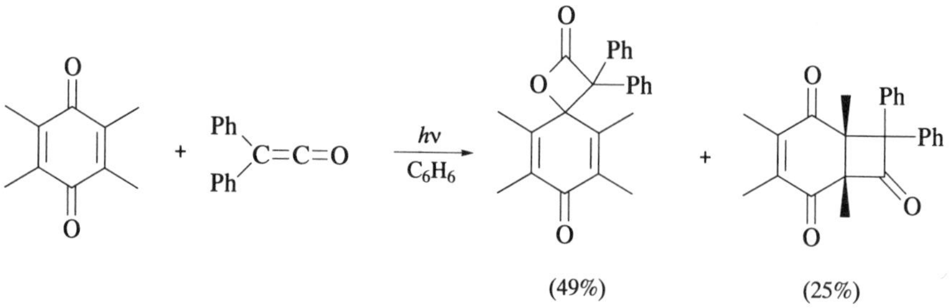

$h\nu$, C_6H_6

(49%) (25%)

(Eq. 88)

Ketene–benzoquinone cycloadditions are quite subject to steric influences. The less-hindered lactone is the sole product reported from toluquinone and diphenylketene (Eq. 89),[152] and 2,6-dimethylquinone and dimethylketene like-

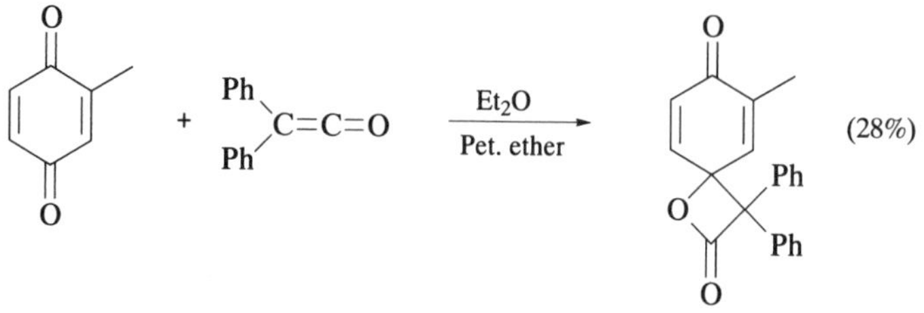

Et_2O, Pet. ether

(28%) (Eq. 89)

wise afford a single lactone (Eq. 90).[165] No examples of double cycloaddition to benzoquinones have been reported.

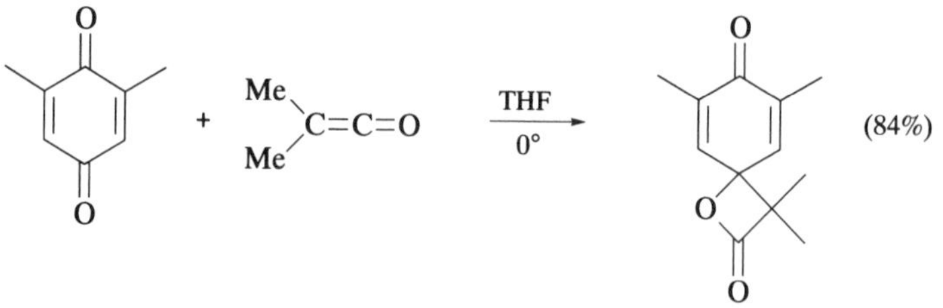

THF, 0°

(84%) (Eq. 90)

Stereochemistry of Carbonyl Cycloadditions. The question of stereochemistry arises in the reactions of aldehydes or unsymmetrical ketones with monosubstituted or unsymmetrically disubstituted ketenes. Unfortunately, many such cycloadditions were described in the older literature wherein the stereochemistry of the product was not determined. There do not appear to be any published examples in which the stereochemistry of a ketone–ketene cycloaddition has been proven.

In aldehyde cycloadditions to monosubstituted and unsymmetrically disubstituted ketenes, mixtures of *cis* and *trans* isomers are generally obtained. In many cases the ratio of *cis* to *trans* products is close to 1; the range is typically from 0.25 to 1.6. Thus *o*-chlorobenzaldehyde and bromomethylketene give a 1:1 mixture of *cis* and *trans* lactones in 50% yield (Eq. 91),[151] and chloral reacts with

$$o\text{-ClC}_6\text{H}_4\text{CHO} + (\text{Br})(\text{Me})\text{C=C=O} \xrightarrow{\text{C}_6\text{H}_{14}} \text{2-oxetanone (H, C}_6\text{H}_4\text{Cl-}o\text{; Br)} \quad (50\%)$$

(Eq. 91)

chloroketene to afford a 40% yield of products wherein the *cis/trans* ratio is 1.6 (Eq. 92).[150] Isopropylketene and chloral give a *cis/trans* ratio of 0.9 for the oxe-

$$\text{CCl}_3\text{CHO} + (\text{Cl})(\text{H})\text{C=C=O} \xrightarrow{\text{Et}_2\text{O}} \text{2-oxetanone (H, CCl}_3\text{; H, Cl)} \quad (40\%)$$

cis/trans = 1.60-1.64

(Eq. 92)

tanone products.[166] In the reaction of monohalo and haloalkylketenes with aldehydes, the *cis* products are formed in slightly greater amounts.[150,151,166]

The most extensive set of stereochemical data relating to ketene–aldehyde cycloadditions deals with silyl- and germylketenes.[147,167] Trialkylsilylketenes add to simple aldehydes in the presence of boron trifluoride etherate to give mixtures in which the *cis* stereochemistry predominates. For example, benzaldehyde and trimethylsilylketene give lactones with a *cis/trans* ratio of 2; with isobutyraldehyde the ratio is 1.5 (Eq. 93).[147]

$$\text{PhCHO} + (\text{TMS})(\text{H})\text{C=C=O} \xrightarrow[\text{-50°}]{\text{BF}_3\text{•Et}_2\text{O}} \text{2-oxetanone (H, Ph; H, TMS)} \quad (65\%)$$

cis/trans = 2.0

(Eq. 93)

When α-haloaldehydes undergo cycloaddition with silylketenes the stereochemical result contrasts with the poor selectivity, favoring slightly the *cis* product with simple aliphatic aldehydes. In halogenated systems the *trans* isomer predominates and is often the exclusive product. Bromal and chloromethyldimethylsilylketene give solely *trans* product (Eq. 94); a similar result is obtained with chloral and triethylsilyketene (Eq. 95). Surprisingly, the stereoselectivity of these cycloadditions is decreased at lower temperatures: bromal and trimethyl-

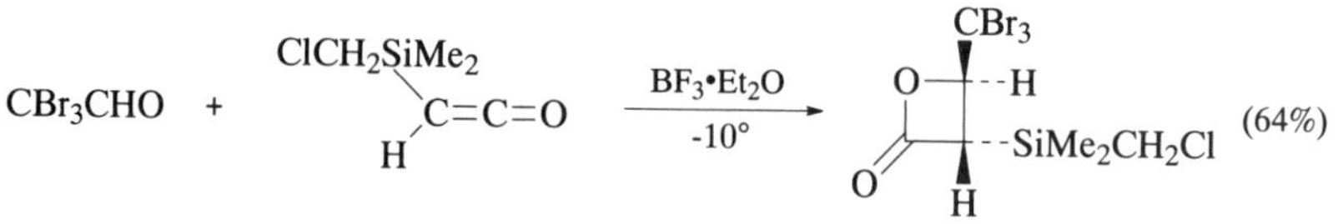

(Eq. 94)

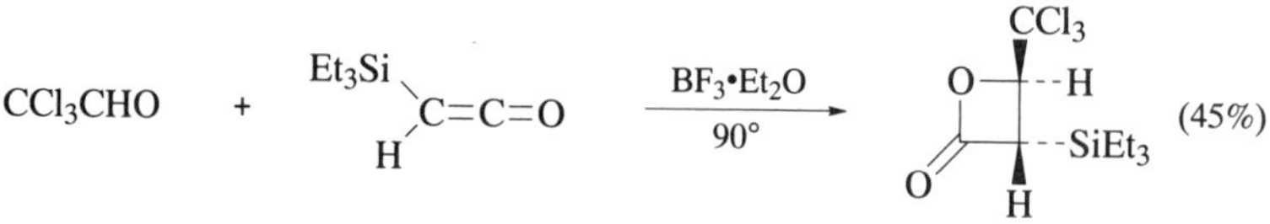

(Eq. 95)

silylketene give an 80/20 *trans*/*cis* mixture at $-50°$, in contrast to 100% *trans* at $-10°$ (Eq. 96).[147]

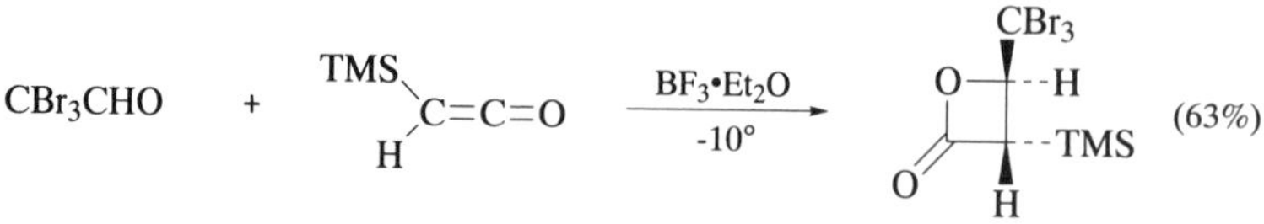

(Eq. 96)

There are obviously many knowledge gaps regarding the stereochemistry of ketene–carbonyl [2+2] cycloadditions. The silylketene–haloaldehyde system is the only one reported to give consistently good stereoselectivity. But outside this system there have been few attempts to define systematically the effects of solvent, temperature, catalyst, method of ketene generation, and steric bulk on the stereochemistry of addition;[151] and the stereochemistry of the lactones formed by ketone–ketene cycloaddition is virtually unexplored.

Chiral Syntheses. The utility of substituted β-lactones as intermediates in synthesis has led to the development of methods for asymmetric induction in their preparation via ketene cycloadditions. The first successful results in this area came from reactions of halogenated aldehydes and ketones with ketenes generated from aryloxy- or dihaloacetyl chlorides and (−)-brucine or (−)-*N*,*N*,α-trimethylphenethylamine as dehydrohalogenation reagents. Thus acetyl chloride and chloral, in the presence of (−)-*N*,*N*,α-trimethylphenethylamine give a 72% yield of (*R*)-(−)-β-trichloromethyl-β-propiolactone of 18.7% optical purity (Eq. 97).[168,169] Other

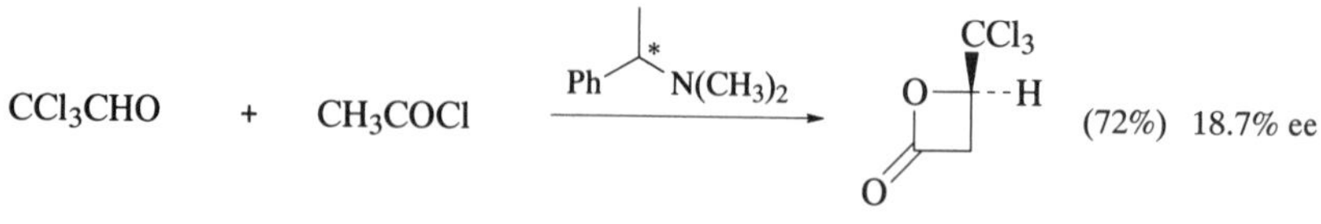

(Eq. 97)

related systems give similar results, and the requirement for a molar equivalent of the chiral base makes the method one of limited synthetic value.

A significant breakthrough in this area came with the announcement that the cycloaddition of ketene with chloral carried out at −50° in the presence of 1–2 mol % of quinidine gives an 89% yield of β-lactone having 98% ee of the *R* configuration (Eq. 98). When quinine is used as the catalyst, (*S*)-(+)-lactone is

$CCl_3CHO + CH_2{=}C{=}O$ → (Quinidine, PhMe, -50°) → β-lactone (CCl_3, H) (89%) 98% ee; (Quinine, PhMe, -50°) → β-lactone (H, CCl_3) (—) 76% ee

(Eq. 98)

obtained with 76% ee; either enantiomer can be recrystallized to optical purity.[170] It should be noted that the original assignment of the *S* configuration to the ketene–chloral adduct obtained with quinidine catalysis was in error; the *R* configuration is in fact obtained with quinidine.[171]

A large number of chiral amine catalysts has been examined, and the cinchona alkaloids are the most effective. A plausible mechanism for the chiral induction, based on formation of an intermediary ketene–amine complex, has been described.[170] The method has been explored in some detail, and excellent results are obtained with both α-halogenated aldehydes and ketones.[171] This method has been applied to the synthesis of, among others, (*R*)- and (*S*)-malic acids,[170] (*R*)- and (*S*)-citramalic acids (Eq. 99),[172] and (*S*)-methyl-3-hydroxyalkanoates (Eq. 100).[173]

Me(CCl_3)C=O + $CH_2{=}C{=}O$ → (Quinidine, PhMe, -25°) → β-lactone (CCl_3) → (several steps) → CO_2H–CH_2–C(OH)–CO_2H (—)

(Eq. 99)

The acid- and base-catalyzed oxetanone ring-opening reactions used in these sequences proceed with inversion of configuration at the β carbon. This result led to early confusion about the correct absolute configuration of the oxetanones. In another application of asymmetric ketene–aldehyde cycloaddition, the (*S*)-ketene–trichloroacetone adduct was converted into a dioxolane, which has utility in the synthesis of 25,26-dihydroxycholecalciferol (Eq. 101).[174]

Optically active polyesters have also been prepared from the β-lactones obtained from ketenes and haloaldehydes under cinchona alkaloid catalysis.[175,176]

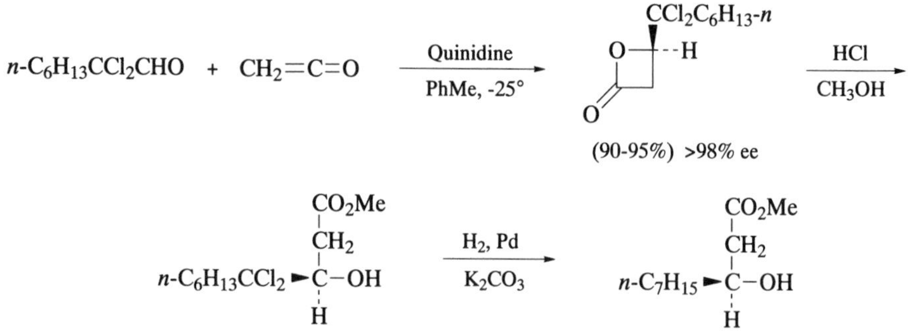

(Eq. 100)

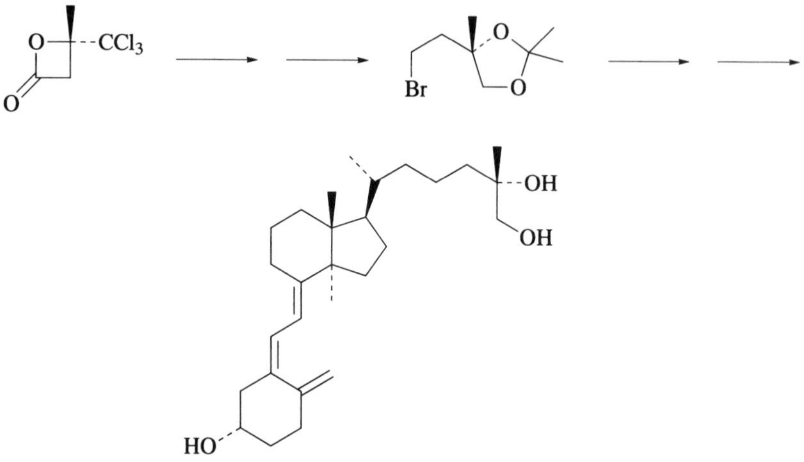

(Eq. 101)

Thiocarbonyl Compounds. Diphenylketene undergoes [2+2] cycloaddition to thiones, thioesters, and thioamides to provide thietanone derivatives in good yield. Thioxanthone and diphenylketene give an adduct in quantitative yield upon irradiation, but other thiones such as thiobenzophenone react upon warming to ca. 60° (Eqs. 102 and 103).[177,178]

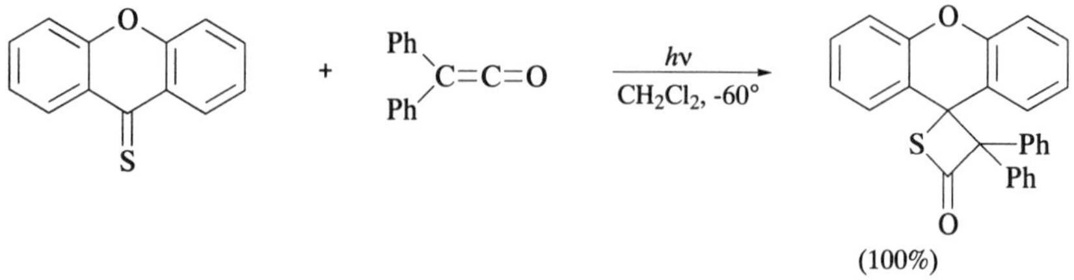

(Eq. 102)

$$\mathrm{Ph_2C{=}S} + \mathrm{Ph_2C{=}C{=}O} \xrightarrow{60^\circ} \text{2,2,4,4-tetraphenylthietan-3-one} \quad (74\%)$$

(Eq. 103)

Thioester cycloadditions are exemplified by the reaction of methyl dithiobenzoate and diphenylketene (Eq. 104); thioamides react similarly.[177,179]

$$\mathrm{PhC(=S)SMe} + \mathrm{Ph_2C{=}C{=}O} \longrightarrow \text{4-SMe-2,2,4-triphenylthietan-3-one} \quad (61\%)$$

(Eq. 104)

There appear to be no reported examples of cycloadditions of ketenes other than diphenylketene to thiono compounds.

Preparation of Olefins by Reaction of Ketenes with Carbonyl and Thiocarbonyl Compounds. The β-lactones formed by [2+2] cycloaddition of ketenes to aldehydes and ketones (and the corresponding thietanones from thiocarbonyl compounds) can be thermally decarboxylated to olefins (Eq. 105).

$$\mathrm{R^1R^2C{=}C{=}O} + \mathrm{R^3R^4C{=}O} \longrightarrow \left[\beta\text{-lactone}\right] \xrightarrow{-\,CO_2} \mathrm{R^1R^2C{=}CR^3R^4}$$

(Eq. 105)

Since the intermediate lactones need not be isolated and the decarboxylation generally requires a temperature in the range 80–180°, the olefin synthesis in its simplest form consists of heating a ketene and carbonyl compound in an inert solvent until carbon dioxide evolution ceases. Lewis acid catalysts of the type used in the preparation of β-lactones are occasionally employed, but it appears that ketenes commonly react with carbonyl compounds without catalysts at the elevated temperatures necessary for decarboxylation. This olefin synthesis provides a convenient alternative to the Wittig and other similar olefinations; the lack of byproducts and ease of workup argue for wider employment of the ketene-based process.

The simplest examples of aldehydes reacting with ketene have been recorded in the patent literature. Thus it is claimed that vinylnorbornene is produced in 94% yield from the cyclopentadiene–acrolein adduct and ketene at 170–190°

(Eq. 106).[180] Gas-phase reaction of hexafluoroacetone with ketene generated in situ from acetic anhydride provides hexafluoroisobutylene (Eq. 107).[181]

CHO + $CH_2{=}C{=}O$ $\xrightarrow[- CO_2]{170\text{-}190°}$ (94%)

(Eq. 106)

CF_3, CF_3 =O + $CH_2{=}C{=}O$ $\xrightarrow[- CO_2]{>290°}$ CF_3, CF_3 (82%)

(Eq. 107)

The stereochemistry of the olefins produced via β-lactone decarboxylation has not been extensively studied, since most examples involve the use of symmetrical ketenes or carbonyl compounds. However, it is clear that the stereochemistry of the intermediate lactone dictates the final olefin stereochemistry. Thus when benzaldehyde and trimethylsilylketene react in the presence of boron trifluoride at $-50°$, both *cis* and *trans* lactones are produced. The isolated *trans* lactone decarboxylates at 50° to give the *E* olefin (13% based on aldehyde), whereas the *cis* lactone decarboxylates at 150° to give a 75% yield of the *Z* olefin (Eq. 108).[167]

TMS(H)C=C=O + PhCHO $\xrightarrow[-50°]{BF_3}$ (O, Ph, TMS) + (O, Ph, TMS)

50° → H, Ph, TMS, H (13%)

150-160° → TMS, Ph, H, H (75%)

(Eq. 108)

In a series of reactions of substituted benzaldehydes with various halocyanoketenes, *Z* olefins are formed exclusively. The stereocontrol is attributed to formation of a single β-lactone. Yields range from 8 to 92%, with the higher yields coming from aldehydes bearing electron-donating substituents as in the preparation of the olefin in Eq. 109.[182]

$p\text{-MeOC}_6\text{H}_4\text{CHO} + (\text{Br})(\text{NC})\text{C=C=O} \xrightarrow[80^\circ]{\text{C}_6\text{H}_6} p\text{-MeOC}_6\text{H}_4\text{CH=C(CN)Br}$ (—)

(Eq. 109)

Carbonyl cycloaddition–elimination to yield olefins is the predominant process even in substrates bearing olefinic and acetylenic functionality, as is seen in the preparation of a complex fulvene in 86% yield (Eq. 110).[183] Furthermore, the

O; C≡C—C≡C; + COCl; C_6H_6, Et_3N, 80°; C≡C—C≡C (86%)

(Eq. 110)

presence of such potentially nucleophilic sites as divalent sulfur does not lead to side reactions (Eq. 111).[184]

Cl; Cl; S; O; + $1\text{-C}_{10}\text{H}_7(\text{Ph})\text{C=C=O}$; C_6H_6, 80°; Cl; Cl; S; Ph; $C_{10}H_7$-1 (74%)

(Eq. 111)

There is at least one example of selective olefin formation from the aldehyde group of a keto aldehyde and dimethylketene,[162] and a number of intramolecular olefin syntheses proceeding via ketene–carbonyl cycloadditions have been described. The preparation of the naphthofuran of Eq. 112[185] is typical; we are aware of no extensions of this technology to the synthesis of larger ring systems.

CHO; O; CO_2H; Et; 1. p-TsCl; 2. Et_3N, C_6H_6, 80°; Et; O (74%)

(Eq. 112)

The elimination of carbonyl sulfide from unisolated intermediate thietanones also provides an olefin synthesis; most recorded examples use thioester substrates (Eq. 113).[186]

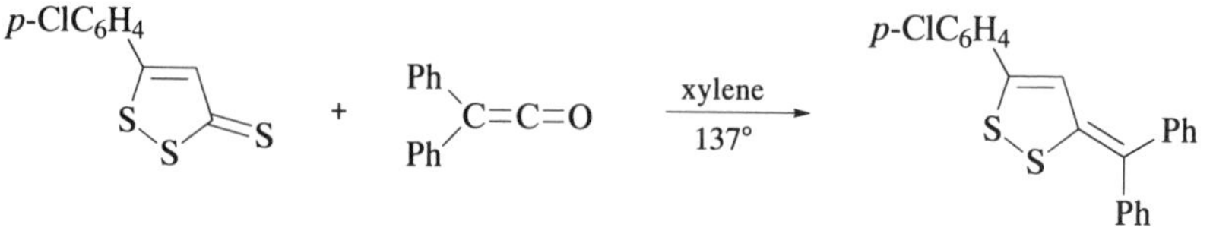

p-ClC$_6$H$_4$ ... + Ph$_2$C=C=O —(xylene, 137°)→ p-ClC$_6$H$_4$ ... Ph, Ph (75%)

(Eq. 113)

Problematical side reactions occur frequently in olefin syntheses attempted with cycloheptatriene or tropone substrates, as in the addition of the ketene from cycloheptatrienyl 7-carbonyl chloride to cycloheptatrienone; only traces of heptafulvalene are recovered and the major product is a lactone (Eq. 114).[187]

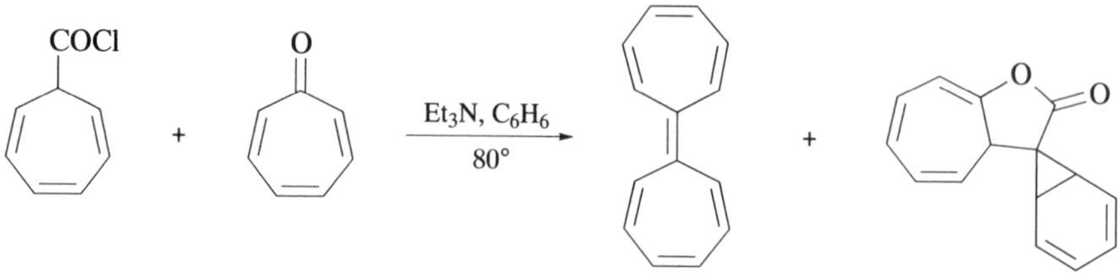

COCl + O —(Et$_3$N, C$_6$H$_6$, 80°)→ (1%) + (80%)

(Eq. 114)

Isocyanates. Ketene and alkyl- or aryl-substituted ketenes undergo [2+2] cycloaddition to alkyl and aryl isocyanates, usually at elevated temperatures, to give good yields of azetidine-2,4-diones (malonimides). The reaction of *p*-phenylenediisocyanate with diethylketene in toluene at 180° to give a bis(malonimide) is typical (Eq. 115).[188] Sulfonyl isocyanates react similarly,[189,190] but cycloadditions of haloketenes to isocyanates do not appear to have been reported.

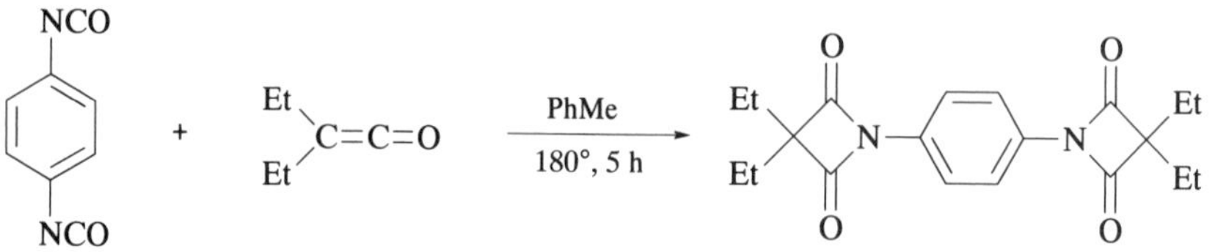

NCO, NCO + Et$_2$C=C=O —(PhMe, 180°, 5 h)→ (91%)

(Eq. 115)

Carbodiimides. The reaction of ketenes with carbodiimides yields 4-imino-2-azetidinones (Eq. 116). The cycloadditions take place at ambient or moderately

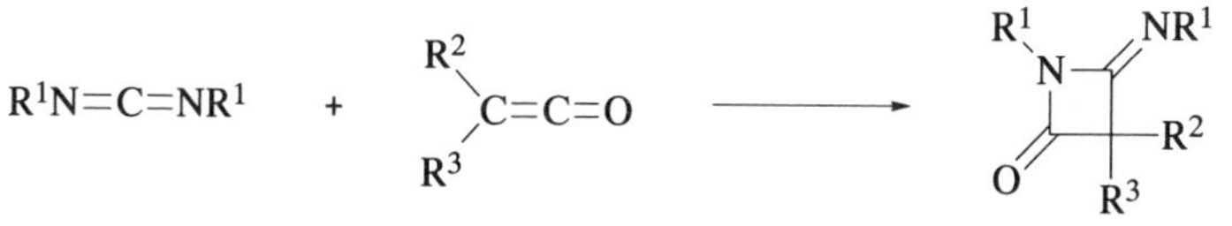

$$R^1N{=}C{=}NR^1 + R^2R^3C{=}C{=}O \longrightarrow \text{(Eq. 116)}$$

elevated temperatures in nonpolar solvents; yields are generally good, and the process tolerates a wide variety of ketene substituents. Steric influences are not very important; the preparation of a tri-*tert*-butyllactam from di-*tert*-butylcarbodiimide and *tert*-butylcyanoketene proceeds in 88% yield (Eq. 117).[191]

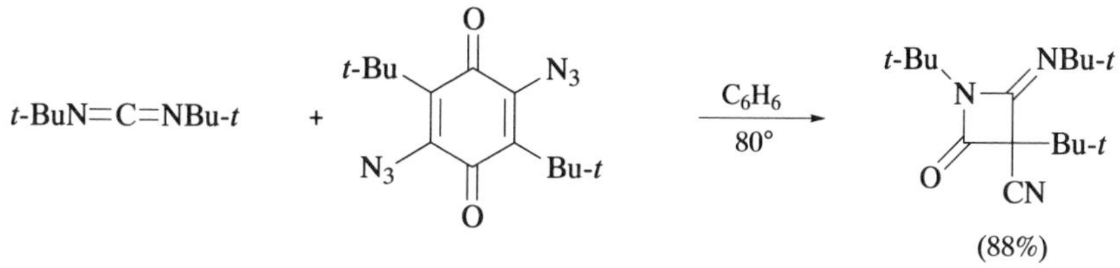

t-BuN=C=NBu-t + (2,5-di-t-Bu-3,6-diazido-1,4-benzoquinone) —(C_6H_6, 80°)→ (88%)

(Eq. 117)

Examples have been reported in which the initial cycloadduct undergoes further reaction with the ketene to provide a spirocyclic product, but this potentially troublesome side reaction can be circumvented by careful avoidance of excess ketene. Thus cyclic pentamethylenecarbodiimide reacts with one equivalent of diphenylketene to give an iminolactam; in the presence of excess ketene, the spirocyclic bis(adduct) is formed in good yield (Eq. 118).[192]

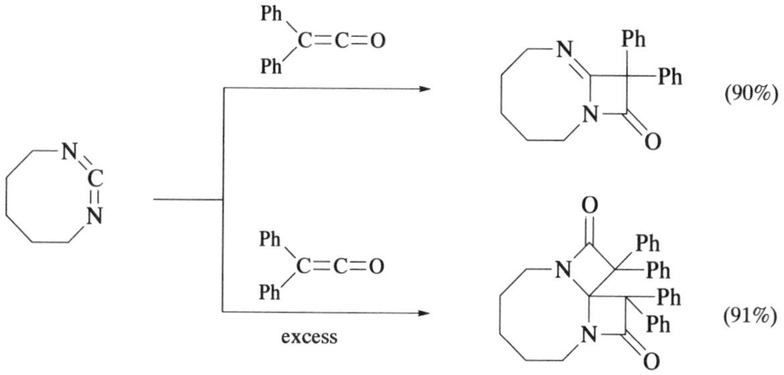

Ph₂C=C=O → (90%); Ph₂C=C=O excess → (91%)

(Eq. 118)

Reaction of dicyclohexylcarbodiimide with the vinylketene formed by ring opening of the cyclobutenone in Eq. 119 gives only the product of [2+2] cycloaddition, and overreaction of the ketene with the product is not reported.[114]

A few cycloadditions of ketenes to unsymmetrical carbodiimides have been investigated; poor to moderate yields of single products are reported, but the composition of the entire product mixture remains undetermined.[193] In the reaction of optically active carbodiimides such as dimenthylcarbodiimide with the

$C_6H_{11}N{=}C{=}NC_6H_{11}$ + [2-chloro-2-cyano-3,4-diphenylcyclobutenone] $\xrightarrow[80°, 16\ h]{C_6H_6}$ [β-lactam product] (76%)

(Eq. 119)

prochiral methylphenylketene, inseparable mixtures of optically active products are obtained. However, when dimenthylcarbodiimide is allowed to react with the ketene as in Eq. 120, an optically active lactam is formed in moderate yield.[194]

$RN{=}C{=}NR$ + (t-Bu)(EtO_2C)$C{=}C{=}O$ ⟶ [β-lactam product] (64%)

R = *l*-menthyl

(Eq. 120)

***N*-Sulfinylamines.** Ketenes undergo facile [2+2] cycloaddition to *N*-sulfinylamines to yield 1,2-thiazetidin-3-one-1-oxides as exemplified by the reaction of cyclohexylsulfinylamine with bis(trifluoromethyl)ketene to give an 86% yield of product (Eq. 121).[195]

$C_6H_{11}N{=}S{=}O$ + $(CF_3)_2C{=}C{=}O$ $\xrightarrow{-30°}$ [thiazetidinone oxide product] (86%)

(Eq. 121)

The process is usually carried out at subambient to ambient temperatures in a wide range of solvents, but fluorenylideneketene reacts with *m*-chlorophenyl-*N*-sulfinylamine only after heating at 60° for 72 hours (Eq. 122).[196] The cycloadducts are generally stable, one exception being the phenylsulfinylamine–ketene adduct, which was prepared and trapped by ring opening with an amine at −78°.[197]

m-$ClC_6H_4N{=}S{=}O$ + [fluorenylideneketene] $\xrightarrow{60°, 72\ h}$ [spiro thiazetidinone oxide product] (87%)

(Eq. 122)

The related dialkylsulfurdiimides also give [2+2] reactions with ketenes to form cyclic products (Eq. 123), but diarylsulfurdiimides fail to react with haloketenes.[198]

t-BuN=S=NBu-t + $CHCl_2COCl$ $\xrightarrow{Et_3N}$ (91%)

(Eq. 123)

Nitroso Compounds. Aryl nitroso compounds react with diarylketenes at ambient or lower temperatures to give product mixtures indicative of poor regioselectivity of cycloaddition. The resulting 1,2-oxazetidin-3-ones are generally stable, but the isomeric 1,2-oxazetidin-4-ones decompose to give carbon dioxide and an imine, which is immediately trapped by additional ketene to give an azetidin-2-one (Eq. 124).[199]

ArN=O + $Ar_2C=C=O$ ⟶ + $\xrightarrow{-CO_2}$

$Ar_2C=NAr$ $\xrightarrow{Ar_2C=C=O}$

(Eq. 124)

It was at one time thought that the nature of the substitution on the aryl nitroso substrate controlled the regiochemistry and degree of concertedness of the cycloaddition,[200] but more recent results indicate that while oxazetidin-3-ones are the major products, both isomers are generally formed in an apparently near-concerted process.[199] Thus nitrosobenzene and diphenylketene react in chloroform at 25° to give 60% of 1,2-oxazetidin-3-one and 13% of the β-lactam (Eq. 125).[199] Products from nitroso compounds and alkyl- or haloketenes have not been reported.

PhN=O + $Ph_2C=C=O$ $\xrightarrow[25°]{CHCl_3}$ (60%) + (13%)

(Eq. 125)

Azo Compounds. Ketenes and azo compounds undergo [2+2] cycloaddition to yield 1,2-diazetidin-3-ones. Most examples of this process involve symmetrical azobenzenes, which give cycloadducts when irradiated in the presence of ketenes (Eq. 126).[201] It is well established that only the *cis* isomers of azobenzenes undergo the ketene cycloaddition process; the reaction of an isolated *cis*-azobenzene with diphenylketene in the absence of irradiation illustrates the point (Eq. 127).[202]

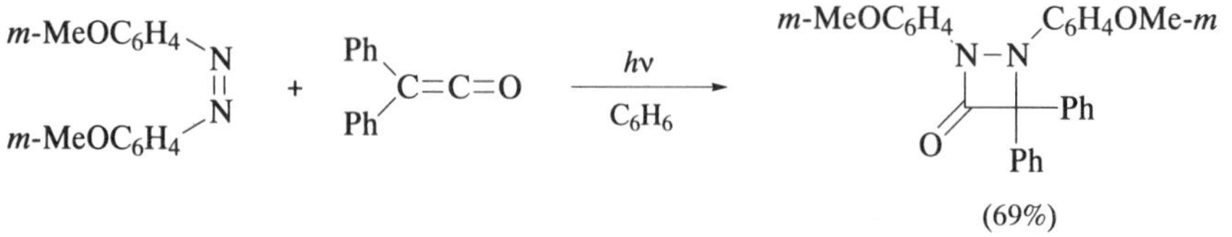

(69%)

(Eq. 126)

PhMe
30°

(80%)

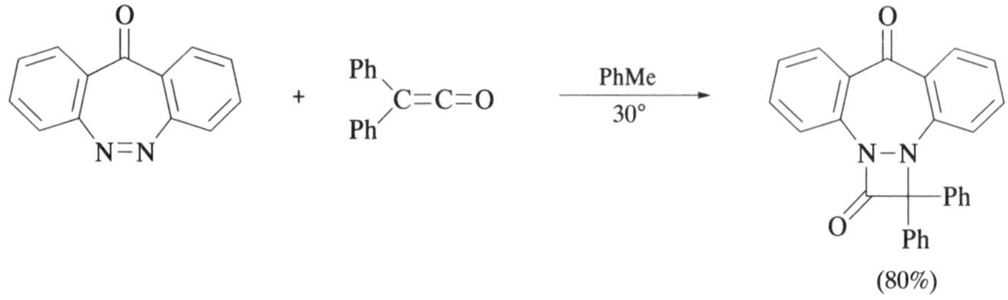

(Eq. 127)

The use of unsymmetrical azobenzenes in the cycloaddition process generally yields mixtures of regioisomeric products as illustrated by the formation of both isomers from 3-methylazobenzene and ketene (Eq. 128).[203] However, some degree of steric control of regiochemistry may be possible, since the *o*-methyl analog gives a single diazetidinone with ketene (Eq. 129).[203]

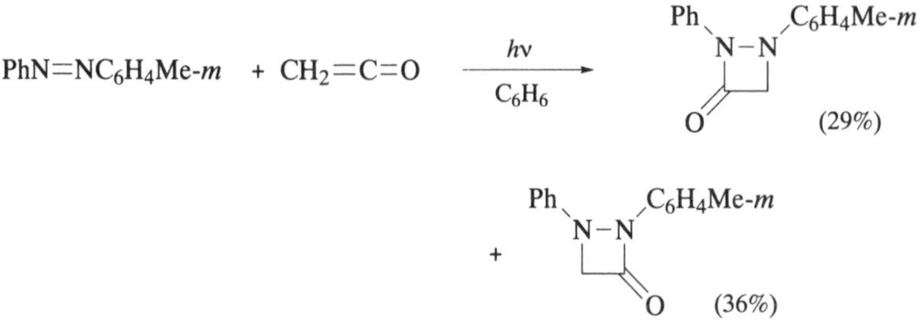

$PhN{=}NC_6H_4Me\text{-}m + CH_2{=}C{=}O \xrightarrow[C_6H_6]{h\nu}$

(29%) + (36%)

(Eq. 128)

$$\text{PhN{=}NC}_6\text{H}_4\text{Me-}o + \text{CH}_2\text{{=}C{=}O} \xrightarrow[\text{C}_6\text{H}_6]{h\nu} \text{(1-Ph-2-(}o\text{-MeC}_6\text{H}_4\text{)-1,2-diazetidin-3-one)} \quad (81\%)$$

(Eq. 129)

In contrast to the azobenzenes, azodicarboxylates react with diphenylketene to give cycloadducts without irradiation (Eq. 130).[204] An azoketone also reacts

$$\text{EtO}_2\text{CN{=}NCO}_2\text{Et} + \text{Ph}_2\text{C{=}C{=}O} \xrightarrow{\text{Pet. ether}} \text{(1,2-bis(CO}_2\text{Et)-4,4-diphenyl-1,2-diazetidin-3-one)} \quad (70\%)$$

(Eq. 130)

with diphenylketene in the dark, but both the orientational selectivity and [2+2]/[4+2] selectivity are poor, with four products being formed (Eq. 131).[205]

$$\text{PhN{=}NCOMe} + \text{Ph}_2\text{C{=}C{=}O} \longrightarrow \text{(1-Ph-2-COMe-4,4-diphenyl-1,2-diazetidin-3-one) (60\%)} + \text{(1-MeCO-2-Ph-4,4-diphenyl-1,2-diazetidin-3-one) (26\%)} + \text{(2-methyl-4-Ph-6,6-diphenyl-1,3,4-oxadiazin-5-one) (10\%)} + \text{(1-Ph-2-COMe-4,4-diphenyl-1,2-dihydrocinnolin-3(4H)-one) (4\%)}$$

(Eq. 131)

Ketene-azo cycloaddition can occur as a secondary process following other ketene cycloadditions. Thus azoquinones react with diphenylketene to give an unobserved cycloadduct which immediately undergoes a [2+2] reaction at the azo linkage to give the final product (Eq. 132).[206]

[3+2] Cycloadditions

Both the carbonyl and the carbon–carbon double bond of ketenes react with 4-electron, 3-center bonds. The retention of stereochemistry by both *cis*- and *trans*-di-*tert*-butylthiocarbonyl ylide in the reaction with diphenylketene is the single case where this criterion for the concertedness of bond formation has been tested (Eq. 133).[207]

(75%)

(Eq. 132)

R^1 = t-Bu, R^2 = H (—)
R^1 = H, R^2 = t-Bu (—)

(Eq. 133)

The azomethine ylide gives different products with ketene and diphenylketene (Eq. 134).[208]

(—)

(Eq. 134)

(—)

Alkyl- and arylnitrones generally react with ketenes by an ionic pathway to give oxazolidinones (Eq. 135);[209] Table XXVII lists one exception to this rule.

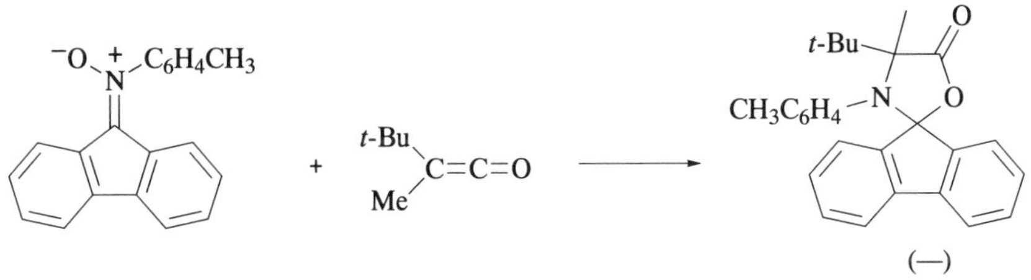

(Eq. 135)

Nitrile oxides react with ketenes, and both isoxazolinone[210,211] and oxazolinone[212] products have been reported. The structural assignment of an isoxazolinone to the product from the reaction of diphenylketene with *tert*-butylnitrile oxide has been challenged.[212] The support for the isoxazolinone structure was based only on the infrared spectrum,[210] while the oxazolinone structure was based on more extensive analysis.[212]

An azimine reacts with diphenylketene to give a rearranged product (Eq. 136).[213] Trapping experiments suggested the intermediacy of a formal [3+2] adduct.

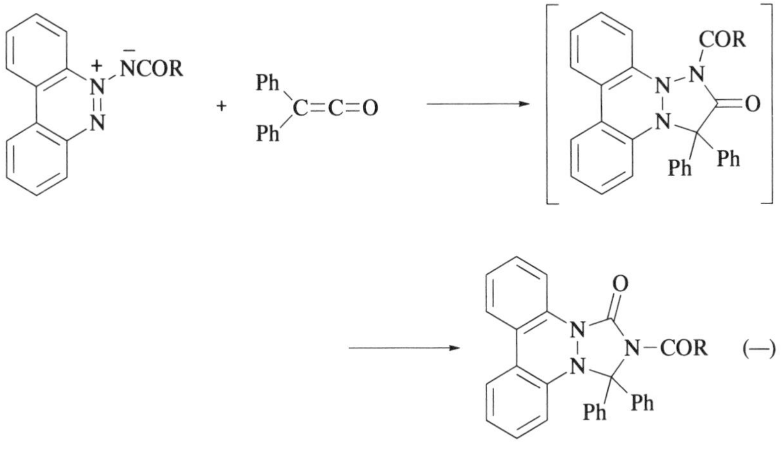

(Eq. 136)

Diazomethane reacts with ketene,[214–217] methylketene,[72] and dimethylketene,[218] as well as with silyl and germanyl ketenes,[219,220] to give cyclopropanones with no indication of a [3+2] process. A rearranged adduct containing two molecules of bis(trifluoromethyl)ketene has been reported.[221] Monoaliphatic and aromatic α-diazoketones give five-membered ring products by an ionic mechanism (Eq. 137).[222,223]

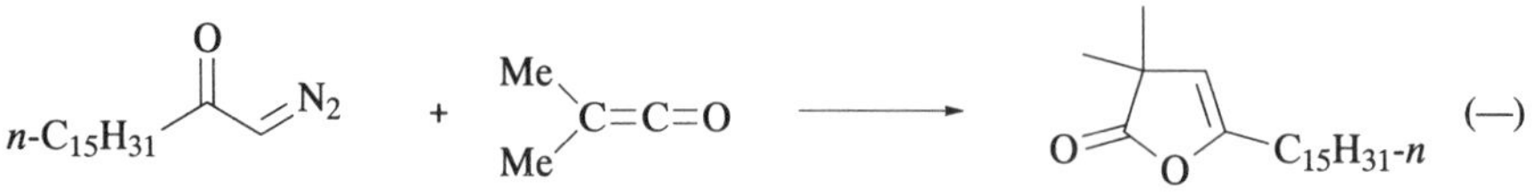

(—)

(Eq. 137)

The reaction of *n*-butyl azide with two molecules of diphenylketene is hypothesized to proceed by way of an α-lactam intermediate.[224,225] The reaction product of two moles of diphenylketene with trimethylsilyl azide is tetraphenylsuccinimide.[226] In both cases, an ionic stepwise mechanism, rather than a cycloaddition, was proposed.

Polyesters and ketones are the isolated products of the reaction of ozone with disubstituted ketenes (Eq. 138).[227–229] The reaction is proposed, by inference from the products, to proceed via a [3+2] addition followed by rearrangements.

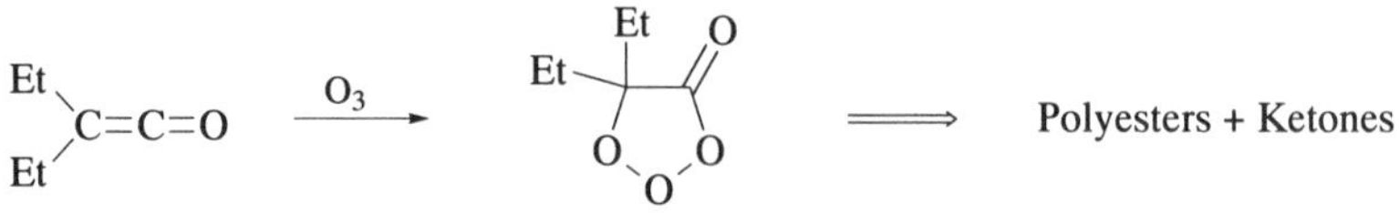

Polyesters + Ketones

(Eq. 138)

Other polar reactants can add to ketenes to give products based on a five-membered ring, as exemplified by the 3-oxidopyridinium betaine in Eq. 139.[230]

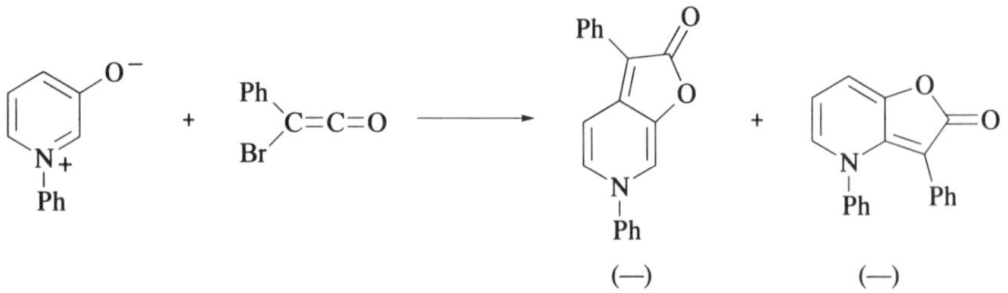

(—) (—)

(Eq. 139)

These processes are probably stepwise ionic reactions leading to cyclic products, rather than true cycloadditions.

[4+2] Cycloaddition of Ketenes

Dienes. The [4+2] cycloaddition of ketenes to simple dienes and dienic enol ethers to give pyran derivatives has not been developed to the point of general preparative utility. While butadiene reacts with bis(trifluoromethyl)ketene to provide a 90% yield of pyran (Eq. 140),[231,232] the reaction of cycloheptatriene

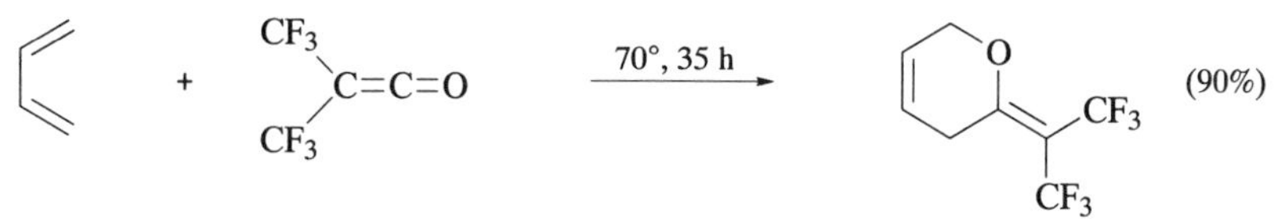

(90%)

(Eq. 140)

and diphenylketene to give a bicyclo[3.2.2] product proceeds in only 11% yield (Eq. 141).[233]

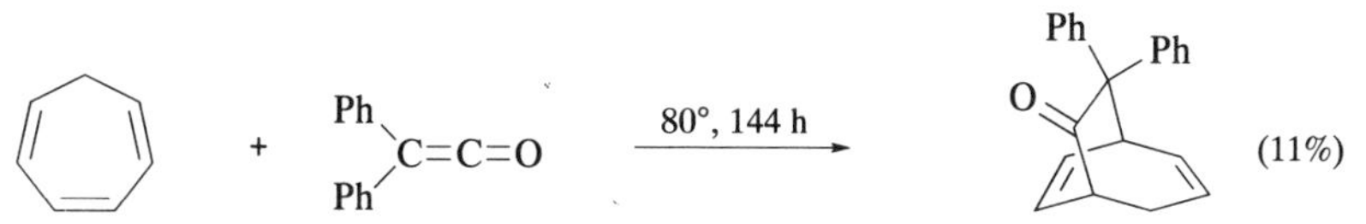

(11%)

(Eq. 141)

Prediction of product structure from such cycloadditions is complicated by the fact that either the ketene carbonyl group or olefinic linkage can serve as the two-electron component, and both modes of addition are seen in many reactions. Thus 3,3,4,4,5,5-hexamethyl-1,2-bis(methylene)cyclopentaene and diphenylketene give equal amounts of two cycloadducts as shown in Eq. 142.[234]

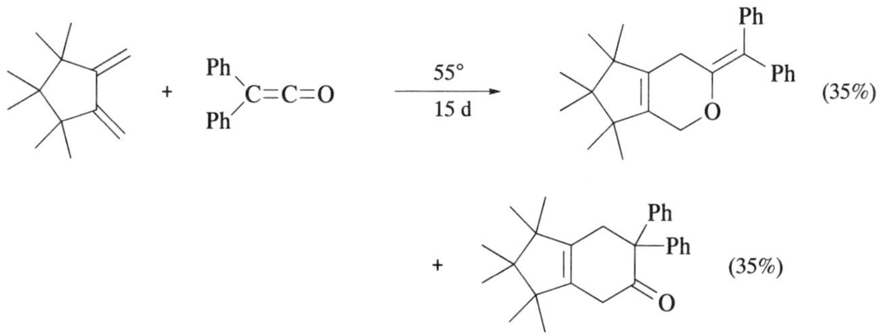

(35%)

(35%)

(Eq. 142)

Competition between [2+2] and [4+2] cycloaddition often gives rise to mixtures of cyclobutanone and pyran products, as is the case with dichloroketene and the bis(trimethylsilyl)diene of Eq. 143.[235,236] It is also known that diene geometry can control the outcome of ketene cycloadditions: the (*Z*)-thioether of Eq. 144 and diphenylketene give cyclobutanone, whereas the pyran is formed when the (*E*)-thioether is employed (Eq. 145).[237]

Moderate yields of [4+2] cycloadducts are obtained with haloketenes and silyloxydienes. The initial product is hydrolyzed in situ to give the isolable pyranone (Eq. 146).[238]

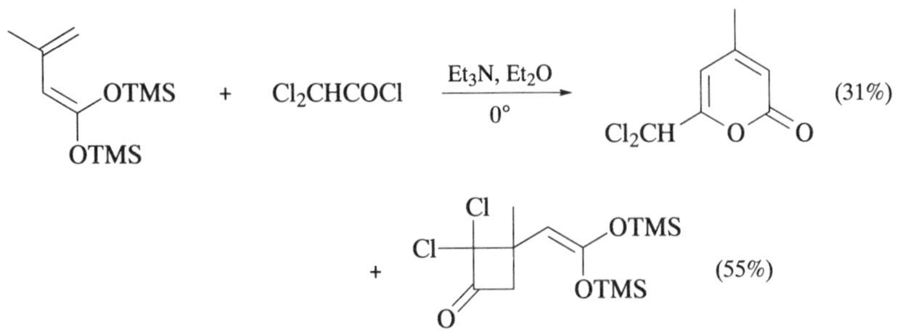

OTMS OTMS + $Cl_2CHCOCl$ $\xrightarrow[0°]{Et_3N,\ Et_2O}$ Cl_2CH O O (31%)

+ Cl Cl O OTMS OTMS (55%)

(Eq. 143)

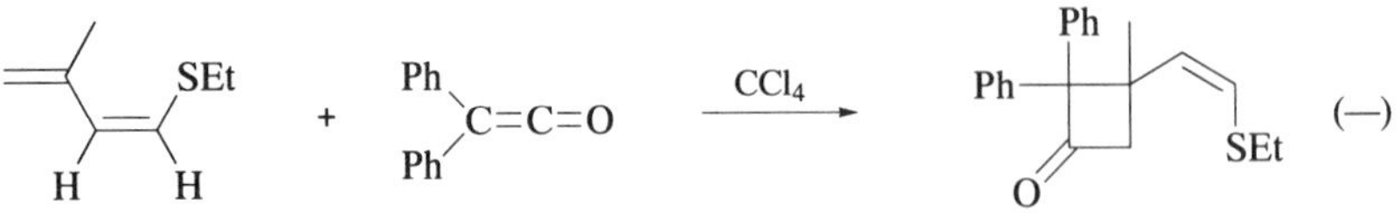

SEt H H + Ph Ph C=C=O $\xrightarrow{CCl_4}$ Ph Ph O SEt (—)

(Eq. 144)

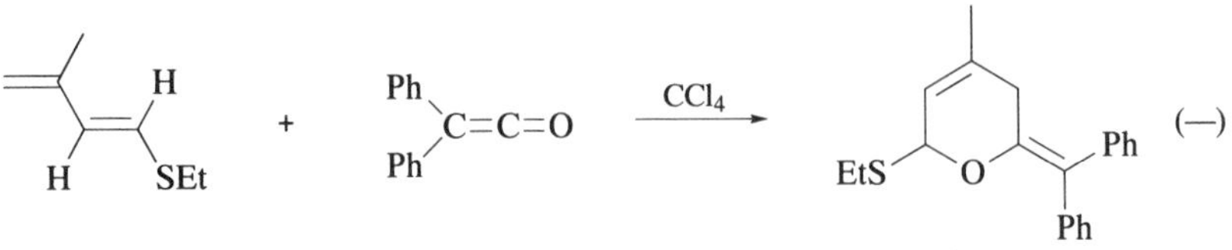

H H SEt + Ph Ph C=C=O $\xrightarrow{CCl_4}$ EtS O Ph Ph (—)

(Eq. 145)

OTMS H OTMS + $ClCH_2COCl$ $\xrightarrow{Et_3N}$ [OTMS ClCH O OTMS]

$\xrightarrow{H^+}$ O $ClCH_2$ O (55%)

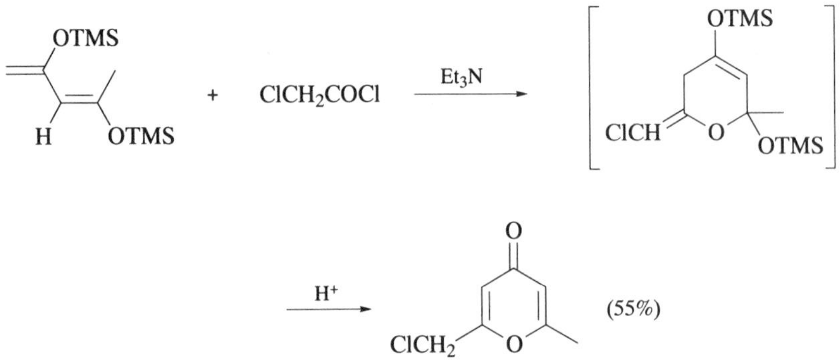

(Eq. 146)

Preparation of phenols by cycloaddition of ketenes to 2-pyranones followed by decarboxylation of the initial bicyclic product generally gives a poor yield (Eq. 147).[239,240]

$$\text{(pyrone)} + NCCH_2CO_2Et \xrightarrow[122\ h]{195\text{-}200^\circ} \text{(pyridinol)}\quad (13\%)$$

(Eq. 147)

Azadienes. Imines of α,β-unsaturated carbonyl compounds (1-azadienes) participate in [4+2] cycloaddition reactions with ketenes to produce dihydropyridones. The other possible regiochemistry of addition, leading to 3-ketodihydropyridines, is not seen. Although competing [2+2] cycloaddition to the imine moiety of azadienes occasionally leads to product mixtures containing β-lactams (Eq. 148), the [4+2] process usually proceeds in acceptable yield at ambient or slightly elevated temperatures in the absence of catalysts.

(—)

(Eq. 148)

Reactions of this type have been studied with three distinct classes of substrates. Linear azadienes formed from cinnamaldehydes and anilines react with ketene and haloketenes to give 3,4-dihydro-2-pyridones in yields typically around 65% (Eq. 149).[241,242]

$$\text{PhCH=CHCH=NC}_6\text{H}_4\text{Me-}p + Cl_2CHCOCl \xrightarrow{Et_3N} \text{(pyridone)}\quad (67\%)$$

(Eq. 149)

In the case of *tert*-butylcyanoketene generated from di-*tert*-butyldiazidoquinone in the presence of cinnamaldehyde-*tert*-butylimine, the *cis*-[4+2] adduct (52%) is accompanied by both *cis*- and *trans*-2-azetidinones (17% and 29%, respectively) (Eq. 150).[23] The ratio of [2+2] to [4+2] adducts in reactions of this type is controlled by the steric bulk on both the ketene and the azadiene; *tert*-butyl-cyanoketene and cinnamaldehyde *p*-methoxyaniline imine give only *trans*-azetidinone (Eq. 151).[23]

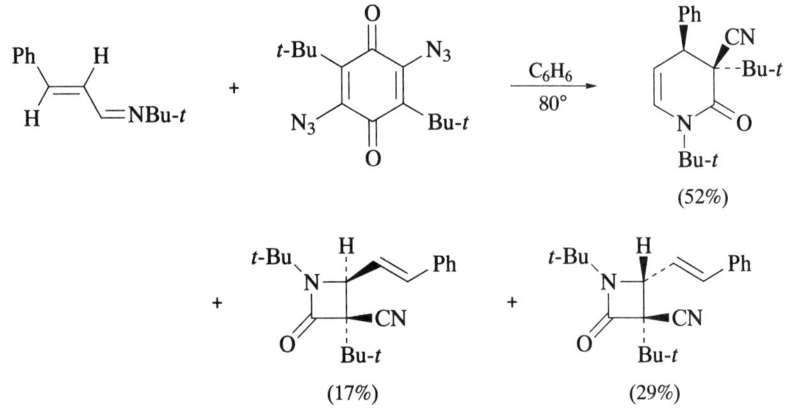

(Eq. 150)

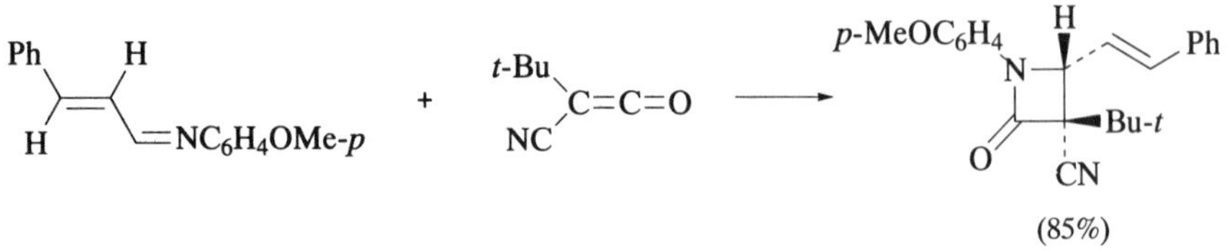

(Eq. 151)

A second class of ketene–azadiene cycloadditions involves 2-styryl-4,5-dihydrothiazoles (Eq. 152), which react with diphenylketene to give thiazolopyridones in good yield.[243]

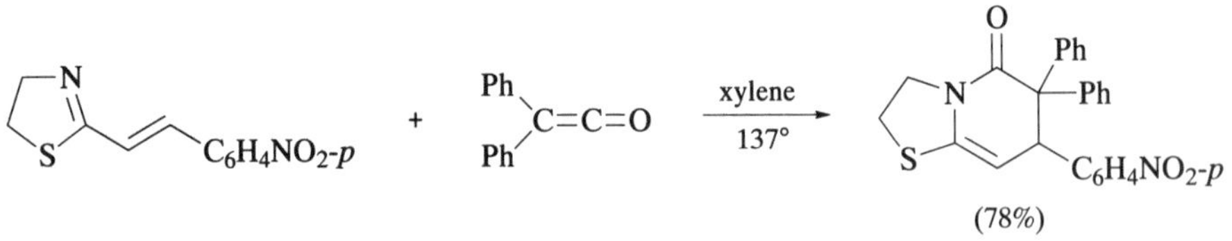

(Eq. 152)

The third group of 1-azadiene-ketene [4+2] cycloadditions utilizes imines of chromone-3-aldehydes, which with dihalo- and haloarylketenes give good yields of pyrido(4,5-*b*)chromones (Eq. 153). Elimination of HCl in situ provides the additional unsaturation in the products.[244]

It would appear that considerable scope for further development lies with [4+2] cycloadditions of ketenes to azadienes.

Amidines. The addition of ketenes to amidines (1,3-diazadienes) results in [4+2] cycloaddition to form 5,6-dihydro-4-pyrimidones. Reaction of the amidine

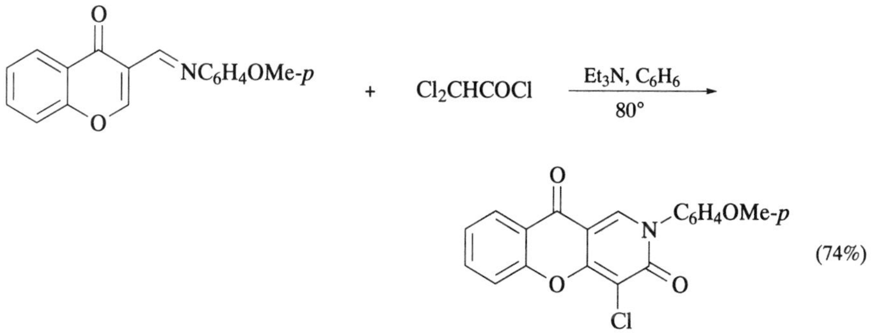

+ $Cl_2CHCOCl$ $\xrightarrow[80°]{Et_3N,\ C_6H_6}$ (74%)

(Eq. 153)

of Eq. 154 with diphenylketene, which results in a 52% yield of pyrimidone, is an example.[245]

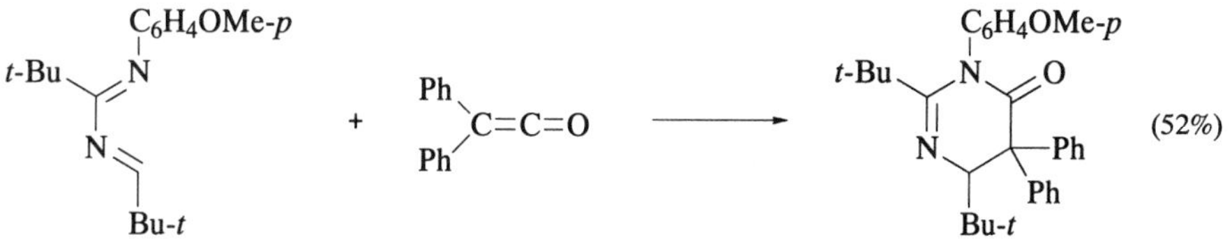

(52%)

(Eq. 154)

Cycloadditions of this type generally proceed in good yield; the orientation of addition is exclusively that leading to 4-pyrimidones. The alternative regiochemistry that would give dihydropyrimidine-5-ones has not been reported. Most recorded examples of this synthetically useful cycloaddition involve either ketenes or amidines bearing groups which may be eliminated to give 4-pyrimidones as the observed products. This process is typified by the reaction of furfuraldehyde 2-aminopyridineimine with dichloroketene, from which a 78% yield of a pyrido(1,2-*b*)pyrimidone is obtained (Eq. 155).[246]

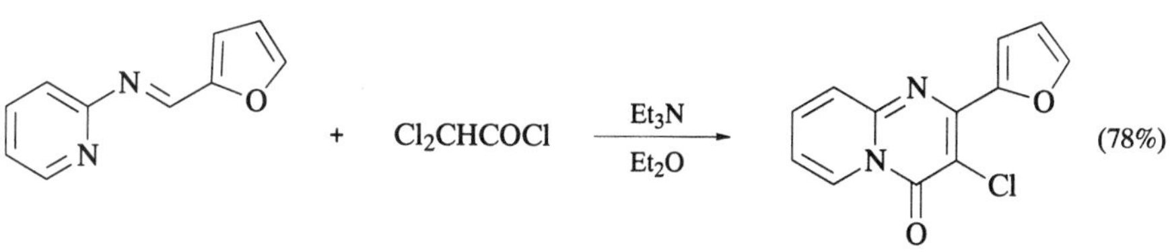

+ $Cl_2CHCOCl$ $\xrightarrow[Et_2O]{Et_3N}$ (78%)

(Eq. 155)

Few competing processes have been reported in these reactions. The only reference to byproducts arising from [2+2] cycloadditions is in the reaction of dichloroketene with imines derived from 2-aminopyridine: a 7% yield of β-lactam accompanied the 45% yield of pyrimidone (Eq. 156).[247]

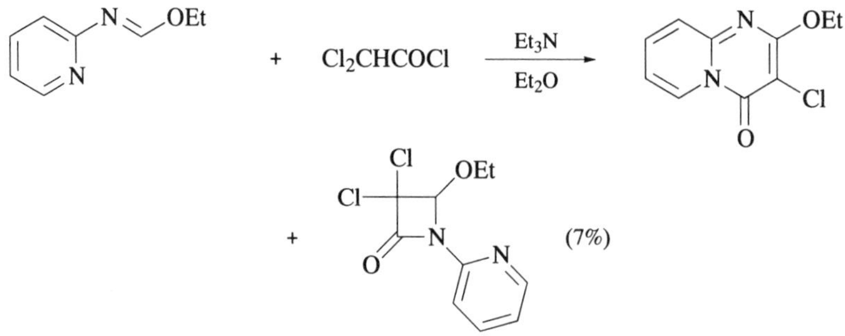

(Eq. 156)

Although amidines made from 2-aminopyridines are used in most examples of this class of ketene cycloadditions, other 2-aminoheterocycles such as the selenadiazole in Eq. 157 provide good yields of the predicted products.[248]

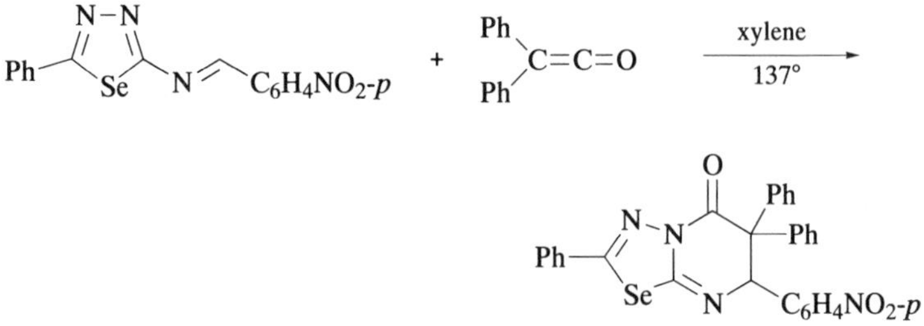

(Eq. 157)

It should also be noted that unusual amidines such as imine–carbodiimides (Eq. 158) react with diphenylketene to give [4+2] adducts unaccompanied by 4-iminoazetidin-2-one byproducts.[249]

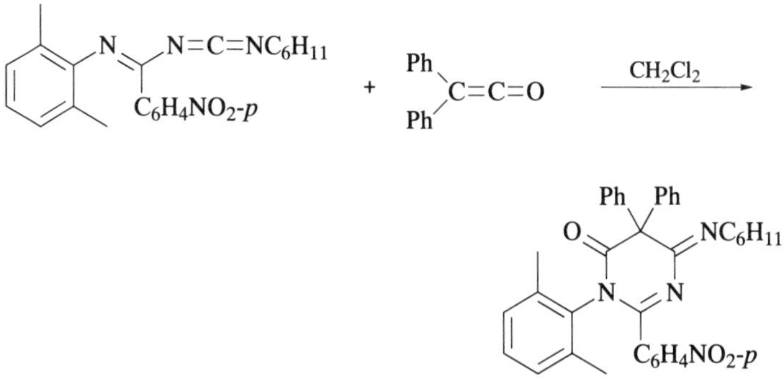

(Eq. 158)

***o*-Quinones.** *o*-Quinones undergo [4+2] cycloadditions with ketenes to give moderate to good yields of benzodioxenones. The addition of *p*-methoxyphenylketene to tetrabromo-*o*-quinone to give a 73% yield of dioxenone (Eq. 159) is typical.[250]

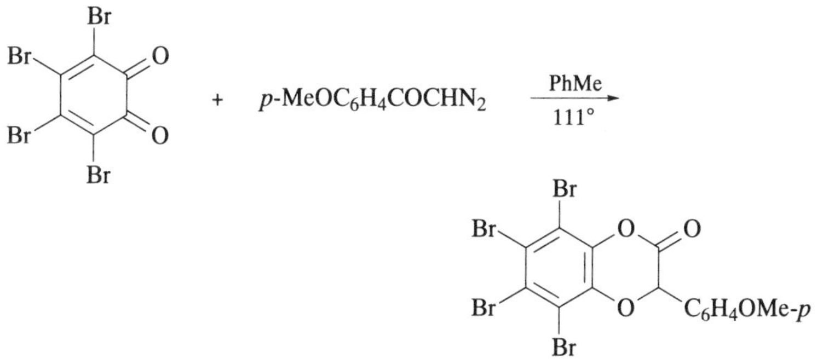

+ *p*-$MeOC_6H_4COCHN_2$ $\xrightarrow[111°]{PhMe}$ (73%) C_6H_4OMe-*p*

(Eq. 159)

Several reactions involving double cycloaddition of tetrahalo-*o*-quinones to bis(ketenes) have been reported (Eq. 160).[250]

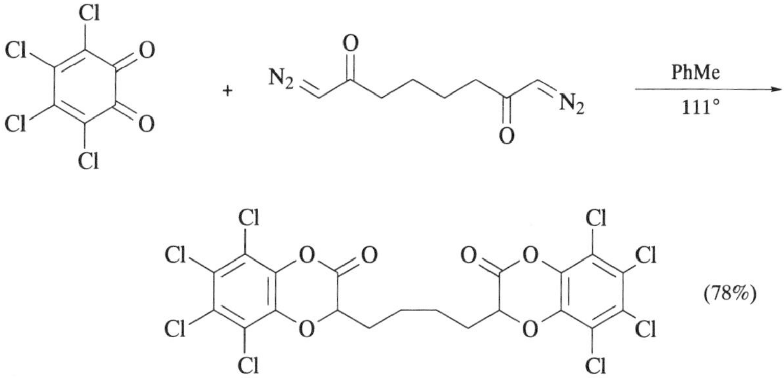

+ N_2 ... N_2 $\xrightarrow[111°]{PhMe}$ (78%)

(Eq. 160)

Analogous reactions with *o*-quinonimines give benzopiperidones (Eq. 161).[251] Some similar examples involve unsymmetrical quinonimines as in Eq. 162, and yield a mixture of the two possible cycloadducts.[251]

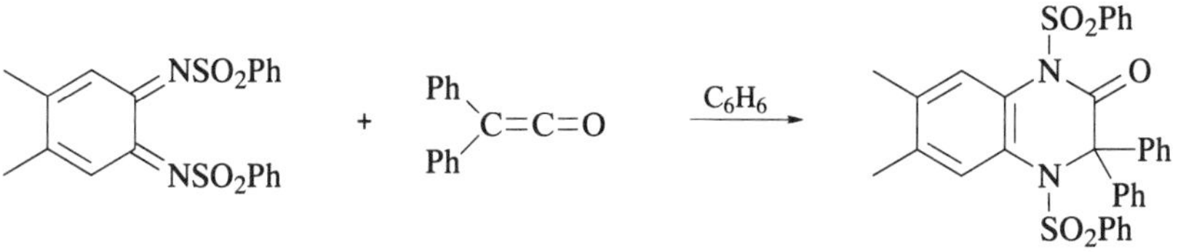

NSO_2Ph / NSO_2Ph + $Ph_2C{=}C{=}O$ $\xrightarrow{C_6H_6}$ (90%)

(Eq. 161)

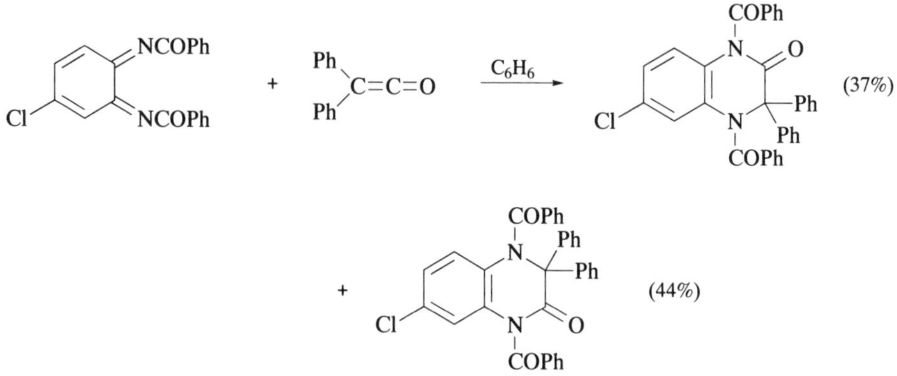

(37%)

(44%)

(Eq. 162)

There appear to be no published examples of the [4+2] cycloaddition of ketenes to α-diketones.

***α,β*-Unsaturated Ketones.** Simple acyclic enones have seldom been reported to undergo [4+2] cycloadditions with ketenes; [2+2] addition to the carbonyl group to yield β-lactones or olefins is the general result of such reactions. The reaction of the enone in Eq. 163 with diphenylketene to give a 36% yield of product is one of the higher-yielding examples.[252]

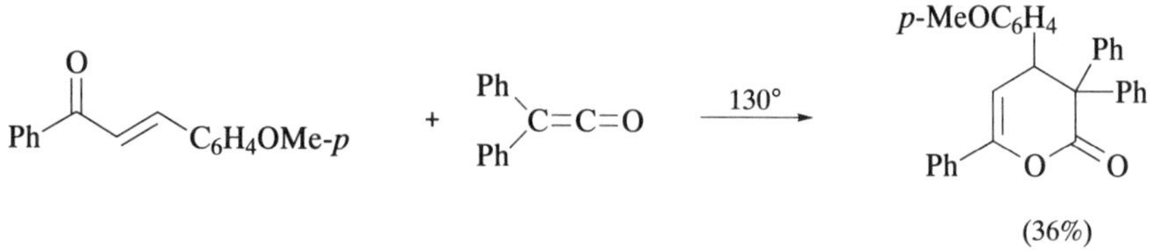

(36%)

(Eq. 163)

In the reactions of benzylidenindanone and acenaphthone with diphenylketene, mixtures of cycloadducts are formed (Eq. 164).[253,254]

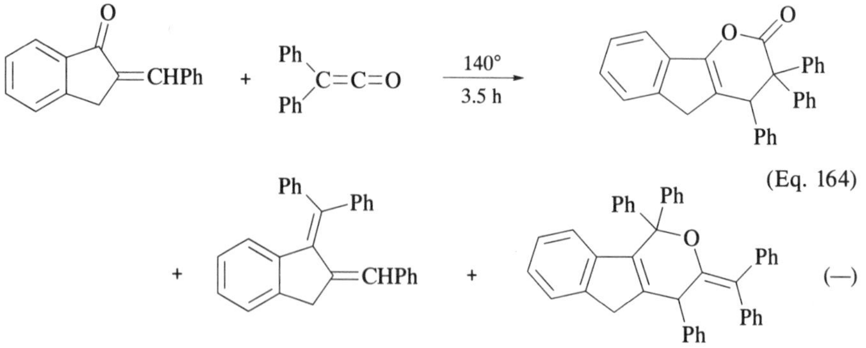

(Eq. 164)

(—)

In contrast to these results, the interaction of ketenes with β-amino and β-alkoxy-α,β-unsaturated ketones (enaminoketones, vinylogous amides, and esters) gives rise to synthetically useful yields of lactones (2,3-dihydro-2-pyranones) arising from [4+2] cycloaddition. Well over 100 examples are recorded; the process, typified by the reaction of the enaminoketone in Eq. 165 with

Me N Ph O + $Cl_2CHCCOCl$ $\xrightarrow{Et_3N}$ Me Ph—N Cl Cl O O

(92%)

(Eq. 165)

dichloroketene to give a lactone (92%),[255] can be carried out with various ketenes under mild conditions. Dichloroketene, ketene, and diphenylketene are most frequently employed.

The scope of this cycloaddition can best be appreciated through examination of the examples in Eqs. 166–168.[256–258]

O CHO O + Ph Ph C=C=O $\xrightarrow{130°}$ O O O O Ph Ph

(60%)[256]

(Eq. 166)

CF_3 CF_3 N Ph O O + CH_2=C=O $\xrightarrow{-70°}$ CF_3 CF_3 N Ph O O O

(60%)[257]

(Eq. 167)

S Pr-*i* N Pr-*i* O + $CHCl_2COCl$ $\xrightarrow{Et_3N}$ *i*-Pr N Pr-*i* Cl Cl S O O

(66%)[258]

(Eq. 168)

In the addition of ketene to β,β-diamino-α,β-unsaturated ketones, elimination of amine occurs in situ and the products are pyranones as illustrated in Eq. 169.[259]

$$+ \; CH_2{=}C{=}O \xrightarrow{Et_2O} \quad (51\%)$$

(Eq. 169)

α,β-Unsaturated Thiones. In exact analogy to the enaminoketone cycloadditions described above, β-amino-α,β-unsaturated thiones (vinylogous thioamides) react with ketenes to give good yields of thiapyranones (Eq. 170).[260,261]

$$p\text{-}ClC_6H_4 \; NMe_2 + CH_2{=}C{=}O \xrightarrow[1\ h]{Me_2CO} p\text{-}ClC_6H_4 \; S \; O \quad (78\%)$$

(Eq. 170)

In all reported examples of this cycloaddition, in situ elimination of either amine from the enethione substrate or chlorine from dichloroketene gives rise to a fully unsaturated thiapyranone. Equation 171 shows that reaction of an

$$p\text{-}ClC_6H_4 \; NMe_2 \xrightarrow[Et_3N,\ C_6H_6]{Cl_2C{=}C{=}O} p\text{-}ClC_6H_4,\ NMe_2,\ Cl \; (41\%)$$

$$p\text{-}ClC_6H_4 \; NMe_2 \xrightarrow[Et_3N,\ C_6H_6]{Ph(H)C{=}C{=}O} p\text{-}ClC_6H_4,\ Ph \; (38\%)$$

(Eq. 171)

α,β-unsaturated thione with a dichloroketene give chlorine elimination; the same thione reacts with phenylketene to afford deaminated product.[260]

A few examples of the [4+2] cycloaddition of diphenylketene to simple unsaturated thiones have been reported; the preparation of the thiolactone in Eq. 172 (51%) is typical.[262]

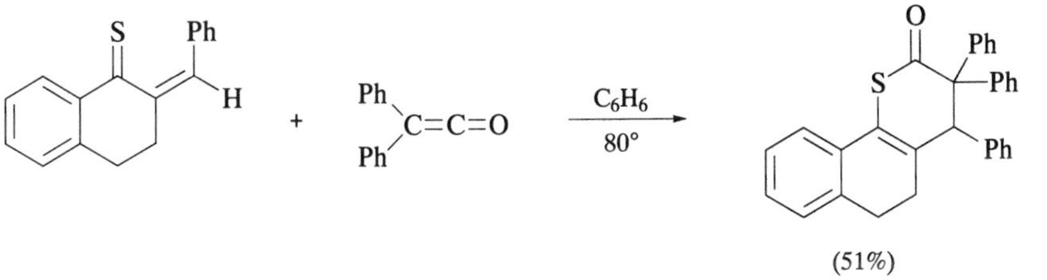

C_6H_6, 80°

(51%)

(Eq. 172)

Acyl Isocyanates and Isothiocyanates. The reaction of ketenes with acyl isocyanates yields 1,3-oxazin-4,6-diones, as illustrated by the reaction of trichloroacetyl isocyanate and dimethylketene as in Eq. 173.[188] The analogous thiazinediones are formed when ketenes react with thioacyl isocyanates, as in the reaction of thiobenzoyl isocyanate with diphenylketene which gives 78% yield (Eq. 174).[263] These reactions are carried out in ether or aromatic solvents in the absence of catalysts. In no case has the isolation of azetidinedione byproducts been reported.

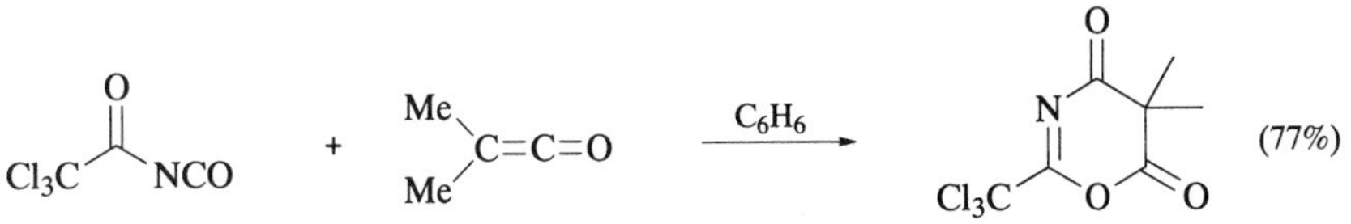

C_6H_6 (77%)

(Eq. 173)

Et_2O (78%)

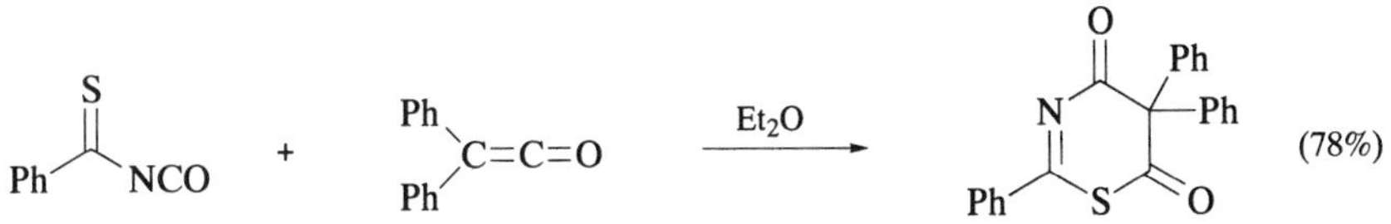

(Eq. 174)

Acyl isothiocyanates exhibit similar chemistry with diphenylketene, yielding 1,3-thiazine-4-thion-6-ones (Eq. 175).[264,265]

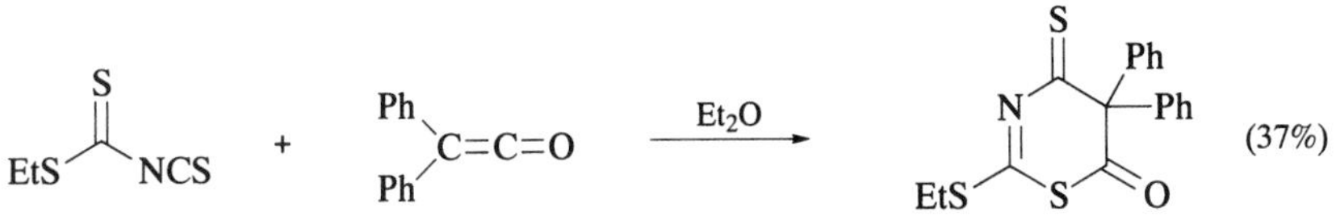

Et_2O (37%)

(Eq. 175)

Vinyl isocyanate and isothiocyanate undergo [4+2] addition to triphenylphosphorane ketene (Eq. 176) to yield the corresponding phosphoranes.[266] Reaction of these isocyanates with the more common ketenes has not been reported.

$Ph_3P{=}C{=}C{=}O$ → (NCO) (98%); (NCS) (94%)

(Eq. 176)

Unsaturated and Acyl Azo Compounds. Diethyl azodicarboxylate and diphenylketene undergo a [4+2] cycloaddition, but the initial 1,3,4-oxadiazin-5-one product reacts with a second mole of ketene to give the final bicyclic product (Eq. 177).[267]

$EtO_2CN{=}NCO_2Et + Ph_2C{=}C{=}O$ →

(—)

(Eq. 177)

The unsaturated azo compound of Eq. 178 gives a 72% yield of diazine with diphenylketene,[268] but other unsaturated azo compounds give mixtures of [4+2] and [2+2] products (Eq. 179).[268]

(72%)

(Eq. 178)

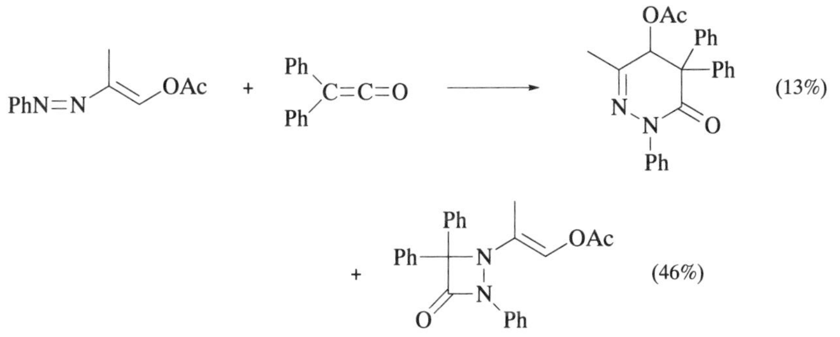

(13%)

(46%)

(Eq. 179)

Thioacyl Imines. Ketene and phenylketene react with thioacyl imines to give moderate yields of 1,3-thiazin-6-ones, as illustrated by the reaction of ketene with the imine of Eq. 180 to give a single adduct (40%); elimination of a mole of dimethylamine occurs spontaneously to give this product.[269]

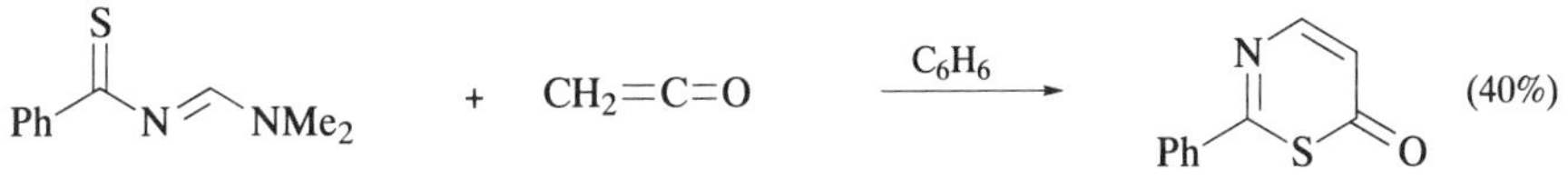

$CH_2{=}C{=}O$ $\xrightarrow{C_6H_6}$ (40%)

(Eq. 180)

Another example of this reaction type is the production of a spirocyclic product (80%) from diphenylketene and the imine of Eq. 181.[270,271]

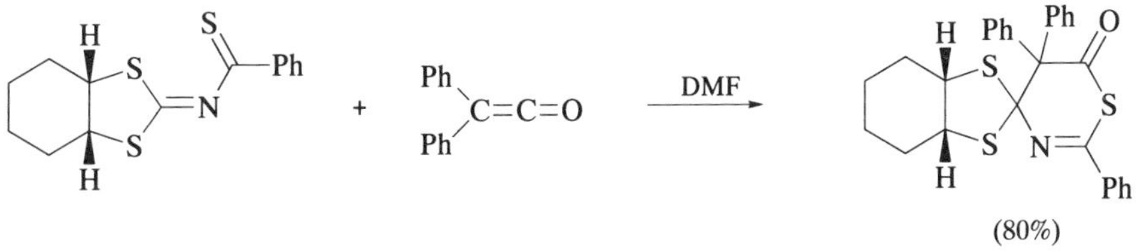

DMF (80%)

(Eq. 181)

Mesoionic Compounds. In a rare example of what appears to be a 1,4-dipolar cycloaddition, 6-oxo-3,6-dihydro-1-pyridinium-4-olates react with the carbonyl moiety of various ketenes to give good yields of the bicyclic heterocycles (Eq. 182).[272] The extension of this reaction to other dipolarophiles has not been reported, but the process proceeds as illustrated with a wide variety of ketenes.

[4+2] Cycloadditions of Acyl- and Vinyl Ketenes to Olefins. Acyl- and vinylketenes undergo facile [4+2] cycloadditions with olefins bearing either

+ $CH_2{=}C{=}O$ $\xrightarrow[20°]{MeCN}$ (83%)

(Eq. 182)

electron-donating or electronegative substituents. Reactions with simple alkenes have not been reported. The products are cyclohexenones (from vinylketenes) and dihydropyran-4-ones (from acylketenes).

Electron-poor olefins such as maleic anhydride and dimethyl fumarate react with the silylvinylketene (Eq. 183) to give good yields of 2-cyclohexenones.[273]

TMS C=C=O

$CHCl_3$, 40°, 12 h — TMS (89%)

(Eq. 183)

CO_2Me / MeO_2C — PhMe, 95°, 38 h — TMS CO_2Me CO_2Me (62%)

The stereochemistry of the products is that expected for typical Diels–Alder reactions. These promising results have not thus far been extended to other unsaturated ketenes.

Both acyl- and vinylketenes undergo [4+2] addition to enamines and enol ethers. Yields are generally good, as in the reaction of the acylketene from the dioxinone of Eq. 184 with butyl vinyl ether.[274,275]

$\xrightarrow{150°}$ [ ] $\xrightarrow{\text{OBu-}n}$ OBu-*n* (75%)

(Eq. 184)

The same ketene reacts with cyclohexenylmorpholine to give a [4+2] product in 32% yield; the initially formed cycloadduct undergoes elimination of morpholine at the high temperature needed to generate the ketene (Eq. 185).[274]

150°

(32%)

(Eq. 185)

The formation of [2+2] cycloaddition byproducts has not been reported in the reaction of acylketenes with enamines, but when vinylketenes are allowed to react with enamines, mixtures of [2+2] and [4+2] cycloadducts sometimes occur. Thus while isobutenylmorpholine and the ketene from 3-methylbutenoyl chloride give only a [4+2] product (67%), use of the isomeric 2-methylketene leads to formation of both products (Eq. 186).[276]

COCl

Et_3N, CH_2Cl_2

(67%)

COCl

(23%) (3%)

(Eq. 186)

[4+2] Cycloadditions of Acyl- and Vinylketenes to Acetylenes. A few examples of this cycloaddition have been recorded. Vinyltrimethylsilylketene and electron-poor acetylenes such as methyl propiolate give moderate yields of phenols (Eq. 187).[273] A single example of an acylketene–benzyne cycloaddition is in the literature.[277]

TMS(vinyl)C=C=O + HC≡CCO_2Me —PhMe, 95°, 63 h→ 2-hydroxy-3-TMS-benzoic acid methyl ester (45%)

(Eq. 187)

The other known acylketene–acetylene [4+2] cycloadditions all involve ynamines. Thus the ketene from the amido acid chloride in Eq. 188 reacts with 1-(diethylamino)propyne to give a 70% yield of the substituted 4-pyranone.[278]

Et_2NC(O)CH(Ph)COCl + MeC≡$CNEt_2$ —Et_3N→ 4-pyranone (Me, Ph, Et_2N, NEt_2) (70%)

(Eq. 188)

A number of organometallic ynamines undergo [4+2] reaction with bis(carboethoxy)ketene to give pyranones (Eq. 189).[279] It would appear that the area of

Me_3SnC≡CN(Me)Ph + $(EtO_2C)_2C$=C=O —Et_2O, 25°→ 4-pyranone (Me_3Sn, CO_2Et, Ph(Me)N, OEt) (—)

(Eq. 189)

ketene–acetylene [4+2] cycloadditions is worthy of further exploration, for the reported examples of this reaction demonstrate the efficient assembly of highly functionalized phenols and heterocycles.

[4+2] Cycloaddition of Acyl- and Vinylketenes to Aldehydes and Ketones. No examples of [4+2] cycloaddition of vinylketenes to aldehydes or ketones are in the literature, but a number of acyl ketenes do undergo such addition to give good yields of 1,3-dioxin-4-ones (Eq. 190). Byproducts arising from [2+2] additions do not appear in these reactions.

The greatest number of reported examples involve generation of the acylketene from adipoyl chloride in the presence of a carbonyl compound such as cyclohexa-

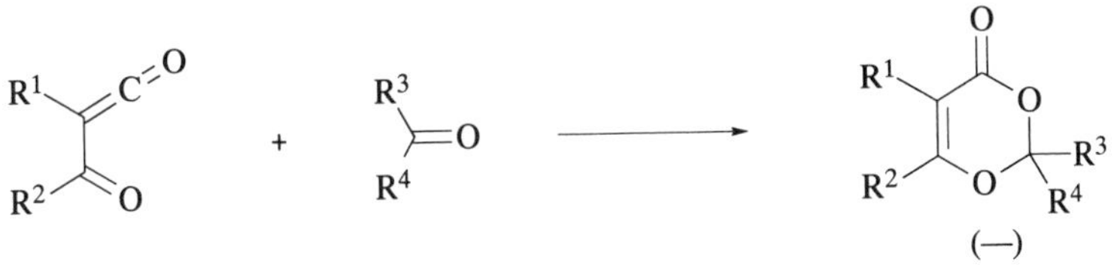

(Eq. 190)

none;[280] the same product is obtained in similar yield when the acylketene is generated from 2-diazocyclohexane-1,3-dione (Eq. 191).[280,281]

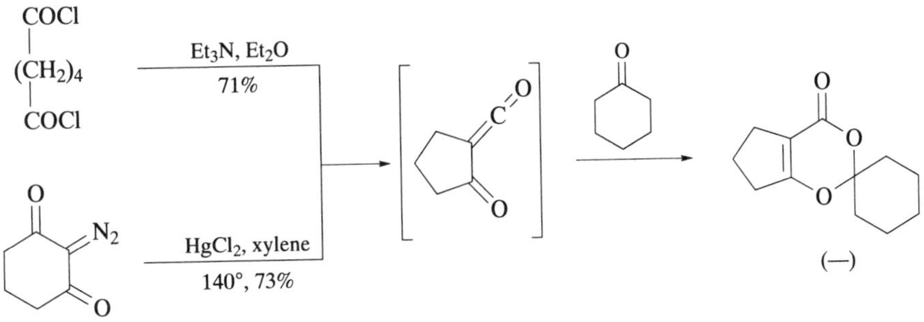

(Eq. 191)

It should be noted that the 1,3-dioxinone products of this cycloaddition are themselves capable of undergoing retro [4+2] fragmentation to regenerate the acylketene, which in the absence of trapping agents will dimerize or decompose. Thus the commercially available 2,2,6-trimethyl-4*H*-1,3-dioxin-4-one reverts to acetone and acetylketene above 100°; in the presence of cyclohexanone, a [4+2] adduct is formed in good yield. If no trapping agent is present, the dimer dehydroacetic acid is formed (Eq. 192).[282]

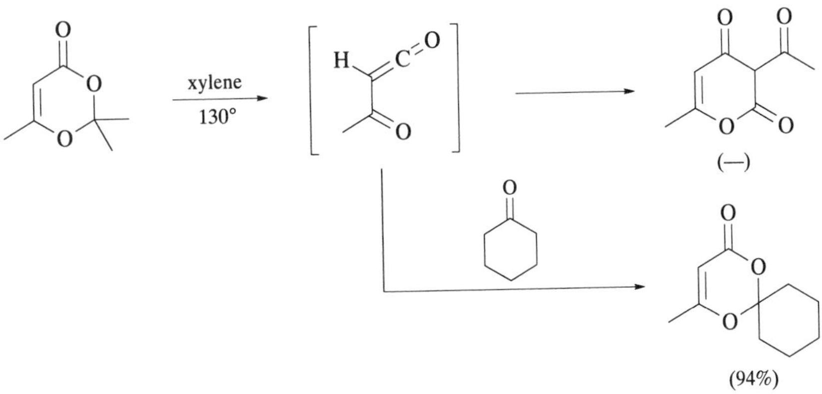

(Eq. 192)

A single example of trapping an iminoketene with a benzaldehyde to give the benzoxazine ring is known (Eq. 193).[277]

$$\text{iminoketene} + p\text{-MeOC}_6\text{H}_4\text{CHO} \xrightarrow{248^\circ} \text{benzoxazinone (NH, } \text{C}_6\text{H}_4\text{OMe-}p\text{) (38\%)}$$

(Eq. 193)

[4+2] Cycloaddition of Acylketenes to Nitriles and Cyanates. Nitriles, cyanamides, and cyanates participate in facile [4+2] cycloaddition reactions with acylketenes to yield derivatives of 1,3-oxazin-4-ones (Eq. 194).

$$R^1, R^2\text{-acylketene} + R^3CN \longrightarrow 1,3\text{-oxazin-4-one } (R^1, R^2, R^3)$$ (Eq. 194)

The requisite acylketenes are usually generated by cycloreversion of 1,3-dioxin-4-ones; the acetone byproduct is removed by distillation in the presence of the nitrile trapping agent. Thus the acetone–diketene adduct, when heated in the presence of dimethylcyanamide, provides an 85% yield of oxazinone (Eq. 195).[123,274]

$$\text{dioxinone} + Me_2NCN \xrightarrow{120^\circ} \text{oxazinone } (NMe_2) \ (85\%)$$

(Eq. 195)

Similarly, a fused dioxinone and m-chlorophenyl cyanate react in boiling xylene to give a [4+2] adduct (57%) (Eq. 196).[123,274]

$$\text{fused dioxinone} + m\text{-ClC}_6\text{H}_4\text{OCN} \xrightarrow{140^\circ} \text{oxazinone } (\text{OC}_6\text{H}_4\text{Cl-}m) \ (57\%)$$

(Eq. 196)

Yields of 1,3-oxazin-4-ones are generally greatest with cyanates and cyanamides. Most nitrile examples involve substituted benzonitriles and give yields in the range 30–50%. Thus *p*-tolunitrile and acylketene precursor at 140° give a 35% yield of the 2-aryl-1,3-oxazin-4-one (Eq. 197).[274]

+ *p*-MeC_6H_4CN —140°→ (35%) C_6H_4Me-*p*

(Eq. 197)

The interesting chlorocarbonylketene from benzylmalonyl dichloride (Eq. 198) reacts with benzonitriles to give modest yields of 6-chloro-1,3-oxazin-4-ones.[283]

COCl, COCl, Ph —120°→ [Ph, C=O, Cl, O] —PhCN→ Ph, N, Cl, O, Ph (40%)

Ph⁄⁄CN → Ph, N, Cl, O, Ph (5%)

(Eq. 198)

This ketene, combined with cinnamonitrile, gives a low yield of cycloadduct (Eq. 198). This is the sole example of a [4+2] cycloaddition of an acylketene to a nitrile other than a benzonitrile.[283]

[4+2] Cycloadditions of Acyl- and Vinylketenes to Heterocumulenes. Although no examples have been published of the cycloaddition of a vinylketene with a heterocumulene, there exist over 100 examples of [4+2] addition of acylketenes to isocyanates, isothiocyanates, carbodiimides, and *N*-sulfinylamines. These reactions lead to generally good yields of highly substituted 1,3-oxazines, 1,3-thiazines, and 1,2,3-oxathiazines.

The isocyanate system has been explored in greatest detail. When 1,3-dioxin-4-ones undergo cycloreversion to acylketenes in the presence of isocyanates, 1,3-oxazin-2,4-diones are produced in generally good yields. Typical examples include the generation of acetylketene from trimethyldioxinone in the presence of an isocyanate (Eq. 199) to give the [4+2] product in 75% yield;[274] an analogous reaction is shown in Eq. 200.[274]

(Eq. 199)

(Eq. 200)

The requisite acylketenes can also be generated by pyrolysis of malonyl dichlorides, as in the preparation of a bis(oxazindione) in 88% yield (Eq. 201).[283]

(Eq. 201)

Isocyanate trapping of the acylketenes generated by Wolff rearrangement of 2-bromo-1,3-cyclohexanedione generally gives modest yields of [4+2] adducts; phenyl isocyanate provides only 25% yield.[284] When the 1,3-dioxin-4-one route to this ketene is used, the yield is 61% (Eq. 202).[274]

(Eq. 202)

Isothiocyanate–acylketene reactions proceed in analogy to the isocyanate examples, but yields are generally poor. Thus phenyl isothiocyanate provides only a 29% yield of the [4+2] adduct (Eq. 203).[274]

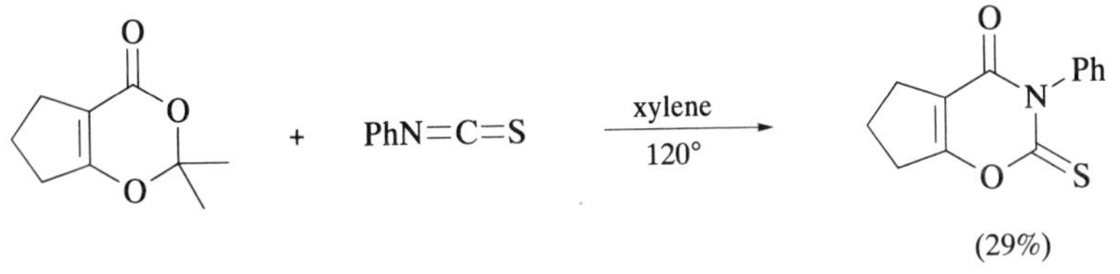

+ PhN=C=S —xylene, 120°→ (29%)

(Eq. 203)

When a dioxinone is heated in the presence of carbodiimides (Eq. 204), excellent yields of 2-imino-1,3-oxazin-4-ones result.[274]

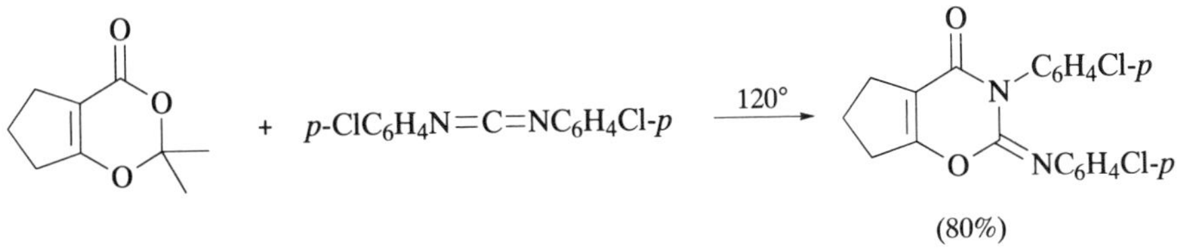

+ p-ClC$_6$H$_4$N=C=NC$_6$H$_4$Cl-p —120°→ (80%)

(Eq. 204)

Interestingly, two multifunctional thioacylketenes react cleanly with diisopropylcarbodiimide (Eqs. 205 and 206). In both reactions the sole reported products

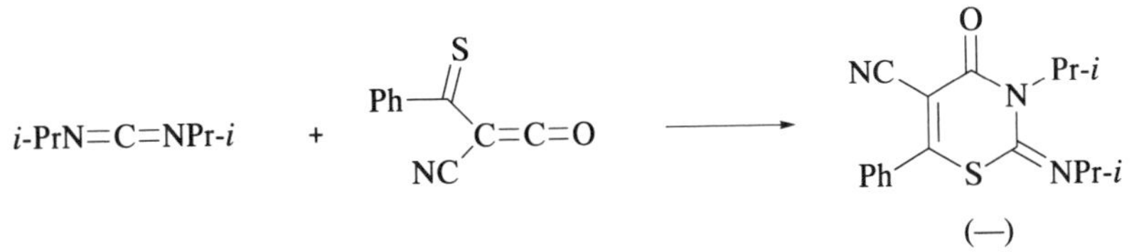

i-PrN=C=NPr-i + → (—)

(Eq. 205)

i-PrN=C=NPr-i + → (—)

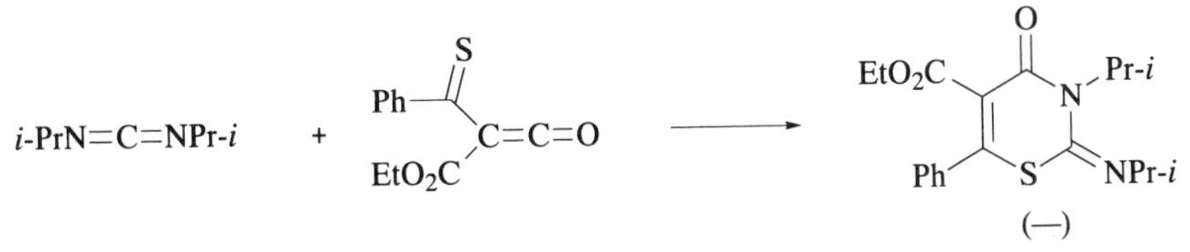

(Eq. 206)

arise from [4+2] addition to the thioacylketene functionality; the other possible modes of reaction involving cyanoketene and carboethoxyketene groups are not seen.[285]

Decomposition of 2-diazo-1,3-diketones in the presence of *N*-sulfinylamines leads to modest yields of 1,2,3-oxathiazin-4-ones. Higher yields of this heterocyclic system are obtained when acylketenes are generated by thermal decarbonylation of dihydrofuran-2,3-diones: compare the reactions of dibenzoyldiazomethane and naphthodihydrofurandione with *N*-phenyl-*N*-sulfinylamine to produce the corresponding cycloadducts in 38% and 74% yields, respectively (Eq. 207).[286,287]

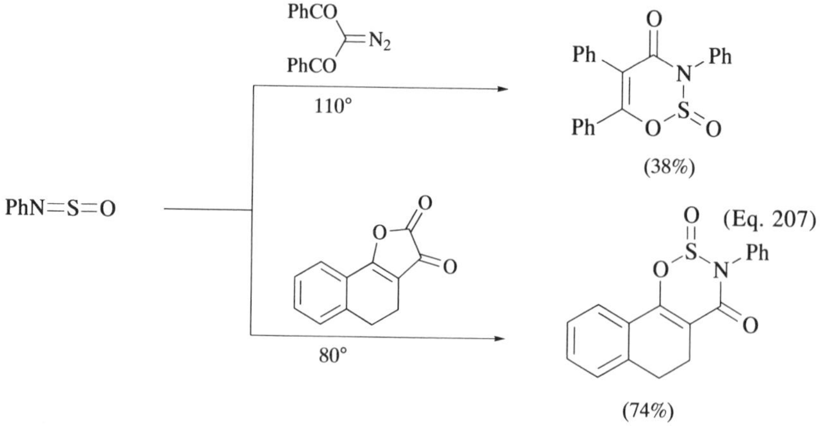

(Eq. 207)

[4+2] Cycloaddition of Acyl- and Vinylketenes to Imines and Azo Compounds. The methyl(dimethylvinyl)ketene formed upon irradiation of a parent diazoketone reacts with a diazacyclopentadiene to give the [4+2] adduct in 46% yield (Eq. 208). No products arising from cycloaddition to the olefin or carbonyl

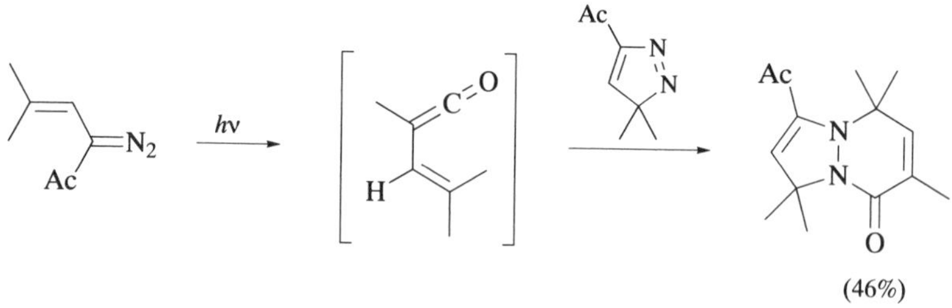

(Eq. 208)

moieties of the diazo reactant are reported.[288] This is the sole literature example of a vinylketene-azo cycloaddition.

Imines undergo [4+2] cycloadditions with acylketenes but the reaction of diphenylketene with phenylbenzoazete to give the [4+2] adduct (30%) is the only reported reaction which could be considered a vinylketene–imine [4+2] cycloaddition (Eq. 209).[289]

Benzalaniline and an acylketene (Eq. 210) yield a mixture of [4+2] and [2+2] adducts.[281,290] In all other reported cases, [4+2] cycloaddition occurs without

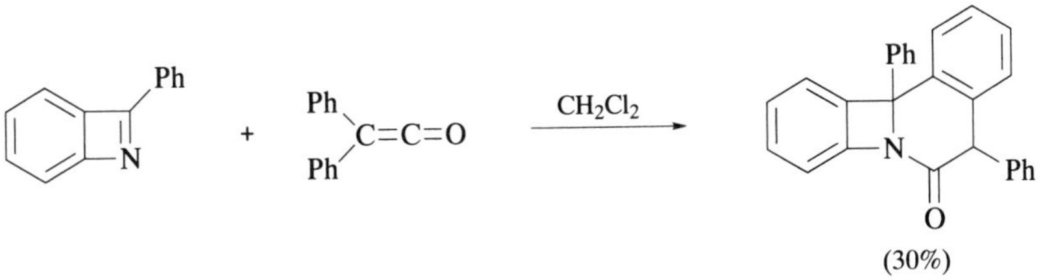

(30%)

(Eq. 209)

(25%) (17%)

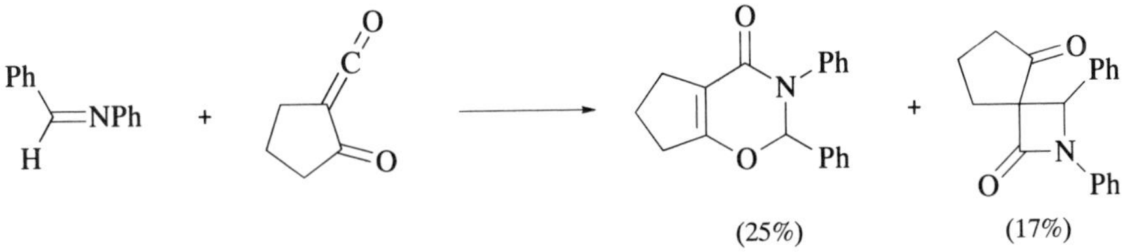

(Eq. 210)

competing [2+2] reaction. The benzodiazepine of Eq. 211, for instance, gives a 72% yield of adduct when heated with dibenzoyldiazomethane, and double cy-

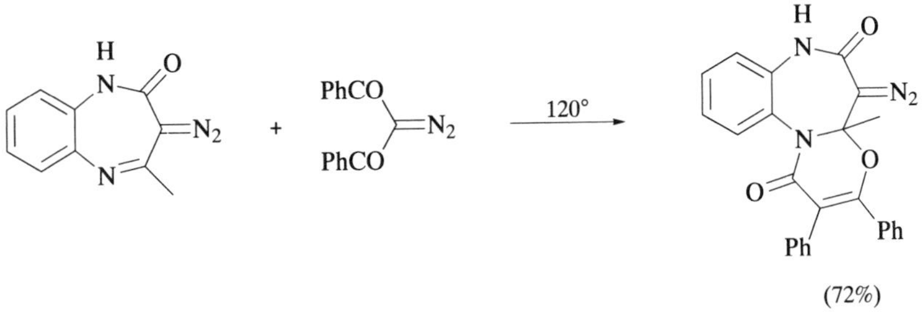

(72%)

(Eq. 211)

cloaddition of the same ketene to a suitable diazepine (Eq. 212) leads to a reasonable yield of adduct.[291]

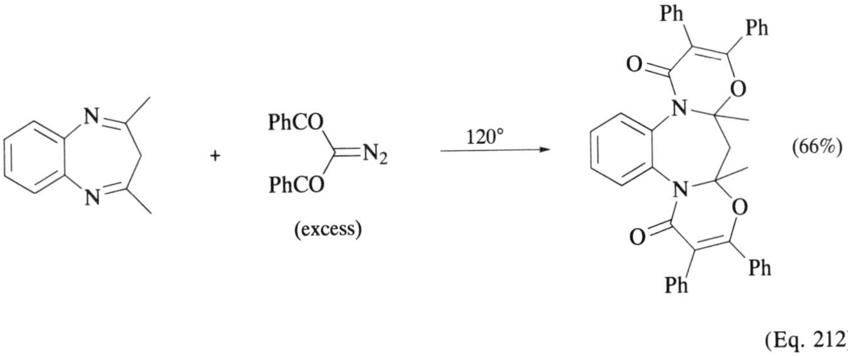

(excess)

(66%)

(Eq. 212)

EXPERIMENTAL PROCEDURES

General

Cycloadditions with reactants that are liquids at room temperature are best performed by simply mixing the two reactants without solvent. If one of the reactants is gaseous, it is more convenient to use a solvent. The progress of the reaction can be estimated by disappearance of the characteristic yellow color of the ketene, by loss of the band at about 2100 cm^{-1} in the infrared spectrum, or by ^1H NMR spectroscopy. The product is often separated from the dimer and other byproducts by chromatography or distillation. Ketene, monoalkylketenes, and dimethylketene are usually allowed to react at or below room temperature, whereas the higher molecular weight ketenes can be heated to temperatures above 100°. The ketene is usually used in excess when dimerization is a major side reaction. Dichloroketene is generated and allowed to react in situ, generally in the presence of halide salts, as described below. The success of the reaction is often determined by the relative rates of cycloaddition and dimerization of the ketene. Although polar solvents and catalysts accelerate the cycloaddition, they are not of general utility since they also accelerate dimerization.

Safety

In general, cycloadditions with ketenes do not require facilities and special training not already possessed by those likely to work with such compounds. Good ventilation is essential, because concentrated ketenes, especially those of low molecular weight, are highly and exothermically reactive, especially with themselves, and have a sharp, overpowering odor. Alone among the ketenes (to our knowledge), dimethylketene forms a dangerously explosive addition product with oxygen.[292] Dimethylketene can be used on a large scale with appropriate precautions; in situ generation from the dimethylketene acylal of dimethylmalonic acid is a practical way to avoid this problem.[293,294] Although oxidation of most other ketenes,[295] if it takes place at all, does not appear to generate explosive products, ketenes should be stored under nitrogen and precautions should be taken that are appropriate for compounds that form peroxides. Ketene itself can be stored neat for days at −78°[74] with some dimerization, but the monoalkylketenes, which can be prepared in concentrated form by pyrolysis of anhydrides, are exceedingly reactive. The dimerization of concentrated methylketene, even at −78°, may occur with such vigor as to vaporize the unreacted ketene with enough force to partially disassemble the apparatus.[110]

Preparation of Ketenes

Although early workers prepared ketene by the pyrolysis of acetone on the hot wire of a "ketene lamp,"[296] ketene is now best prepared in the laboratory by cracking commercially available diketene at atmospheric pressure in a hot tube, as described in a detailed procedure.[74] Ketene is prepared on an industrial scale by the pyrolysis of acetic acid. Mono- or dialkylketenes are made by dehydrohalogenation of an acid chloride with triethylamine in diethyl ether[297] or by dehalogenation of an α-halo acid halide with zinc or zinc–copper couple, as described below. The zinc dehalogenation method is applicable only to additions to olefins that are not susceptible to cationic polymerization. The zinc halide etherate formed in the reaction catalyzes the polymerization of olefins such as ethyl vinyl ether, styrene, furan, enol ethers, cyclopentadiene, and other conjugated dienes.[29] Various techniques described below have proven useful in sequestering the zinc salts and minimizing this problem. The more volatile alkylketenes are best prepared by pyrolyzing the corresponding anhydride in a hot tube, followed by bulb-to-bulb distillation to separate the ketene from carboxylic acid and uncracked anhydride.[56,74] Dichloroketene, which is generated and used in situ, is prepared either by dehydrohalogenation of dichloroacetyl chloride or by reduction of trichloroacetyl chloride. Superior results are obtained, especially with hindered olefins, if phosphorus oxychloride is used to sequester the zinc chloride.[298] A more recent procedure recommends the use of 1,2-dimethoxyethane to suppress rearrangement of sensitive adducts of allylic ethers.[299] Ultrasonic irradiation is claimed to accelerate dichloroketene reactions, resulting in shorter reaction times, better yields, and the ability to use ordinary instead of activated zinc. The detailed preparation of ketene,[74] diphenylketene,[39] and *tert*-butylcyanoketene[300,301] have been described. Less common, but valuable in specific cases, are ketene preparations by the pyrolysis of esters[302] (especially isopropylidene malonates;[293,303,304] thermal[305] or photochemical[306] ring opening of cyclobutenones; photorearrangement of 1-silyl-1,2-diones to yield silyloxyketenes;[307] pyrolysis of trimethylsilyloxyketene acetals;[308] photochemical cycloreversion of 9,10-bridged anthracenes;[309] cyanoketenes by the pyrolysis of 2,5-diazido-1,4-benzoquinones[23,301] or 4-azido-2-furanones;[310–312] and the Wolff rearrangement of α-diazoketones.[313–315]

2,2-Dimethyl-3-octylcyclobutanone (Preparation of Dimethylketene from the Dimethylketenacylal of Dimethylmalonic Acid).[316] A mixture of 528 g (4 moles) of dimethylmalonic acid and 1623 g (16 moles) of acetic anhydride was heated slowly to boiling at a pressure of about 10 torr. The acetic acid formed was removed continuously over an efficient fractionating column. The removal of the theoretical amount of acetic acid from the reaction mixture was followed by spontaneous evolution of CO_2, as indicated by a rise in pressure. The evolution of CO_2 was allowed to continue at atmospheric pressure (vented apparatus) and was complete within a few hours. Removal of the excess acetic anhydride by vacuum distillation afforded 300 g (80%) of a crystalline residue of the dimethylketenacylal of dimethylmalonic acid; mp 80° (from petroleum ether).

A mixture of 9.2 g (0.05 mol) of the dimethylketenacylal of dimethylmalonic acid, 50 g (0.36 mol) of 1-decene, and 100 mg of potassium carbonate was heated slowly to 130°. The carbon dioxide formed was removed through a reflux condenser. The reaction was complete after 6 hours. Distillation of the mixture in vacuo afforded 14 g of 2,2-dimethyl-3-octylcyclobutanone (67% yield based on the dimethylketenacylal); bp 88° (1 mm).

2,2-Dichloro-3,3,4-trimethylcyclobutanone (Dichloroketene via Zinc–Copper Dehalogenation of Trichloroacetyl Chloride in the Presence of Triethyl Phosphite).[317] To a mixture of 10.0 g (155 mmol) of zinc–copper couple and 5.96 g (85 mmol) of 2-methyl-2-butene in 75 mL of dry ether, stirred under argon at room temperature, was added over 45 minutes to a solution of 24.44 g (134.4 mmol) of trichloroacetyl chloride and 20.56 g (134.1 mmol) of phosphorus oxychloride in 75 mL of dry ether. The mixture was stirred overnight after which the ether solution was separated from the precipitated zinc chloride and added to hexane, and the resulting mixture was partially concentrated under reduced pressure in order to complete precipitation of the chloride. The supernatant layer was decanted and washed successively with a cold aqueous solution of sodium bicarbonate, water, and brine and then dried over anhydrous sodium sulfate. Evaporation of the solvent under reduced pressure followed by distillation gave 13.35 g (87%) of 2,2-dichloro-3,3,4-trimethylcyclobutanone, bp 75° (2 mm); IR 1805, 1460, 1180, 870, 805, 745, 685 cm^{-1}.

7,7-Dichlorobicyclo[3.2.0]hept-2-en-6-one [Dichloroketene (via Dehydrohalogenation of Dichloroacetyl Chloride with Triethylamine) and Cyclopentadiene].[318,319] To a vigorously stirred solution of 375 g (5.7 mol) of freshly distilled cyclopentadiene and 143.5 g (0.97 mol) of dichloroacetyl chloride was added 100 g (0.99 mol) of triethylamine over a period of 1.5 hours. After stirring for 15 hours under nitrogen, the mixture was filtered. Distillation afforded 124.3 g (72%) of 7,7-dichlorobicyclo[3.2.0]hept-2-en-6-one, bp 60–65° (2.5 mm); ^1H NMR ($CDCl_3$) δ 2.30–2.90 (m, 2H), 3.90–4.35 (m, 2H), 5.65–6.10 (m, 2H).

Bicyclo[3.2.0]hept-2-en-6-one. (Dechlorination of a 2,2-Dichlorocyclobutanone with Zinc/Acetic Acid).[318,319] To a vigorously stirred suspension of 261.7 g (4.00 mol) of zinc dust and 400 mL of glacial acetic acid was added dropwise at room temperature 124.3 g (0.70 mol) of 7,7-dichlorobicyclo[3.2.0]hept-2-en-6-one. After the addition was complete, the temperature was raised to and maintained at 70° for 40 minutes. The mixture was then cooled and treated with ether, and the zinc residue was removed by filtration. The ethereal layer was washed with a saturated solution of sodium carbonate to remove acetic acid and then dried with magnesium sulfate. Distillation afforded 61.5 g (81%) of bicyclo[3.2.0]hept-2-en-6-one, bp 60° (ca. 15 mm); ^1H NMR ($CDCl_3$) δ 2.30–2.80 (m, 3H), 3.10–3.55 (m, 2H), 3.65–3.95 (m, 1H), 5.60–5.85 (m, 2H).

3-Ethoxy-2,2-dimethylcyclobutanone (Cycloaddition of Dimethylketene, Prepared from Isobutyric Anhydride, to a Vinyl Ether).[320,321] CAUTION: Dimethylketene reacts readily with oxygen to form an explosive peroxide, and it dimerizes readily to a solid product which may plug passageways in an experimental apparatus.

Isobutyric anhydride was passed under a slow stream of nitrogen into an electrically heated 16 × 76 mm glass tube. The vaporized anhydride was conducted to a Vycor glass tube, 15 mm in diameter and 46 cm long, heated electrically with Nichrome alloy ribbon (1100 W). Temperatures were measured with a thermowell which extended through the entire length of the Vycor pyrolysis chamber. Gaseous products from the pyrolysis chamber passed through an efficient water-jacketed copper condenser, and thence through two glass cyclone separators. The residual vapors were then passed through a trap held at 50°, and conducted to a cold condenser and receiver to collect the dimethylketene, bp 34°. With the pyrolysis tube at 500–525° and a nitrogen stream flowing at 1.5 cubic feet per hour, isobutyric anhydride was passed through the system at the rate of 2400 mL per hour. During a period of 45 minutes, 1710 g of anhydride was introduced and 144 g of dimethylketene was collected, corresponding to a conversion of 19%.[320] (The apparatus described for the cracking of diketene[74] is more clearly described than the one in the patent procedure above, and has been successfully used by the authors to prepare dimethylketene.)

To 960 g (13.3 mol) of ethyl vinyl ether, stirred at room temperature under a nitrogen atmosphere, 600 g (8.6 mol) of dimethylketene was added over a period of 4 hours. The mixture was stirred for several hours. Distillation through a 12-inch Vigreux column gave 315 g (4.4 mol) of unchanged ethyl vinyl ether and 975 g (80%) of 3-ethoxy-2,2-dimethylcyclobutanone, bp 82–183° (38 mm); ^{1}H NMR: δ 3.82 (t, 1H), 3.47 (q, 2H), 3.08 (d, 1H), 2.98 (d, 1H), 1.2 (t, 3H), 1.12 (s, 6H).[321]

6-Methyl-2-oxa-1-phenyl-3,4-benzobicyclo[3.2.0]heptan-7-one (Intramolecular Cycloaddition).[322] A solution of (*o*-propenylphenoxy)phenylacetic acid (1.2 g, 4.5 mmol) in benzene (50 mL) was added over 5 hours through a syringe to a solution of triethylamine (2.3 g, 22.5 mmol) and *p*-toluenesulfonyl chloride (1.7 g, 9 mmol) in benzene (50 mL) at reflux. After the addition was complete, the mixture was heated at reflux for 6 hours. Upon cooling, the mixture was washed with water (3 × 50 mL) and concentrated to about 30 mL. This concentrate was stirred with 3% aqueous sodium hydroxide solution (250 mL) for 10 hours to remove excess *p*-toluenesulfonyl chloride. The benzene layer was dried with magnesium sulfate and filtered, and the benzene evaporated under reduced pressure. The residue was purified by column chromatography on silica gel (3% ethyl acetate in hexane) to give 0.9 g (83%) of 6-methyl-2-oxa-1-phenyl-3,4-benzobicyclo[3.2.0]heptan-7-one as a white solid, mp 95–96°; IR (neat) 1783, 1611, 1592 cm^{-1}. Anal. Calcd. for $C_{17}H_{14}O_2$: C, 81.58; H, 5.64. Found: C, 81.40; H, 5.61.

4,4-Dichloro-3-*n*-pentyl-2-cyclobuten-1-one (Cycloaddition of Dichloroketene to an Alkyne).[113] In a flame-dried, 100-mL three-necked flask equipped with argon atmosphere, stirrer, reflux condenser, and constant pressure addition funnel was placed 0.40 g (18 mmol) of activated zinc, 0.576 g (6 mmol) of 1-heptyne, and 50 mL of anhydrous ether. To this stirred mixture was added dropwise over 1 hour a solution of 1.79 g (12 mmol) of phosphorus oxychloride (freshly distilled from potassium carbonate), trichloroacetyl chloride (12 mmol), and 10 mL of anhydrous ether. The mixture was then stirred at reflux for 4 hours and the residual zinc removed by filtration on a pad of Celite. The ether solution was washed with water, 5% sodium bicarbonate solution, and brine, and dried over potassium carbonate. After removal of ether under reduced pressure, the product was purified by bulb-to-bulb distillation at 100° bath temperature (0.1 mm), to give 1.08 g (90%) of the title compound as a clear oil. IR v_{max} (neat) 1800, 1585 cm^{-1}; ^{1}H NMR ($CDCl_3$) δ 6.12 (m, 1H, J = 2 Hz), 2.7 (t, 2H, J = 6 Hz), 2.0–0.7 (m, 9H). Anal. Calcd. for $C_9H_{13}Cl_2O$: C, 52.19; H, 5.85. Found: C, 52.10; H, 5.79.

4-Ethyl-4-butyl-3-ethoxy-2-cyclobuten-1-one ([2+2] Cycloaddition of a Dialkylketene to an Acetylenic Ether).[124] To a stirred solution of 51 g (0.73 mol) of ethoxyacetylene in 150 mL of hexane at room temperature was added 92 g (0.73 mol) of butylethylketene. Stirring was continued for 6 hours at room temperature, then the solution was heated under reflux overnight. Distillation through a 10-in. Vigreux column gave 73 g (51%) of pure product boiling at 88° (0.8 mm), n_D^{20} 1.4665.

4-[2-(Ethylthio)propyl]-2-oxetanone ([2+2] Cycloaddition of Ketene to an Aldehyde).[323] A solution of 38.6 g (0.288 mol) of 3-(ethylthio)butyraldehyde and 1.5 mL of boron trifluoride etherate in 350 mL of ether was stirred at room temperature. Ketene gas was added through a frit at such a rate that the reaction temperature could be maintained close to 20° by an external cold water bath. After 1.5 hours IR analysis of an aliquot showed all aldehyde to be consumed and a strong β-lactone carbonyl band to be present at 1818 cm^{-1}. The mixture was purged with nitrogen, washed with 25 mL of saturated aqueous sodium bicarbonate solution, dried over anhydrous magnesium sulfate, and the ether removed under reduced pressure. Kugelrohr distillation (75–80°, 1.0 mm) gave 36.0 g (72%) of pure product. IR v_{max} (neat) 1818, 1110, 815 cm^{-1}; MS m/z 174 (calc, 174). Anal. Calcd. for $C_8H_{14}O_2S$: C, 55.1; H, 8.09. Found: C, 54.9; H, 8.14.

(*R*)-4-(1,1-Dichloroethyl)-2-oxetanone (Asymmetric Induction in [2+2] Cycloaddition of Ketene to an Aldehyde).[171] A 250-mL three-neck flask was equipped with thermometer, ketene inlet tube, and stirrer and charged with a solution of 389 mg (1.2 mmol) of quinidine and 13.4 g (105 mmol) of 2,2-dichloropropionaldehyde in 150 mL of toluene. The mixture was stirred at −25° and about 1 equivalent of ketene was bubbled into the mixture at approximately 30 mmol/hour. After completion of the ketene addition, the mixture was warmed to room temperature and the catalyst was removed by washing with 3 × 30 mL of 4 N HCl. The mixture was dried over magnesium sulfate and toluene was evaporated under reduced pressure, leaving a residue that was purified by flash column chromatography (dichloromethane elution) and bulb-to-bulb distillation (90°, 0.5 mm) to give 16.2 g (96%) of product; $[\alpha]_{578}^{RT}$ + 19.7°. Crystallization from methylcyclohexane gave 12.5 g (77%) of optically pure product, $[\alpha]_{578}^{RT}$ + 21.5° (c = 1, cyclohexane), mp 51.1–51.2°. IR v_{max} (neat) 1845 cm^{-1}; ^{1}H NMR ($CDCl_3$) δ 2.2 (3H, s), 3.6 (2H, d), 4.6 (1H, t). Anal. Calcd. for $C_5H_6O_2Cl_2$: C, 35.53; H, 3.58; Cl, 41.95. Found: C, 35.36; H, 3.57; Cl, 41.43.

α,α-Dichloro-β,β-bis(carboethoxy)propiolactone ([2+2] Cycloaddition of Dichloroketene to a Ketone).[324] A solution of 27 g (0.15 mol) of diethyl mesoxalate and 45.5 g (0.32 mol) of dichloroacetyl chloride in 100 mL of absolute ether was stirred at 10° during the dropwise addition of 31.2 g (0.31 mol) of triethylamine. After 1 hour the solution was filtered and the filtrate stripped of ether. The residue was distilled to give 32.5 g (76%) of product of bp 120° (1.5 mm). IR υ_{max} 1875 cm^{-1}. Anal. Calcd. for $C_9H_{10}Cl_2O_6$: C, 37.9; H, 3.8; Cl, 24.9. Found: C, 37.5; H, 3.6; Cl, 25.4.

3,3-Dimethyl-2-(2-methyl-3*H*-naphtho[1,8-*bc*]thiophene-3-ylidene)butyronitrile (Olefin Synthesis by Reaction of a Ketene with a Ketone).[325] A solution of 0.4 g (2 mmol) of 2-methyl-3*H*-naphtho[1,8-*bc*]thiophen-3-one and 0.302 g (1 mmol) of 2,5-diazido-3,6-di-*tert*-butyl-1,4-benzoquinone in 10 mL of dry benzene was heated at reflux in an argon atmosphere for 1 hour. The solvent was removed and the residue was chromatographed on a silica gel column with petroleum ether (bp 60–80°)/benzene (2:1). The major product recrystallized from hexane to give 0.209 g (75%) of product; mp 134°. IR υ_{max} (KBr) 3030, 2970, 2860, 2160 cm^{-1}. ^1H NMR $CDCl_3$): δ 1.57 (s, 9H), 2.72 (s, 3H), 7.07 (d, 1H, $J = 10$ Hz), 7.45 (d, 1H, $J = 10$ Hz), 7.31–7.84 (m, 2H). Anal. Calcd. for $C_{18}H_{17}NS$: C, 77.37; H, 6.13; N, 5.01; S, 11.47. Found: C, 77.17; H, 6.11; N, 5.06; S, 11.53.

***N,N'*-Phenylenebis(2,2-diethylmalonamide) ([2+2] Cycloaddition of a Ketene to an Isocyanate).**[188] A solution of 8 g (0.05 mol) of *p*-phenylene diisocyanate and 14.7 g (0.15 mol) of diethylketene in 100 mL of toluene was heated in a sealed tube at 180° for 18 hours. The reaction mixture was then evaporated to dryness and the solid crude product was washed with cyclohexane to give 16.3 g (91%) of product; mp 130° from ethanol. IR (CCl_4) υ_{max} 1748 cm^{-1}.

1,2-Di(*m*-methoxyphenyl)-4,4-diphenyl-1,2-diazetidinone ([2+2] Cycloaddition of a Ketene to an Azo Compound).[201] A solution of 0.61 g (2.5 mmol) of 3,3'-dimethoxyazobenzene and 0.51 g (2.5 mmol) of diphenylketene in 100 mL of benzene was irradiated for 0.3 hour through Pyrex with a General Electric Uviarc UA-3 lamp. Removal of benzene left a glassy residue that was extracted with four 50-mL portions of boiling hexane. Recrystallization of the residue from ethanol gave 0.72 g (69%) of product; mp 103–106°. IR υ_{max} 1780 cm^{-1}. Anal. Calcd. for $C_{28}H_{24}N_2O_3$: C, 77.04; H, 5.54. Found: C, 76.92; H, 5.55.

5-Diphenylmethylene-2,4-di-*tert*-butyl-1,3-oxathiolane ([3+2] Cycloaddition to Diphenylketene).[326] A solution of 1.80 g (9 mmol) of *trans*-2,5-di-*tert*-butyl-1,3,4-thiadiazoline and diphenylketene (2.10 g, 10.8 mmol) in 20 mL of methylcyclohexane was heated at reflux for 5 hours. After removal of solvent, the crude product was chromatographed over 100 g of silica gel using benzene as eluant to afford 3.11 g (94%) of 5-diphenylmethylene-2,4-di-*tert*-butyl-1,3-oxathiolane, mp 120–123°; IR (KBr) 1630, 1030, 1010, 700, 695 cm^{-1}. Anal. Calcd. for $C_{24}H_{30}OS$: C, 78.63; H, 8.27; S, 8.75. Found: C, 78.64; H, 8.26; S, 8.81.

3,6-Dihydro-2-[trifluoro-1-(trifluoromethyl)ethylidene]-2*H*-pyran ([4+2] Cycloaddition of Ketene to a Diene).[231] A mixture of 10.5 g (59 mmol) of bis(trifluoromethyl)ketene, 3.2 g (59 mmol) of 1,4-butadiene, and 0.01 g of hydroquinone was heated at 70° in a sealed ampule for 35 hours. Distillation of the reaction mixture yielded 12.5 g (90%) of the pyran product, bp 76–78° (15 mm). IR v_{max} (neat) 1628 cm^{-1}; ^{19}F NMR (CCl$_4$) δ 56.5 (q, $J = 9.9$ Hz), 59.8 (q, $J = 9.9$ Hz). ^{1}H NMR (CCl$_4$) δ 3.25 (m, 2H), 4.74 (m, 2H), 6.10 (m, 2H). Anal. Calcd. for $C_8H_6F_6O$: C, 41.4; H, 2.58. Found: C, 41.7; H, 2.64.

2-(*p*-Methoxyphenyl)-4-chloro-8-methyl[1]benzopyrano[3,2-*c*]pyridin-3,1(2*H*)-dione ([4+2] Cycloaddition of a Ketene to a 1-Azadiene).[244] To a solution of 0.789 g (3 mmol) of 6-methyl-3-(*p*-methoxyphenyliminomethyl)chromone and 0.91 g (9 mmol) of triethylamine in 200 mL of dry benzene heated at reflux was added a solution of 0.88 g (6 mmol) of dichloroacetyl chloride in benzene. The mixture was cooled and filtered, and the filtrate was retained. The residue was slurried with water to remove triethylamine hydrochloride and the resulting product was dried and added to that obtained by evaporating the benzene filtrate. The combined crude product was recrystallized from anisole to give 0.79 g (72%) of the title compound, mp 318°. Anal. Calcd. for $C_{20}H_{14}ClNO_4$: C, 65.3; H, 3.8; N, 3.8. Found: C, 65.7; H, 3.5; N, 3.8.

3-Chloro-2-ethoxy-9-methylpyrido[1,2-*a*]pyrimidin-4(4*H*)-one ([4+2] Cycloaddition of a Ketene to an Amidine).[247] A solution of 3.54 g (24 mmol) of dichloroacetyl chloride in 19 mL of dry 1,2-dimethoxyethane was added dropwise to a solution of 3.28 g (20 mmol) of ethyl *N*-(3-methyl-2-pyridyl)formimidate and 4.85 g (48 mmol) of triethylamine in 40 mL of dry dimethoxyethane with stirring at −15°. The mixture was then stirred for 5 hours at room temperature. Removal of the solvent under vacuum gave a residue that was dissolved in 100 mL of dichloromethane, washed with water, dried over sodium sulfate, and concentrated under vacuum. The residue was subjected to chromatography over 150 g of silica gel; elution with 2:1 hexane:ethyl acetate gave 3.85 g (80%) of the title compound, mp 165–166° (recrystallized from benzene). IR v_{max} (CHCl$_3$) 1675, 1635 cm^{-1}; ^{1}H NMR (CDCl$_3$) δ 1.50 (t, 3H), 4.60 (q, 2H), 8.97 (1H, dd). Anal. Calcd. for $C_{11}H_{11}ClN_2O_2$: C, 55.35; H, 4.65; N, 11.74. Found: C, 55.36; H, 4.60; N, 11.71.

3,3-Diphenyl-4-methoxy-3,4,5,6-tetrahydro-2*H*-naphtho[2,1-*e*]pyran-2-one ([4+2] Cycloaddition of a Ketene to an Enone).[327] A mixture of 1.1 g (5.67 mmol) of freshly distilled diphenylketene and 1.07 g (5.67 mmol) of β-methoxymethylene)-α-tetralone was heated at 82° until the IR spectrum of the mixture disclosed loss of the 2100 cm^{-1} ketene band. The mixture was allowed to cool to about 50° and 5–7 mL of petroleum ether was added. The resulting crystalline product was recrystallized from benzene/hexane to give 1.78 g (82%) of the pyranone, mp 161–162°. IR v_{max} (neat) 1760, 1600 cm^{-1}; ^{1}H NMR (CDCl$_3$) δ 2.30–2.85 (br, 4H), 3.10 (s, 3H), 4.20 (s, 1H), 7.21 (s, 10H). Anal. Calcd. for $C_{26}H_{22}O_3$: C, 81.68; H, 5.76. Found: C, 81.50; H, 5.81.

3-(*p*-Chlorophenyl)-3,4-dihydro-6-methyl-2*H*-1,3-oxazine-2,4-dione ([4+2] Cycloaddition of an Acylketene to an Isocyanate).[274] A stirred mixture of 28.4 g (0.2 mol) of 2,2,4-trimethyl-1,3-dioxin-4-one and 38.3 g (0.25 mol) of *p*-chlorophenyl isocyanate was heated at 120–130° for 15–20 minutes while the acetone formed was removed by distillation. The cooled residue was triturated with petroleum ether and the crude product was recrystallized from methanol to give 40.4 g (85%) of product; mp 215–217°. Anal. Calcd. for $C_{11}H_8ClNO_3$: C, 55.7; H, 3.4; N, 5.9; Cl, 14.9. Found: C, 55.7, H, 3.4; N, 6.2; Cl, 14.7.

TABULAR SURVEY

The search of published literature for ketene cycloadditions extends to the end of 1988, and a few later papers are cited. The patent literature search covers the same period, but the authors have included examples only from patents which in our experience or judgment ought to be reproducible.

There are a few entries in the tables for which it could be argued that the involvement of a free ketene reactant is unproven or even unlikely. Our basis for including such entries is that the reaction products are predictable if one assumes a ketene intermediate; whether or not a free ketene has a finite existence under the reaction conditions is of secondary importance to most preparative chemists.

A number of reaction starting material and product structures show a particular geometry of olefinic substitution. The reader should be aware that the original literature does not in every case rigorously establish such geometry; thus the original papers should be read critically by readers for whom this is an important point.

The tables follow the order of discussion in the Scope and Limitations section. Within each table, compounds reacting with ketenes are listed according to increasing carbon number, and increasing hydrogen number within a given carbon number. Carbon(s) and hydrogens residing in ester, ether, silyl ether, alkylamine, and other pendant groups are counted except in those cases where the availability of a series of closely homologous reactants allows creation of a subtable within a table. Yields are given in parentheses; a dash in the appropriate column indicates that yield(s) or reaction conditions were not reported.

The following abbreviations appear in the tables:

Ac	acetyl
Bn	benzyl
Bu	butyl
DMF	*N,N*-dimethylformamide
Et	ethyl
Me	methyl
Ph	phenyl
Pht	*o*-phthalyl
Pr	propyl
TBDMS	*tert*-butyldimethylsilyl
TMS	trimethylsilyl
Ts	*p*-toluenesulfonyl

TABLE I. KETENE DIMERS

Ketene or Ketene Source	Conditions	Product(s) and Yield(s) (%)	Refs.
C_2			
$Cl_2CHCOBr$	Zn, Et_2O or octane	(—)	7
$Cl_2CHCOCl$	Et_3N, n-C_6H_{14}, acenaphthalene	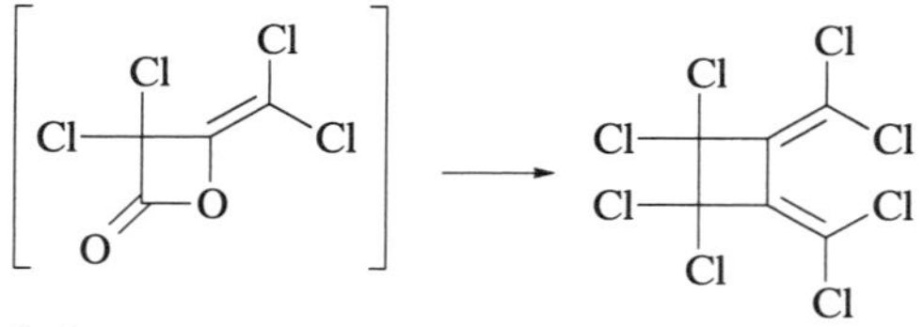 (—)	63
$BrCH_2COBr$	Et_3N, Et_2O, -78°	(—)	328
$ClCH_2COCl$	Et_3N, Et_2O	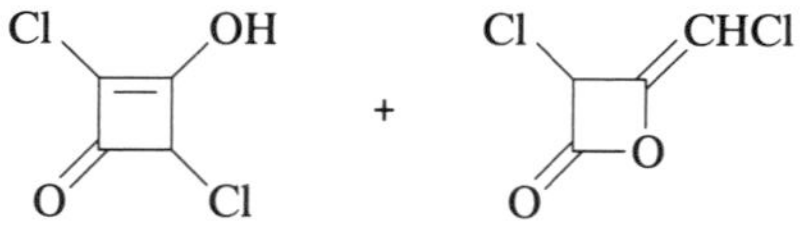 + (—)	329
$ClCH_2COBr$	Et_3N, Et_2O, -78°	(—)	328
FCH_2COCl	Et_3N, Et_2O, -78°	(—)	328
$F_5SCH_2CO_2H$	P_4O_{10}, 60-160°	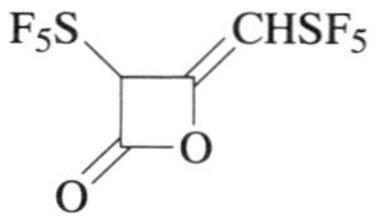 (—)	330
CH_2=C=O	(—)	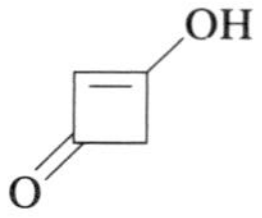 (4-5) + 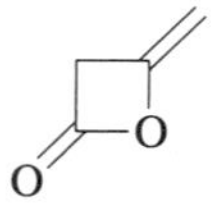 (88-90)	59

TABLE I. KETENE DIMERS (*Continued*)

Ketene or Ketene Source	Conditions	Product(s) and Yield(s) (%)	Refs.
XCH_2COX	Et_3N, Et_2O	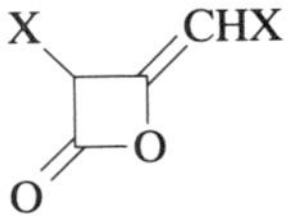 X, CHX; O, O; X = Br (40); X = Cl (50)	331
C_3			
MeCHBrCOBr	Zn, $ZnCl_2$, Et_2O, -5°	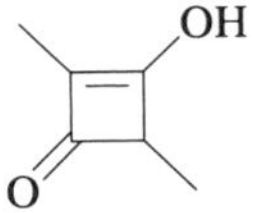 OH; O (5)	332, 333
	Zn, THF, -30°	" (88)	334
EtCOCl	Et_3N, Et_2O	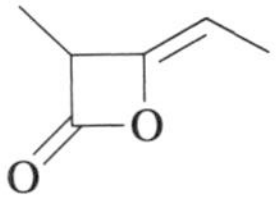 O; O (74)	331
$(EtCO)_2O$	Pyrolysis	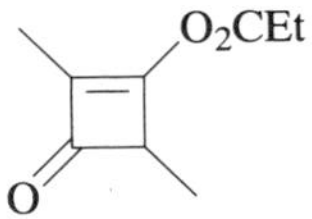 O_2CEt; O (76) 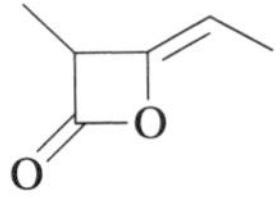 + O; O (13)	335
C_4			
$(CF_3)_2C{=}C{=}O$	—	(—)	65, 336
	Et_2NO	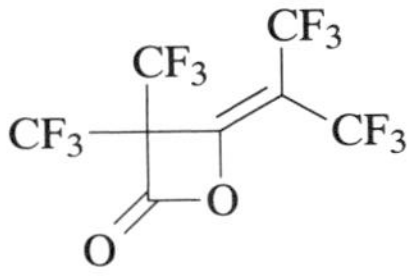 CF_3, CF_3, CF_3, CF_3; O; O (60)	66
	Et_3NF	" (91)	65

TABLE I. KETENE DIMERS (*Continued*)

Ketene or Ketene Source	Conditions	Product(s) and Yield(s) (%)	Refs.
	Et_3N	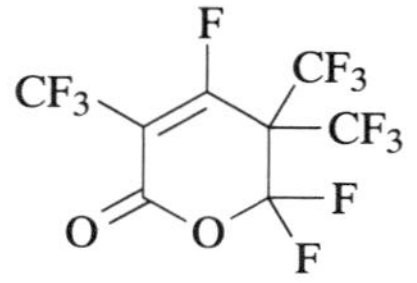 CF_3, F, CF_3, CF_3, F, F, O, O (79)	67
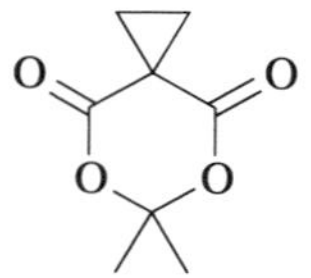 O, O, O, O	500°, 0.05 mm	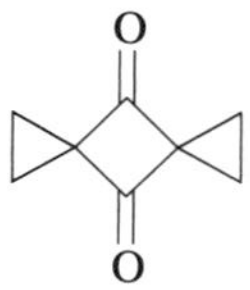 O, O (45)	304
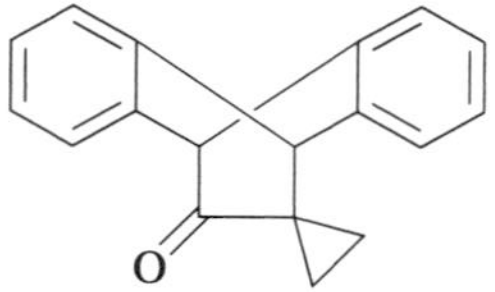 O	*h*ν (Corex), pentane	" (38)	309
CH_2=CHCH=C=O	rt	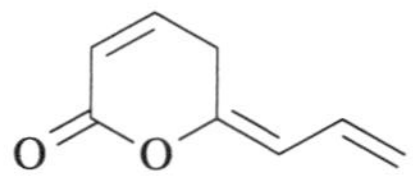 O, O (70)	337
EtCHBrCOBr	Zn, THF, -50°	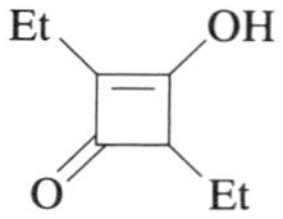 Et, OH, O, Et (19)	334, 332
n-PrCOCl	Et_3N, ligroin	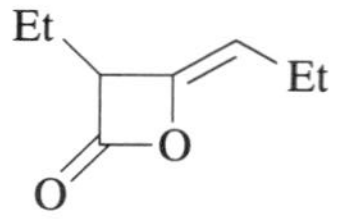 Et, Et, O, O (70)	331, 338

TABLE I. KETENE DIMERS (*Continued*)

Ketene or Ketene Source	Conditions	Product(s) and Yield(s) (%)	Refs.
$(n\text{-PrCO})_2O$	Pyrolysis	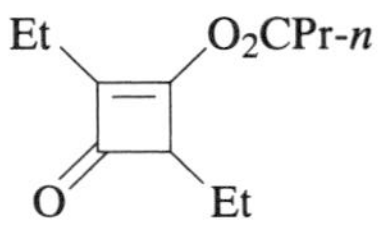 Et, O_2CPr-*n*, Et (40) 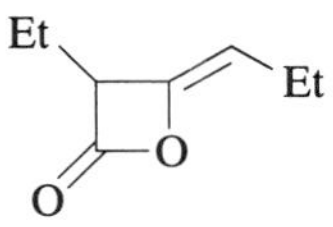 + Et, Et, O, O (10)	335
$Me_2CHCOCl$	Et_3N, CS_2	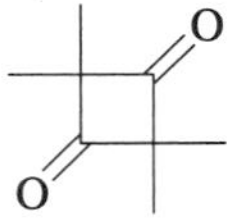 O, O **I** (18)	339
$Me_2C{=}C{=}O$	Various solvents	**I** (99)	40, 340, 341
	$P(OEt)_3$	**I** (4) 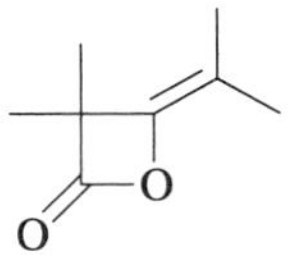 + O, O (93)	342
	$AlCl_3$, PhMe, -56°	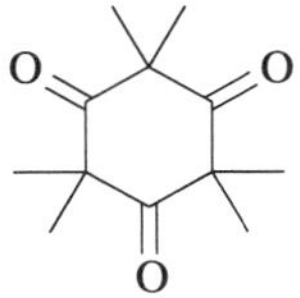 O, O, O (54)	62
C_5			
$Me_2C{=}CHCOCl$	Me_3N, hexane	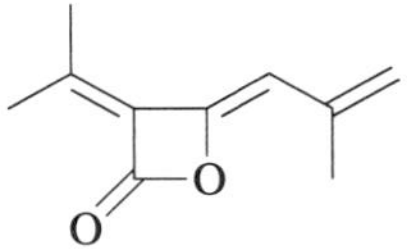 O, O (60)	343

TABLE I. KETENE DIMERS (*Continued*)

Ketene or Ketene Source	Conditions	Product(s) and Yield(s) (%)	Refs.
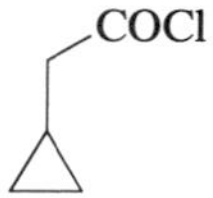 COCl	Et_3N	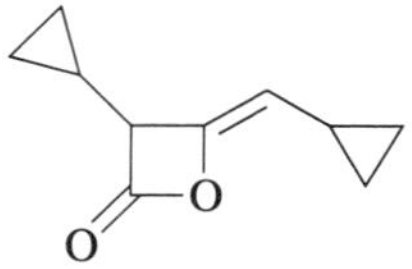 O O (—)	344
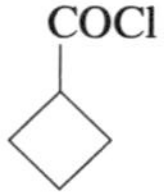 COCl	Et_3N, C_6H_6, Et_2O	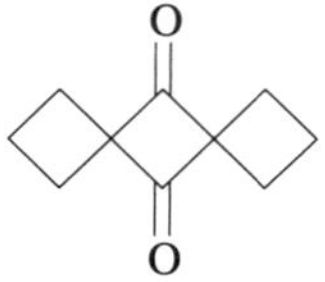 O O (70)	345
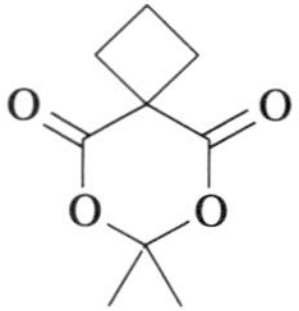 O O O O	430°, 0.003 mm	" (100)	304
n-BuCOCl	Et_3N, ligroin	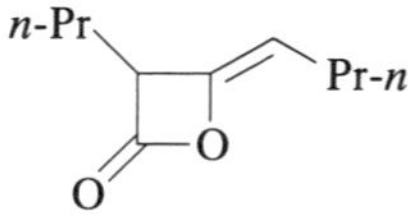 *n*-Pr Pr-*n* O O (93)	331
i-BuCOCl	Et_3N, Et_2O	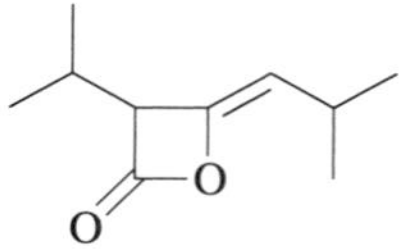 O O **I** (56)	331

TABLE I. KETENE DIMERS (*Continued*)

Ketene or Ketene Source	Conditions	Product(s) and Yield(s) (%)	Refs.
$(i\text{-PrCO})_2O$	Pyrolysis	**I** (9) 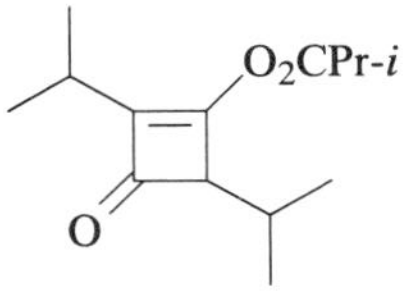 + [structure: O_2CPr-*i*] (57)	335
C_6			
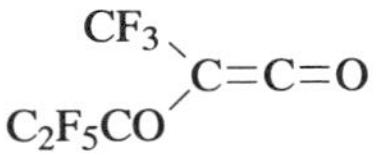 $CF_3(C_2F_5CO)C{=}C{=}O$	CsF (cat.), tetraglyme	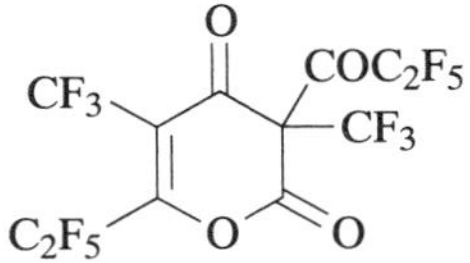 [structure: CF_3, C_2F_5, COC_2F_5, CF_3] (90)	346
	CsF (molar), tetraglyme	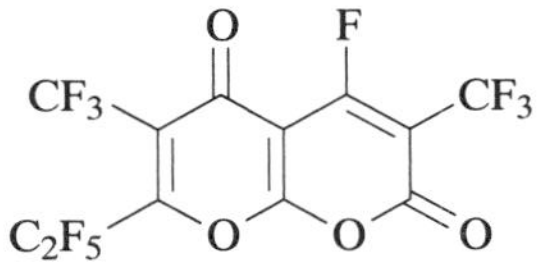 [structure: CF_3, C_2F_5, F, CF_3] (47)	346
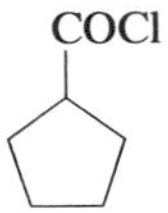 [structure: cyclopentyl-COCl]	Et_3N, Et_2O	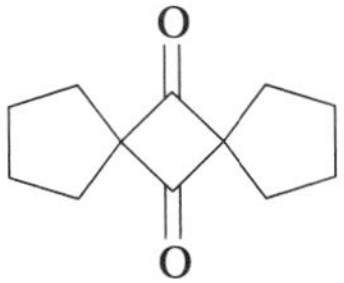 [structure] (85)	345
$t\text{-BuCH}_2COCl$	Et_3N	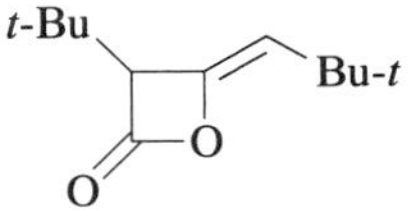 [structure: *t*-Bu, Bu-*t*] (30)	347

TABLE I. KETENE DIMERS (*Continued*)

Ketene or Ketene Source	Conditions	Product(s) and Yield(s) (%)	Refs.
$Et_2C{=}C{=}O$	25°, 48 d	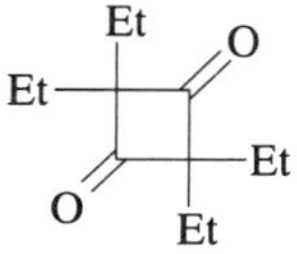 (83)	341, 348
n-$C_5H_{11}COCl$	Et_3N, ligroin	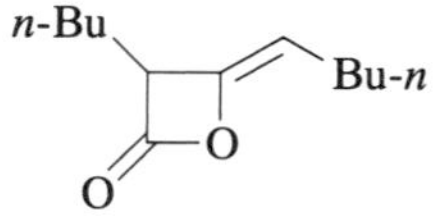 (40)	331
$ClOC(CH_2)_4COCl$	Et_3N, Et_2O	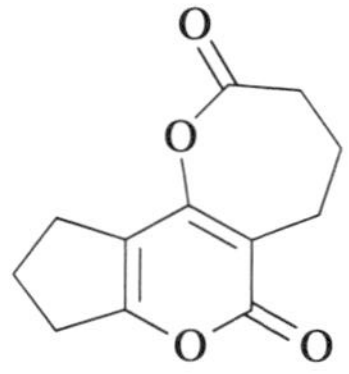 (4)	349
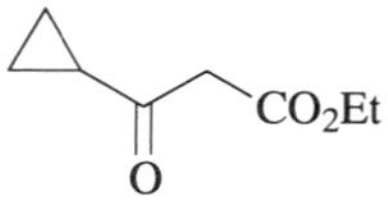 CO_2Et	410°	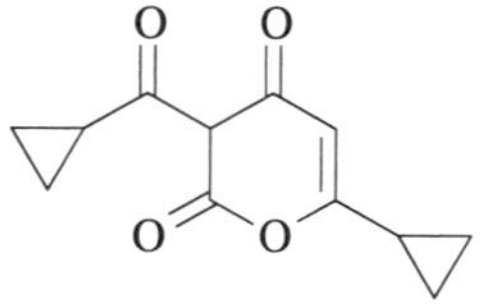 (79)	350
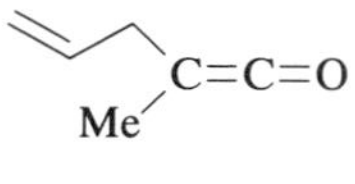 Me C=C=O	30°, 4 d	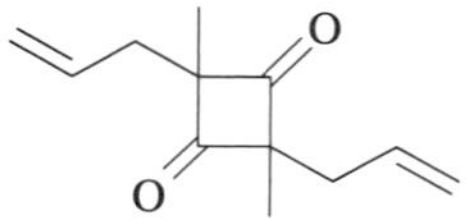 (78)	341

TABLE I. KETENE DIMERS (*Continued*)

Ketene or Ketene Source	Conditions	Product(s) and Yield(s) (%)	Refs.
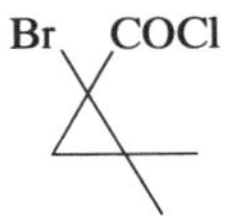 Br, COCl (cyclopropane)	$ZnCl_2$	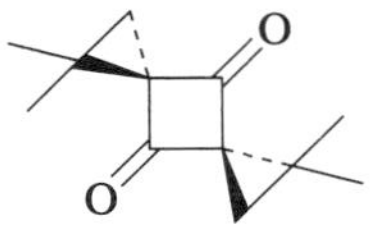 O, O (45)	351
$TMS(CH_2)_2COCl$	Et_3N, hexane	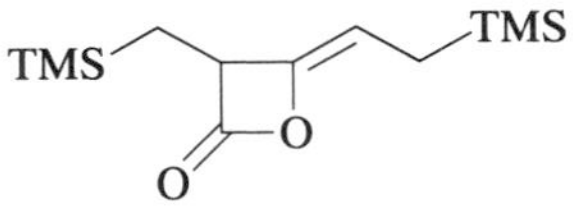 TMS, TMS, O, O (85)	352
	C_6H_6	(—)	353, 68
C_7			
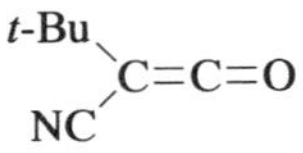 t-Bu, NC, C=C=O	Et_3N, C_6H_6	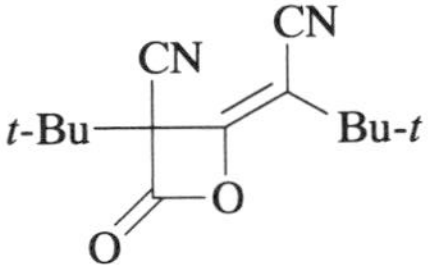 CN, CN, t-Bu, Bu-t, O, O (—)	68
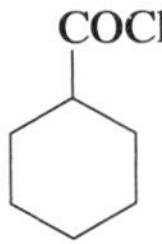 COCl (cyclohexane)	Et_3N, Et_2O	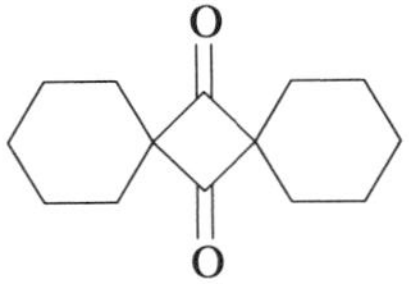 O, O (82)	345, 354, 355
C_8			
BnCOCl	Et_3N, Et_2O	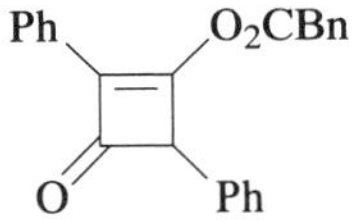 Ph, O_2CBn, O, Ph (67)	356, 71
PhCHClCOCl	Zn, Et_2O	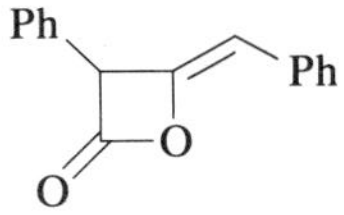 Ph, Ph, O, O (14)	357

TABLE I. KETENE DIMERS (*Continued*)

Ketene or Ketene Source	Conditions	Product(s) and Yield(s) (%)	Refs.
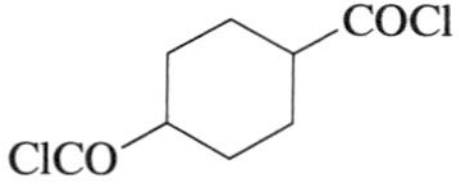 ClCO, COCl	Et_3N, C_6H_6	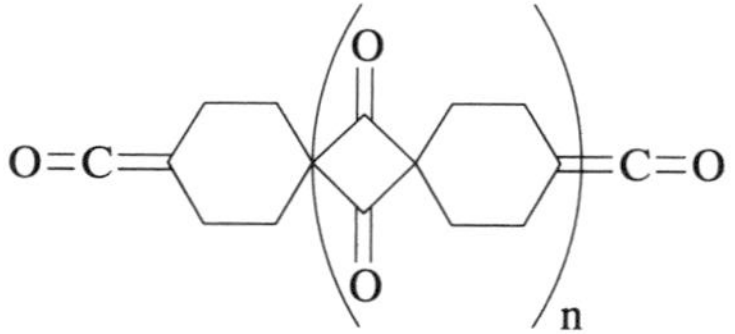 O=C=, =C=O, O, O, n; n = 1, 2, 3, 4 (—)	358
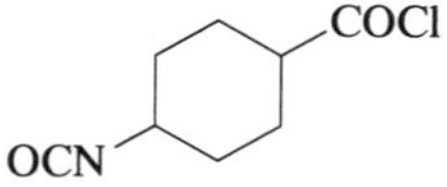 OCN, COCl	PhNHMe, dioxane, 80°	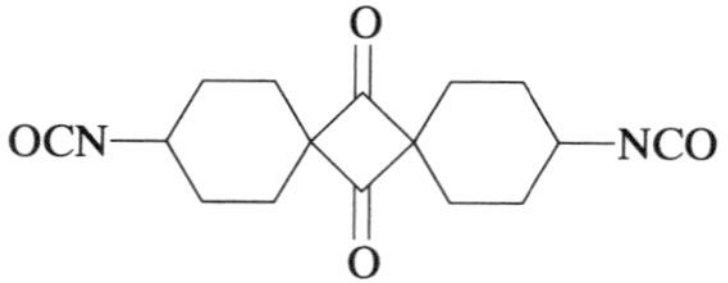 OCN—, —NCO, O, O (74)	359
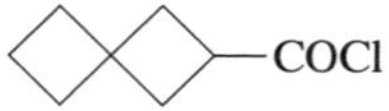 —COCl	Et_3N, C_6H_6	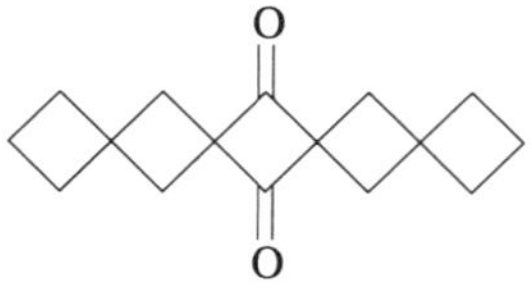 O, O (77)	360
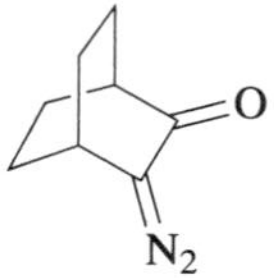 O, N_2	Cu, C_6H_6	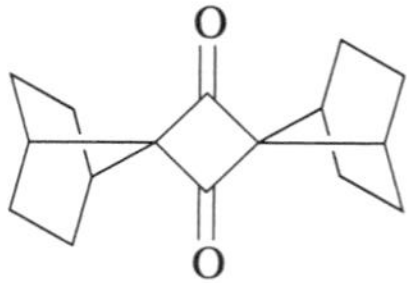 O, O (46)	313

TABLE I. KETENE DIMERS (*Continued*)

Ketene or Ketene Source	Conditions	Product(s) and Yield(s) (%)	Refs.
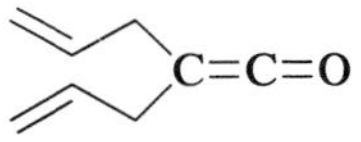 C=C=O	30°, 11 d	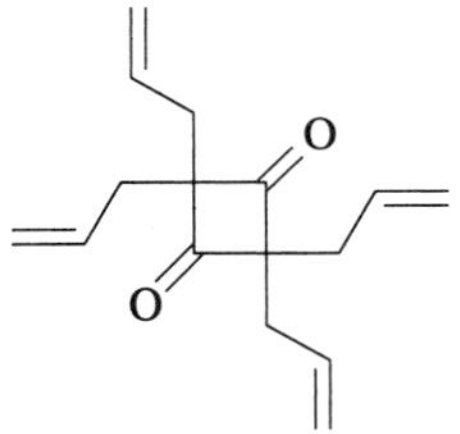 O O (—)	341
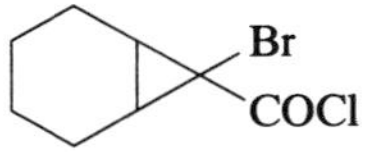 Br COCl	Zn-Ag	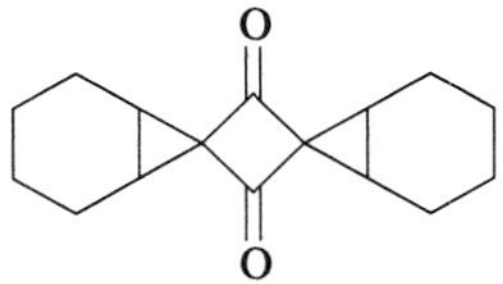 O O (20)	351
$ClOC(CH_2)_6COCl$	Et_3N, Et_2O	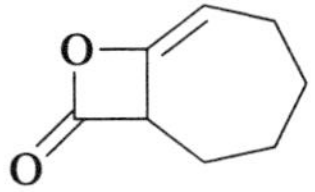 O O (—)	361
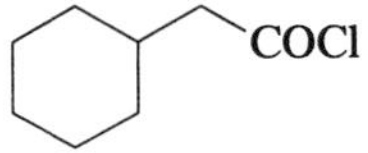 COCl	Et_3N	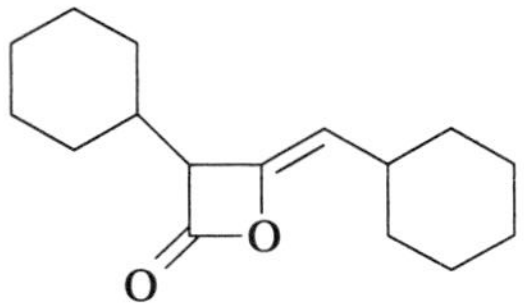 O O (40)	355
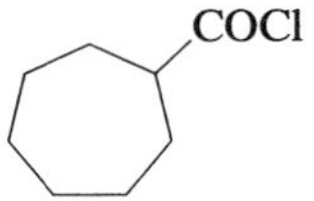 COCl	Et_3N, C_6H_6	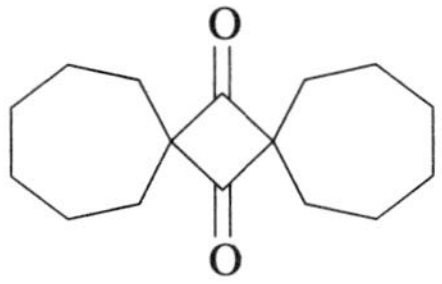 O O (8)	362

TABLE I. KETENE DIMERS (*Continued*)

Ketene or Ketene Source	Conditions	Product(s) and Yield(s) (%)	Refs.
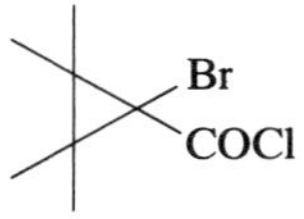 Br, COCl	Zn-Ag	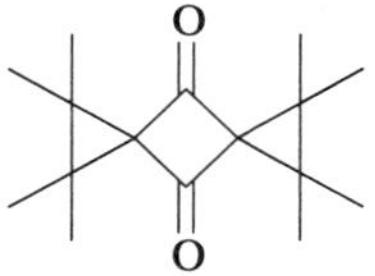 O, O (45)	351
n-$C_7H_{15}COCl$	Et_3N, Et_2O	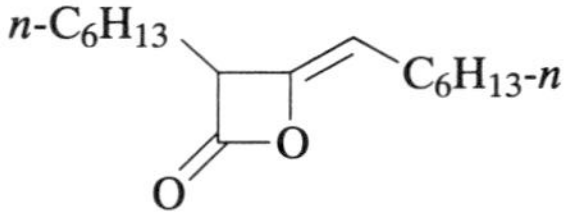 n-C_6H_{13}, C_6H_{13}-n, O, O (56)	331
n-Pr, n-Pr C=C=O	25°, 77 d	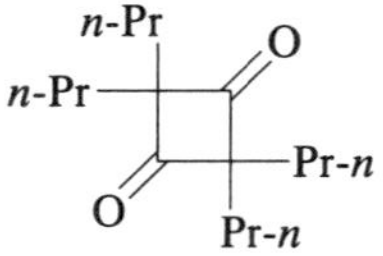 n-Pr, n-Pr, O, O, Pr-n, Pr-n (10)	341
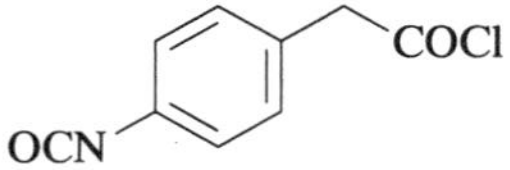 OCN, COCl	Et_3N, Et_2O	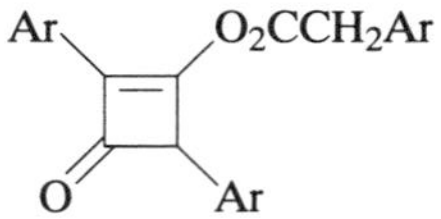 Ar, O_2CCH_2Ar, O, Ar (35)	359
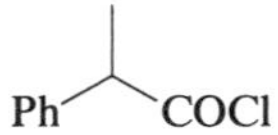 Ph, COCl	Et_3N, 100°	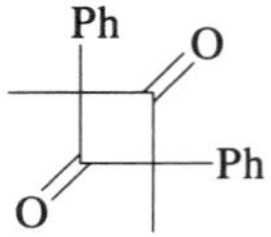 Ph, O, O, Ph *cis*, (10) *trans*, (46)	363
C_9			
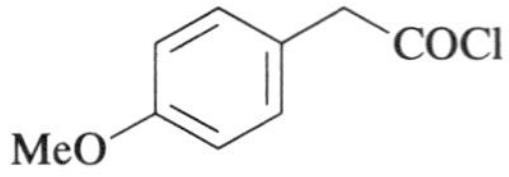 MeO, COCl	Et_3N	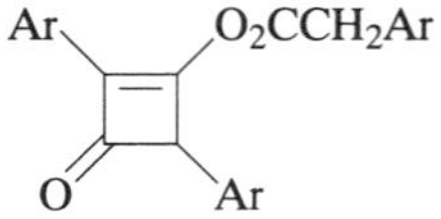 Ar, O_2CCH_2Ar, O, Ar (40) 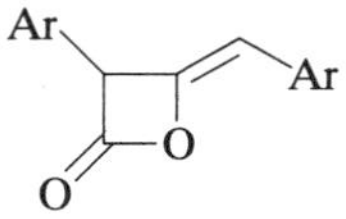 + Ar, Ar, O, O (5)	71, 364

TABLE I. KETENE DIMERS (*Continued*)

Ketene or Ketene Source	Conditions	Product(s) and Yield(s) (%)	Refs.
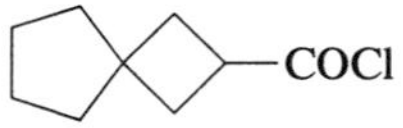 COCl	Et_3N, C_6H_6	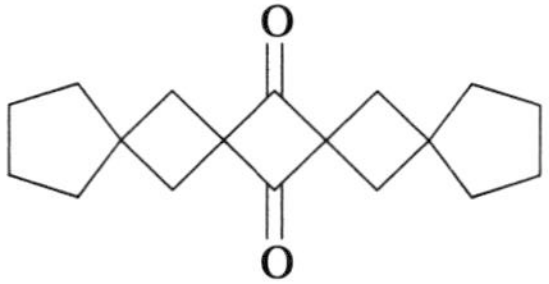 O O (42)	360
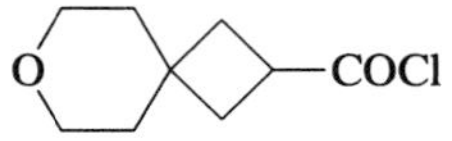 O COCl	Et_3N, C_6H_6	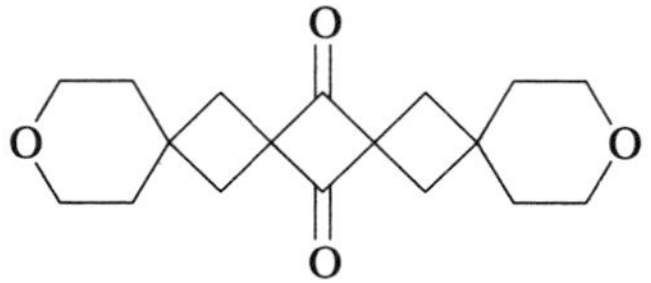 O O O O (54)	365
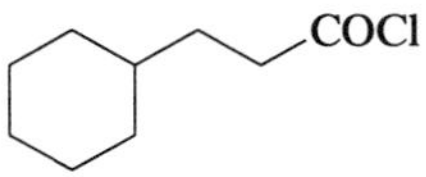 COCl	Et_3N	Liquid dimer (18)	355
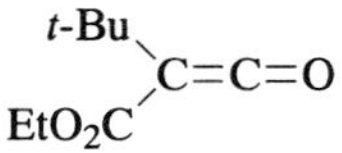 t-Bu(EtO$_2$C)C=C=O	rt, 2 min	No dimerization	69
C_{10}			
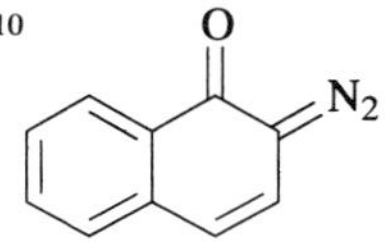 O N_2	Xylene, reflux	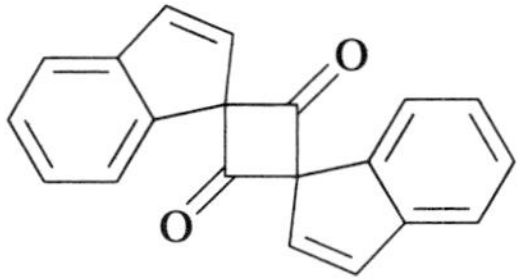 O O (—)	366
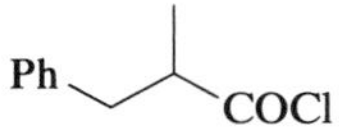 Ph COCl	Et_3N, 100°	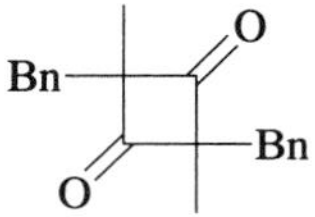 Bn O O Bn (19) *trans:cis* = 40:60	363

TABLE I. KETENE DIMERS (*Continued*)

Ketene or Ketene Source	Conditions	Product(s) and Yield(s) (%)	Refs.
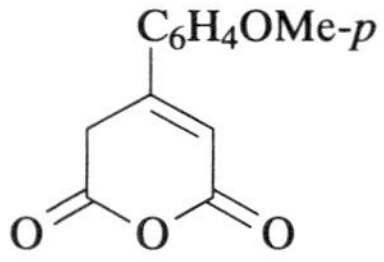 C_6H_4OMe-*p*; O, O, O	$PhCO_2Na$, $PhCO_2H$, 140-150°	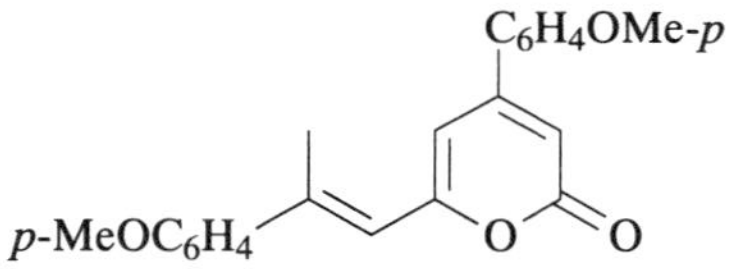 C_6H_4OMe-*p*; *p*-$MeOC_6H_4$; O, O (26)	367
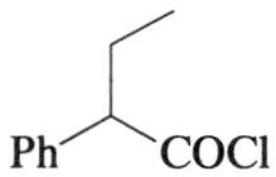 Ph, COCl	Et_3N, 100°	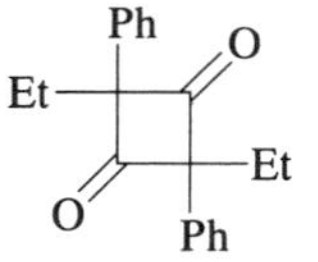 Ph, O, Et, Et, O, Ph (28) *trans*:*cis* = 38:62	363
Bn, Me, C=C=O	Et_3N, 100°	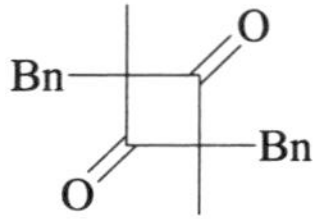 Bn, O, Bn, O (—)	341
 MeO_2C, COCl	Et_3N, 100°	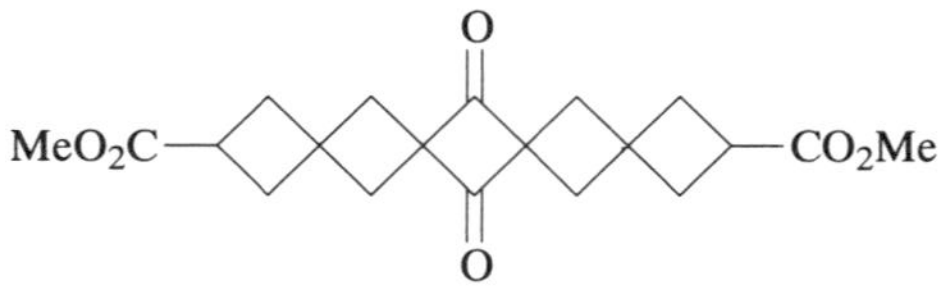 MeO_2C, O, O, CO_2Me (50)	368
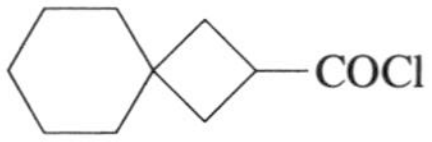 COCl	Et_3N, C_6H_6	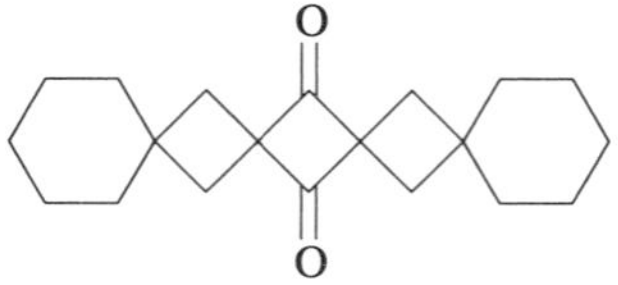 O, O (65)	360

TABLE I. KETENE DIMERS (*Continued*)

Ketene or Ketene Source	Conditions	Product(s) and Yield(s) (%)	Refs.
 C_6H_{11}(CH$_2$)$_3$COCl	Et_3N	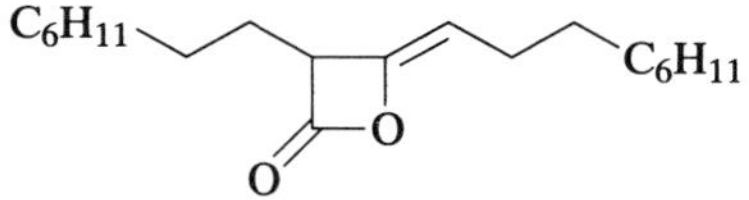 C_6H_{11}, C_6H_{11} (25)	355
C_{11}			
 Pht(CH$_2$)$_3$COCl	Et_3N	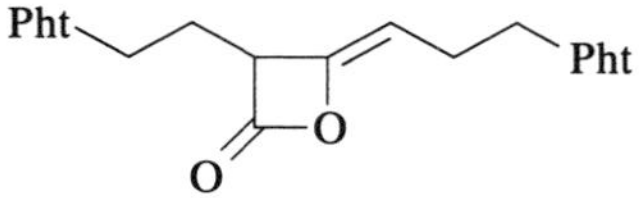 Pht, Pht (—)	369
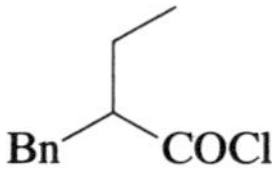 Bn, COCl	Et_3N, 100°	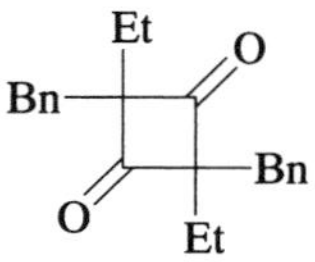 Et, Bn, O, Bn, O, Et (14) *trans*:*cis* = 42:58	363
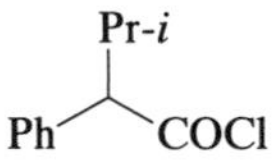 Pr-*i*, Ph, COCl	Me_3N, 100°	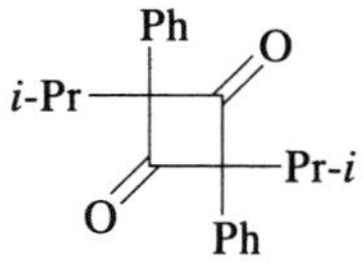 Ph, *i*-Pr, O, Pr-*i*, O, Ph (9)	363
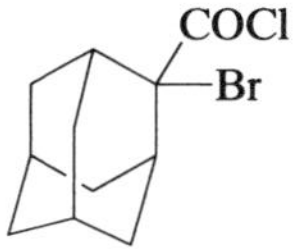 COCl, Br	Zn, Et_2O	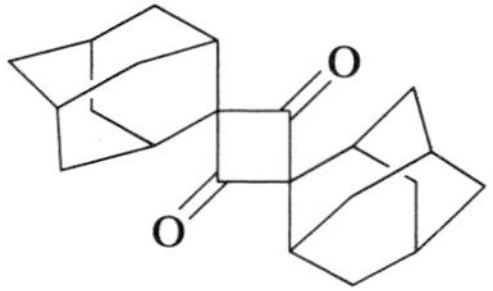 O, O (79)	370

TABLE I. KETENE DIMERS (*Continued*)

Ketene or Ketene Source	Conditions	Product(s) and Yield(s) (%)	Refs.
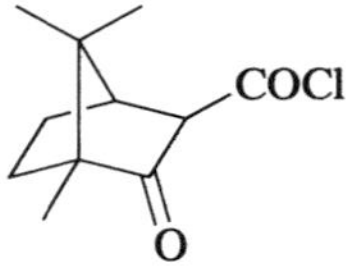 COCl	Et_3N, Et_2O	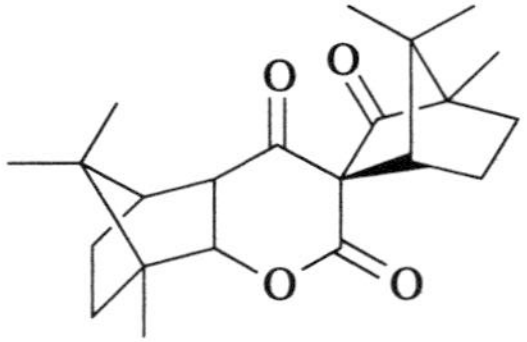 (87)	371
		+ 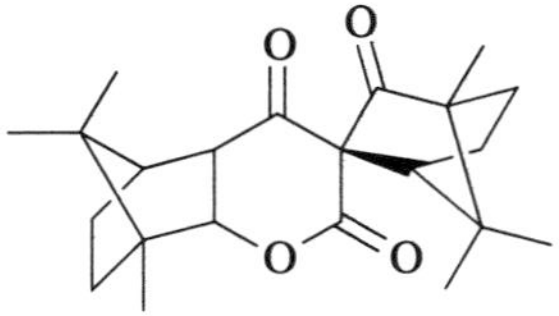 (—)	
	Quinoline , Et_2O	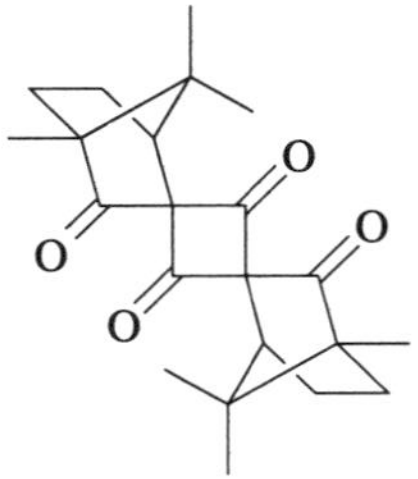 (—)	372
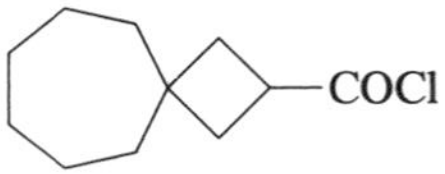 COCl	Et_3N, C_6H_6	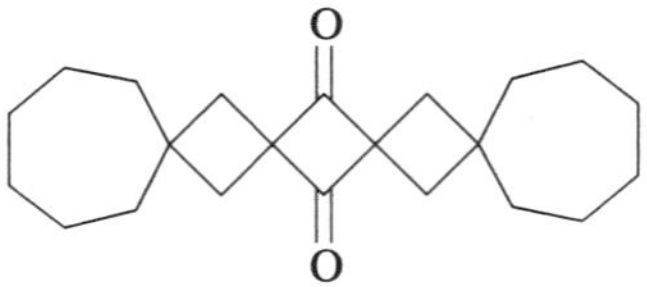 (50)	362

TABLE I. KETENE DIMERS (*Continued*)

Ketene or Ketene Source	Conditions	Product(s) and Yield(s) (%)	Refs.
$C_6H_{11}(CH_2)_4COCl$	Et_3N	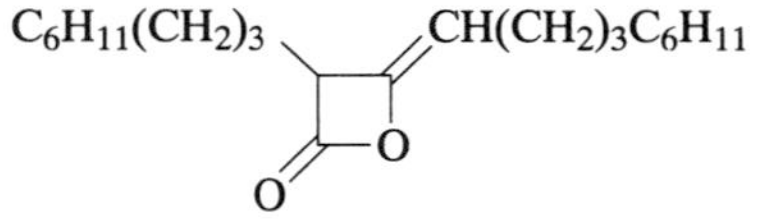 $C_6H_{11}(CH_2)_3$, $CH(CH_2)_3C_6H_{11}$, O, O (36)	355
C_{12}			
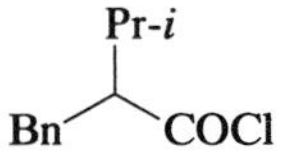 Pr-*i*, Bn, COCl	Et_3N, 100°	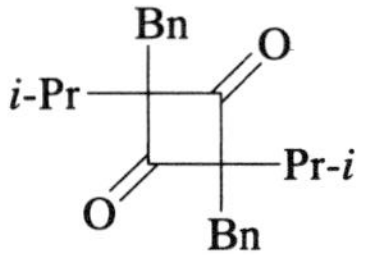 Bn, O, *i*-Pr, Pr-*i*, O, Bn (3) *trans*:*cis* = 36:64	363
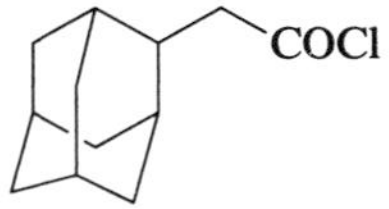 COCl	Et_3N, Et_2O	—	370
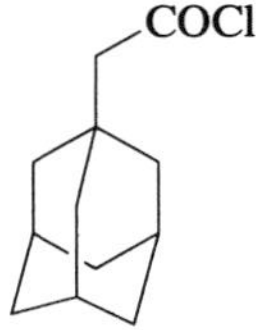 COCl	Et_3N, Et_2O	—	370
 COCl	Et_3N, C_6H_6	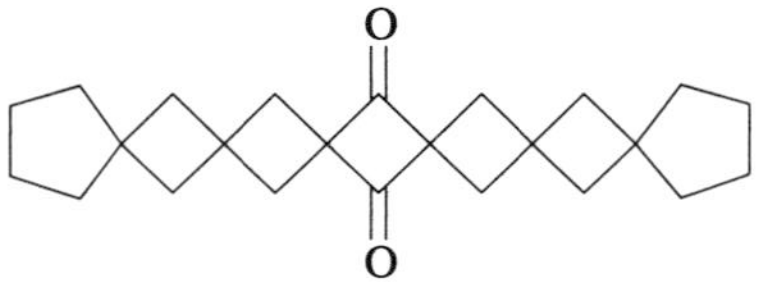 O, O (61)	360

TABLE I. KETENE DIMERS (*Continued*)

Ketene or Ketene Source	Conditions	Product(s) and Yield(s) (%)	Refs.
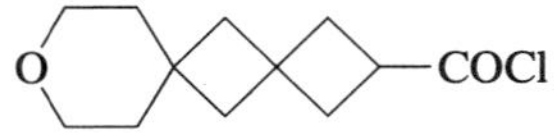 —COCl	Et_3N, C_6H_6	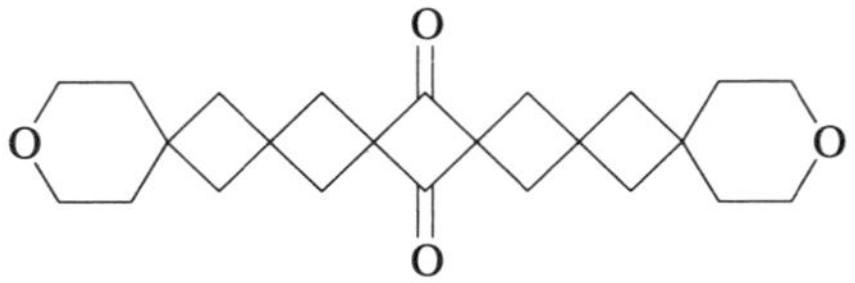 (50)	365
n-$C_{11}H_{23}COCl$	Et_3N, Et_2O	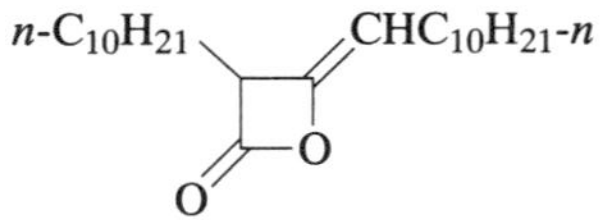 n-$C_{10}H_{21}$, $CHC_{10}H_{21}$-n, O, O (100)	331
C_{13}			
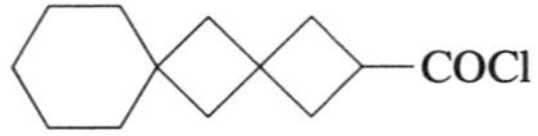 —COCl	Et_3N, C_6H_6	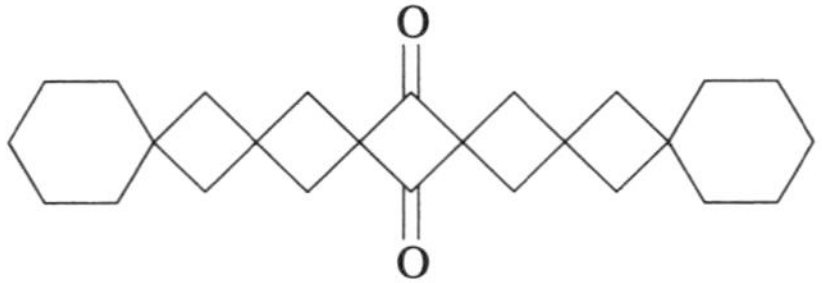 (38)	360
C_{14}			
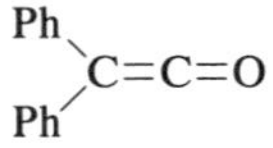 $Ph_2C=C=O$	PhCOCl, 100°	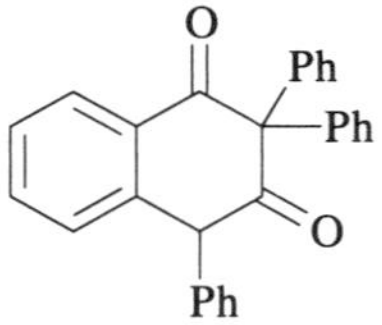 O, Ph, Ph, O, Ph (—)	61
	200°, 6 h	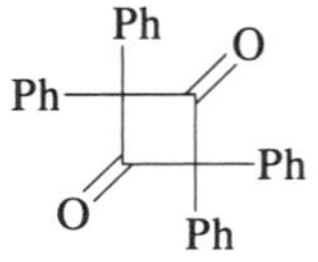 Ph, Ph, O, O, Ph, Ph (8) 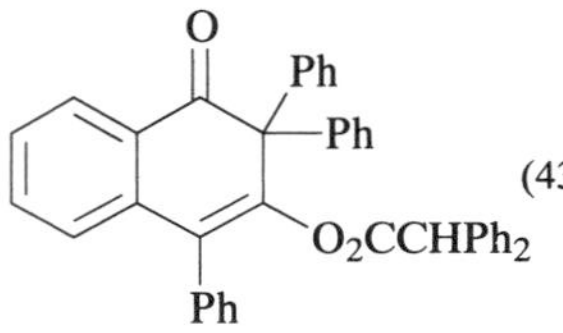 + O, Ph, Ph, O_2CCHPh_2, Ph (43)	61, 373

TABLE I. KETENE DIMERS (*Continued*)

Ketene or Ketene Source	Conditions	Product(s) and Yield(s) (%)	Refs.
	rt, 1-methylimidazole (N, N, Me)	[structure: O, Ph, Ph, N, N, Me, Ph, Ph, O] + [structure: Ph, Ph, Ph, Ph, O, O] (6-30)	374, 375
[structure]—COCl	Et_3N, C_6H_6	[structure: O, O] (55)	362
n-$C_{13}H_{27}COCl$	Et_3N, ligroin	[structure: n-$C_{12}H_{25}$, $CHC_{12}H_{25}$-n, O, O] (—)	331
[structure: Ph, Bn, COCl]	Et_3N, 100°	[structure: Ph, O, Bn, Ph, O, Bn] (18)	376
[structure: O]—COCl	Et_3N, C_6H_6	[structure: O, O, O, O] (18)	365

TABLE I. KETENE DIMERS (*Continued*)

Ketene or Ketene Source	Conditions	Product(s) and Yield(s) (%)	Refs.
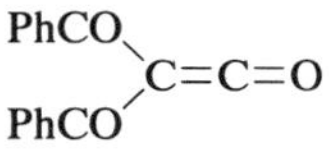 $(PhCO)_2C=C=O$	> -75°	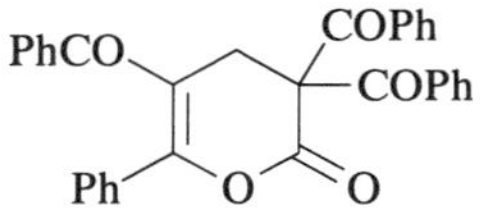 PhCO, Ph, O, O, COPh, COPh (—)	377
C_{16}			
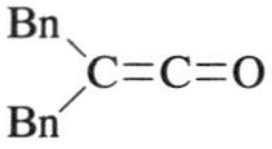 $Bn_2C=C=O$	25°, 24 h	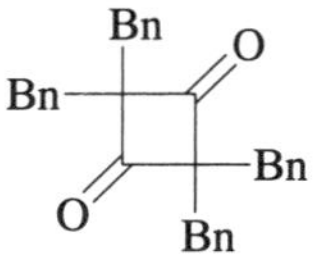 Bn, Bn, O, O, Bn, Bn (—)	341, 378
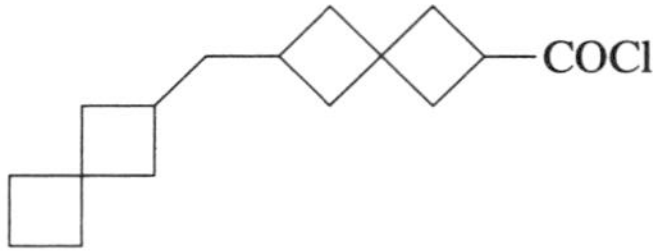 COCl	Et_3N, C_6H_6	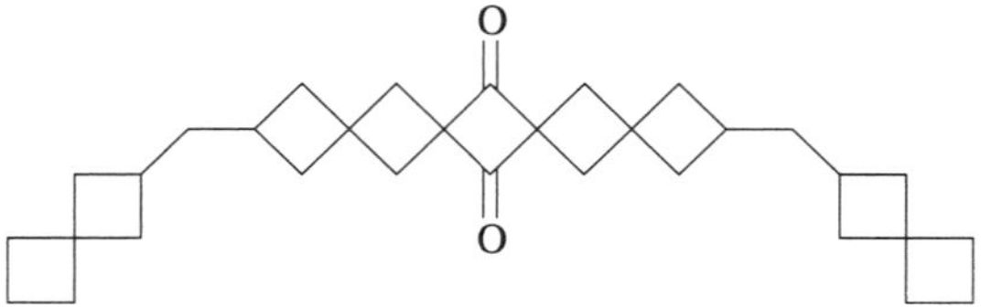 O, O (20)	360
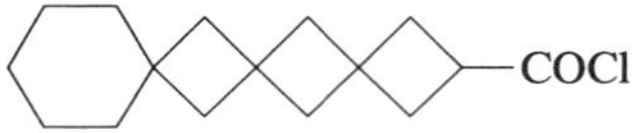 COCl	Et_3N, C_6H_6	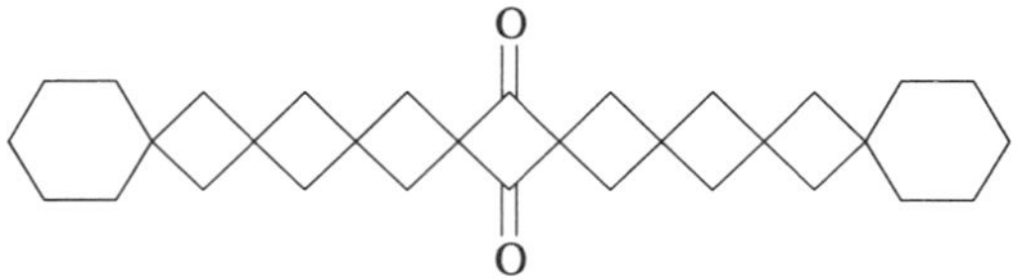 O, O (53)	360

TABLE I. KETENE DIMERS (*Continued*)

Ketene or Ketene Source	Conditions	Product(s) and Yield(s) (%)	Refs.
$n\text{-}C_{15}H_{31}COCl$	Et_3N, Et_2O	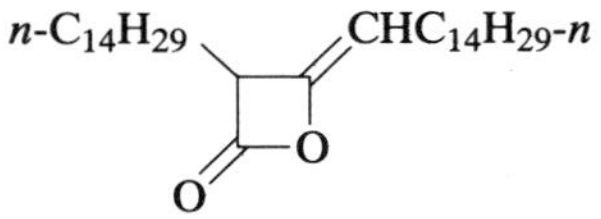 $n\text{-}C_{14}H_{29}$, $CHC_{14}H_{29}\text{-}n$, O, O (very high)	379
C_{18}			
$n\text{-}C_{17}H_{35}COCl$	Et_3N, C_6H_6	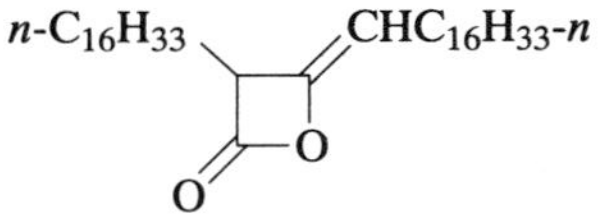 $n\text{-}C_{16}H_{33}$, $CHC_{16}H_{33}\text{-}n$, O, O (90)	331
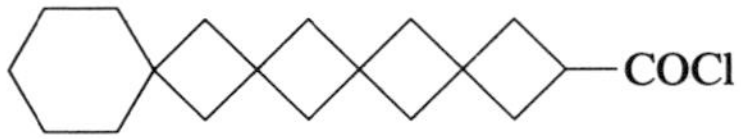 —COCl	Et_3N, C_6H_6	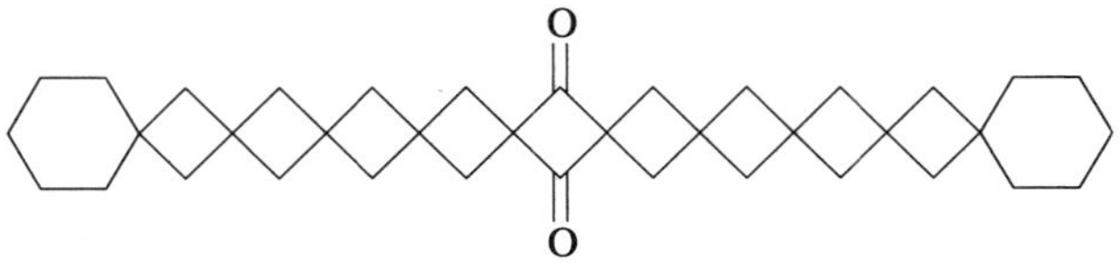 O, O (28)	360

TABLE II. MIXED KETENE DIMERS

Ketene or Ketene Source	Ketene or Ketene Source	Conditions	Product(s) and Yield(s) (%)	Refs.
C_2				
$CD_2=C=O$	$CH_2=C=O^{18}$	0°	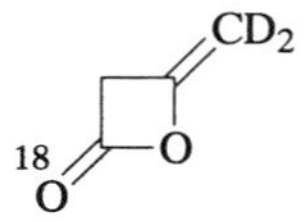 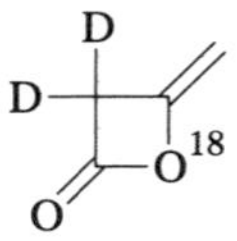 [structure: 4-(CD$_2$)-oxetan-2-one, ^{18}O carbonyl] + [structure: 3,3-D$_2$-4-methylene-oxetan-2-one, ^{18}O ring] (—)	380
$CH_2=C=O$	$(CF_3)_2C=C=O$	-78°	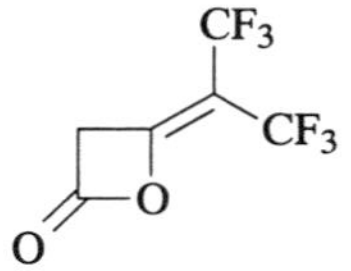 [structure: 4-(bis(CF$_3$)methylene)oxetan-2-one] (89)	336
	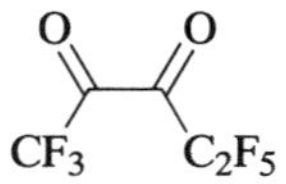 $CF_3COCOC_2F_5$	Et_2O, -20°	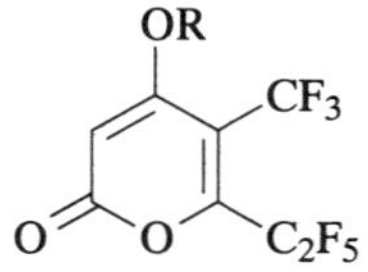 [structure: 4-OR-5-CF$_3$-6-C$_2$F$_5$-2-pyranone] R = H, Ac (67)	346
	$t\text{-Bu}(NC)C=C=O$	PhMe	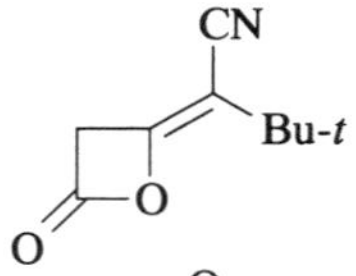 [structure: 4-(CN)(Bu-t)methylene-oxetan-2-one] (36)	43
	$Ph_3P=C=C=O$	C_6H_6, rt	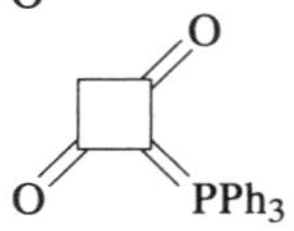 [structure: 2-PPh$_3$-ylidene-cyclobutane-1,3-dione] (—)	381
C_3				
$Me(H)C=C=O$	$(CF_3)_2C=C=O$	-80°	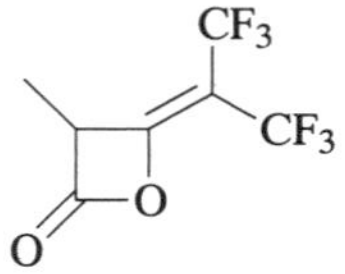 [structure: 3-methyl-4-(bis(CF$_3$)methylene)oxetan-2-one] (61)	336

TABLE II. MIXED KETENE DIMERS (*Continued*)

Ketene or Ketene Source	Ketene or Ketene Source	Conditions	Product(s) and Yield(s) (%)	Refs.
$Et^{14}COCl$	$n\text{-}C_7H_{15}COCl$	Et_3N, Et_2O, rt	(structure, 14, 14) (25) + $n\text{-}C_6H_{13}$ (structure, 14) (8) + (structure, 14) $C_6H_{13}\text{-}n$ (17) + $n\text{-}C_6H_{13}$ (structure) $C_6H_{13}\text{-}n$ (24)	382
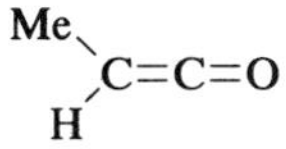 $Me(H)C=C=O$	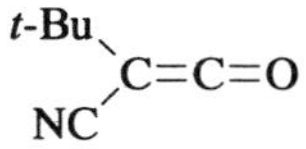 $t\text{-}Bu(NC)C=C=O$	PhMe	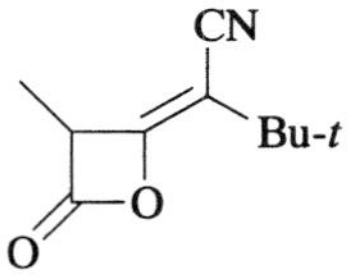 (structure) CN, Bu-*t* (49)	41,43
MeCHClCOCl	$Me_2CHCOCl$	Et_3N, C_6H_6	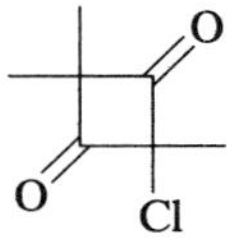 (structure) Cl (34)	73
	$Et_2CHCOCl$	Et_3N, C_6H_6, reflux 4 d	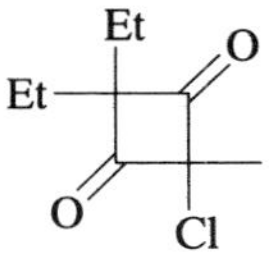 (structure) Et, Et, Cl (63)	383, 73

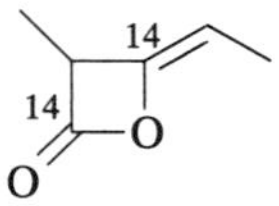

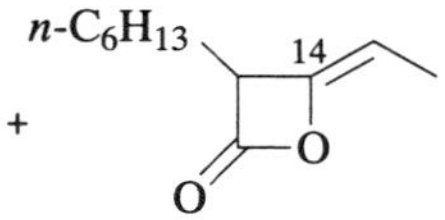

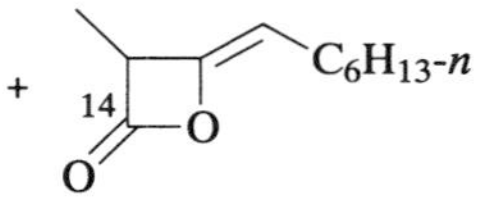

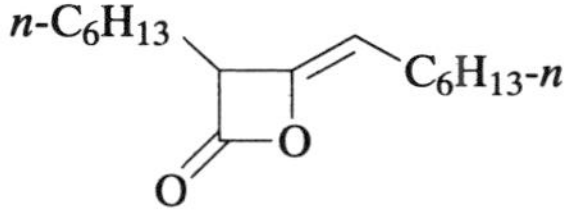

TABLE II. MIXED KETENE DIMERS (*Continued*)

Ketene or Ketene Source	Ketene or Ketene Source	Conditions	Product(s) and Yield(s) (%)	Refs.
	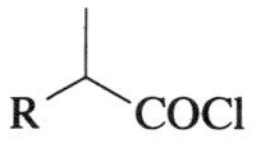 R–CH(Me)–COCl	Et_3N, C_6H_6	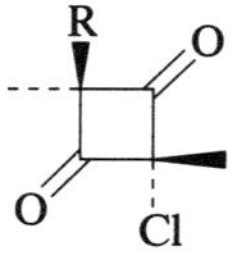 I 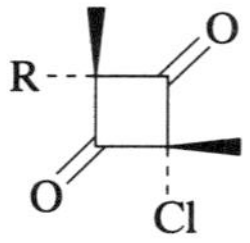 + II (structures: R, O, O, Cl) R / I + II / I:II *n*-Pr / (42) / 1:1 *i*-Pr / (57) / 1:1	347
	t-$BuCH_2COCl$	Et_3N, C_6H_6	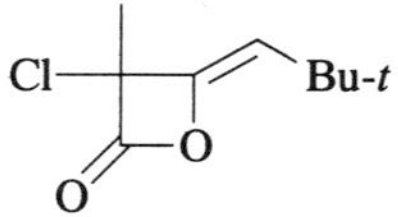 (structure: Cl, Bu-*t*, O, O) (32)	347
	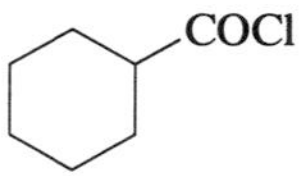 cyclohexyl-COCl	Et_3N, C_6H_6	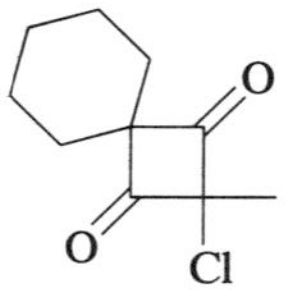 (structure: O, O, Cl) (62)	73, 383
C_4				
$Me_2CHCOCl$	EtCHClCOCl	Et_3N, C_6H_6	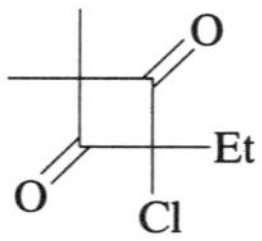 (structure: O, O, Et, Cl) (41)	73

TABLE II. MIXED KETENE DIMERS (*Continued*)

Ketene or Ketene Source	Ketene or Ketene Source	Conditions	Product(s) and Yield(s) (%)	Refs.
$Me_2C=C=O$	$(CF_3)_2C=C=O$	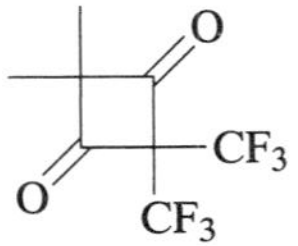	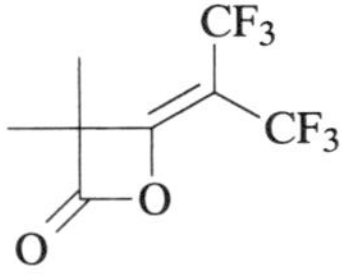 2,2-dimethyl-4,4-bis(CF$_3$)cyclobutane-1,3-dione + 3,3-dimethyl-4-[(CF$_3$)$_2$C=]oxetan-2-one	336
		Cyclohexane	(81) + (0)	
		EtOAc	(19) + (24)	
$Me_2CHCOCl$	i-PrCHClCOCl	Et_3N, C_6H_6, rt	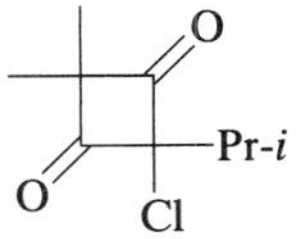 2,2-dimethyl-4-chloro-4-Pr-i-cyclobutane-1,3-dione (40)	73
	t-BuCHClCOCl	Et_3N, C_6H_6	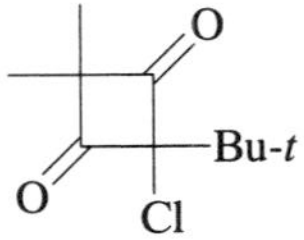 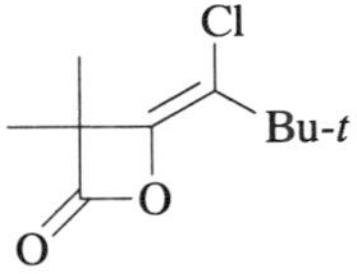 2,2-dimethyl-4-chloro-4-Bu-t-cyclobutane-1,3-dione (56) + 3,3-dimethyl-4-[Cl(Bu-t)C=]oxetan-2-one (13)	73
$Me_2C=C=O$	t-Bu(NC)C=C=O	PhMe	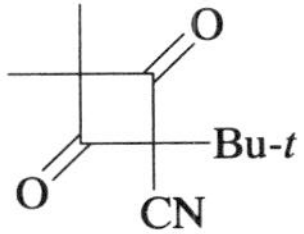 2,2-dimethyl-4-cyano-4-Bu-t-cyclobutane-1,3-dione (38)	43,41
EtCHClCOCl	$Et_2CHCOCl$	Et_3N, C_6H_6	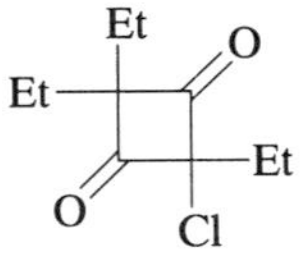 2,2,4-triethyl-4-chlorocyclobutane-1,3-dione (65)	383, 73

TABLE II. MIXED KETENE DIMERS (*Continued*)

Ketene or Ketene Source	Ketene or Ketene Source	Conditions	Product(s) and Yield(s) (%)	Refs.
	t-$BuCH_2COCl$	Et_3N, C_6H_6	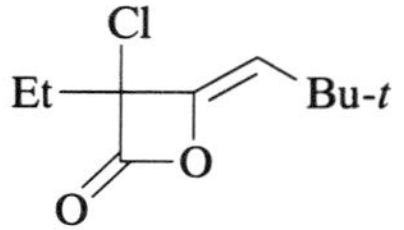 (10)	347
	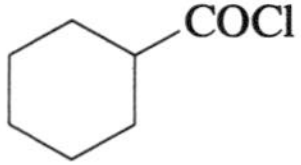	Et_3N, C_6H_6	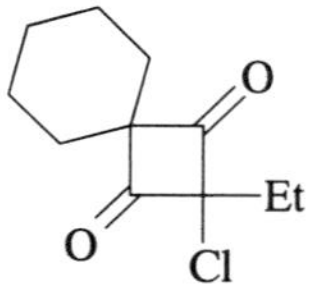 (35)	73
$(CF_3)_2C=C=O$	n-Bu(Et)C=C=O	Hexane	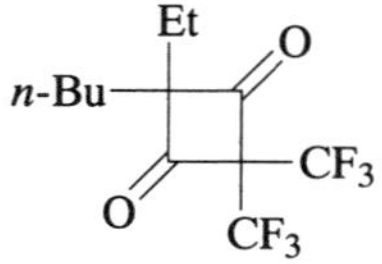 (—)	336
	$Ph_2C=C=O$	100°, 1 week	(—)	336
	TMS(H)C=C=O	-78°	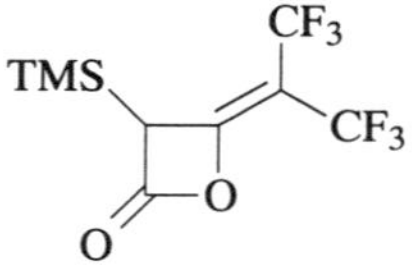 (—)	384
C_5				
Et(Me)C=C=O	t-Bu(NC)C=C=O	PhMe	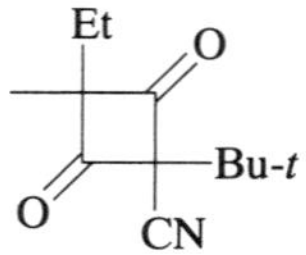 (44)	43

TABLE II. MIXED KETENE DIMERS (*Continued*)

Ketene or Ketene Source	Ketene or Ketene Source	Conditions	Product(s) and Yield(s) (%)	Refs.
C_6				
$Et_2CHCOCl$	t-BuCHClCOCl	Et_3N, C_6H_6	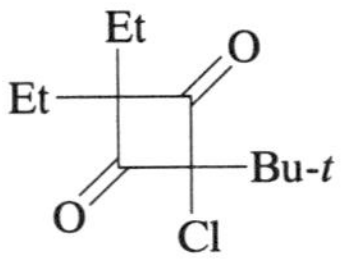 (43)	73
t-$BuCH_2COCl$	t-BuCHXCOCl	Et_3N, C_6H_6	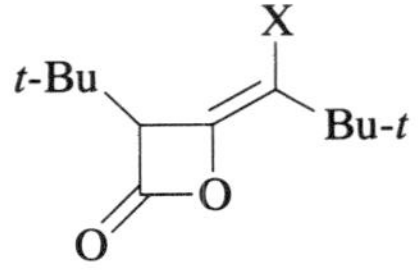 X = Cl (20) X = Br (10)	347
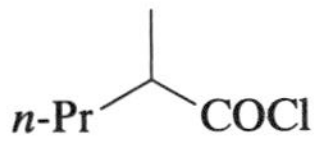 n-Pr … COCl	t-BuCHClCOCl	Et_3N, C_6H_6	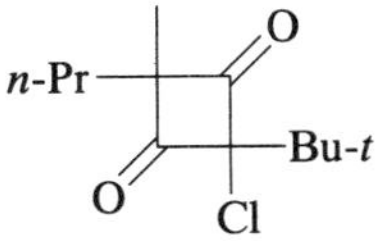 (48)	347
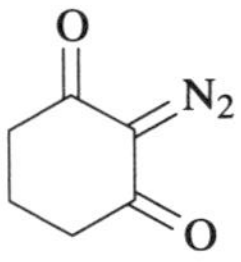	Ph, C=C=O, Ph	Xylene, 140°	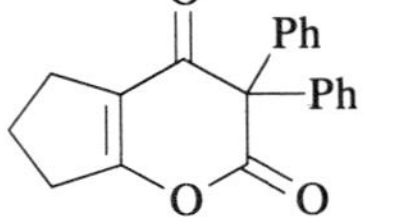 (—)	385
t-BuCHBrCOCl	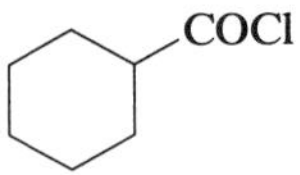 COCl	Et_3N, C_6H_6	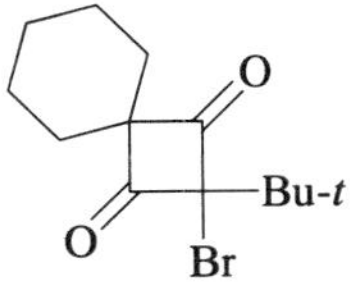 (36)	73
Ph, C=C=O, Ph	$Ph_3P{=}C{=}C{=}O$	C_6H_6, rt	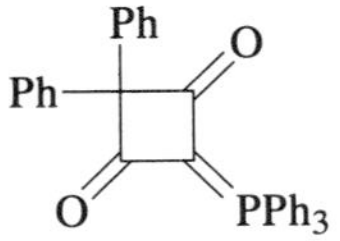 (—)	381

TABLE III. [2+2] CYCLOADDITION OF KETENES TO ACYCLIC OLEFINS

Reactant	Ketene or Ketene Source	Conditions	Product(s) and Yield(s) (%)	Refs.
C_2				
$CH_2{=}CH_2$	$(CF_3)_2C{=}C{=}O$	130-200°	(—)	84
	O, O, O, O	140°	O (60)	316
	$Ph_2C{=}C{=}O$	85-90°, 12 d	Ph, Ph, O (60)	386
C_3				
$MeO(F)C{=}CF_2$	$CF_3(C_2F_5CO)C{=}C{=}O$	100°	O, F, F, F, MeO, O, CF_3, C_2F_5 (73)	346
$MeCH{=}CH_2$	$(CF_3)_2C{=}C{=}O$	150°, 8 h	CF_3, CF_3, O (**I**) + CF_3, CF_3, O (**II**); **I** + **II** (19); **I**:**II** = 2.3	84

TABLE III. [2+2] CYCLOADDITION OF KETENES TO ACYCLIC OLEFINS (*Continued*)

Reactant	Ketene or Ketene Source	Conditions	Product(s) and Yield(s) (%)	Refs.
	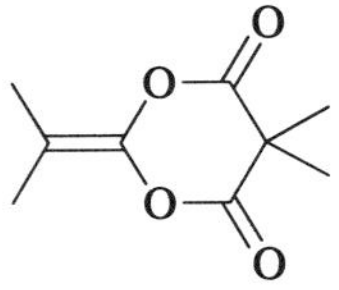 [structure]	K_2CO_3 (cat.), PhMe, 130°	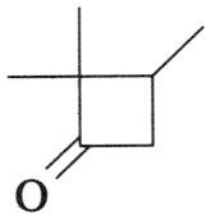 [structure] (72)	316
	$CF_3(C_2F_5CO)C{=}C{=}O$	150°, 8 h	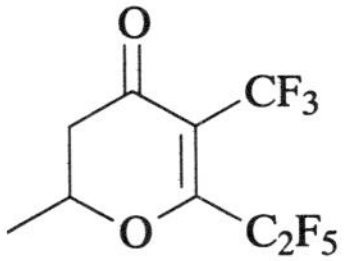 [structure: O, CF_3, C_2F_5] (20)	346
	$Ph_2C{=}C{=}O$		[structure: Ph, Ph, O]	386
		C_6H_6, 110-115°, 2 d	(43)	
		rt, 8 min	(97)	
C_4 $Me_2C{=}CH_2$	$(CF_3)_2C{=}C{=}O$	100°	[structure: CF_3, O, CF_3] (86)	84, 387
	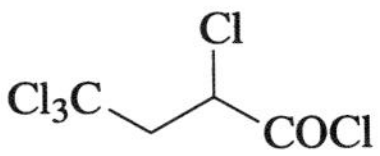 $Cl_3CCH_2CH(Cl)COCl$	Et_3N, cyclohexane	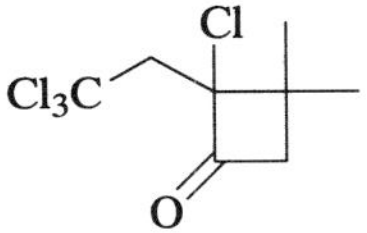 [structure: Cl_3C, Cl, O] (67)	388, 389

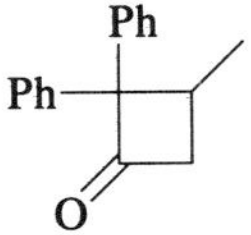

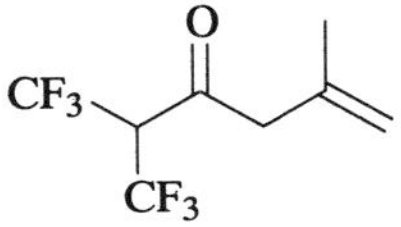

TABLE III. [2+2] CYCLOADDITION OF KETENES TO ACYCLIC OLEFINS (*Continued*)

Reactant	Ketene or Ketene Source	Conditions	Product(s) and Yield(s) (%)	Refs.
	$Me_2C{=}C{=}O$	140°	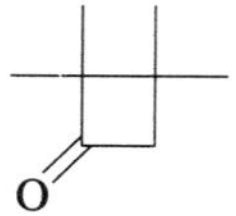 (65)	390, 37
	$EtOCH_2COCl$	Et_3N	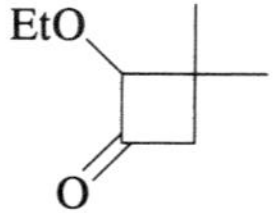 EtO (30)	391
	t-Bu(NC)C=C=O	C_6H_6, 78°	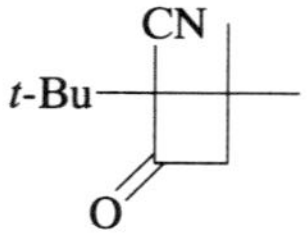 t-Bu, CN (44) 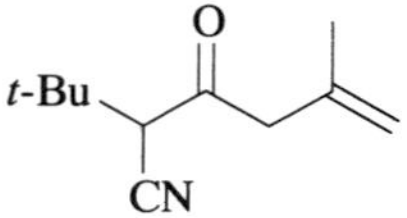 + t-Bu, CN, O (22)	392
	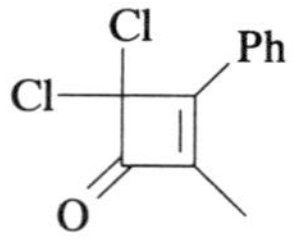 Cl, Cl, Ph, O	$h\nu$, C_6H_6	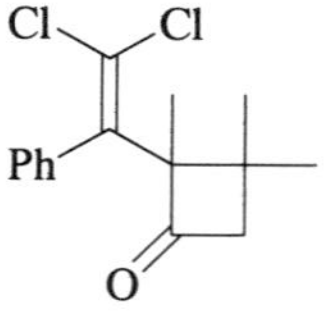 Cl, Cl, Ph, O (40)	306
	$Ph_2C{=}C{=}O$	70°, 8 d	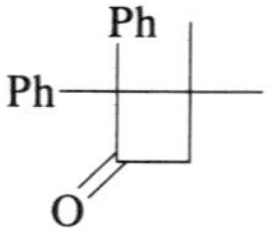 Ph, Ph, O (66)	386

TABLE III. [2+2] CYCLOADDITION OF KETENES TO ACYCLIC OLEFINS (*Continued*)

Reactant	Ketene or Ketene Source	Conditions	Product(s) and Yield(s) (%)	Refs.
cis-MeCH=CHMe	Cl_3CCOCl	Zn, $POCl_3$, Et_2O	2,2-dichloro-3,4-dimethylcyclobutanone (79)	82
trans-MeCH=CHMe	Cl_3CCOCl	Zn, $POCl_3$, Et_2O	2,2-dichloro-3,4-dimethylcyclobutanone (80)	82
cis-MeCH=CHMe	MeCH(Br)COCl	Et_3N, hexane	2-bromo-2,3,4-trimethylcyclobutanone (22)	393
trans-MeCH=CHMe	$(CF_3)_2C{=}C{=}O$	100°, overnight	(—)	84
cis-MeCH=CHMe	$Me_2C{=}C{=}O$	100°	2,2,3,4-tetramethylcyclobutanone (—)	37
	$Me_2C{=}C{=}O$	C_6H_6, 105°, 8 h	2,2,3,4-tetramethylcyclobutanone **I** (94) *cis*:*trans* = 97.5:2.5	294

TABLE III. [2+2] CYCLOADDITION OF KETENES TO ACYCLIC OLEFINS (*Continued*)

Reactant	Ketene or Ketene Source	Conditions	Product(s) and Yield(s) (%)	Refs.
(E)-MeCH=CHMe	2-isopropylidene-5,5-dimethyl-1,3-dioxane-4,6-dione	C_6H_6, 105°, 3 h, K_2CO_3	**I** (70) *cis:trans* = 95:5	294
	"	C_6H_6, 105°, 8 h	**I** (10) *cis:trans* = 40:60	294
	"	C_6H_6, 105°, 3 h, K_2CO_3	**I** (32) *cis:trans* = 19:81	294
	$Me_2C{=}C{=}O$	100°	**I** (—)	37
(Z)-MeCH=CHMe	$EtOCH_2COCl$	Et_3N	2-EtO-3,4-dimethylcyclobutanone (45)	391
(E)-MeCH=CHMe	$EtOCH_2COCl$	Et_3N	2-EtO-3,4-dimethylcyclobutanone (31)	391
(Z)-MeCH=CHMe	$(CF_3)(C_2F_5CO)C{=}C{=}O$	95°, 310 h	2,3-dihydro-2,3-dimethyl-5-CF_3-6-C_2F_5-pyran-4-one (10) + $MeCH{=}C(Me)COCH(CF_3)COC_2F_5$ (1)	346

TABLE III. [2+2] CYCLOADDITION OF KETENES TO ACYCLIC OLEFINS (*Continued*)

Reactant	Ketene or Ketene Source	Conditions	Product(s) and Yield(s) (%)	Refs.
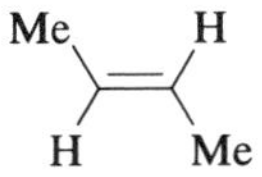 Me, H / H, Me (trans)	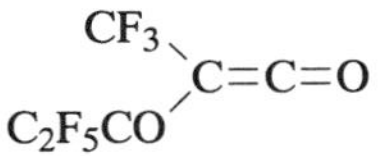 $CF_3(C_2F_5CO)C{=}C{=}O$	95°, 310 h	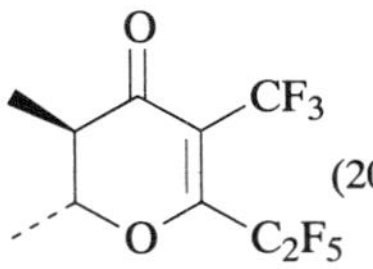 O, CF_3, C_2F_5 (20) 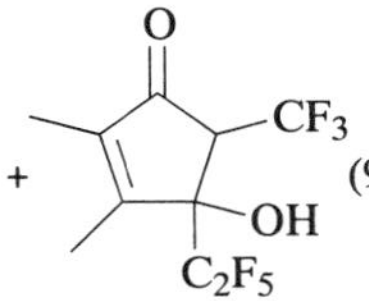 + O, CF_3, OH, C_2F_5 (9) 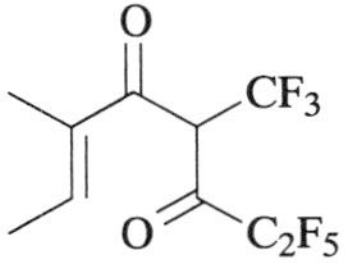 + O, CF_3, O, C_2F_5 (10)	346
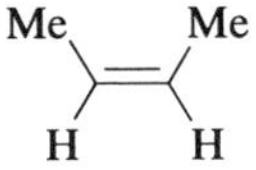 Me, Me / H, H (cis)	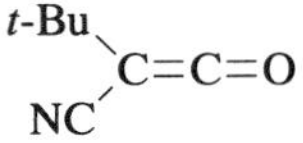 t-Bu(NC)C=C=O	C_6H_6, 78°	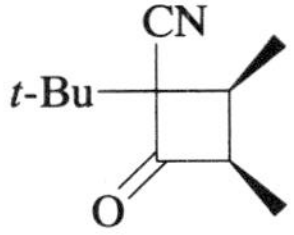 t-Bu, CN, O (—)	394
MeCH=CHMe	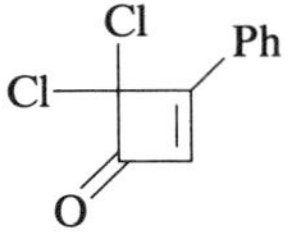 Cl, Cl, Ph, O (cyclobutenone)	$h\nu$, C_6H_6	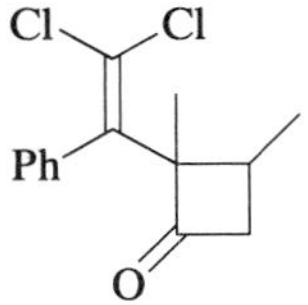 Cl, Cl, Ph, O (30)	306
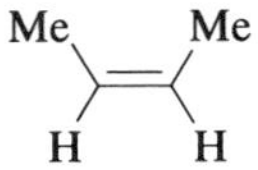 Me, Me / H, H (cis)	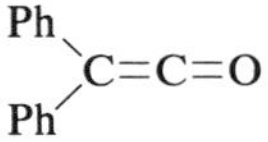 $Ph_2C{=}C{=}O$	90-95°, 3 d	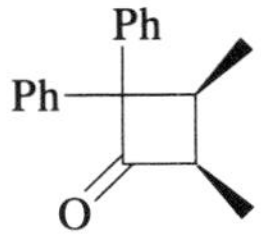 Ph, Ph, O (96)	386, 395

TABLE III. [2+2] CYCLOADDITION OF KETENES TO ACYCLIC OLEFINS (*Continued*)

Reactant	Ketene or Ketene Source	Conditions	Product(s) and Yield(s) (%)	Refs.
(E)-MeCH=CHMe	$Ph_2C=C=O$	90-95°, 3 min	2,2-diphenyl-3,4-dimethylcyclobutanone (46)	386, 395
$EtCH=CH_2$	$(CF_3)_2C=C=O$	100°, 37 d	2,2-bis(CF$_3$)-3-ethylcyclobutanone (10) + $(CF_3)_2CHCOCH_2CH=CHCH_3$ (17) + $(CF_3)_2C=C(O_2CCH(CF_3)_2)CH_2CH=CHCH_3$ (34)	84
$MeSCH_2CH=CH_2$	Cl_3CCOCl	Zn, $MeCH(OMe)_2$, reflux 60 h	(—)	299
C$_5$				
$CH_2=CHCH_2CH=CH_2$	$Ph_2C=C=O$	98°, 2 weeks	(—)	396
vinylcyclopropane	Cl_3CCOCl	Zn, Et_2O	2,2-dichloro-3-cyclopropylcyclobutanone (58)	78
methylenecyclobutane	Cl_3CCOCl	Zn, Et_2O	2,2-dichlorospiro[3.3]heptan-1-one (60)	397, 398

TABLE III. [2+2] CYCLOADDITION OF KETENES TO ACYCLIC OLEFINS (*Continued*)

Reactant	Ketene or Ketene Source	Conditions	Product(s) and Yield(s) (%)	Refs.
$Me_2C{=}CHMe$	$EtCHClCOCl$	Et_3N, C_6H_{14}	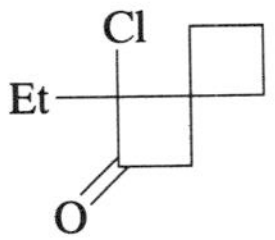 Cl, Et, O (35)	399
	$MeCHClCOCl$	Et_3N, C_6H_{14}	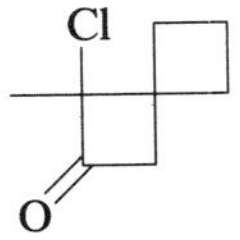 Cl, O (30)	399
	Cl_3CCOCl	Zn, Et_2O, $POCl_3$	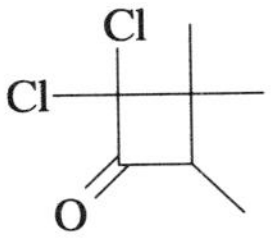 Cl, Cl, O (87)	82, 317
	Cl_3CCOCl	Zn, Et_2O	" (73)	400
	Cl(NC)C=C=O	PhMe, 110°, 1.75 h	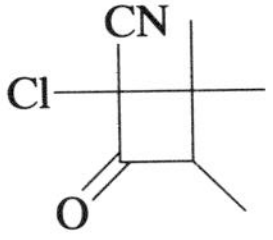 CN, Cl, O (80)	312
	$Me_2C{=}C{=}O$	100°	(—)	37
	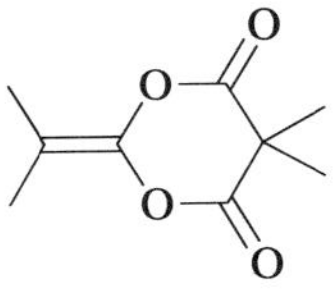 O, O, O, O	PhMe, 110°, K_2CO_3	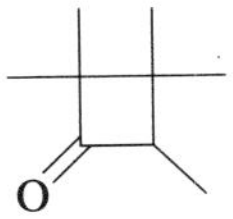 O (30)	293

TABLE III. [2+2] CYCLOADDITION OF KETENES TO ACYCLIC OLEFINS (*Continued*)

Reactant	Ketene or Ketene Source	Conditions	Product(s) and Yield(s) (%)	Refs.
	t-Bu(NC)C=C=O	C_6H_6, 5°	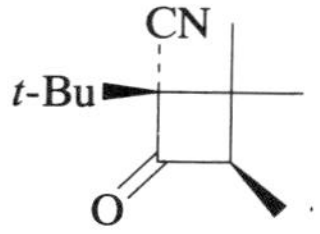 t-Bu, CN (86)	401
	n-BuC≡C(NC)C=C=O	C_6H_6, 80°	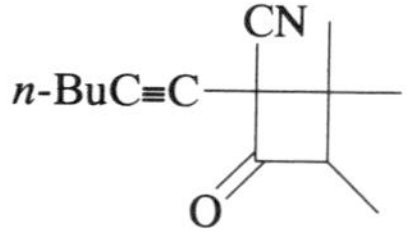 n-BuC≡C, CN (45)	402
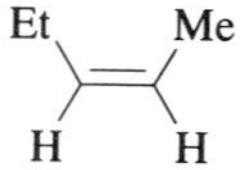 Et(H)C=C(Me)(H)	Me(Me)C=C=O	100°	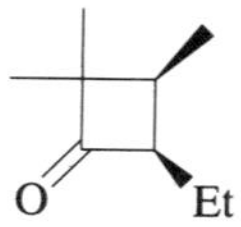 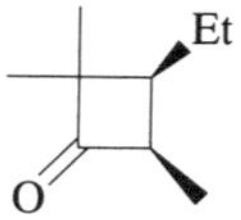 Et I + Et II (—) I:II = 51:49	37
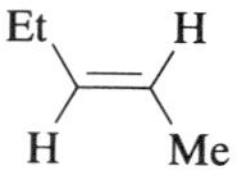 Et(H)C=C(H)(Me)	Me(Me)C=C=O	100°	(—)	37
n-PrCH=CH$_2$	Cl_3CCOCl	Zn, Et_2O	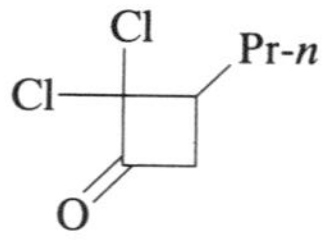 Cl, Cl, Pr-n (58)	400, 403
	MeCOCOTMS	hν, C_5H_{10}	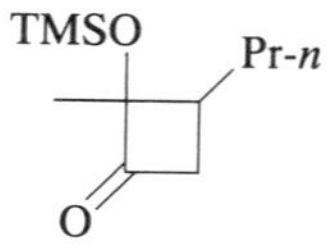 TMSO, Pr-n (—)	307

TABLE III. [2+2] CYCLOADDITION OF KETENES TO ACYCLIC OLEFINS (*Continued*)

Reactant	Ketene or Ketene Source	Conditions	Product(s) and Yield(s) (%)	Refs.
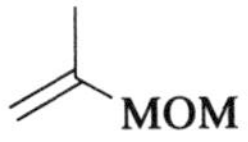 MOM	Cl_3CCOCl	Zn, Et_2O	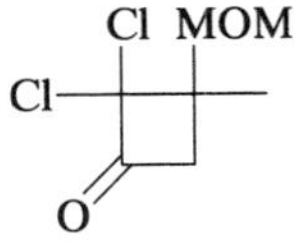 Cl, Cl, MOM (56)	404, 299
 NMe_2	*n*-Bu, Et C=C=O	180°	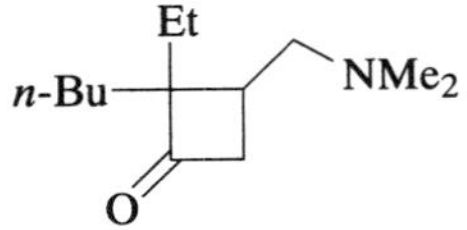 *n*-Bu, Et, NMe_2 (32)	405
C_6 	$Cl_2RCCOCl$	Zn, Et_2O	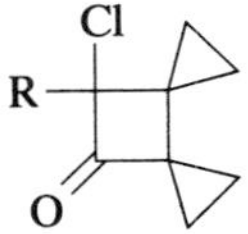 R, Cl; R = H (63); R = Cl (59)	406, 407
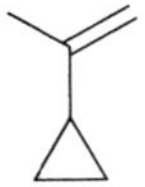	Cl_3CCOCl	Zn, Et_2O	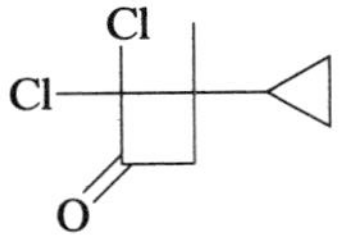 Cl, Cl (56-60)	78
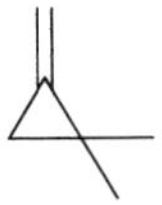	Me, Me C=C=O	120°	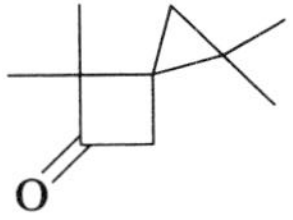 (36) 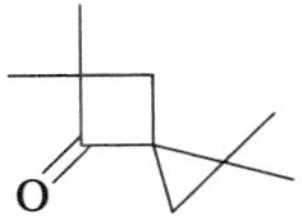 + (29)	408
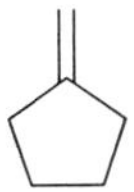	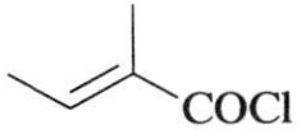 COCl	Et_3N	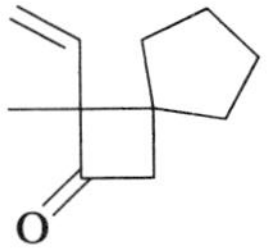 (79)	409

TABLE III. [2+2] CYCLOADDITION OF KETENES TO ACYCLIC OLEFINS (*Continued*)

Reactant	Ketene or Ketene Source	Conditions	Product(s) and Yield(s) (%)	Refs.
	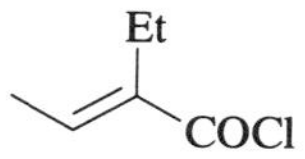 Et, COCl	Et_3N	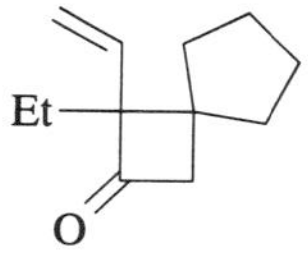 Et (67)	409
	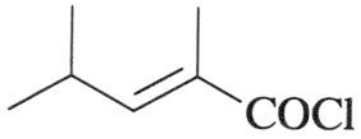 COCl	Et_3N	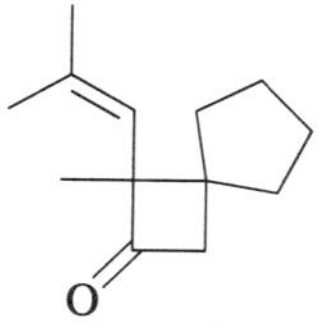 (20)	409
	Ph, Ph, C=C=O	100°	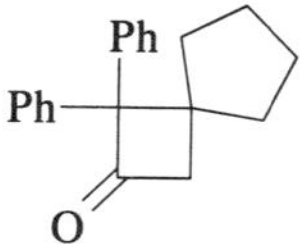 Ph, Ph, O (—)	410
n-BuCH=CH_2	CF_3, CF_3, C=C=O	150°, 8 h	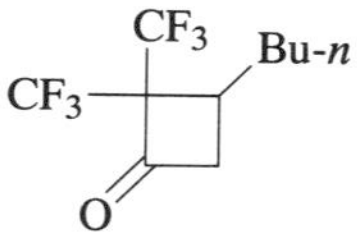 CF_3, CF_3, Bu-n, O (12) 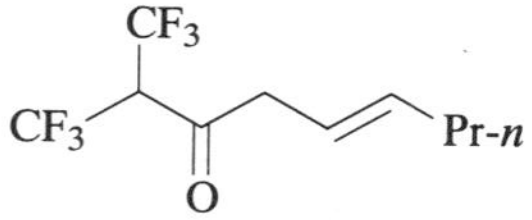 + CF_3, CF_3, O, Pr-n (8)	84
	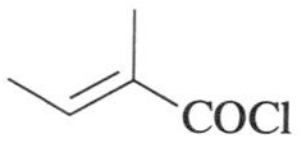 COCl	Et_3N, $CHCl_3$	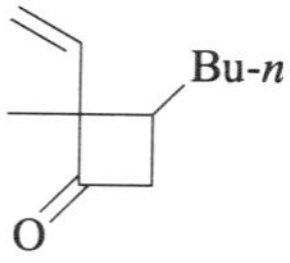 Bu-n, O (24)	411

TABLE III. [2+2] CYCLOADDITION OF KETENES TO ACYCLIC OLEFINS (*Continued*)

Reactant	Ketene or Ketene Source	Conditions	Product(s) and Yield(s) (%)	Refs.
	$Ph_2C{=}C{=}O$	100°	2,2-diphenyl-3-(Bu-*n*)cyclobutanone (87-91)	386
cis-EtCH=CHEt	Cl(NC)C=C=O	PhMe, 103°, 1.75 h	2-Cl-2-CN-3,4-diethylcyclobutanone (93)	312
trans-EtCH=CHEt	Cl(NC)C=C=O	PhMe, 103°, 1.75 h	2-Cl-2-CN-3,4-diethylcyclobutanone (67)	312
cis-EtCH=CHEt	*n*-BuC≡C(NC)C=C=O	C_6H_6, 80°	2-(*n*-BuC≡C)-2-CN-3,4-diethylcyclobutanone (48) + isomer (12)	402
trans-EtCH=CHEt	*n*-BuC≡C(NC)C=C=O	C_6H_6, 80°	2-(*n*-BuC≡C)-2-CN-3,4-diethylcyclobutanone (38) + isomer (38)	402

TABLE III. [2+2] CYCLOADDITION OF KETENES TO ACYCLIC OLEFINS (*Continued*)

Reactant	Ketene or Ketene Source	Conditions	Product(s) and Yield(s) (%)	Refs.
$CH_2{=}CEt_2$	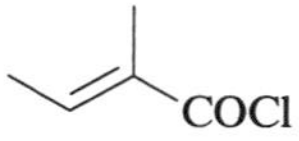 [structure] COCl	Et_3N	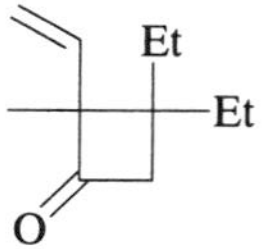 [structure: Et, Et, O] (12)	409
$Me_2C{=}CMe_2$	Cl_3CCOCl	Zn, Et_2O, $POCl_3$	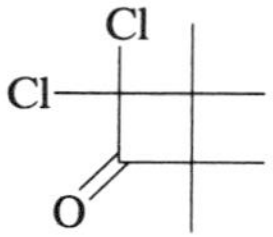 [structure: Cl, Cl, O] (61)	82, 298
	Cl_3CCOCl	Zn, Et_2O	" (75)	400
	$Cl_2CHCOCl$	Zn, Et_2O	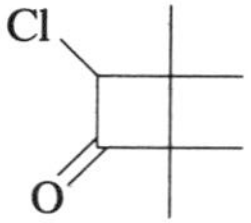 [structure: Cl, O] (55)	400
	Cl(NC)C=C=O	PhMe, 103°, 1.75 h	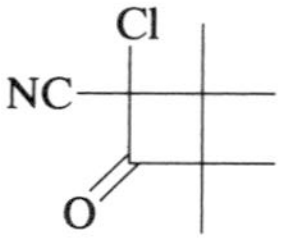 [structure: Cl, NC, O] (74)	312
	$(CF_3)_2C{=}C{=}O$	100°, 45 d	(—)	84
	$Me_2C{=}C{=}O$	100°	(—)	37

TABLE III. [2+2] CYCLOADDITION OF KETENES TO ACYCLIC OLEFINS (*Continued*)

Reactant	Ketene or Ketene Source	Conditions	Product(s) and Yield(s) (%)	Refs.
	$EtOCH_2COCl$	Et_3N	EtO, O (43)	391
	n-BuC≡C(NC)C=C=O	C_6H_6, 80°	n-BuC≡C, CN, O (50)	402
	Ph_2C=C=O	(—)	(—)	412
Me(Me)C=C(MOM)H	Cl_3CCOCl	Zn, $MeCH(OMe)_2$, reflux 40-60 h	Cl, Cl, O, MOM (76)	299, 404
	Cl(NC)C=C=O	PhMe	NC, Cl, O, MOM (15)	404
TMS (allyl)	RClHCCOCl	Et_3N, C_6H_{14}	Cl, R, CH_2TMS, O R = Cl (54) R = Me (62)	352

TABLE III. [2+2] CYCLOADDITION OF KETENES TO ACYCLIC OLEFINS (*Continued*)

Reactant	Ketene or Ketene Source	Conditions	Product(s) and Yield(s) (%)	Refs.
C_7				
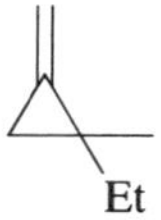 Et	$Me_2C{=}C{=}O$	120°	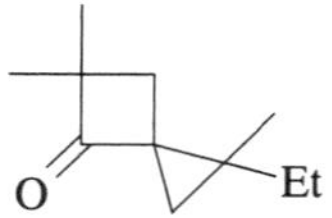 O, Et (29) 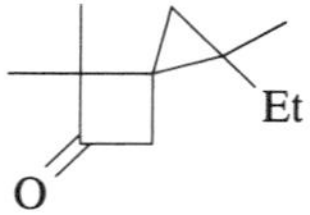 + O, Et (36)	408
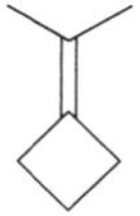	$Cl_2C{=}C{=}O$	—	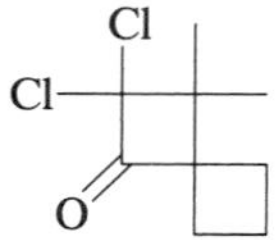 Cl, Cl, O (—)	413
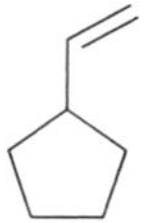	$Ph_2C{=}C{=}O$	100°	(—)	410
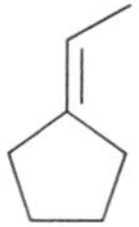	$Ph_2C{=}C{=}O$	100°, 12 h	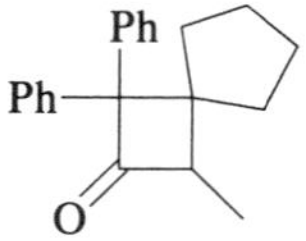 Ph, Ph, O (—)	410
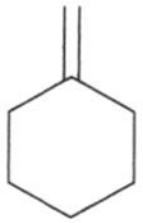	$Cl_2CHCOCl$	Et_3N	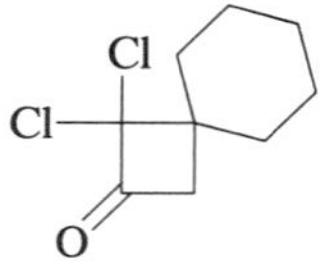 Cl, Cl, O (55)	414, 415
	MeCHClCOCl	Et_3N, C_6H_{14}	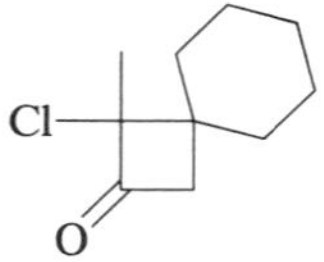 Cl, O (60)	399, 416

TABLE III. [2+2] CYCLOADDITION OF KETENES TO ACYCLIC OLEFINS (*Continued*)

Reactant	Ketene or Ketene Source	Conditions	Product(s) and Yield(s) (%)	Refs.
	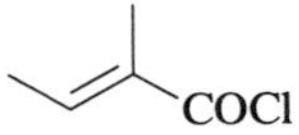 COCl	Et_3N	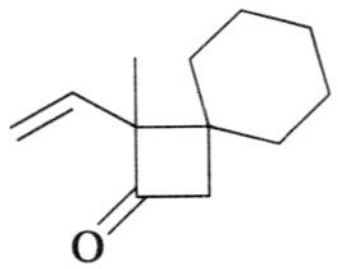 O (42)	409
	Ph, Ph C=C=O	100°	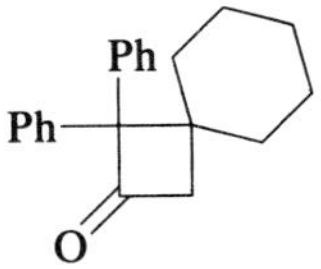 Ph, Ph, O (—)	410
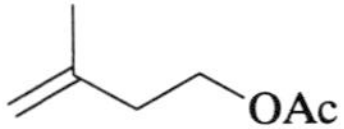 OAc	Cl_3CCOCl	Zn-Cu, Et_2O	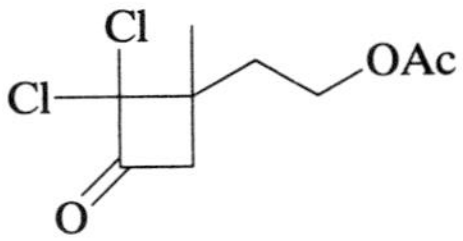 Cl, Cl, OAc, O (68)	417
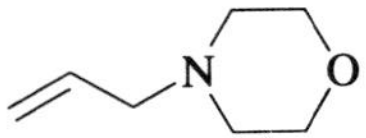 N, O	*n*-Bu, Et C=C=O	180°, 8 h	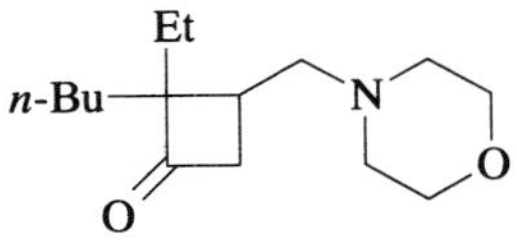 *n*-Bu, Et, N, O, O (28)	405
n-$C_5H_{11}CH{=}CH_2$	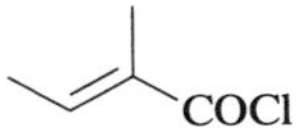 COCl	Et_3N	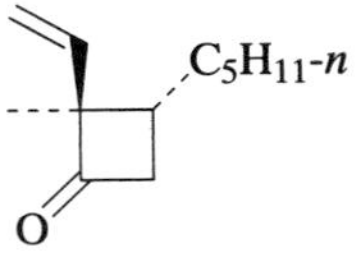 C_5H_{11}-*n*, O (28) 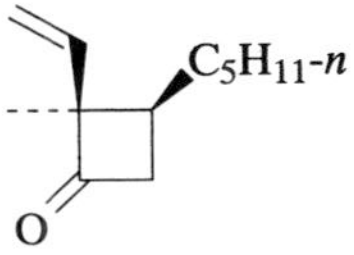 + C_5H_{11}-*n*, O (12)	409

TABLE III. [2+2] CYCLOADDITION OF KETENES TO ACYCLIC OLEFINS (*Continued*)

Reactant	Ketene or Ketene Source	Conditions	Product(s) and Yield(s) (%)	Refs.
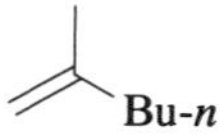 Bu-*n*	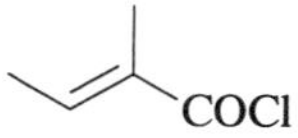 COCl	Et_3N	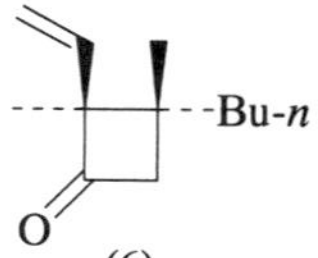 Bu-*n* (6) 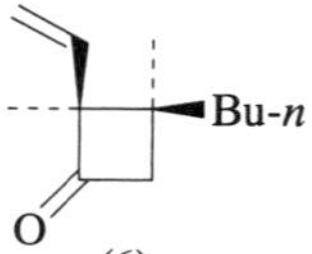 + Bu-*n* (6)	409
C_8				
	$Ph_2C=C=O$	70°, 6 h	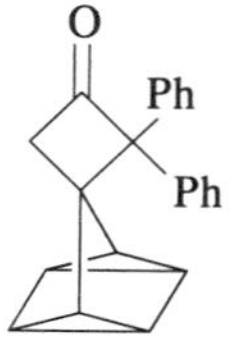 Ph, Ph (80)	418
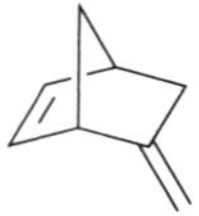	Cl_3CCOCl	Zn, Et_2O	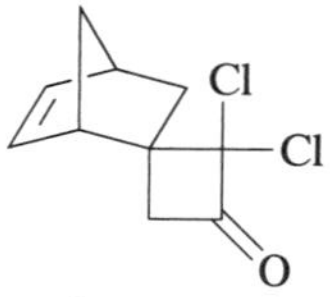 Cl, Cl (60)	400
	MeCHClCOCl	Et_3N, C_6H_{14}	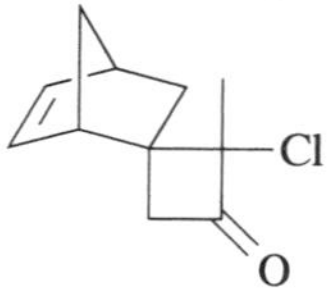 Cl (65)	399
	$Me_2CHCOCl$	Et_3N, C_6H_{14}	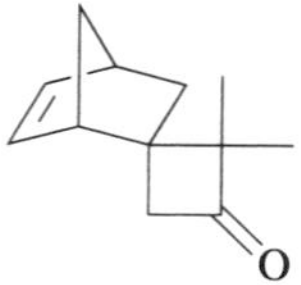 + 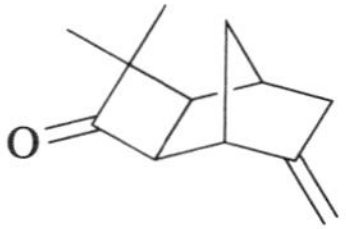 (10)	399

TABLE III. [2+2] CYCLOADDITION OF KETENES TO ACYCLIC OLEFINS (*Continued*)

Reactant	Ketene or Ketene Source	Conditions	Product(s) and Yield(s) (%)	Refs.
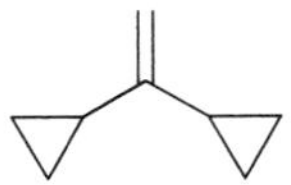	Cl_3CCOCl	Zn, Et_2O	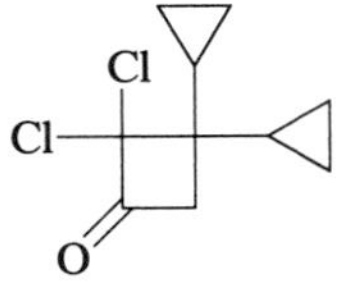 Cl, Cl, O (56-60)	419
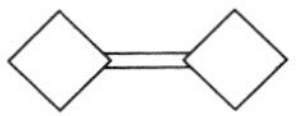	Cl_3CCOCl	Zn-Cu	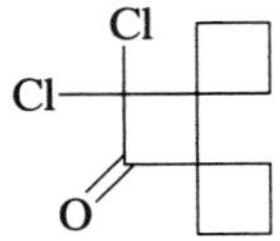 Cl, Cl, O (—)	420
	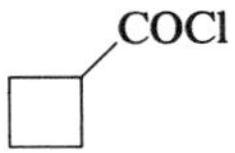 COCl	Et_3N	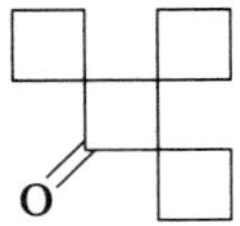 O (54)	421
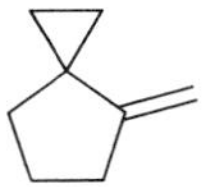	$Cl_2CHCOCl$	Et_3N, C_5H_{12}	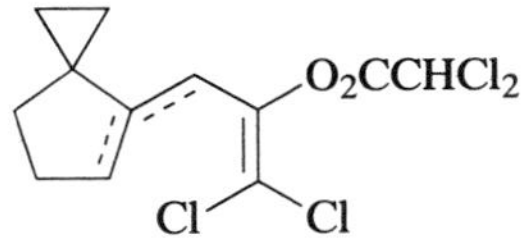 O_2CCHCl_2, Cl, Cl (30)	422
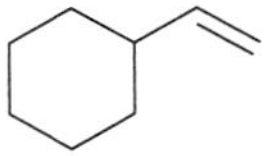	Ph, Ph, C=C=O	100°	(—)	422
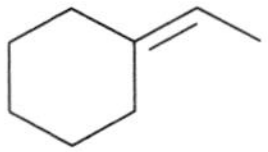	Ph, Ph, C=C=O	100°	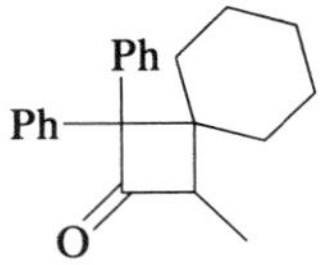 Ph, Ph, O (—)	410

TABLE III. [2+2] CYCLOADDITION OF KETENES TO ACYCLIC OLEFINS (*Continued*)

Reactant	Ketene or Ketene Source	Conditions	Product(s) and Yield(s) (%)	Refs.
N-allylpiperidine	n-Bu(Et)C=C=O	180°	2-n-Bu-2-Et-3-(piperidinomethyl)cyclobutanone (22)	405
C_9				
Allyl–OPh	Cl_3CCOCl	Zn, Et_2O	2,2-dichloro-3-(phenoxymethyl)cyclobutanone (47)	404
	Cl_3CCOCl	$MeCH(OMe)_2$, Zn, reflux 40-60 h	" (86)	299
Allyl–SPh	Cl_3CCOCl	$MeCH(OMe)_2$, Zn, reflux 40-60 h	2,2-dichloro-3-(phenylthiomethyl)cyclobutanone (33)	299
C_{10}				
R^1,R^1,R^2,R^2-cyclopropylidene=2,2-dimethylcyclopropylidene	$Me_2C=C=O$	—	dispiro cyclobutanone; R^1 = Me, R^2 = H (—); R^1 = H, R^2 = Me (—)	423

TABLE III. [2+2] CYCLOADDITION OF KETENES TO ACYCLIC OLEFINS (*Continued*)

Reactant	Ketene or Ketene Source	Conditions	Product(s) and Yield(s) (%)	Refs.
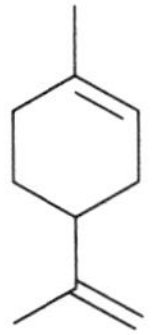	Cl_3CCOCl	Zn, Et_2O, ultrasound	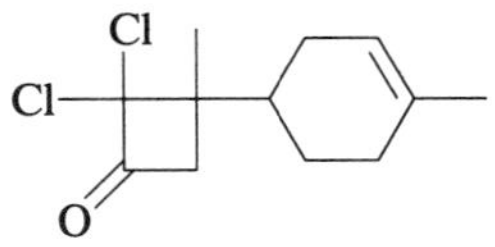 Cl, Cl, O (80)	424
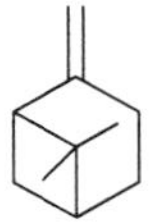	MeCHClCOCl	Et_3N, C_6H_{14}	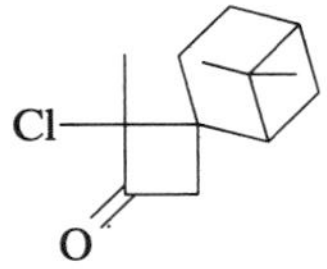 Cl, O (73)	416, 399
R^2, O, R^1	Cl_3CCOCl	Zn, Et_2O	Cl, Cl, O, O, R^2, R^1 **I** + R^1, R^2, O, O, Cl, Cl **II**	404
			R^1 / R^2 / **I + II** / **I:II**: Me / H / (76) / 3:7; H / Me / (68) / 15:85	
n-$C_8H_{17}CH{=}CH_2$	Cl_3CCOCl	Zn-Cu, Et_2O, $POCl_3$	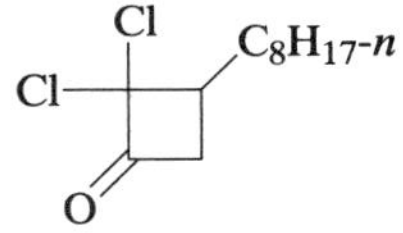 Cl, Cl, O, C_8H_{17}-n (88)	82,83

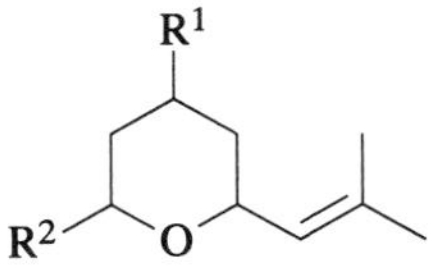

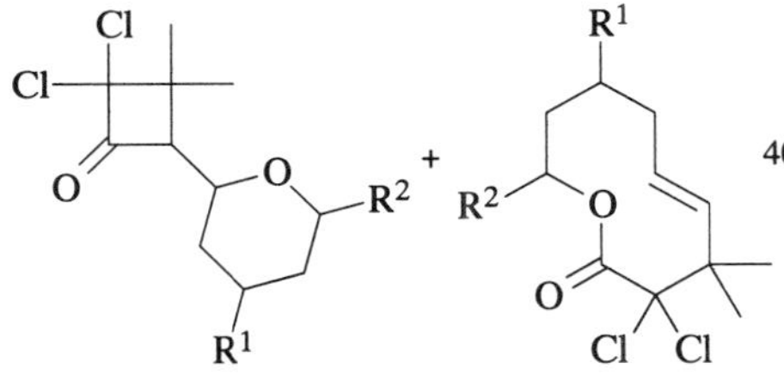

R^1	R^2	**I + II**	**I:II**
Me	H	(76)	3:7
H	Me	(68)	15:85

TABLE III. [2+2] CYCLOADDITION OF KETENES TO ACYCLIC OLEFINS (*Continued*)

Reactant	Ketene or Ketene Source	Conditions	Product(s) and Yield(s) (%)	Refs.
	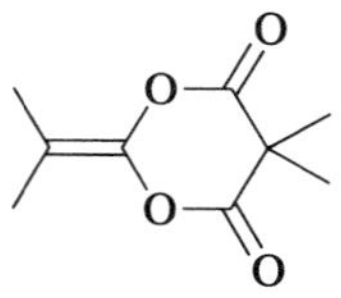	PhMe, 110°, K_2CO_3 (cat.)	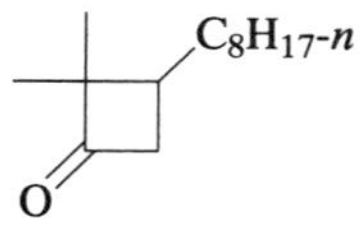 C_8H_{17}-*n* (67)	293
C_{11} 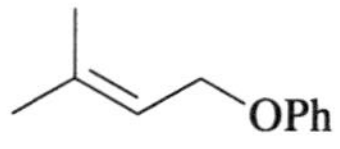 OPh	Cl_3CCOCl	Zn, $POCl_3$, glyme, reflux 40-60 h	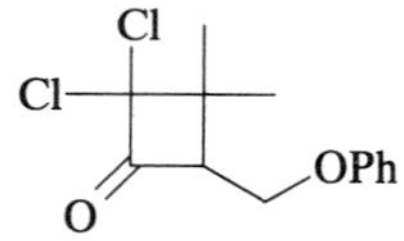 Cl Cl OPh O (65)	299
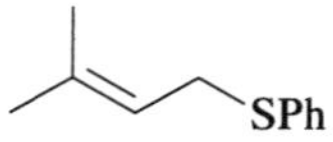 SPh	Cl_3CCOCl	Zn, Et_2O	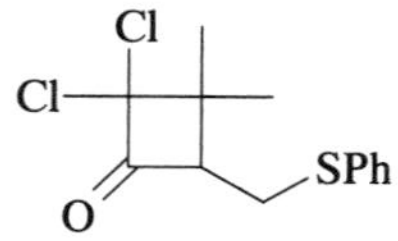 Cl Cl SPh O (19) 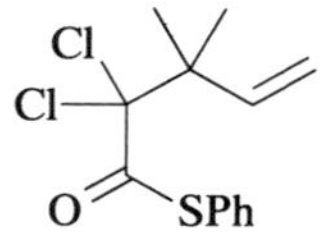 **I** + Cl Cl O SPh (26)	404
	Cl_3CCOCl	Zn, $POCl_3$, glyme, reflux 40-60 h	" (66)	299
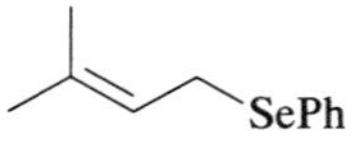 SePh	Cl_3CCOCl	Zn, Et_2O	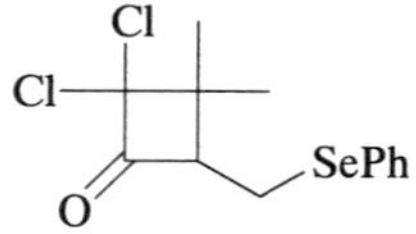 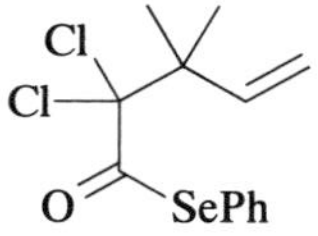 Cl Cl SePh O + Cl Cl O SePh	404

TABLE III. [2+2] CYCLOADDITION OF KETENES TO ACYCLIC OLEFINS (*Continued*)

Reactant	Ketene or Ketene Source	Conditions	Product(s) and Yield(s) (%)	Refs.
 Bu-*t*	$Cl_2CHCOCl$	Et_3N, C_6H_{14}	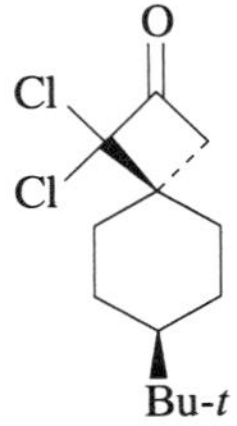 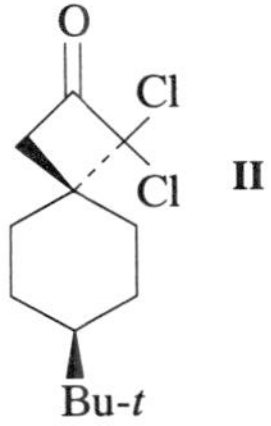 O, Cl, Cl, Bu-*t* **I** + O, Cl, Cl, Bu-*t* **II** **I + II** (50), **I:II** = 4:1	414
C_{14} 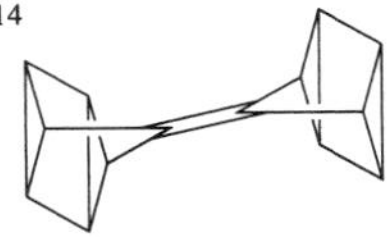	Ph, Ph C=C=O	—	(—)	425
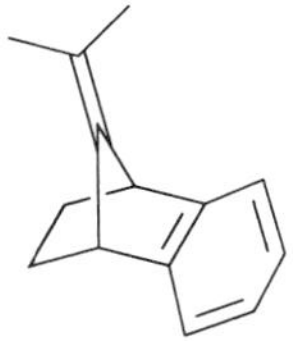	Cl_3CCOCl	Zn-Cu, Et_2O, $POCl_3$	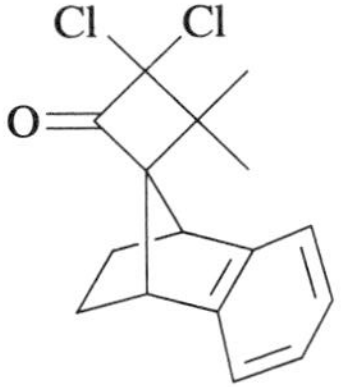 Cl, Cl, O (40) 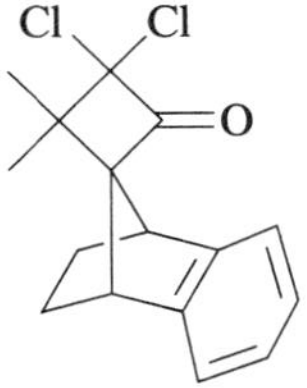 + Cl, Cl, O (13)	426
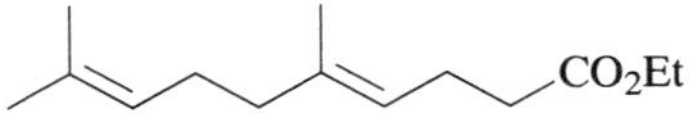 CO_2Et	Cl_3CCOCl	Zn-Cu, Et_2O, $POCl_3$	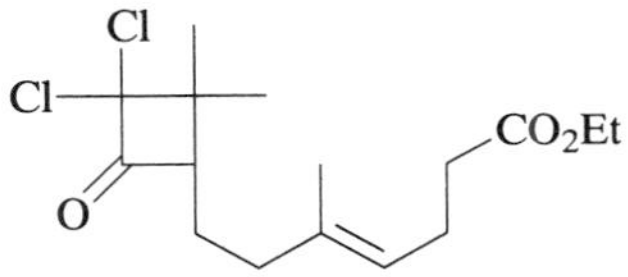 Cl, Cl, O, CO_2Et (75)	75

TABLE IV. [2+2] CYCLOADDITION OF KETENES TO CYCLIC OLEFINS

Reactant	Ketene or Ketene Source	Conditions	Product(s) and Yield(s) (%)	Refs.
C_4				
	$(CF_3)_2C=C=O$	C_6H_6	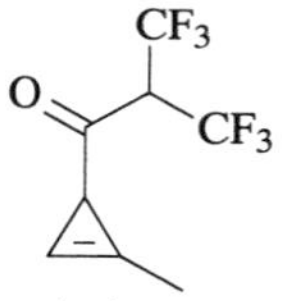 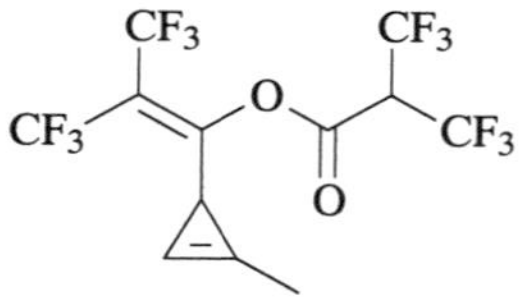 (—) + (—)	79
	$t\text{-Bu}(NC)C=C=O$	C_6H_6, 5°	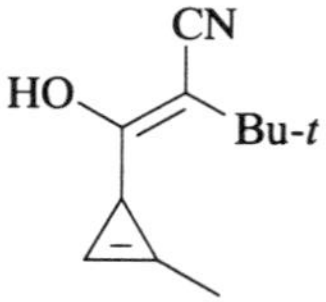 (73) 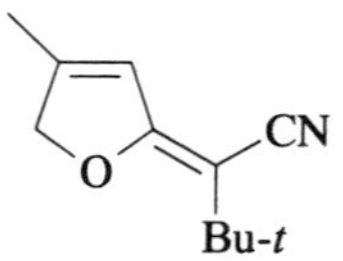 + (17)	79
	Cl_3CCOCl	Zn, Et_2O	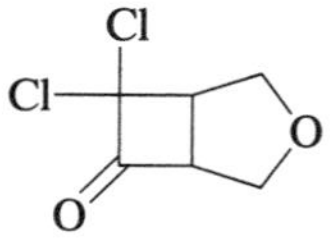 (40)	427, 404, 428
	$s\text{-Bu}(Et)C=C=O$	180°	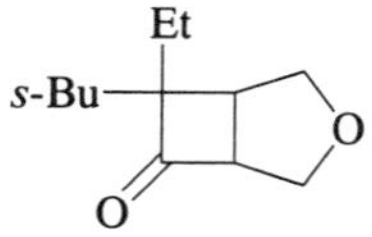 (30)	321
C_5				
	$t\text{-Bu}(NC)C=C=O$	25°, 5 h	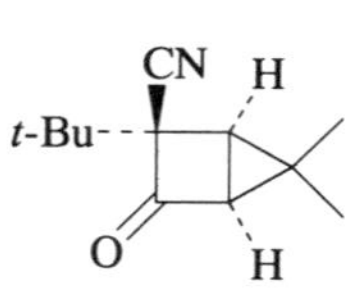 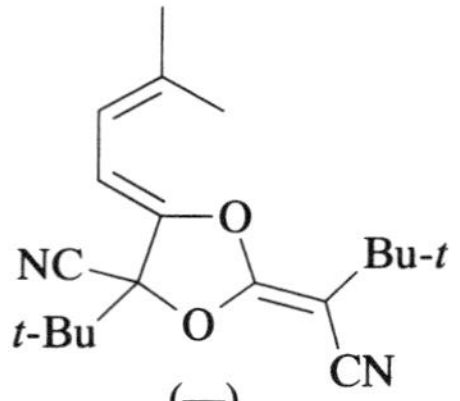 (—) + (—)	429

TABLE IV. [2+2] CYCLOADDITION OF KETENES TO CYCLIC OLEFINS (*Continued*)

Reactant	Ketene or Ketene Source	Conditions	Product(s) and Yield(s) (%)	Refs.
cyclopentene	$Cl_2CHCOCl$	Et_3N, C_5H_{12}	(67)	5, 430, 431
	Cl_3CCOCl	Zn, Et_2O, ultrasound	" (70)	424
	MeCHBrCOCl	Et_3N, Et_2O	(—)	432
	MeCHClCOCl	Et_3N, Et_2O	(77)	416, 432
	$(CF_3)_2C{=}C{=}O$	175°, 8 h	(—)	84
	$EtOCH_2COCl$	Et_3N	(56)	391

TABLE IV. [2+2] CYCLOADDITION OF KETENES TO CYCLIC OLEFINS (*Continued*)

Reactant	Ketene or Ketene Source	Conditions	Product(s) and Yield(s) (%)	Refs.
	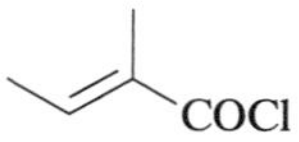 COCl	Et_3N	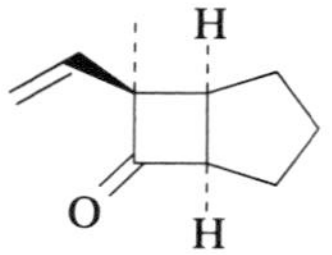 I 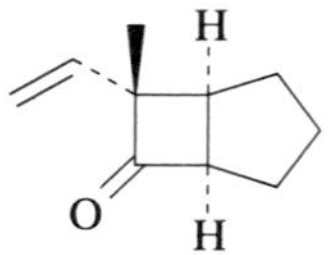 + II; **I + II** (28), **I:II** = 3:7	409
	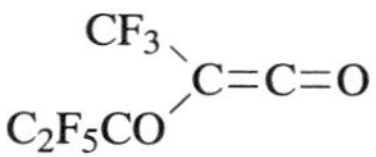 CF_3, C_2F_5CO C=C=O	180°, 8 h	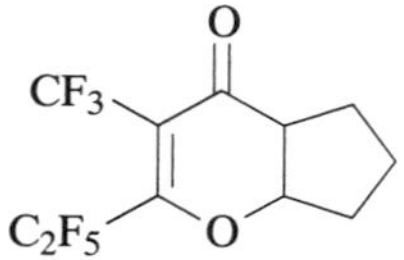 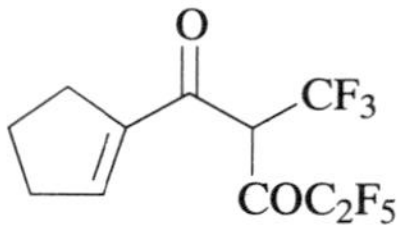 CF_3, C_2F_5 (—) + O, CF_3, COC_2F_5 (—)	346
	PhCHClCOCl	Et_3N	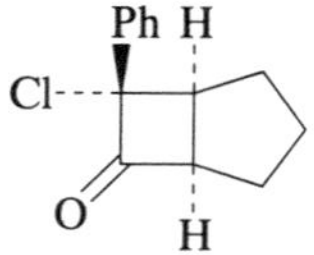 Ph H Cl O H (—)	433
	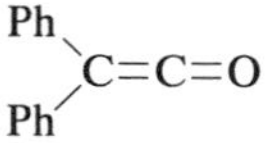 Ph, Ph C=C=O	MeCN, 100°, 9 d	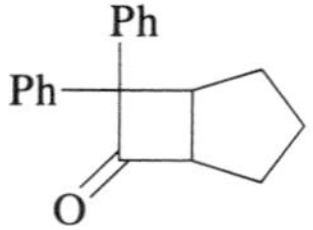 Ph Ph O (62)	386
C_6 	$Cl_2CHCOCl$	Et_3N, Et_2O	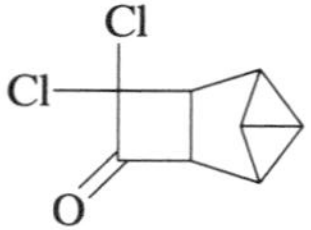 Cl Cl O (70)	434, 435

TABLE IV. [2+2] CYCLOADDITION OF KETENES TO CYCLIC OLEFINS (*Continued*)

Reactant	Ketene or Ketene Source	Conditions	Product(s) and Yield(s) (%)	Refs.
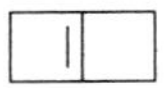	$CH_2=C=O$	1. CH_2Cl_2, -78° 2. Al_2O_3	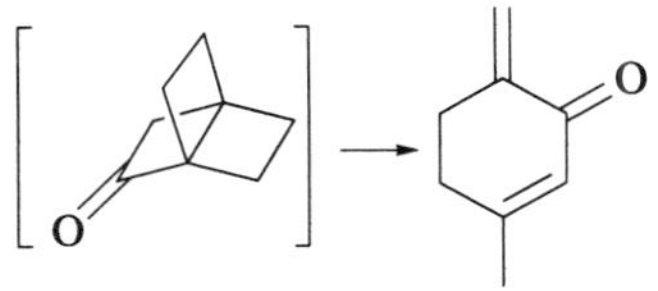 [O] → O (—)	436
	$(CF_3)_2C=C=O$	25°	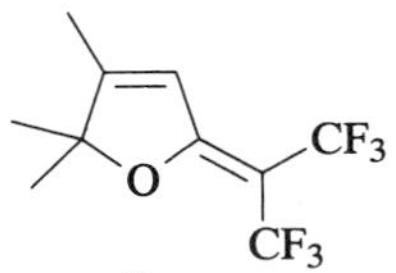 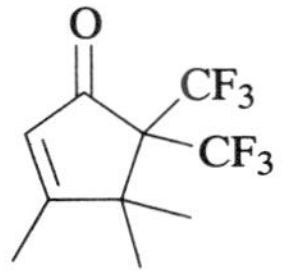 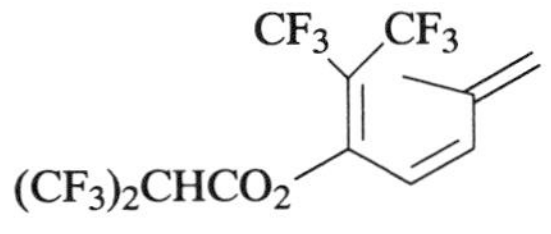 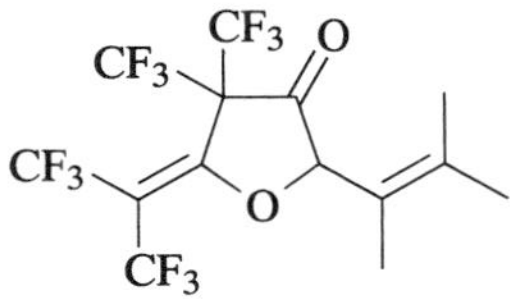 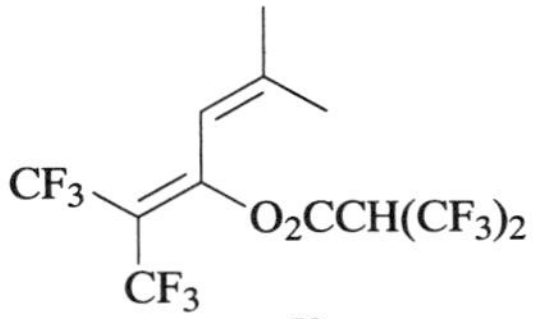 CF_3 CF_3 O **I** + O CF_3 CF_3 **II** + CF_3 CF_3 $(CF_3)_2CHCO_2$ **III** + CF_3 CF_3 O CF_3 CF_3 O **IV** + CF_3 CF_3 $O_2CCH(CF_3)_2$ **V**	437

Solvent	I	II	III	IV	V
Cyclohexane	(63)	(27)	(5)	(1)	(4)
$Cl(CH_2)_2Cl$	(61)	(9)	(11)	(18)	(1)
MeCN	(21)	(1)	(0)	(72)	(6)

TABLE IV. [2+2] CYCLOADDITION OF KETENES TO CYCLIC OLEFINS (*Continued*)

Reactant	Ketene or Ketene Source	Conditions	Product(s) and Yield(s) (%)	Refs.
	t-Bu(NC)C=C=O	C_6H_6, 25°, 15 min	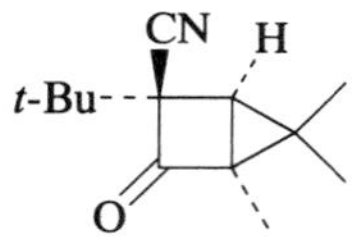 (28) 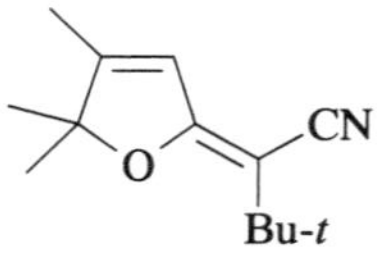 + (7) 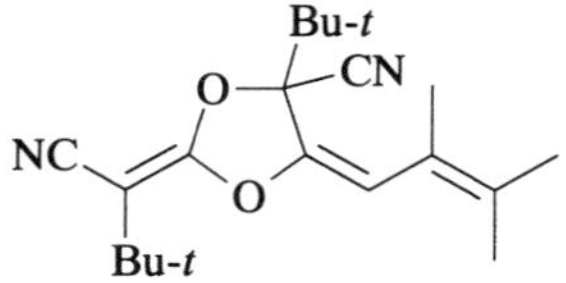 + (40)	429
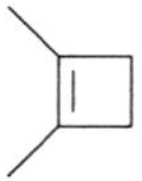	Cl_3CCOCl	Zn, Et_2O	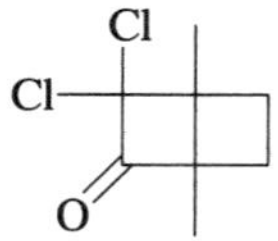 (17) + 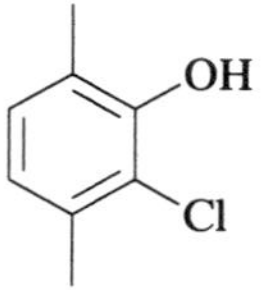 (3)	438
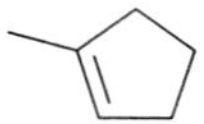	Cl_3CCOBr	Zn, Et_2O	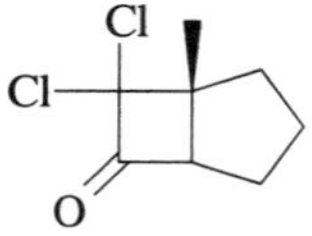 (18)	439, 440
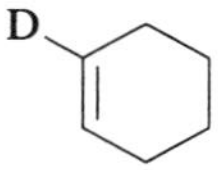	Ph_2C=C=O	100°, 10 d	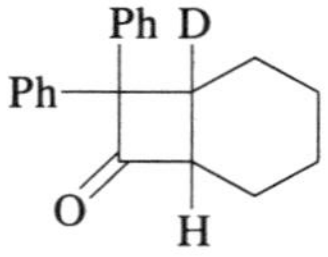 (—) + 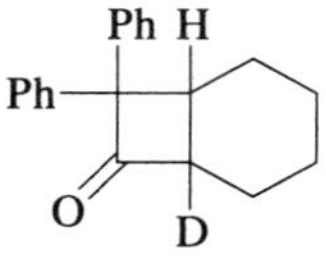 (—)	53

TABLE IV. [2+2] CYCLOADDITION OF KETENES TO CYCLIC OLEFINS (*Continued*)

Reactant	Ketene or Ketene Source	Conditions	Product(s) and Yield(s) (%)	Refs.
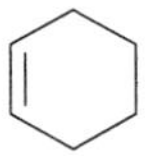	$Cl_2CHCOCl$	Et_3N	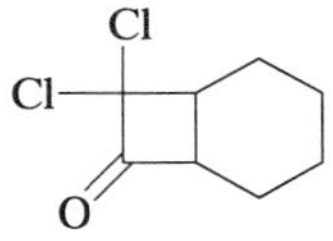 (57)	431, 441
	Cl_3CCOBr	Zn, Et_2O	" (70)	400, 439
	Cl_3CCOCl	Zn, Et_2O	" (90)	400
	Cl_3CCOCl	Zn-Cu, Et_2O	" (52)	403
	MeCHBrCOCl	Et_3N	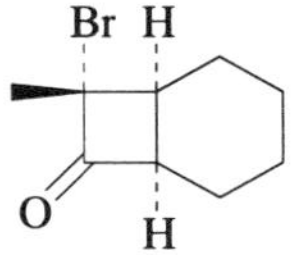 (—)	432
	MeCHClCOCl	Et_3N, C_6H_{14}	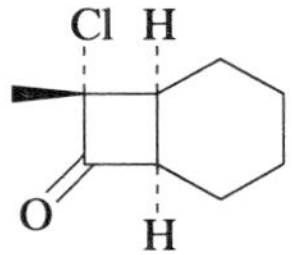 **I** + 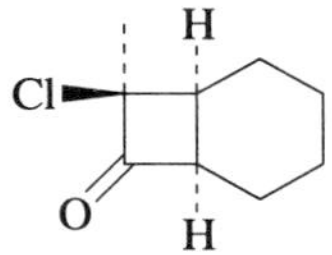 **II** **I** + **II** (26), **I**:**II** = 1:5.5	442, 416
	Cl(NC)C=C=O	PhMe, reflux	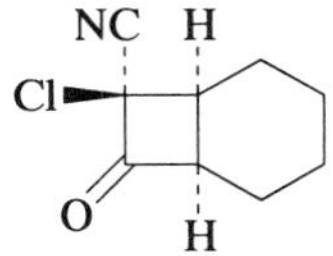 (64)	443, 312

TABLE IV. [2+2] CYCLOADDITION OF KETENES TO CYCLIC OLEFINS (*Continued*)

Reactant	Ketene or Ketene Source	Conditions	Product(s) and Yield(s) (%)	Refs.
	$(CF_3)_2C{=}C{=}O$	100°, 7 weeks	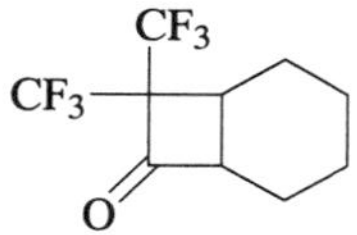 CF_3, CF_3, O (—)	84
	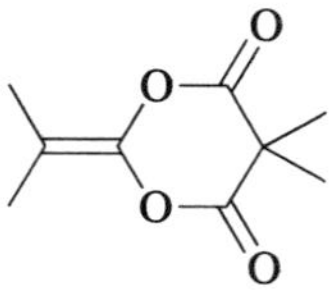 O, O, O, O	PhMe, 110°, K_2CO_3 (cat.)	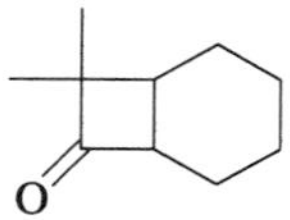 O (32)	293
	$EtOCH_2COCl$	Et_3N	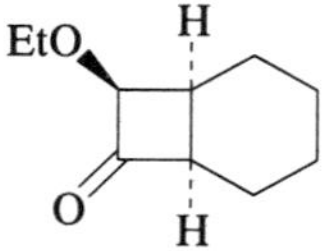 EtO, H, H, O (46) 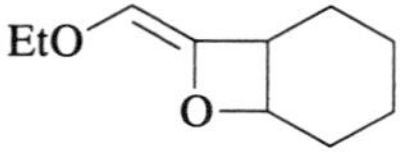 + EtO, O (18) 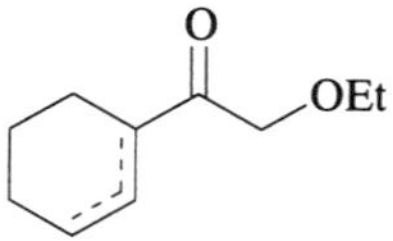 + O, OEt (18)	391
	$CF_3(C_2F_5CO)C{=}C{=}O$	95°, 22 d	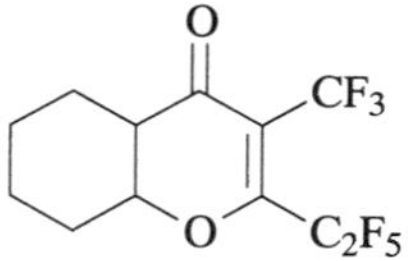 O, CF_3, O, C_2F_5 (13) 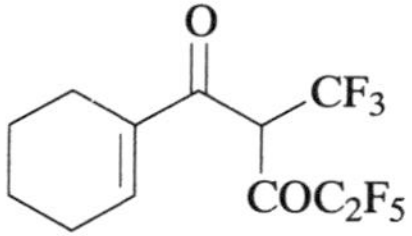 + O, CF_3, COC_2F_5 (10)	346

TABLE IV. [2+2] CYCLOADDITION OF KETENES TO CYCLIC OLEFINS (*Continued*)

Reactant	Ketene or Ketene Source	Conditions	Product(s) and Yield(s) (%)	Refs.
	MeCOCOTMS	$h\nu$, C_5H_{10}	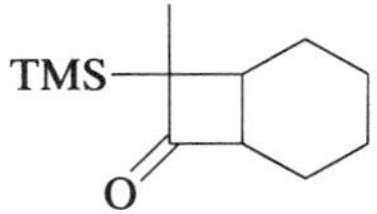 TMS, O (—)	307
	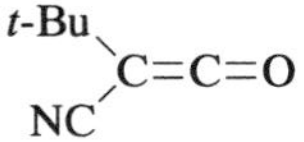 t-Bu(NC)C=C=O	C_6H_6, reflux, 3 h	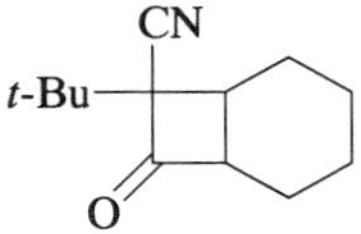 CN, t-Bu, O (63)	353
	PhCHClCOCl	Et_3N	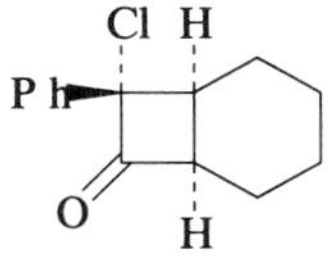 Cl H, Ph, O, H (70)	433
	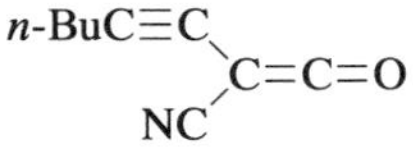 n-BuC≡C(NC)C=C=O	C_6H_6, reflux	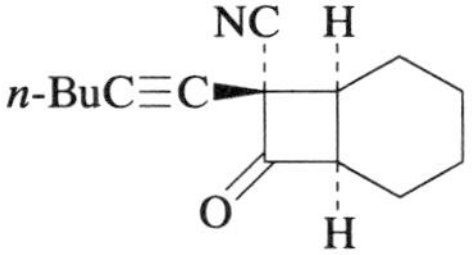 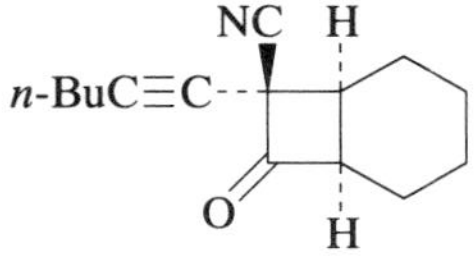 NC H, n-BuC≡C, O, H **I** + NC H, n-BuC≡C, O, H **II** **I** + **II** (44), **I**:**II** = 4:1	402

TABLE IV. [2+2] CYCLOADDITION OF KETENES TO CYCLIC OLEFINS (*Continued*)

Reactant	Ketene or Ketene Source	Conditions	Product(s) and Yield(s) (%)	Refs.
	$Ph_2C{=}C{=}O$	100°, 10 d	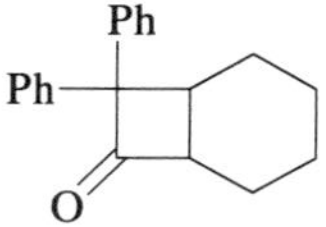 (60)	444, 412, 445
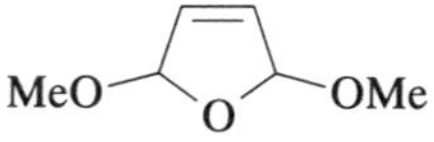 2,5-dimethoxy-2,5-dihydrofuran (MeO, OMe)	Cl_3CCOCl	Zn, Et_2O	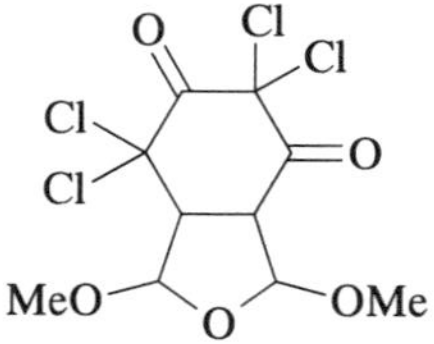 (—)	427
C_7				
norbornadiene	$Cl_2CHCOCl$	Et_3N	(25)	431, 441
	$(CF_3)_2C{=}C{=}O$	100°	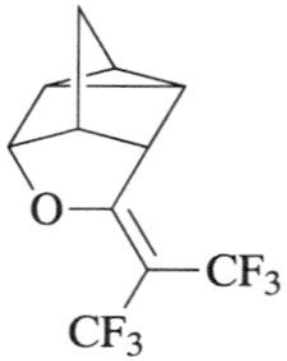 (81)	84
	$CF_3(C_2F_5CO)C{=}C{=}O$	Phenothiazine, 60°, 2 h	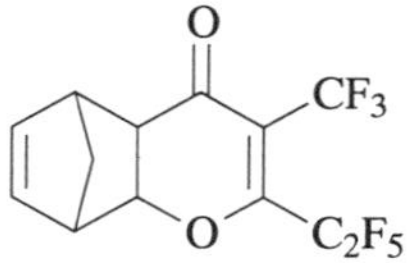 (—)	346

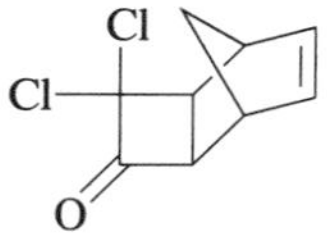

TABLE IV. [2+2] CYCLOADDITION OF KETENES TO CYCLIC OLEFINS (*Continued*)

Reactant	Ketene or Ketene Source	Conditions	Product(s) and Yield(s) (%)	Refs.
	RCHXCOCl	Et_3N, solvent	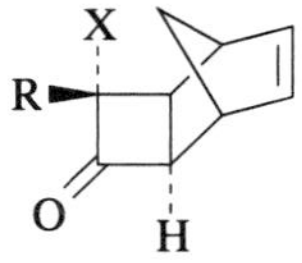 I, 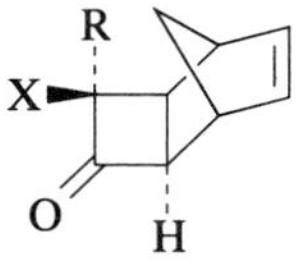 II (R, X; O; H)	446
	t-Bu(NC)C=C=O	C_6H_6, 78°	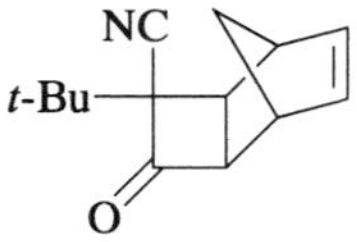 t-Bu, NC, O (26) 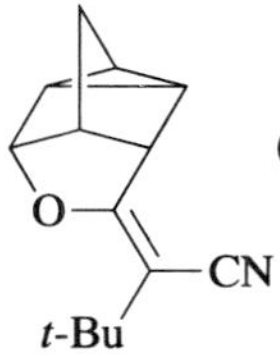 + O, CN, t-Bu (19)	392, 447
	$Ph_2C=C=O$	C_6H_6, 65°, 12 d	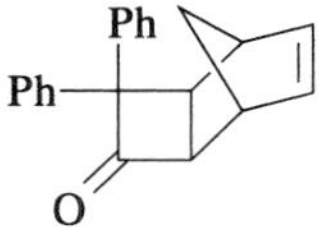 Ph, Ph, O (60)	418

Solvent	R	X	I + II	I:II
C_6H_{12}	Me	Cl	(16)	100:0
MeCN	Me	Cl	(7)	40:60
C_6H_{12}	Me	Br	(17)	100:0
MeCN	Me	Br	(5)	0:100
C_6H_{12}	Et	Cl	(14)	100:0
MeCN	Et	Cl	(8)	44:56
C_6H_{12}	*i*-Pr	Cl	(6)	100:0
MeCN	*i*-Pr	Cl	(1)	100:0
C_6H_{12}	*t*-Bu	Br	No 1:1 adduct	
MeCN	*t*-Bu	Br	No 1:1 adduct	

TABLE IV. [2+2] CYCLOADDITION OF KETENES TO CYCLIC OLEFINS (*Continued*)

Reactant	Ketene or Ketene Source	Conditions	Product(s) and Yield(s) (%)	Refs.
 (norbornene)	$Cl_2CHCOCl$	Et_3N, C_6H_{12}	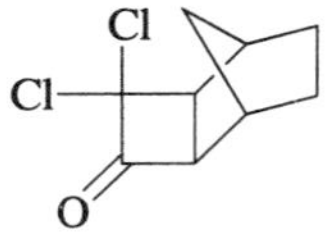 Cl, Cl– (structure) (10)	441
	Cl_3CCOCl	Zn, Et_2O	" (64)	400
	"	Zn, Et_2O, $POCl_3$	" (70)	298
	"	Zn, Et_2O, ultrasound	" (75)	424
	MeCHClCOCl	Et_3N, MeCN	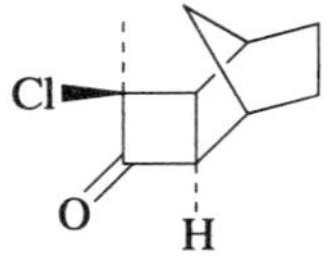 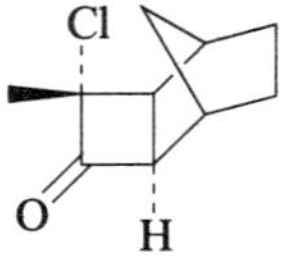 Cl (structure) **I** + Cl (structure) **II** **I** + **II** (5), **I**:**II** = 1:1	446
	MeCHClCOCl	Et_3N, C_6H_{12}	**II** (12)	446
	$(CF_3)_2C{=}C{=}O$	175°, 8 h	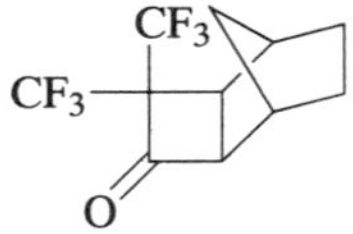 CF_3, CF_3 (structure) (97)	84
	$(CF_3)(C_2F_5CO)C{=}C{=}O$	60°, 16 h	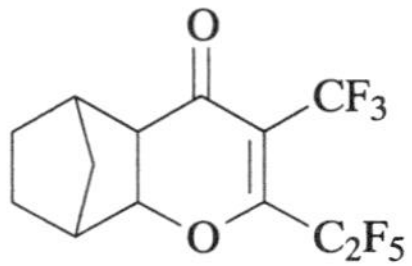 CF_3, C_2F_5 (structure) (67)	346

TABLE IV. [2+2] CYCLOADDITION OF KETENES TO CYCLIC OLEFINS (*Continued*)

Reactant	Ketene or Ketene Source	Conditions	Product(s) and Yield(s) (%)	Refs.
	$Ph_2C{=}C{=}O$	C_6H_6, reflux 14 d	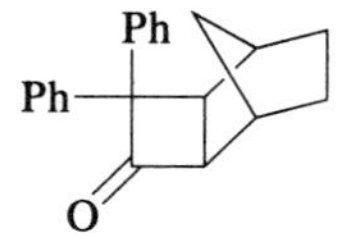 **I** (84)	448
		Excess Olefin, 70°, 2 d	**I** (45-68) + 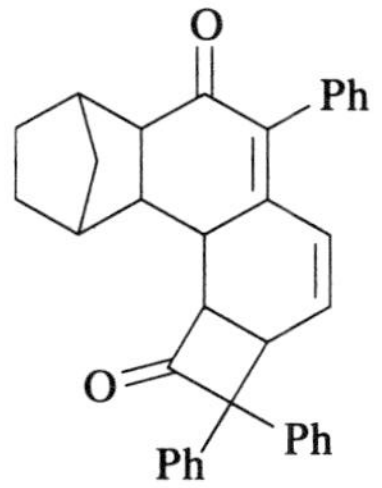 (10-16)	448
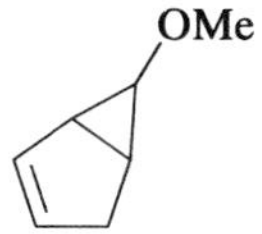	$Cl_2CHCOCl$	Et_3N, C_5H_{12}	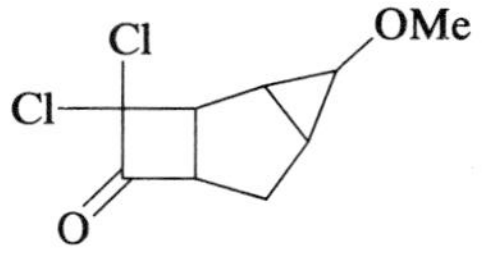 (80)	89
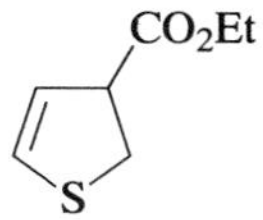	$Cl_2CHCOCl$	Et_3N, CCl_4	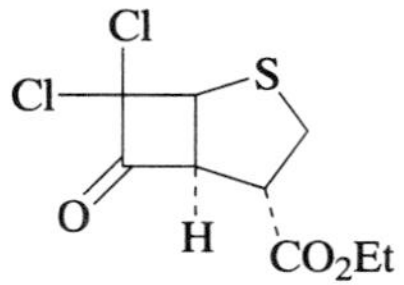 (33)	449

TABLE IV. [2+2] CYCLOADDITION OF KETENES TO CYCLIC OLEFINS (*Continued*)

Reactant	Ketene or Ketene Source	Conditions	Product(s) and Yield(s) (%)	Refs.
	$Cl_2CHCOCl$	Et_3N	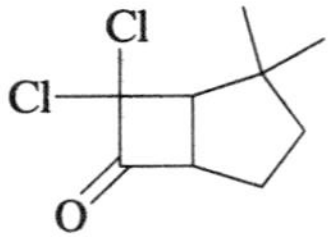 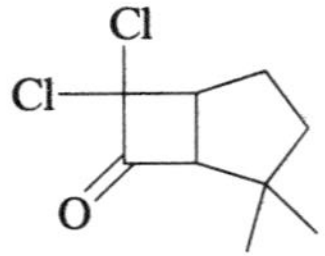 Cl, Cl ... O **I** + Cl, Cl ... O **II** **I** + **II** (51), **I**:**II** = 5:1	450, 451
	t-Bu(NC)C=C=O	C_6H_6, 78°	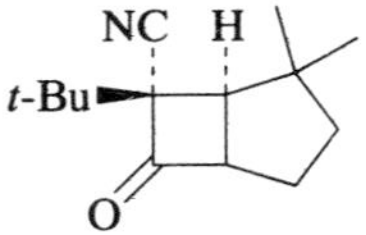 t-Bu, NC, H ... O (—)	450
	Ph_2C=C=O	100°, 11 d	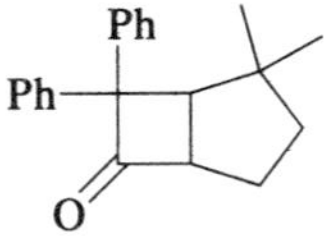 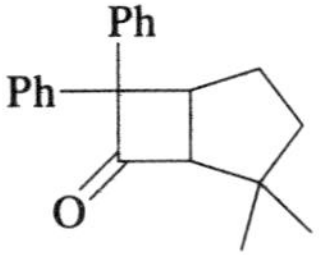 Ph, Ph ... O **I** + Ph, Ph ... O **II** **I** + **II** (90), **I**:**II** = 5.7:1	450
	MeCHClCOCl	Et_3N	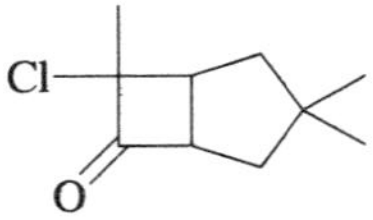 Cl ... O (—)	452
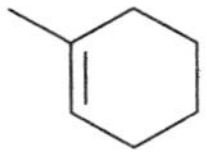	Cl_3CCOCl	Zn-Cu, Et_2O, $POCl_3$	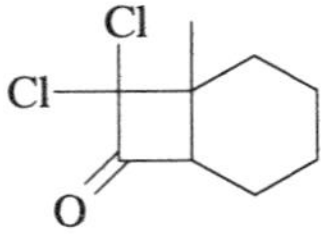 Cl, Cl ... O (83)	82,83

TABLE IV. [2+2] CYCLOADDITION OF KETENES TO CYCLIC OLEFINS (*Continued*)

Reactant	Ketene or Ketene Source	Conditions	Product(s) and Yield(s) (%)	Refs.
	Cl_3CCOCl	Zn, Et_2O, $POCl_3$	" (79)	298
	"	Zn, Et_2O	" (80)	400, 439, 440
	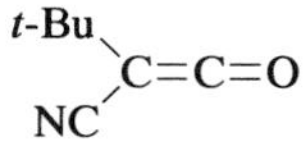 t-Bu(NC)C=C=O	C_6H_6, 78°	(—)	55
	Cl_3CCOCl	Zn-Cu, Et_2O, $POCl_3$	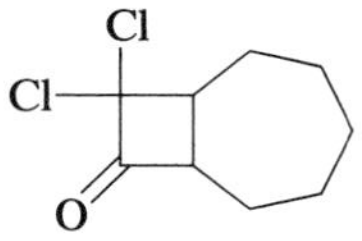 (85)	83, 453
	MeCHClCOCl	Et_3N, C_6H_{14}	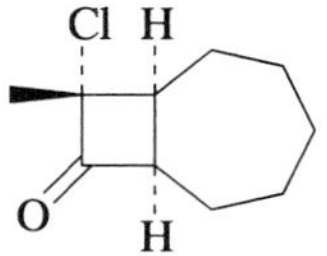 (66)	416
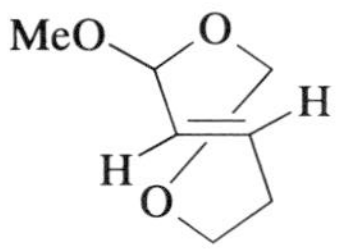	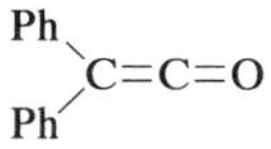 $Ph_2C=C=O$	-45 to 60°	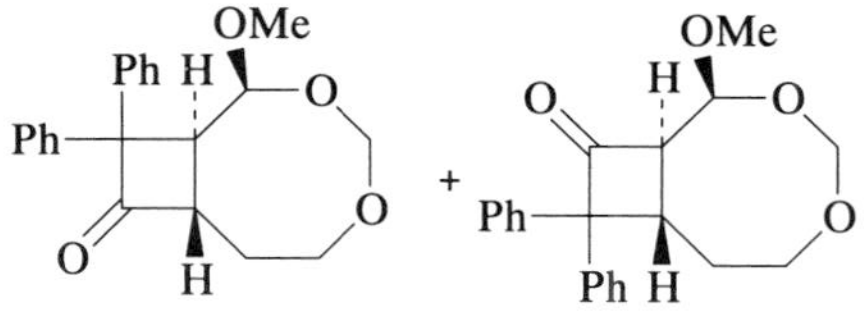 I + II (—), I:II = 7:1	454

TABLE IV. [2+2] CYCLOADDITION OF KETENES TO CYCLIC OLEFINS (*Continued*)

Reactant	Ketene or Ketene Source	Conditions	Product(s) and Yield(s) (%)	Refs.
C_8				
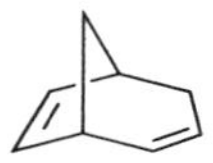	Cl_3CCOCl	Zn, Et_2O	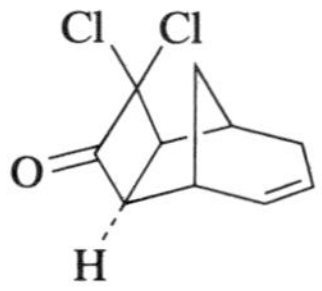 Cl Cl, O, H (14) 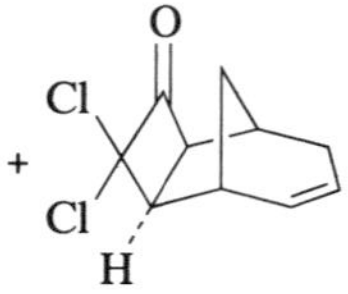 + O, Cl, Cl, H (70)	455
	t-Bu, NC, C=C=O	C_6H_6, 78°	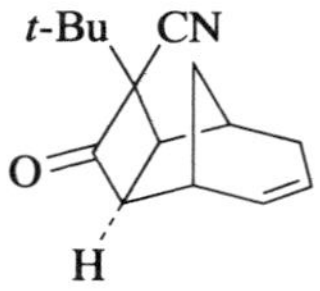 t-Bu CN, O, H (—)	394
	Cl_3CCOCl	Zn, Et_2O	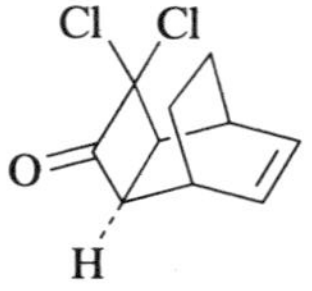 Cl Cl, O, H (86)	76
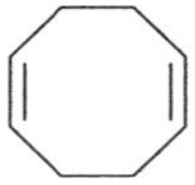	$Cl_2CHCOBr$	Et_3N, C_6H_{14}	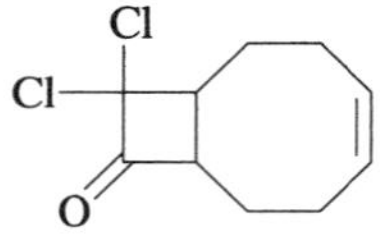 Cl, Cl, O (53)	403
	Cl_3CCOCl	Zn, Et_2O, ultrasound	" (70)	424
	Cl_3CCOCl	Zn-Cu, Et_2O	" (60)	456, 453

TABLE IV. [2+2] CYCLOADDITION OF KETENES TO CYCLIC OLEFINS (*Continued*)

Reactant	Ketene or Ketene Source	Conditions	Product(s) and Yield(s) (%)	Refs.
	MeCHClCOCl	Et_3N, C_6H_{14}	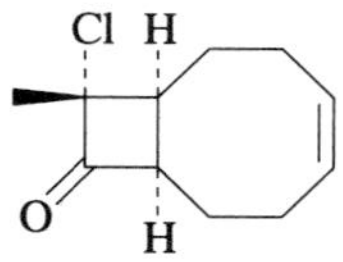 (50)	416
	Cl(NC)C=C=O	PhMe, reflux	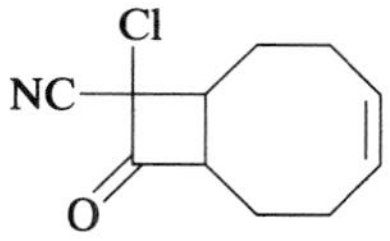 (78)	443
	$Me_2C=C=O$	—	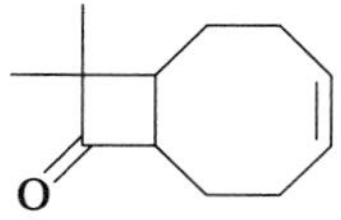 (55)	457
	$Ph_2C=C=O$	60°, 1 h	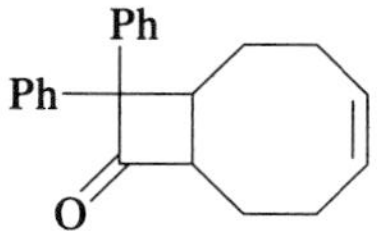 (—)	458
	Cl_3CCOCl	Zn, Et_2O	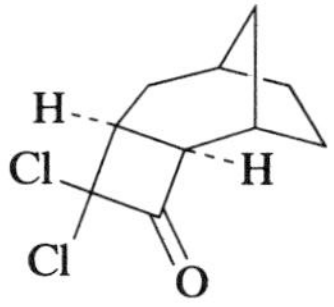 (34) + 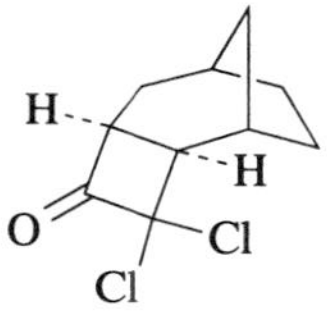 (34)	455
	Cl_3CCOCl	Zn, Et_2O	(—)	76

TABLE IV. [2+2] CYCLOADDITION OF KETENES TO CYCLIC OLEFINS (*Continued*)

Reactant	Ketene or Ketene Source	Conditions	Product(s) and Yield(s) (%)	Refs.
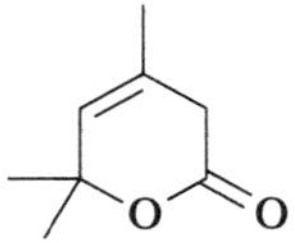	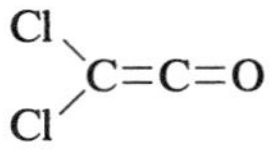 $Cl_2C{=}C{=}O$	—	(—)	459
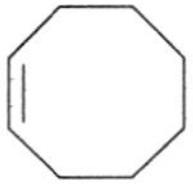	$Cl_2CHCOCl$	Et_3N	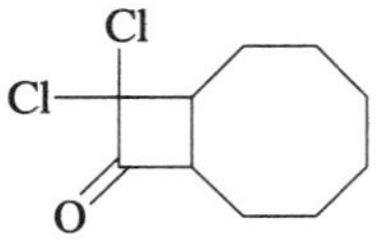 Cl, Cl (50)	57
	MeCOCl	Et_3N	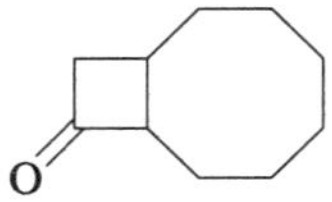 (27)	406
	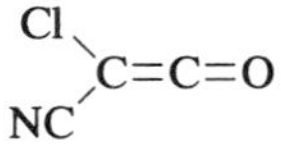 Cl(NC)C=C=O	PhMe, reflux	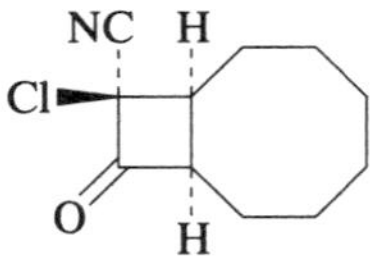 NC, H, Cl, H (72)	443
	MeCHClCOCl	Et_3N, C_6H_{14}, 25°	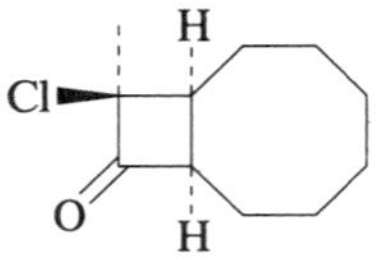 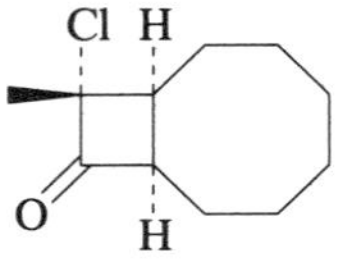 Cl, H, H **I** + Cl, H, H **II** **I** + **II** (15), **I**:**II** = 5:1	442
	MeCHClCOCl	Et_3N, C_6H_{14}, 25°	**II** (83)	416

TABLE IV. [2+2] CYCLOADDITION OF KETENES TO CYCLIC OLEFINS (*Continued*)

Reactant	Ketene or Ketene Source	Conditions	Product(s) and Yield(s) (%)	Refs.
	EtCOCl	Et_3N	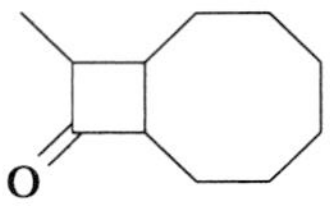 (17)	460
	$Me_2C{=}C{=}O$	70-75°, 12 h	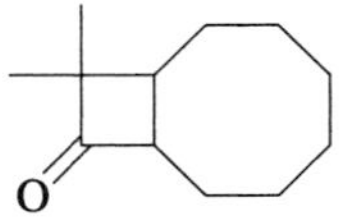 (77)	457
	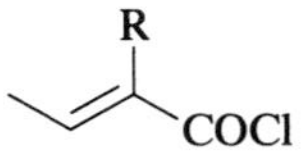 R, COCl	Et_3N	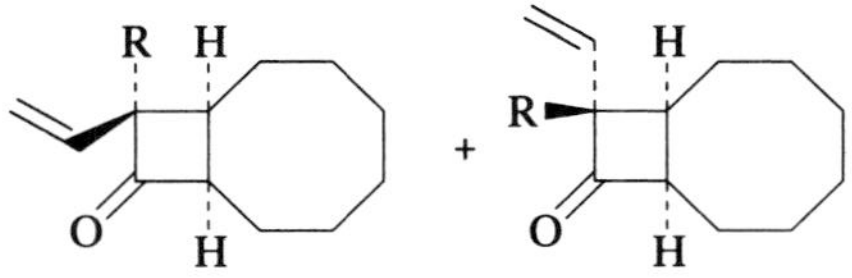 R, H, H, O (I) + R, H, H, O (II) R = Me, **I** + **II** (60), **I**:**II** = 1:3 R = Et, **I** + **II** (52), **I**:**II** = 15:85	409
	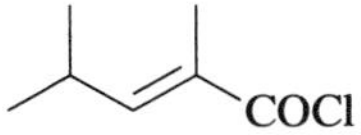 COCl	Et_3N	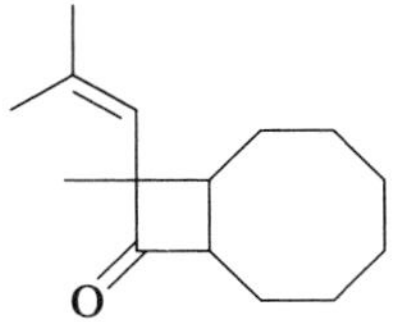 O (33)	409
	$Cl_2CHCOCl$	Et_3N	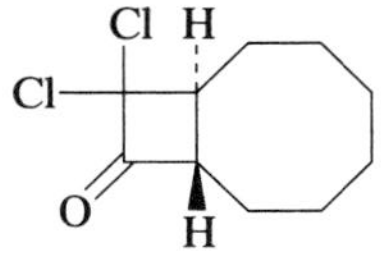 Cl, Cl, H, H, O (100)	57

TABLE IV. [2+2] CYCLOADDITION OF KETENES TO CYCLIC OLEFINS (*Continued*)

Reactant	Ketene or Ketene Source	Conditions	Product(s) and Yield(s) (%)	Refs.
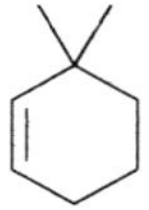	$Cl_2CHCOCl$	Et_3N	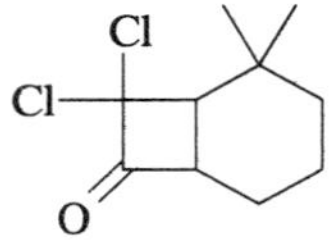 I 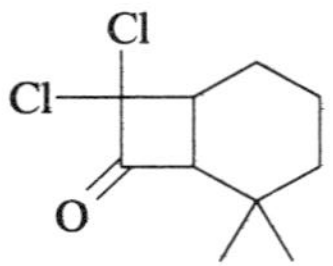 + II; I + II (54), I:II = 1:2.6	450
	$Cl_2C{=}C{=}O$	—	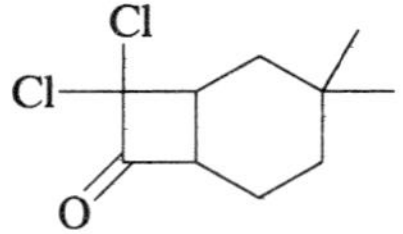 (—)	451
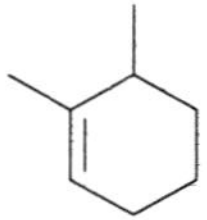	$Cl_2C{=}C{=}O$	—	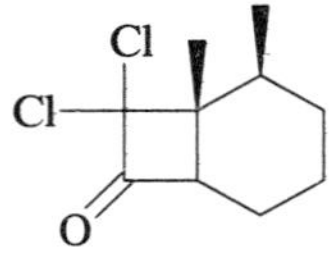 (89)	461
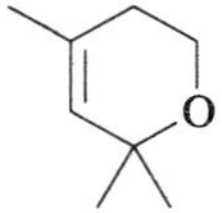	Cl_3CCOCl	Zn, Et_2O, $POCl_3$	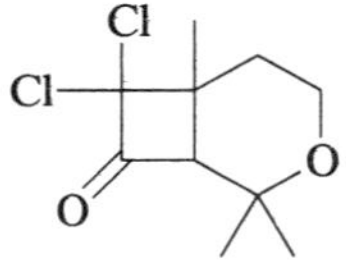 I 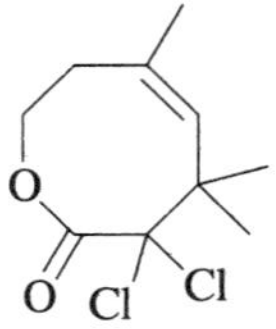 + II	462
	Cl_3CCOCl	Zn, glyme	" (50-60)	462
C_9				
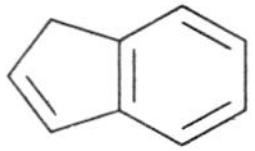	$Cl_2CHCOCl$	Et_3N	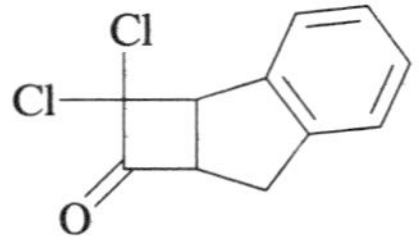 (41)	431, 463, 441

TABLE IV. [2+2] CYCLOADDITION OF KETENES TO CYCLIC OLEFINS (*Continued*)

Reactant	Ketene or Ketene Source	Conditions	Product(s) and Yield(s) (%)	Refs.
	Cl_3CCOBr	Zn, Et_2O	" (60-70)	439
	Cl_3CCOCl	Zn, Et_2O, ultrasound	" (80)	424
	Cl_3CCOCl	Zn, Et_2O, $POCl_3$	" (81)	298
	MeCHClCOCl	Et_3N, C_6H_{14}	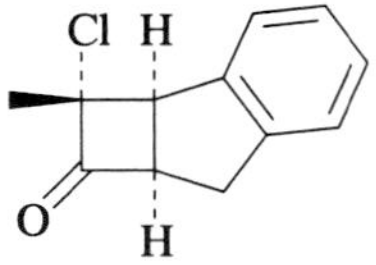 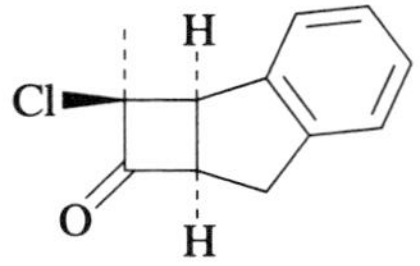 Cl H, O, H **I** + Cl, H, O, H **II** **I** + **II** (78), **I**:**II** = 5:1	46
	$Me_2C{=}C{=}O$	—	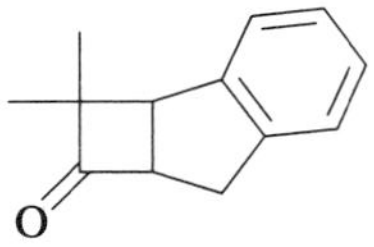 O (64)	457
	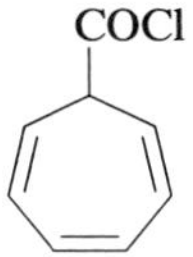 COCl (cycloheptatrienyl)	Et_3N, Et_2O	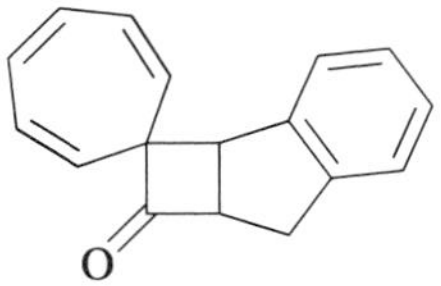 O (24)	464
	$Ph_2C{=}C{=}O$	120°, 4 h	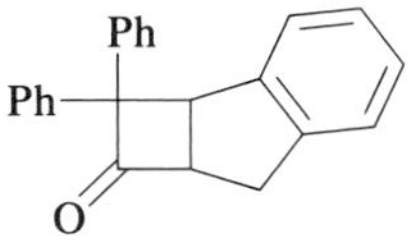 Ph, Ph, O (62)	465

TABLE IV. [2+2] CYCLOADDITION OF KETENES TO CYCLIC OLEFINS (*Continued*)

Reactant	Ketene or Ketene Source	Conditions	Product(s) and Yield(s) (%)	Refs.
	Cl_3CCOCl	Zn, Et_2O	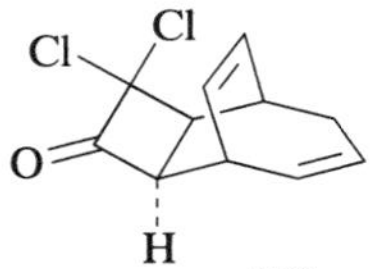 Cl, Cl, O, H (12) 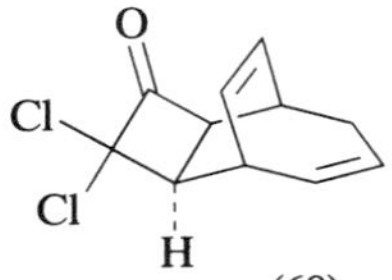 + O, Cl, Cl, H (69)	455
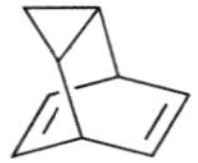	Cl_3CCOCl	Zn, Et_2O	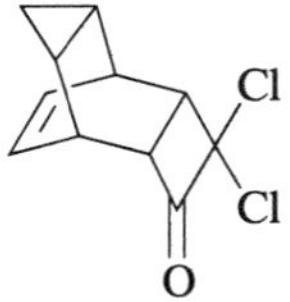 Cl, Cl, O (74) 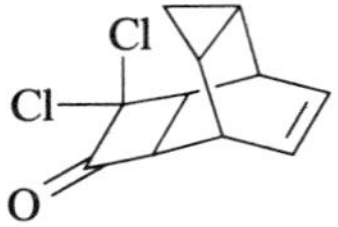 + Cl, Cl, O (12)	76
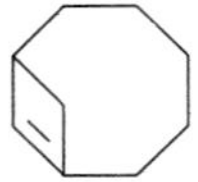	$Ph_2C=C=O$	rt, 2 h	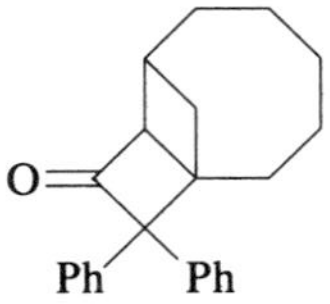 O, Ph, Ph (95)	77
	$Ph_2C=C=O$	C_6H_6, rt	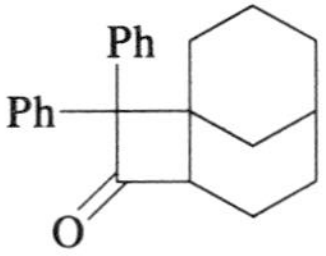 Ph, Ph, O (77)	466
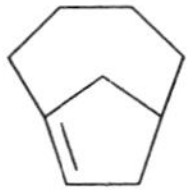	$Ph_2C=C=O$	C_6H_6, rt	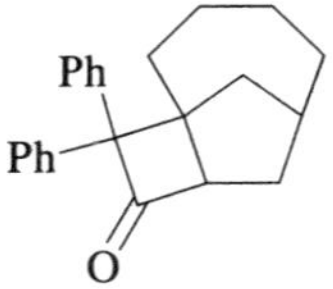 Ph, Ph, O (97)	466

TABLE IV. [2+2] CYCLOADDITION OF KETENES TO CYCLIC OLEFINS (*Continued*)

Reactant	Ketene or Ketene Source	Conditions	Product(s) and Yield(s) (%)	Refs.
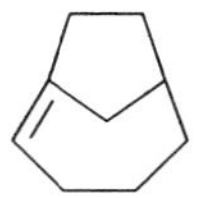	$Ph_2C{=}C{=}O$	C_6H_6, rt	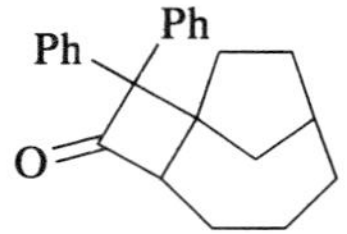 (83)	466
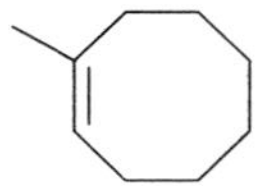	$Ph_2C{=}C{=}O$	C_6H_6, 120°, 72 h	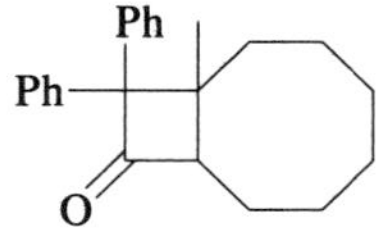 (78)	466
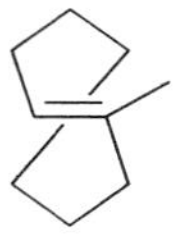	$Ph_2C{=}C{=}O$	C_6H_6, rt, 10 min	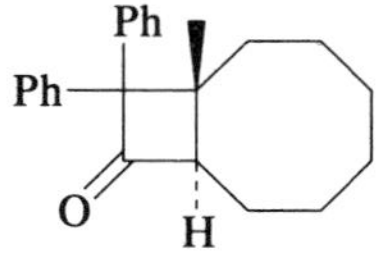 (87)	466
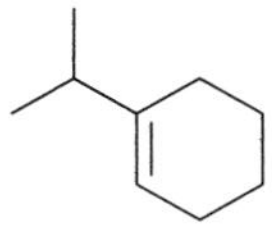	Cl_3CCOBr	Zn, Et_2O	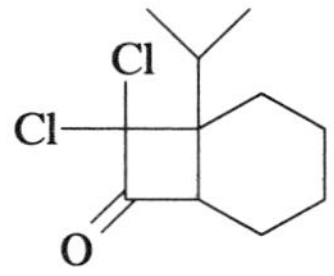 (—)	439
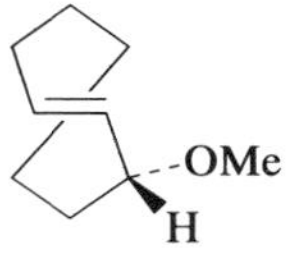 OMe, H	$Ph_2C{=}C{=}O$	-45 to 60°	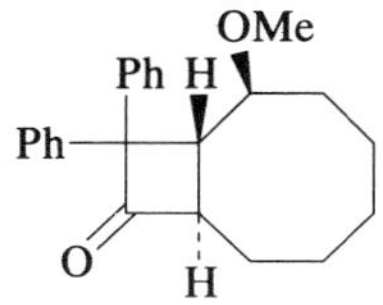 I 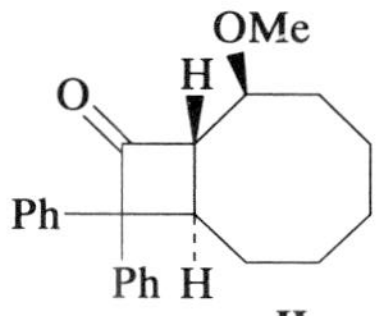 + II (—), I:II = 1:5	454

TABLE IV. [2+2] CYCLOADDITION OF KETENES TO CYCLIC OLEFINS (*Continued*)

Reactant	Ketene or Ketene Source	Conditions	Product(s) and Yield(s) (%)	Refs.
C_{10}				
	Cl_3CCOBr	Zn, Et_2O	Cl, Cl (71)	439
	Cl_3CCOBr	Zn, Et_2O	Cl, Cl (56)	439
	Cl_3CCOCl	Zn, Et_2O	Cl, Cl (95)	455
	$Cl_2C{=}C{=}O$	—	Cl, Cl (60)	329
	$Ph_2C{=}C{=}O$	$h\nu$, Pyrex, C_6H_6	Ph, Ph (49) + Ph, Ph (25)	153

TABLE IV. [2+2] CYCLOADDITION OF KETENES TO CYCLIC OLEFINS (*Continued*)

Reactant	Ketene or Ketene Source	Conditions	Product(s) and Yield(s) (%)	Refs.
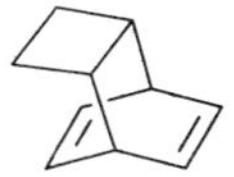	Cl_3CCOCl	Zn, Et_2O	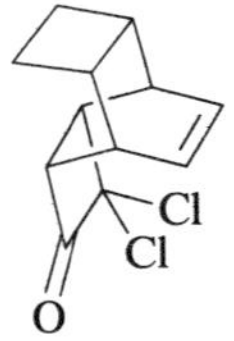 Cl, Cl, O (14) 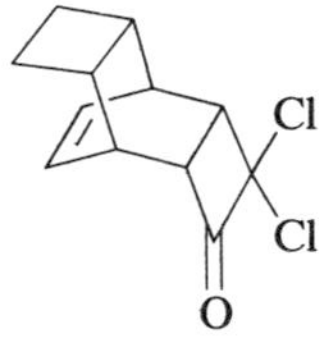 + Cl, Cl, O (62)	76
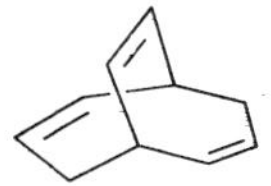	Cl_3CCOCl	Zn, Et_2O	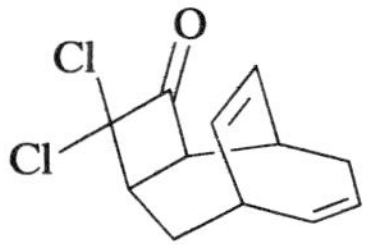 Cl, Cl, O (86)	455
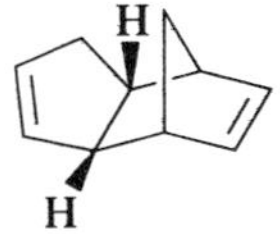 H, H	$Cl_2CHCOCl$	Et_3N	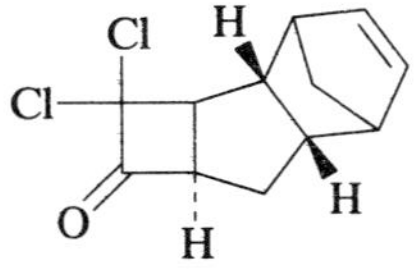 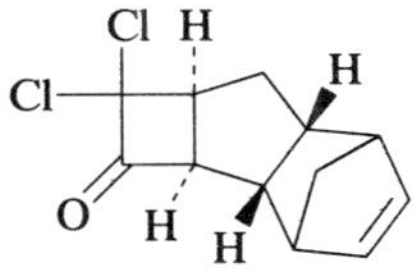 Cl, Cl, H, H, H, O **I** + Cl, Cl, H, H, H, H, O **II**; **I + II** (86), **I:II** = 19:81	467, 431, 468
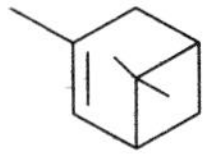	Cl_3CCOCl	Zn, Et_2O	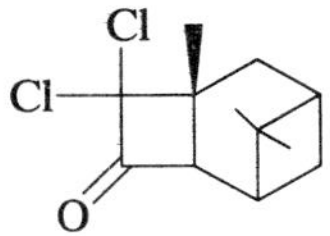 Cl, Cl, O (16)	400
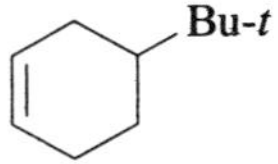 Bu-*t*	Cl_3CCOBr	Zn	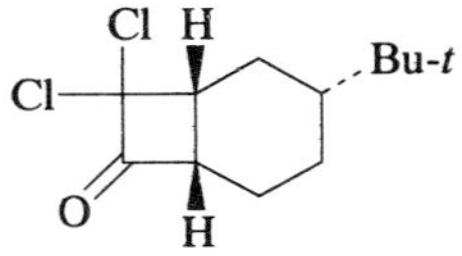 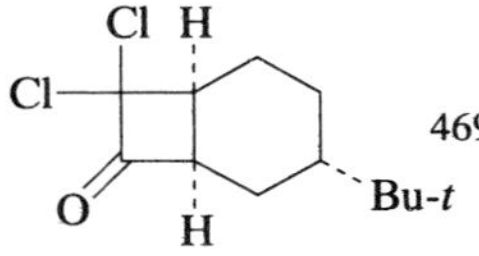 Cl, Cl, H, H, O, Bu-*t* (—) + Cl, Cl, H, H, O, Bu-*t* (—)	469

TABLE IV. [2+2] CYCLOADDITION OF KETENES TO CYCLIC OLEFINS (*Continued*)

Reactant	Ketene or Ketene Source	Conditions	Product(s) and Yield(s) (%)	Refs.
C_{11}				
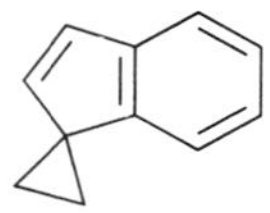	$Cl_2CHCOCl$	Et_3N, C_6H_6	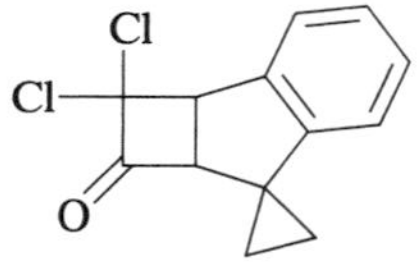 Cl, Cl, O (53)	470
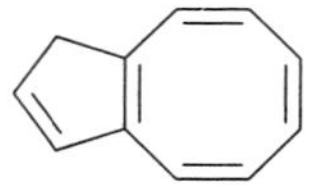	$Cl_2CHCOCl$	Et_3N, C_6H_6	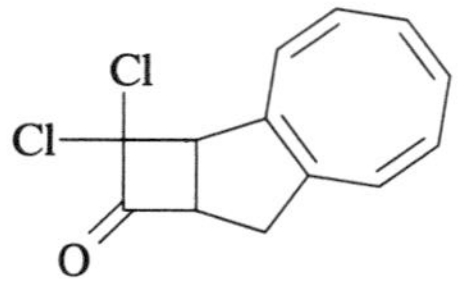 Cl, Cl, O (89)	471
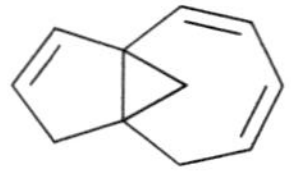	Cl_3CCOCl	Zn-Cu, Et_2O, $POCl_3$	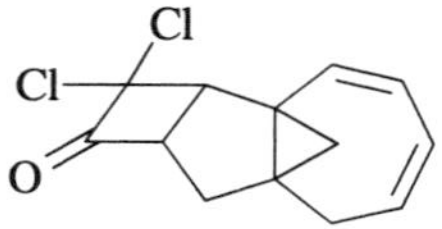 Cl, Cl, O (55)	472
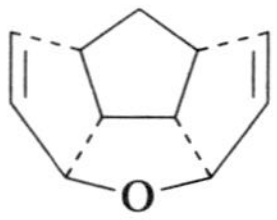 O	Cl_3CCOCl	Zn-Cu, Et_2O	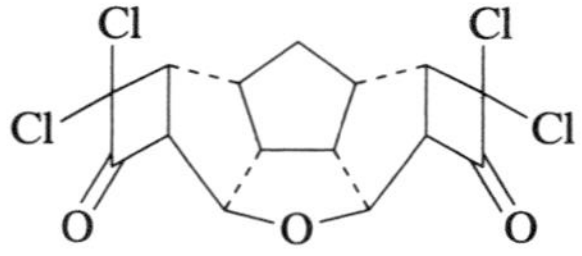 Cl, Cl, Cl, Cl, O, O, O (35-40)	473, 474, 424
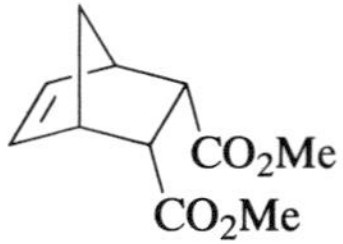 CO_2Me, CO_2Me	Cl_3CCOCl	Zn-Cu, Et_2O, $POCl_3$	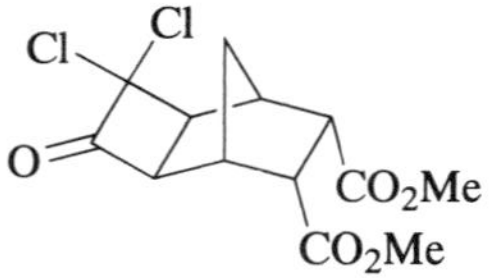 Cl, Cl, O, CO_2Me, CO_2Me (65)	475

TABLE IV. [2+2] CYCLOADDITION OF KETENES TO CYCLIC OLEFINS (*Continued*)

Reactant	Ketene or Ketene Source	Conditions	Product(s) and Yield(s) (%)	Refs.
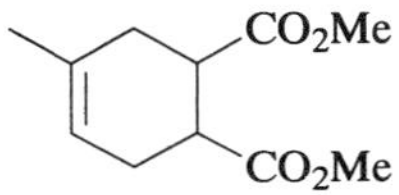 CO_2Me, CO_2Me	Cl_3CCOCl	Zn-Cu, Et_2O, $POCl_3$	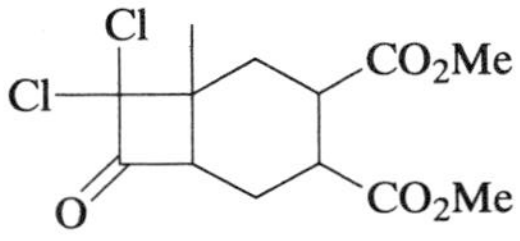 Cl, Cl, O, CO_2Me, CO_2Me (65)	476
	$R_2C{=}C{=}O$	C_6H_6, rt	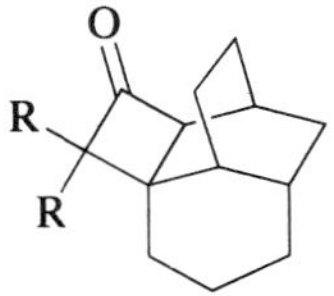 R, R, O; R = H (25); R = Ph (38)	477, 478
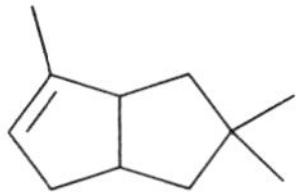	Cl_3CCOCl	Zn-Cu, Et_2O, $POCl_3$	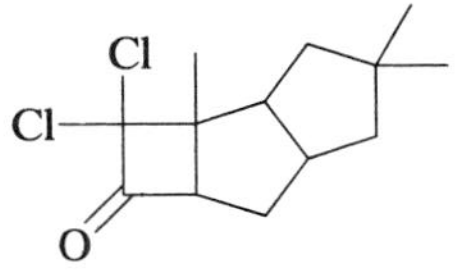 Cl, Cl, O (60)	452
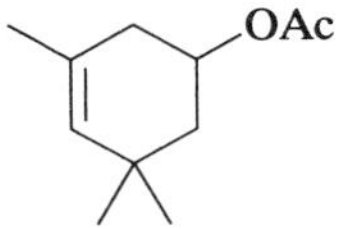 OAc	Cl_3CCOCl	Zn-Cu, Et_2O, $POCl_3$	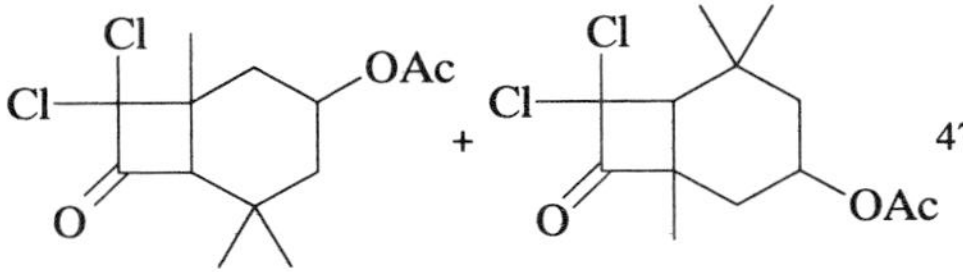 Cl, Cl, O, OAc **I** + Cl, Cl, O, OAc **II**; **I** + **II** (63), **I**:**II** = 1.8:1	476

TABLE IV. [2+2] CYCLOADDITION OF KETENES TO CYCLIC OLEFINS (*Continued*)

Reactant	Ketene or Ketene Source	Conditions	Product(s) and Yield(s) (%)	Refs.
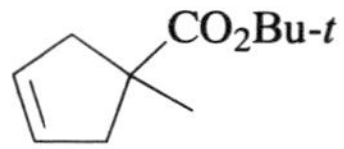 1-methyl-3-cyclopentene-1-carboxylic acid t-butyl ester (CO_2Bu-*t*)	Cl_3CCOCl	Zn-Cu, Et_2O, $POCl_3$	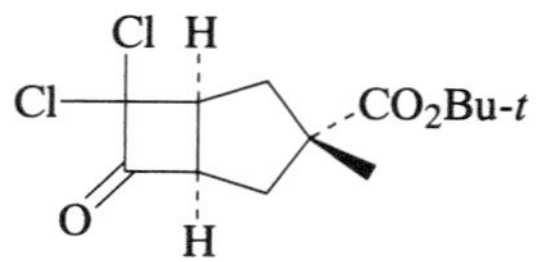 Cl, Cl, H, H, O, CO_2Bu-*t* (50) 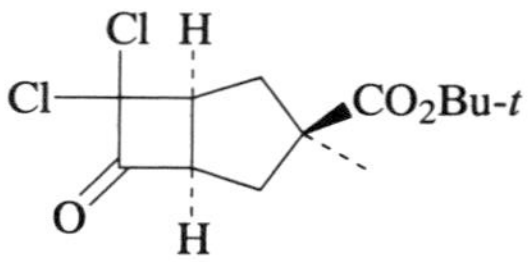 + Cl, Cl, H, H, O, CO_2Bu-*t* (18)	479
C_{12}				
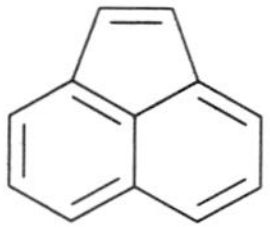 acenaphthylene	$Cl_2CHCOCl$	Et_3N, PhMe	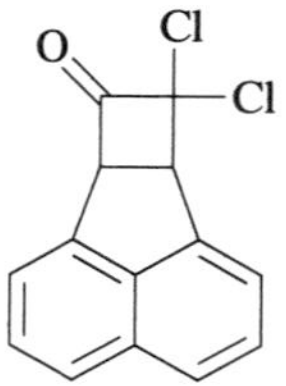 O, Cl, Cl (10)	480, 481, 63
	Cl_3CCOCl	Zn-Cu, Et_2O, $POCl_3$	" (50-60)	482
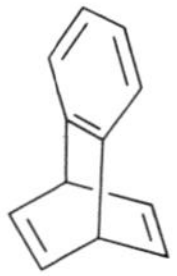 benzobarrelene	Cl_3CCOCl	Zn-Cu, Et_2O	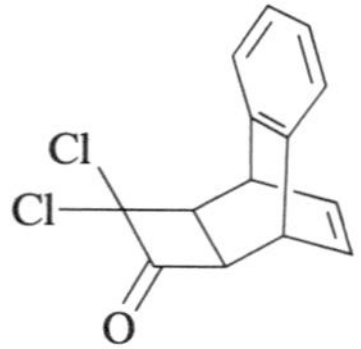 Cl, Cl, O (89)	76

TABLE IV. [2+2] CYCLOADDITION OF KETENES TO CYCLIC OLEFINS (*Continued*)

Reactant	Ketene or Ketene Source	Conditions	Product(s) and Yield(s) (%)	Refs.
	t-Bu(NC)C=C=O	C_6H_6, 78°	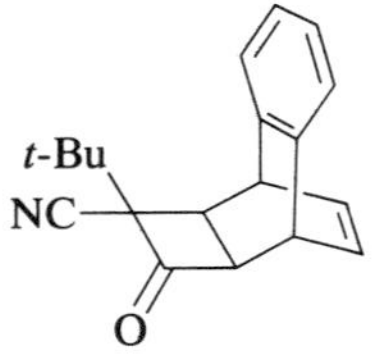 t-Bu, NC (33) 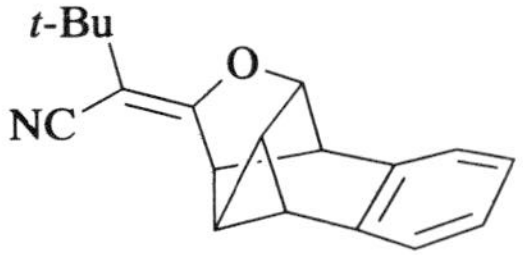 + t-Bu, NC, O (66)	394
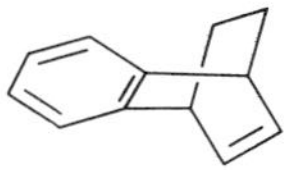	Cl_3CCOCl	Zn, Et_2O	(—)	76
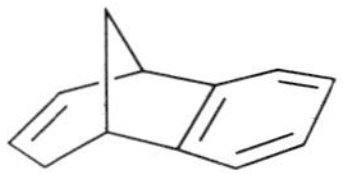	Cl_3CCOCl	Zn, Et_2O	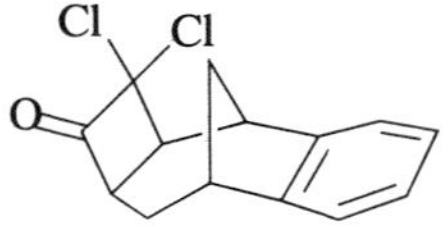 Cl, Cl, O (77)	455
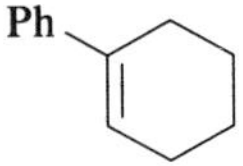 Ph	Cl_3CCOBr	Zn, Et_2O	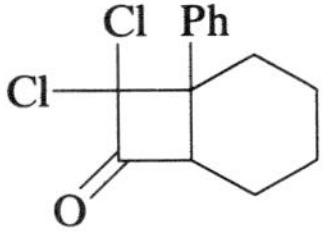 Cl, Cl, Ph, O (61)	439, 440
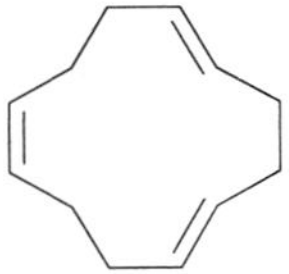	Cl_3CCOCl	Zn, Et_2O, ultrasound	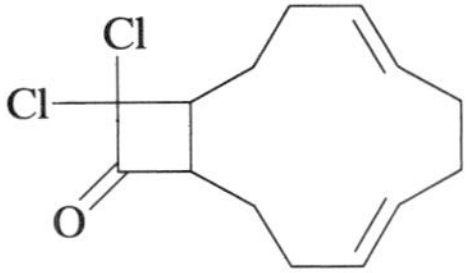 Cl, Cl, O (90)	424

TABLE IV. [2+2] CYCLOADDITION OF KETENES TO CYCLIC OLEFINS (*Continued*)

Reactant	Ketene or Ketene Source	Conditions	Product(s) and Yield(s) (%)	Refs.
O, O	Cl, Cl, C=C=O	—	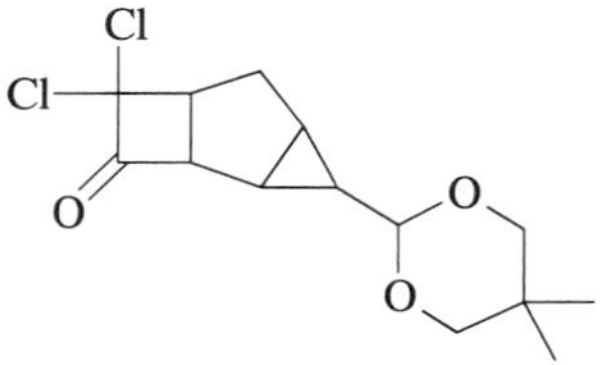 Cl, Cl, O, O, O (81)	483
C_{13}				
	Cl_3CCOCl	Zn-Cu, Et_2O, $POCl_3$	Cl, Cl, O (—)	484
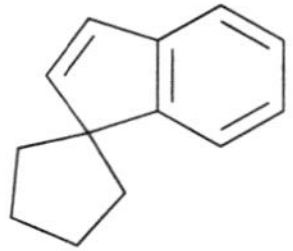	$Cl_2CHCOCl$	Et_3N, C_6H_6	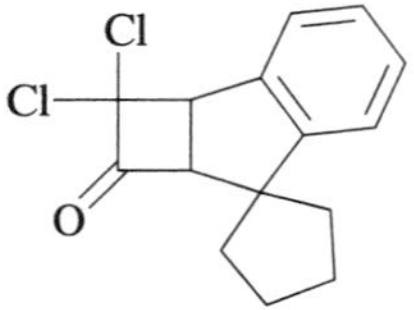 Cl, Cl, O (58)	470
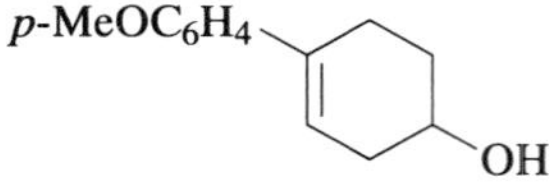 *p*-$MeOC_6H_4$, OH	Cl_3CCOBr	Zn-Cu, Et_2O	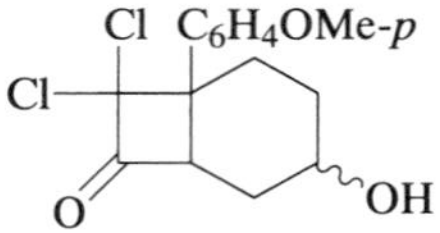 Cl, Cl, C_6H_4OMe-*p*, O, OH (56)	485

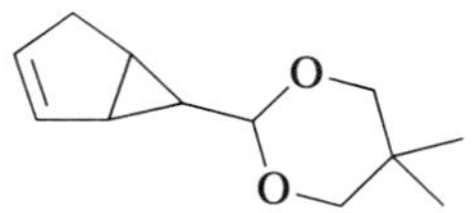

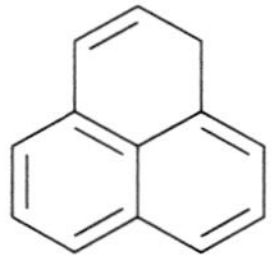

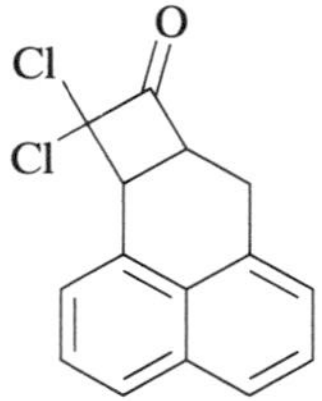

TABLE IV. [2+2] CYCLOADDITION OF KETENES TO CYCLIC OLEFINS (*Continued*)

Reactant	Ketene or Ketene Source	Conditions	Product(s) and Yield(s) (%)	Refs.
C_{14}				
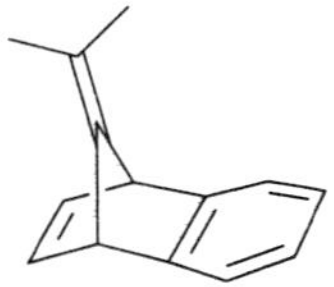	Cl_3CCOCl	Zn-Cu, Et_2O, $POCl_3$	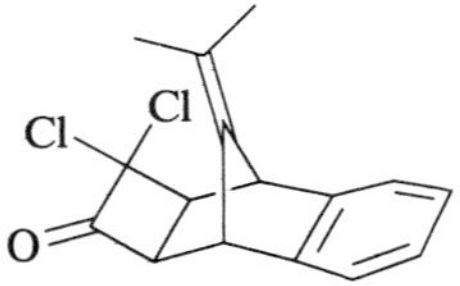 Cl, Cl, O (27) 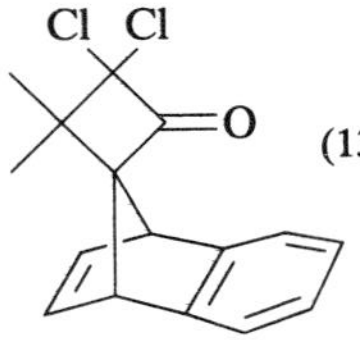 + Cl Cl, O (13) 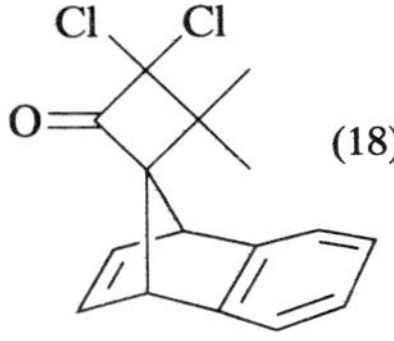 + Cl Cl, O (18)	426
C_{15}				
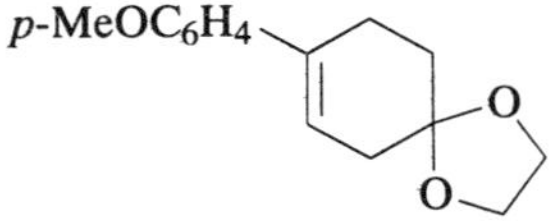 p-$MeOC_6H_4$, O, O	Cl_3CCOBr	Zn-Cu, Et_2O	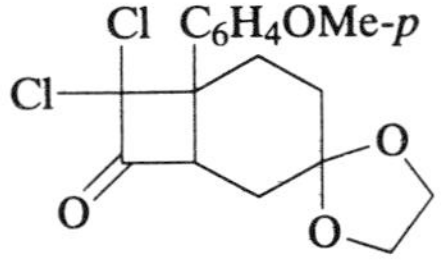 Cl C_6H_4OMe-p, Cl, O, O, O (—)	485
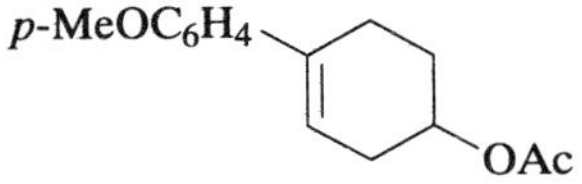 p-$MeOC_6H_4$, OAc	Cl_3CCOBr	Zn-Cu, Et_2O	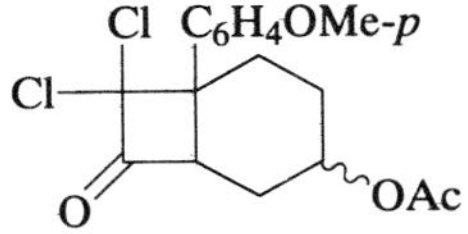 Cl C_6H_4OMe-p, Cl, O, OAc (85)	485
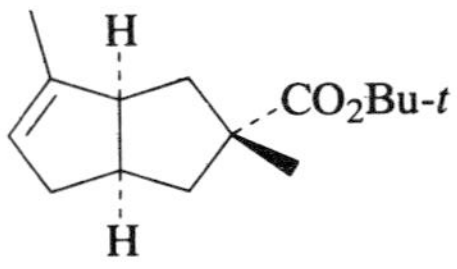 H, H, CO_2Bu-t	Cl_3CCOCl	Zn-Cu, $POCl_3$	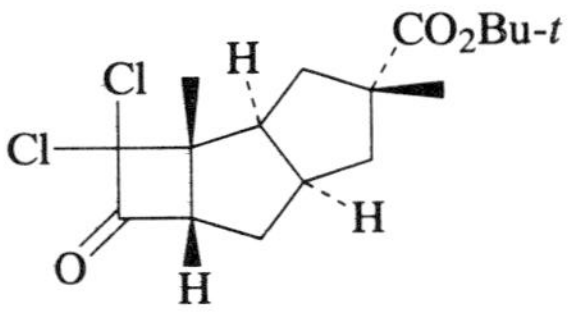 CO_2Bu-t, Cl, H, Cl, O, H, H (68)	479

TABLE IV. [2+2] CYCLOADDITION OF KETENES TO CYCLIC OLEFINS (*Continued*)

Reactant	Ketene or Ketene Source	Conditions	Product(s) and Yield(s) (%)	Refs.
C_{16}				
OBz	Cl_3CCOCl	Zn, Et_2O, $POCl_3$	Cl, Cl, OBz, O **I** + Cl, Cl, OBz, O **II** **I** + **II** (61), **I**:**II** = 1.85:1	476
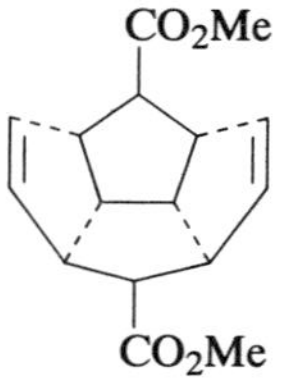 CO_2Me, CO_2Me	Cl_3CCOCl	Zn-Cu, $POCl_3$	(—)	486
C_{19}				
Ph, Ph, O	Ph(Ph)C=C=O	130°, 11 d	Ph, Ph H, H, Ph, Ph, O, O, H, H (—) + O, H H, Ph, Ph, O, Ph, Ph H, H (—)	487, 445
C_{21}				
MeO, MeO, OBn	Cl_3CCOBr	Zn-Cu, Et_2O	Cl Ar, Cl, O, OBn (80)	485

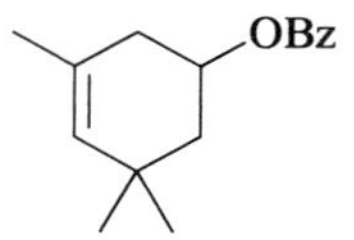

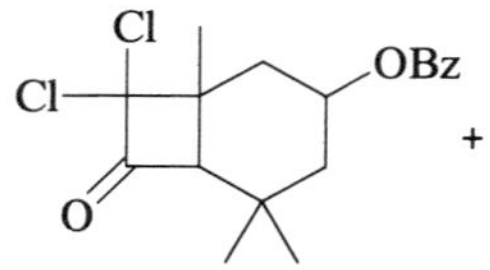

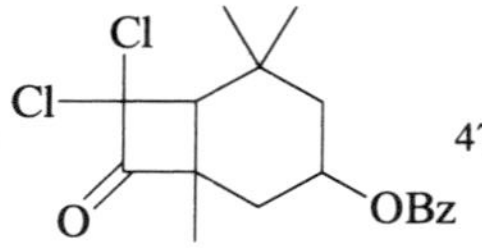

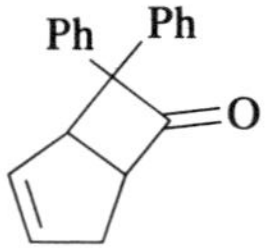

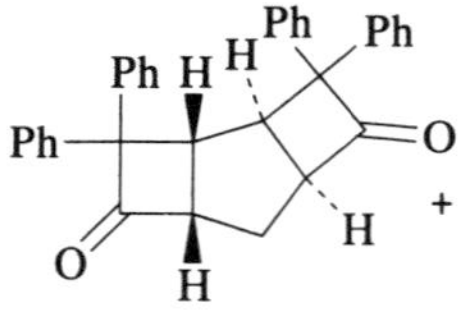

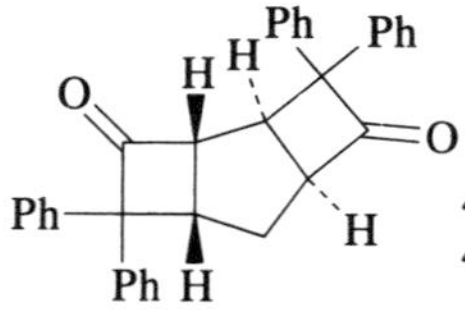

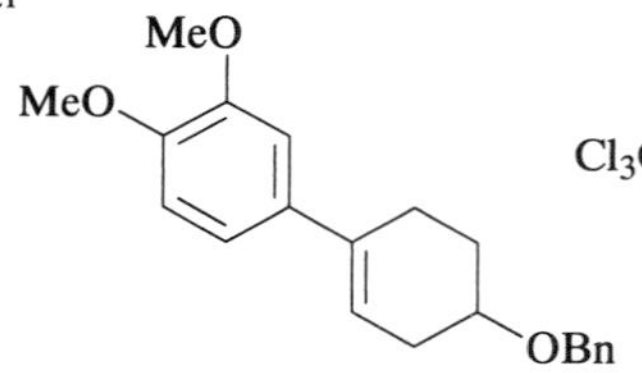

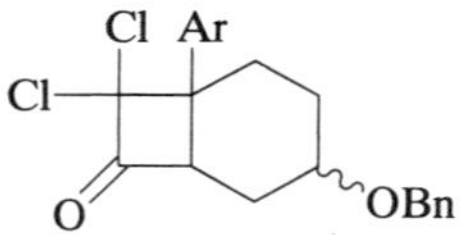

TABLE IV. [2+2] CYCLOADDITION OF KETENES TO CYCLIC OLEFINS (*Continued*)

Reactant	Ketene or Ketene Source	Conditions	Product(s) and Yield(s) (%)	Refs.
C_{22}				
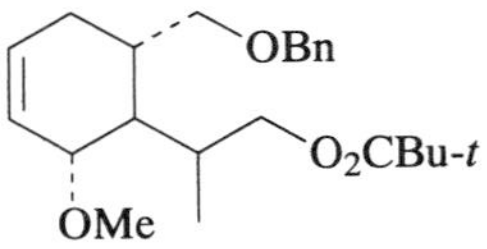 OBn, O_2CBu-*t*, OMe	Cl_3CCOCl	Zn-Cu, Et_2O	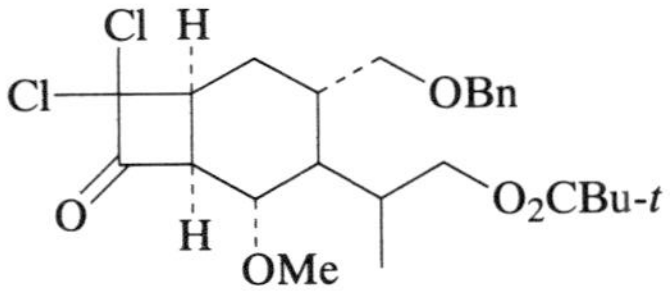 Cl, Cl, H, H, O, OBn, O_2CBu-*t*, OMe (44)	488
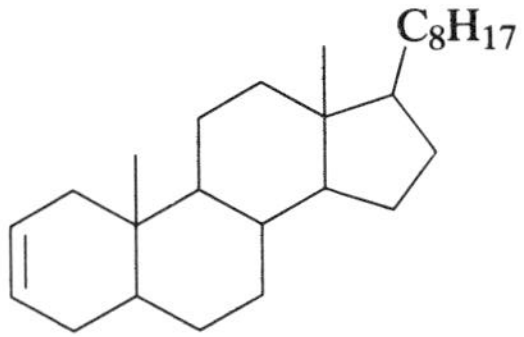 C_8H_{17}	Cl_3CCOBr	Zn, Et_2O	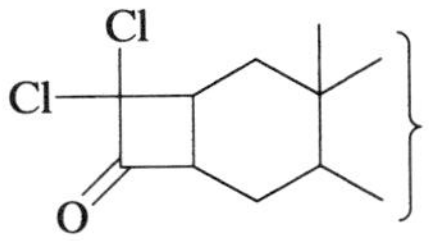 Cl, Cl, O (52) 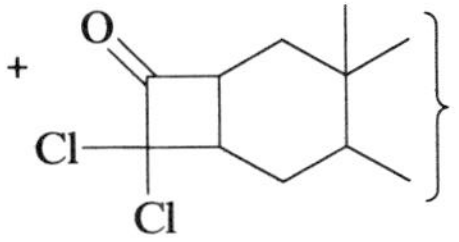 + O, Cl, Cl (4)	489, 82, 298, 490
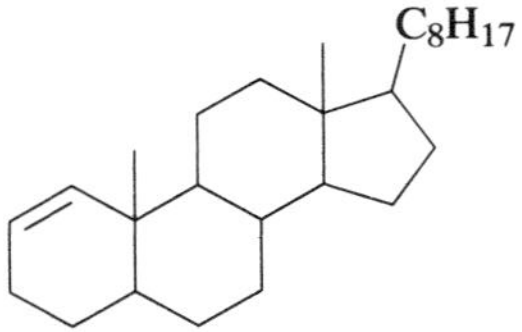 C_8H_{17}	Cl_3CCOBr	Zn, Et_2O	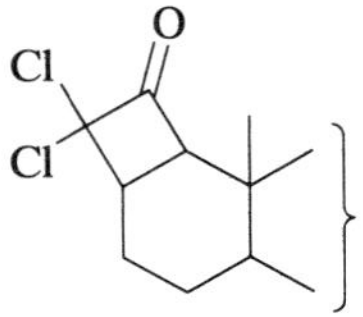 Cl, Cl, O (58)	489
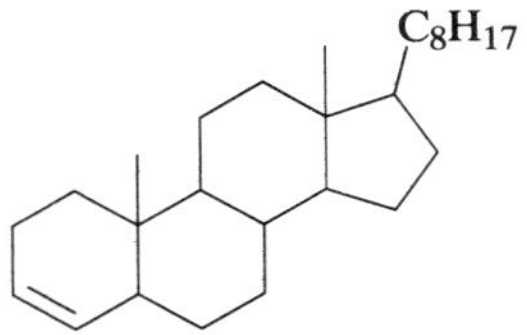 C_8H_{17}	Cl_3CCOBr	Zn, Et_2O	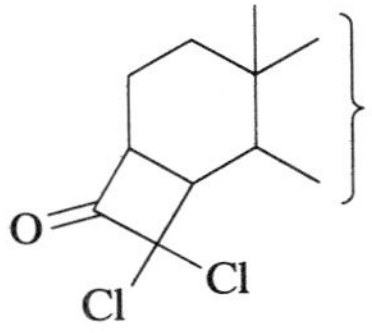 O, Cl, Cl (54)	489

TABLE V. [2+2] CYCLOADDITION OF KETENES TO ACYCLIC DIENES

Reactant	Ketene or Ketene Source	Conditions	Product(s) and Yield(s) (%)	Refs.
C_4				
2-chloro-1,3-butadiene	$Ph_2C{=}C{=}O$	CCl_4	2,2-diphenyl-3-chloro-3-vinylcyclobutanone **I** + 2,2-diphenyl-3-(1-chlorovinyl)cyclobutanone **II**; **I + II** (—), **I:II** = 4:1	491
1,3-butadiene	$Cl_2CHCOCl$	Et_3N, C_6H_{12}	2,2-dichloro-3-vinylcyclobutanone (—)	431
	$CH_2{=}C{=}O$	100°, 2 h	3-vinylcyclobutanone (—)	87
	$(CF_3)_2C{=}C{=}O$	100°, 60 h	2-[bis(trifluoromethyl)methylene]-3,6-dihydro-2H-pyran (78)	84, 231, 85
	$Me_2C{=}C{=}O$	100°, 30 min	2,2-dimethyl-3-vinylcyclobutanone (—)	492
	2-isopropylidene-5,5-dimethyl-1,3-dioxane-4,6-dione	K_2CO_3 (cat.) PhMe, 140°	" (50)	293

TABLE V. [2+2] CYCLOADDITION OF KETENES TO ACYCLIC DIENES (*Continued*)

Reactant	Ketene or Ketene Source	Conditions	Product(s) and Yield(s) (%)	Refs.
	n-Bu(Et)C=C=O	150°	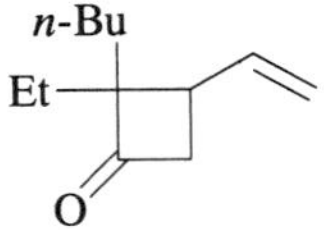 n-Bu, Et, O (45)	493
	Ph(Ph)C=C=O	THF, C_7H_{16}, 30°	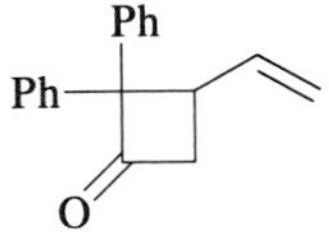 Ph, Ph, O (55)	491, 396
C_5 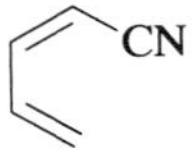 CN	Ph(Ph)C=C=O	4 weeks	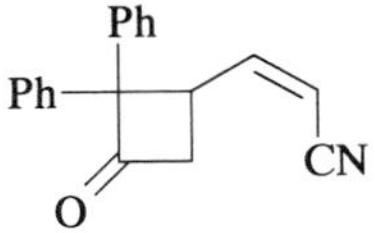 Ph, Ph, O, CN (89)	494, 36, 495, 496
	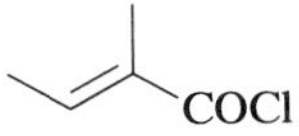 COCl	Et_3N	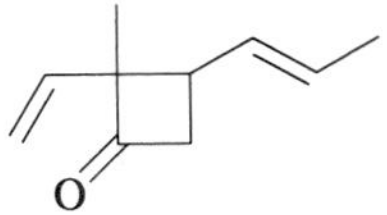 O (—)	305
	n-Bu(Et)C=C=O	180°	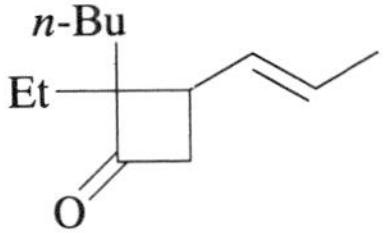 n-Bu, Et, O (64)	493
	Ph(Ph)C=C=O	20°, 4 d	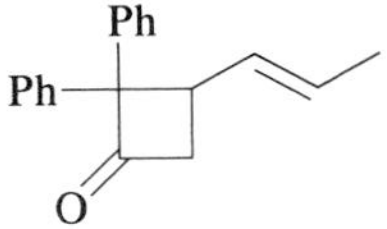 Ph, Ph, O (95)	36, 495, 494, 496, 412

TABLE V. [2+2] CYCLOADDITION OF KETENES TO ACYCLIC DIENES (*Continued*)

Reactant	Ketene or Ketene Source	Conditions	Product(s) and Yield(s) (%)	Refs.
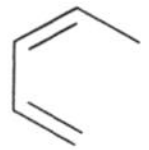	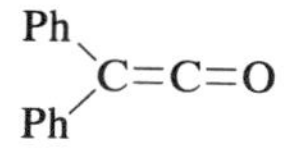 Ph, Ph C=C=O	20°, 4 d	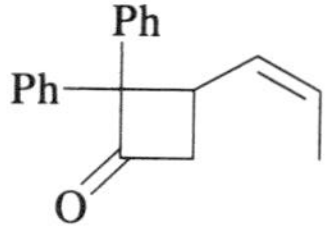 Ph, Ph, O (99)	494, 36, 495
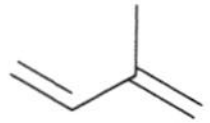	Cl_3CCOCl	Zn, Et_2O, $POCl_3$	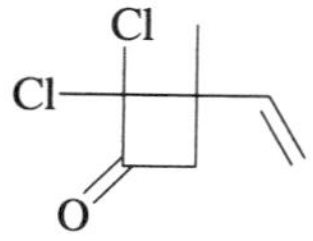 Cl, Cl, O (62) 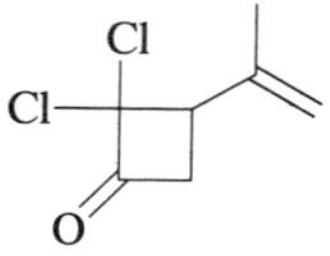 + Cl, Cl, O (19)	497
	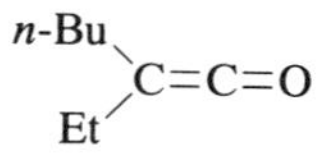 *n*-Bu, Et C=C=O	150°, 4 h	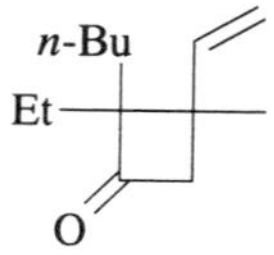 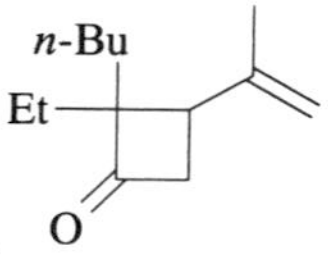 *n*-Bu, Et, O **I** + *n*-Bu, Et, O **II** **I** + **II** (45)	493
	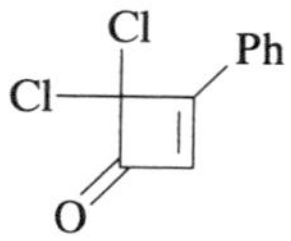 Cl, Cl, Ph, O	*h*ν, C_6H_6	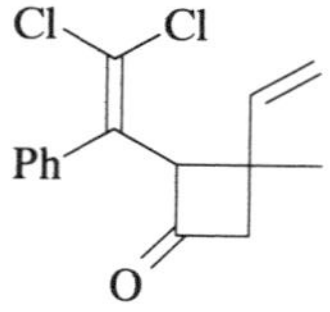 Cl, Cl, Ph, O (70)	306
	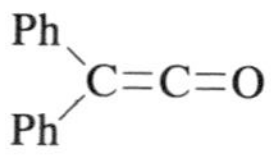 Ph, Ph C=C=O	16 d	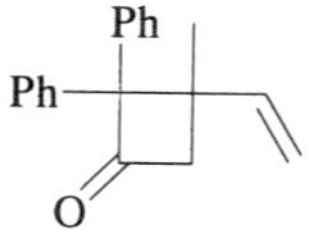 Ph, Ph, O (64) 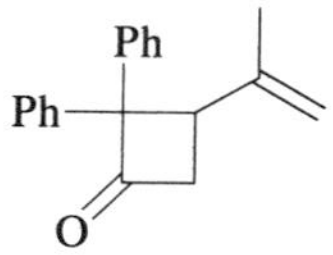 + Ph, Ph, O (28)	494, 491, 495, 36

TABLE V. [2+2] CYCLOADDITION OF KETENES TO ACYCLIC DIENES (*Continued*)

Reactant	Ketene or Ketene Source	Conditions	Product(s) and Yield(s) (%)	Refs.
C_6				
OAc	$Ph_2C=C=O$	42°	Ph, Ph, O, OAc (—)	495, 496
	"	100°, 10 h	Adduct, mp 146°	396
	"	14 weeks	Ph, Ph, O (86)	36, 494, 396, 412
	$Me_2C=C=O$	125°	O (60)	498
OEt	$Ph_2C=C=O$	C_6H_6	Ph, Ph, O, OEt (—) + EtO, O, Ph, Ph (—)	496, 495
OEt	$Ph_2C=C=O$	100°, 16 h	OEt, O, Ph, Ph (—)	496

TABLE V. [2+2] CYCLOADDITION OF KETENES TO ACYCLIC DIENES (*Continued*)

Reactant	Ketene or Ketene Source	Conditions	Product(s) and Yield(s) (%)	Refs.
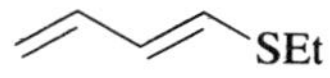 SEt	Ph(Ph)C=C=O	42°	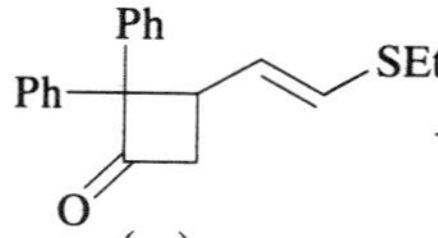 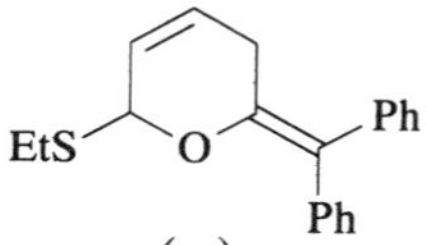 Ph, Ph, O, SEt (—) + EtS, O, Ph, Ph (—)	495
C_7				
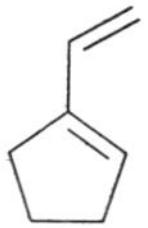	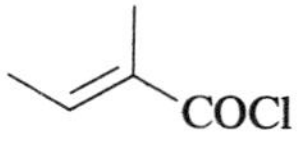 COCl	Et_3N	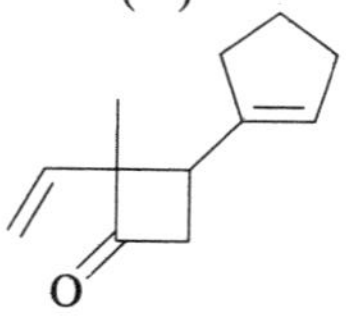 O (—)	305
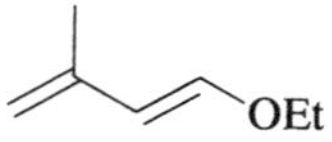 OEt	Ph(Ph)C=C=O	—	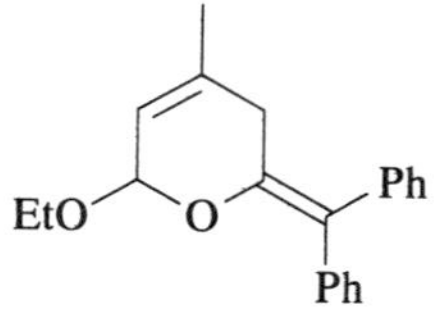 EtO, O, Ph, Ph (—)	496, 495
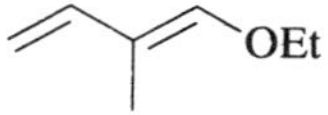 OEt	Ph(Ph)C=C=O	—	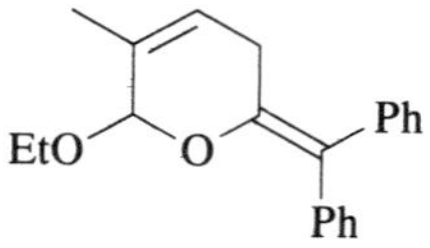 EtO, O, Ph, Ph (—)	496, 495
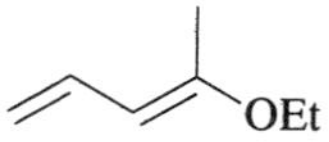 OEt	Ph(Ph)C=C=O	—	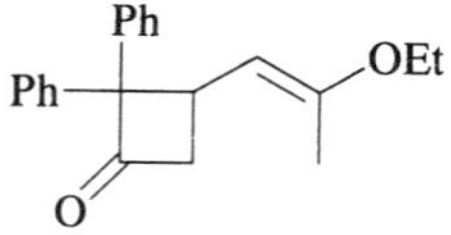 Ph, Ph, O, OEt (—)	496, 495

TABLE V. [2+2] CYCLOADDITION OF KETENES TO ACYCLIC DIENES (*Continued*)

Reactant	Ketene or Ketene Source	Conditions	Product(s) and Yield(s) (%)	Refs.
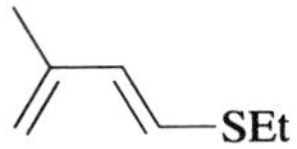 SEt	Ph, Ph C=C=O	42°	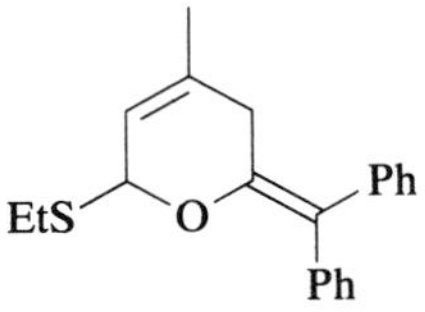 EtS, O, Ph, Ph (—)	496, 495
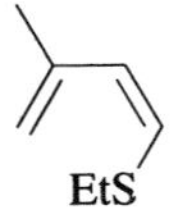 EtS	Ph, Ph C=C=O	42°	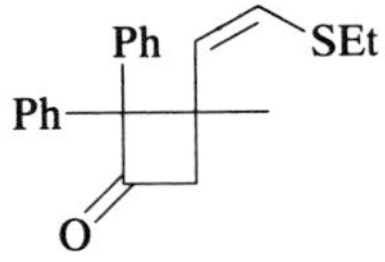 Ph, Ph, SEt, O (—)	496, 495
 OTMS	$Cl_2CHCOCl$	Et_3N, C_6H_{14}	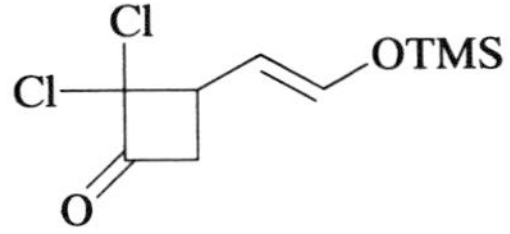 Cl, Cl, OTMS, O (71)	499
	MeCHClCOCl	Et_3N, C_6H_{14}	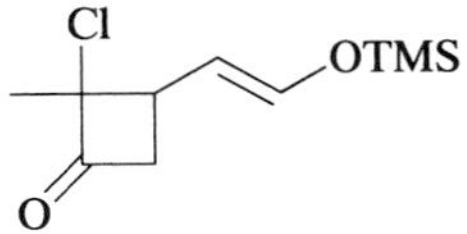 Cl, OTMS, O (66)	499
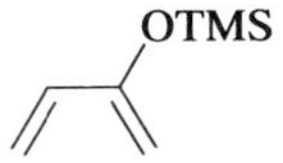 OTMS	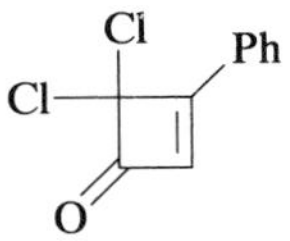 Cl, Cl, Ph, O	$h\nu$, C_6H_6	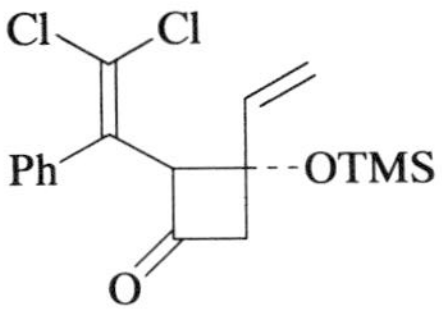 Cl, Cl, Ph, OTMS, O (50)	306

TABLE V. [2+2] CYCLOADDITION OF KETENES TO ACYCLIC DIENES (*Continued*)

Reactant	Ketene or Ketene Source	Conditions	Product(s) and Yield(s) (%)	Refs.
C_8				
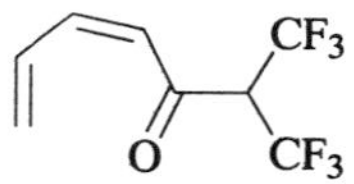	$(CF_3)_2C=C=O$	100°, 16 h	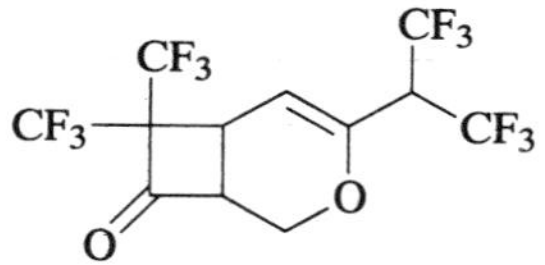 (—)	84
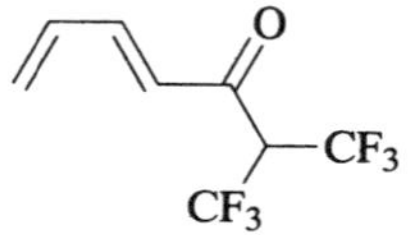	$(CF_3)_2C=C=O$	100°	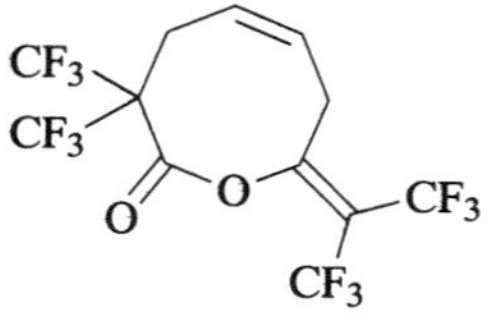 (—)	85
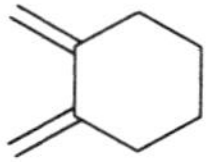	Cl_3CCOCl	Zn, Et_2O, $POCl_3$	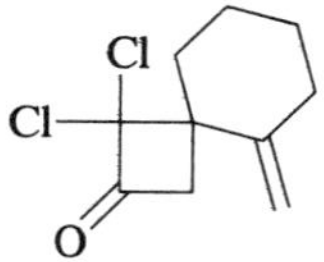 (61)	500
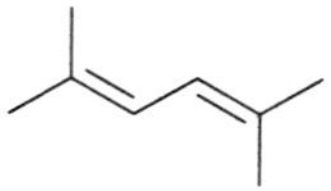	Cl_3CCOCl	Zn, Et_2O	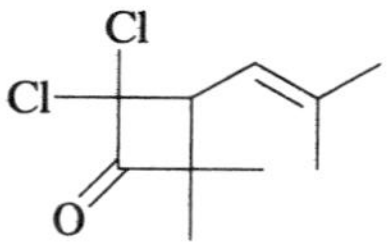 (30) + 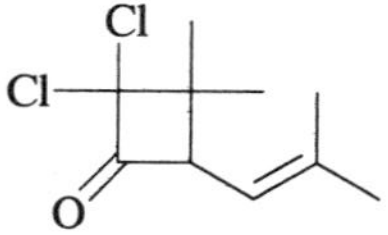 (13)	400
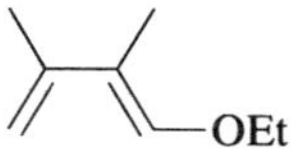	$Ph_2C=C=O$	42°	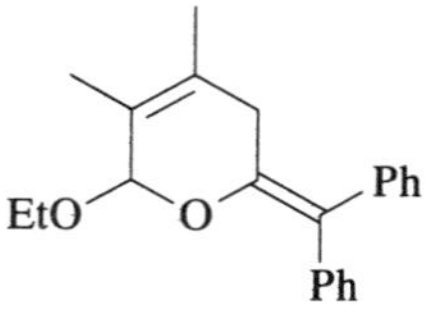 (—)	495

TABLE V. [2+2] CYCLOADDITION OF KETENES TO ACYCLIC DIENES (*Continued*)

Reactant	Ketene or Ketene Source	Conditions	Product(s) and Yield(s) (%)	Refs.
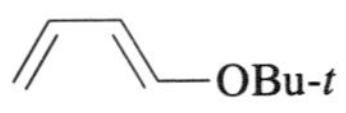 OBu-*t*	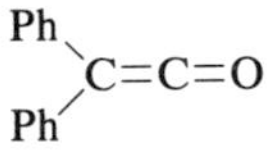 Ph, Ph C=C=O	C_6H_6, 48 h	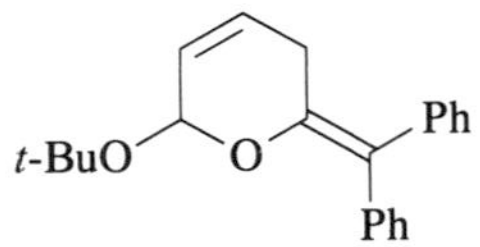 *t*-BuO, O, Ph, Ph (—)	86
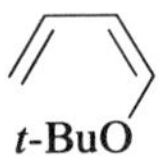 *t*-BuO	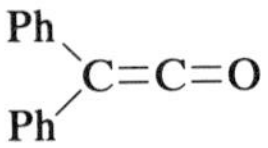 Ph, Ph C=C=O	C_6H_6, 48 h	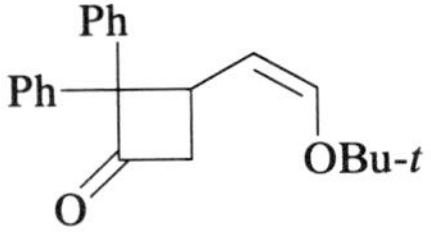 Ph, Ph, O, OBu-*t* (—)	86
C_9				
i-Pr, OEt	Ph, Ph C=C=O	42°	Pr-*i*, EtO, O, Ph, Ph (—)	495
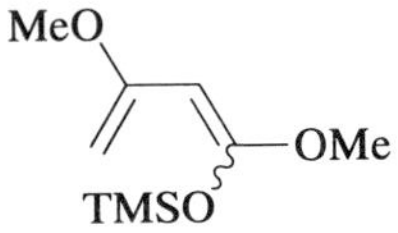 MeO, OMe, TMSO	$Cl_2CHCOCl$	Et_3N, Et_2O	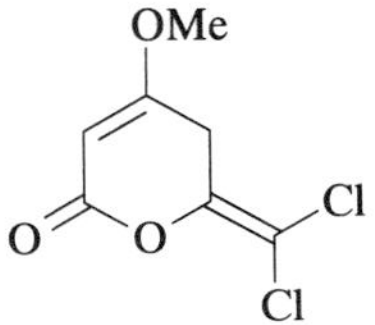 OMe, O, O, Cl, Cl (30) +	235
			O, OMe, Cl, Cl, CO_2Me (34)	

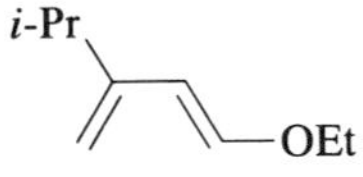

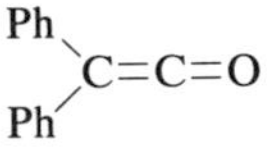

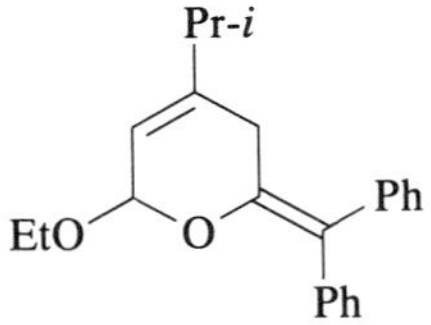

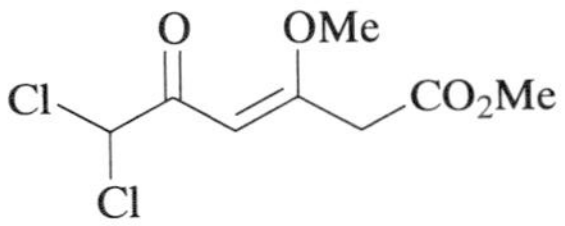

TABLE V. [2+2] CYCLOADDITION OF KETENES TO ACYCLIC DIENES (*Continued*)

Reactant	Ketene or Ketene Source	Conditions	Product(s) and Yield(s) (%)	Refs.
	$ClCH_2COCl$	Et_3N, Et_2O	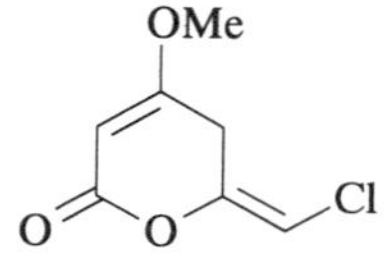 [structure: OMe, O, O, Cl] (23) 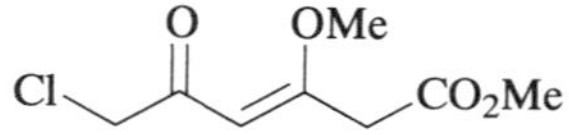 + [structure: Cl, O, OMe, CO_2Me]	235
	[structure: Ph, Ph, C=C=O]	1. Et_2O 2. MeOH, HCl	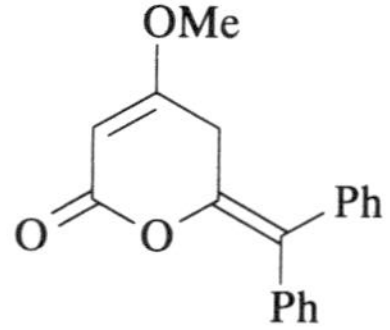 [structure: OMe, O, O, Ph, Ph] (56)	235
C_{10} 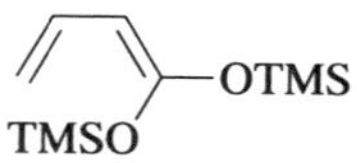 [structure: TMSO, OTMS]	[structure: R, R, C=C=O] R = Cl, Ph	—	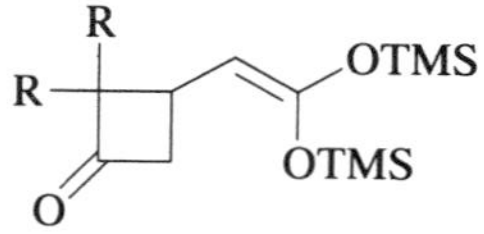 [structure: R, R, O, OTMS, OTMS] (—)	235
C_{11} 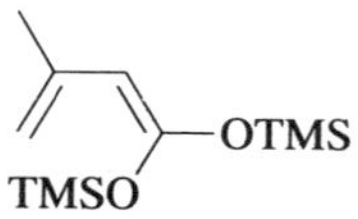 [structure: TMSO, OTMS]	$Cl_2CHCOCl$	Et_3N, Et_2O	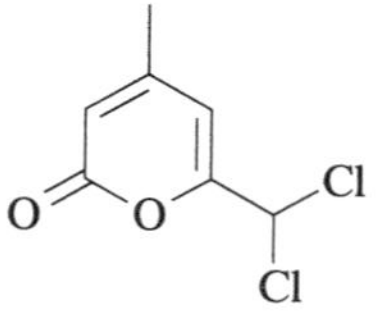 [structure: O, O, Cl, Cl] (31) 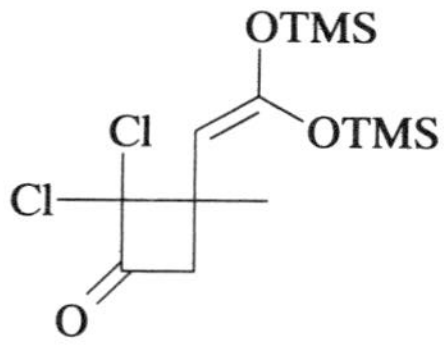 + [structure: Cl, Cl, O, OTMS, OTMS] (55)	235

TABLE V. [2+2] CYCLOADDITION OF KETENES TO ACYCLIC DIENES (*Continued*)

Reactant	Ketene or Ketene Source	Conditions	Product(s) and Yield(s) (%)	Refs.
C_{13} 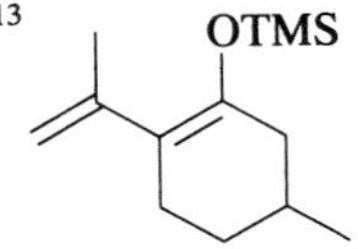 OTMS	R(R)C=C=O R = Cl, Me	Et_3N	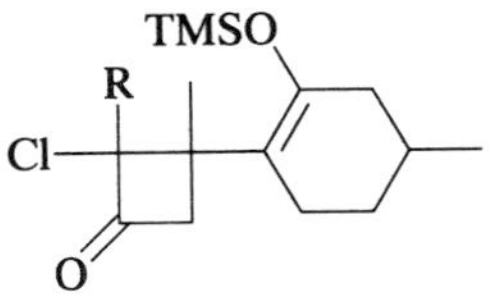 TMSO, R, Cl, O R = Cl (58) R = Me (49)	499
C_{16} 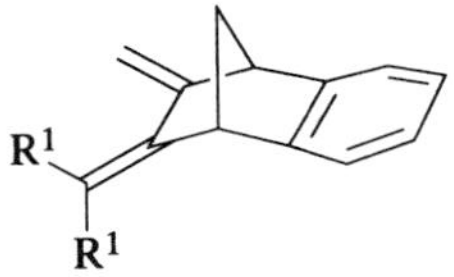 R^1, R^1	$R_2ClCCOCl$	Zn, Et_2O, $POCl_3$	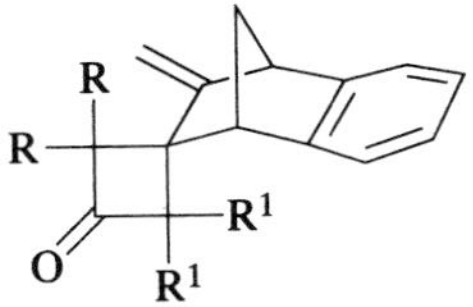 R, R, O, R^1, R^1 R = Cl, R^1 = H (43) R = H, R^1 = Cl (14)	501
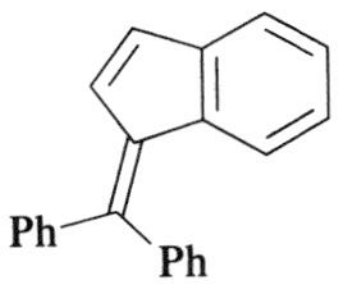 Ph, Ph	$Cl_2CHCOCl$	Et_3N, C_6H_{14}	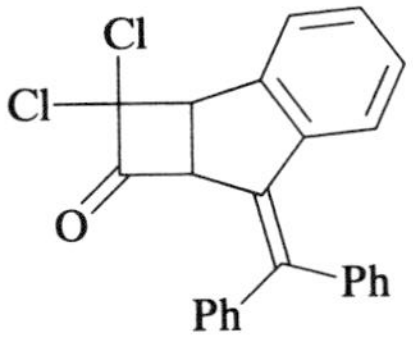 Cl, Cl, O, Ph, Ph (22)	502, 503

TABLE VI. [2+2] CYCLOADDITION OF KETENES TO CYCLIC DIENES

Reactant	Ketene or Ketene Source	Conditions	Product(s) and Yield(s) (%)	Refs.
C_5 (cyclopentadiene)	$Br_2CHCOCl$	Et_3N, C_6H_{14}	Br, Br, O (58)	504
	$Cl_2CHCOCl$	Et_3N	Cl, Cl, O (72)	319, 505, 6, 5, 318, 506, 83, 507, 508
	$BrCH_2COBr$	Et_3N, Et_2O	Br, H, H, O (5)	509, 328
	$ClCH_2COCl$	Et_3N	Cl, H, H, H, O *endo* + Cl, H, H, H, O *exo*; *endo*:*exo* = >97:<3	38, 510, 509, 511, 512
	FCH_2COCl	Et_3N, Et_2O, -78°	F, H, H, O (40)	509, 328, 46

TABLE VI. [2+2] CYCLOADDITION OF KETENES TO CYCLIC DIENES (*Continued*)

Reactant	Ketene or Ketene Source	Conditions	Product(s) and Yield(s) (%)	Refs.
	$CH_2=C=O$	rt	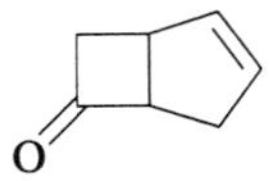 (—)	513-515
	Cl(NC)C=C=O	PhMe, 103°, 1.75 h	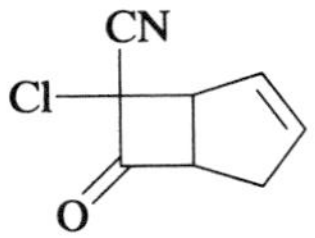 Cl, CN, O (30)	312
	MeCHBrCOCl	Et_3N	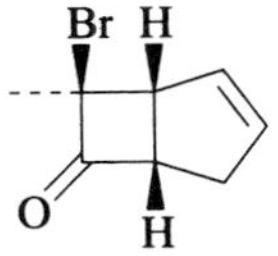 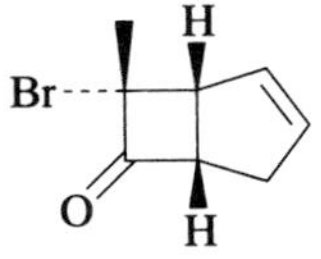 Br H, O, H **I** + Br, H, O, H **II** **I** + **II** (63), **I**:**II** = 56:44	38, 516-518
	MeCHClCOCl	Et_3N, solvent	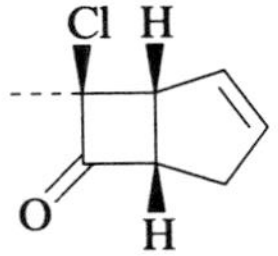 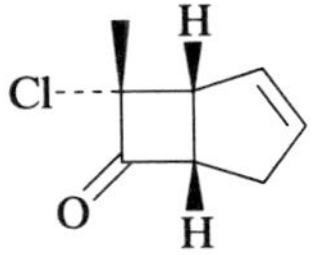 Cl H, O, H **I** + Cl, H, O, H **II**	

Solvent	**I** + **II**	**I**:**II**	Ref.
C_6H_{14}	(75)	81:19	516
C_6H_{14}	(67)	80:20	38
C_5H_{12}	(88)	73:27	511
Et_3N	(32)	67:33	517
$CHCl_3$	(40)	62:38	517
MeCN	(62)	37:63	516

TABLE VI. [2+2] CYCLOADDITION OF KETENES TO CYCLIC DIENES (*Continued*)

Reactant	Ketene or Ketene Source	Conditions	Product(s) and Yield(s) (%)	Refs.
	EtCOCl	Et_3N, C_6H_{14}	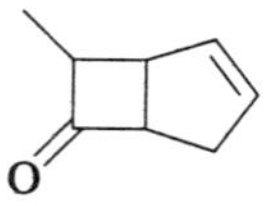 (20) *endo*:*exo* = 98:2	38, 509, 519
	$MeOCH_2COCl$	Et_3N	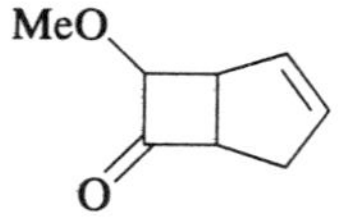 MeO (9) *endo*:*exo* = >95:<5	38
	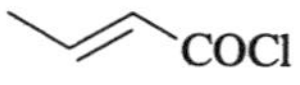 COCl	Et_3N, $CHCl_3$	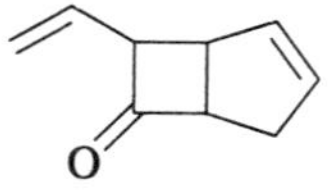 (38) *endo*:*exo* = 82:18	411, 305
	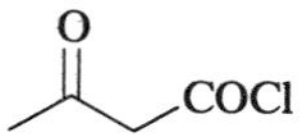 O COCl	Et_3N	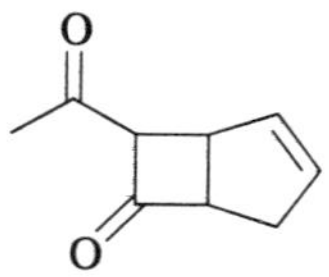 O (56) *endo*:*exo* = 94:6	520
	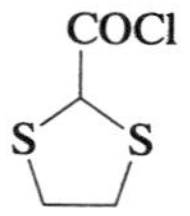 COCl S S	Et_3N	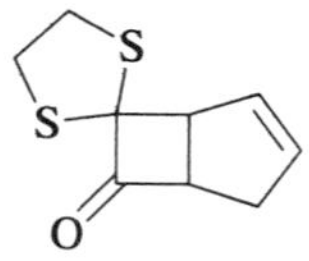 S S (84)	521

TABLE VI. [2+2] CYCLOADDITION OF KETENES TO CYCLIC DIENES (*Continued*)

Reactant	Ketene or Ketene Source	Conditions	Product(s) and Yield(s) (%)	Refs.
	R, Et, COCl R = Cl, Br	Et_3N, solvent	R H, Et, O, H **I** + Et H, R, O, H **II** R / Solvent / I + II / I:II Br / C_6H_{14} / (70) / 62:38 Br / MeCN / (54) / 21:79 Cl / C_6H_{14} / (77) / 84:16 Cl / C_5H_{12} / (70) / 89:11 Cl / MeCN / (77) / 52:48	516, 518, 511, 510
	COCl	Et_3N, $CHCl_3$	H, O, H **I** + H, O, H **II** **I** + **II** (59), **I**:**II** = 92:8	411
	n-PrCOBr	Et_3N, CCl_4	Et, H, O, H (34)	509
	Me, Me, C=C=O	rt, 16 h	O (77)	38,522-524

TABLE VI. [2+2] CYCLOADDITION OF KETENES TO CYCLIC DIENES (*Continued*)

Reactant	Ketene or Ketene Source	Conditions	Product(s) and Yield(s) (%)	Refs.
	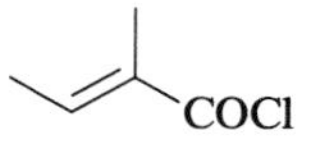 COCl	Et_3N, $CHCl_3$	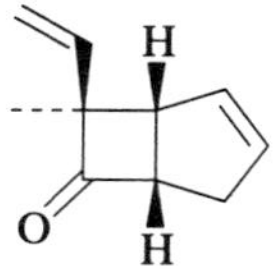 O, H, H (51) 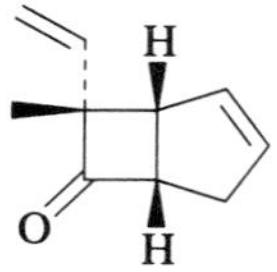 + O, H, H (27)	38
	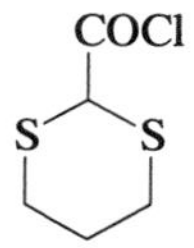 COCl, S, S	Et_3N, Et_2O	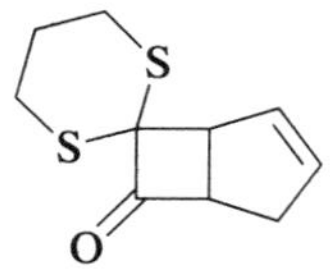 S, S, O (70)	525
	i-PrCHXCOCl	Et_3N, solvent	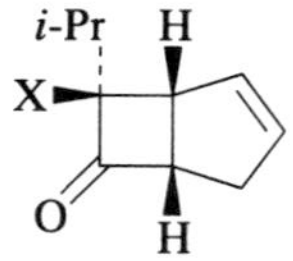 *i*-Pr, H, X, O, H **I** + *i*-Pr, H, X, O, H **II** 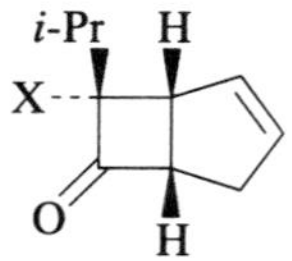	
	n-BuCOCl	Et_3N	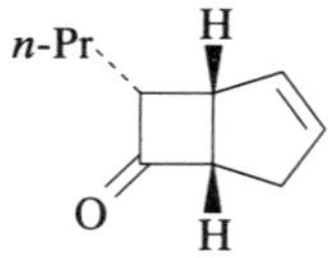 *n*-Pr, H, O, H (—)	519

Solvent	X	I	II	Ref.
C_6H_{14}	Br	(55)	(21)	516
MeCN	Br	(17)	(29)	516
C_6H_{14}	Cl	(65)	(6)	516
—	Cl	(55)	(3)	510
C_5H_{12}	Cl	(55)	(3)	511
MeCN	Cl	(40)	(37)	516

TABLE VI. [2+2] CYCLOADDITION OF KETENES TO CYCLIC DIENES (*Continued*)

Reactant	Ketene or Ketene Source	Conditions	Product(s) and Yield(s) (%)	Refs.
	i-BuCOCl	Et_3N, $CHCl_3$	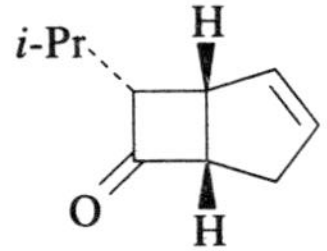 *i*-Pr, H, H, O (37)	509, 519
	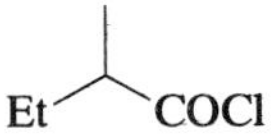 Et, COCl	Et_3N	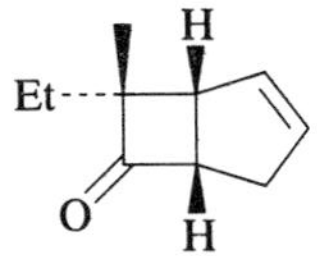 Et, H, H, O (43) 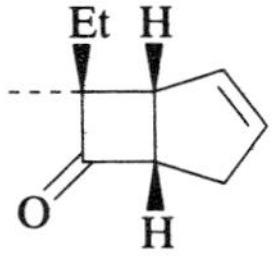 + Et, H, H, O (29)	38
	t-BuCHBrCOCl	Et_3N, solvent	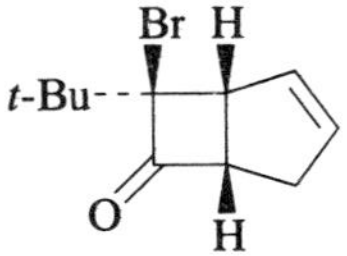 *t*-Bu, Br, H, H, O MeCN (54) C_6H_{14} (11)	516
	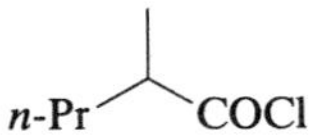 *n*-Pr, COCl	Et_3N, MeCN	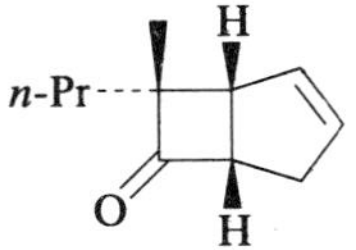 *n*-Pr, H, H, O (47) 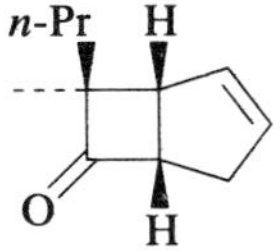 + *n*-Pr, H, H, O (34)	516, 38
	t-$BuCH_2COCl$	Et_3N	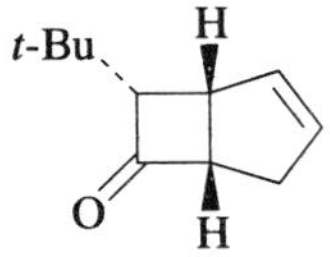 *t*-Bu, H, H, O (22)	526, 519

TABLE VI. [2+2] CYCLOADDITION OF KETENES TO CYCLIC DIENES (*Continued*)

Reactant	Ketene or Ketene Source	Conditions	Product(s) and Yield(s) (%)	Refs.
	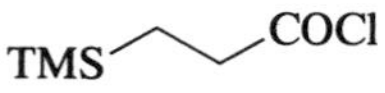 TMS–CH2CH2–COCl	Et_3N, C_6H_{14}	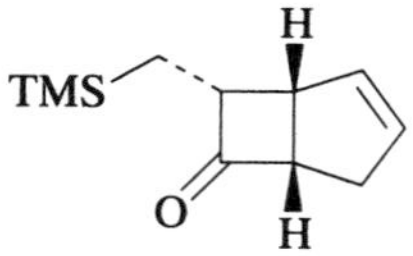 TMS, H, H, O (65)	352
	MeCOCOTMS	$h\nu$, C_5H_{12}	TMSO, O (—)	307
	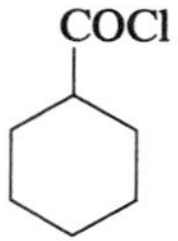 COCl (cyclohexyl)	Et_3N, C_6H_6	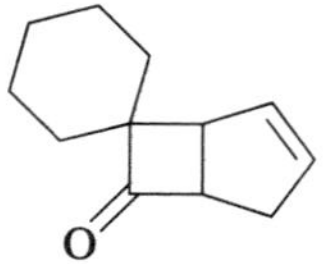 O (65)	527
	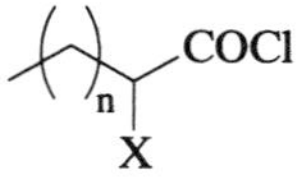 COCl, n, X	Et_3N, C_6H_{14}	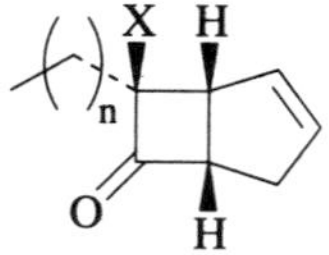 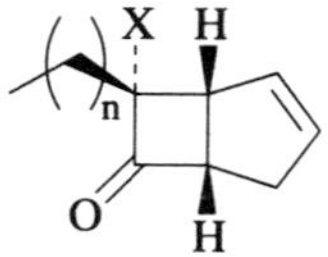 X, H, n, O, H **I** + X, H, n, O, H **II**	528
			n / X / I / II: 2 / Br / (43) / (27); 3 / Br / (25) / (15); 2 / Cl / (50) / (5)	
	PhCHXCOCl	Et_3N	Ph, X, H, O, H — X: Br (53)	529
			Cl (95)	511
			H (26)	38
			Me (85)	529

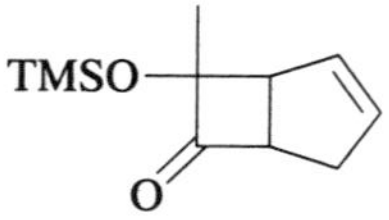

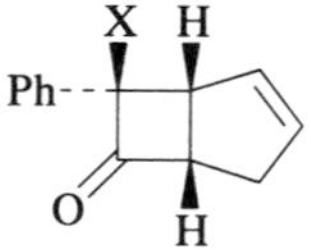

TABLE VI. [2+2] CYCLOADDITION OF KETENES TO CYCLIC DIENES (*Continued*)

Reactant	Ketene or Ketene Source	Conditions	Product(s) and Yield(s) (%)	Refs.
	$PhSCH_2COCl$	Et_3N, C_6H_{14}	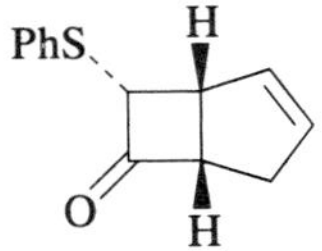 PhS, H, H, O (—)	521
	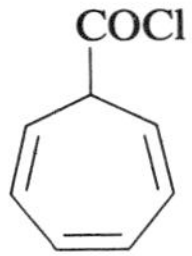 COCl	Et_3N, Et_2O	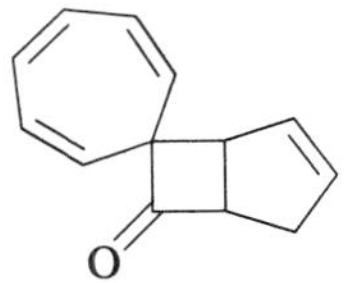 O (60)	424
	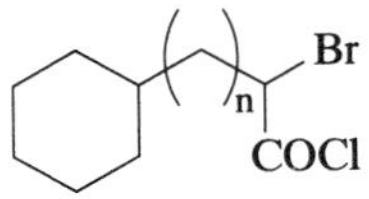 Br, n, COCl	Et_3N, C_6H_{14}	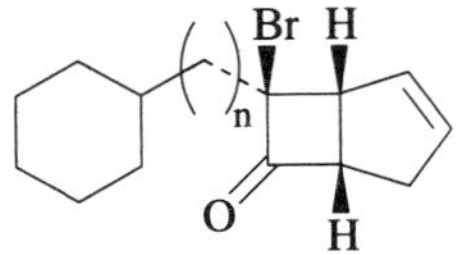 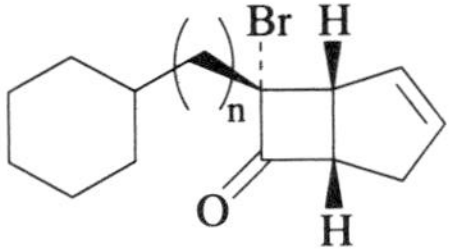 Br, H, n, O, H **I** + Br, H, n, O, H **II** n / **I** / **II**: 0 (62) (20); 1 (59) (24); 2 (66) (26)	528
	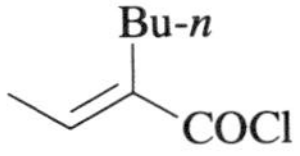 Bu-*n*, COCl	Et_3N, $CHCl_3$	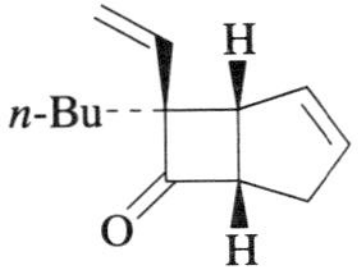 *n*-Bu, H, O, H (54) 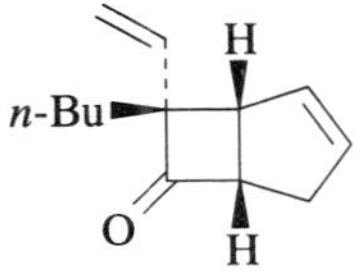 + *n*-Bu, H, O, H (5)	411

TABLE VI. [2+2] CYCLOADDITION OF KETENES TO CYCLIC DIENES (*Continued*)

Reactant	Ketene or Ketene Source	Conditions	Product(s) and Yield(s) (%)	Refs.
	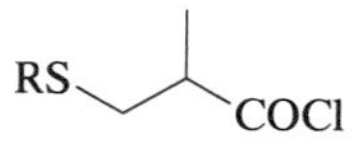 $RSCH_2CH(CH_3)COCl$	Et_3N, C_6H_{12}	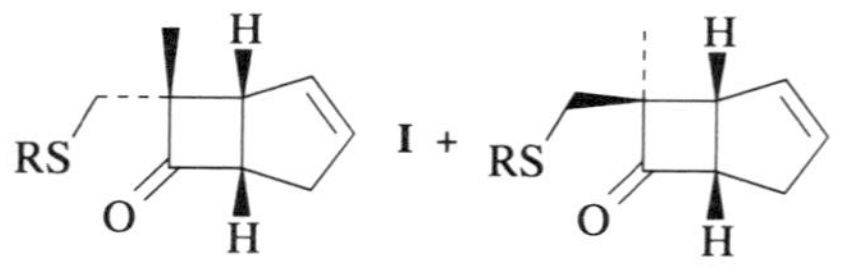 I + II R: Ph, I (56), II (28); *t*-Bu, I (53), II (26)	530, 521
	n-$C_8H_{17}CHBrCOCl$	Et_3N, C_6H_{12}	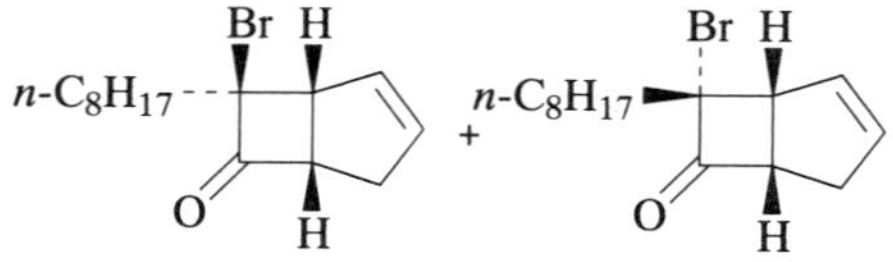 (44) + (21)	528
	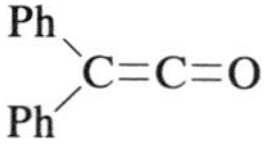 $Ph_2C{=}C{=}O$	C_6H_{14}, rt, 24 h	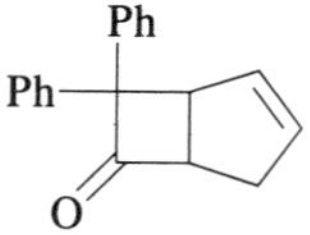 (92)	445, 491, 531, 36, 412, 340, 444, 532
	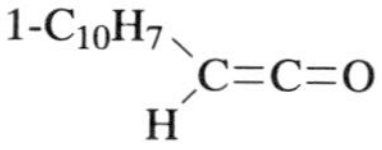 1-$C_{10}H_7CH{=}C{=}O$	—	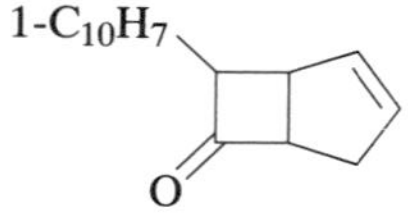 (—)	533
	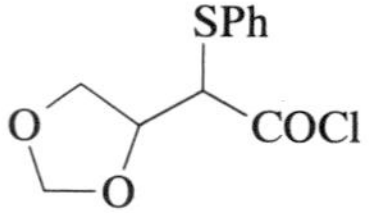 (1,3-dioxolan-4-yl)CH(SPh)COCl	Et_3N, Et_2O	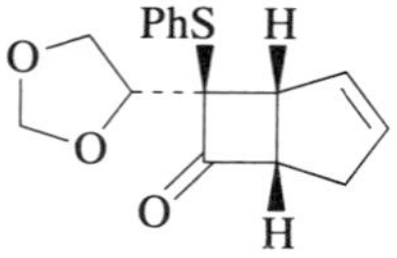 (57)	521

TABLE VI. [2+2] CYCLOADDITION OF KETENES TO CYCLIC DIENES (*Continued*)

Reactant	Ketene or Ketene Source	Conditions	Product(s) and Yield(s) (%)	Refs.
C_6	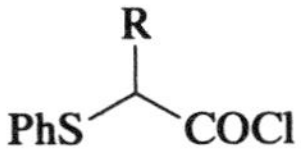 PhS, R, COCl	Et_3N, Et_2O	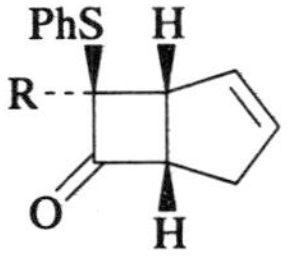 PhS, H, R, O, H; R = n-C_7H_{15} (84); R = n-Bu (82)	521
	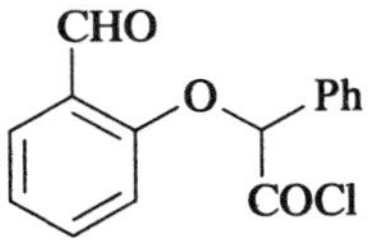 CHO, O, Ph, COCl	Et_3N, C_6H_6	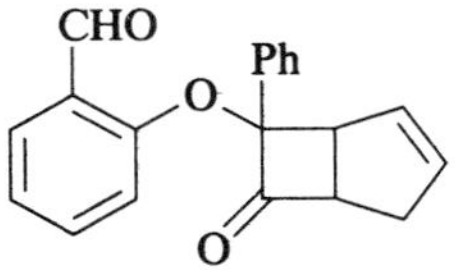 CHO, O, Ph, O (—)	293
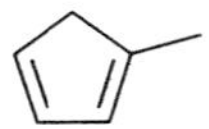	$Cl_2CHCOCl$	Et_3N, C_6H_{14}	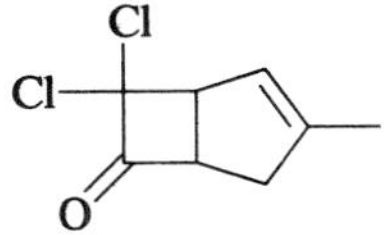 Cl, Cl, O (55)	528
	MeCHBrCOCl	Et_3N, C_6H_{14}	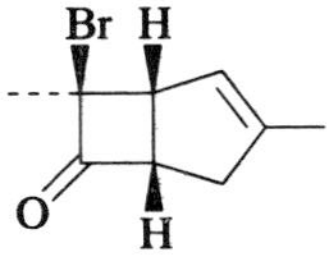 Br, H, O, H (50)	528
	MeCHClCOCl	Et_3N, C_6H_{14}	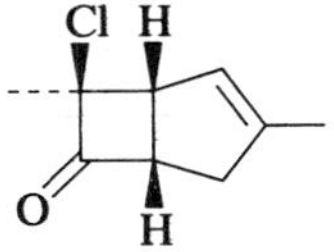 Cl, H, O, H (60) 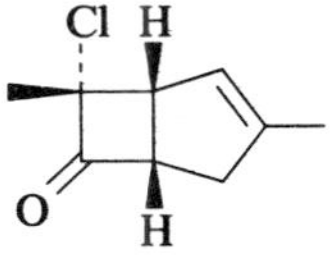 + Cl, H, O, H (15)	528

TABLE VI. [2+2] CYCLOADDITION OF KETENES TO CYCLIC DIENES (*Continued*)

Reactant	Ketene or Ketene Source	Conditions	Product(s) and Yield(s) (%)	Refs.
	$AcOCH_2COCl$	Et_3N	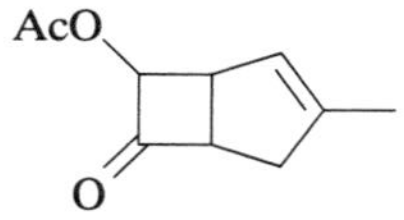 AcO (—)	520
	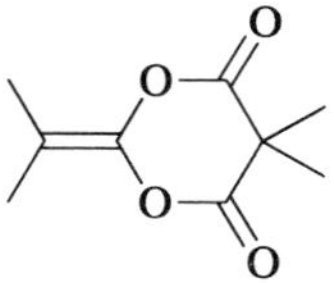 O O O O	K_2CO_3	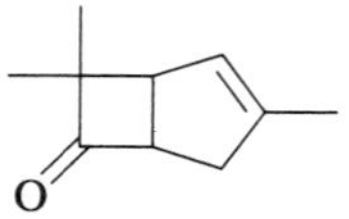 O (55) 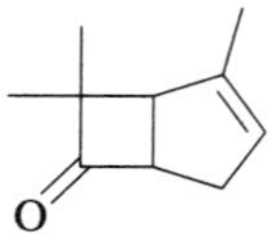 + O (17)	534
	n-PrCOCl	Et_3N, $CHCl_3$	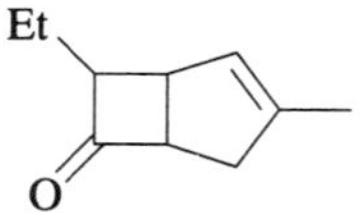 Et O (76) 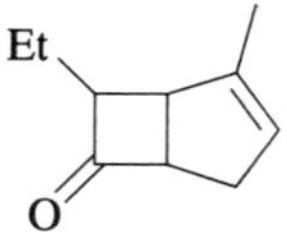 + Et O (5)	534
	n-PrCHClCOCl	Et_3N, C_6H_{14}	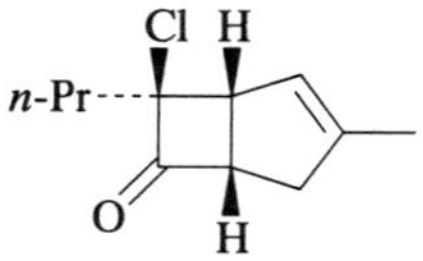 Cl H *n*-Pr O H (55) 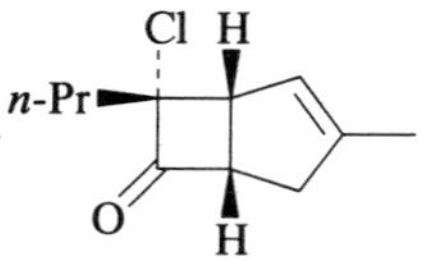 + Cl H *n*-Pr O H (5)	528
	$Cl_2CHCOCl$	Et_3N	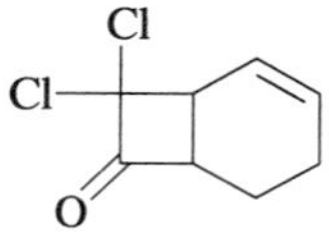 Cl Cl O (60)	431, 441, 88, 535
	$CH_2{=}C{=}O$	PhMe, 100°, 4 d	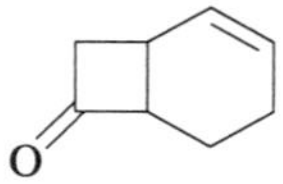 O (—)	514

TABLE VI. [2+2] CYCLOADDITION OF KETENES TO CYCLIC DIENES (*Continued*)

Reactant	Ketene or Ket ene Source	Conditions	Product(s) and Yield(s) (%)	Refs.
	MeCHXCOCl	Et_3N, solvent	[structure] **I** + [structure] **II** Solvent / X / I / II C_6H_{14} / Br / (26) / (14) C_6H_{14} / Cl / (42) / (8) MeCN / Cl / (5) / (40)	442 442 536
	[structure: Cl, COCl]	Et_3N	[structure: Cl, O] (—)	305
	[structure: R^1, R^2, O]	80°, 60 h	[structure: R^1, R^2, O] → [structure: O, R^1, R^2] R^1 = H, R^2 = Me (49) R^1 = H, R^2 = *n*-Bu (33) R^1 = R^2 = Me (91)	305

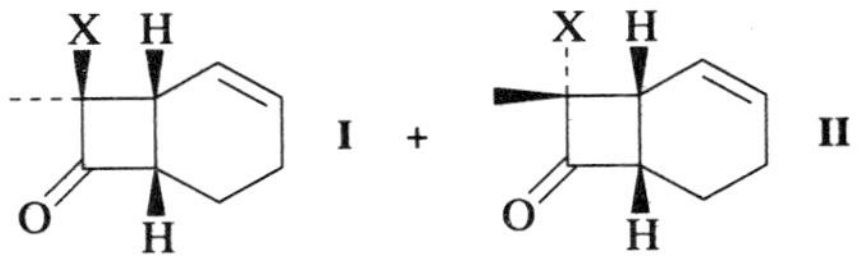

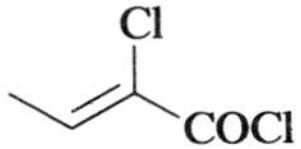

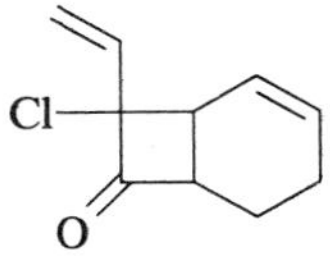

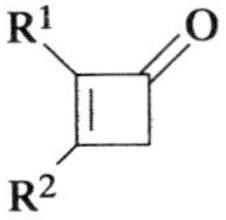

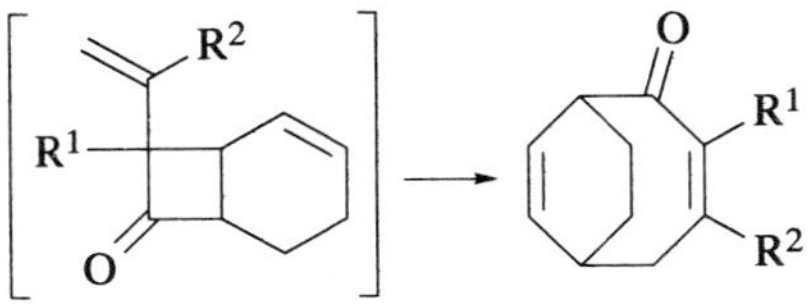

TABLE VI. [2+2] CYCLOADDITION OF KETENES TO CYCLIC DIENES (*Continued*)

Reactant	Ketene or Ketene Source	Conditions	Product(s) and Yield(s) (%)	Refs.
	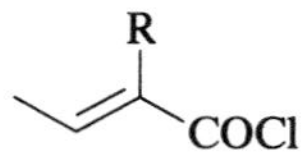 R; COCl	Et_3N, $CHCl_3$	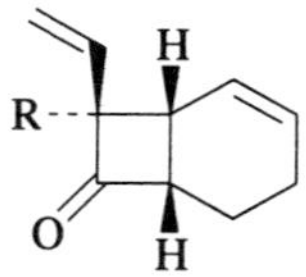 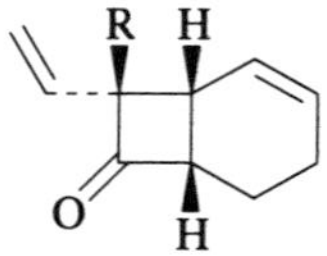 R; H; O; H; I + R; H; O; H; II R: Me, I (37), II (20) R: *n*-Bu, I (41), II (1)	411
	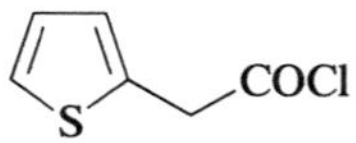 S; COCl	Et_3N, PhMe	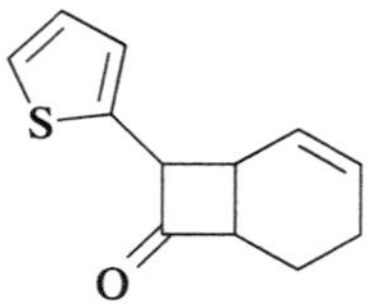 S; O (32)	537
	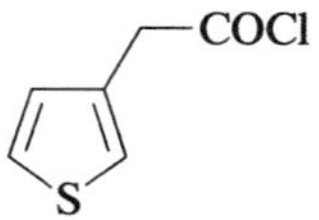 COCl; S	Et_3N, PhMe	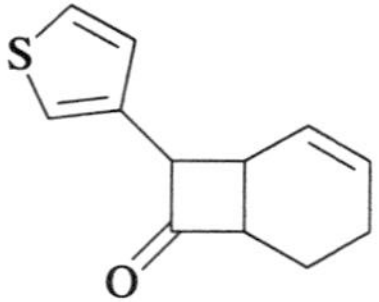 S; O (33)	537
	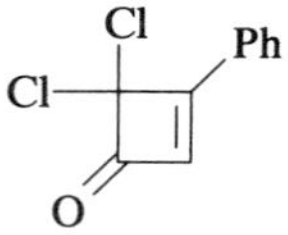 Cl; Cl; Ph; O	$h\nu$, C_6H_6	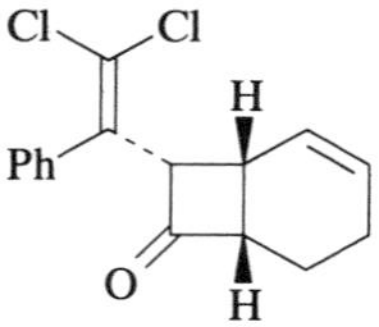 Cl; Cl; Ph; H; O; H (75)	306

TABLE VI. [2+2] CYCLOADDITION OF KETENES TO CYCLIC DIENES (*Continued*)

Reactant	Ketene or Ketene Source	Conditions	Product(s) and Yield(s) (%)	Refs.
	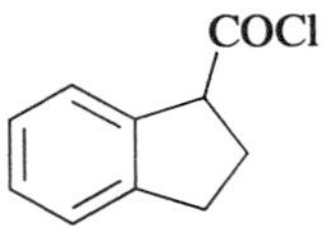 COCl	Et_3N, C_6H_6	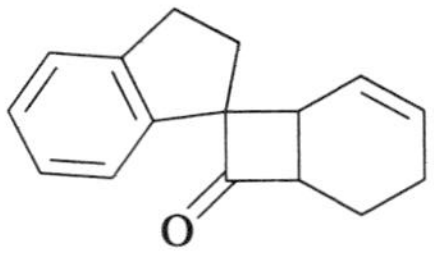 O (99)	538
	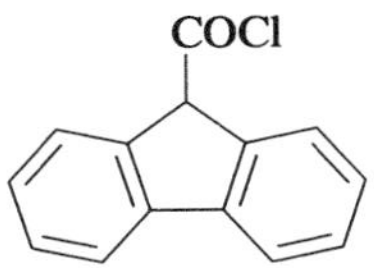 COCl	Et_3N, C_6H_6	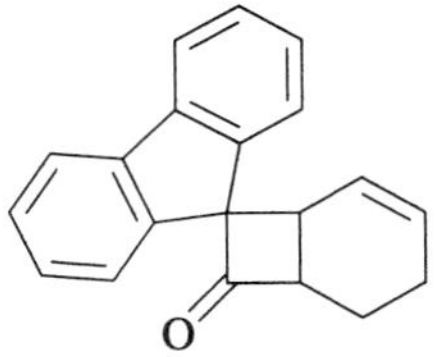 O (62)	538
	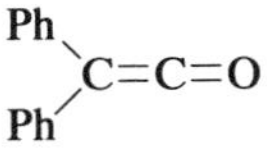 Ph, Ph C=C=O	—	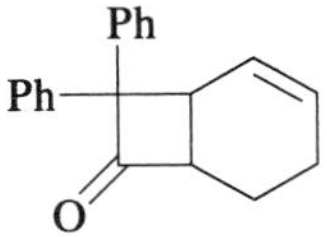 Ph, Ph, O (—)	412
C_7 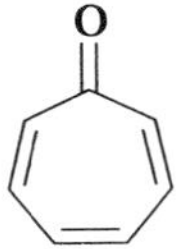 O	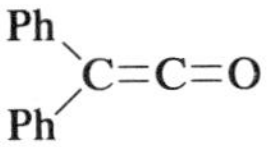 Ph, Ph C=C=O	C_6H_6	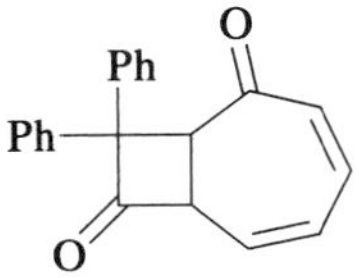 Ph, Ph, O, O (98)	539
	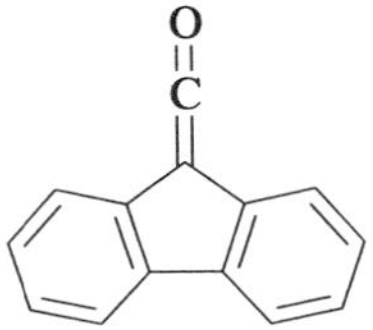 O=C=	—	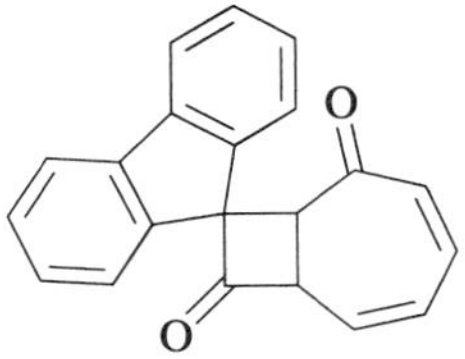 O, O (63)	539

TABLE VI. [2+2] CYCLOADDITION OF KETENES TO CYCLIC DIENES (*Continued*)

Reactant	Ketene or Ketene Source	Conditions	Product(s) and Yield(s) (%)	Refs.
	$(CF_3)_2C=C=O$	100°, 16 h	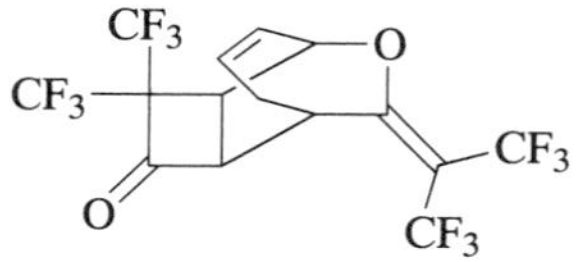 (68)	84
	$Cl_2CHCOCl$	Et_3N, C_6H_{14}	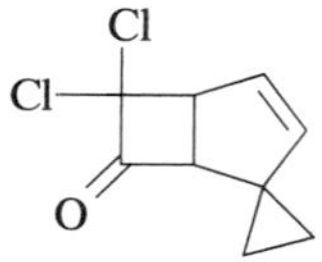 (55)	525, 450
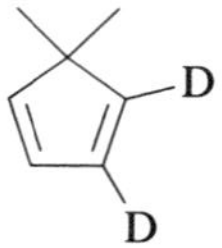 D, D	$Ph_2C=C=O$	C_6H_6, 78°	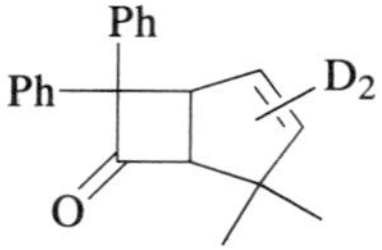 D_2 (—)	52
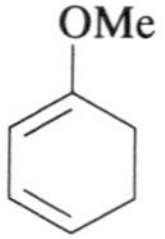 OMe	$Ph_2C=C=O$	42°	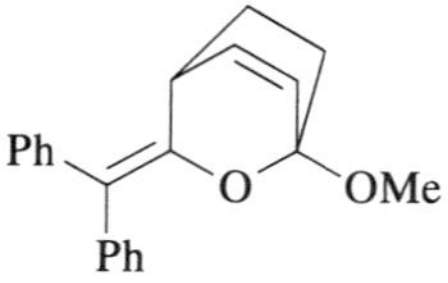 (—)	495
C_8				
	$Cl(NC)C=C=O$	PhMe, 109°	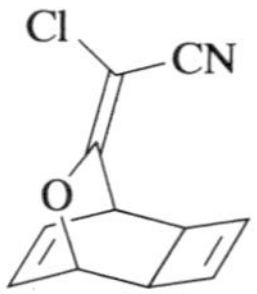 (19)	443

TABLE VI. [2+2] CYCLOADDITION OF KETENES TO CYCLIC DIENES (*Continued*)

Reactant	Ketene or Ketene Source	Conditions	Product(s) and Yield(s) (%)	Refs.
	$(CF_3)_2C=C=O$	100°, 3 d	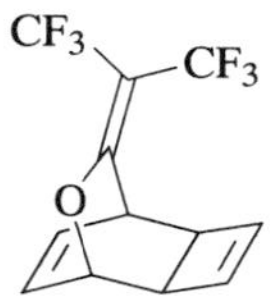 (36)	84
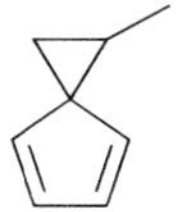	$Cl_2CHCOCl$	Et_3N, C_6H_{14}	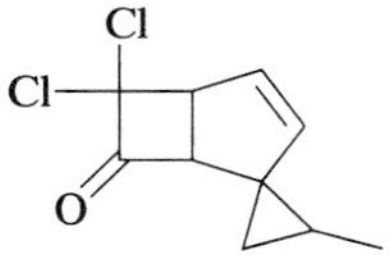 (93)	470
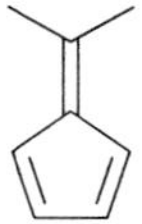	$Cl_2CHCOCl$	Et_3N, C_6H_{14}	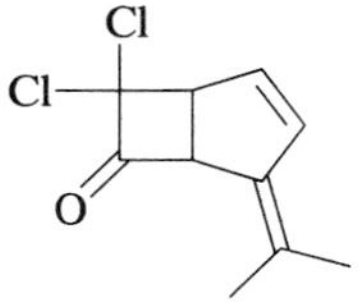 (96)	540
	$ClCH_2COCl$	Et_3N, C_6H_{14}	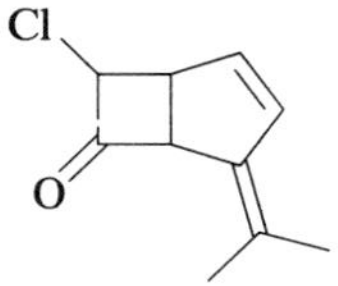 (22)	540
	$AcOCH_2COCl$	Et_3N	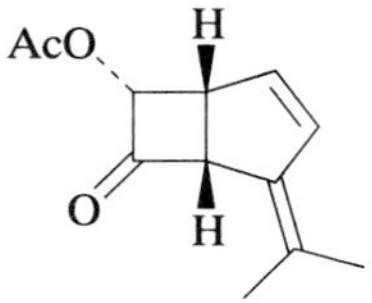 (—)	520

TABLE VI. [2+2] CYCLOADDITION OF KETENES TO CYCLIC DIENES (*Continued*)

Reactant	Ketene or Ketene Source	Conditions	Product(s) and Yield(s) (%)	Refs.
	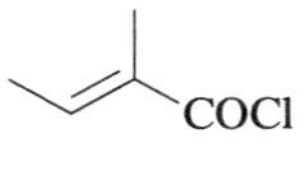 COCl	Et_3N	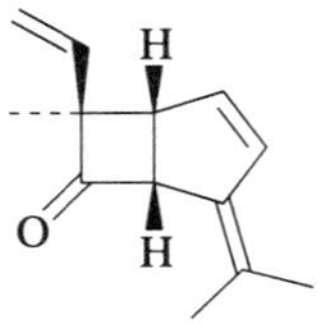 H, H, O (—)	541
	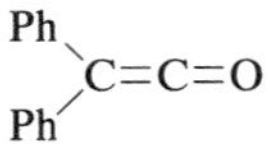 Ph, Ph C=C=O	rt, 6 d	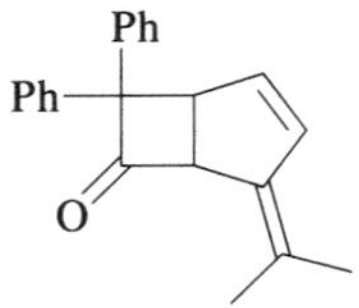 Ph, Ph, O (—)	372
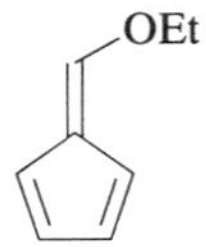 OEt	*i*-PrCOCl	Et_3N, C_6H_{14}	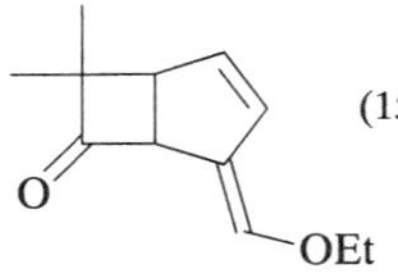 O, OEt (15) 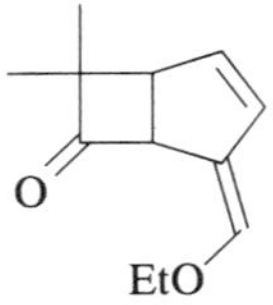 + O, EtO (45)	92
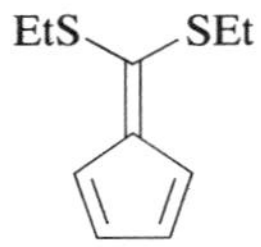 EtS, SEt	$Cl_2CHCOCl$	Et_3N, Et_2O, -10°	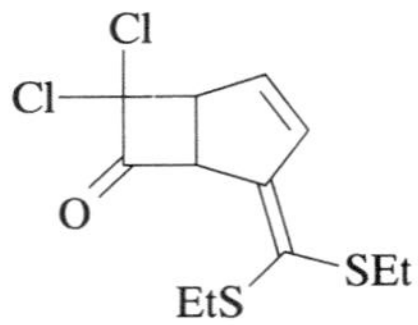 Cl, Cl, O, EtS, SEt (—)	542

TABLE VI. [2+2] CYCLOADDITION OF KETENES TO CYCLIC DIENES (*Continued*)

Reactant	Ketene or Ketene Source	Conditions	Product(s) and Yield(s) (%)	Refs.
 Pr-*i*	$Cl_2CHCOCl$	Et_3N, Et_2O	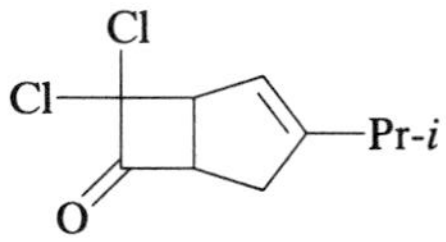 Cl, Cl, O, Pr-*i* (3) 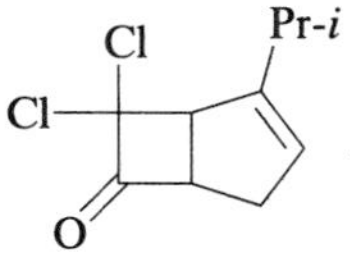 + Cl, Cl, O, Pr-*i* (23) 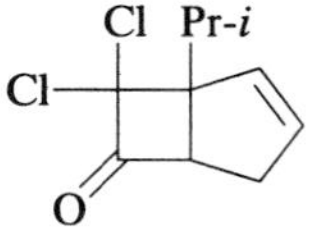 + Cl, Cl, Pr-*i*, O (54)	543
	$Cl_2CHCOCl$	Et_3N, C_5H_{12}	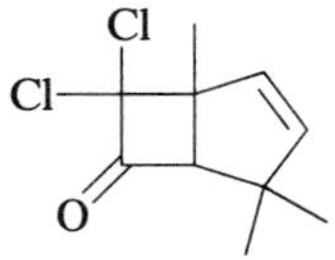 Cl, Cl, O (13) 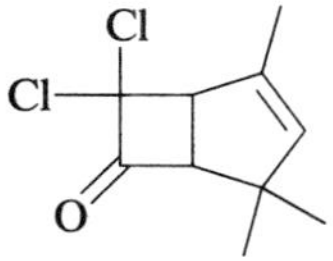 + Cl, Cl, O (13)	459
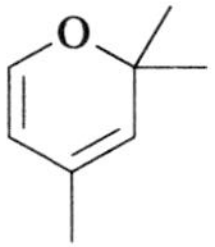 O	Cl_3CCOCl	Zn, $POCl_3$	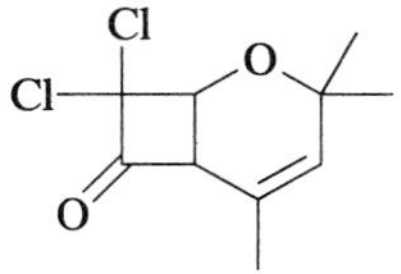 Cl, Cl, O, O (45)	459
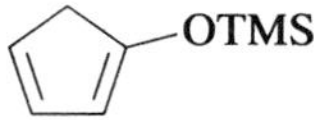 OTMS	RCHClCOCl	Et_3N	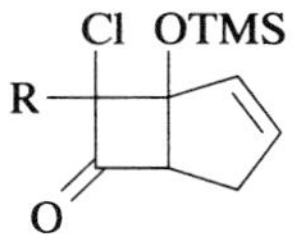 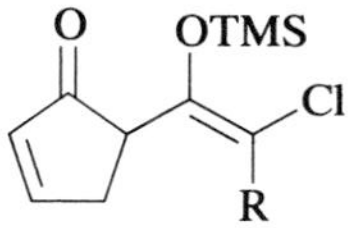 R, Cl, OTMS, O **I** + O, OTMS, Cl, R **II**	499

R	I	II
Cl	(65)	(22)
Me	(82)	(0)

TABLE VI. [2+2] CYCLOADDITION OF KETENES TO CYCLIC DIENES (*Continued*)

Reactant	Ketene or Ketene Source	Conditions	Product(s) and Yield(s) (%)	Refs.
OTMS	$Cl_2CHCOCl$	Et_3N	Cl, Cl, H, H, O, OTMS (69)	544, 545
C_9	$Cl_2CHCOCl$	Et_3N, C_6H_{14}	Cl, Cl, O (55)	470
	$Cl_2CHCOCl$	Et_3N, C_5H_{10}	Cl, Cl, O (53)	450, 451
OTMS	RCHClCOCl	Et_3N	Cl, R, OTMS, O **I** + O, OTMS, Cl, R **II** R / **I** / **II** Cl / (51) / (17) Me / (73) / (0)	499
C_{10} Fe(CO)$_3$	$Ph_2C{=}C{=}O$	C_6H_6, rt	Ph, Ph, O, Fe(CO)$_3$ (25)	546

TABLE VI. [2+2] CYCLOADDITION OF KETENES TO CYCLIC DIENES (*Continued*)

Reactant	Ketene or Ketene Source	Conditions	Product(s) and Yield(s) (%)	Refs.
	R(Cl)C=C=O	C_6H_{14}	R = H (22) R = Cl (84)	547
	i-PrCOCl	Et_3N, CH_2Cl_2	(65)	92
C_{11}				
	$Cl_2CHCOCl$	Et_3N, C_6H_{14}	(46)	471
	$Cl_2CHCOCl$	Et_3N, C_6H_{14}	(70)	548
TMSO OTMS	$Cl_2CHCOCl$	Et_3N, C_6H_{14}	Cl H OTMS, Cl, O, H, OTMS (72)	549

TABLE VI. [2+2] CYCLOADDITION OF KETENES TO CYCLIC DIENES (*Continued*)

Reactant	Ketene or Ketene Source	Conditions	Product(s) and Yield(s) (%)	Refs.
C_{12} 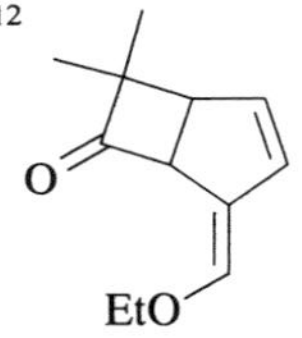 EtO	*i*-PrCOCl	Et_3N, C_6H_{14}	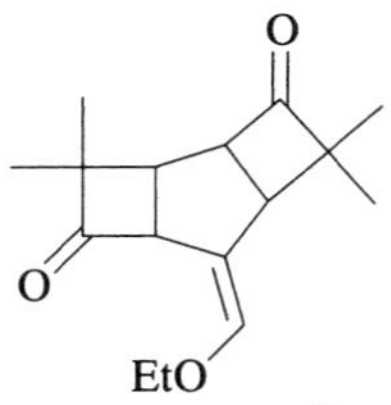 EtO (57)	92
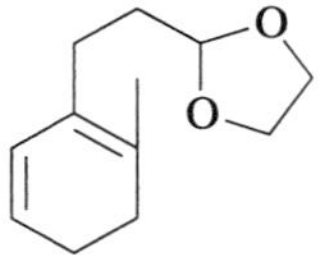 O, O	$Cl_2CHCOCl$	Et_3N, C_6H_{14}	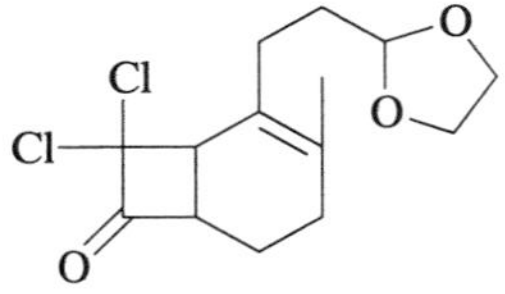 Cl, Cl, O, O, O (—)	550
C_{13} 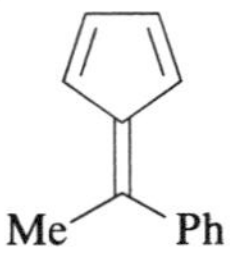 Me, Ph	Ph, Ph C=C=O	C_6H_6	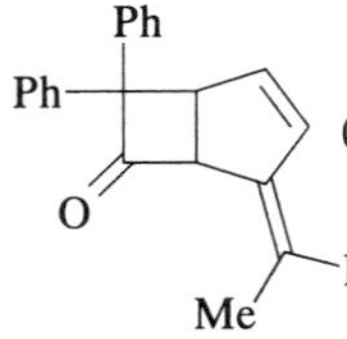 Ph, Ph, O, Me, Ph (42) 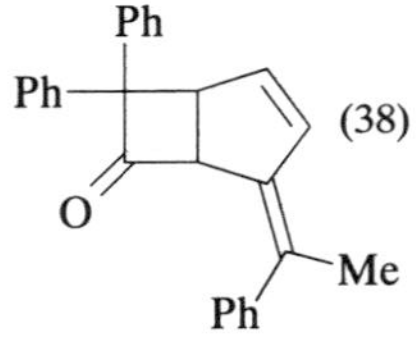 + Ph, Ph, O, Ph, Me (38)	551
C_{14} 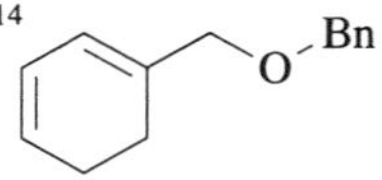 O-Bn	MeCHClCOCl	Et_3N	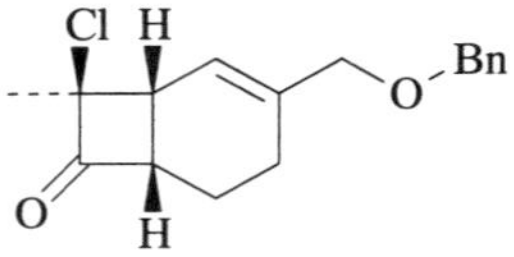 Cl, H, H, O, O-Bn (63) 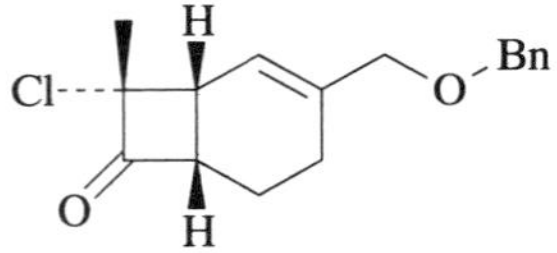 + Cl, H, H, O, O-Bn (23)	552

TABLE VI. [2+2] CYCLOADDITION OF KETENES TO CYCLIC DIENES (*Continued*)

Reactant	Ketene or Ketene Source	Conditions	Product(s) and Yield(s) (%)	Refs.
C_{17}				
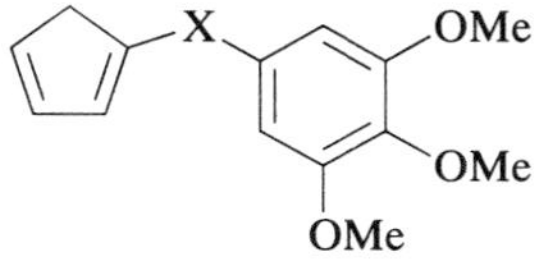 X, OMe, OMe, OMe	$Cl_2CHCOCl$	Et_3N, C_6H_{14}	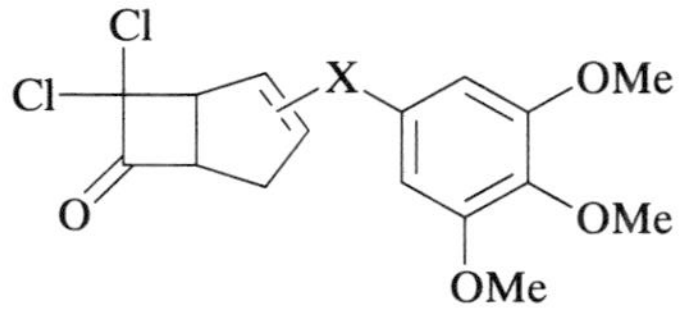 Cl, Cl, O, X, OMe, OMe, OMe X = $CH_2C{\equiv}C$ (32) X = $CH_2CH{=}CH$ (—) X = $(CH_2)_3$ (42)	553
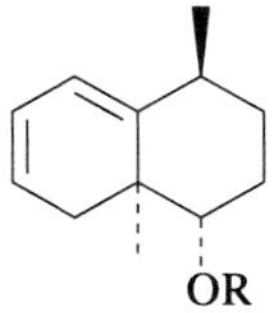 OR	$Cl_2CHCOCl$	Et_3N	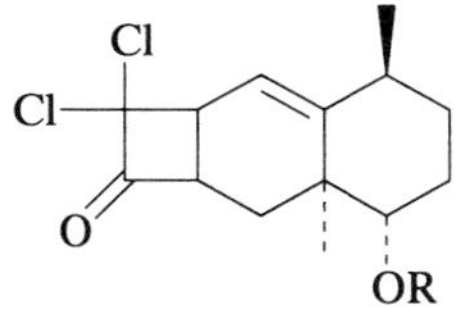 Cl, Cl, O, OR R = OTBDMS (—) R = OBn (—)	554 555
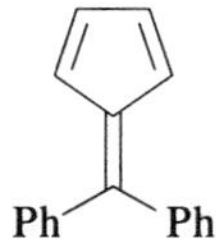 Ph, Ph	$Cl_2CHCOCl$	Et_3N, C_6H_{14}	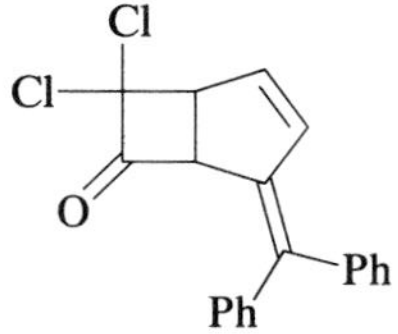 Cl, Cl, O, Ph, Ph (55)	502, 503
	i-PrCOCl	Et_3N, C_6H_{14}	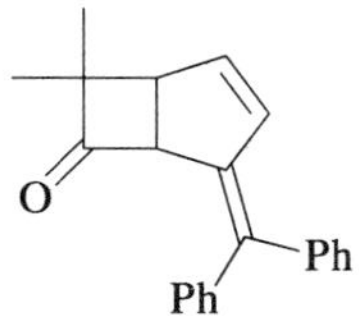 O, Ph, Ph (32)	551

TABLE VI. [2+2] CYCLOADDITION OF KETENES TO CYCLIC DIENES (*Continued*)

Reactant	Ketene or Ketene Source	Conditions	Product(s) and Yield(s) (%)	Refs.
	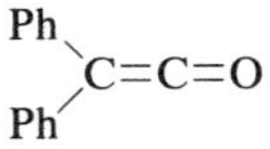 $Ph_2C=C=O$	C_6H_6, rt, 2 d	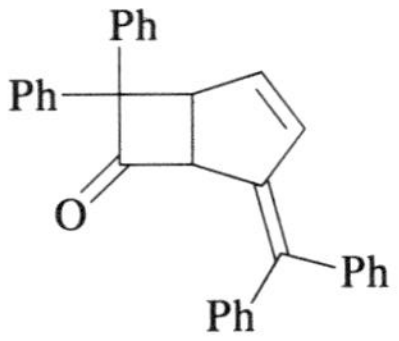 Ph, Ph, O, Ph, Ph (86)	551
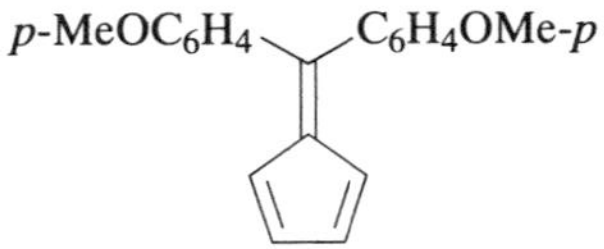 p-$MeOC_6H_4$, C_6H_4OMe-p	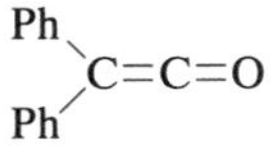 $Ph_2C=C=O$	C_6H_{12}	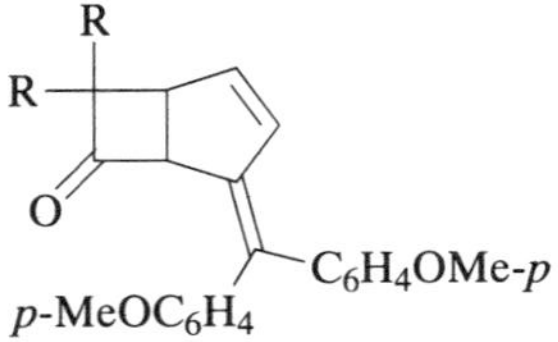 R, R, O, p-$MeOC_6H_4$, C_6H_4OMe-p R = Me (64) R = Ph, (91)	551

TABLE VII. [2+2] CYCLOADDITION OF KETENES TO ARENES

Reactant	Ketene or Ketene Source	Conditions	Ar	R^1	R^2	R^3	R^4	R^5	Yield (%)	Refs.
$R^1(Ar)C=C(R^2)R^3$	$R^4R^5C=C=O$	[a]	p-ClC_6H_4	H	H	H	Ph	Ph	(82)	444
			p-ClC_6H_4	H	H	H	Ph	Ph	(100)	556
			Ph	H	D	D	Ph	Ph	(99)	50
			Ph	D	H	H	Me	Me	(—)	557
			Ph	D	H	H	Ph	Ph	(99)	50
			Ph	D	H	H	Ph	Ph	(—)	51
			Ph	H	D	H	t-Bu	CN	(—)	558
			Ph	H	D	H	Ph	Ph	(—)	51
			Ph	H	H	D	t-Bu	CN	(—)	558
			Ph	H	H	H	Cl	Cl	(88)	83
			Ph	H	H	H	Cl	Cl	(19)	403
			Ph	H	H	H	Cl	Cl	(87)	298
			Ph	H	H	H	Cl	CN	(86)	312
			Ph	H	H	H	CF_3	CF_3	(80)	231
			Ph	H	H	H	CF_3	CF_3	(—)	387
			Ph	H	H	H	CF_3	CF_3	(80)	84
			Ph	H	H	H	Me	Me	(20)	457
			Ph	H	H	H	Me	Me	(40)	293
			Ph	H	H	H	Ph	Ph	(93)	444
			Ph	H	H	H	Ph	Ph	(100)	556
			Ph	H	H	H	Ph	Ph	(94)	386
			Ph	CD_3	H	H	Ph	Ph	(—)	21
			p-$CF_3C_6H_4$	H	H	H	Ph	Ph	(100)	556

Product: cyclobutanone bearing R^4, R^5 (C-4), R^1, Ar (C-3), R^2, R^3 (C-2), C=O.

TABLE VII. [2+2] CYCLOADDITION OF KETENES TO ARENES (*Continued*)

Reactant	Ketene or Ketene Source	Conditions	Product(s) and Yield(s) (%)	Refs.

Ar	R^1	R^2	R^3	R^4	R^5	Yield (%)	Refs.
Ph	Me	H	H	Ph	Ph	(98)	386
Ph	Me	H	H	Ph	Ph	(1)	559
Ph	Me	H	H	Ph	Ph	(98)	81
p-MeC_6H_4	H	H	H	Ph	Ph	(100)	556
p-MeC_6H_4	H	H	H	Ph	Ph	(81)	444
p-$MeOC_6H_4$	H	H	H	Ph	Ph	(84)	444
p-$MeOC_6H_4$	H	H	H	CF_3	CF_3	(79)	133
p-MeC_6H_4	Me	H	H	Cl	Cl	(84)	560
m-MeC_6H_4	Me	H	H	Me	Me	(—)	561

Reactant	Ketene or Ketene Source	Conditions	Product(s) and Yield(s) (%)	Refs.
C_8 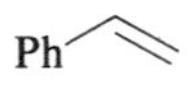 Ph–CH=CH$_2$	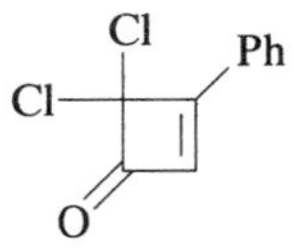 4,4-dichloro-3-phenylcyclobut-2-enone (Cl, Cl, Ph, O)	$h\nu$, C_6H_6	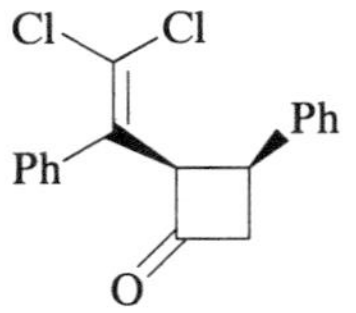 cyclobutanone with Cl$_2$C=C(Ph)– and Ph substituents (90)	306
	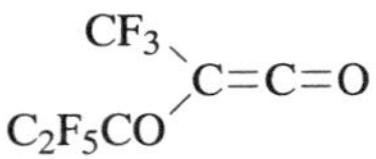 $CF_3(C_2F_5CO)C=C=O$	Phenothiazine, 95°, 15 min	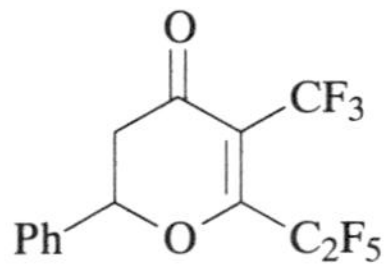 2-Ph-dihydropyran-4-one with CF_3, C_2F_5 (60)	346
C_9 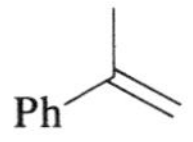 Ph–C(Me)=CH$_2$	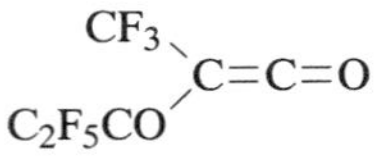 $CF_3(C_2F_5CO)C=C=O$	95°, 60 h	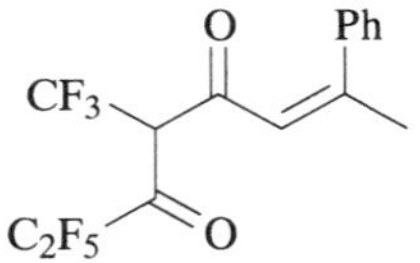 $C_2F_5COCH(CF_3)COCH=C(Ph)Me$ (59)	346

TABLE VII. [2+2] CYCLOADDITION OF KETENES TO ARENES (*Continued*)

Reactant	Ketene or Ketene Source	Conditions	Product(s) and Yield(s) (%)	Refs.
	$(CF_3)_2C{=}C{=}O$	100°, 2 h	$(CF_3)_2CHC(O)CH{=}C(Ph)CH_3$ (96)	84
C_{14}				
$Ph_2C{=}CH_2$	$Ph_2C{=}C{=}O$	150°, 3 d	1,4,4-Triphenyl-3,4-dihydronaphthalen-2-yl diphenylacetate (O_2CCHPh_2) (58)	80, 444, 562

[a] The conditions were different for most entries and can be found by consulting the reference.

TABLE VIII. [2+2] CYCLOADDITION OF KETENES TO ALLENES

Reactant	Ketene or Ketene Source	Conditions	Product(s) and Yield(s) (%)	Refs.
C_3				
$CH_2=C=CH_2$	$(CF_3)_2C=C=O$	165°	No reaction	336
	$(n\text{-Bu})(Et)C=C=O$	—	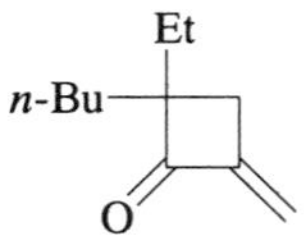	563
	$Ph_2C=C=O$	—	(—)	564
C_5				
$MeCH=C=CHMe$	$(CF_3)_2C=C=O$	100°, 60 h	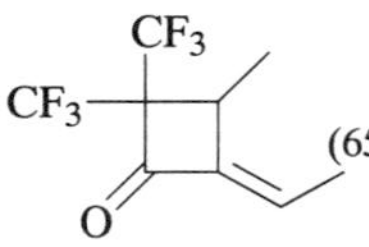 (65) + 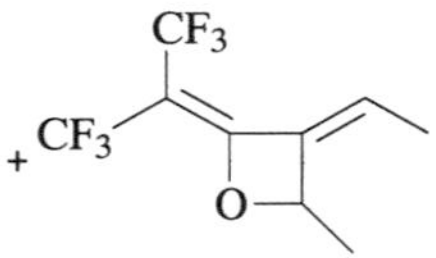 (17)	336
	$Me_2C=C=O$	130°, 2 h	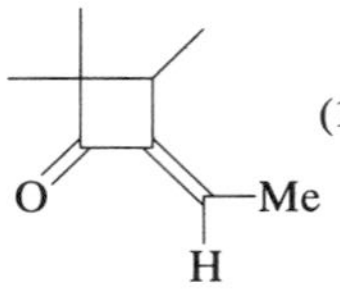 (17) + 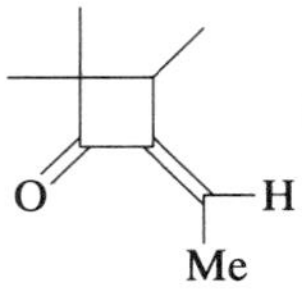 (50)	93, 565-567

TABLE VIII. [2+2] CYCLOADDITION OF KETENES TO ALLENES (*Continued*)

Reactant	Ketene or Ketene Source	Conditions	Product(s) and Yield(s) (%)	Refs.
	t-Bu(NC)C=C=O	25°, 1 h	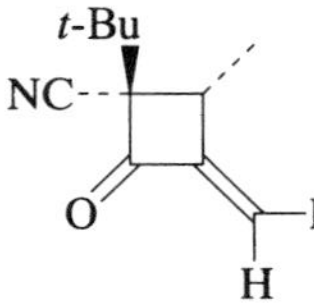 (13) + (2) +	568, 94
			(22) + (23)	
$Me_2C=C=CH_2$	$Me_2C=C=O$	120°, 2 h	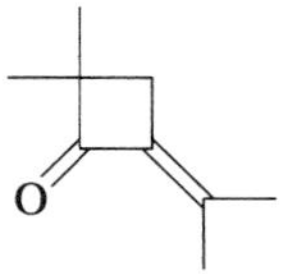 (48) + 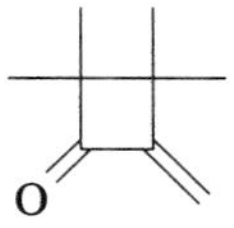 (20)	93, 569
	t-Bu(NC)C=C=O	0°, 1.5 h	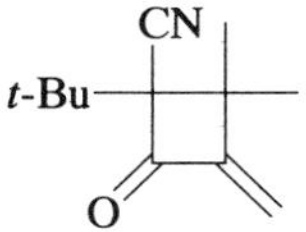 (41) 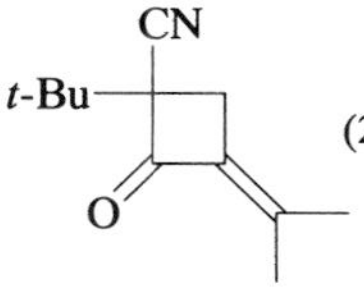 + (21)	568
	Ph(Cl)C=C=O	—	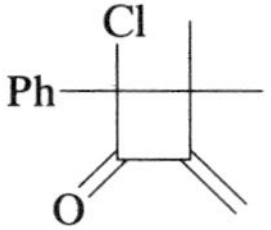 (4) + 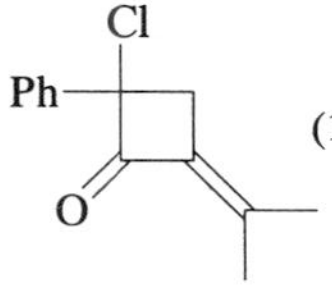 (19)	564

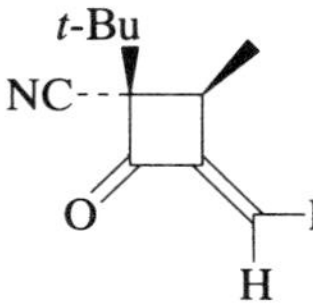

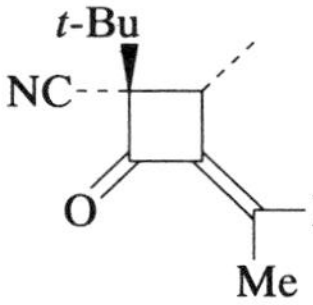

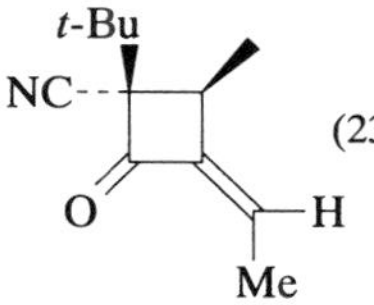

TABLE VIII. [2+2] CYCLOADDITION OF KETENES TO ALLENES (*Continued*)

Reactant	Ketene or Ketene Source	Conditions	Product(s) and Yield(s) (%)	Refs.
	$Ph_2C=C=O$	—	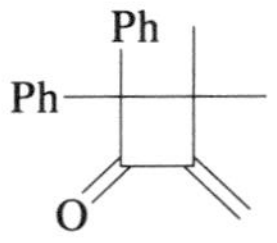 (7) + 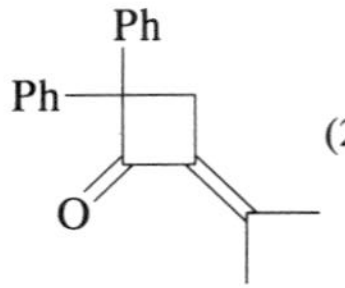 (22)	564
C_6				
$Me_2C=C=C(Me)H$	$Me_2C=C=O$	95°, 1 h	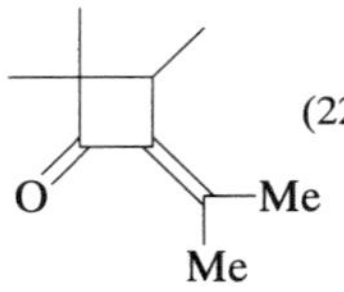 (22) 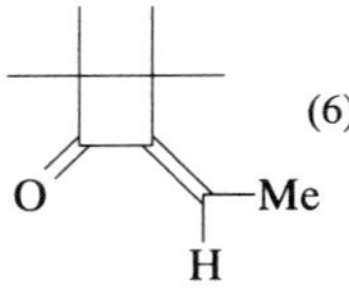 + (6) + (52)	93, 567
C_7				
$Et(Me)C=C=CH_2$	$Me_2C=C=O$	120°, 2 h	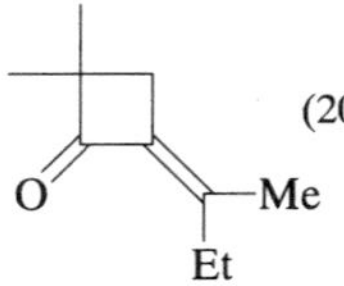 (20) 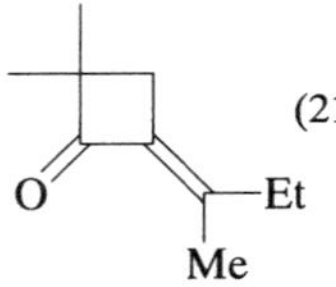 + (21) + (19)	93, 569

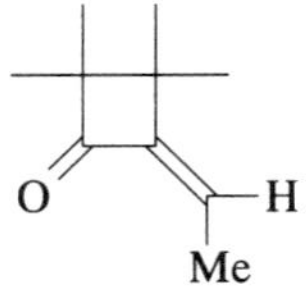

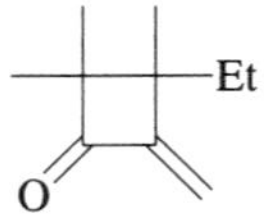

TABLE VIII. [2+2] CYCLOADDITION OF KETENES TO ALLENES (*Continued*)

Reactant	Ketene or Ketene Source	Conditions	Product(s) and Yield(s) (%)	Refs.
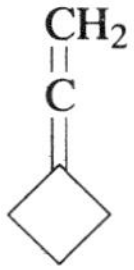 $CH_2=C=$(cyclobutylidene)	$Me_2C=C=O$	50°, 6 h	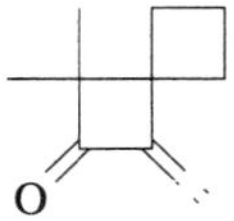 (53) 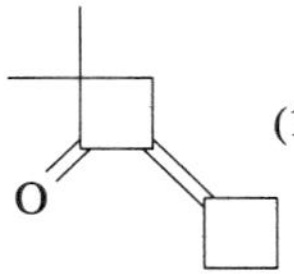 + (17)	93
$Me_2C=C=CMe_2$	$R^1R^2CHCOCl$	Et_3N, C_6H_{14}	R[1] R[2] Yield (%): Me Cl (72); Me Br (65); Et Cl (70); Cl Cl (55); Cl Br (45); Cl H (25); Me H (20); Me Ph (90); Et Ph (90)	570
	$(CF_3)_2C=C=O$	Et_2O	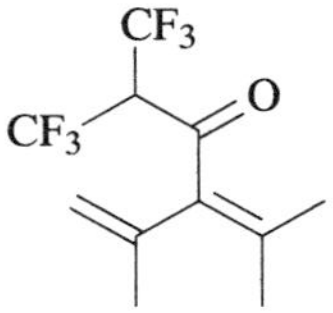 CF_3, CF_3, O (5) + CF_3, CF_3, O, O, CF_3, CF_3 (79)	336
	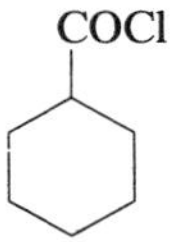 (cyclohexyl)COCl	Et_3N, C_6H_6	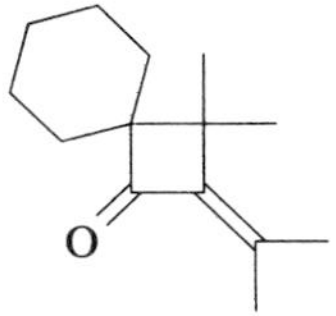 (60)	527

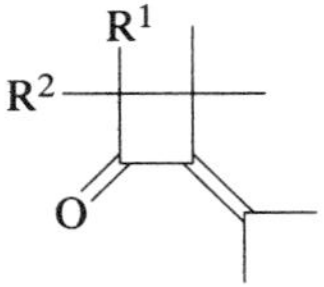

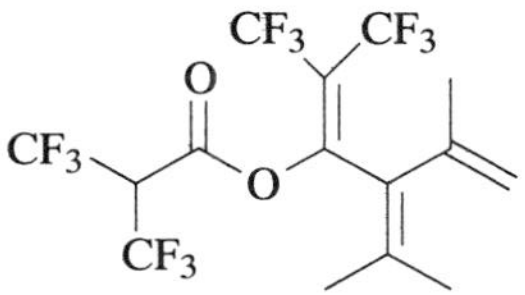

TABLE VIII. [2+2] CYCLOADDITION OF KETENES TO ALLENES (*Continued*)

Reactant	Ketene or Ketene Source	Conditions	Product(s) and Yield(s) (%)	Refs.
	t-Bu(NC)C=C=O	C_6H_6, 78°	(77)	568
	n-Bu(Et)C=C=O	rt, 4 h	(48)	563
	$Ph_2C=C=O$	rt, 4 h	(—)	563
Et(Me)C=C=C(Me)H	$Me_2C=C=O$	95°, 1 h	(10) + (42) + (2) + (16)	93

TABLE VIII. [2+2] CYCLOADDITION OF KETENES TO ALLENES (*Continued*)

Reactant	Ketene or Ketene Source	Conditions	Product(s) and Yield(s) (%)	Refs.
C_8				
t-Bu(Me)C=C=CH_2	t-Bu(NC)C=C=O	C_6H_6, 78°	(16) + (1) + (6) + (4)	571
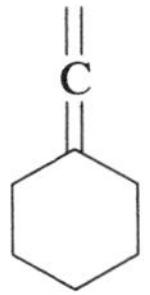 (cyclohexylidene)C=C	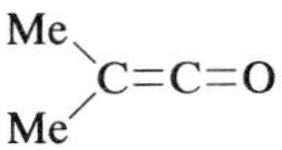 Me_2C=C=O	120°, 2 h	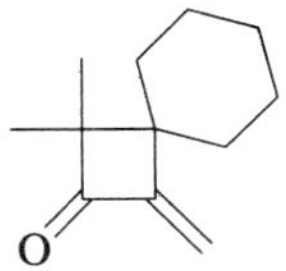 (34) 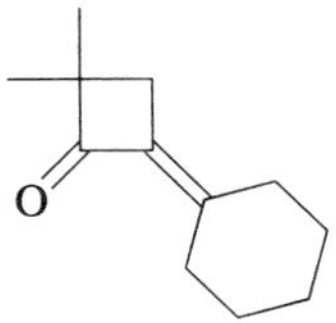 + (36)	93, 569
C_9				
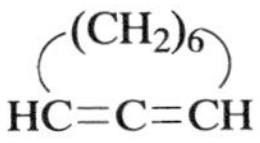 HC=C=CH with $(CH_2)_6$ ring	RCHClCOCl	Et_3N, C_6H_{14}	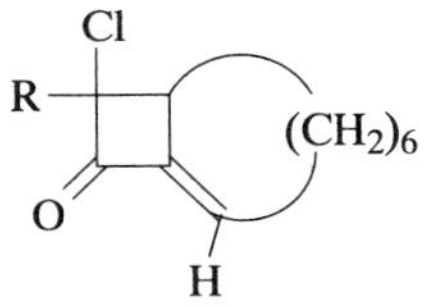 R = Cl (75) R = Me (72)	570

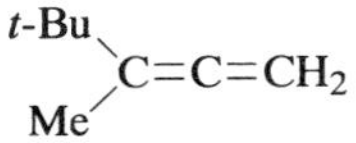

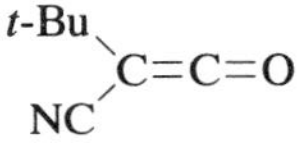

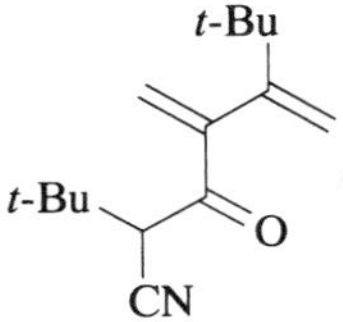

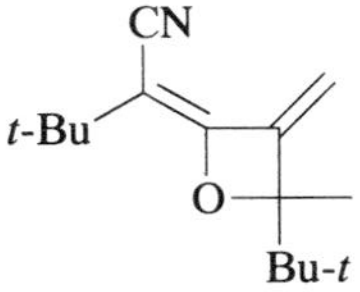

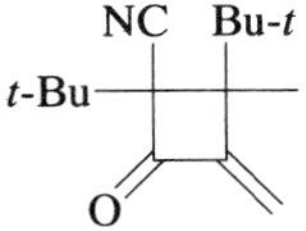

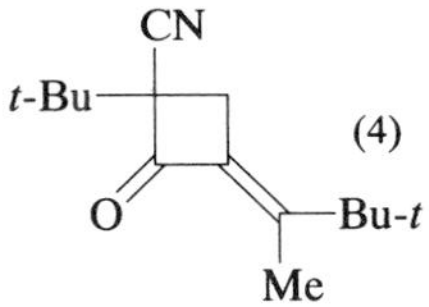

TABLE VIII. [2+2] CYCLOADDITION OF KETENES TO ALLENES (*Continued*)

Reactant	Ketene or Ketene Source	Conditions	Product(s) and Yield(s) (%)	Refs.
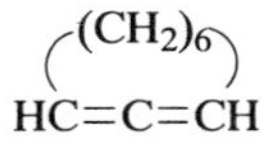 $(CH_2)_6$ HC=C=CH $[\alpha]^{20}_D$ = -7.6°	$Me_2C=C=O$	—	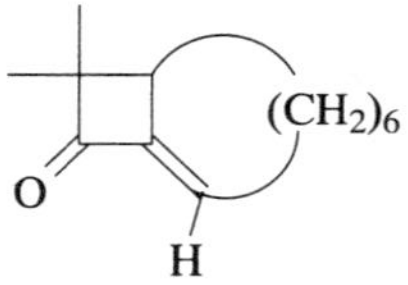 $(CH_2)_6$ O H (—) $[\alpha]^{20}_D$ = +2.2°	566, 572
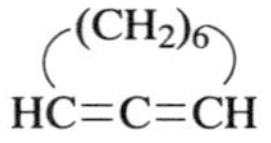 $(CH_2)_6$ HC=C=CH $[\alpha]^{25}_D$ = +20.43°	t-Bu(NC)C=C=O	C_6H_6	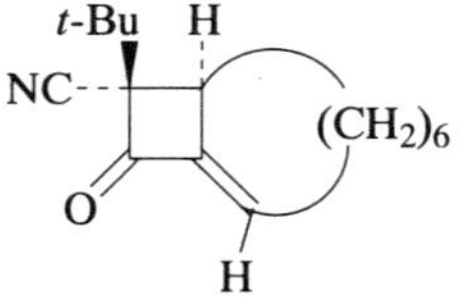 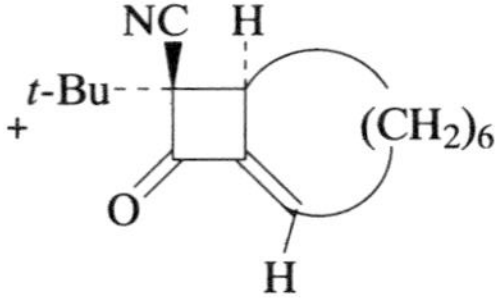 t-Bu H NC $(CH_2)_6$ O H $[\alpha]^{25}_D$ = +29.68° + NC H t-Bu $(CH_2)_6$ O H $[\alpha]^{25}_D$ = -29.41°	573
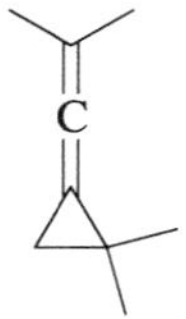 C	$Me_2C=C=O$	50°, 6 h	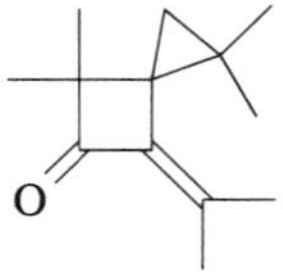 O (100)	93, 423
C_{10}				
$(Me_2C=C=CH)_2$	$Ph_2CHCOCl$	Et_3N, C_6H_{14}	Ph Ph H C=C=C Me Me O (76) + Ph Ph O (<4)	574

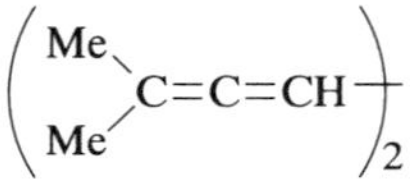

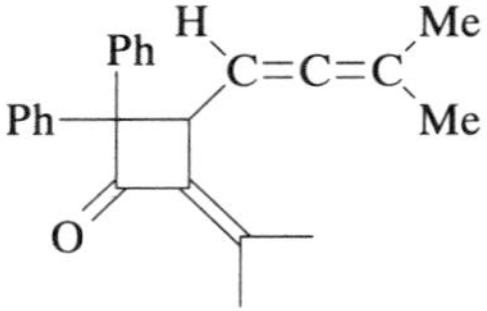

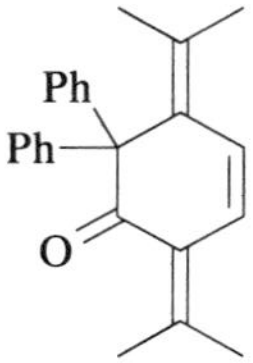

TABLE VIII. [2+2] CYCLOADDITION OF KETENES TO ALLENES (*Continued*)

Reactant	Ketene or Ketene Source	Conditions	Product(s) and Yield(s) (%)	Refs.
	$Cl_2CHCOCl$	Et_3N, Et_2O	Cl, Cl, H, C=C=C, Me, Me, O (35)	574
H, C=C=C, H $[\alpha]^{21}_D$ = +22.4°	Me, C=C=O, Me	80°	O, H (92) $[\alpha]^{23}_D$ = -6.5°	575, 566
$(CH_2)_7$ HC=C=CH	Me, C=C=O, Me	50°, 6 h	O, $(CH_2)_7$, H (100)	93
C_{11}				
t-Bu, C=C=C, Bu-*t*, H, H	Me, C=C=O, Me	125°, 2 h	O, Bu-*t*, Bu-*t*, H (15) + O, Bu-*t*, H, Bu-*t* (10)	93
$(CH_2)_8$ HC=C=CH	Me, C=C=O, Me	120°, 4 h	O, $(CH_2)_8$, H (32) + O, H, $(CH_2)_8$ (40)	93

TABLE VIII. [2+2] CYCLOADDITION OF KETENES TO ALLENES (*Continued*)

Reactant	Ketene or Ketene Source	Conditions	Product(s) and Yield(s) (%)	Refs.
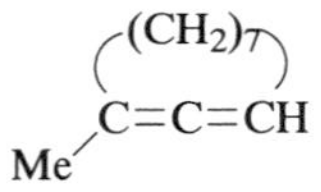 Me–C=C=CH (ring closed by $(CH_2)_7$)	$Me_2C=C=O$	40°, 2 h	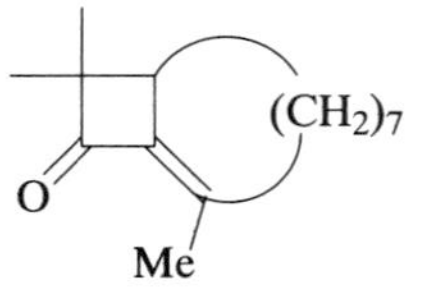 (59) 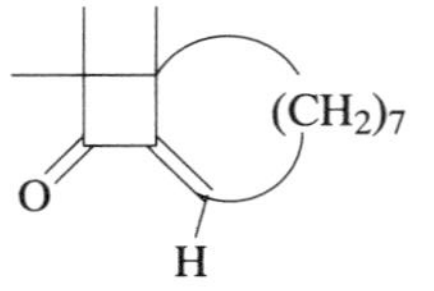 + (26)	93, 570
C_{13}				
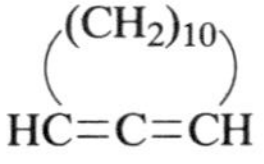 HC=C=CH (ring closed by $(CH_2)_{10}$) $[\alpha]^{20}_D$ = +4.4°	$Me_2C=C=O$	—	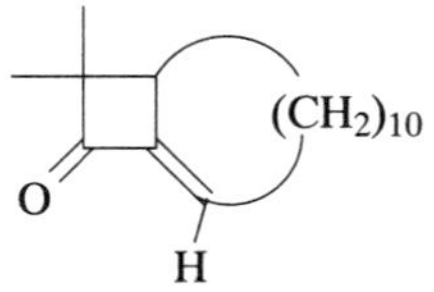 $[\alpha]^{20}_D$ = -0.6° 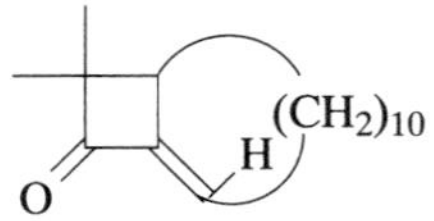 + $[\alpha]^{20}_D$ = -0.7°	566
C_{15}				
Ph(H)C=C=C(H)Ph $[\alpha]_D$ = -365°	t-Bu(NC)C=C=O	C_6H_6, 78°	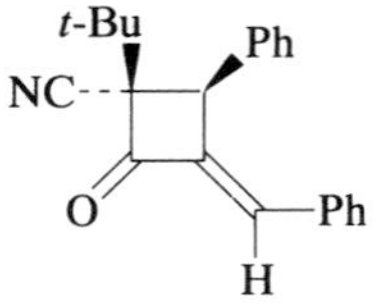 (38) $[\alpha]_D$ = -21.4° 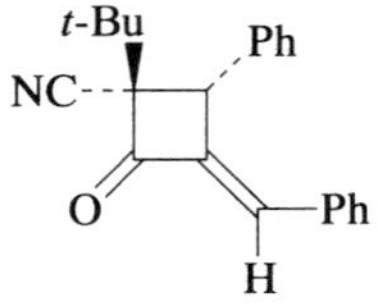 + (11) $[\alpha]_D$ = -213°	95

TABLE IX. [2+2] CYCLOADDITION OF KETENES TO ENAMINES

Reactant	Ketene or Ketene Source	Conditions	Product(s) and Yield(s) (%)	Refs.
C_2				
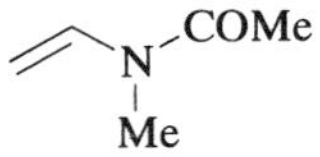 $CH_2{=}CH{-}N(Me){-}COMe$	n-Bu(Et)C=C=O	MeCN	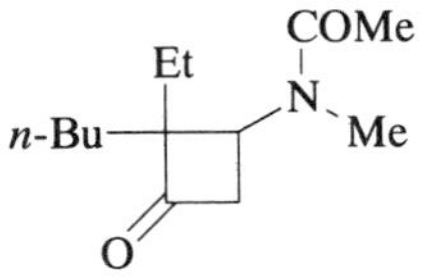 (structure: n-Bu, Et, O, N(Me)COMe) (62)	100
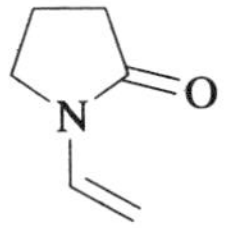 (structure: N, O)	$CH_2{=}C{=}O$	MeCN, reflux	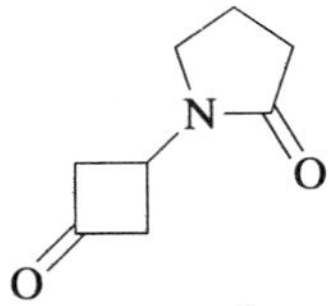 (structure: N, O, O) (—)	100
	$Me_2C{=}C{=}O$	Et_2O, rt	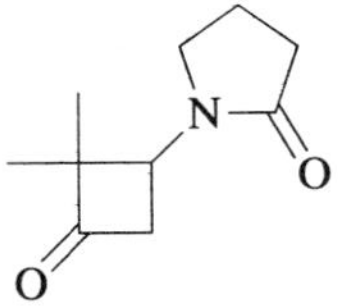 (structure: N, O, O) (30)	405, 100
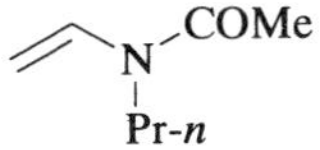 $CH_2{=}CH{-}N(Pr\text{-}n){-}COMe$	$Me_2C{=}C{=}O$	Et_2O, rt	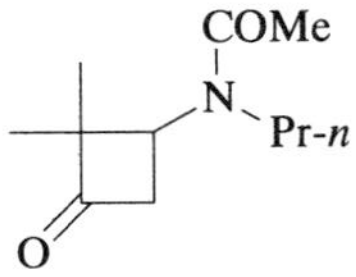 (structure: O, N(Pr-n)COMe) (48)	405, 100
	n-Bu(Et)C=C=O	180°	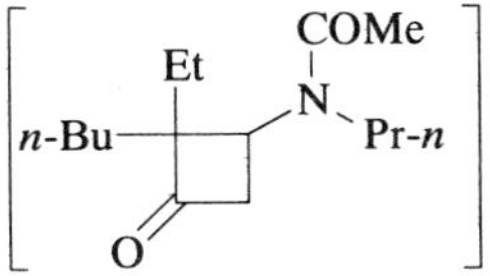 [(structure: n-Bu, Et, O, N(Pr-n)COMe)] (—)	405

TABLE IX. [2+2] CYCLOADDITION OF KETENES TO ENAMINES (*Continued*)

Reactant	Ketene or Ketene Source	Conditions	Product(s) and Yield(s) (%)	Refs.
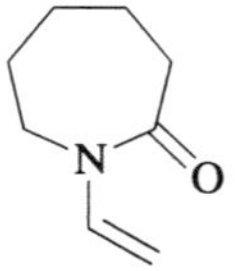 N, O	Me, Me, C=C=O	MeCN	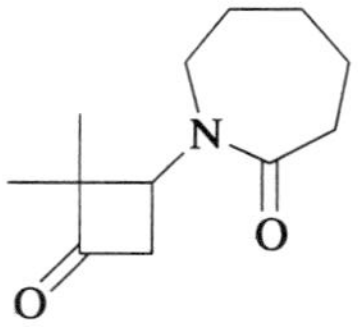 N, O, O (—)	100
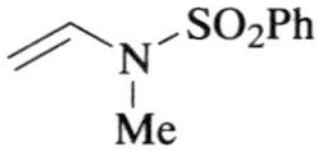 N, SO_2Ph, Me	R^1, R^2, C=C=O	Et_2O, rt	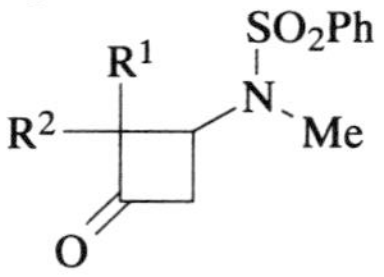 R^1, R^2, SO_2Ph, N, Me, O $R^1 = R^2 = $ Me (24) R^1 = Et, R^2 = *i*-Bu (85)	405
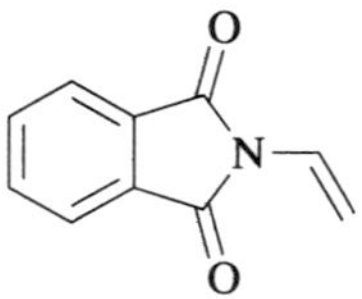 O, N, O	Me, Me, C=C=O	Et_2O, rt	(—)	405
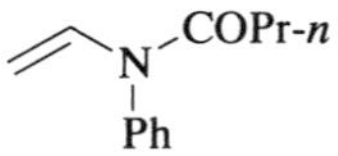 N, COPr-*n*, Ph	*n*-C_8H_{17}, *n*-C_8H_{17}, C=C=O	MeCN, reflux	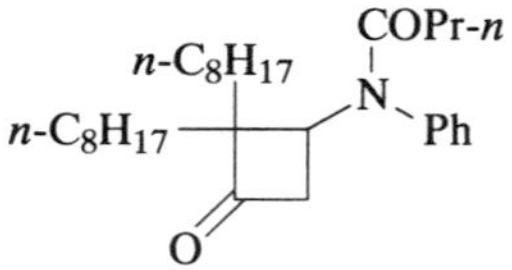 *n*-C_8H_{17}, *n*-C_8H_{17}, COPr-*n*, N, Ph, O (—)	100
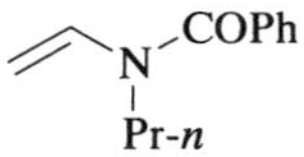 N, COPh, Pr-*n*	Me, Me, C=C=O	C_6H_6	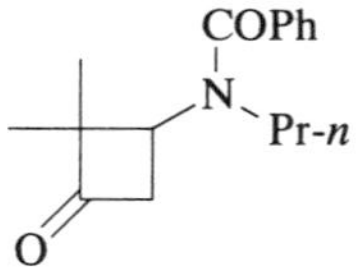 COPh, N, Pr-*n*, O (86)	405, 100

TABLE IX. [2+2] CYCLOADDITION OF KETENES TO ENAMINES (*Continued*)

Reactant	Ketene or Ketene Source	Conditions	Product(s) and Yield(s) (%)	Refs.
$CH_2{=}CH{-}N(Me)COC_8H_{17}$-*n*	*n*-C_8H_{17}(H)C=C=O	MeCN, reflux	(—)	100
$CH_2{=}CH{-}N(Bu\text{-}n)COC_6H_4Me$-*p*	cyclohexylidene C=O	MeCN, reflux	(—)	100
C_3				
1-(1-propenyl)piperidine	$CH_2{=}C{=}O$	Et_2O, -40°	(14)	576
$MeCH{=}CH{-}N(Pr\text{-}n)COMe$	$Ph_2C{=}C{=}O$	MeCN, reflux	(99)	100, 405
1,3-dipiperidino-1-propene	$R_2C{=}C{=}O$	C_6H_{14}	R = H (—) R = Me (—)	405

TABLE IX. [2+2] CYCLOADDITION OF KETENES TO ENAMINES (*Continued*)

Reactant	Ketene or Ketene Source	Conditions	Product(s) and Yield(s) (%)	Refs.
Et–CH=CH–N(Me)COMe	$(p\text{-MeOC}_6\text{H}_4)_2$C=C=O	MeCN, reflux	p-MeOC$_6$H$_4$, p-MeOC$_6$H$_4$, N(Me)COMe, Et, O (cyclobutanone) (—)	100
1-(1-butenyl)-X-heterocycle (N, X)	R_2C=C=O	[a]	[cyclobutanone: R, R, N-heterocycle (X), Et, O]	
			R = H, X = O, Yield (%) (—)	98,577
			R = H, X = CH_2, Yield (%) (—)	576
			R = Me, X = CH_2, Yield (%) (—)	577
Me_2C=CH–NMe_2	R^1R^2C=C=O	[a]	cyclobutanone (R^1, R^2, NMe_2, O) **I** + lactone (R^1, R^2, NMe_2, O, O, =CR^1R^2) **II**	
			R^1 = H, R^2 = H, Yield **I** (%) (—), Yield **II** (%) (—)	577
			R^1 = Me, R^2 = Me, Yield **I** (%) (64), Yield **II** (%) (—)	578, 577
			R^1 = Me, R^2 = Me, Yield **I** (%) (—), Yield **II** (%) (32)	579
			R^1 = Et, R^2 = Et, Yield **I** (%) (—), Yield **II** (%) (12)	579, 578
			R^1 = Et, R^2 = n-Bu, Yield **I** (%) (—), Yield **II** (%) (—)	579, 578
			R^1 = Me, R^2 = Et, Yield **I** (%) (—), Yield **II** (%) (—)	578

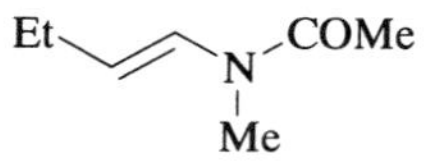

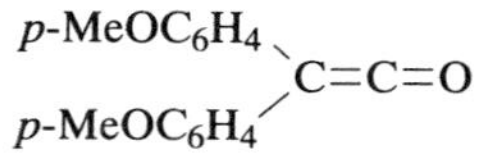

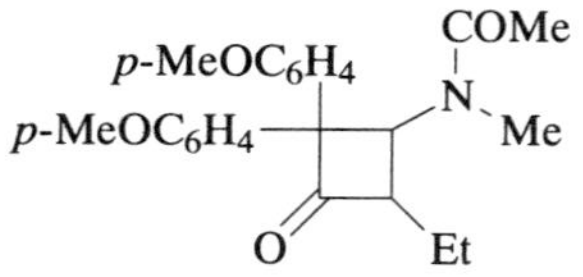

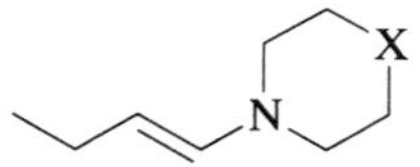

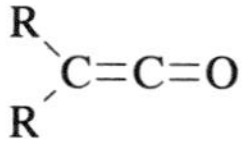

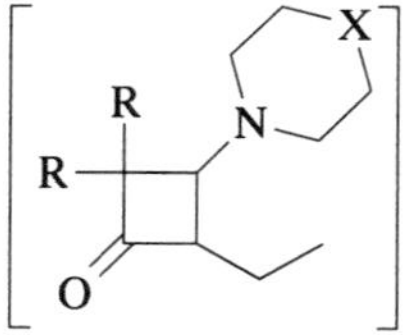

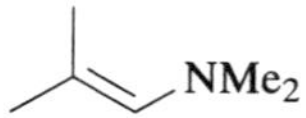

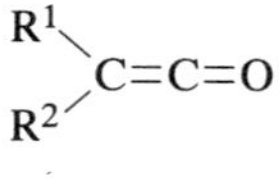

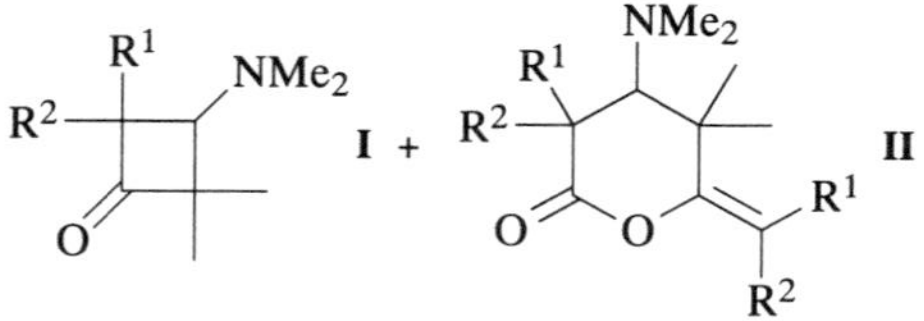

TABLE IX. [2+2] CYCLOADDITION OF KETENES TO ENAMINES (*Continued*)

Reactant	Ketene or Ketene Source	Conditions	Product(s) and Yield(s) (%)			Refs.
1-(2-methylprop-1-enyl) cyclic amine (ring containing X)	$R^1R^2C{=}C{=}O$	a	aminocyclobutanone + δ-lactone			
			R^1	R^2	X	
			H	H	—	576
			Me	Me	—	580, 581
			Me	H	—	582
			Et	*n*-Bu	—	582
			Br	CH_2=CH	—	97
			Ph	Ph	—	99
			H	H	O	576, 98
			Me	H	O	582
			Cl	H	O	582
			PhO	H	O	582
			MeO	H	O	582
			Ph	H	O	582
			Me	Me	O	582, 578, 64
			Et	Et	O	582, 578
			Ph	Ph	O	582, 580, 99
			Me	Et	O	578
			CH_2=CH	H	O	97
			CH_2=C(Me)	H	O	97
			CH_2=CH	Br	O	97
			CH_2=C(Me)	Br	O	97
			H	H	CH_2	576

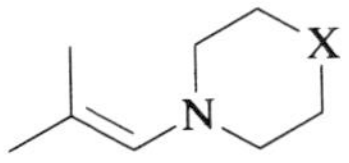

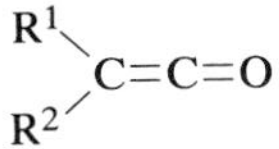

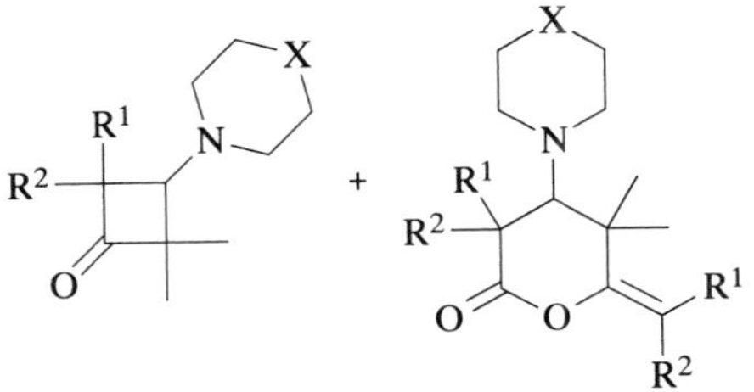

TABLE IX. [2+2] CYCLOADDITION OF KETENES TO ENAMINES (*Continued*)

Reactant	Ketene or Ketene Source	Conditions	Product(s) and Yield(s) (%)	Refs.

R^1	R^2	X	Refs.
Me	Me	CH_2	578-580, 583
Et	Et	CH_2	578
–$(CH_2)_5$–		CH_2	578
PhO	H	CH_2	582
i-Pr	H	CH_2	582
Bu	H	CH_2	582
n-$C_{16}H_{33}$	H	CH_2	582
Me	Me	NMe	578

COCl

Et_3N

X N O I + O N X II — 97

X	Solvent	Temp.	Yield **I** (%)	Yield **II** (%)
—	CH_2Cl_2	rt	(0)	(53)
—	C_6H_{14}	70°	(28)	(28)
O	CH_2Cl_2	rt	(9)	(17)
O	C_6H_{14}	70°	(24)	(2)

N N — Me C=C=O Me — — O N N O (—) — 578

TABLE IX. [2+2] CYCLOADDITION OF KETENES TO ENAMINES (*Continued*)

Reactant	Ketene or Ketene Source	Conditions	Product(s) and Yield(s) (%)	Refs.
C_5				
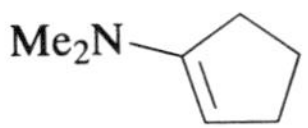 Me_2N– (1-cyclopentenyl)	$CH_2{=}C{=}O$	Et_2O, rt, 4 h	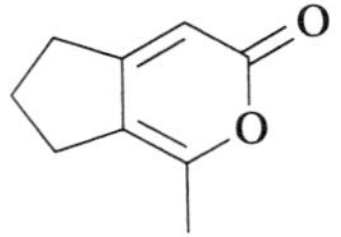 (20)	98
C_6				
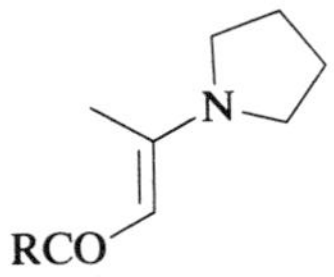 RCO, N (pyrrolidinyl)	$CH_2{=}C{=}O$	Et_2O, rt, 4 h	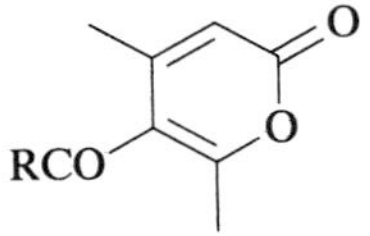 RCO, O, O; R = Me (52); R = MeO (35)	98
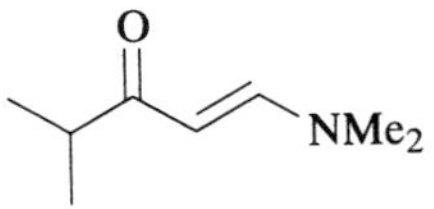 O, NMe_2	$R^1R^2C{=}C{=}O$	C_6H_6, rt, 2 h	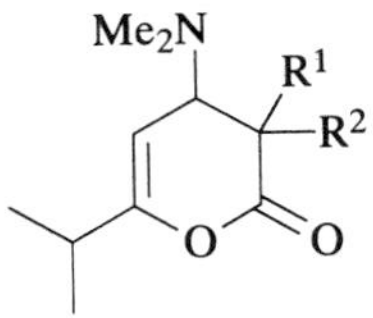 Me_2N, R^1, R^2, O, O; $R^1 = R^2$ = Me (86); R^1 = Et, R^2 = *n*-Bu (52)	584
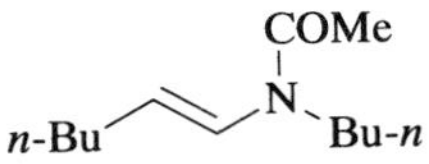 *n*-Bu, N(COMe)Bu-*n*	Ph(Me)C=C=O	MeCN, reflux	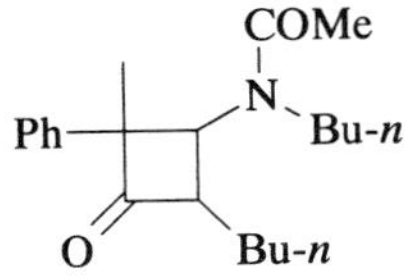 Ph, N(COMe)Bu-*n*, Bu-*n*, O (—)	100
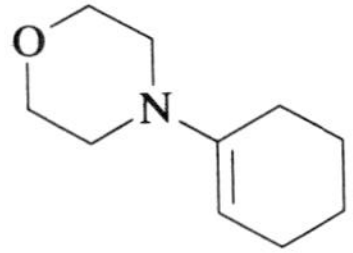 O, N (morpholinyl), 1-cyclohexenyl	$CH_2{=}C{=}O$	Et_2O, -40°, 2 h	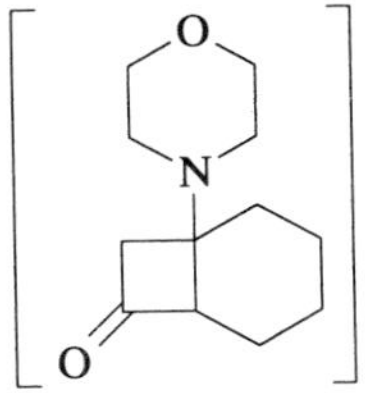 [O, N, O] (—)	576, 585, 98

TABLE IX. [2+2] CYCLOADDITION OF KETENES TO ENAMINES (*Continued*)

Reactant	Ketene or Ketene Source	Conditions	Product(s) and Yield(s) (%)	Refs.
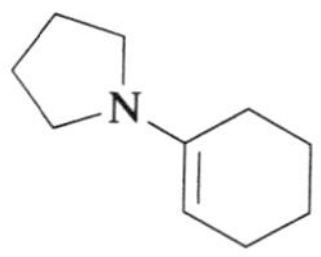 N	$CH_2{=}C{=}O$	Et_2O, rt, 4 h	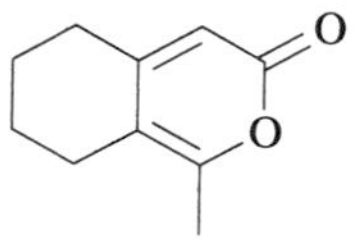 O, O (35)	98
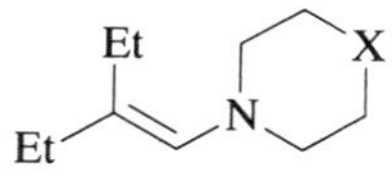 Et, Et, N, X	$CH_2{=}C{=}O$	Et_2O, -40°, 2 h	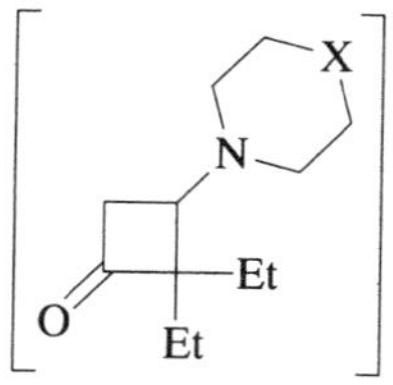 [X, N, Et, Et, O] (—) X = —, O, CH_2	576
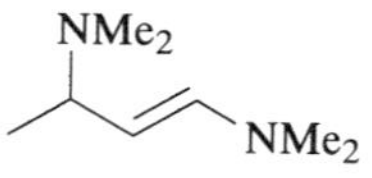 NMe_2, NMe_2	Me, Me, $C{=}C{=}O$	C_6H_{14}	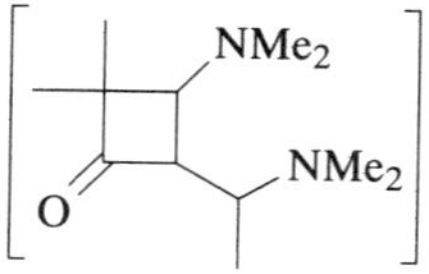 [NMe_2, NMe_2, O] (—)	405
C_7				
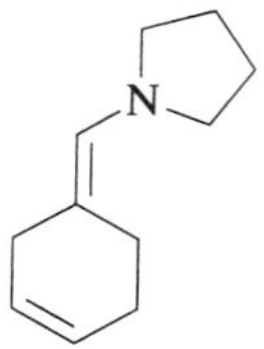 N	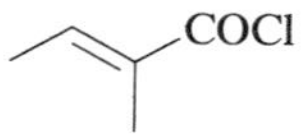 COCl	Et_3N, CH_2Cl_2, rt	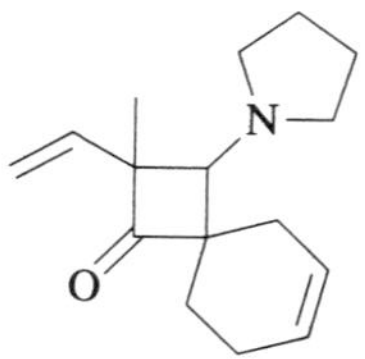 N, O (83)	97

TABLE IX. [2+2] CYCLOADDITION OF KETENES TO ENAMINES (*Continued*)

Reactant	Ketene or Ketene Source	Conditions	Product(s) and Yield(s) (%)	Refs.
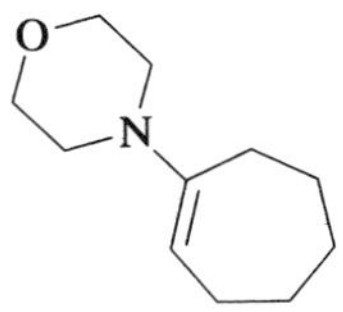	RCH_2COCl	Et_3N	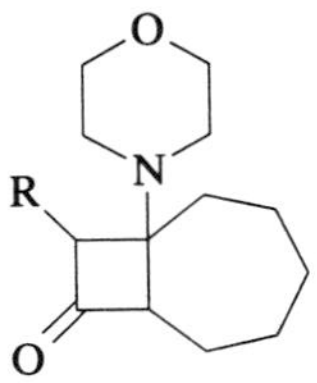 R = H (41) R = Me (29)	460
C_8	MeCOCl	Et_3N, Et_2O	(—)	460, 586
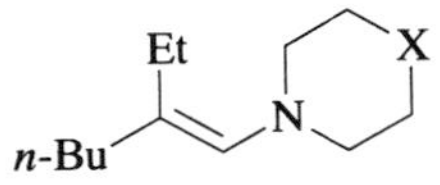 X; Et; *n*-Bu	$CH_2{=}C{=}O$	Et_2O, -40°, 30 min	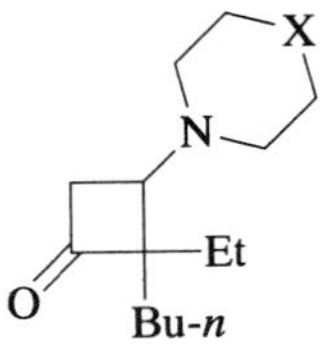 X; Et; Bu-*n*; (—) X = —, O, CH_2	576
C_9 $(CH_2)_x$	MeCOCl	Et_3N, Et_2O	$(CH_2)_x$ (—) X = 7-11	586

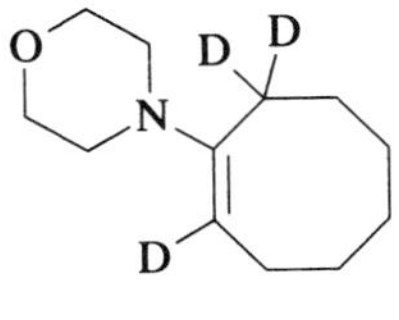

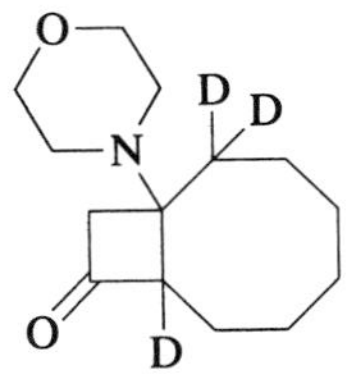

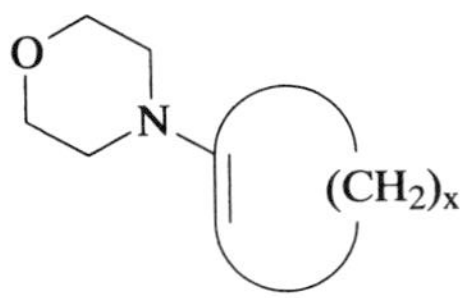

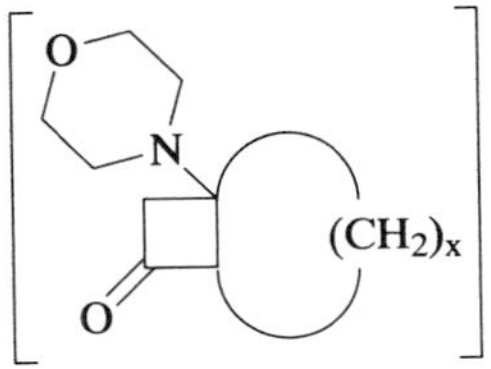

TABLE IX. [2+2] CYCLOADDITION OF KETENES TO ENAMINES (*Continued*)

Reactant	Ketene or Ketene Source	Conditions	Product(s) and Yield(s) (%)	Refs.
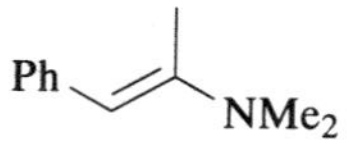 Ph, NMe_2	$CH_2{=}C{=}O$	Et_2O, rt, 4 h	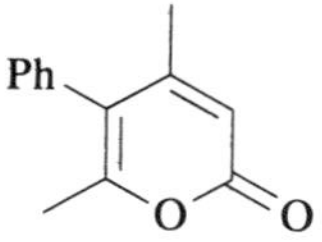 Ph, O, O (34)	98
C_{10}				
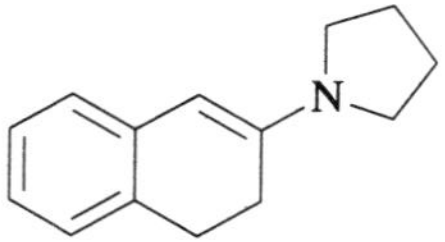 N	$CH_2{=}C{=}O$	Et_2O, rt, 4 h	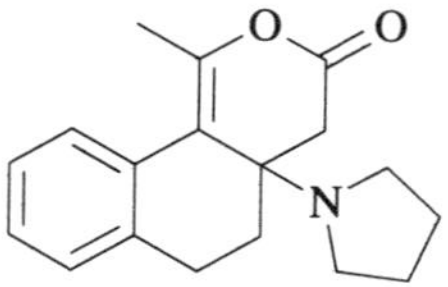 O, O, N (31)	98
C_{12}				
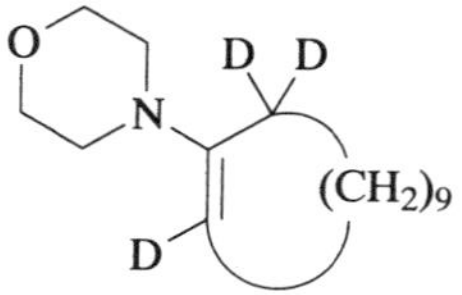 O, N, D D, D, $(CH_2)_9$	EtCOCl	Et_3N, Et_2O	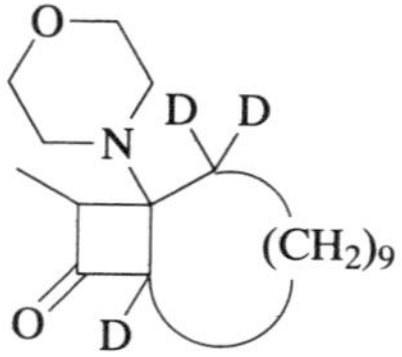 O, N, D D, O, D, $(CH_2)_9$ (—)	460

TABLE IX. [2+2] CYCLOADDITION OF KETENES TO ENAMINES (*Continued*)

Reactant	Ketene or Ketene Source	Conditions	Product(s) and Yield(s) (%)			Refs.
1-morpholino-cyclododecene ($(CH_2)_{10}$ ring)	$R^1R^2C{=}C{=}O$	[a]	2-R^2-2-R^1-1-morpholino-bicyclo (cyclobutanone fused, $(CH_2)_{10}$)			
			R^1	R^2	Yield (%)	
			H	H	(39)	586, 587, 96, 588
			Me	H	(31)	460, 96, 89
			Me	Me	(0)	460
			Me	Cl	(—)	460
			n-Bu	H	(85)[b]	96, 89
			n-$C_{10}H_{21}$	H	(85)[b]	96, 89
			n-$C_{16}H_{33}$	H	(76)[b]	96, 89
			MeO_2CCH_2	H	(65)[b]	96, 89
			$EtO_2C(CH_2)_5$	H	(83)[b]	96, 89
C_{22}						
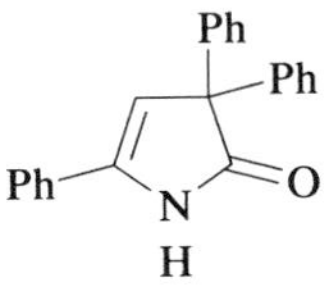 3,3,5-triphenyl-1,3-dihydro-2H-pyrrol-2-one (Ph, Ph, Ph, N–H, =O)	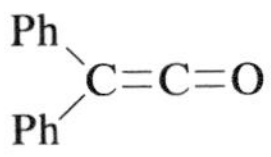 $Ph_2C{=}C{=}O$	60°, 30 min	bicyclic pyrrolidinone–pyranone product (H, N, Ph, Ph, Ph, O, O, Ph, Ph, $=CPh_2$)	(—)		589

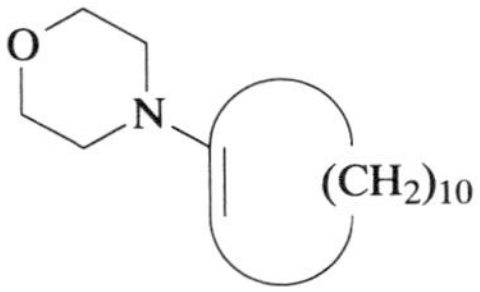

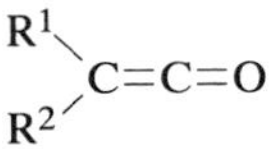

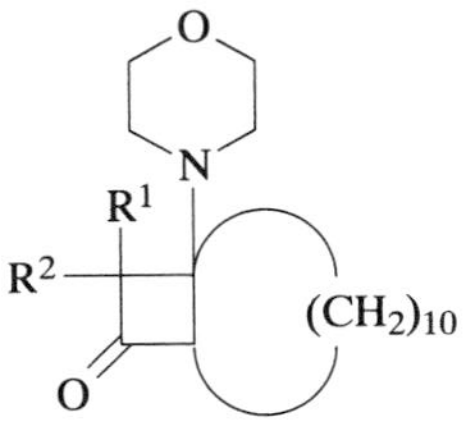

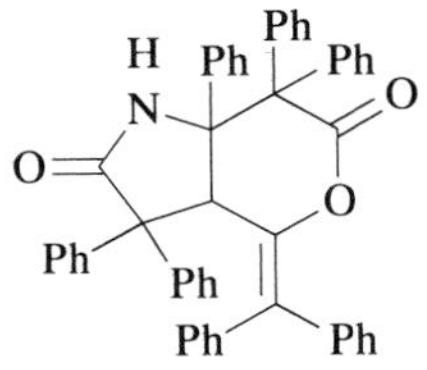

TABLE IX. [2+2] CYCLOADDITION OF KETENES TO ENAMINES (*Continued*)

Reactant	Ketene or Ketene Source	Conditions	Product(s) and Yield(s) (%)	Refs.
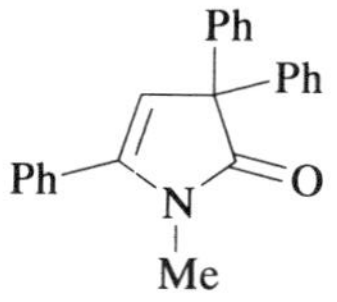 [structure: 1-Me-3,3,5-triphenyl-pyrrol-2-one]	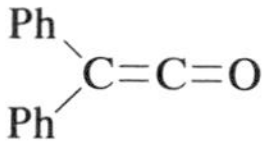 $Ph_2C{=}C{=}O$	rt, 5 d	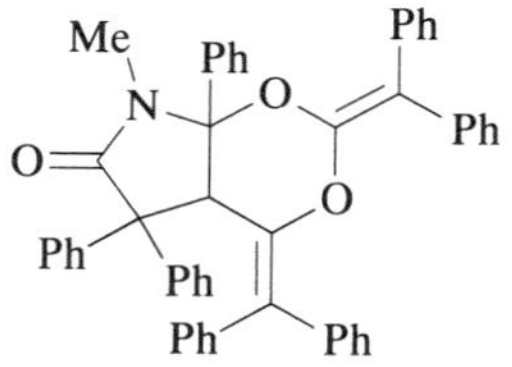 [structure] (—)	589

[a] The conditions were different for most entries and can be found by consulting the reference.
[b] The yield is of the 1,3-cyclotetradecanedione after hydrolysis.

TABLE X. [2+2] CYCLOADDITION OF KETENES TO ENOL ETHERS

Reactant	Ketene or Ketene Source	Conditions	Product(s) and Yield(s) (%)	Refs.
C_3				
$CH_2=CHOMe$	$CH_2=C=O$	100°, 4 h	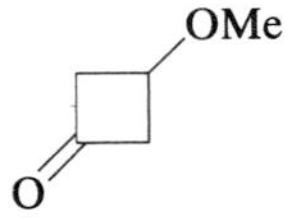 OMe, O (—)	590
C_4				
$(CH_2=CH)_2O$	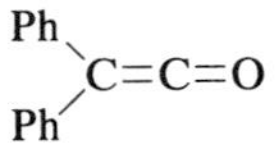 Ph, Ph, C=C=O	C_6H_{14}, rt, 7 d	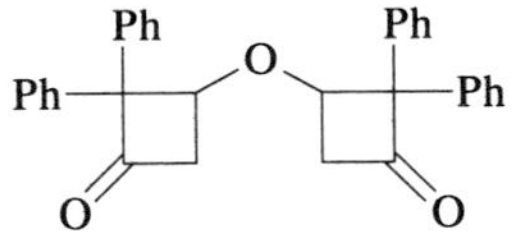 Ph, Ph, O, Ph, Ph, O, O (75)	591
 O	$Cl_2CHCOCl$	Et_3N, Et_2O	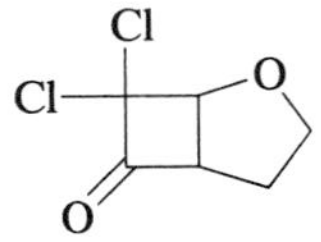 Cl, Cl, O, O (56)	428
	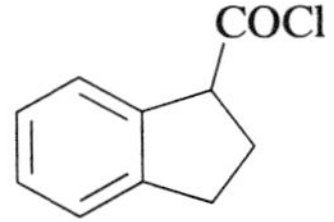 COCl	Et_3N, Et_2O	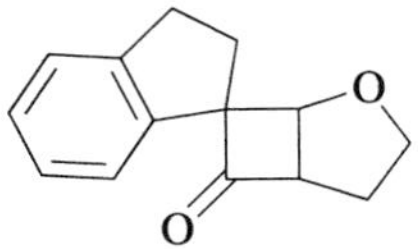 O, O (89)	592
	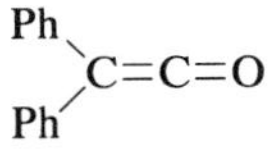 Ph, Ph, C=C=O	0°	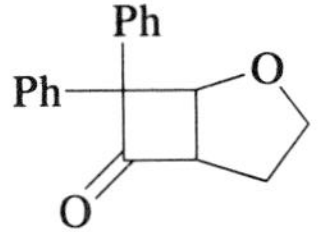 Ph, Ph, O, O (99)	103
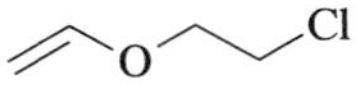 O, Cl	$CH_2=C=O$	MeCN, $ZnCl_2$, 50°	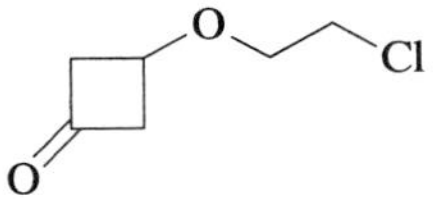 O, Cl, O (20)	101

TABLE X. [2+2] CYCLOADDITION OF KETENES TO ENOL ETHERS (*Continued*)

Reactant	Ketene or Ketene Source	Conditions	Product(s) and Yield(s) (%)	Refs.
	$Me_2C=C=O$	rt	2,2-dimethyl-3-(2-chloroethoxy)cyclobutanone (70)	321
MeCH=CHOMe	$Ph_2C=C=O$	rt	2,2-diphenyl-3-methoxy-4-methylcyclobutanone (—)	593
CH_2=CHOEt	$R^1R^2C=C=O$	a	2-R^1-2-R^2-3-ethoxycyclobutanone	
			R^1 = Cl, R^2 = Cl (45)	403, 329
			R^1 = H, R^2 = H (30)	101, 590
			R^1 = Br, R^2 = Me (39)	442
			R^1 = Me, R^2 = Me (80)	321, 523
			R^1 = H, R^2 = OEt (85)	394, 101
			R^1 = H, R^2 = C(Me)=CH_2 (23)	343
			R^1 = H, R^2 = CH_2OTMS (60)	352
			R^1 = CN, R^2 = *t*-Bu (98)	55
			R^1 = Et, R^2 = *n*-Bu (81)	321
			R^1 = H, R^2 = PhC=CCl_2 (80)	306
			R^1 = Ph, R^2 = Ph (82)	103, 591
	$(CF_3)_2C=C=O$	C_6H_{14}, rt, 16 h	[2-(bis(trifluoromethyl)methylene)-4-ethoxyoxetane] → $(CF_3)_2$CHC(O)CH=CHOEt	133, 387, 231

TABLE X. [2+2] CYCLOADDITION OF KETENES TO ENOL ETHERS (*Continued*)

Reactant	Ketene or Ketene Source	Conditions	Product(s) and Yield(s) (%)	Refs.
C_5 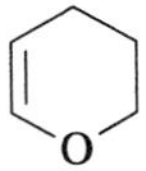 3,4-dihydro-2H-pyran	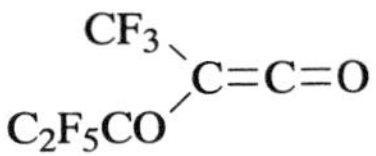 $CF_3(C_2F_5CO)C=C=O$	Et_2O, 25°	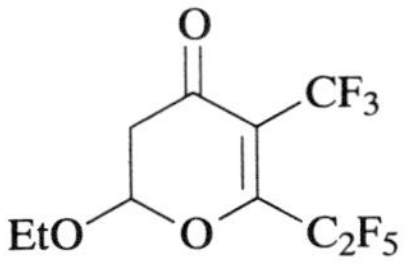 2-EtO-5-CF_3-6-C_2F_5-2,3-dihydro-4H-pyran-4-one (75)	346
	$Cl(NC)C=C=O$	PhMe, 103°	$ClCH(CN)CO$-(5-yl-3,4-dihydro-2H-pyran) (32)	312
	$R^1R^2C=C=O$	*a*	7-R^1-7-R^2-2-oxabicyclo[4.2.0]octan-8-one	
			R^1 = Cl, R^2 = Cl (35)	431, 441
			R^1 = H, R^2 = H (45)	101, 591
			R^1 = Br, R^2 = Me (50)	442
			R^1 = Cl, R^2 = Me (40)	442
			R^1 = Me, R^2 = Me (80)	321, 457
			R^1 = Me, R^2 = OTMS (—)	307
			R^1R^2 = $-(CH_2)_5-$ (67)	527
			R^1 = H, R^2 = $PhC=CCl_2$ (80)	306
			R^1 = Ph, R^2 = Ph (99)	103, 356
			R^1 = H, R^2 = $(CH_2)_3C(Me)=CH_2$ (50)	594

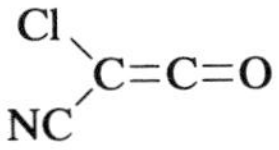

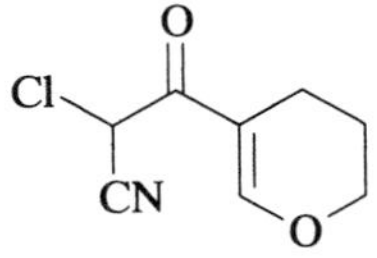

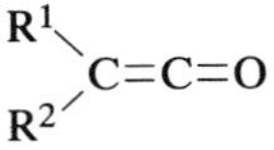

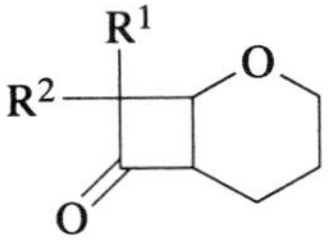

TABLE X. [2+2] CYCLOADDITION OF KETENES TO ENOL ETHERS (*Continued*)

Reactant	Ketene or Ketene Source	Conditions	Product(s) and Yield(s) (%)	Refs.
Et–CH=CH~OMe	$Ph_2C=C=O$	rt	2,2-diphenyl-3-methoxy-4-ethylcyclobutanone (Ph, Ph, OMe, Et, O) (—)	593
$CH_2=C(CH_3)OEt$	t-Bu(NC)C=C=O	C_6H_6, 78°	t-Bu–CH(CN)–C(=O)–CH=C(CH_3)OEt (—)	55
	$Ph_2C=C=O$	rt	[2,2-diphenyl-3-ethoxy-3-methylcyclobutanone (Ph, Ph, OEt, O)] → Ph–CH(Ph)–C(=O)–CH=C(CH_3)OEt (96)	103
cis-EtO–CH=CH–CH_3	$R^1R^2C=C=O$	*a*	2-R[1]-2-R[2]-3-ethoxy-4-methylcyclobutanone (R[1], R[2], OEt, O)	
			R[1] Me, R[2] Me, Yield (64)	321
			R[1] Ph, R[2] Me, Yield (100)	42
			R[1] Ph, R[2] Et, Yield (100)	42
			R[1] Ph, R[2] *n*-Pr, Yield (100)	42
			R[1] Ph, R[2] *i*-Pr, Yield (100)	42
			R[1] Ph, R[2] *t*-Bu, Yield (100)	42
			R[1] Ph, R[2] Ph, Yield (—)	593,45

TABLE X. [2+2] CYCLOADDITION OF KETENES TO ENOL ETHERS (*Continued*)

Reactant	Ketene or Ketene Source	Conditions	Product(s) and Yield(s) (%)	Refs.
(E)-1-ethoxypropene (OEt)	R(Ph)C=C=O	[a]	2-R-2-Ph-3-OEt-4-methylcyclobutanone; R: Me (100), Et (100), n-Pr (100), i-Pr (100), t-Bu (100), Ph (—)	42, 42, 42, 42, 42, 45
vinyl–O–CH$_2$–OEt	n-Bu(Et)C=C=O	180°	2-n-Bu-2-Et-3-(OCH$_2$OEt)cyclobutanone (34)	321
vinyl–OTMS	Cl_3CCOCl	Zn, Et_2O	2,2-Cl$_2$-3-OTMS-cyclobutanone (84)	104
C_6				
vinyl–OBu-n	Me$_2$C=C=O	rt	2,2-Me$_2$-3-OBu-n-cyclobutanone (65)	321
	Ph$_2$C=C=O	C_6H_6, rt, 3 h	2,2-Ph$_2$-3-OBu-n-cyclobutanone (99)	103
vinyl–OBu-i	Me$_2$C=C=O	rt	2,2-Me$_2$-3-OBu-i-cyclobutanone (54)	321

TABLE X. [2+2] CYCLOADDITION OF KETENES TO ENOL ETHERS (*Continued*)

Reactant	Ketene or Ketene Source	Conditions	Product(s) and Yield(s) (%)	Refs.
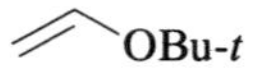 OBu-*t*	Me_2CO	Ketene lamp	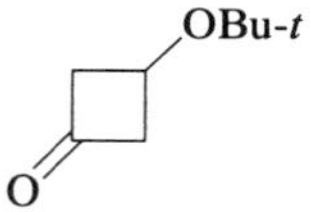 OBu-*t*, O (60)	595, 590
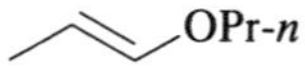 OPr-*n*	CF_3, C_2F_5CO, C=C=O	$CHCl_3$, 25°	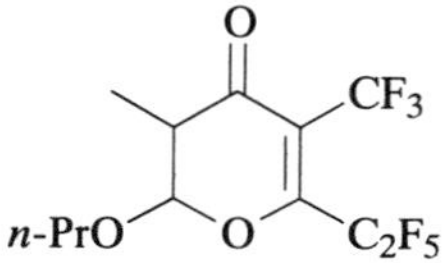 O, CF_3, *n*-PrO, O, C_2F_5 (65)	346
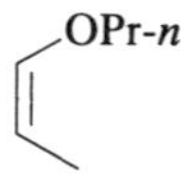 OPr-*n*	CF_3, CF_3, C=C=O	C_6H_{14}, -50°	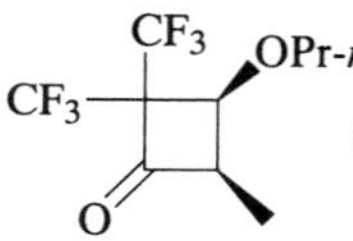 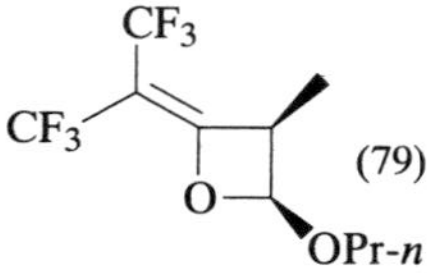 CF_3, CF_3, OPr-*n*, O (20) + CF_3, CF_3, O, OPr-*n* (79)	133
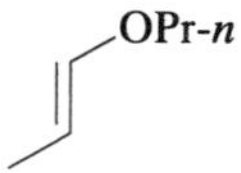 OPr-*n*	CF_3, CF_3, C=C=O	0°, 2 h	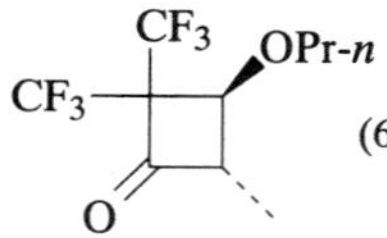 CF_3, CF_3, OPr-*n*, O (63) 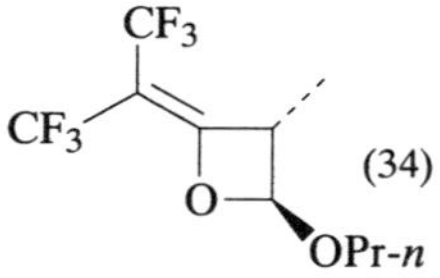 + CF_3, CF_3, O, OPr-*n* (34)	133
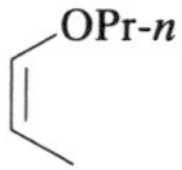 OPr-*n*	Me, Me, C=C=O	Et_2O, rt, 1 h	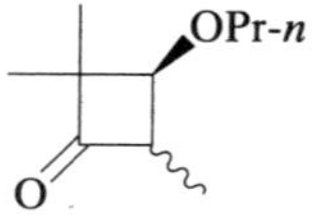 OPr-*n*, O (89) *cis*:*trans* = 98.8:1.2	47, 395, 596
	Ph, Ph, C=C=O	Et_2O, rt, 1 h	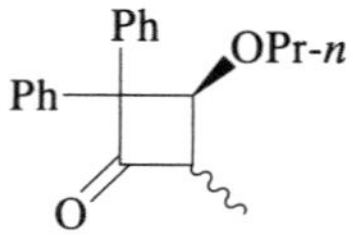 Ph, Ph, OPr-*n*, O (99) *cis*:*trans* = 96.2:3.8	47, 395, 596

TABLE X. [2+2] CYCLOADDITION OF KETENES TO ENOL ETHERS (*Continued*)

Reactant	Ketene or Ketene Source	Conditions	Product(s) and Yield(s) (%)	Refs.
OPr-*n*	Me(Me)C=C=O	Et_2O, rt, 1 h	OPr-*n*, O (60) *cis*:*trans* = 0.8:99.2	47, 596, 395
	Ph(Ph)C=C=O	Et_2O, rt, 1 h	Ph, Ph, OPr-*n*, O (86) *cis*:*trans* = 2.5:97.5	47, 596, 597
Et, OEt *cis*:*trans* = 3:97	Me(Me)C=C=O	C_6H_6	OEt, Et, O (43) *cis*:*trans* = 12:88	598
cis:*trans* = 88:12	Me(Me)C=C=O	C_6H_6	OEt, Et, O (69)	598
	$EtOCH_2COCl$	Et_3N	EtO, OEt, Et, O (45)	101
Et, OEt	Ph(Ph)C=C=O	PhCN, 40°	Ph, Ph, OEt, Et, O (—)	45, 593

TABLE X. [2+2] CYCLOADDITION OF KETENES TO ENOL ETHERS (*Continued*)

Reactant	Ketene or Ketene Source	Conditions	Product(s) and Yield(s) (%)	Refs.
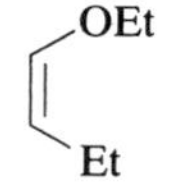 OEt, Et	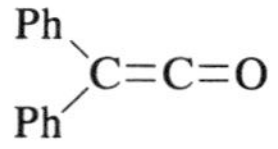 Ph, Ph, C=C=O	PhCN, 40°	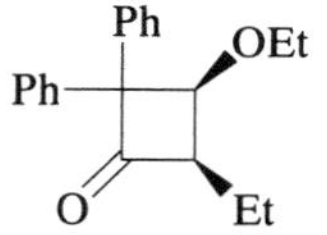 Ph, Ph, OEt, O, Et (—)	45, 593
	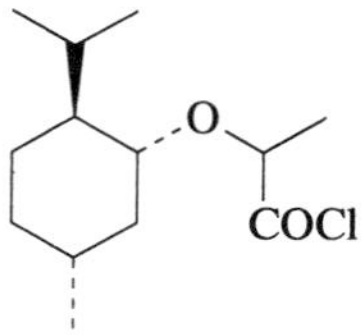 O, COCl	Et_3N, *t*-BuOMe	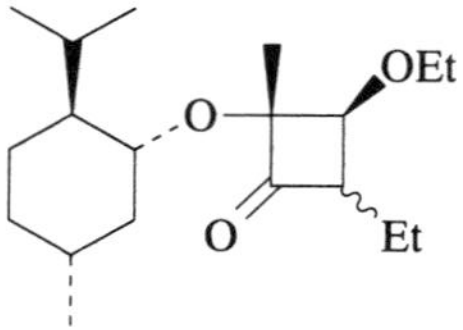 O, OEt, O, Et (65) *cis:trans* =20:80	599
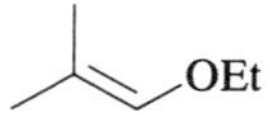 OEt	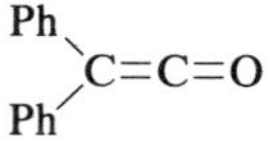 Ph, Ph, C=C=O	MeCN, 1 min	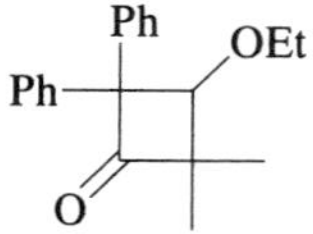 Ph, Ph, OEt, O (43)	103, 45
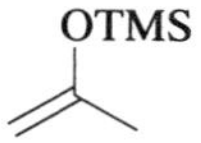 OTMS	Cl_3CCOCl	Zn, Et_2O	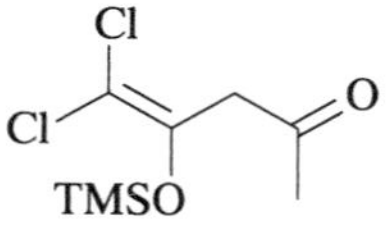 Cl, Cl, TMSO, O (79)	104
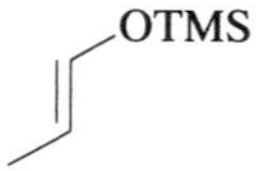 OTMS	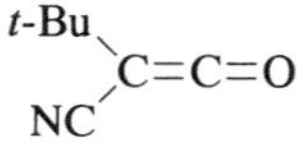 *t*-Bu, NC, C=C=O	C_6H_6, rt	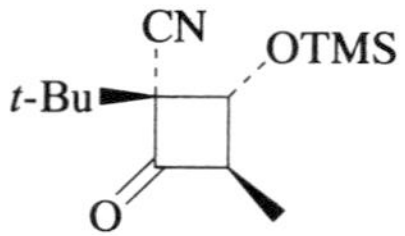 CN, OTMS, *t*-Bu, O (97)	93
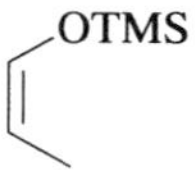 OTMS	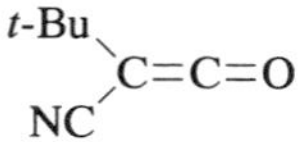 *t*-Bu, NC, C=C=O	C_6H_6, rt	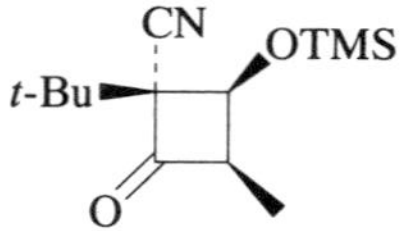 CN, OTMS, *t*-Bu, O (95)	93

TABLE X. [2+2] CYCLOADDITION OF KETENES TO ENOL ETHERS (*Continued*)

Reactant	Ketene or Ketene Source	Conditions	Product(s) and Yield(s) (%)	Refs.
C_7				
MeO (1-methoxycyclohexene)	$CH_2{=}C{=}O$	MeCN, $ZnCl_2$, 60°	MeO (bicyclic cyclobutanone) (60)	101
O, OEt (2-ethoxy-3,4-dihydro-2H-pyran)	Me, Me C=C=O	rt	O, OEt (bicyclic cyclobutanone) (85)	321
OTMS	R^1, R^2 C=C=O	[a]	R^1, R^2, OTMS, O (cyclobutanone)	
			R^1 Cl, R^2 Cl, Yield (%) (92)	600, 104
			R^1 Cl, R^2 Me, Yield (%) (61)	104
			R^1 Cl, R^2 Ph, Yield (%) (—)	104
			R^1 CN, R^2 *t*-Bu, Yield (%) (92)	40
OTMS	Me, H C=C=O	THF	OTMS, O (cyclobutanone) (44)	601
Et, OTMS	Cl_3CCOCl	Zn, Et_2O	Cl, Cl, OTMS, Et, O (cyclobutanone) (82)	104
C_8				
O, NHCOPr-*i*	Me, Me C=C=O	rt	O, NHCOPr-*i*, O (cyclobutanone) (80)	321

TABLE X. [2+2] CYCLOADDITION OF KETENES TO ENOL ETHERS (*Continued*)

Reactant	Ketene or Ketene Source	Conditions	Product(s) and Yield(s) (%)	Refs.
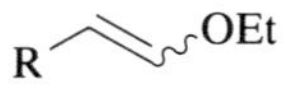 R–CH=CH–OEt; R = *i*-Pr, *t*-Bu	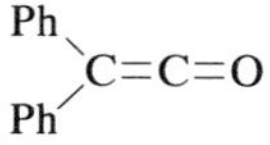 $Ph_2C=C=O$	PhCN, 40°	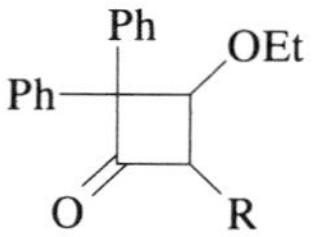 (—)	45
TMSO-cyclopentene	$R^1R^2C=C=O$	*a*	R^1 Cl, R^2 Cl (77)	104, 600
			R^1 Cl, R^2 Me (79)	104, 416
			R^1 Cl, R^2 Ph (80)	104
			R^1 PhO, R^2 Me (79)	104
			R^1 H, R^2 PhO (68)	104
			R^1 H, R^2 MeO (67)	104
			R^1 H, R^2 Cl (53)	104
$CH_2=CH$–OTBDMS	$R^1R^2C=C=O$	*a*	R^1 Cl, R^2 Me (80)	602
			R^1 Et, R^2 Et (56)	102
			R^1 Ph, R^2 Ph (86)	102
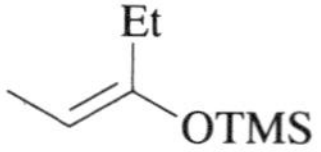 $CH_3CH=C(Et)$OTMS	Cl_3CCOCl	Zn, Et_2O	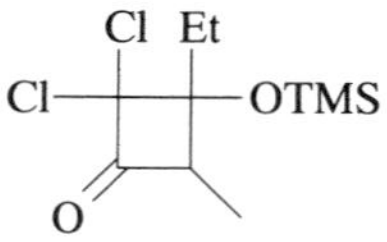 (80)	104
$CH_2=CH$–$OSiEt_3$	$Ph_2C=C=O$	—	(100)	603

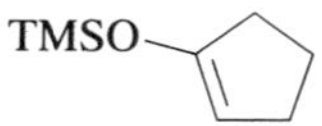

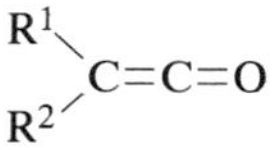

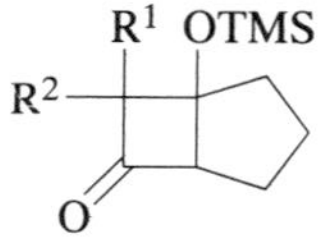

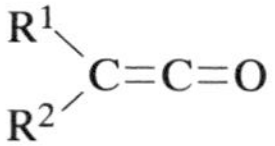

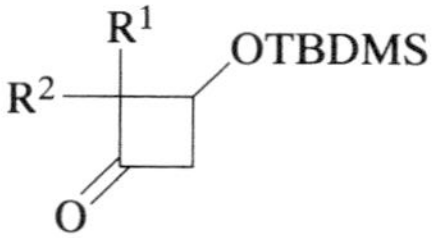

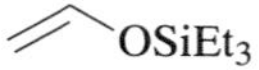

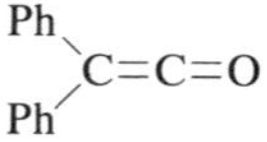

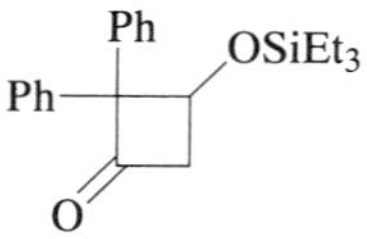

TABLE X. [2+2] CYCLOADDITION OF KETENES TO ENOL ETHERS (*Continued*)

Reactant	Ketene or Ketene Source	Conditions	Product(s) and Yield(s) (%)	Refs.
C_9				
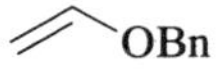 OBn	$CH_2{=}C{=}O$	100°, 4 h	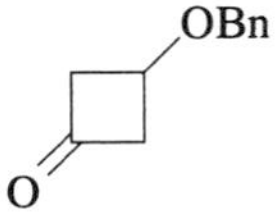 OBn, O (20)	590
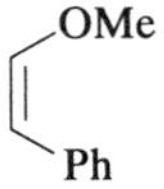 OMe, Ph	Ph, Ph C=C=O	rt, 10 weeks	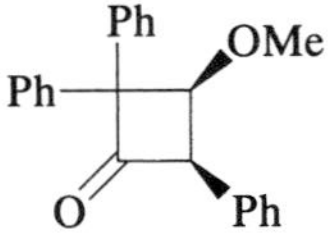 Ph, Ph, OMe, Ph, O (95)	103
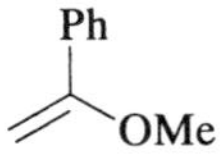 Ph, OMe	Ph, Ph C=C=O	rt	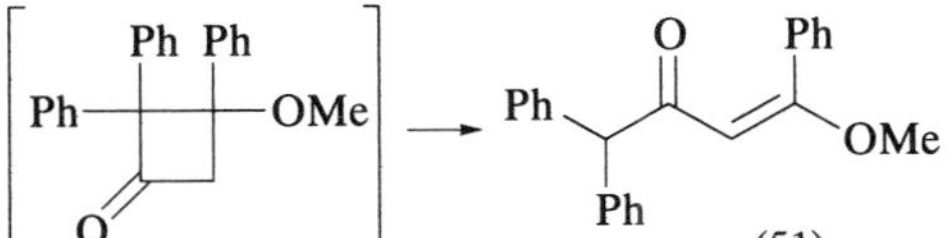 [Ph, Ph, Ph, OMe, O] → Ph, Ph, O, Ph, OMe (51)	103
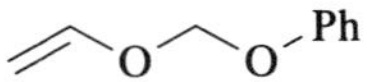 O, O, Ph	*n*-Bu, Et C=C=O	180°	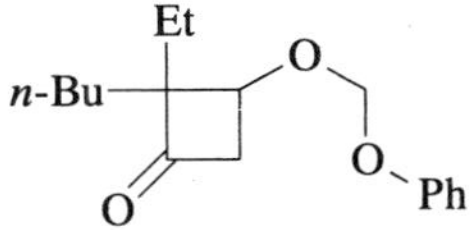 Et, *n*-Bu, O, O, Ph, O (45)	321
 OC_6H_4OMe-*p*	Me, Me C=C=O	rt	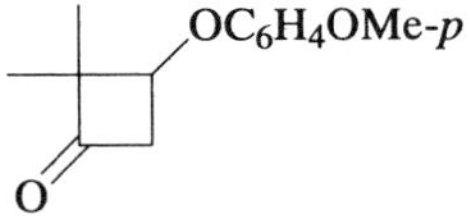 OC_6H_4OMe-*p*, O (46)	321
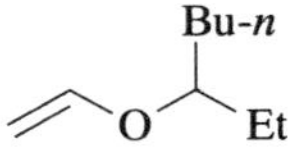 O, Bu-*n*, Et	Me, Me C=C=O	rt	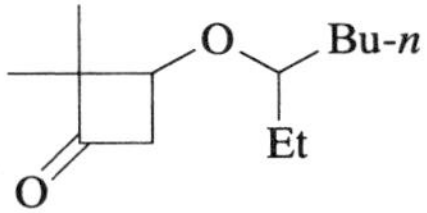 O, Bu-*n*, Et, O (56)	321

TABLE X. [2+2] CYCLOADDITION OF KETENES TO ENOL ETHERS (*Continued*)

Reactant	Ketene or Ketene Source	Conditions	Product(s) and Yield(s) (%)	Refs.
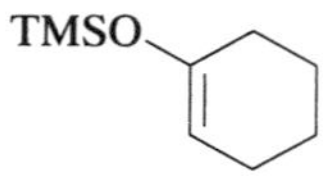 TMSO (cyclohexenyl)	Cl_3CCOCl	Zn-Cu, Et_2O	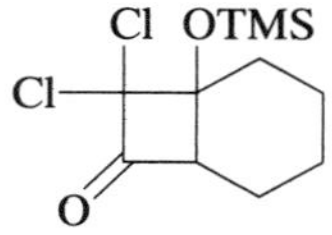 Cl, Cl, OTMS, O (81)	600
	MeCHClCOCl	Et_3N, C_6H_{14}	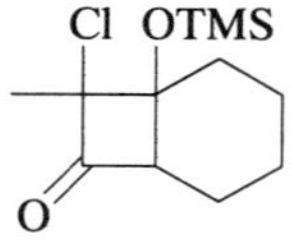 Cl, OTMS, O (20)	416
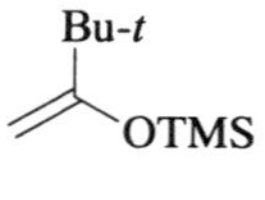 Bu-*t*, OTMS	Cl_3CCOCl	Zn, Et_2O	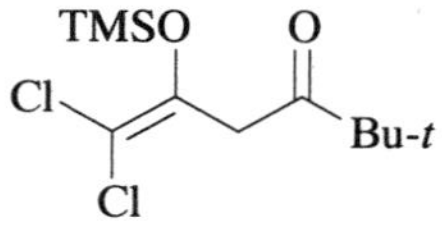 TMSO, O, Cl, Cl, Bu-*t* (82)	104, 600
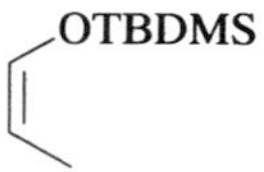 OTBDMS	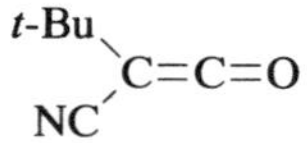 *t*-Bu, NC C=C=O	C_6H_6, rt, 32 d	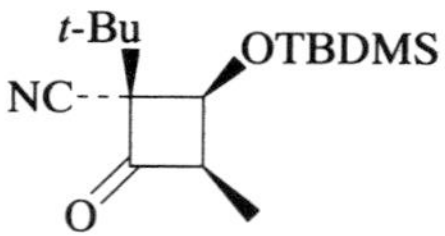 *t*-Bu, NC, OTBDMS, O (5)	40
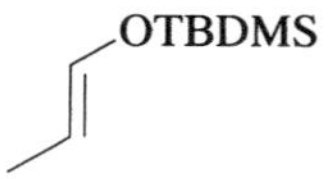 OTBDMS	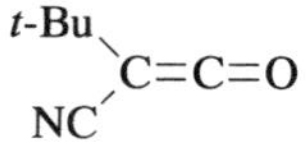 *t*-Bu, NC C=C=O	C_6H_6, 12 h	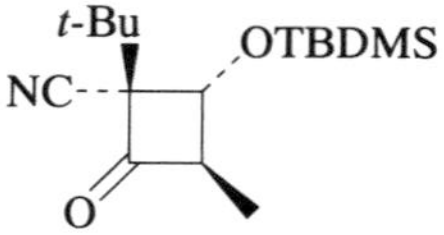 *t*-Bu, NC, OTBDMS, O (95)	40
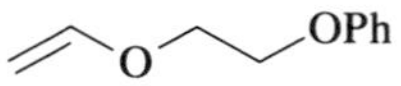 O, OPh	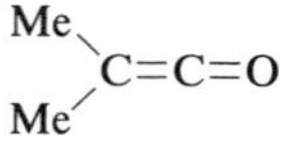 Me, Me C=C=O	rt	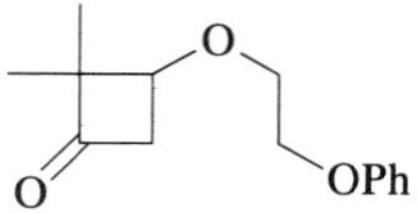 O, O, OPh (67)	321

TABLE X. [2+2] CYCLOADDITION OF KETENES TO ENOL ETHERS (*Continued*)

Reactant	Ketene or Ketene Source	Conditions	Product(s) and Yield(s) (%)	Refs.
C_{10}				
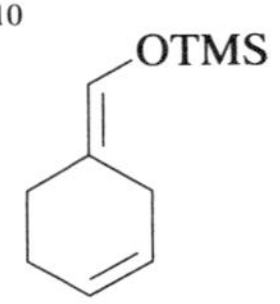 OTMS	RCHClCOCl	Et_3N, C_6H_{14}	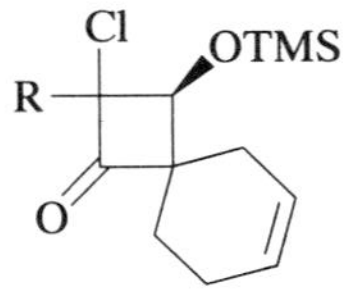 Cl, R, OTMS, O; R = Cl (70), R = Me (49), R = Ph (—)	104
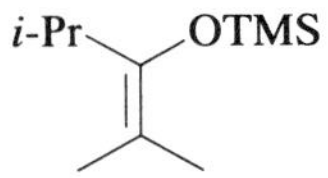 *i*-Pr, OTMS	Cl_3CCOCl	Zn, Et_2O	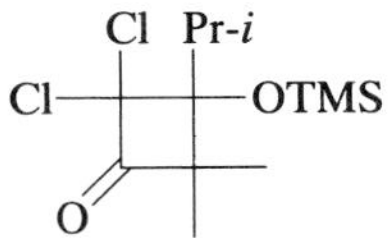 Cl, Cl, Pr-*i*, OTMS, O (88)	400
C_{11}				
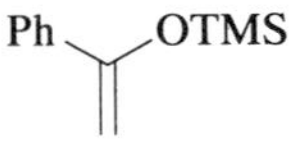 Ph, OTMS	Cl_3CCOCl	Zn-Cu, Et_2O	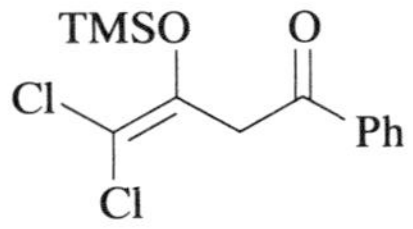 TMSO, O, Cl, Cl, Ph (—)	600
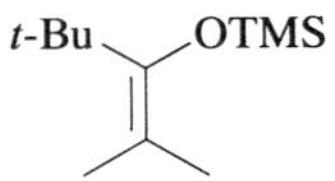 *t*-Bu, OTMS	Cl_3CCOCl	Zn, Et_2O	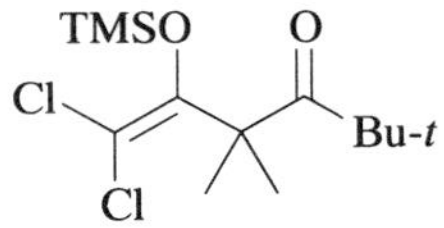 TMSO, O, Cl, Cl, Bu-*t* (66)	104
C_{12}				
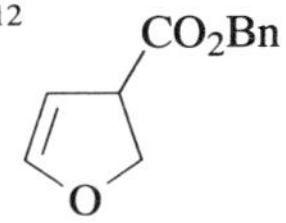 CO_2Bn, O	$Cl_2CHCOCl$	Et_3N	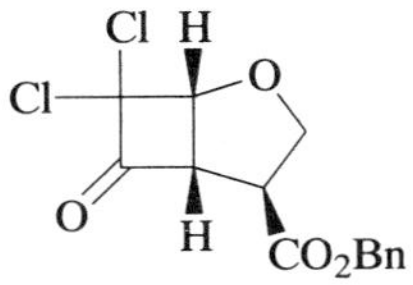 Cl, Cl, H, O, H, O, CO_2Bn (59)	236

TABLE X. [2+2] CYCLOADDITION OF KETENES TO ENOL ETHERS (*Continued*)

Reactant	Ketene or Ketene Source	Conditions	Product(s) and Yield(s) (%)	Refs.
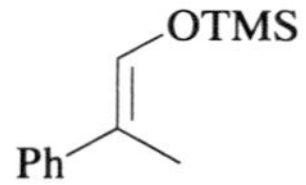 OTMS, Ph	Cl_3CCOCl	Zn-Cu, Et_2O	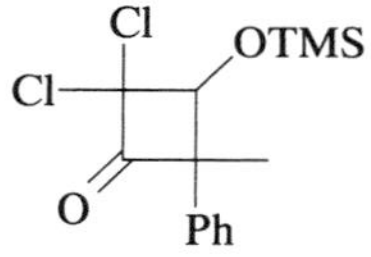 Cl, Cl, OTMS, O, Ph (86)	600, 104
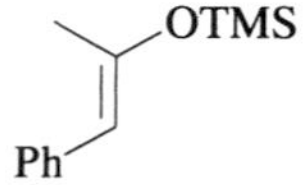 OTMS, Ph	Cl_3CCOCl	Zn-Cu, Et_2O	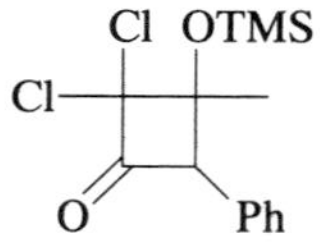 Cl, Cl, OTMS, O, Ph (59)	600
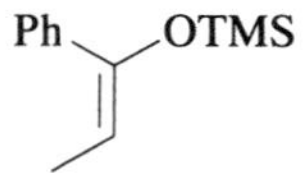 Ph, OTMS	Cl_3CCOCl	Zn-Cu, Et_2O	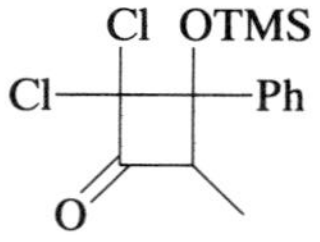 Cl, Cl, OTMS, Ph, O (94)	600
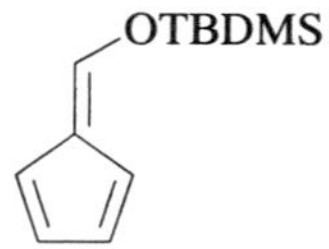 OTBDMS	$Cl_2CHCOCl$	Et_3N, C_6H_{12}, rt	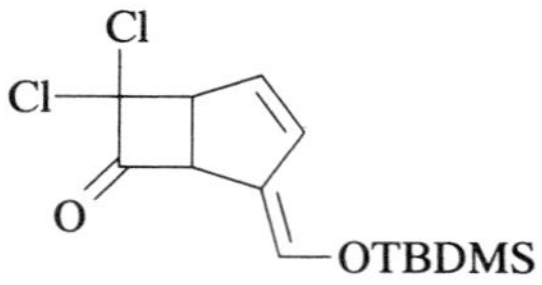 Cl, Cl, O, OTBDMS (34)	604
OTBDMS, Bu-*t*	Ph, Ph, C=C=O	rt	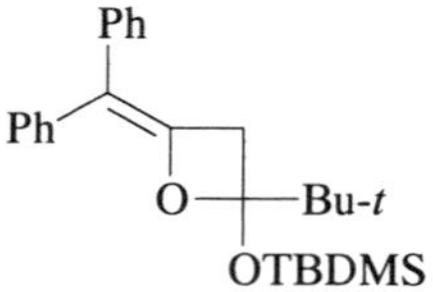 Ph, Ph, O, Bu-*t*, OTBDMS (41) +	102
			TBDMSO, O, Ph, Ph, Bu-*t* (45)	

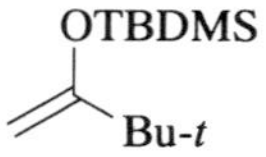

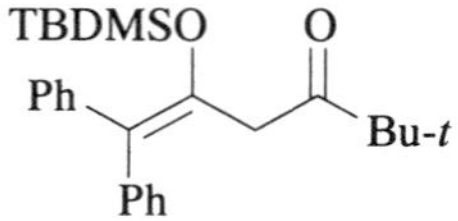

TABLE X. [2+2] CYCLOADDITION OF KETENES TO ENOL ETHERS (*Continued*)

Reactant	Ketene or Ketene Source	Conditions	Product(s) and Yield(s) (%)	Refs.
C_{13}				
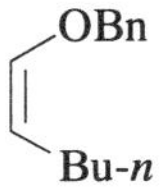 OBn, Bu-*n* (structure)	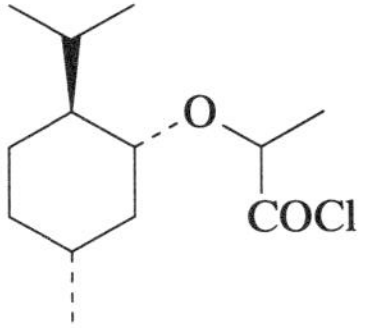 O, COCl (structure)	Et_3N, *t*-BuOMe	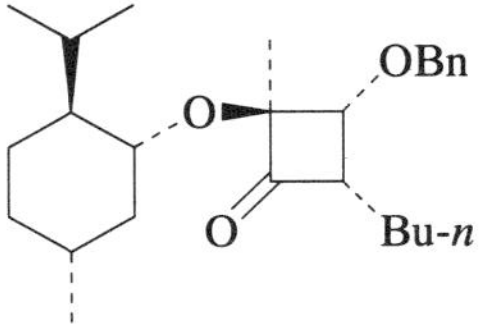 O, OBn, O, Bu-*n* (structure) (50)	599
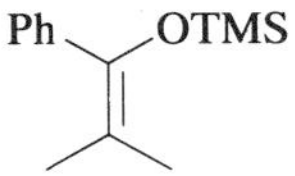 Ph, OTMS (structure)	Cl_3CCOCl	Zn, Et_2O	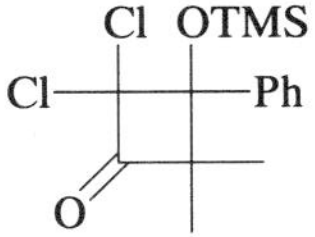 Cl, OTMS, Cl, Ph, O (structure) (71)	104
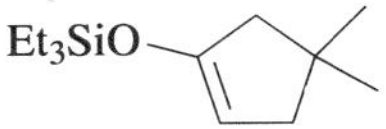 Et_3SiO (structure)	$Cl_2CHCOCl$	Et_3N, C_5H_{12}	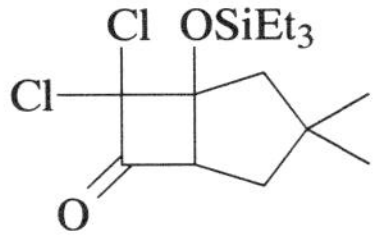 Cl, $OSiEt_3$, Cl, O (structure) (83)	605, 606
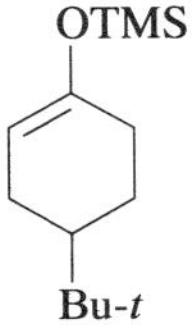 OTMS, Bu-*t* (structure)	Cl_3CCOCl	Zn-Cu, Et_2O	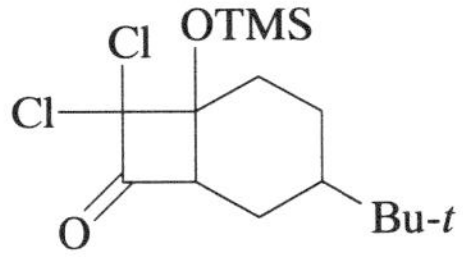 Cl, OTMS, Cl, O, Bu-*t* (structure) (95)	590

TABLE X. [2+2] CYCLOADDITION OF KETENES TO ENOL ETHERS (*Continued*)

Reactant	Ketene or Ketene Source	Conditions	Product(s) and Yield(s) (%)					Refs.
OTBDMS, Ar	$R^1R^2C{=}C{=}O$	[a]	I: R^1, OTBDMS, R^2, Ar, O; II: TBDMSO, O, R^1, R^2, Ar					
			Ar	R^1	R^2	I	II	
			p-$O_2NC_6H_4$	Cl	Cl	(—)	(—)	600
			p-ClC_6H_4	Et	Et	**I** + **II** = (80)		102
			p-$MeOC_6H_4$	Et	Et	**I** + **II** = (77)		102
			Ph	Et	Et	**I** + **II** = (76)		102
			p-ClC_6H_4	Ph	Ph	(—)	(99)	102
			p-$MeOC_6H_4$	Ph	Ph	(—)	(6)	102
			Ph	Ph	Ph	(—)	(76)	102
C_{16}								
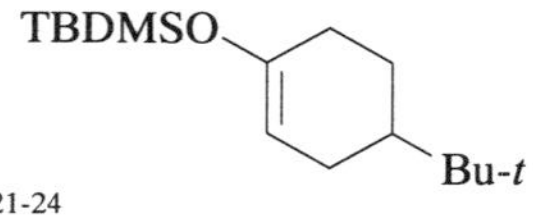 TBDMSO, Bu-*t*	Cl_3CCOCl	Zn-Cu, Et_2O	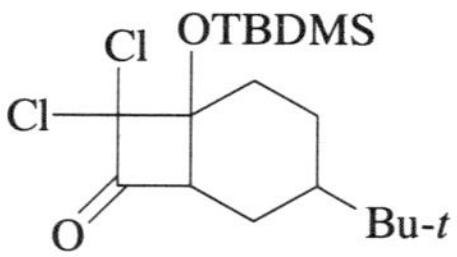 Cl, OTBDMS, Cl, O, Bu-*t* (92)					600
$C_{21\text{-}24}$								
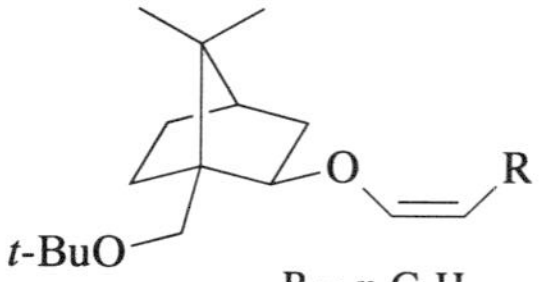 *t*-BuO, O, R; R = *n*-C_5H_{11}, $Ph(CH_2)_2$, *n*-C_8H_{17}	Cl_3CCOCl	Zn-Cu, Et_2O	Cl, Cl, O, O, R, OBu-*t*					

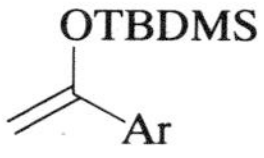

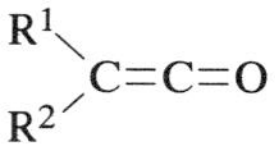

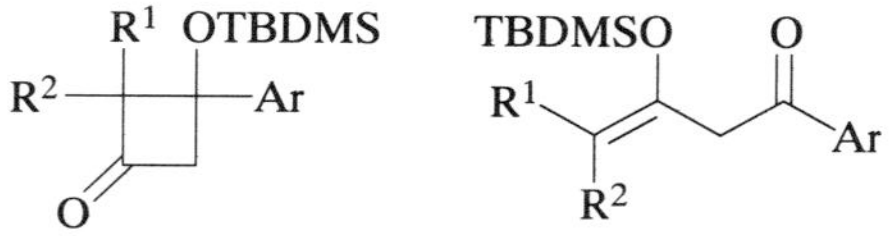

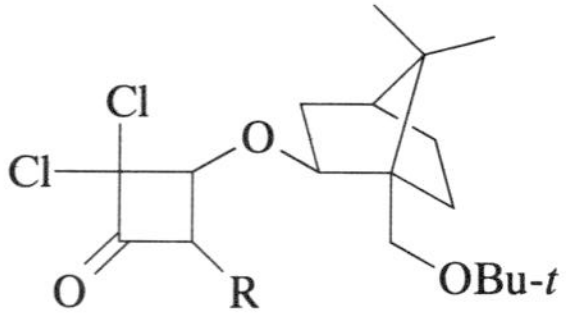

[a] The conditions for most entries are different, and can be found by consulting the reference.

TABLE XI. [2+2] CYCLOADDITION OF KETENES TO ENOL CARBOXYLATES

Reactant	Ketene or Ketene Source	Conditions	Product(s) and Yield(s) (%)	Refs.
$CH_2{=}CHO_2CR^1$	$R^2R^3C{=}C{=}O$	[a]	2-R^2-2-R^3-3-(O_2CR^1)cyclobutanone	

R^1	R^2	R^3	Yield (%)	Refs.
H	$-(CH_2)_6-$		(—)	608
CCl_3	Me	Me	(31)	608
Me	CN	*t*-Bu	(97)	55
Me	Et	*n*-Bu	(36)	608,321
Me	Et	*s*-Bu	(34)	321
Me	Et	*i*-Bu	(30)	608
Me	Ph	Ph	(72)	608,321
CH=CHMe	Me	Ph	(—)	608
CH_2=CMe	Ph	Ph	(72)	608
t-Bu	H	H	(—)	608
Ph	Ph	Ph	(44)	608
$C_{11}H_{23}$	Ph	Ph	(—)	608

Reactant	Ketene or Ketene Source	Conditions	Product(s) and Yield(s) (%)	Refs.
C_4				
$CH_2{=}CHOAc$	$(CF_3)_2C{=}C{=}O$	100°	$(CF_3)_2CHC(O)CH{=}CHOAc$ (40) + 2,2-bis(CF_3)-3-OAc-cyclobutanone (—) + 3-[$(CF_3)_2C{=}$]-2-OAc-oxetane (—)	133

TABLE XI. [2+2] CYCLOADDITION OF KETENES TO ENOL CARBOXYLATES (*Continued*)

Reactant	Ketene or Ketene Source	Conditions	Product(s) and Yield(s) (%)	Refs.
	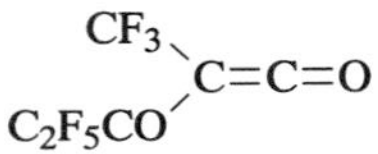 $CF_3(C_2F_5CO)C{=}C{=}O$	Phenothiazine, 95°, 15 min	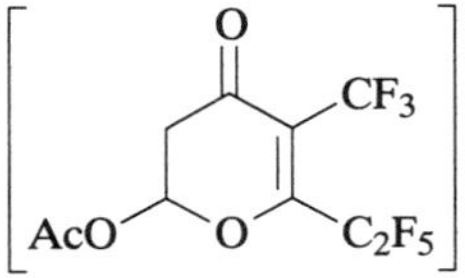 [AcO, O, O, CF_3, C_2F_5 pyranone] (—)	346
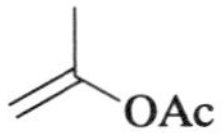 OAc (isopropenyl acetate)	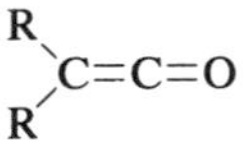 $R_2C{=}C{=}O$	rt, 2 d	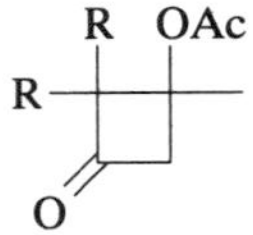 R, R, OAc, O cyclobutanone; R = Ph (—); R = n-C_8H_{17} (—); R = CN, t-Bu (—)	608
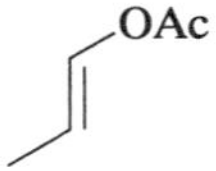 OAc (propenyl acetate)	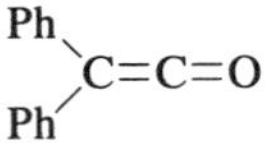 $Ph_2C{=}C{=}O$	rt, 2 d	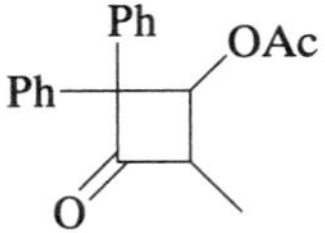 Ph, Ph, OAc, O cyclobutanone (—)	608
C_9 O_2CPh (vinyl benzoate)	$(CF_3)_2C{=}C{=}O$	100°, 16 h	CF_3, CF_3, O_2CPh, O cyclobutanone (42) + CF_3, CF_3, O, O_2CPh oxetane (34)	133, 609
C_{10} O_2CPh (isopropenyl benzoate)	t-Bu(NC)C=C=O	C_6H_6, 78°	t-Bu, $PhCO_2$, CN, O (—)	55

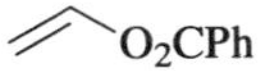

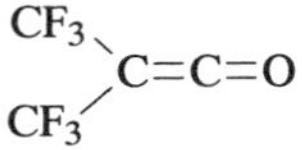

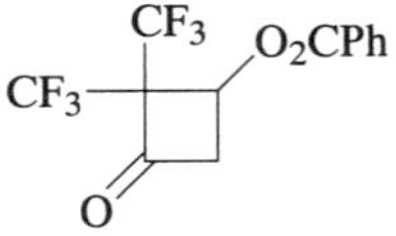

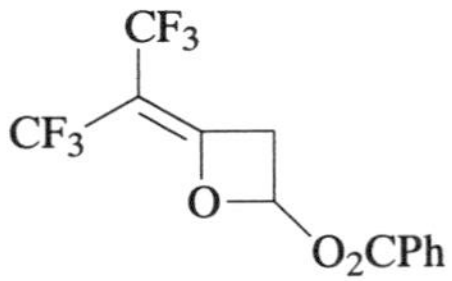

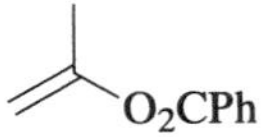

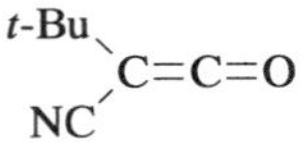

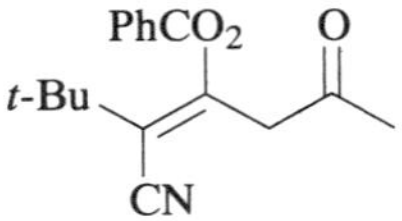

[a] The conditions for most entries are different and can be found by consulting the reference.

TABLE XII. [2+2] CYCLOADDITION OF KETENES TO POLYOXYGENATED OLEFINS

Reactant	Ketene or Ketene Source	Conditions	Product(s) and Yield(s) (%)	Refs.
C_3				
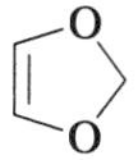 O, O	Ph, Ph C=C=O	MeCN, rt, 6 d	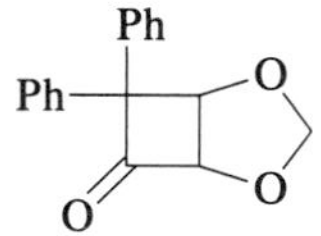 Ph, Ph, O, O, O (52)	103
C_4				
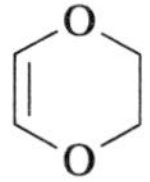 O, O	Ph, Ph C=C=O	45 d	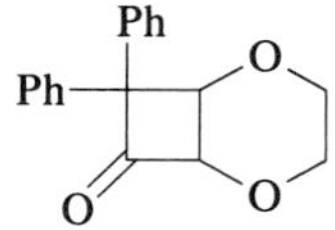 Ph, Ph, O, O, O (80)	103
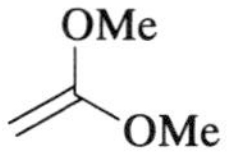 OMe, OMe	Ph, Ph C=C=O	80°, 30 min	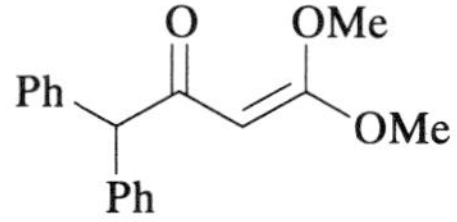 O, OMe, Ph, OMe, Ph (57)	610
	MeCHClCOCl	Et_3N	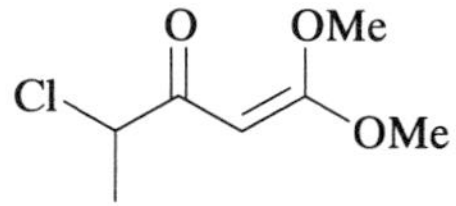 O, OMe, Cl, OMe (85)	611
C_5				
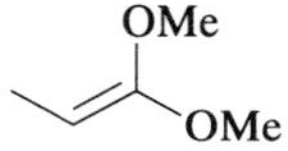 OMe, OMe	CH_2=C=O	MeCN, $ZnCl_2$, 25°	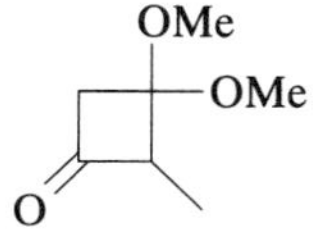 OMe, OMe, O (35)	101
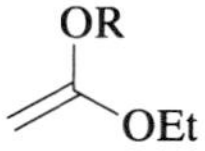 OR, OEt	TMS, H C=C=O	90°, 2 h	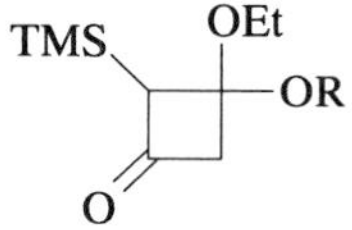 TMS, OEt, OR, O R = Me, Et (50)	612

TABLE XII. [2+2] CYCLOADDITION OF KETENES TO POLYOXYGENATED OLEFINS (*Continued*)

Reactant	Ketene or Ketene Source	Conditions	Product(s) and Yield(s) (%)	Refs.
C_6				
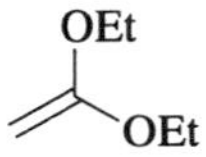 OEt, OEt	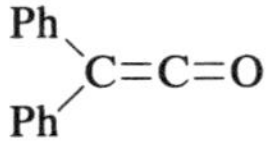 Ph, Ph C=C=O	C_5H_{12}, -50°	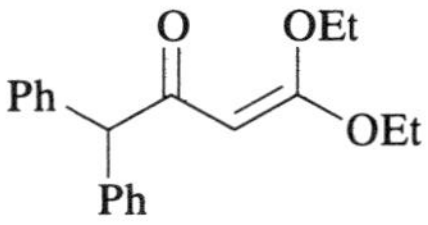 O, OEt, Ph, OEt, Ph (99)	103
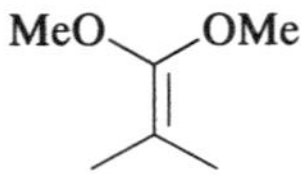 MeO, OMe	$Cl_2CHCOCl$	Et_3N	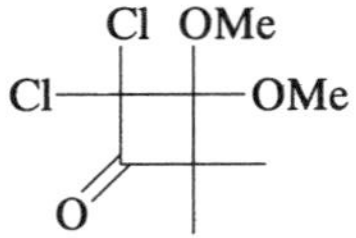 Cl, OMe, Cl, OMe, O (82)	611
	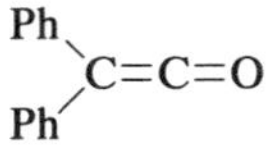 Ph, Ph C=C=O	100°	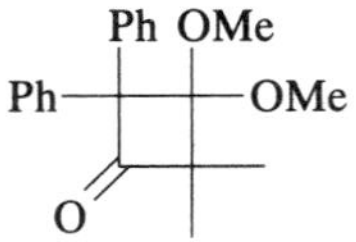 Ph, OMe, Ph, OMe, O (36)	105
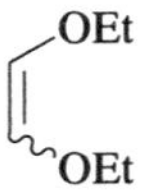 OEt, OEt	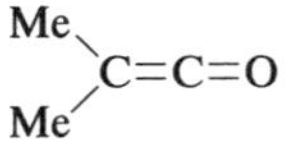 Me, Me C=C=O	rt	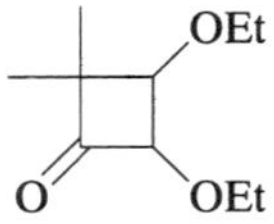 OEt, O, OEt (20)	321
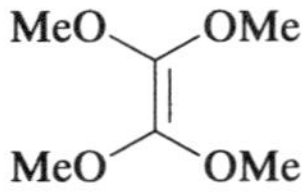 MeO, OMe, MeO, OMe	CH_2=C=O	MeCN, $ZnCl_2$, rt	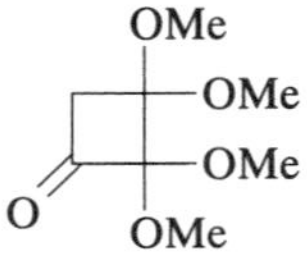 OMe, OMe, OMe, O, OMe (40)	101
	CH_2=C=O	rt, 4 h	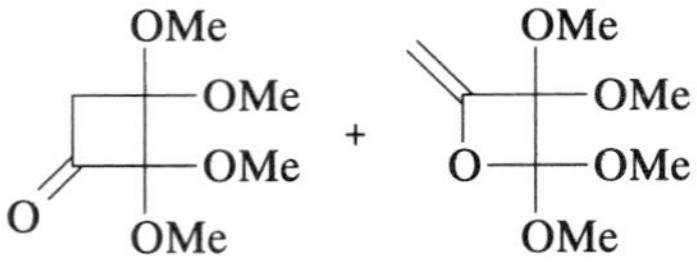 OMe, OMe, OMe, O, OMe + OMe, OMe, O, OMe, OMe (total 72)	106, 613

TABLE XII. [2+2] CYCLOADDITION OF KETENES TO POLYOXYGENATED OLEFINS (*Continued*)

Reactant	Ketene or Ketene Source	Conditions	Product(s) and Yield(s) (%)	Refs.
	RCH_2COCl	Et_3N, C_6H_{14}	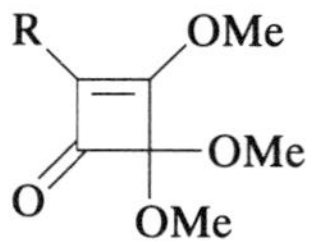 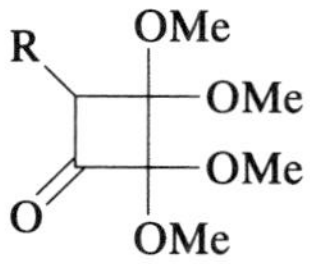 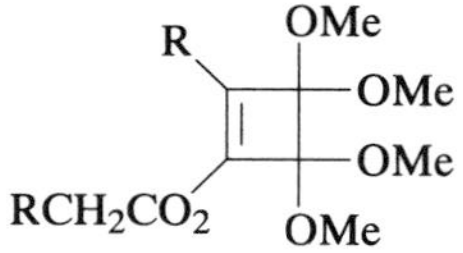 R, OMe, OMe, OMe, O **I** + R, OMe, OMe, OMe, OMe, O **II** + R, OMe, OMe, OMe, OMe, RCH_2CO_2 **III** R = *t*-Bu, **I + II + III** (83) R = CH_2CO_2Me, **I + II + III** (34)	614
	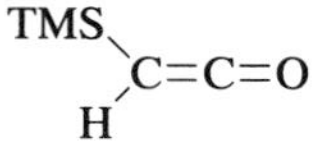 TMS, H, C=C=O	90°, 2 h	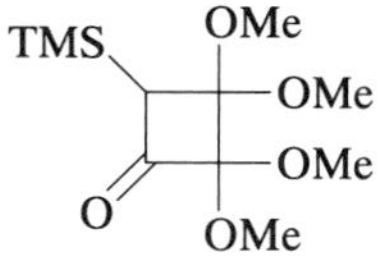 TMS, OMe, OMe, OMe, OMe, O (65)	615
	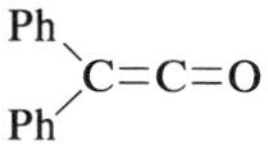 Ph, Ph, C=C=O	Dioxane, 60-70°	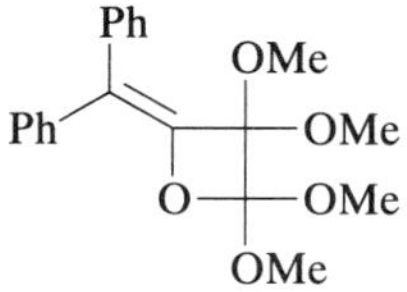 Ph, Ph, OMe, OMe, OMe, OMe, O (45)	106
C_7 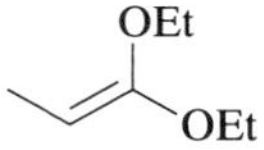 OEt, OEt	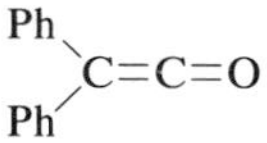 Ph, Ph, C=C=O	80°, 30 min	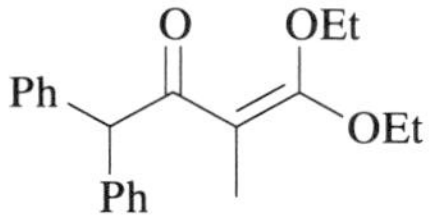 Ph, Ph, O, OEt, OEt (72)	610

TABLE XII. [2+2] CYCLOADDITION OF KETENES TO POLYOXYGENATED OLEFINS (*Continued*)

Reactant	Ketene or Ketene Source	Conditions	Product(s) and Yield(s) (%)	Refs.
C_8				
MeO, OMe (cyclopentylidene)	$Ph_2C=C=O$	100°, 12 h	Ph, Ph, OMe, OMe, O (35)	105
MeO, OTMS	MeO, OTMS	200°, 4 h	OTMS, CO_2Me (75)	308
C_9				
MeO, OTMS, OTMS	RCHClCOCl	Et_3N, C_6H_{14}	R, OTMS, CO_2TMS, Cl, OMe; R = H (80), R = Cl (87), R = Me (85), *cis:trans* = 1:1	611
C_{10}				
Ph, OMe, OMe	$Ph_2C=C=O$	90-95°, 7 h	Ph, O, OMe, OMe, Ph, Ph (40)	616, 610
MeO, OTMS (cyclopentylidene)	MeO, OTMS (cyclopentylidene)	200°, 4 h	OTMS, CO_2Me (75)	308
TMSO, TMSO	X_3CCOCl	Zn, Et_2O	X, X, OTMS, OTMS, O; X = F (34), X = Cl (55), X = Br (45)	617

TABLE XII. [2+2] CYCLOADDITION OF KETENES TO POLYOXYGENATED OLEFINS (*Continued*)

Reactant	Ketene or Ketene Source	Conditions	Product(s) and Yield(s) (%)	Refs.
MeO, OTMS (olefin)	RCHClCOCl	Et_3N, C_6H_{14}	OTMS, R, Cl, CO_2Me; R = H (80), R = Cl (80), R = Me (82)	611
C_{11}				
TMSO, OTMS (cyclohexylidene)	TMSO, OMe (cyclohexylidene)	200°, 4 h	OTMS, CO_2Me (75)	308
TMSO, TMSO (cyclopentene)	X_3CCOCl	Zn, Et_2O	X, X, OTMS, O, OTMS; X = F (47), X = Cl (75), X = Br (74)	617
EtO, OEt, EtO, OEt	$CH_2{=}C{=}O$	0-5°, 14 h	OEt, OEt, OEt, OEt, O (88) + O, CO_2Et, EtO, OEt (6)	613

TABLE XII. [2+2] CYCLOADDITION OF KETENES TO POLYOXYGENATED OLEFINS (*Continued*)

Reactant	Ketene or Ketene Source	Conditions	Product(s) and Yield(s) (%)	Refs.
	RCH_2COCl	Et_3N, C_6H_{14}	2-R-3,3,4,4-tetraethoxycyclobutanone **I** + 1-R-2-(RCH_2CO_2)-3,3,4,4-tetraethoxycyclobutene **II** + 4-R-3,4-... 2-R-3-OEt-4,4-diethoxycyclobut-2-enone **III** R: Me, **I + II + III** (55); *t*-Bu (81); *n*-C_6H_{13} (78); C_6H_{11} (75); Bn (85); Cl_2C=CH (62); Ph (60); *p*-$MeOC_6H_4$ (78); *p*-ClC_6H_4 (71)	614
	TMS(H)C=C=O	90°, 2 h	2-TMS-3,3,4,4-tetraethoxycyclobutanone (68)	615
$(TMSO)_2C$=CH(OTMS)	RCHClCOCl	Et_3N, C_6H_{14}	RC(Cl)=C(OTMS)CH(OTMS)CO_2TMS R = H (80) R = Cl (85) R = Me (85)	611

TABLE XII. [2+2] CYCLOADDITION OF KETENES TO POLYOXYGENATED OLEFINS (*Continued*)

Reactant	Ketene or Ketene Source	Conditions	Product(s) and Yield(s) (%)	Refs.
C_{12}				
Ph, OMe, OTMS	Ph, OMe, OTMS	200°, 4 h	TMSO, Ph, CO_2Me, Ph (85)	308
TMSO, TMSO	X_3CCOCl	Zn, Et_2O	X, X, OTMS, O, OTMS; X = F (55), X = Cl (65), X = Br (82)	617
C_{14}				
n-PrO, OPr-n, n-PrO, OPr-n	TMS, H, C=C=O	—	TMS, OPr-n, OPr-n, OPr-n, O, OPr-n (—)	615
C_{18}				
n-BuO, OBu-n, n-BuO, OBu-n	CH_2=C=O	0-5°, 14 h	OBu-n, OBu-n, OBu-n, O, OBu-n (83)	613
C_{24}				
Ph, O, OMe, OMe, Ph, Ph	Ph, Ph, C=C=O	80-85°, 1 h	Ph OMe, Ph, OMe, Ph, O, O, Ph, Ph (—)	616

TABLE XIII. INTRAMOLECULAR CYCLOADDITIONS

Reactant	Conditions	Product(s) and Yield(s) (%)	Refs.
C_6			
$R^2R^1C{=}CR^3CHR^4CHR^5OCHR^6COCl$	Et_3N, C_6H_6, reflux	I + II	

R^1	R^2	R^3	R^4	R^5	R^6	Yield I	Yield II	Refs.
H	H	H	H	H	H	(16)	(0)	618, 619
H	H	Me	H	H	H	(72)	(0)	618, 619
Me	Me	H	H	H	H	(0)	(52)	618, 619
H	H	Me	H	H	Me	(50)	(0)	620
H	H	Me	H	Me	H	(73)	(0)	620, 619
H	H	Me	Me	H	H	(63)	(0)	620, 619
H	Et	H	H	H	H	(47)	(19)	619
H	–$(CH_2)_4$–		H	H	H	(58)	(5)	620

Reactant	Conditions	Product(s) and Yield(s) (%)	Refs.
C_7			
$R^1CH{=}CHCHR^2NHCHR^3CHR^4CO_2Et$	400°	bicyclic cyclobutanone-pyrroline	

R^1	R^2	R^3	R^4	Yield	Refs.
H	H	Me	H	(46)	302, 621
H	H	Me	Me	(62)	302
H	Me_2	H	H	(67)	302
Me	H	Me	Me	(46)	302
H	H	Ph	H	(45)	621
H	H	Me	Ph	(60)	302
H	H	–$(CH_2)_3$–		(32)	302
H	H	–$(CH_2)_4$–		(33)	302

Reactant	Conditions	Product(s) and Yield(s) (%)	Refs.
$CH_2{=}CHCH_2NHCH{=}C(CN)CO_2Et$	400°	3-cyanopyrrole (30)	302

TABLE XIII. INTRAMOLECULAR CYCLOADDITIONS (*Continued*)

Reactant	Conditions	Product(s) and Yield(s) (%)	Refs.
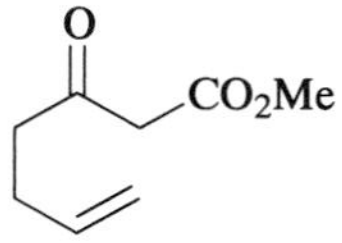 5-hexenoyl ketoester (CO_2Me)	550°, 10^{-2} torr	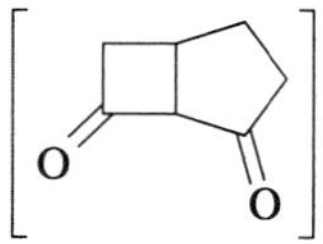 [bicyclic dione] (—)	622, 623
 epoxy bicyclic ketone	$h\nu$, C_6H_6	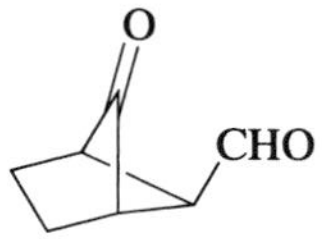 bicyclic keto aldehyde (CHO) (—)	624
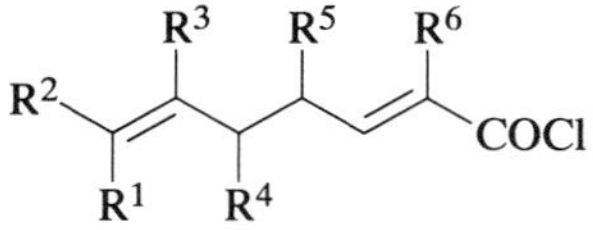 dienoyl chloride (R¹–R⁶, COCl)	Et_3N, PhMe, reflux	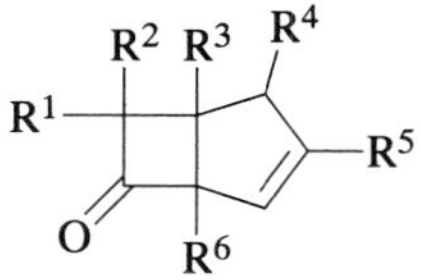 I 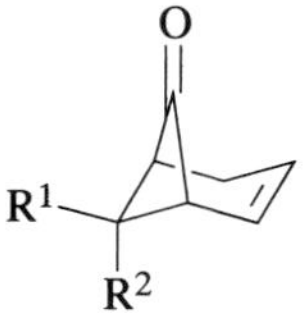 + II	

R^1	R^2	R^3	R^4	R^5	R^6	Yield I	Yield II	Refs.
H	H	H	H	H	H	(48)	(0)	625, 107
H	H	H	H	H	Me	(48)	(0)	107
H	Me	H	H	H	H	(38)	(9)	107
Me	Me	H	H	H	H	(0)	(41)	107
Et	H	H	H	H	H	(46)	(5)	625, 107
H	H	H	H	$CH_2CH{=}CH_2$	H	(83)	(0)	626
H	H	Me	H	$CH_2C(Me){=}CH_2$	H	(84)	(0)	626
H	H	H	$CH_2OTBDPS$	H	H	(39)	(0)	107
H	H	H	CH_2OMe	H	H	(40)	(0)	107

TABLE XIII. INTRAMOLECULAR CYCLOADDITIONS (*Continued*)

Reactant	Conditions	Product(s) and Yield(s) (%)	Refs.
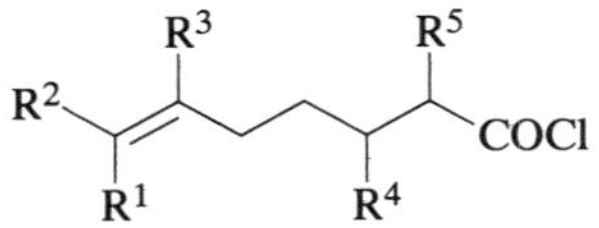 R^2, R^1, R^3, R^4, R^5 ... COCl	Et_3N, C_6H_6, reflux	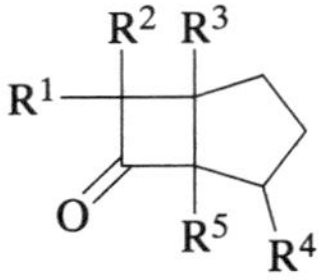 I 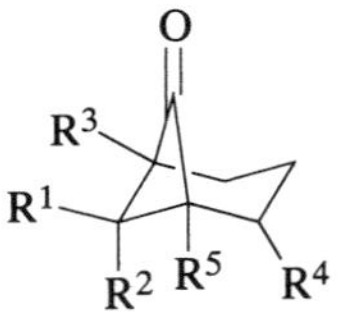 + II	

R^1	R^2	R^3	R^4	R^5	Yield I	Yield II	Refs.
H	H	H	H	Cl	(68)	(0)	627
H	H	H	H	H	(9)	(0)	626
H	H	Me	H	H	(—)	(0)	628, 594, 626
H	H	Ph	H	H	(—)	(0)	628, 594
H	H	H	H	$CH=CH_2$	(51)	(0)	625
H	$-(CH_2)_3-$		H	H	(50)	(0)	107, 629
H	$-(CH_2)_3-$		H	Me	(45)	(0)	630
H	Et	H	H	H	(26)	(14)	627
Me	Me	H	Me	Cl	(0)	(55)	627
H	H	H	H	3-furyl	(71)	(0)	631
H	H	H	H	Ph	(74)	(0)	631
H	H	H	H	1-cyclohexenyl	(68)	(0)	631
H	H	H	H	2-naphthyl	(88)	(0)	631

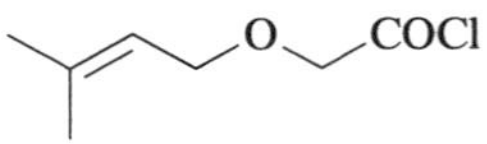

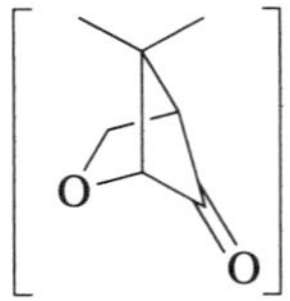

Reactant	Conditions	Product(s) and Yield(s) (%)	Refs.
O ... COCl	Et_3N, C_6H_6	[bicyclic ketone] (—)	620

TABLE XIII. INTRAMOLECULAR CYCLOADDITIONS (*Continued*)

Reactant	Conditions	Product(s) and Yield(s) (%)	Refs.
C_8			
[structure]	$h\nu$, Et_2O	[structure] (—)	632
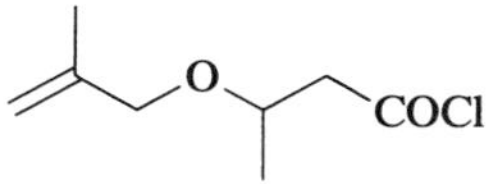 [structure] COCl	Et_3N, CH_2Cl_2	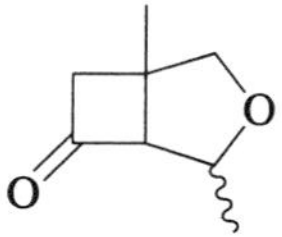 [structure] (70)	633
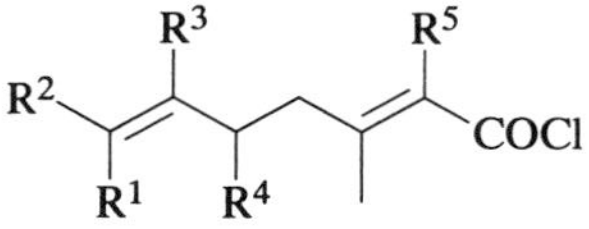 [structure] COCl	Et_3N, PhMe	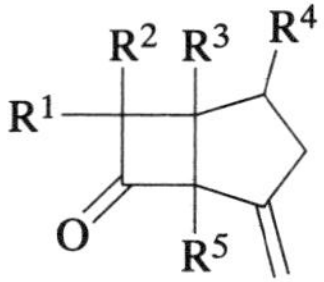 I, 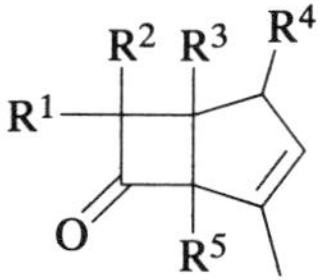 II, 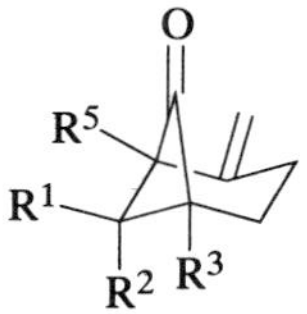 III	

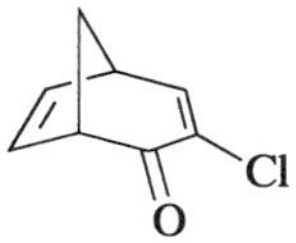

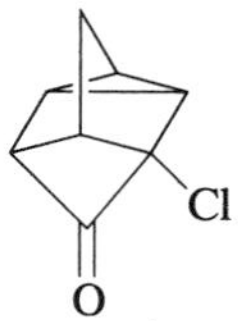

R^1	R^2	R^3	R^4	R^5	Yield **I**	Yield **II**	Yield **III**	Refs.
H	H	H	H	H	(41)	(0)	(0)	625
H	H	H	H	H	(48)	(12)	(0)	110
H	H	Me	H	H	(43)	(0)	(0)	625, 107, 110
Me	Me	H	H	H	(0)	(0)	(45)	109, 107, 634-636
CH_2OMe	Me	H	H	H	(0)	(0)	(30)	109
H	H	H	H	Me	(27)	(0)	(9)	637
H	H	H	H	$CH_2C{\equiv}CEt$	(38)	(0)	(7)	637
H	H	H	Me	H	(39)	(0)	(0)	110, 107
$Me_2C{=}CH(CH_2)_2$	Me	H	H	H	(0)	(0)	(43)	638, 109
Me	$Me_2C{=}CH(CH_2)_2$	H	H	H	(0)	(0)	(39)	109
$EtCH{=}CH(CH_2)_7$	Me	H	H	H	(0)	(0)	(30)	634
Me	$EtCH{=}CH(CH_2)_7$	H	H	H	(0)	(0)	(38)	634

TABLE XIII. INTRAMOLECULAR CYCLOADDITIONS (*Continued*)

Reactant	Conditions	Product(s) and Yield(s) (%)	Refs.
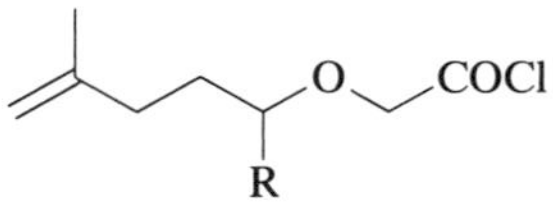 COCl, O, R	Et_3N, C_6H_6, reflux	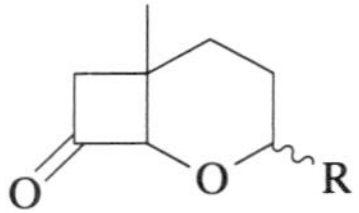 O, O, R; R = H (62); R = Me (58)	620, 619
C_9 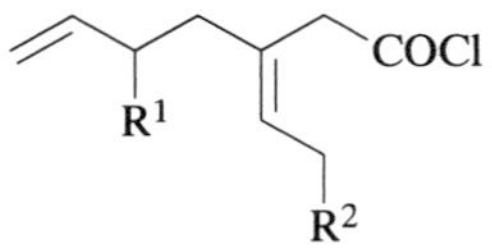 COCl, R^1, R^2	Et_3N, PhMe, reflux	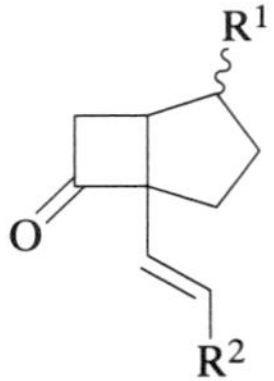 R^1, O, R^2; R^1 = H, R^2 = H (50); R^1 = Me, R^2 = H (52); R^1 = H, R^2 = Et (52)	108
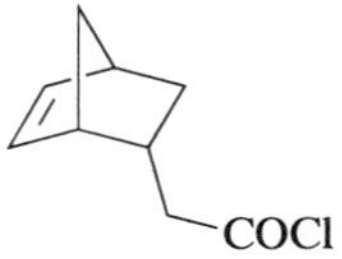 COCl	Et_3N, C_6H_6, reflux	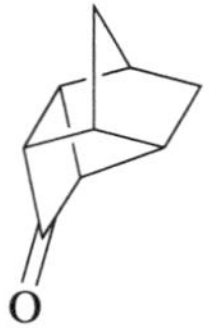 O (61)	639
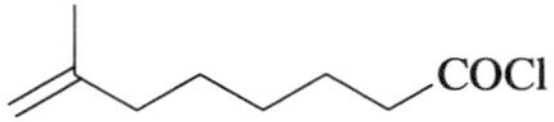 COCl	Et_3N	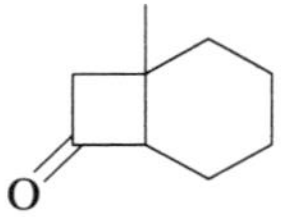 O (3)	626
C_{10} 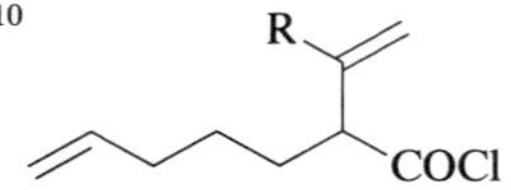 R, COCl	Et_3N, PhMe, reflux	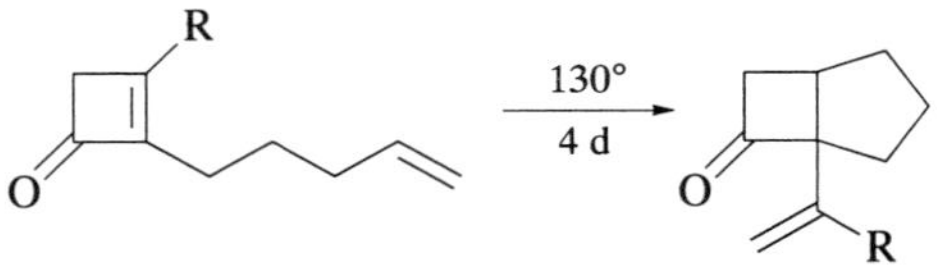 R, O; R = Me (56); R = Et (87) —130°, 4 d→ O, R; R = Me (76); R = Et (75)	108

TABLE XIII. INTRAMOLECULAR CYCLOADDITIONS (*Continued*)

Reactant	Conditions	Product(s) and Yield(s) (%)	Refs.
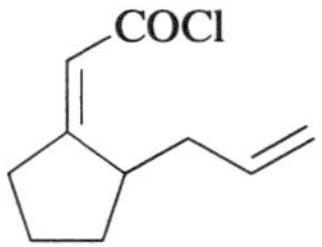 COCl	Et_3N	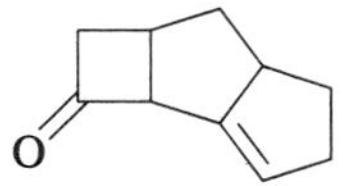 (62)	640
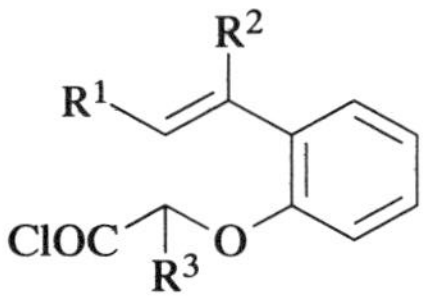 R^1, R^2, ClOC, R^3	Et_3N, C_6H_6, reflux	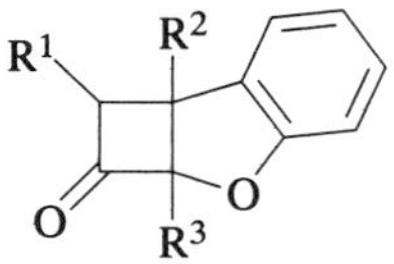 R^1 R^2 R^3 Yield	
		H H H (70)	619, 620
		Me H H (72)	618
		H H Me (60)	618
		Me H Me (76)	618
		Me H Et (71)	618
		H Ph H (88)	618
		Me H Ph (85)	618, 322
		H Ph Me (84)	618
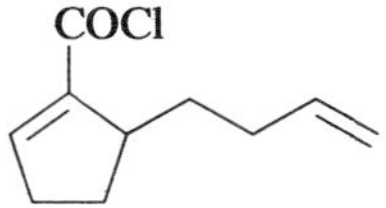 COCl	*i*-Pr_2NEt, DMAP, PhMe, reflux	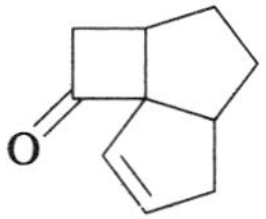 (78)	629
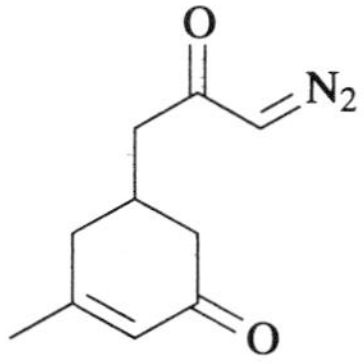 O, N_2, O	*hν*, C_6H_6	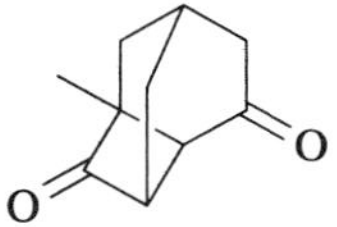 (24)	641
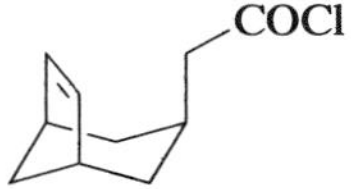 COCl	Et_3N	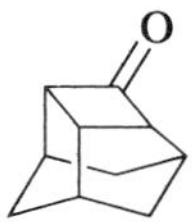 (54-76)	642

TABLE XIII. INTRAMOLECULAR CYCLOADDITIONS (*Continued*)

Reactant	Conditions	Product(s) and Yield(s) (%)	Refs.
COCl	$Cl_2CHCHCl_2$, reflux or Et_3N	O (1) + O (10)	643
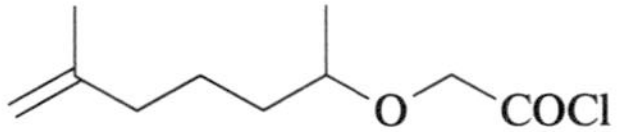 O COCl	Et_3N, C_6H_6, reflux	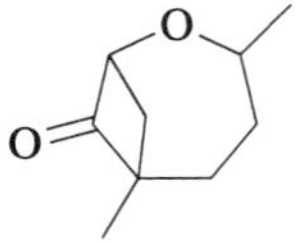 O O (30)	619, 620
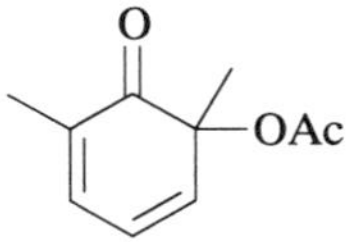 O OAc	*hν*	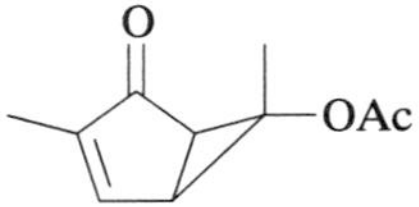 O OAc (36-43)	644
O OAc R^1 R^2	*hν*	O OAc R^1 R^2 **I** + O R^2 R^1 OAc **II**	644
		R^1 H, R^2 H: Yield **I** (45), Yield **II** (27)	
		R^1 Me, R^2 H: Yield **I** (57), Yield **II** (45)	
		R^1 Me, R^2 Me: Yield **I** (42), Yield **II** (30)	
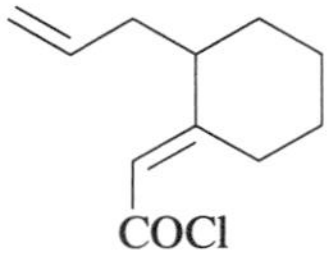 C_{11} COCl	Et_3N, PhMe, reflux	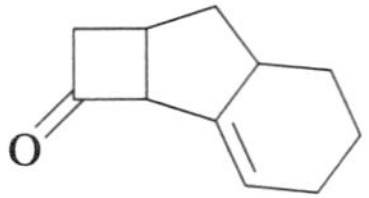 O (33) + O (25)	110

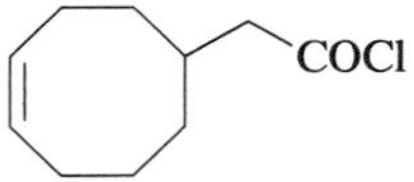

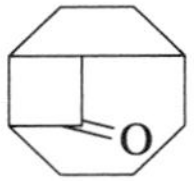

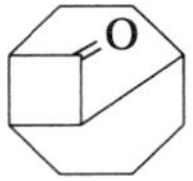

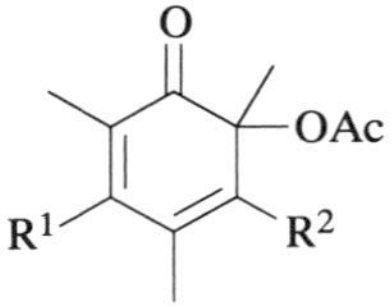

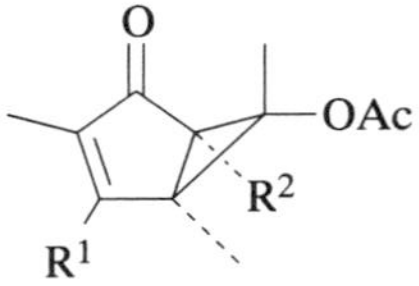

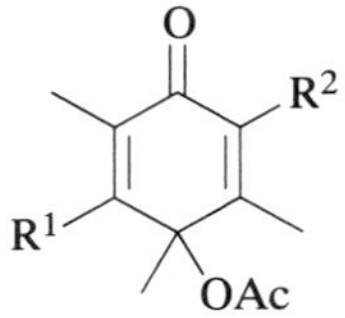

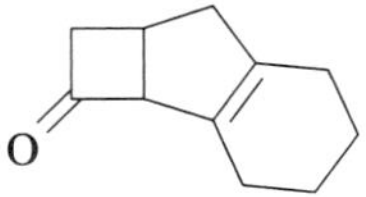

TABLE XIII. INTRAMOLECULAR CYCLOADDITIONS (*Continued*)

Reactant	Conditions	Product(s) and Yield(s) (%)	Refs.
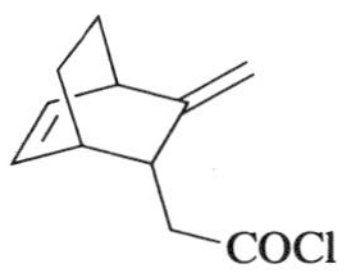 COCl	Et_3N, C_6H_6, reflux	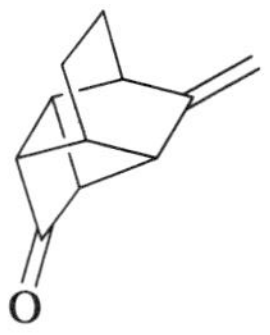 O (44)	645
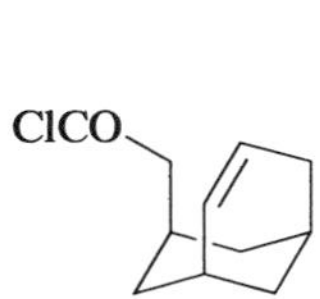 ClCO	Et_3N, C_6H_6, reflux	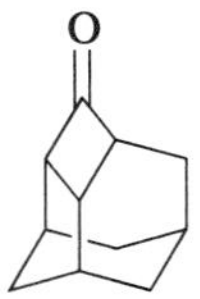 O (46)	646
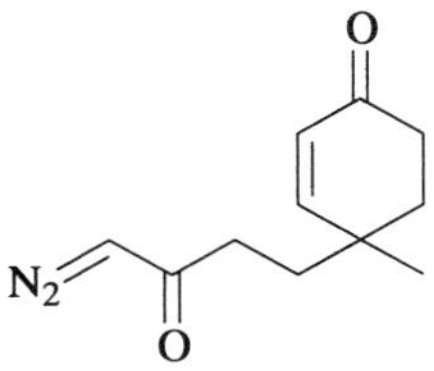 O, N_2, O	$h\nu$, C_6H_{12}, dioxane	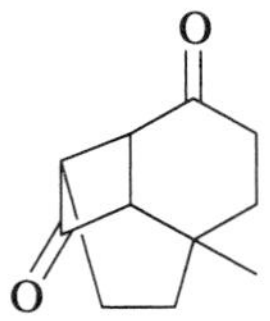 O, O (32)	315, 641
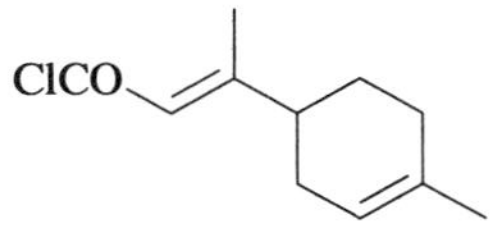 ClCO	Et_3N, PhMe, reflux	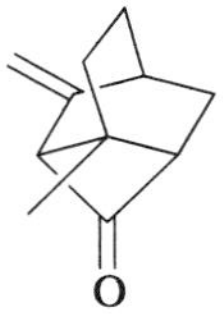 O (59)	647

TABLE XIII. INTRAMOLECULAR CYCLOADDITIONS (*Continued*)

Reactant	Conditions	Product(s) and Yield(s) (%)	Refs.
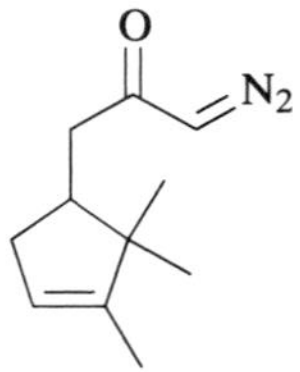 N_2, O	$h\nu$, C_5H_{12}	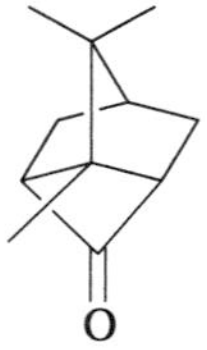 O (79)	648
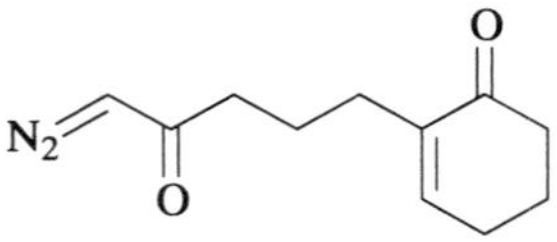 N_2, O, O	$h\nu$, C_6H_{12}, dioxane	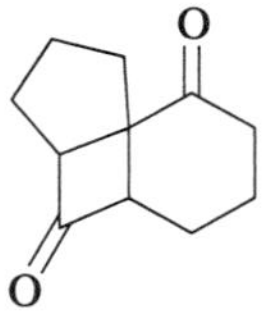 O, O (39)	315
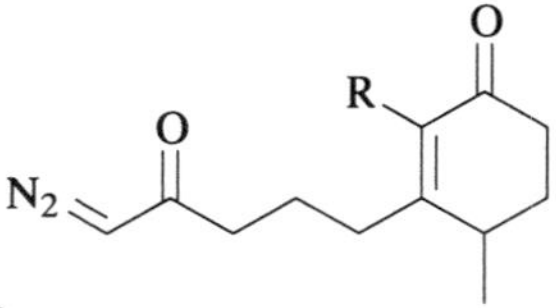 N_2, O, R, O	$h\nu$, C_6H_{12}, dioxane	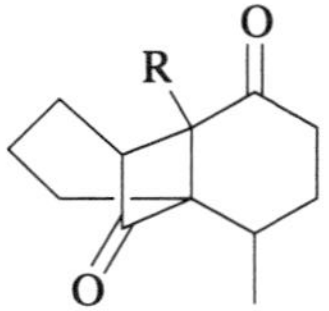 R, O, O; R = H (34); R = Me (30)	315, 649
C_{12}			
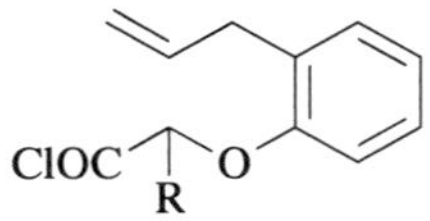 ClOC, R, O	Et_3N, C_6H_6, reflux	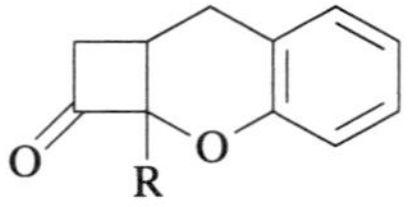 O, R, O; R = Me (43); R = Et (49)	618, 322
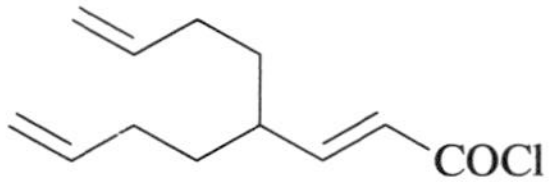 COCl	Et_3N	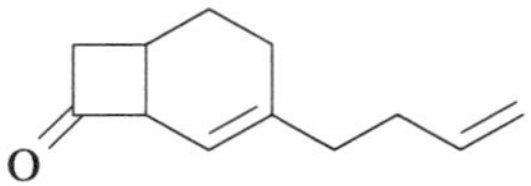 O (8)	626

TABLE XIII. INTRAMOLECULAR CYCLOADDITIONS (*Continued*)

Reactant	Conditions	Product(s) and Yield(s) (%)	Refs.
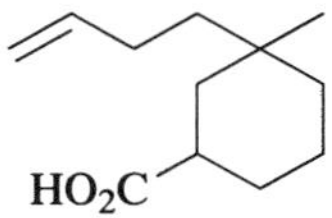 HO_2C	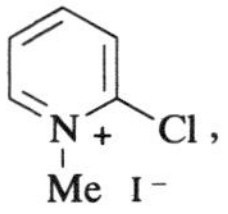 2-chloro-N-methylpyridinium iodide (N+ Me I^-, Cl), Et_3N, MeCN	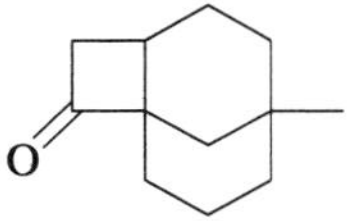 O (47)	650
C_{13} O, R, O, N_2	$h\nu$, C_6H_{12}, dioxane	O, R, O; R = H (30); R = Me (30)	315
C_{14} O, O, N_2	$h\nu$, C_6H_6, dioxane	O, O (40)	641, 649
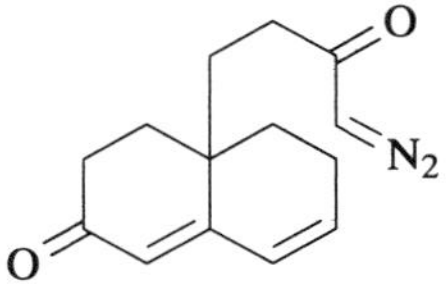 O, O, N_2	$h\nu$, C_6H_{12}, dioxane	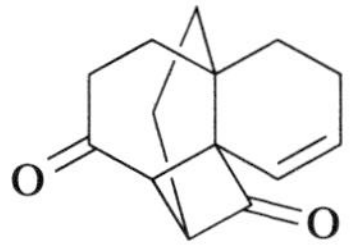 O, O (29)	315
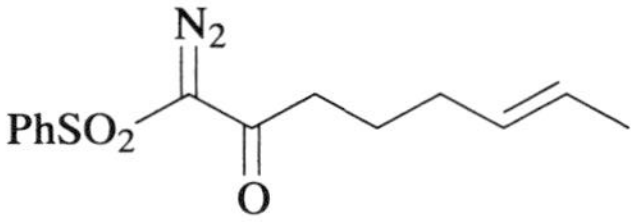 $PhSO_2$, N_2, O	174°, n-$C_{10}H_{21}$	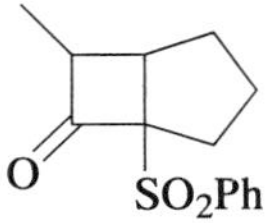 O, SO_2Ph (31)	651

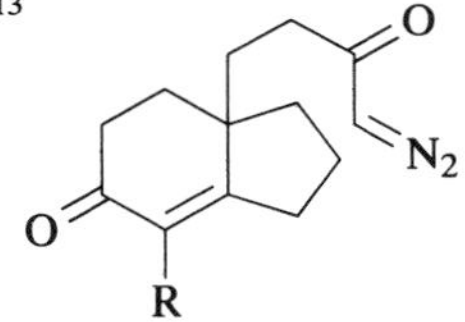

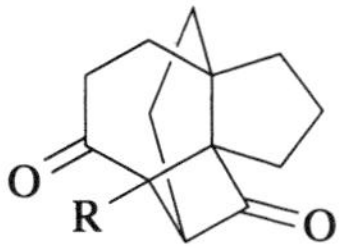

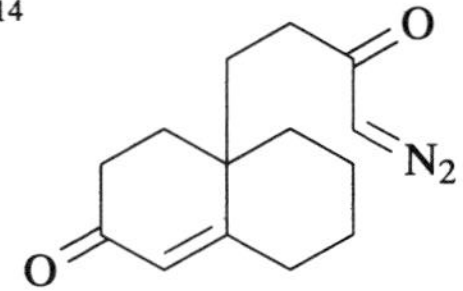

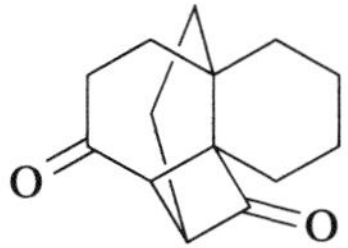

TABLE XIII. INTRAMOLECULAR CYCLOADDITIONS (*Continued*)

	Reactant	Conditions	Product(s) and Yield(s) (%)	Refs.
	COCl	Et_3N, PhMe, reflux	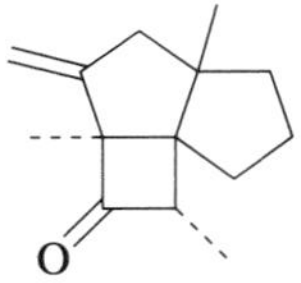 (31) + O, Et (6)	647
C_{15}	O, R^2, R^1, O, COCl, Ph	Et_3N	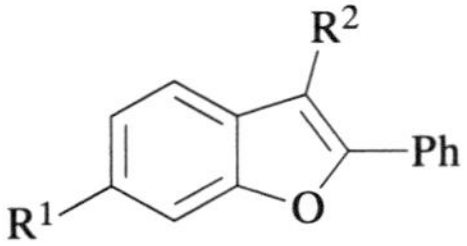 R^2, Ph, R^1, O; $R^1 = R^2 = H$ (75); R^1 = MeO, R^2 = Ph (82)	322
	O, NOH	1. NH_2Cl, THF 2. *hν*, Et_2O	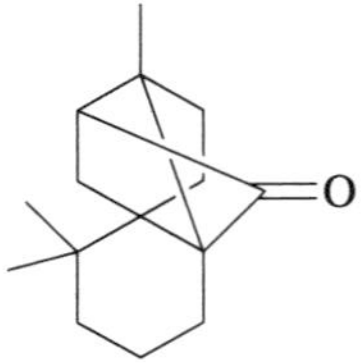 O (90)	652
	OHC	Distill in vacuo	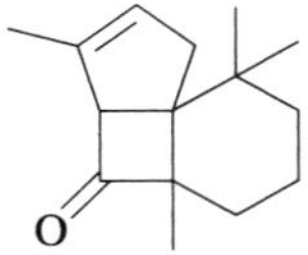 O (50)	653

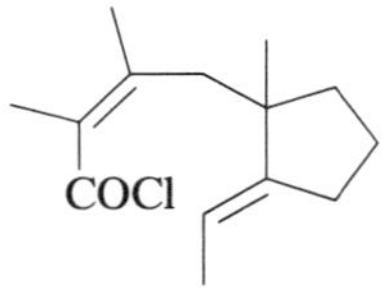

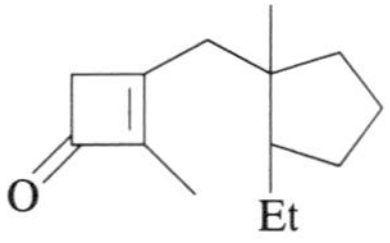

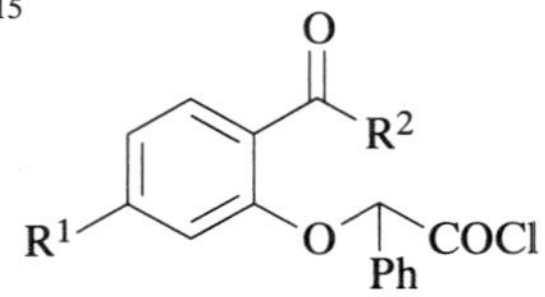

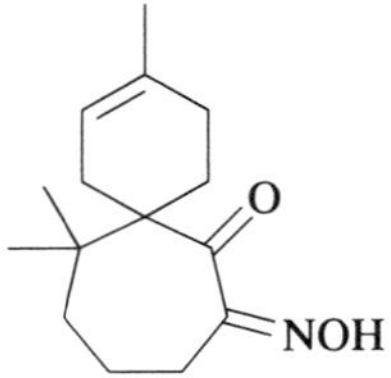

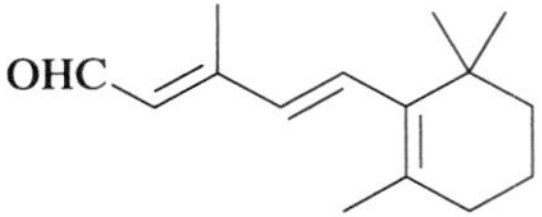

TABLE XIII. INTRAMOLECULAR CYCLOADDITIONS (*Continued*)

Reactant	Conditions	Product(s) and Yield(s) (%)	Refs.
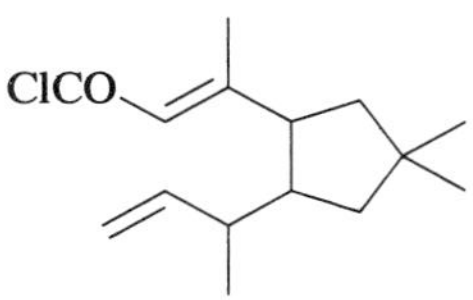 ClCO	i-Pr_2NEt, 110°	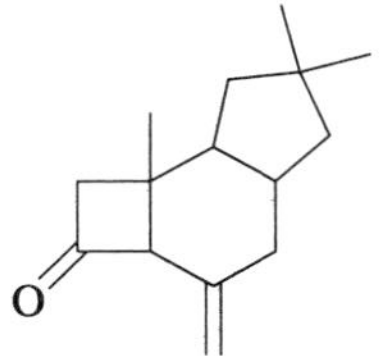 (57)	654
C_{16} 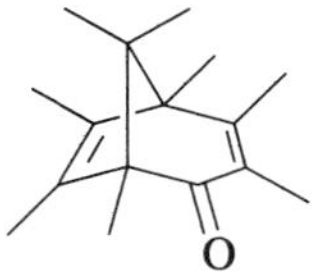	$h\nu$, MeOH	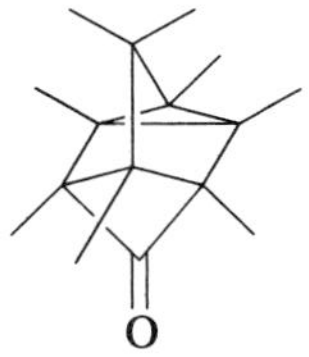 (100)	655
C_{18} 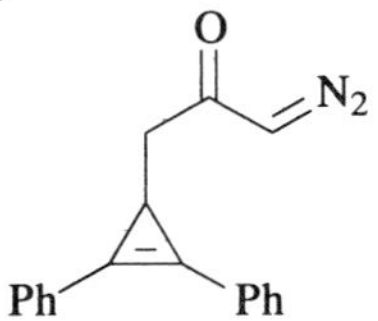 N_2, Ph, Ph	$h\nu$, THF	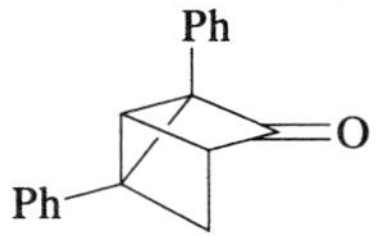 Ph, Ph (—)	656
C_{23} 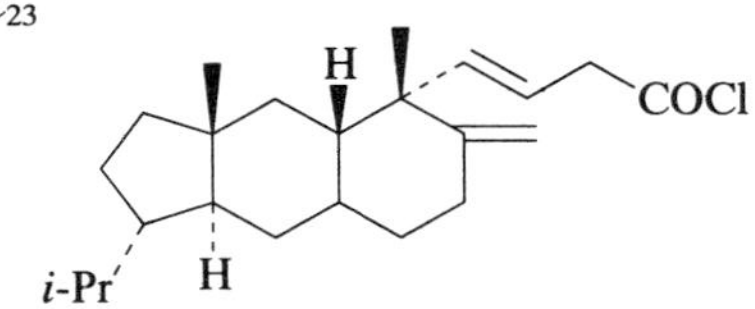 H, COCl, i-Pr, H	Et_3N, C_6H_6, reflux	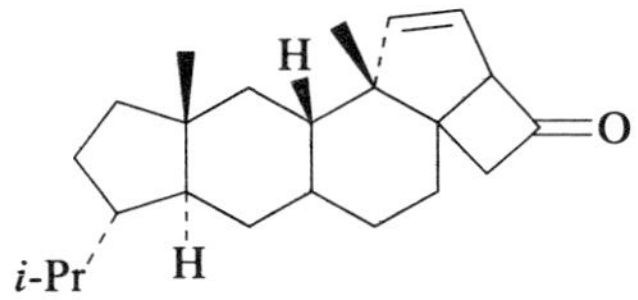 H, i-Pr, H (80)	657

TABLE XIII. INTRAMOLECULAR CYCLOADDITIONS (*Continued*)

Reactant	Conditions	Product(s) and Yield(s) (%)	Refs.
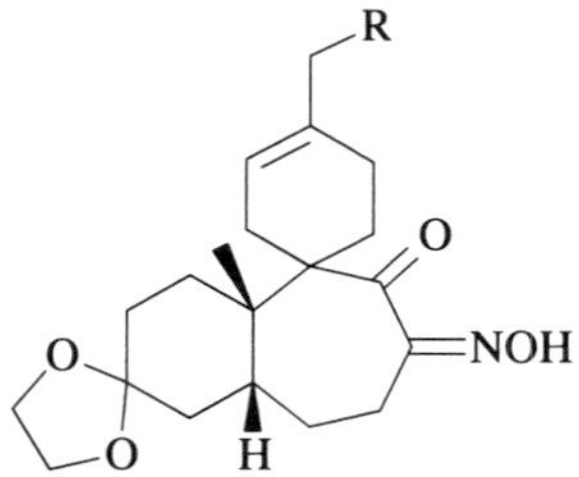 R, O, =NOH, O, O, H	1. NH_2Cl, THF 2. *h*ν, Et_2O, -75°	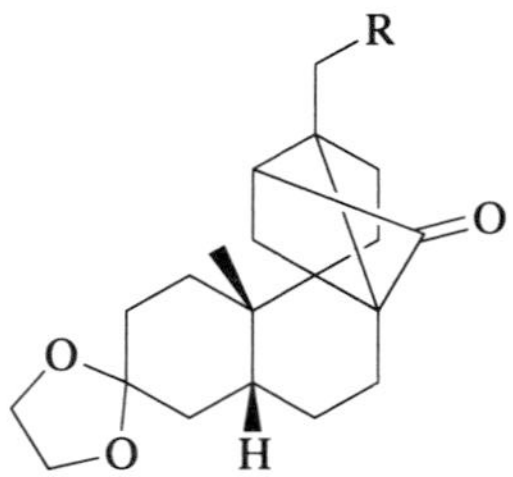 R, O, O, O, H R = H (80) R = TMS (60)	658, 659, 652

TABLE XIV. [2+2] CYCLOADDITION OF KETENES TO ALKYNES

Reactant	Ketene or Ketene Source	Conditions	Product(s) and Yield(s) (%)	Refs.
C_4				
$MeC\equiv CMe$	Cl_3CCOCl	Zn-Cu, Et_2O 7 min	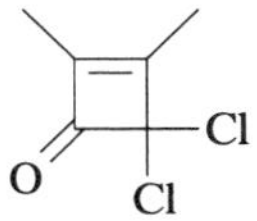 **I** (60) 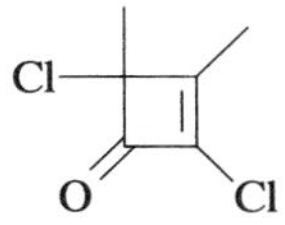 + **II** (2.5)	116
		Zn-Cu, $POCl_3$, Et_2O, 14 h	**I** (85)	111, 116
	$Cl_2CHCOCl$	Et_3N, 12-15°	**I** (12)	660
	$(CF_3)_2C=C=O$	150°	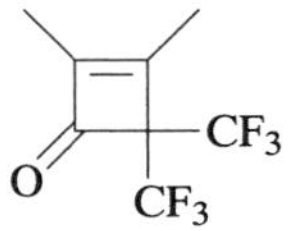	661
	"	100°, 6 h	" (65)	661
C_5				
$HC\equiv CCH=CHOMe$	$Me_2C=C=O$	MeCN	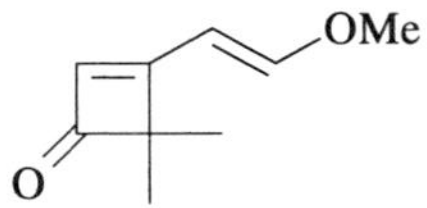 (38)	115
	"	Et_2O, rt	" (65)	662
	$(n\text{-Bu})(Et)C=C=O$	MeCN, 82°	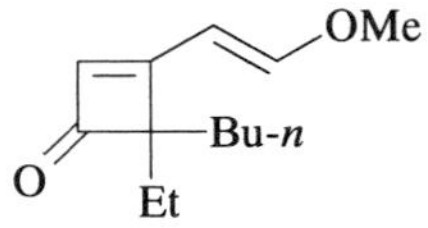 (39)	662

TABLE XIV. [2+2] CYCLOADDITION OF KETENES TO ALKYNES (*Continued*)

Reactant	Ketene or Ketene Source	Conditions	Product(s) and Yield(s) (%)	Refs.
$HC\equiv CC(Me)=CH_2$	Cl_3CCOCl	Zn-Cu, $POCl_3$, Et_2O, 14 h	2,2-dichloro-3-isopropenylcyclobut-3-enone (45)	111, 113
C_6				
$HC\equiv CBu$-*n*	"	"	3-Bu-*n*-2,2-dichlorocyclobut-3-enone (65-77)	111-113
	"	Zn-Cu, Et_2O, 8 min	" (51)	116
	$(CF_3)_2C=C=O$	150°, 8 h	3-Bu-*n*-2,2-bis(trifluoromethyl)cyclobut-3-enone (11.5)	661
$HC\equiv CBu$-*t*	Cl_3CCOCl	Zn-Cu, $POCl_3$, Et_2O, 10°, 14 h	3-Bu-*t*-2,2-dichlorocyclobut-3-enone (80)	112
	2,5-diazido-3,6-di-*t*-butyl-1,4-benzoquinone	C_6H_6	2,3-di-Bu-*t*-2-cyanocyclobut-3-enone (40-80)	663

TABLE XIV. [2+2] CYCLOADDITION OF KETENES TO ALKYNES (*Continued*)

Reactant	Ketene or Ketene Source	Conditions	Product(s) and Yield(s) (%)	Refs.
EtC≡CEt	Cl_3CCOCl	Zn-Cu, $POCl_3$, Et_2O, 14 h	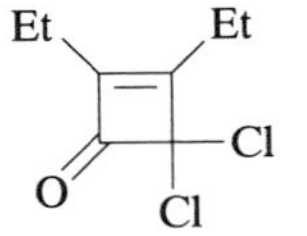 **I** (30) 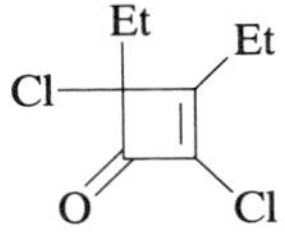 + **II** (56)	116
	"	"	**I** (57-62)	111, 113
	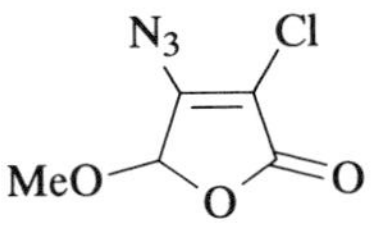 MeO, N_3, Cl furanone	PhMe, 103°	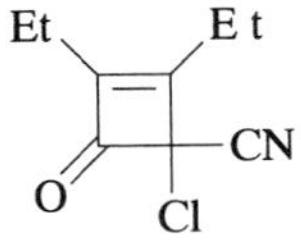 Et, Et, CN, Cl cyclobutenone (61)	114
	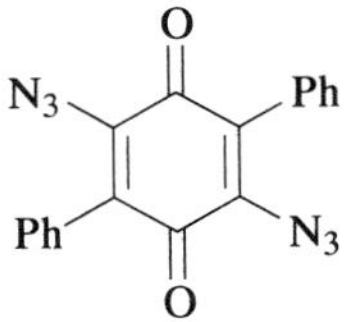 N_3, Ph, Ph, N_3 benzoquinone	CCl_4, 77°	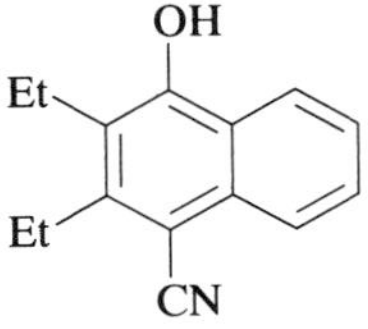 OH, Et, Et, CN naphthalene (33)	117
C_7				
HC≡C$(CH_2)_3$OAc	Cl_3CCOCl	Zn-Cu, $POCl_3$, Et_2O	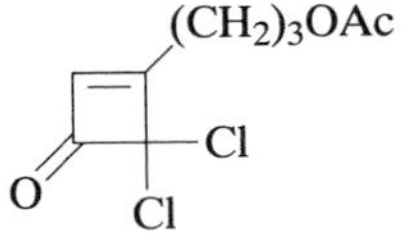 $(CH_2)_3$OAc, Cl, Cl cyclobutenone (80)	112
HC≡CC_5H_{11}-*n*	"	"	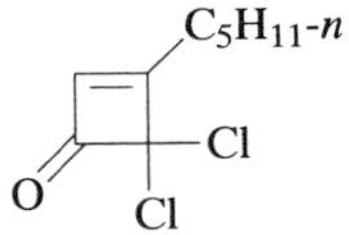 C_5H_{11}-*n*, Cl, Cl cyclobutenone (70)	113

TABLE XIV. [2+2] CYCLOADDITION OF KETENES TO ALKYNES (*Continued*)

Reactant	Ketene or Ketene Source	Conditions	Product(s) and Yield(s) (%)	Refs.
C_8				
$HC\equiv CC_6H_4Cl$-*p*	$(CF_3)_2C=C=O$	90°, 16 h	4,4-bis(CF_3)-3-(C_6H_4Cl-*p*)cyclobut-2-enone (62)	661
	"	150°	" (—)	661
$HC\equiv CPh$	Cl_3CCOCl	Zn-Cu, $POCl_3$, Et_2O	4,4-dichloro-3-phenylcyclobut-2-enone (75)	112, 113
	$(CF_3)_2C=C=O$	100-150°	4,4-bis(CF_3)-3-phenylcyclobut-2-enone (79-80)	148, 231, 387, 661
	2,5-diazido-3,6-di-*t*-butyl-1,4-benzoquinone (N_3, Bu-*t*, *t*-Bu, N_3)	C_6H_6	4-Bu-*t*-4-CN-3-phenylcyclobut-2-enone (40-80)	663
	2,5-diazido-3,6-diphenyl-1,4-benzoquinone (N_3, Ph, Ph, N_3)	CCl_4, 77°	4-cyano-3-phenyl-1-naphthol (OH, Ph, CN) (67)	117

TABLE XIV. [2+2] CYCLOADDITION OF KETENES TO ALKYNES (*Continued*)

Reactant	Ketene or Ketene Source	Conditions	Product(s) and Yield(s) (%)	Refs.
	Ph, Ph C=C=O	rt	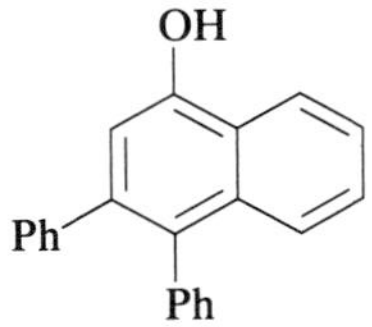 OH, Ph, Ph (81)	118
n-PrC≡CPr-*n*	Cl_3CCOCl	Zn-Cu, $POCl_3$, Et_2O, 10°, 4-26 h	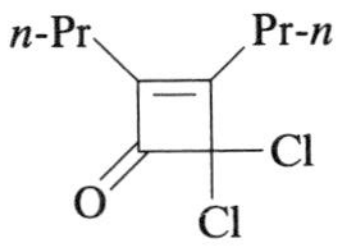 *n*-Pr, Pr-*n*, O, Cl, Cl (—)	112
C_9				
MeC≡CPh	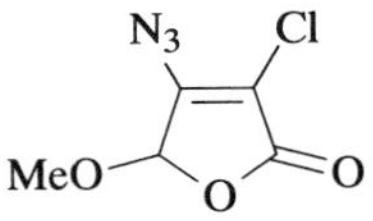 N_3, Cl, MeO, O, O	PhMe, 103°	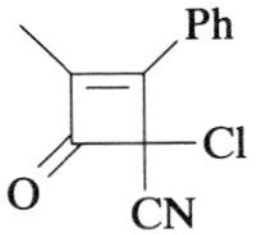 Ph, O, Cl, CN (84)	114
HC≡CC_6H_4Me-*p*	CF_3, CF_3 C=C=O	150°	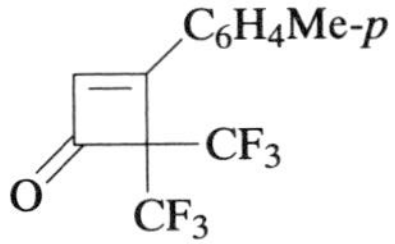 C_6H_4Me-*p*, O, CF_3, CF_3 (—)	661
	Ph, Ph C=C=O	rt, 36 h	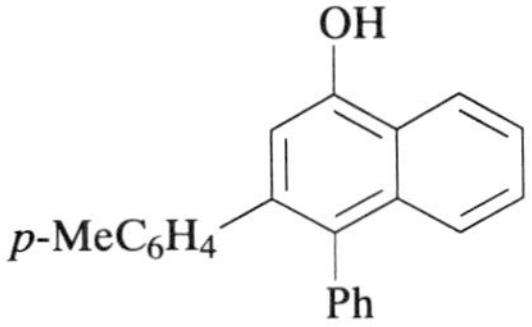 OH, *p*-MeC_6H_4, Ph (77)	664

TABLE XIV. [2+2] CYCLOADDITION OF KETENES TO ALKYNES (*Continued*)

Reactant	Ketene or Ketene Source	Conditions	Product(s) and Yield(s) (%)	Refs.
C_{10}				
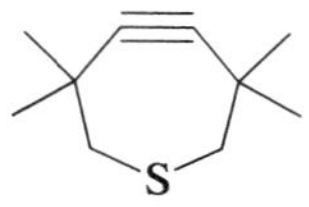 [structure]	$Cl_2CHCOCl$	Et_3N, C_5H_{12}	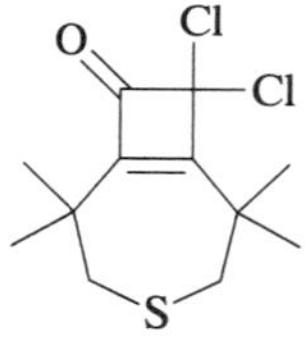 [structure] (80)	60
t-BuC≡CBu-t	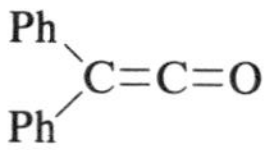 $Ph_2C{=}C{=}O$	175°, 5.5 h	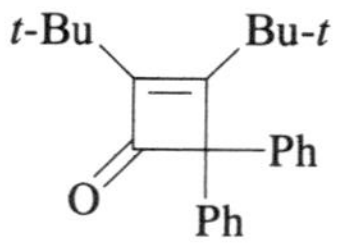 [structure] (47)	665
C_{12}				
t-BuC≡CPh	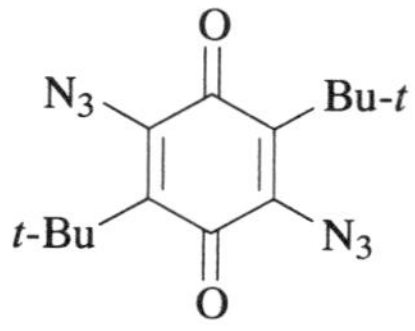 [structure]	C_6H_6	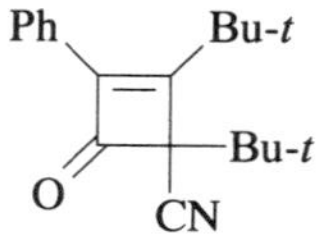 [structure] (40-80)	663
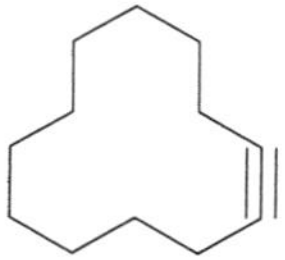 [structure]	Cl_3CCOCl	Zn-Cu, $POCl_3$, Et_2O, 10°, 4-26 h	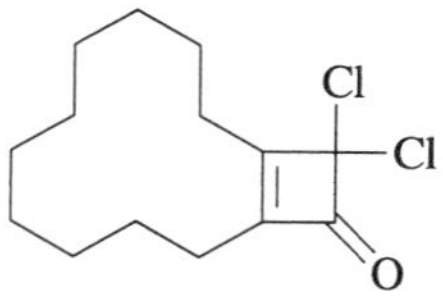 [structure] (—)	112
C_{14}				
PhC≡CPh	Cl_3CCOCl	Zn-Cu, $POCl_3$, Et_2O	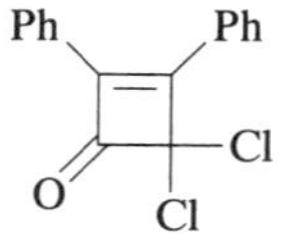 [structure] (32-45)	111, 113

TABLE XIV. [2+2] CYCLOADDITION OF KETENES TO ALKYNES (*Continued*)

Reactant	Ketene or Ketene Source	Conditions	Product(s) and Yield(s) (%)	Refs.
	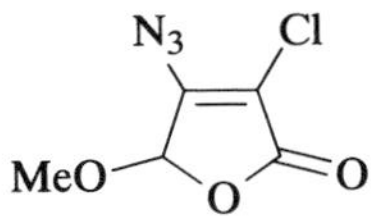 N_3, Cl, MeO, O, O	PhMe, 103°	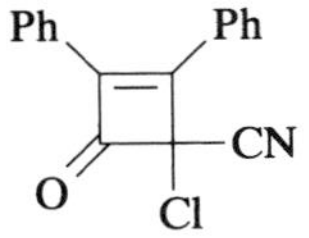 Ph, Ph, CN, O, Cl (77)	114
	$(CF_3)_2C{=}C{=}O$	150-200°	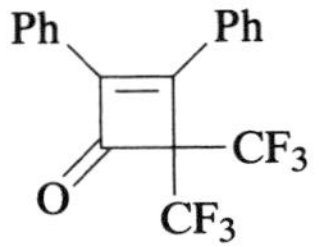 Ph, Ph, CF_3, O, CF_3 (95)	661
	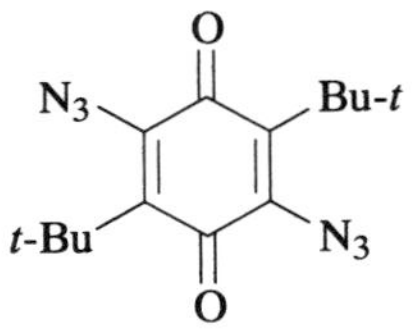 O, N_3, Bu-*t*, *t*-Bu, N_3, O	C_6H_6	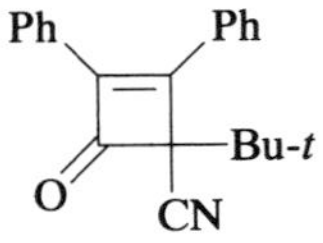 Ph, Ph, Bu-*t*, O, CN (40-80)	663
	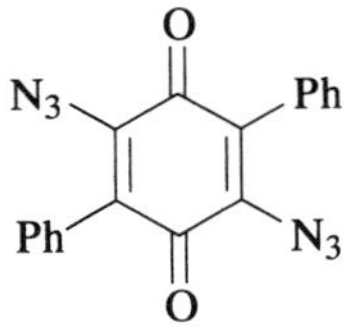 O, N_3, Ph, Ph, N_3, O	CCl_4, 77°	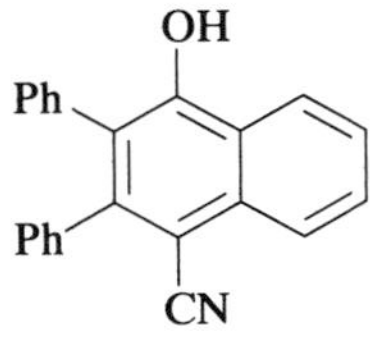 OH, Ph, Ph, CN (41)	117
	$Ph_2C{=}C{=}O$	70-80°, 3 d	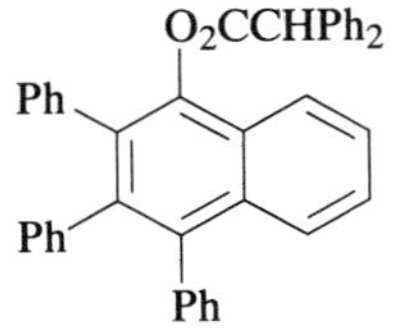 O_2CCHPh_2, Ph, Ph, Ph (82)	119

TABLE XIV. [2+2] CYCLOADDITION OF KETENES TO ALKYNES (*Continued*)

Reactant	Ketene or Ketene Source	Conditions	Product(s) and Yield(s) (%)	Refs.
C_{16} 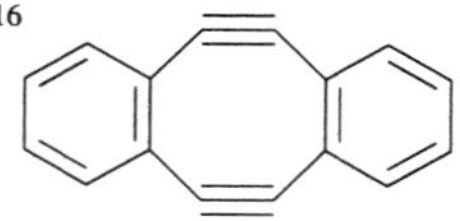	$Cl_2CHCOCl$	Et_3N, C_5H_{12}, 20°, 12 h	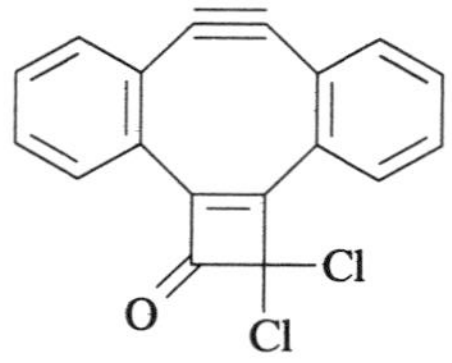 (42)	666
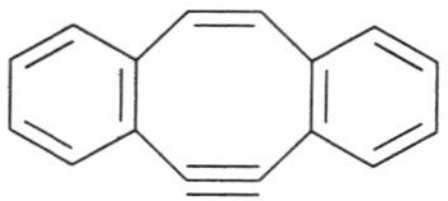	$Cl_2CHCOCl$	Et_3N, C_5H_{12}, 20°, 12 h	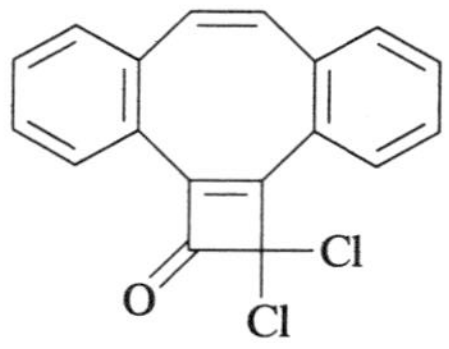 (17)	666

TABLE XV. [2+2] CYCLOADDITION OF KETENES TO ACETYLENIC ETHERS

Reactant	Ketene or Ketene Source	Conditions	Product(s) and Yield(s) (%)	Refs.
C_3				
HC≡COMe	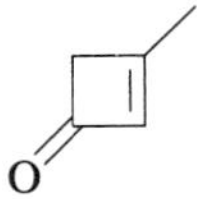 O	C_6H_6, 80-160° 26 h	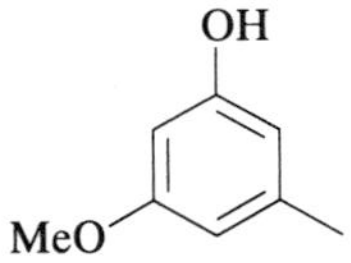 OH, MeO (71)	131
C_4				
HC≡COEt	CH_2=C=O	CH_2Cl_2, 0°	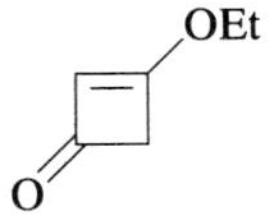 OEt, O (30-31)	126, 127
	$Cl_2CHCOCl$	Et_3N	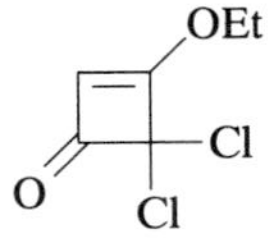 OEt, Cl, Cl, O (—)	667
	Me_2C=C=O	MeCN	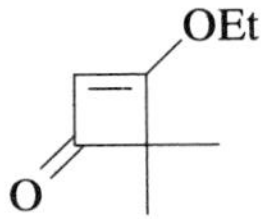 OEt, O (57-80)	123-125, 668
	i-C_3H_7COCl	Et_3N, Et_2O, 0°	" (65-66)	126, 127
	$Me_2CBrCOBr$	Zn, EtOAc, 0°	" (58)	128
	$(CF_3)_2C$=C=O	-80°	$(CF_3)_2C$=C=$CHCO_2Et$ (39)	133

TABLE XV. [2+2] CYCLOADDITION OF KETENES TO ACETYLENIC ETHERS (*Continued*)

Reactant	Ketene or Ketene Source	Conditions	Product(s) and Yield(s) (%)	Refs.
	$Et_2CHCOCl$	Et_3N, 0°	OEt, Et, Et, O (10-30)	126
	$C_5H_9CH_2COCl$	Et_3N, 0°	OEt, C_5H_9, O (10-30)	126
	$C_6H_{11}CH_2COCl$	Et_3N, 0°	OEt, C_6H_{11}, O (10-30)	126
	n-Bu, Et, C=C=O	C_6H_{14}, 25-40°	OEt, Bu-n, Et, O (70)	123,124
	N_3, Ph, Ph, N_3, O, O	CCl_4, 77°	OH, OEt, CN (30)	117

TABLE XV. [2+2] CYCLOADDITION OF KETENES TO ACETYLENIC ETHERS (*Continued*)

Reactant	Ketene or Ketene Source	Conditions	Product(s) and Yield(s) (%)	Refs.
	4-Cl-4-CN-2-R^1-3-R^2-cyclobutenone	C_6H_6, 40°	6-Cl-6-CN-2-Me-3-EtO-4-R^1-5-R^2-cyclohexa-2,4-dienone: R^1 = Me, R^2 = Ph (33); R^1 = Ph, R^2 = Ph (69); R^1 = Et, R^2 = Et (85)	114
	$Ph_2C{=}C{=}O$	C_6H_6, rt	3-EtO-4-Ph-1-naphthol (35)	132
	$R^1R^2C{=}C{=}O$	$MeNO_2$, -18 to 20°	3-OEt-6-R^3-8a-R^4-azulen-1(8aH)-one, I + II	

R^1	R^2	I: R^3	I: R^4	I: Yield	II: R^3	II: R^4	II: Yield	I:II	Refs.
Ph	Ph	H	Ph	(35)	—	—	(—)	—	132, 135-138
Ph	p-ClC_6H_4	H	p-ClC_6H_4	(—)	Cl	Ph	(—)	2.3	134
Ph	p-BrC_6H_4	H	p-BrC_6H_4	(—)	Br	Ph	(—)	2.4	134
Ph	p-$MeOC_6H_4$	H	p-$MeOC_6H_4$	(—)	MeO	Ph	(—)	50	134
Ph	p-MeC_6H_4	H	p-MeC_6H_4	(—)	Me	Ph	(—)	0.2	134
Ph	p-$EtO_2CC_6H_4$	H	p-$EtO_2CC_6H_4$	(—)	EtO_2C	Ph	(—)	22.2	134

TABLE XV. [2+2] CYCLOADDITION OF KETENES TO ACETYLENIC ETHERS (*Continued*)

Reactant	Ketene or Ketene Source	Conditions	Product(s) and Yield(s) (%)	Refs.
	$(Ph)_2CHCOCl$	Et_3N, 0°	[structure: OEt, Ph, Ph, O] (—)	126
	$(o\text{-}MeC_6H_4)_2C{=}C{=}O$	$MeNO_2$, 0°	[structure: OEt, o-MeC_6H_4, O] (—)	134
MeC≡COMe	$Ph_2C{=}C{=}O$	C_6H_6, 60°	[structure: OMe, Ph, Ph, O] (24) + [structure: OH, MeO, Ph] (8)	132
MeC≡CXMe	[structure: R^3, R^3, R^2, R^1, O]	C_6H_6 or PhMe, 80-160°, 2.5-21 h	[structure: OH, R^3, R^2, R^1, MeX]	131
	X = O, R^1 = H, R^2 = Me, R^3 = Cl		(65)	
	X = O, R^1 = H, R^2 = Me, R^3 = H		(92)	
	X = O, R^1 = Me, R^2 = Me, R^3 = H		(61)	
	X = S, R^1 = Me, R^2 = Me, R^3 = H		(86)	
	X = S, R^1 = H, R^2 = n-Bu, R^3 = H		(82)	

TABLE XV. [2+2] CYCLOADDITION OF KETENES TO ACETYLENIC ETHERS (*Continued*)

Reactant	Ketene or Ketene Source	Conditions	Product(s) and Yield(s) (%)	Refs.
C_5				
MeC≡COEt	$CH_2{=}C{=}O$	$MeNO_2$, 0°, 10 d	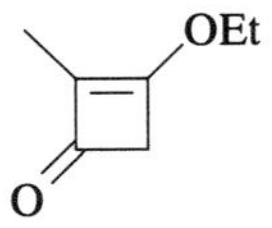 OEt, O (15)	669
	MeC≡COEt	140-150°	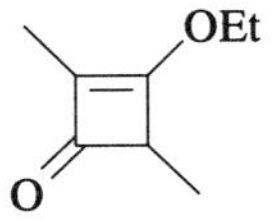 OEt, O (71)	129
	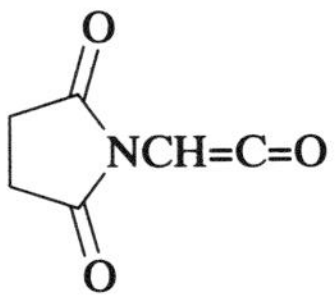 O, NCH=C=O, O	—	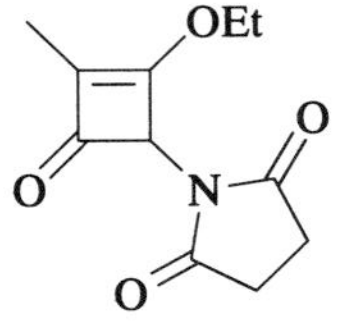 OEt, O, N, O, O (70)	670
	Ph, Ph, C=C=O	C_6H_6, rt	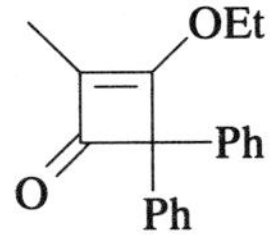 OEt, O, Ph, Ph (80)	129, 132
	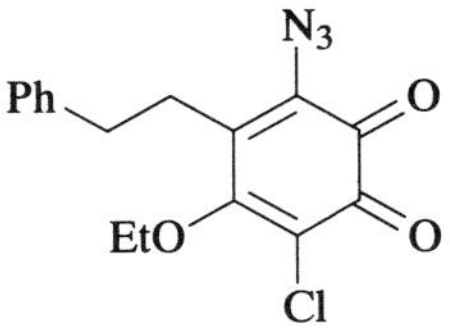 Ph, N_3, O, O, EtO, Cl	C_6H_6, 66°	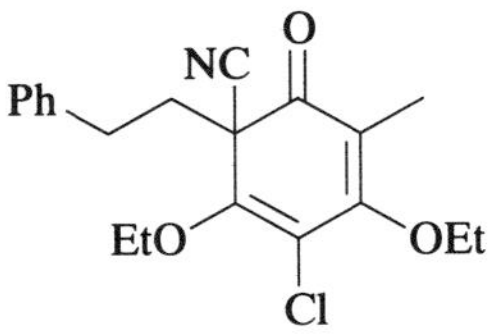 Ph, NC, O, EtO, OEt, Cl (49)	671
EtC≡COMe	$CH_2{=}C{=}O$	$MeNO_2$, 0°	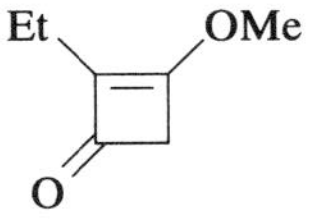 Et, OMe, O (30)	672

TABLE XV. [2+2] CYCLOADDITION OF KETENES TO ACETYLENIC ETHERS (*Continued*)

Reactant	Ketene or Ketene Source	Conditions	Product(s) and Yield(s) (%)	Refs.
C_6				
HC≡COBu-*t*	HC≡COBu-*t*	30°, 16 h	OBu-*t*, O (95)	673
EtC≡COEt	CH_2=C=O	$MeNO_2$, 0°	Et, OEt, O (35)	672
C_7				
MeC≡COBu-*t*	CH_2=C=O	$MeNO_2$, 0°	OBu-*t*, O (51)	672
	MeC≡COBu-*t*	80-90°	OBu-*t*, O (—)	674
C_8				
EtC≡COBu-*t*	CH_2=C=O	$MeNO_2$, 0°	Et, OBu-*t*, O (52)	672
	EtC≡COBu-*t*	80-90°	Et, OBu-*t*, O, Et (—)	674

TABLE XV. [2+2] CYCLOADDITION OF KETENES TO ACETYLENIC ETHERS (*Continued*)

Reactant	Ketene or Ketene Source	Conditions	Product(s) and Yield(s) (%)	Refs.
n-BuC≡COEt	CH_2=C=O	$MeNO_2$, 0°	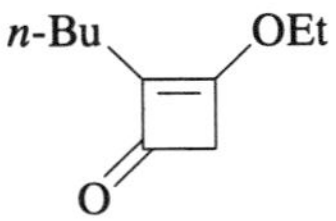 n-Bu, OEt, O (34)	672
C_9				
n-C_5H_{11}C≡COEt	CH_2=C=O	$MeNO_2$, 0°	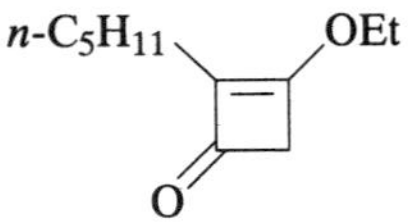 n-C_5H_{11}, OEt, O (56)	672
	n-C_5H_{11}C≡COEt	120-130°	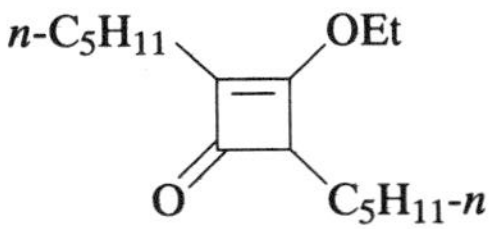 n-C_5H_{11}, OEt, O, C_5H_{11}-n (83)	129
n-C_6H_{13}C≡COMe	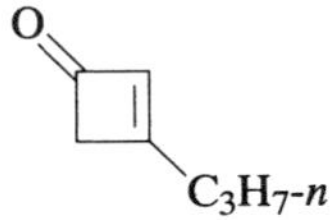 O, C_3H_7-n	C_6H_6, 80-160°, 22 h	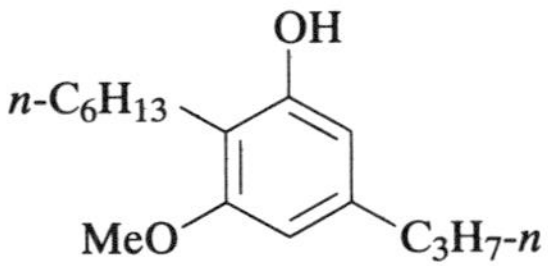 OH, n-C_6H_{13}, MeO, C_3H_7-n (82)	675, 131
	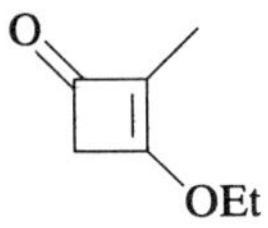 O, OEt	C_6H_6, 80-160°, 22 h	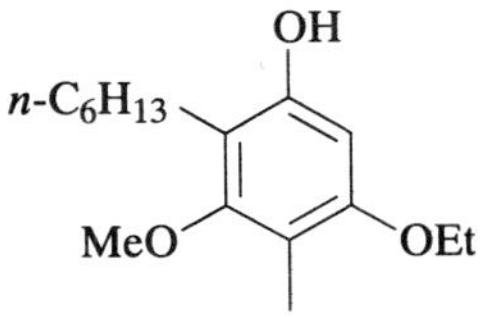 OH, n-C_6H_{13}, MeO, OEt (71)	675, 131
	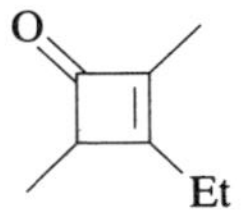 O, Et	C_6H_6, 80-160°, 22 h	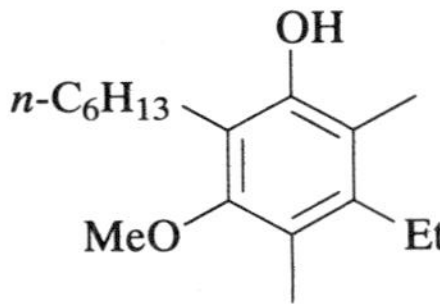 OH, n-C_6H_{13}, MeO, Et (33)	675, 131

TABLE XV. [2+2] CYCLOADDITION OF KETENES TO ACETYLENIC ETHERS (*Continued*)

Reactant	Ketene or Ketene Source	Conditions	Product(s) and Yield(s) (%)		Refs.
C_{10}					
t-BuOC≡COBu-*t*	*t*-BuOC≡COBu-*t*	C_6H_6, 80°	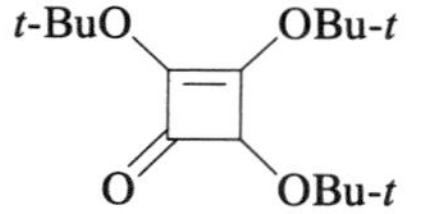 *t*-BuO, OBu-*t*, O, OBu-*t*	(100)	130
n-C_6H_{13}C≡COEt	$Me_2CBrCOBr$	Zn, EtOAc, 0°	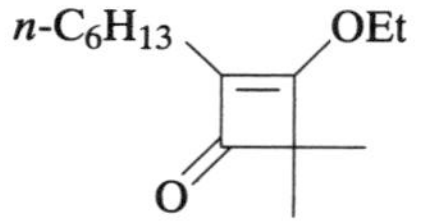 *n*-C_6H_{13}, OEt, O	(68)	128
C_{16}					
n-$C_{12}H_{25}$C≡COEt	$Me_2CBrCOBr$	Zn, EtOAc, 0°	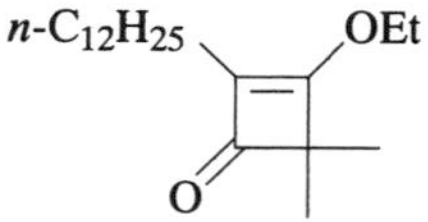 *n*-$C_{12}H_{25}$, OEt, O	(—)	128
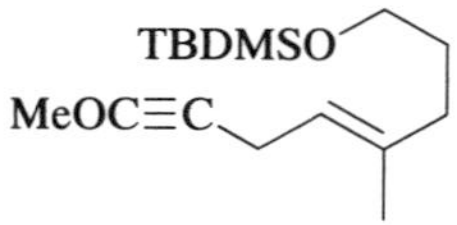 TBDMSO, MeOC≡C	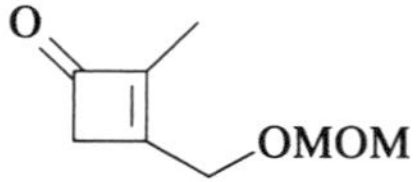 O, OMOM	C_6H_6, 120°, 14 h	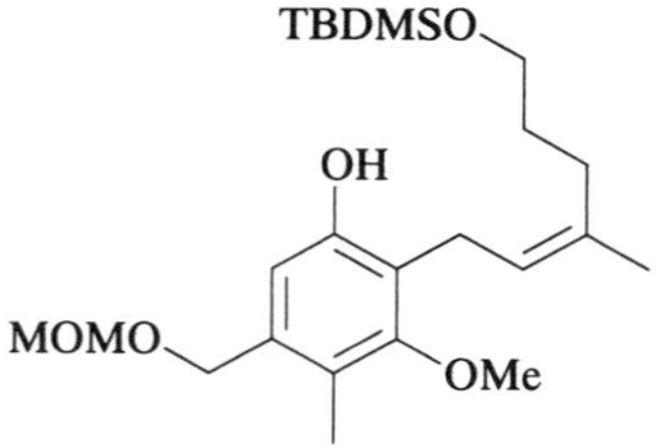 TBDMSO, OH, MOMO, OMe	(73)	676
C_{18}					
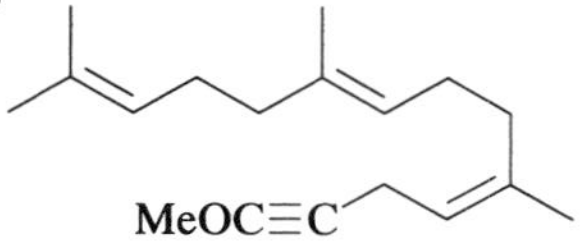 MeOC≡C	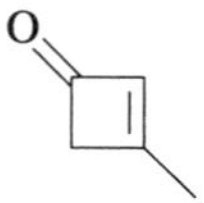 O	C_6H_6, 80°, 4 h	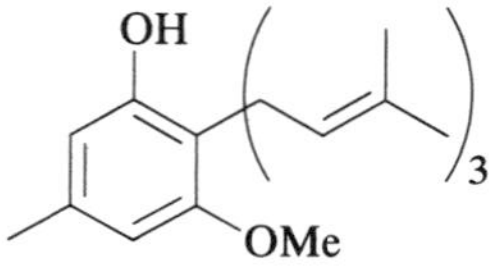 OH, OMe, 3	(65)	131

TABLE XVI. [2+2] CYCLOADDITION OF KETENES TO YNAMINES

Reactant	Ketene or Ketene Source	Conditions	Product(s) and Yield(s) (%)	Refs.
C_7				
$MeC{\equiv}CNEt_2$	$CH_2{=}C{=}O$	Et_2O, -50°	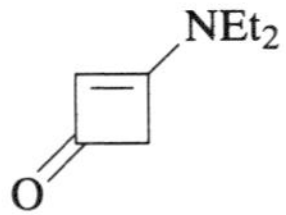 3-(diethylamino)cyclobut-2-enone (7)	139
	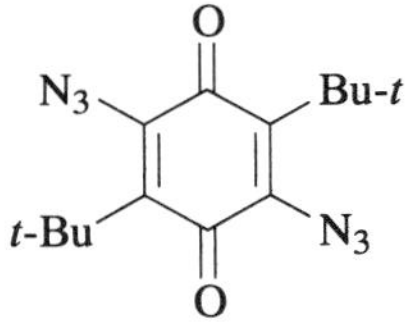 2,5-diazido-3,6-di-*t*-butyl-1,4-benzoquinone	C_6H_6, 80°	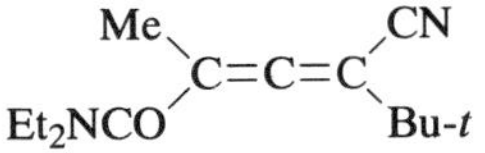 $Et_2NCO(Me)C{=}C{=}C(CN)Bu\text{-}t$ (53)	142
	$R^1R^2CHCOCl$	Et_3N	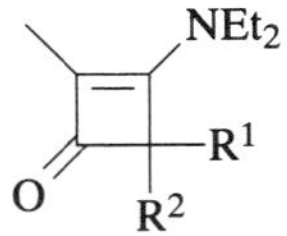 I 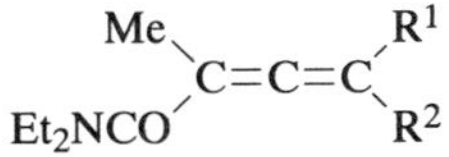 (2-methyl-3-diethylamino-4-R^1-4-R^2-cyclobut-2-enone) + II $Et_2NCO(Me)C{=}C{=}CR^1R^2$	
	R^1 / R^2		I / II	
	Me / CO_2Et		(3) / (66)	278
	Me / CO_2NEt_2		(3) / (95)	278
	Ph / CO_2Et		(32) / (32)	278
	Ph / H		(37) / (9)	140, 677
	$p\text{-}ClC_6H_4COCHN_2$	$h\nu$, C_6H_6	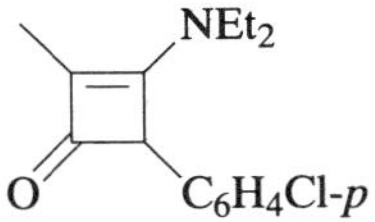 2-methyl-3-diethylamino-4-($C_6H_4Cl\text{-}p$)cyclobut-2-enone (15) 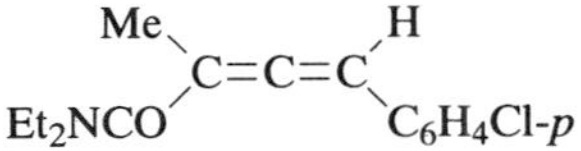 + $Et_2NCO(Me)C{=}C{=}C(H)C_6H_4Cl\text{-}p$ (25)	140

TABLE XVI. [2+2] CYCLOADDITION OF KETENES TO YNAMINES (*Continued*)

Reactant	Ketene or Ketene Source	Conditions	Product(s) and Yield(s) (%)	Refs.
	p-MeOC$_6$H$_4$COCHN$_2$	$h\nu$, C_6H_6	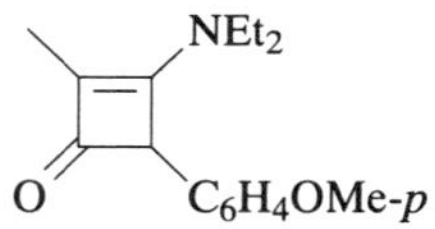 NEt$_2$; O; C$_6$H$_4$OMe-p (42)	140
	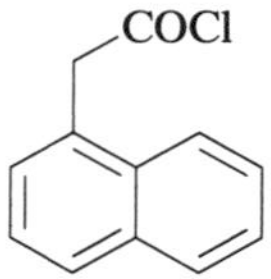 COCl	Et_3N, Et_2O, rt	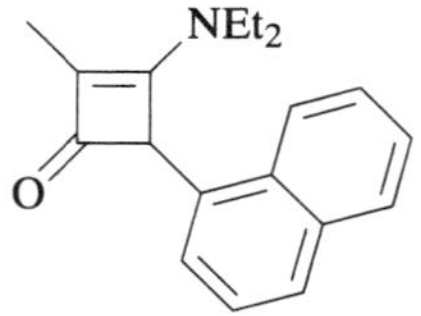 NEt$_2$; O (—)	677
	Ph, Ph C=C=O	C_6H_6, 15°	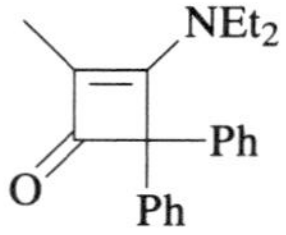 NEt$_2$; Ph; Ph; O **I** (16) 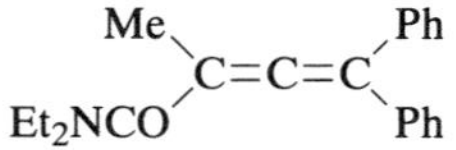 + Me, Et$_2$NCO C=C=C Ph, Ph **II** (57)	140
	Ph, Ph C=C=O	C_6H_6	**I** (11) 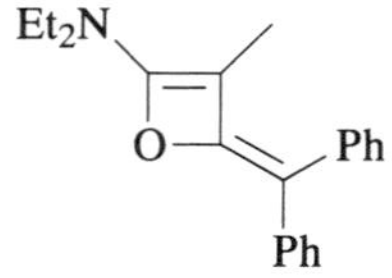 + Et$_2$N; O; Ph; Ph (27)	678
	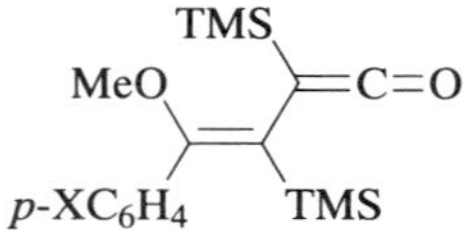 TMS; MeO; C=O; p-XC$_6$H$_4$; TMS	C_6H_{14}, 25°, 3 h	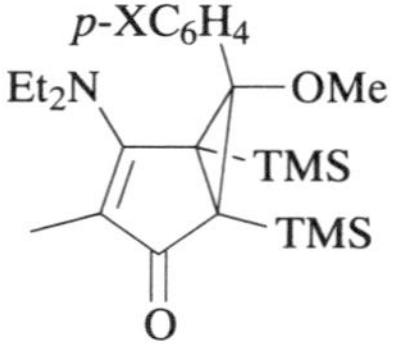 p-XC$_6$H$_4$; Et$_2$N; OMe; TMS; TMS; O — X = H (—)	143
			X = Me (—)	143
			X = CF$_3$ (38)	143, 144

TABLE XVI. [2+2] CYCLOADDITION OF KETENES TO YNAMINES (*Continued*)

Reactant	Ketene or Ketene Source	Conditions	Product(s) and Yield(s) (%)	Refs.
C_8				
$MeCOC{\equiv}CNEt_2$	$Ph_2C{=}C{=}O$	C_6H_6, 15°	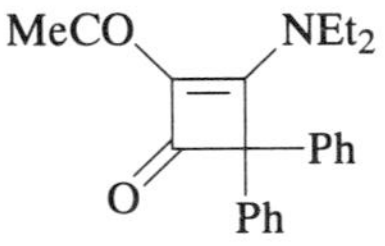 4-acetyl-3-(diethylamino)-2,2-diphenylcyclobut-3-enone (7) 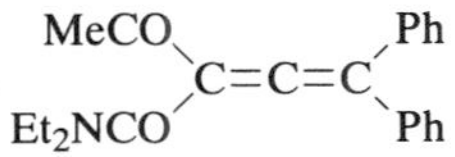 + $(MeCO)(Et_2NCO)C{=}C{=}CPh_2$ (11)	140
$MeO_2CC{\equiv}CNEt_2$	$Ph_2C{=}C{=}O$	C_6H_6	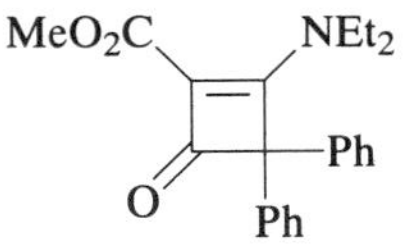 4-(methoxycarbonyl)-3-(diethylamino)-2,2-diphenylcyclobut-3-enone (15) 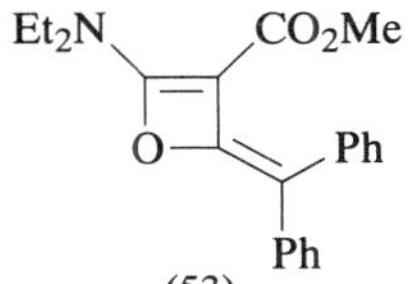 + 2-(diethylamino)-3-(methoxycarbonyl)-4-(diphenylmethylene)-oxete (53)	678
C_9				
$HC{\equiv}CN(Me)Ph$	$CH_2{=}C{=}O$	MeCN	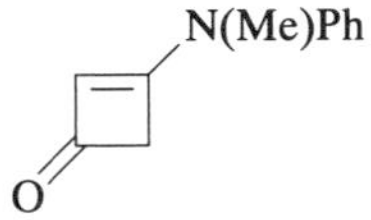 3-[N(Me)Ph]cyclobut-2-enone (7) + $CH_2{=}C{=}CHCON(Me)Ph$ (34)	679
C_{10}				
$MeC{\equiv}CN(Me)Ph$	$CH_2{=}C{=}O$	MeCN	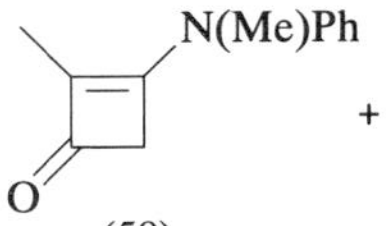 2-methyl-3-[N(Me)Ph]cyclobut-2-enone (50) 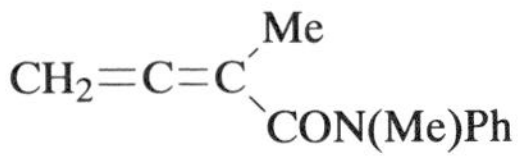 + $CH_2{=}C{=}C(Me)CON(Me)Ph$ (12)	679
	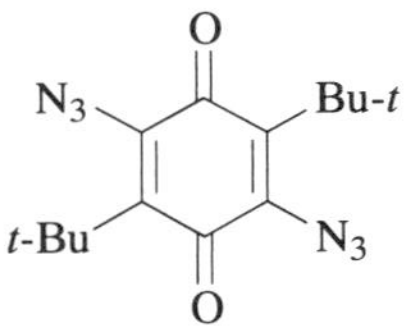 2,5-diazido-3,6-di-t-butyl-1,4-benzoquinone	C_6H_6, 80°	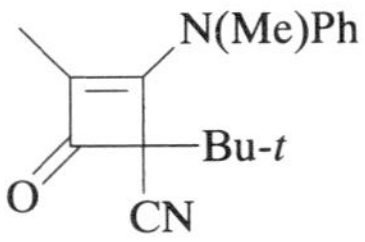 2-methyl-3-[N(Me)Ph]-4-Bu-t-4-CN-cyclobut-2-enone (23)	142

TABLE XVI. [2+2] CYCLOADDITION OF KETENES TO YNAMINES (*Continued*)

Reactant	Ketene or Ketene Source	Conditions	Product(s) and Yield(s) (%)	Refs.
n-$C_6H_{13}C{\equiv}CNMe_2$	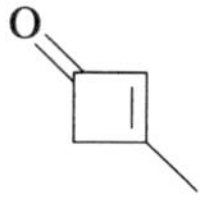 O,	C_6H_6, 150°, 2 h	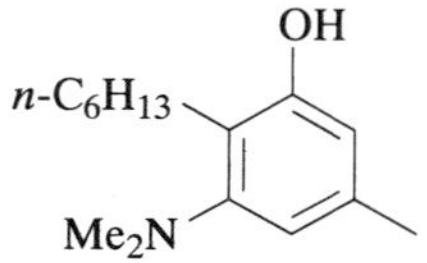 OH, n-C_6H_{13}, Me_2N (74)	131
	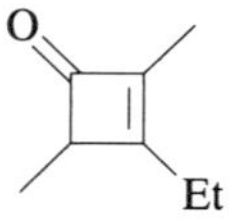 O, Et	C_6H_6, 160°, 3 h	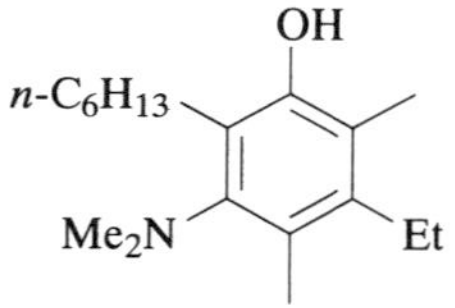 OH, n-C_6H_{13}, Me_2N, Et (83)	131
$Et_2NC{\equiv}CNEt_2$	$Ph_2C{=}C{=}O$	Et_2O, -50°	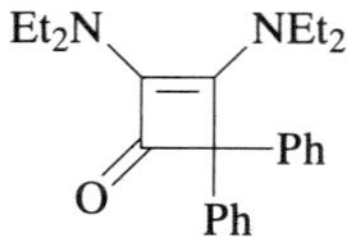 Et_2N, NEt_2, O, Ph, Ph (94)	139
C_{12}				
$PhC{\equiv}CNEt_2$	$CH_2{=}C{=}O$	Et_2O, -50°	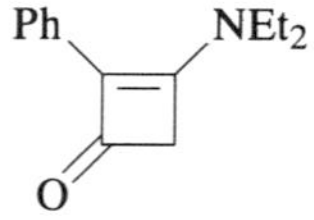 Ph, NEt_2, O (78)	139
	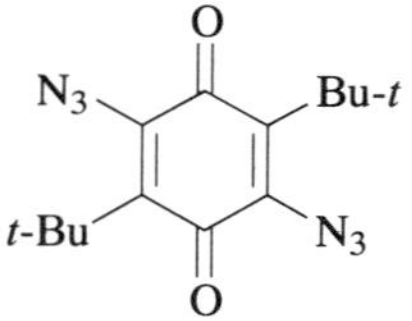 O, N_3, Bu-t, t-Bu, N_3, O	C_6H_6, 80°	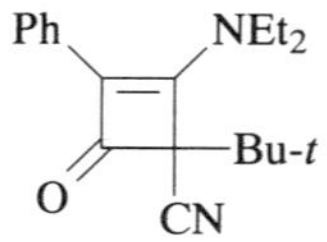 Ph, NEt_2, O, Bu-t, CN (45) 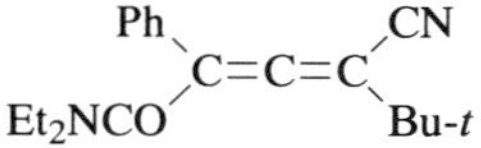 + Ph(CN)... $Et_2NCO(Ph)C{=}C{=}C(CN)Bu$-$t$ (5)	142

TABLE XVI. [2+2] CYCLOADDITION OF KETENES TO YNAMINES (*Continued*)

Reactant	Ketene or Ketene Source	Conditions	Product(s) and Yield(s) (%)	Refs.
	$R^1R^2CHCOCl$	Et_3N, Et_2O, rt	2-Ph-3-NEt_2-4-R^1-4-R^2-cyclobut-2-enone	
	R^1 / R^2			
	H / H		(—)	677
	H / Me		(—)	677
	Me / CO_2Et		(14)	278
	Me / $CONEt_2$		(21)	278
	H / Ph		(—)	677
	Ph / CO_2Et		(65)	278
	Ph / Et		(57)	677
	H / $C_{10}H_7$-1		(—)	677
	$Me_2C{=}C{=}O$	THF, 0°	2-Ph-3-NEt_2-4,4-dimethylcyclobut-2-enone (17)	678
	$Ph_2C{=}C{=}O$	Et_2O, -50°	2-Ph-3-NEt_2-4,4-diphenylcyclobut-2-enone **I** (95)	139
	$Ph_2C{=}C{=}O$	C_6H_6, 15°	**I** (54) + $Et_2NCO(Ph)C{=}C{=}CPh_2$ (14)	140

TABLE XVI. [2+2] CYCLOADDITION OF KETENES TO YNAMINES (*Continued*)

Reactant	Ketene or Ketene Source	Conditions	Product(s) and Yield(s) (%)	Refs.
C_{13}				
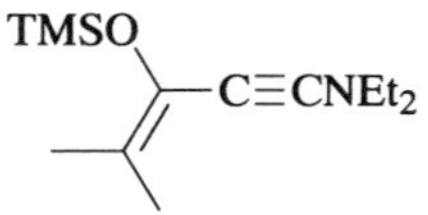 TMSO, C≡CNEt$_2$ (structure)	Me, Me C=C=O	Et_2O	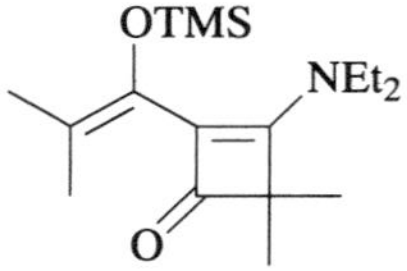 OTMS, NEt$_2$, O (structure) (10)	141
C_{14}				
HC≡CNPh$_2$	CH_2=C=O	MeCN	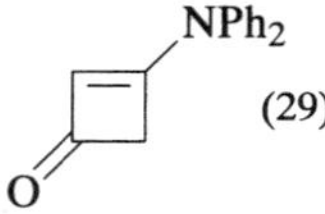 NPh$_2$, O (structure) (29) + CH_2=C=C(CONPh$_2$)H (36)	679
C_{15}				
PhC≡CN(Me)Ph	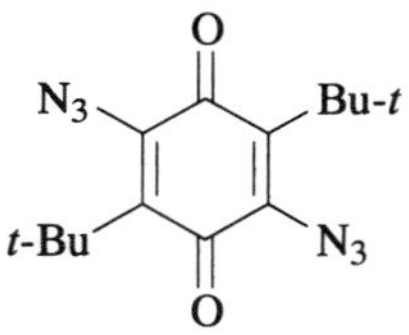 O, O, N$_3$, N$_3$, Bu-*t*, *t*-Bu (structure)	C_6H_6, 80°	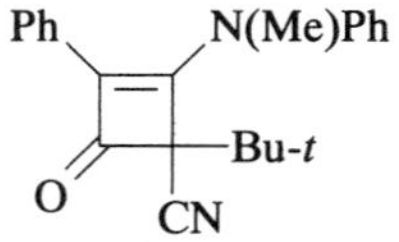 Ph, N(Me)Ph, Bu-*t*, CN, O (structure) (15)	142
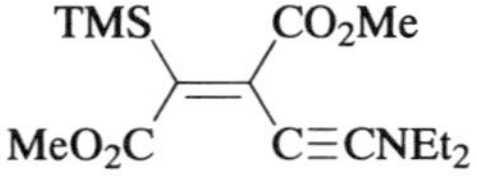 TMS, CO$_2$Me, MeO$_2$C, C≡CNEt$_2$ (structure)	Ph, Ph C=C=O	Et_2O	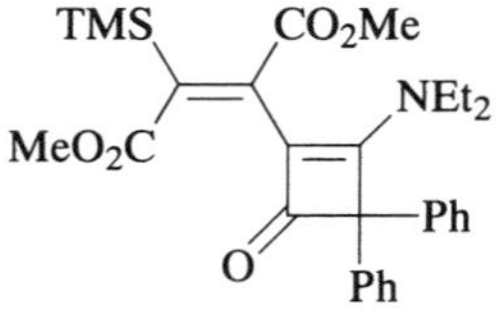 TMS, CO$_2$Me, MeO$_2$C, NEt$_2$, Ph, Ph, O (structure) (—)	680
C_{16}				
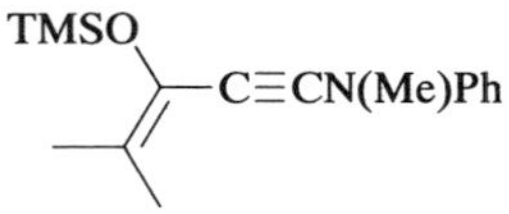 TMSO, C≡CN(Me)Ph (structure)	Me, Me C=C=O	Et_2O	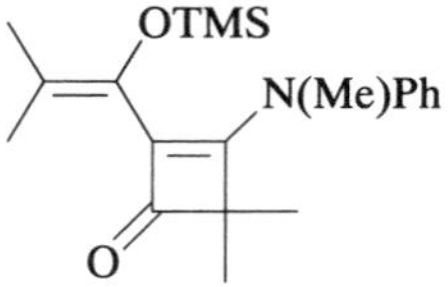 OTMS, N(Me)Ph, O (structure) (38)	141

TABLE XVI. [2+2] CYCLOADDITION OF KETENES TO YNAMINES (*Continued*)

Reactant	Ketene or Ketene Source	Conditions	Product(s) and Yield(s) (%)	Refs.
C_{19}				
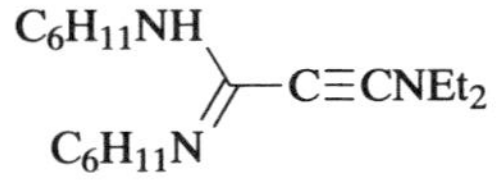 $C_6H_{11}NH$, $C_6H_{11}N$ (amidine) —$C{\equiv}CNEt_2$	$Ph_2C{=}C{=}O$	Et_2O	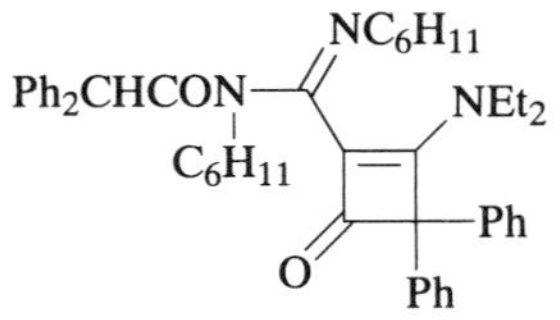 $Ph_2CHCON(C_6H_{11})$—$C({=}NC_6H_{11})$-cyclobutenone (NEt_2, Ph, Ph) (36)	681
$C_{23\text{-}41}$				
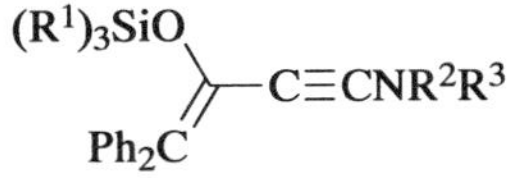 $(R^1)_3SiO(Ph_2C{=})C$—$C{\equiv}CNR^2R^3$	$R^4R^5C{=}C{=}O$	Et_2O	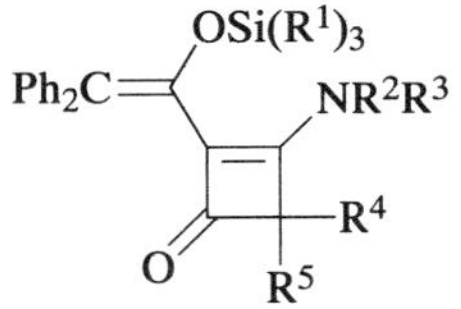 $Ph_2C{=}C(OSi(R^1)_3)$-cyclobutenone (NR^2R^3, R^4, R^5)	141
R^1 R^2 R^3	R^4 R^5			
Me Et Et	Me Me		(60)	
Me Et Et	Me Ph		(37)	
Me Et Et	Ph Ph		(61)	
Me Me Ph	Ph Ph		(68)	
Ph Me Ph	Ph Ph		(43)	
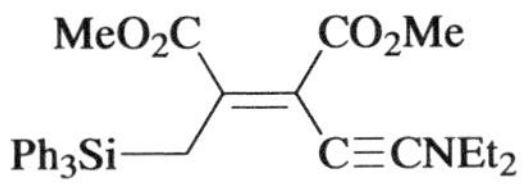 $Ph_3SiCH_2(MeO_2C)C{=}C(CO_2Me)C{\equiv}CNEt_2$	$R_2C{=}C{=}O$	—	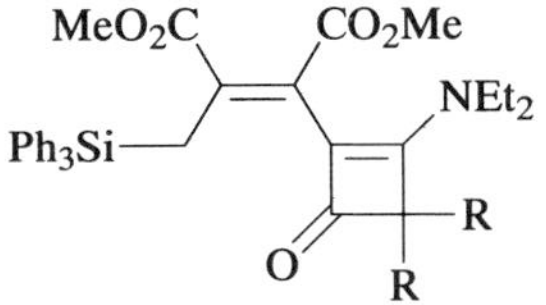 $Ph_3SiCH_2(MeO_2C)C{=}C(CO_2Me)$-cyclobutenone ($NEt_2$, R, R)	682
	R			
	H		(—)	
	Me		(—)	
	Ph		(—)	

TABLE XVII. CYCLOADDITION OF KETENES TO ORGANOMETALLIC ACETYLENES

Reactant	Ketene or Ketene Source	Conditions	Product(s) and Yield(s) (%)	Refs.
C_5				
HC≡CTMS	Cl_3CCOCl	Zn, Et_2O	4,4-dichloro-3-TMS-cyclobutenone (TMS, Cl, Cl, O) (66)	120
	"	Zn, $POCl_3$, Et_2O	" (60)	113
C_6				
MeC≡CTMS	"	Zn, Et_2O	2-Me-3-TMS-4,4-dichlorocyclobutenone (TMS, Cl, Cl, O) **I** (27) + 2-TMS-3-Me-4,4-dichlorocyclobutenone (TMS, Cl, Cl, O) **II** (47)	120
	"	Zn, $POCl_3$, Et_2O	**I** (30) + **II** (30)	113
C_7				
EtOC≡CTMS	"	Zn, $POCl_3$, Et_2O	2-TMS-3-OEt-4,4-dichlorocyclobutenone (TMS, OEt, Cl, Cl, O) (86)	120
	$Cl_2CHCOCl$	Et_3N, Et_2O	" (—)	121
C_9				
$TMSC≡CNEt_2$	$CH_2{=}C{=}O$	MeCN	$CH_2{=}C{=}C(CONEt_2)(TMS)$ (17)	683
C_{10}				
$Et_3SiC≡COEt$	$Cl_2CHCOCl$	Et_3N, Et_2O	2-Et_3Si-3-OEt-4,4-dichlorocyclobutenone (Et_3Si, OEt, Cl, Cl, O) (—)	121

TABLE XVII. CYCLOADDITION OF KETENES TO ORGANOMETALLIC ACETYLENES (*Continued*)

Reactant	Ketene or Ketene Source	Conditions	Product(s) and Yield(s) (%)	Refs.
$Et_3GeC\equiv COEt$	$Cl_2CHCOCl$	Et_3N, Et_2O	2-Et_3Ge-3-OEt-4,4-dichlorocyclobut-2-enone (—)	121
C_{11}				
$PhC\equiv CTMS$	Cl_3CCOCl	Zn, Et_2O	2-TMS-3-Ph-4,4-dichlorocyclobut-2-enone (93)	120
C_{12}				
$TMSC\equiv CN(Me)Ph$	$CH_2=C=O$	MeCN	3-N(Me)Ph-cyclobut-2-enone (15) + $CH_2=C=C(CON(Me)Ph)(TMS)$ (23)	683
C_{17}				
$TMSC\equiv CNPh_2$	$CH_2=C=O$	MeCN	3-NPh_2-cyclobut-2-enone (17) + $CH_2=C=C(CONPh_2)(TMS)$ (39)	683
C_{18}				
$R^1C\equiv CNEt_2$	$R^2R^3C=C=O$	—	2-R^1-3-NEt_2-4-R^2-4-R^3-cyclobut-2-enone **I** + $R^1(Et_2NCO)C=C=CR^2R^3$ **II**	140

R^1	R^2	R^3	I	II
Ph_2P	Ph	Ph	(10)	(0)
$Ph_2P(O)$	H	Ph	(35)	(12)
$Ph_2P(O)$	Ph	Ph	(12)	(0)
$Ph_2P(S)$	Ph	Ph	(49)	(0)
$Ph_2P(NTs)$	Ph	Ph	(6)	(0)

TABLE XVII. CYCLOADDITION OF KETENES TO ORGANOMETALLIC ACETYLENES (*Continued*)

Reactant	Ketene or Ketene Source	Conditions	Product(s) and Yield(s) (%)	Refs.
C_{26}				
$Ph_2AsC\equiv CNPh_2$	$CH_2=C=O$	MeCN	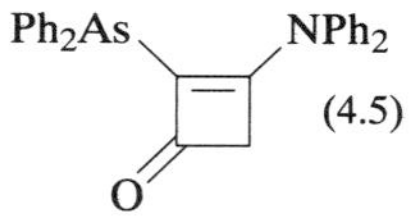 3-(Ph_2As)-2-(NPh_2)cyclobut-2-enone (4.5) + $CH_2=C=C(AsPh_2)(NPh_2)$ (33)	679
C_{27}				
$Ph_3SiC\equiv CN(Me)Ph$	$CH_2=C=O$	MeCN	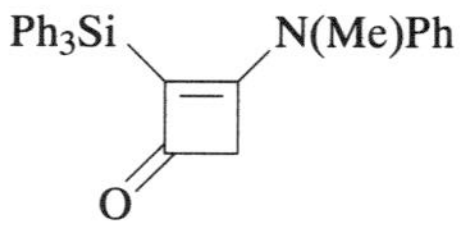 3-(Ph_3Si)-2-(N(Me)Ph)cyclobut-2-enone (6) + $CH_2=C=C(SiPh_3)(CON(Me)Ph)$ (7)	683
	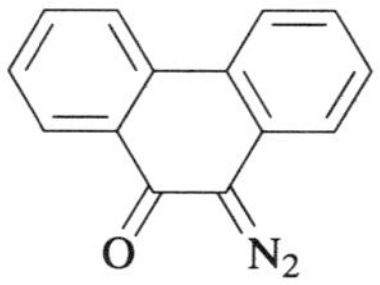 10-diazo-9(10H)-phenanthrenone	$h\nu$, C_6H_6	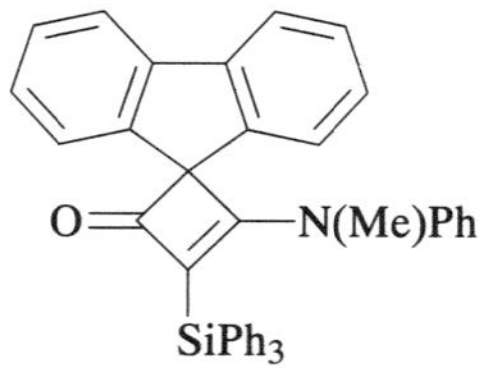 spiro[cyclobutenone-fluorene] with N(Me)Ph and $SiPh_3$ substituents, C=O (19)	141
$Ph_3GeC\equiv CN(Me)Ph$	$CH_2=C=O$	MeCN	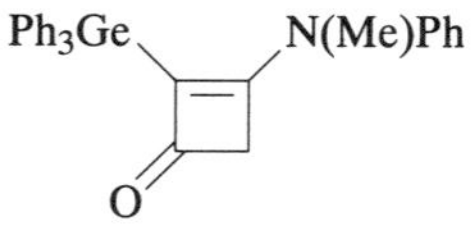 3-(Ph_3Ge)-2-(N(Me)Ph)cyclobut-2-enone (8) + $CH_2=C=C(GePh_3)(CON(Me)Ph)$ (35)	679
C_{32}				
$Ph_3SiC\equiv CNPh_2$	$CH_2=C=O$	MeCN	$CH_2=C=C(SiPh_3)(NPh_2)$ (41)	683

TABLE XVII. CYCLOADDITION OF KETENES TO ORGANOMETALLIC ACETYLENES (*Continued*)

Reactant	Ketene or Ketene Source	Conditions	Product(s) and Yield(s) (%)	Refs.
	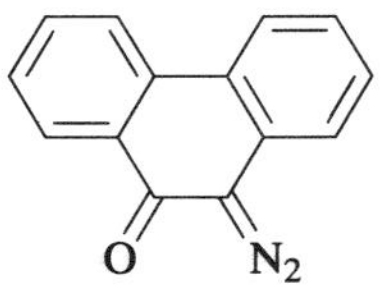 O, N_2	$h\nu$, C_6H_6	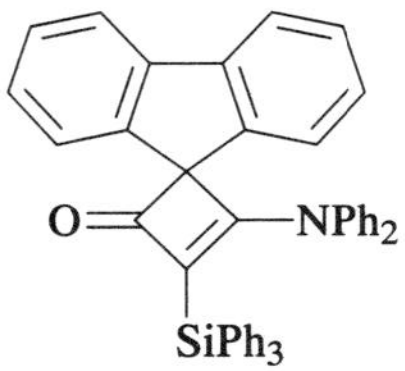 O, NPh_2, $SiPh_3$ (12)	141
$R^1C{\equiv}CNi(PPh_3)(\eta^5\text{-}C_5H_5)$	$R^2R^3C{=}C{=}O$	C_6H_6	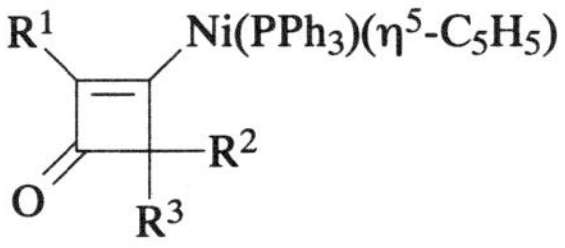 R^1, $Ni(PPh_3)(\eta^5\text{-}C_5H_5)$, O, R^2, R^3	122

R^1	R^2	R^3	Yield (%)
H	H	H	(27)
H	Me	Ph	(72)
H	Ph	Ph	(32)
Me	Ph	Ph	(82)
$CH{=}CH_2$	Ph	Ph	(70)
Ph	H	H	(84)
Ph	Me	H	(68)
Ph	Ph	H	(73)
Ph	Ph	Ph	(78)
C_6H_4Br-*p*	Ph	Ph	(62)
C_6H_4Me-*p*	Ph	Ph	(80)

TABLE XVIII. [2+2] CYCLOADDITION OF KETENES TO ALDEHYDES

Reactant	Ketene or Ketene Source	Conditions	Product(s) and Yield(s) (%)	Refs.
C_1				
HCHO	$CH_2{=}C{=}O$	$AlCl_3$, $ZnCl_2$, EtOAc, 10°	β-propiolactone (—)	684
	$CH_2{=}C{=}O$	$AlCl_3$, Me_2CO, β-propiolactone	" (93)	685, 686
	$(CF_3)_2C{=}C{=}O$	$ZnCl_2$, Et_2O, 150°	3,3-bis(CF_3)-oxetan-2-one (—)	148
	$Me_2C{=}C{=}O$	$AlCl_3$, I_2, glyme, 50°	3,3-dimethyloxetan-2-one (89)	687
	$Me_2C{=}C{=}O$	$HgCl_2$, *i*-PrOAc, 50-55°	" (93)	149
	TMS(H)C=C=O	BF_3•Et_2O, 20°	3-TMS-oxetan-2-one (30)	147
	$R^1R^2C{=}C{=}O$	$ZnCl_2$, *i*-PrOAc, 50°	3-R^1-3-R^2-oxetan-2-one	
	R^1 = Me, R^2 = Et		(72)	149
	R^1 = Et, R^2 = Et		(—)	149
	R^1 = Me, R^2 = *n*-Pr		(—)	149
	R^1 = Et, R^2 = *n*-Bu		(49)	149
	R^1 = *t*-Bu, R^2 = *t*-Bu		(68)	688

TABLE XVIII. [2+2] CYCLOADDITION OF KETENES TO ALDEHYDES (*Continued*)

Reactant	Ketene or Ketene Source	Conditions	Product(s) and Yield(s) (%)	Refs.
C_2				
CF_3CHO	MeCOCl	Et_3N, Et_2O, 0-10°	O, O, CF_3 (20)	168
Cl_3CCHO	$Cl_2CHCOCl$	Et_3N, Et_2O or C_6H_{14}, 0-25°	O, O, CCl_3, Cl, Cl (39-43)	168
	$Cl_2CHCOCl$	(–)-Brucine, CCl_4, -20°	" (–)-isomer (72)	169
	$ClCH_2COCl$	Et_3N, Et_2O or C_6H_{14}	O, O, CCl_3, Cl (40) *cis*:*trans* = 1.60-1.64	150, 166
	MeCOCl	Et_3N, Et_2O	O, O, CCl_3 (69)	168
	MeCOCl	(–)-PhCH(NMe_2)Me, Et_2O, 0°	" (–)-isomer (72)	169
	$CH_2{=}C{=}O$	rt	" (70-92)	175, 689
	$CH_2{=}C{=}O$	$BF_3{\bullet}Et_2O$, CH_2Cl_2, -60°	" (70-92)	690
	$CH_2{=}C{=}O$	(–)-Brucine, $CHCl_3$, -25°	" (+)-isomer (—)	169
	$CH_2{=}C{=}O$	Quinidine, PhMe, -50°	" (89-95) *R*, 98% ee	170, 171, 174

TABLE XVIII. [2+2] CYCLOADDITION OF KETENES TO ALDEHYDES (*Continued*)

Reactant	Ketene or Ketene Source	Conditions	Product(s) and Yield(s) (%)	Refs.
	MeCHBrCOCl	Et_3N, C_6H_{14}, rt	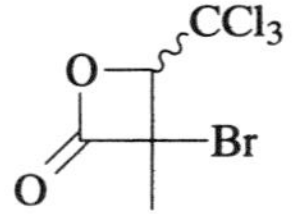 CCl_3, Br (60)	151
	MeCHClCOCl	Et_3N, C_6H_{14}, rt	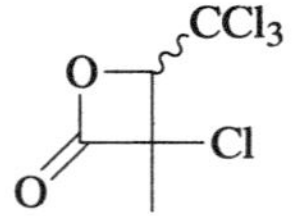 CCl_3, Cl (53)	151
	EtCOCl	Et_3N, C_6H_{14}, rt	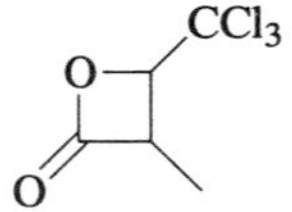 CCl_3 (43) *cis*:*trans* = 1.39	150
	i-PrCOCl	Et_3N, Et_2O	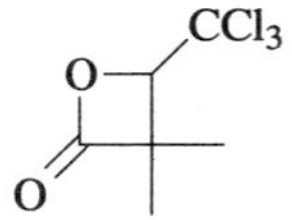 CCl_3 (15-65)	166, 689
	$Me_2C{=}C{=}O$	$BF_3{\bullet}Et_2O$, -15°	" (60)	689
	i-PrCHBrCOCl	Zn, Et_2O	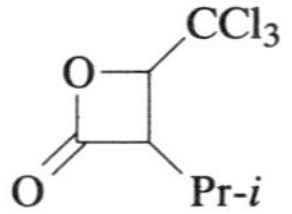 CCl_3, Pr-*i* (25) *cis*:*trans* = 0.9	166
	$ClCH_2SiMe_2(H)C{=}C{=}O$	$BF_3{\bullet}Et_2O$, -50°	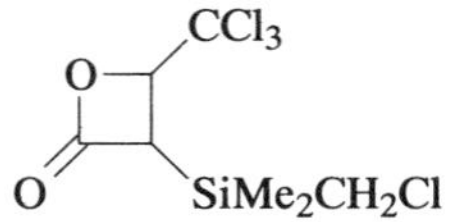 CCl_3, $SiMe_2CH_2Cl$ (51) *cis*:*trans* = 0.25	147

TABLE XVIII. [2+2] CYCLOADDITION OF KETENES TO ALDEHYDES (*Continued*)

Reactant	Ketene or Ketene Source	Conditions	Product(s) and Yield(s) (%)	Refs.
	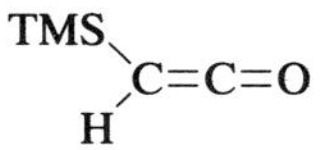 TMS(H)C=C=O	$BF_3 \cdot Et_2O$, 90°	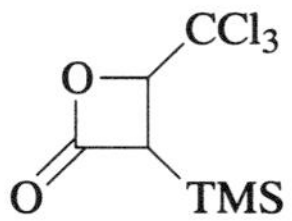 β-lactone (4-CCl₃, 3-TMS) (69) *cis*:*trans* = 0.25	147
	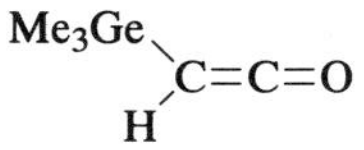 Me_3Ge(H)C=C=O	$BF_3 \cdot Et_2O$, 50°	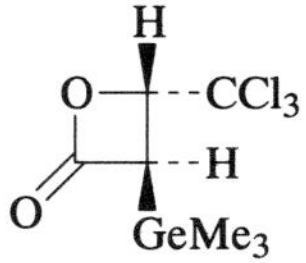 β-lactone (4-CCl₃, 3-GeMe₃) (67)	147
	2,4-$Cl_2C_6H_3OCH_2COCl$	Et_3N, Et_2O, 0-10°	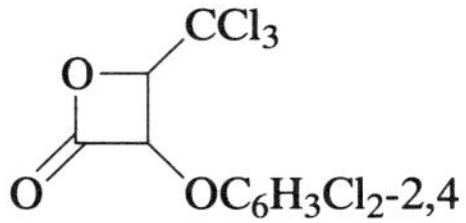 β-lactone (4-CCl₃, 3-$OC_6H_3Cl_2$-2,4) (63)	147
	o-$ClC_6H_4OCH_2COCl$	(–)-Brucine, $CHCl_3$, -25°	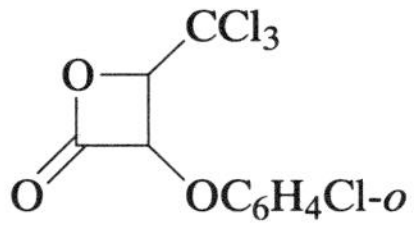 β-lactone (4-CCl₃, 3-OC_6H_4Cl-*o*) (45) (–)-isomer	168, 169
	o-$ClC_6H_4OCH_2COCl$	Et_3N, Et_2O, 0-10°	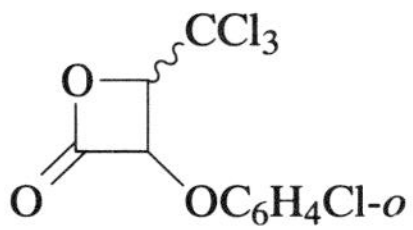 β-lactone (4-CCl₃ wavy, 3-OC_6H_4Cl-*o*) (45)	168
	$C_6H_5OCH_2COCl$	Et_3N, Et_2O, 0-10°	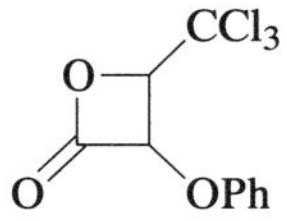 β-lactone (4-CCl₃, 3-OPh) (36)	168
	$C_6H_5OCH_2COCl$	Et_3N, Et_2O, 0-10°	" (61) *cis*:*trans* = 1.7	691

TABLE XVIII. [2+2] CYCLOADDITION OF KETENES TO ALDEHYDES (*Continued*)

Reactant	Ketene or Ketene Source	Conditions	Product(s) and Yield(s) (%)	Refs.
	$C_6H_{11}COCl$	Et_3N, C_6H_6	spiro β-lactone with CCl_3, cycloheptane ring (—)	527
	$Et_3Si(H)C{=}C{=}O$	$BF_3{\bullet}Et_2O$, 90°	β-lactone: H, CCl_3; H, $SiEt_3$ (45)	147
Br_3CCHO	$Cl_2CHCOCl$	Et_3N, Et_2O, 0-10°	β-lactone: CBr_3; Cl, Cl (11)	168
	$ClCH_2SiMe_2(H)C{=}C{=}O$	$BF_3{\bullet}Et_2O$, -10°	β-lactone: CBr_3, H; $SiMe_2CH_2Cl$, H (64)	147
	$TMS(H)C{=}C{=}O$	$BF_3{\bullet}Et_2O$, -10°	β-lactone: CBr_3, H; TMS, H (63)	147
Cl_2CHCHO	$CH_2{=}C{=}O$	Quinidine, PhMe, -25°	β-lactone: $CHCl_2$, H (67)	171

TABLE XVIII. [2+2] CYCLOADDITION OF KETENES TO ALDEHYDES (*Continued*)

Reactant	Ketene or Ketene Source	Conditions	Product(s) and Yield(s) (%)	Refs.
Br_2CHCHO	$CH_2{=}C{=}O$	$BF_3 \cdot Et_2O$, CH_2Cl_2, -70°	(54)	690
MeCHO	$Cl_2CHCOCl$	Et_3N, Et_2O, 10°	(51)	324, 692
	$Cl_2CHCOCl$	Et_3N, C_6H_{14}, 0°	" (—)	693
	$CH_2{=}C{=}O$	$BF_3 \cdot Et_2O$, Et_2O, -20°	(80)	158
	$CH_2{=}C{=}O$	SiO_2/Al_2O_3, 10-15°	" (90)	694, 695
	TMS(H)C=C=O	$BF_3 \cdot Et_2O$, 20°	(64) *cis*:*trans* = 1.5	147
	Me_3Ge(H)C=C=O	$BF_3 \cdot Et_2O$, -78°	(75)	147
C_3				
HC≡CCHO	$CH_2{=}C{=}O$	$BF_3 \cdot Et_2O$, Et_2O, -20°	(40)	158

TABLE XVIII. [2+2] CYCLOADDITION OF KETENES TO ALDEHYDES (*Continued*)

Reactant	Ketene or Ketene Source	Conditions	Product(s) and Yield(s) (%)	Refs.
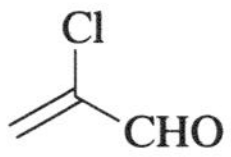 Cl, CHO	$CH_2{=}C{=}O$	BF_3•Et_2O, Et_2O, -20°	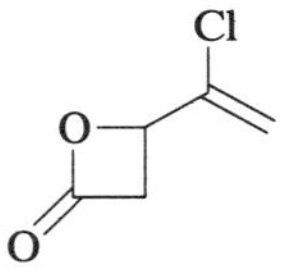 O, O, Cl (45)	158
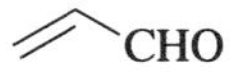 CHO	$CH_2{=}C{=}O$	BF_3•Et_2O, Et_2O, -20°	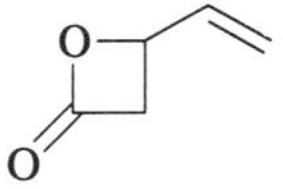 O, O (87)	696
	$CH_2{=}C{=}O$	BF_3•Et_2O, PhMe, -25°	" (72)	697
 O, CHO	$CH_2{=}C{=}O$	$ZnCl_2$, Et_2O, 0-10°	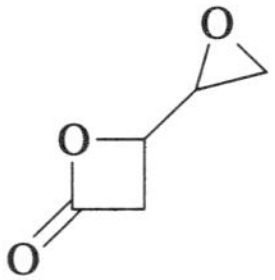 O, O, O (42)	159, 160
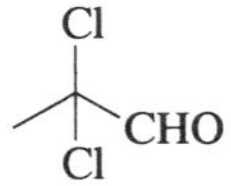 Cl, Cl, CHO	MeCOCl	Et_3N, Et_2O, 0°	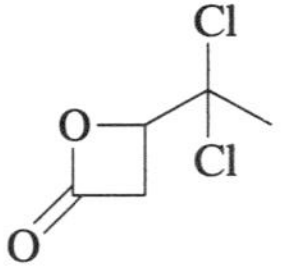 O, O, Cl, Cl (5)	175
	$CH_2{=}C{=}O$	Quinidine, PhMe, -25°	" (95) *R*, 91% ee	171, 175
	$CH_2{=}C{=}O$	Brucine, PhMe, -25°	" (15)	175
EtCHO	$CH_2{=}C{=}O$	BF_3•Et_2O, Et_2O, -20°	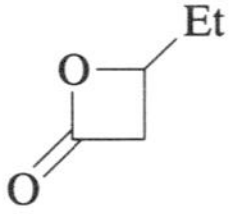 O, O, Et (75)	158

TABLE XVIII. [2+2] CYCLOADDITION OF KETENES TO ALDEHYDES (*Continued*)

Reactant	Ketene or Ketene Source	Conditions	Product(s) and Yield(s) (%)	Refs.
	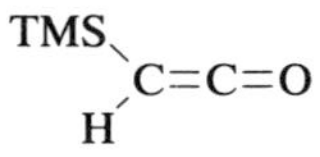 TMS(H)C=C=O	BF_3•Et_2O	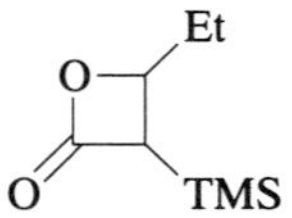 O, Et, TMS (60)	698
	"	BF_3•Et_2O	" (68) *cis*:*trans* = 1.5	147
C_4  CHO	CH_2=C=O	ZnO_2CPr, PhMe	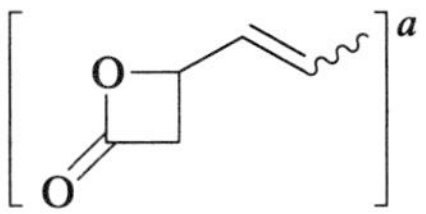 [O, O][a] (—)	164
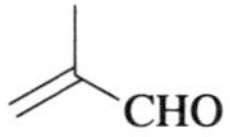 CHO	CH_2=C=O	BF_3•Et_2O, Et_2O, -20°	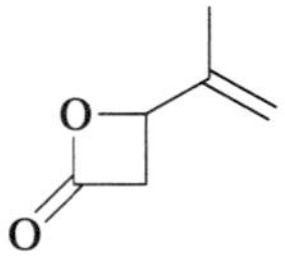 O, O (70)	158
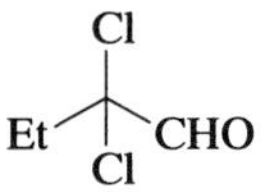 Cl, Et, CHO, Cl	CH_2=C=O	BF_3•Et_2O, Et_2O, 10°	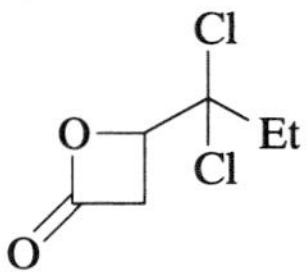 O, O, Cl, Et, Cl (49)	176
	CH_2=C=O	Quinidine, PhMe, -25°	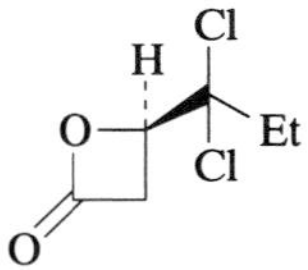 O, O, H, Cl, Et, Cl (72-95) 89-100% ee	171, 173, 176

TABLE XVIII. [2+2] CYCLOADDITION OF KETENES TO ALDEHYDES (*Continued*)

Reactant	Ketene or Ketene Source	Conditions	Product(s) and Yield(s) (%)	Refs.
EtCH(Cl)CHO	$CH_2=C=O$	$BF_3 \cdot Et_2O$, CH_2Cl_2, -60°	4-(1-chloropropyl)oxetan-2-one (60)	690
n-PrCHO	TMS(H)C=C=O	$BF_3 \cdot Et_2O$	4-(Pr-*n*)-3-TMS-oxetan-2-one (54)	698
	"	$BF_3 \cdot Et_2O$, Et_2O, 20°	" (62) *cis:trans* = 1.5	147
i-PrCHO	$Cl_2CHCOCl$	Et_3N, Et_2O, 10-20°	4-(Pr-*i*)-3,3-dichloro-oxetan-2-one (40)	324, 692
	TMS(H)C=C=O	$BF_3 \cdot Et_2O$	4-(Pr-*i*)-3-TMS-oxetan-2-one (56)	698
	"	"	" (90) *cis:trans* = 1.5	147
C_5				
n-PrCCl_2CHO	$CH_2=C=O$	Quinidine, PhMe, -25°	4-(1,1-dichlorobutyl)oxetan-2-one (H, wedge) (66) 84% ee	176

TABLE XVIII. [2+2] CYCLOADDITION OF KETENES TO ALDEHYDES (*Continued*)

Reactant	Ketene or Ketene Source	Conditions	Product(s) and Yield(s) (%)	Refs.
i-BuCHO	TMS(H)C=C=O	BF_3•Et_2O, Et_2O, 20°	[β-lactone: Bu-*i*, TMS] (86) *cis*:*trans* = 1.5	147
C_6				
n-BuCCl_2CHO	CH_2=C=O	Quinidine, PhMe, -25°	[β-lactone: H, CCl_2Bu-*n*] (90-95) >98% ee	173
EtS-branched CHO [structure]	CH_2=C=O	BF_3•Et_2O, Et_2O, 20°	[β-lactone: SEt side chain] (72)	323
C_7				
2,4,5-$Cl_3C_6H_2CHO$	CH_2=C=O	BF_3•Et_2O, PhMe, 0°	[β-lactone: $C_6H_2Cl_3$-2,4,5] (65)	699
o-ClC_6H_4CHO	Br-CH(CH_3)COCl	Et_3N, C_6H_{14}, rt	[β-lactone: C_6H_4Cl-*o*, Br, Me] (50)	151
p-$O_2NC_6H_4CHO$	CH_2=C=O	BF_3•Et_2O, PhMe, 0°	[β-lactone: $C_6H_4NO_2$-*p*] (79)	699

TABLE XVIII. [2+2] CYCLOADDITION OF KETENES TO ALDEHYDES (*Continued*)

Reactant	Ketene or Ketene Source	Conditions	Product(s) and Yield(s) (%)	Refs.
XC_6H_4CHO	$Cl_2CHCOCl$	Et_3N, $ZnCl_2$, Et_2O, 3-25°	[structure: 2-oxetanone, 4-C_6H_4X, 3,3-Cl,Cl]	
X				
H			(55)	161
o-Cl			(53)	161
m-Cl			(59)	161
p-Cl			(59-66)	161, 324, 692
o-NO_2			(85)	161
m-NO_2			(71)	161
p-NO_2			(65)	161
p-CN			(82)	161
o-Me			(37)	161
m-Me			(53)	161
p-Me			(44)	161
m-MeO			(38)	161
p-MeO			(23)	161
m-AcO			(44)	161
p-AcO			(29)	161
PhCHO	$Cl_2CHCOCl$	Et_3N, Et_2O, 10°	[structure: 2-oxetanone, 4-Ph, 3,3-Cl,Cl] (30)	324, 692
	$CH_2{=}C{=}O$	$BF_3•Et_2O$, PhMe, 4 h	[structure: 2-oxetanone, 4-Ph] (95-96)	700, 699

TABLE XVIII. [2+2] CYCLOADDITION OF KETENES TO ALDEHYDES (*Continued*)

Reactant	Ketene or Ketene Source	Conditions	Product(s) and Yield(s) (%)	Refs.
	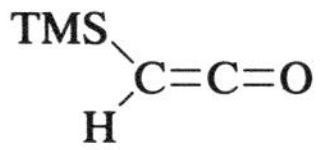 TMS(H)C=C=O	BF_3•Et_2O, -50°	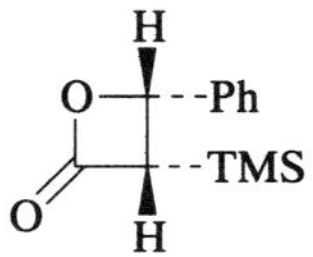 H, O, Ph, TMS, H, O (65)	167
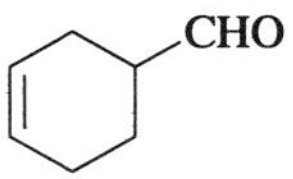 CHO	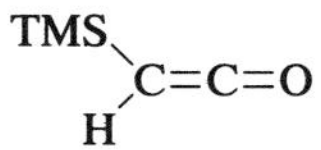 TMS(H)C=C=O	BF_3•Et_2O, -50°	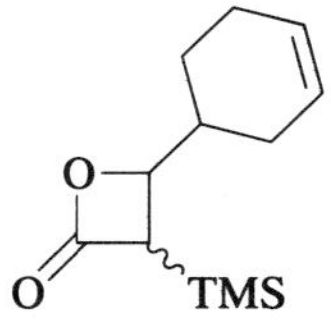 O, O, TMS (65) *cis*:*trans* = 2	698
C_8				
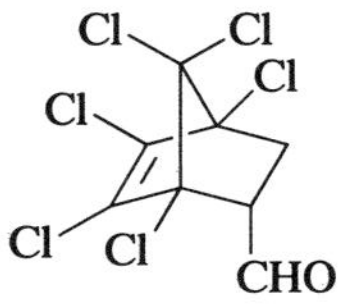 Cl, Cl, Cl, Cl, Cl, Cl, CHO	CH_2=C=O	$ZnCl_2$, EtOAc, -50°	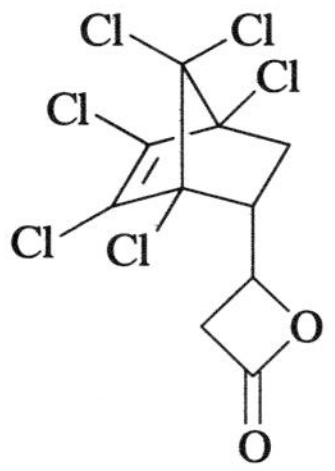 Cl, Cl, Cl, Cl, Cl, Cl, O, O (21)	701
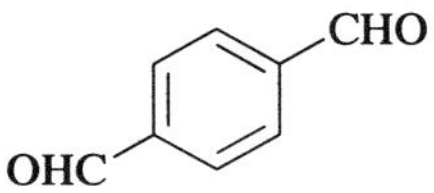 OHC, CHO	$Cl_2CHCOCl$	Et_3N, Et_2O	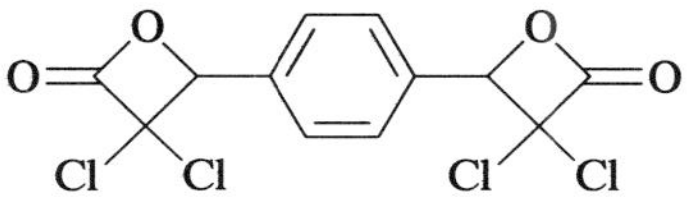 O, O, Cl, Cl, O, O, Cl, Cl (33)	702

TABLE XVIII. [2+2] CYCLOADDITION OF KETENES TO ALDEHYDES (*Continued*)

Reactant	Ketene or Ketene Source	Conditions	Product(s) and Yield(s) (%)	Refs.
(norbornenyl)CHO	$CH_2{=}C{=}O$	ZnCl, 80°	(—)	703
$n\text{-}C_6H_{13}CCl_2CHO$	$CH_2{=}C{=}O$	Quinidine, PhMe, -25°	H, Cl, Cl, $C_6H_{13}\text{-}n$ (90-95) >98% ee	173
$n\text{-}C_7H_{15}CHO$	$CH_2{=}C{=}O$	—	$C_7H_{15}\text{-}n$ (>84)	704
	$CH_2{=}C{=}O$	$BF_3{\bullet}Et_2O$, THF, -60 to -70°	" (—)	705
n-Bu(Et)CHCHO	$CH_2{=}C{=}O$	$BF_3{\bullet}Et_2O$, THF, -60 to -70°	Et, Bu-n (50)	706, 705
$PhCCl_2CHO$	$CH_2{=}C{=}O$	Quinidine, PhMe, -25°	H, Cl, Cl, Ph (89) 90% ee	171

TABLE XVIII. [2+2] CYCLOADDITION OF KETENES TO ALDEHYDES (*Continued*)

Reactant	Ketene or Ketene Source	Conditions	Product(s) and Yield(s) (%)	Refs.
C_9				
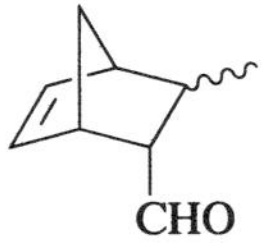 CHO	$CH_2{=}C{=}O$	$AlCl_3$, $ZnCl_2$, EtOAc, <20°	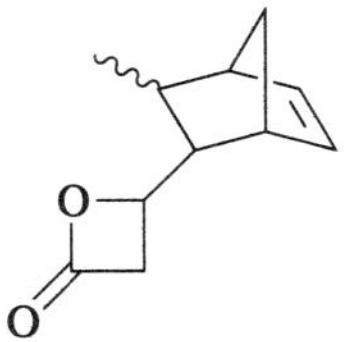 O, O (—)	703
C_{10}				
$CH_2{=}CH(CH_2)_7CHO$	$CH_2{=}C{=}O$	BF_3•Et_2O, THF, -70 to -60°	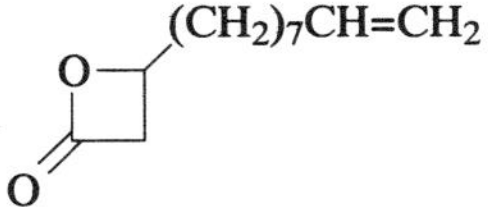 $(CH_2)_7CH{=}CH_2$ O, O (—)	705
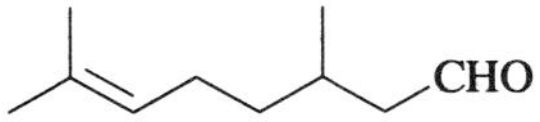 CHO	$CH_2{=}C{=}O$	BF_3, THF	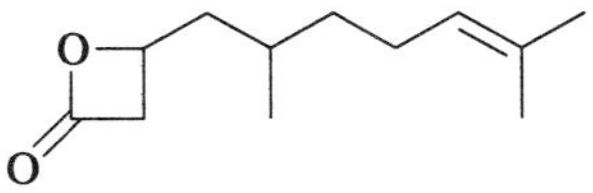 O, O (49)	707
C_{11}				
n-$C_{10}H_{21}CHO$	$CH_2{=}C{=}O$	BF_3•Et_2O, THF, -70 to -60°	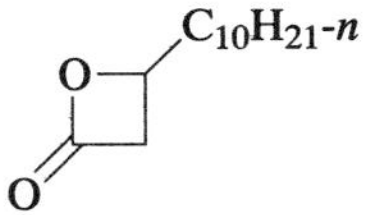 $C_{10}H_{21}$-n O, O (—)	705
C_{12}				
n-$C_{11}H_{23}CHO$	$CH_2{=}C{=}O$	BF_3•Et_2O, THF, -70 to -60°	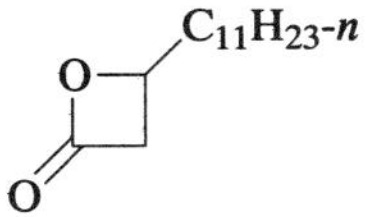 $C_{11}H_{23}$-n O, O (95.6)	705
C_{15}				
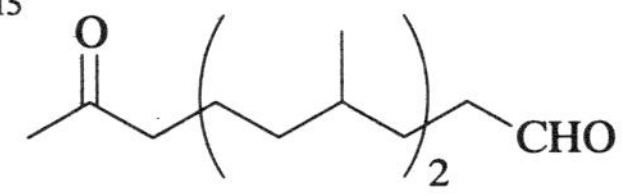 O, 2, CHO	Me, Me $C{=}C{=}O$	—	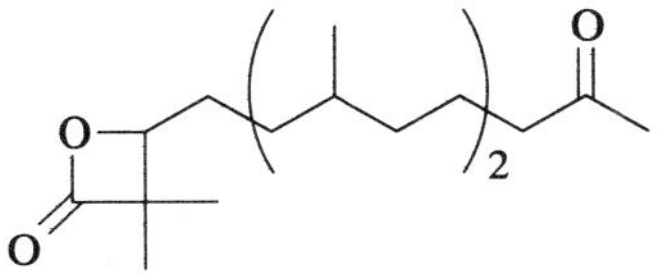 O, O, 2, O (>82)	162

[a] This product polymerizes.

TABLE XIX. [2+2] CYCLOADDITION OF KETENES TO KETONES

Reactant	Ketene or Ketene Source	Conditions	Product(s) and Yield(s) (%)	Refs.
C_3				
$CO(CN)_2$	$R_2C{=}C{=}O$	—	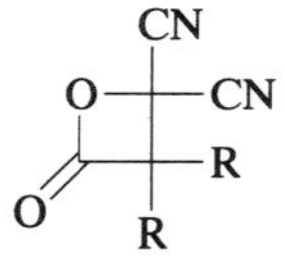 2-oxetanone: O–C(CN)(CN)–C(R)(R)–C(=O)	
	R	Solvent		
	H	C_6H_6	(81.6)	145, 708
	Me	Et_2O	(86)	145
	Et	C_6H_{14}	(—)	145
	n-Pr	C_6H_{14}	(80)	145
$CO(CF_3)_2$	$CH_2{=}C{=}O$	Et_2O, -78°	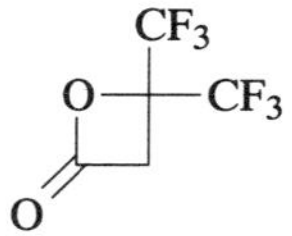 2-oxetanone: O–C(CF_3)(CF_3)–CH_2–C(=O) (96)	146
	$Ph_2C{=}C{=}O$	KF, MeCN	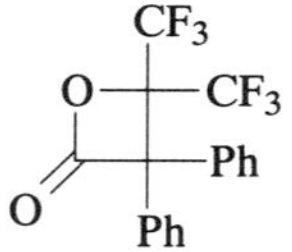 2-oxetanone: O–C(CF_3)(CF_3)–C(Ph)(Ph)–C(=O) (59)	709
$CO(CF_2Cl)_2$	*o*-$ClC_6H_4CH_2COCl$	(–)-Brucine, $CHCl_3$, -30°	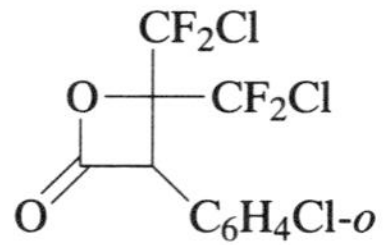 2-oxetanone: O–C(CF_2Cl)(CF_2Cl)–CH(C_6H_4Cl-*o*)–C(=O) (—) (+)-isomer	169

TABLE XIX. [2+2] CYCLOADDITION OF KETENES TO KETONES (*Continued*)

Reactant	Ketene or Ketene Source	Conditions	Product(s) and Yield(s) (%)	Refs.
	$R^1R^2CHCOCl$		2-oxetanone: 4,4-bis(CF_2Cl), 3-R^1, 3-R^2	
	R^1 / R^2			
	H / Cl	Et_3N, C_6H_{14}, rt	(52)	151
	Me / Cl	Et_3N, C_6H_{14}, rt	(55)	151
	Me / Br	Et_3N, C_6H_{14}, rt	(60)	151
	H / o-ClC_6H_4	(–)-Brucine, $CHCl_3$ -35°	(—), racemic	169
	H / o-ClC_6H_4	(–)-PhCH(Me)NMe₂, CCl_4, -30°	(—), (–)-isomer	169
$CO(CCl_3)_2$	MeCOCl	Et_3N, Et_2O, 0°	2-oxetanone: 4,4-bis(CCl_3) (6)	168
	$CH_2{=}C{=}O$	190°	" (80-85)	689,710, 711
	o-$ClC_6H_4CH_2COCl$	(–)-PhCH(Me)NMe₂, CCl_4, -20°	2-oxetanone: 4,4-bis(CCl_3), 3-C_6H_4Cl-o (—)	169
CH_3COCF_3	$CH_2{=}C{=}O$	BF_3 45Et_2O, (i-Pr)$_2$O, -70°	2-oxetanone: 4-Me, 4-CF_3 (57)	690

TABLE XIX. [2+2] CYCLOADDITION OF KETENES TO KETONES (*Continued*)

Reactant	Ketene or Ketene Source	Conditions	Product(s) and Yield(s) (%)	Refs.
	$CH_2{=}C{=}O$	$BF_3{\cdot}Et_2O$, Et_2O, 0°	O, CF_3, O (72)	712
	$CH_2{=}C{=}O$	Quinidine, PhMe, -25°	CF_3, O, O (83)	171, 172
$CO(CH_2F)_2$	$CH_2{=}C{=}O$	$ZnCl_2$, Et_2O, 25-30°	CH_2F, O, CH_2F, O (95)	713
$CO(CH_2Cl)_2$	$CH_2{=}C{=}O$	$BF_3{\cdot}Et_2O$, CH_2Cl_2, -30°	CH_2Cl, O, CH_2Cl, O (85)	690
	Cl_3CCOCl	Zn, Et_2O, rt	CH_2Cl, O, CH_2Cl, Cl, O, Cl (23)	151
O, Cl	$CH_2{=}C{=}O$	$BF_3{\cdot}Et_2O$, $CHCl_3$, -30°	O, CH_2Cl, O (80)	690
	$CH_2{=}C{=}O$	$BF_3{\cdot}Et_2O$, Et_2O, -15°	" (75)	714

TABLE XIX. [2+2] CYCLOADDITION OF KETENES TO KETONES (*Continued*)

Reactant	Ketene or Ketene Source	Conditions	Product(s) and Yield(s) (%)	Refs.
	Cl_3CCOCl	Zn, Et_2O, rt	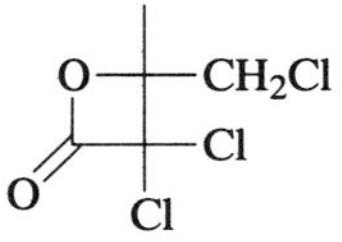 O, O, CH_2Cl, Cl, Cl (15)	151
$COMe_2$	$F_2BrCCOCl$	Zn, Me_2CO, -10°	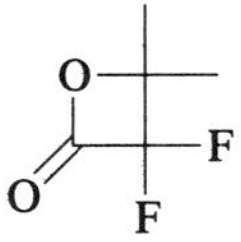 O, O, F, F (50)	715
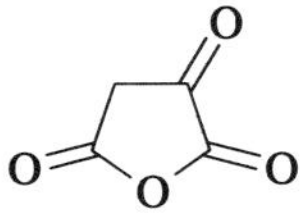 O, O, O, O	$CH_2{=}C{=}O$	$ZnCl_2$, 0°	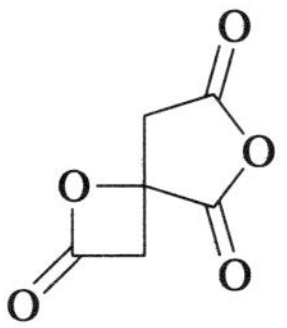 O, O, O, O, O, O (—)	716
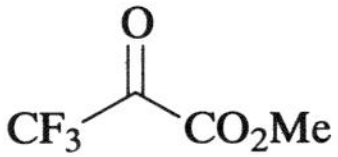 O, CF_3, CO_2Me	$CH_2{=}C{=}O$	Et_2O, -70°	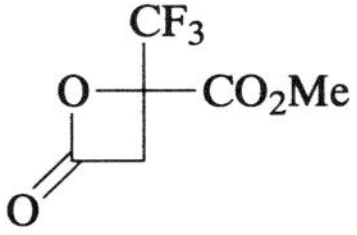 CF_3, O, O, CO_2Me (36)	717
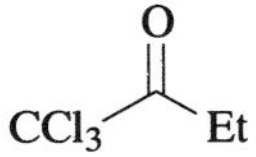 O, CCl_3, Et	$CH_2{=}C{=}O$	Quinidine, PhMe, -25°	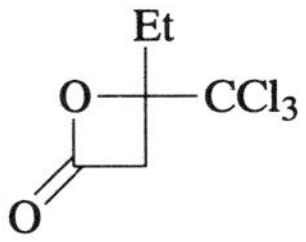 Et, O, O, CCl_3 (1-2)	171
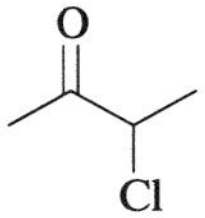 O, Cl	$CH_2{=}C{=}O$	$BF_3{\bullet}Et_2O$, Et_2O, -15°	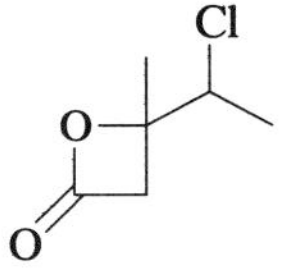 Cl, O, O (84)	714

TABLE XIX. [2+2] CYCLOADDITION OF KETENES TO KETONES (*Continued*)

Reactant	Ketene or Ketene Source	Conditions	Product(s) and Yield(s) (%)	Refs.
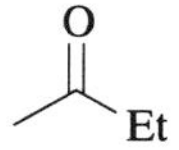 O, Et	Cl_3CCOCl	Zn, Et_2O, rt	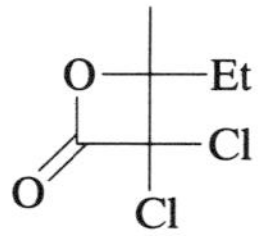 O, Et, Cl, O, Cl (35)	151
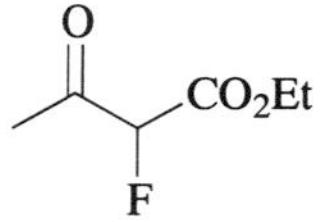 O, CO_2Et, F	$CH_2{=}C{=}O$	$ZnCl_2$, $CHCl_3$, 5°	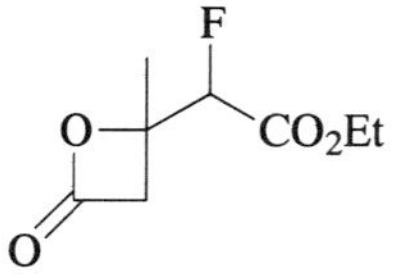 F, O, CO_2Et, O (45)	713
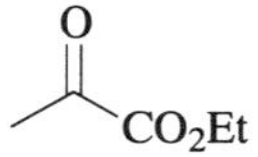 O, CO_2Et	$Cl_2CHCOCl$	Et_3N, Et_2O, 10°	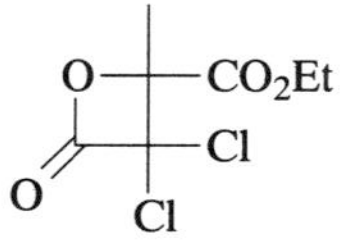 O, CO_2Et, Cl, O, Cl (33)	692
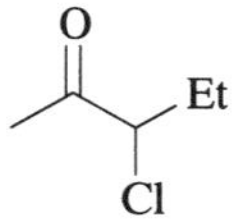 O, Et, Cl	$CH_2{=}C{=}O$	$BF_3{\bullet}Et_2O$, Et_2O, -15°	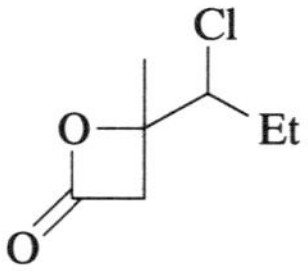 Cl, O, Et, O (—)	714
C_5 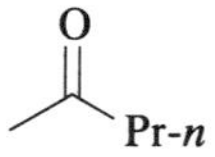 O, Pr-*n*	Cl_3CCOCl	Zn, Et_2O, rt	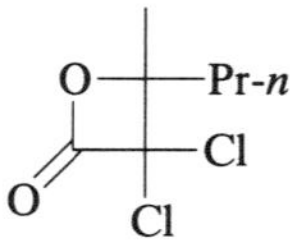 O, Pr-*n*, Cl, O, Cl (35)	151
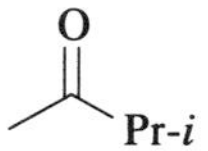 O, Pr-*i*	Cl_3CCOCl	Zn, Et_2O, rt	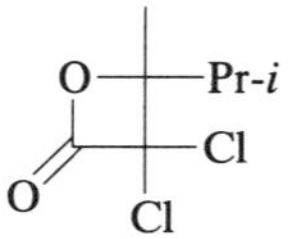 O, Pr-*i*, Cl, O, Cl (20)	151

TABLE XIX. [2+2] CYCLOADDITION OF KETENES TO KETONES (*Continued*)

Reactant	Ketene or Ketene Source	Conditions	Product(s) and Yield(s) (%)	Refs.
	$CH_2{=}C{=}O$	$BF_3{\cdot}Et_2O$, CCl_4, 0°	(2-oxetanone, 4-Me, 4-Pr-*i*) (—)	156
(2,5-cyclohexadiene-1,4-dione, R^1, R^2, R^3, R^4)	$Ph_2C{=}C{=}O$	Et_2O, pet. ether	(spiro β-lactone, Ph, Ph, R^1, R^2, R^3, R^4)	
R^1 = Cl, R^2 = Cl, R^3 = Cl, R^4 = Cl			(38)	153
R^1 = Cl, R^2 = Cl, R^3 = Cl, R^4 = H			(61)	152
R^1 = H, R^2 = Cl, R^3 = Cl, R^4 = H			(70)	152
R^1 = Cl, R^2 = H, R^3 = Cl, R^4 = H			(56)	152
R^1 = H, R^2 = Br, R^3 = Br, R^4 = H			(78)	152
R^1 = H, R^2 = Cl, R^3 = H, R^4 = H			(74)	152
R^1 = H, R^2 = H, R^3 = H, R^4 = H			(57)	718
(1,4-benzoquinone)	$Me_2C{=}C{=}O$	THF, 0°	(spiro β-lactone, gem-dimethyl) (64)	165

TABLE XIX. [2+2] CYCLOADDITION OF KETENES TO KETONES (*Continued*)

Reactant	Ketene or Ketene Source	Conditions	Product(s) and Yield(s) (%)	Refs.
O; O	Cl_3CCOCl	Zn-Cu, Et_2O	O; Cl; Cl; O; O (87)	155
OEt; Cl; O; Cl	$Cl_2CHCOCl$	Et_3N, C_6H_{14}, 0°	OEt; O; Cl; Cl; O; Cl; Cl (—)	693
MeO_2C; O; CO_2Me	$CH_2{=}C{=}O$	$ZnCl_2$, 0°	CO_2Me; O; CO_2Me; O (—)	716
O; CO_2Et; F	$CH_2{=}C{=}O$	$ZnCl_2$, $CHCl_3$	O; CO_2Et; O; F (45)	713
O	$CH_2{=}C{=}O$	$BF_3{\bullet}Et_2O$, Et_2O, 0°	O; O (90.5)	719, 720
$^{13}CH_3CO^{13}CH_2CH_2OAc$	$CH_2{=}C{=}O$	$BF_3{\bullet}Et_2O$, Et_2O, -30°	$^{13}CH_3$; O; $^{13}CH_2CH_2OAc$; O (—)	721

TABLE XIX. [2+2] CYCLOADDITION OF KETENES TO KETONES (*Continued*)

Reactant	Ketene or Ketene Source	Conditions	Product(s) and Yield(s) (%)	Refs.
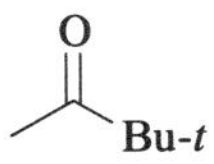 O, Bu-*t*	$CH_2{=}C{=}O$	$BF_3{\bullet}Et_2O$, CCl_4, 0°	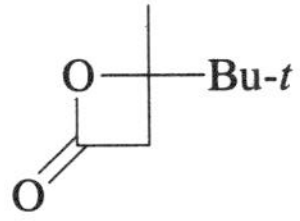 O, Bu-*t*, O (67)	156
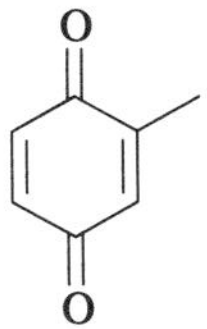 O, O	Ph, Ph, $C{=}C{=}O$	Et_2O, pet. ether	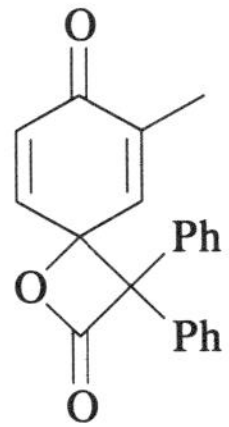 O, O, Ph, Ph, O (28)	152
o-ClC_6H_4CHO	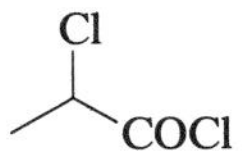 Cl, COCl	Et_3N, C_6H_{14}, rt	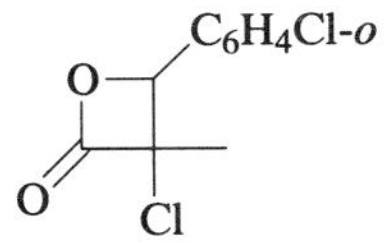 C_6H_4Cl-*o*, O, O, Cl (45)	151
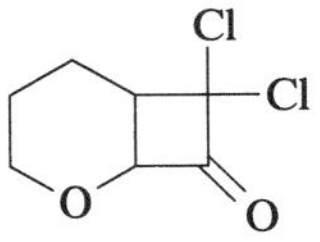 Cl, Cl, O, O	$Cl_2CHCOCl$	Et_3N, C_6H_{14}, rt	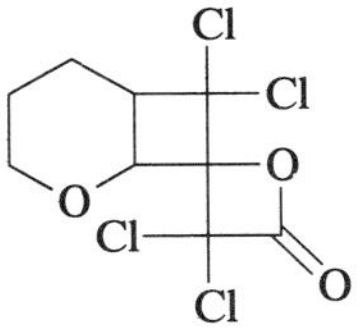 Cl, Cl, O, O, Cl, Cl, O (—)	693
$CO(CO_2Et)_2$	$Cl_2CHCOCl$	Et_3N, Et_2O, 0°	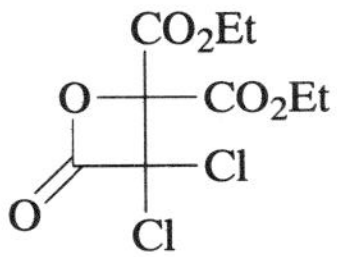 CO_2Et, O, CO_2Et, Cl, O, Cl (76)	324

TABLE XIX. [2+2] CYCLOADDITION OF KETENES TO KETONES (*Continued*)

	Reactant	Ketene or Ketene Source	Conditions	Product(s) and Yield(s) (%)	Refs.
	2,2,3,3-tetramethylcyclopropanone	$Me_2C{=}C{=}O$	CH_2Cl_2, -78°, 6 d	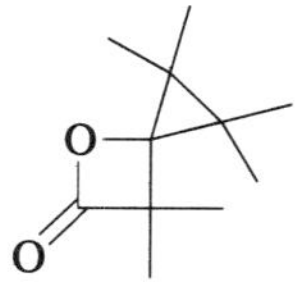 (—)	157
C_8 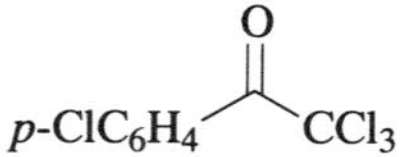	$p\text{-ClC}_6H_4COCCl_3$	$CH_2{=}C{=}O$	Quinidine, PhMe, -25°	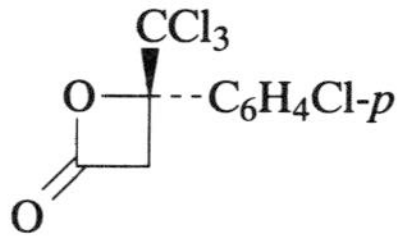 β-lactone, CCl_3, $C_6H_4Cl\text{-}p$ (68) 90% ee	171
	$p\text{-O}_2NC_6H_4COCCl_3$	$CH_2{=}C{=}O$	Quinidine, PhMe, -25°	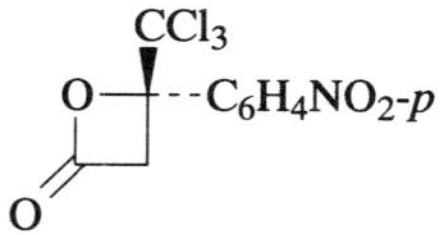 β-lactone, CCl_3, $C_6H_4NO_2\text{-}p$ (95) 89% ee	171
	$ClCH_2COPh$	Cl_3CCOCl	Zn, Et_2O, rt	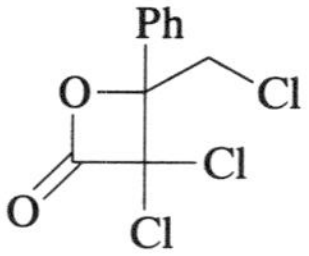 β-lactone, Ph, CH_2Cl, Cl, Cl (20)	151
	2,6-dimethyl-1,4-benzoquinone	$Me_2C{=}C{=}O$	THF, 0°	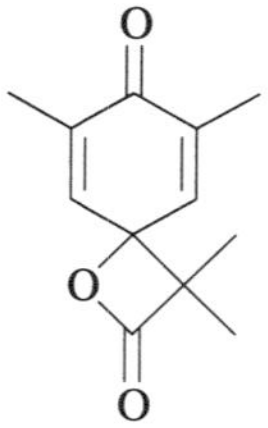 spiro β-lactone (84)	165

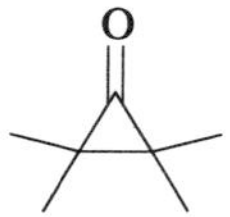

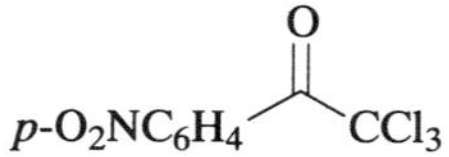

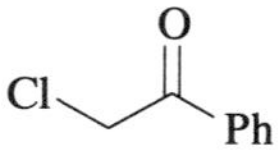

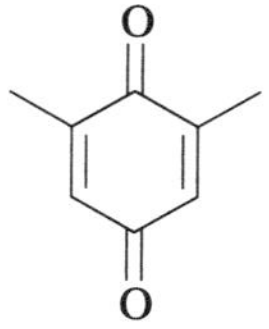

TABLE XIX. [2+2] CYCLOADDITION OF KETENES TO KETONES (*Continued*)

Reactant	Ketene or Ketene Source	Conditions	Product(s) and Yield(s) (%)	Refs.
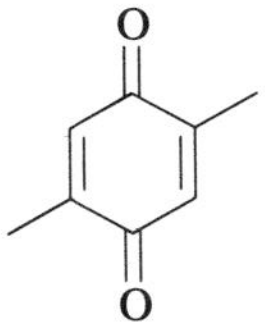 2,5-dimethyl-1,4-benzoquinone	$Ph_2C{=}C{=}O$	Et_2O, pet. ether	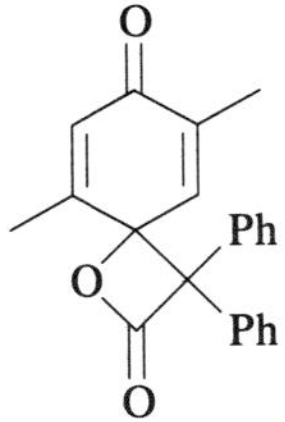 (36)	152
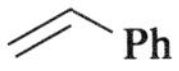 $CH_2{=}CHPh$	$Cl_2CHCOCl$	Et_3N, C_6H_{14}, 0°	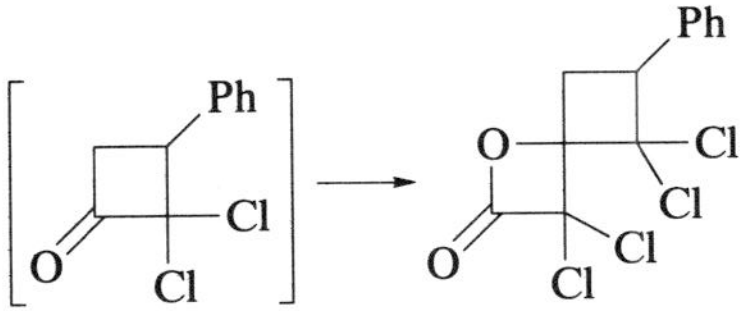 [2,2-dichloro-3-phenylcyclobutanone] → spiro β-lactone with Cl, Cl, Cl, Cl, Ph (—)	693
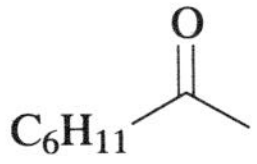 $C_6H_{11}COCH_3$	$CH_2{=}C{=}O$	$BF_3{\bullet}Et_2O$, CCl_4, 0°	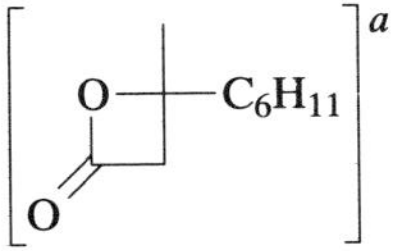 [β-lactone, C_6H_{11}][a] (—)	156
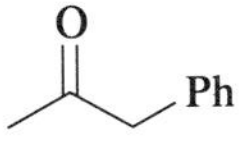 CH_3COCH_2Ph	Cl_3CCOCl	Zn, Et_2O, rt	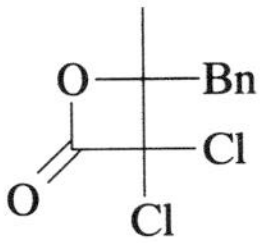 β-lactone, Bn, Cl, Cl (40)	151
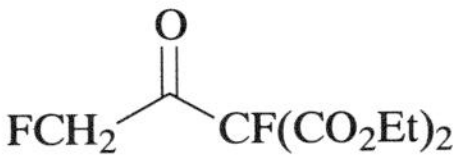 $FCH_2COCF(CO_2Et)_2$	$CH_2{=}C{=}O$	$ZnCl_2$, Et_2O, 25°	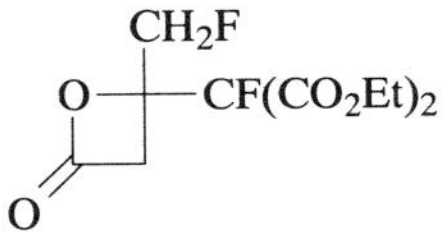 β-lactone, CH_2F, $CF(CO_2Et)_2$ (24)	713

TABLE XIX. [2+2] CYCLOADDITION OF KETENES TO KETONES (*Continued*)

Reactant	Ketene or Ketene Source	Conditions	Product(s) and Yield(s) (%)	Refs.
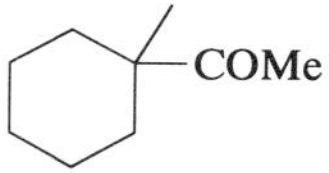 1-methylcyclohexyl–COMe	$CH_2{=}C{=}O$	$BF_3 \cdot Et_2O$, CCl_4, 0°	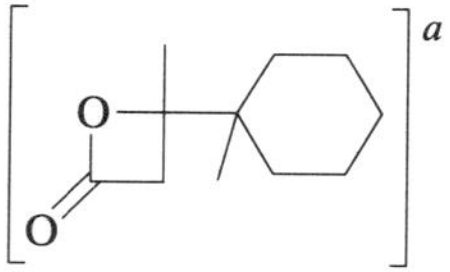 [oxetan-2-one product] [a] (—)	156
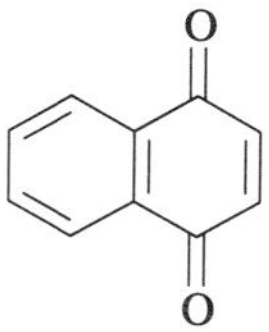 1,4-naphthoquinone	$Me_2C{=}C{=}O$	THF, 0°	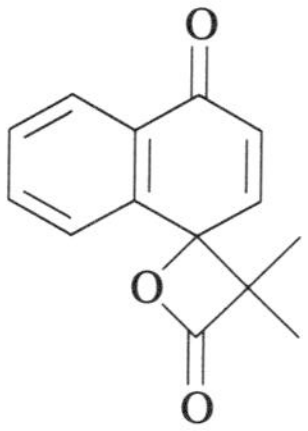 [spiro-β-lactone] (70)	165
	$Ph_2C{=}C{=}O$	Et_2O, pet. ether	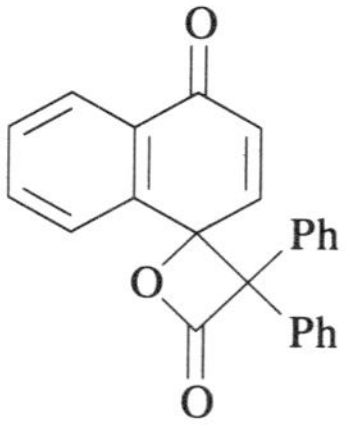 [spiro-β-lactone, Ph, Ph] (43)	152
C_{10} 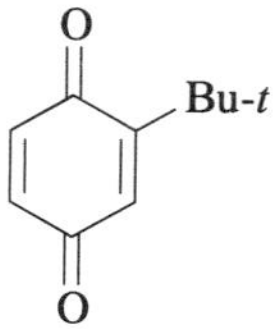 2-Bu-*t*-1,4-benzoquinone	$Me_2C{=}C{=}O$	THF, 0°	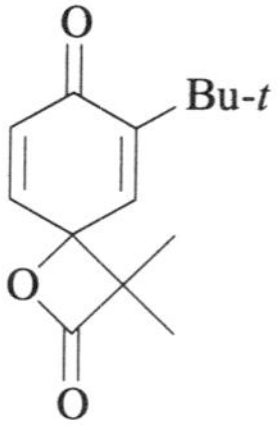 [spiro-β-lactone, Bu-*t*] (—)	165

TABLE XIX. [2+2] CYCLOADDITION OF KETENES TO KETONES (*Continued*)

Reactant	Ketene or Ketene Source	Conditions	Product(s) and Yield(s) (%)	Refs.
$PhCOCO_2Et$	$Cl_2CHCOCl$	Et_3N, Et_2O, 10°	β-lactone (Ph, CO_2Et, Cl, Cl) (38)	692
duroquinone	$Ph_2C=C=O$	$h\nu$, C_6H_6	cyclobutanone adduct (Ph, Ph) (25) + spiro β-lactone (Ph, Ph) (49)	153
C_{11} bicyclic ketone with CCl_3CO_2	Cl_3CCOCl	Zn, Et_2O	spiro β-lactone (CCl_3CO_2, Cl, Cl) (16)	722
C_{14} anthraquinone	$Ph_2C=C=O$	$h\nu$, C_6H_6	spiro β-lactone (Ph, Ph) (53)	153

TABLE XIX. [2+2] CYCLOADDITION OF KETENES TO KETONES (*Continued*)

Reactant	Ketene or Ketene Source	Conditions	Product(s) and Yield(s) (%)	Refs.
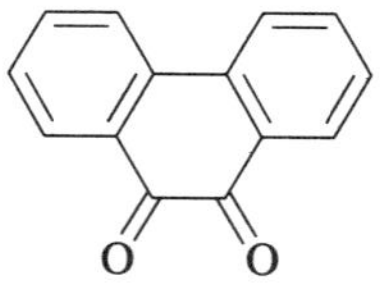 O O	Me, Me C=C=O	$ZnCl_2$, THF	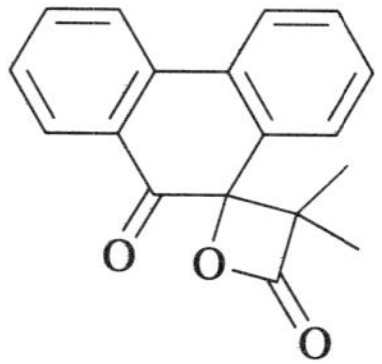 O O O (62)	154
C_{15} 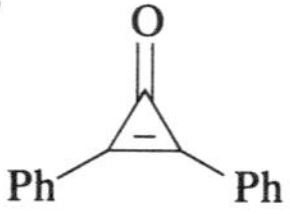 O Ph Ph	CF_3, CF_3 C=C=O	C_6H_6, 25°, 5 d	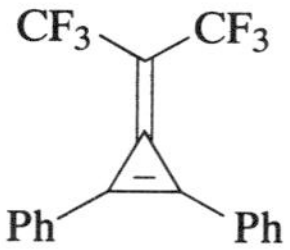 CF_3 CF_3 Ph Ph (79)	723
C_{16} 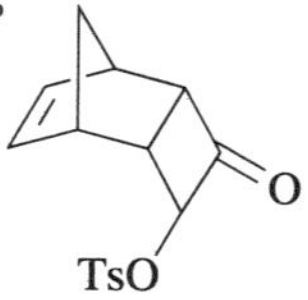 O TsO	Cl_3CCOCl	Zn, Et_2O	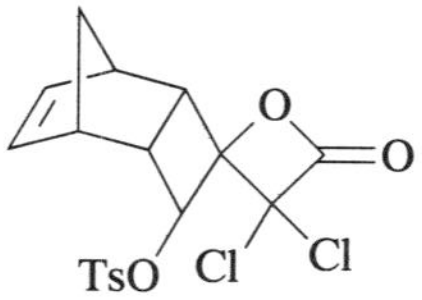 O O TsO Cl Cl (26)	722

TABLE XIX. [2+2] CYCLOADDITION OF KETENES TO KETONES (*Continued*)

Reactant	Ketene or Ketene Source	Conditions	Product(s) and Yield(s) (%)	Refs.
C_{17}				
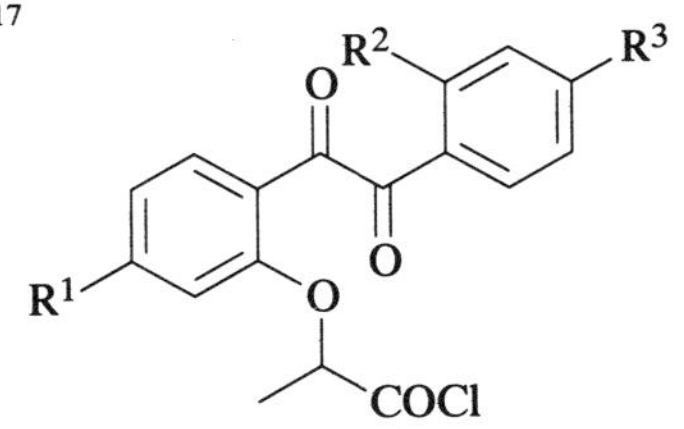	Intramolecular	Et_3N, C_6H_6, 50°	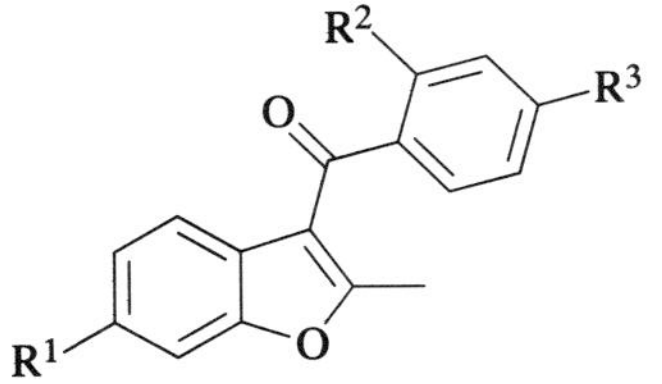 I + 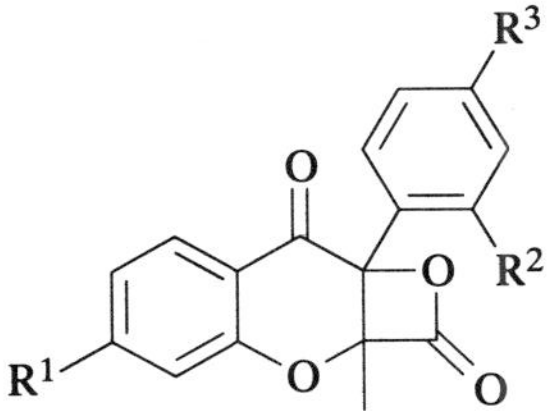 II	724

R^1	R^2	R^3	I	II
H	H	H	(63)	(trace)
H	OMe	H	(24)	(49)
OMe	OMe	OMe	(0)	(57)

[a] The product was not isolated.

TABLE XX. [2+2] CYCLOADDITION OF KETENES TO THIOCARBONYL COMPOUNDS

Reactant	Ketene or Ketene Source	Conditions	Product(s) and Yield(s) (%)	Refs.
C_3				
$(CF_3)_2C{=}S$	$Ph_2C{=}C{=}O$	KF, MeCN	2,2-bis(CF_3)-3,3-diphenylthietan-4-one (59)	709
C_8				
PhC(=S)SMe	$Ph_2C{=}C{=}O$	—	2-Ph-2-SMe-3,3-diphenylthietan-4-one (61)	179
C_9				
N-methyl-3-thioxophthalimide (S, NMe, O)	$Ph_2C{=}C{=}O$	$h\nu$, CH_2Cl_2, -60°	spiro thietanone (Ph, Ph, NMe, O) (59)	177
N-methyl-dithiophthalimide (S, NMe, S)	$Ph_2C{=}C{=}O$	$h\nu$, CH_2Cl_2, -60°	spiro thietanone (Ph, Ph, NMe, S) (42)	177
C_{10}				
PhC(=S)SCH$_2$CH=CH$_2$	$Ph_2C{=}C{=}O$	—	2-Ph-2-(SCH$_2$CH=CH$_2$)-3,3-diphenylthietan-4-one (61)	179

TABLE XX. [2+2] CYCLOADDITION OF KETENES TO THIOCARBONYL COMPOUNDS (*Continued*)

Reactant	Ketene or Ketene Source	Conditions	Product(s) and Yield(s) (%)	Refs.
C_{13}				
xanthene-9-thione (C=S)	$Ph_2C{=}C{=}O$	$h\nu$, CH_2Cl_2, -60°	spiro[xanthene-9,2'-thietan]-4'-one, 3',3'-Ph (100)	177
$Ph_2C{=}S$	$Ph_2C{=}C{=}O$	60°	3,3,4,4-tetraphenylthietan-2-one (S, Ph, Ph, Ph, Ph, O) (74)	178
	$Ph_2C{=}C{=}O$	Et_2O	" (80-90)	725
p-$MeOC_6H_4C({=}S)C_6H_4OMe$-$p$	$Ph_2C{=}C{=}O$	Et_2O	4,4-bis(C_6H_4OMe-p)-3,3-diphenylthietan-2-one (—)	725
	$Ph_2C{=}C{=}O$	60°	" (34)	178

TABLE XXI. OLEFINS FROM REACTION OF KETENES WITH CARBONYL AND THIOCARBONYL COMPOUNDS

Reactant	Ketene or Ketene Source	Conditions	Product(s) and Yield(s) (%)	Refs.
C_2				
Cl_3CCHO	MeCHClCOCl	1. Et_3N 2. Pyrolysis, 2 mm	$CH_2=C=CHCCl_3$ (40-50)	726
	EtCHClCOCl	"	$MeCH=C=CHCCl_3$ (40-50)	726
	n-PrCHClCOCl	"	$EtCH=C=CHCCl_3$ (40-50)	726
C_3				
$(CF_3)_2C=C=O$	$CH_2=C=O$	550°, 14 s	$(CF_3)_2C=CH_2$ (26)	727
	$CH_2=C=O$	290-340°	" (82-100)	181
C_4				
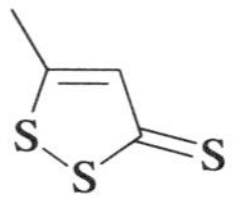 S S S	$Ph_2C=C=O$	Xylene, 137°	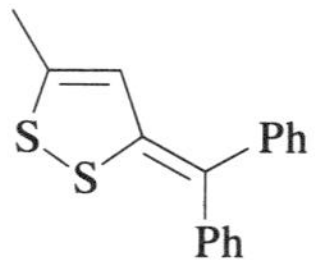 S S Ph Ph (10)	186
C_5				
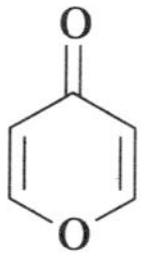 O O	$Ph_2C=C=O$	150°	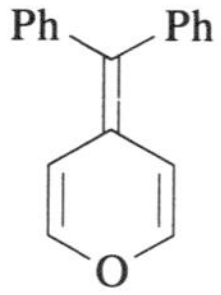 Ph Ph O (50)	728
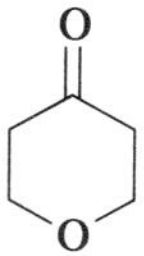 O O	$Ph_2C=C=O$	—	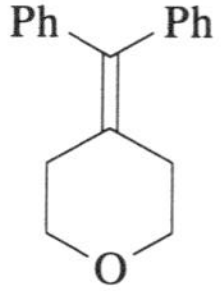 Ph Ph O (29)	539, 729

TABLE XXI. OLEFINS FROM REACTION OF KETENES WITH CARBONYL AND THIOCARBONYL COMPOUNDS (*Continued*)

Reactant	Ketene or Ketene Source	Conditions	Product(s) and Yield(s) (%)	Refs.
C_6 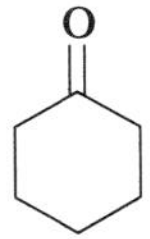 Cyclohexanone	$CH_2{=}C{=}O$	BF_3•Et_2O, 0-2°, 110°	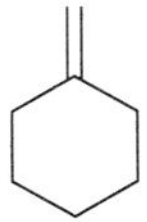 Methylenecyclohexane (60)	730
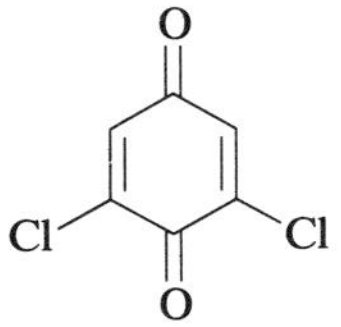 2,6-Dichloro-1,4-benzoquinone	$Ph_2C{=}C{=}O$	100°, 24 h	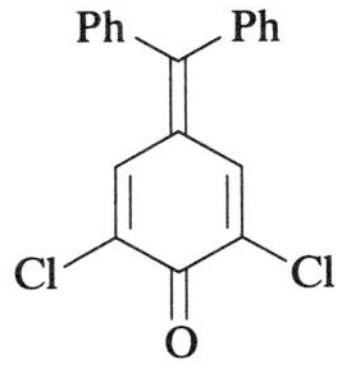 4-(Diphenylmethylene)-2,6-dichlorocyclohexa-2,5-dienone (—)	152
	$Ph_2C{=}C{=}O$	130°, 1.5 h	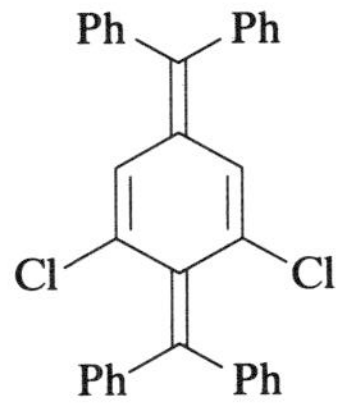 2,5-Dichloro-1,4-bis(diphenylmethylene)cyclohexa-2,5-diene (—)	152
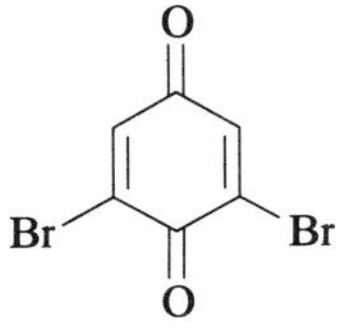 2,6-Dibromo-1,4-benzoquinone	$Ph_2C{=}C{=}O$	Xylene, 140°	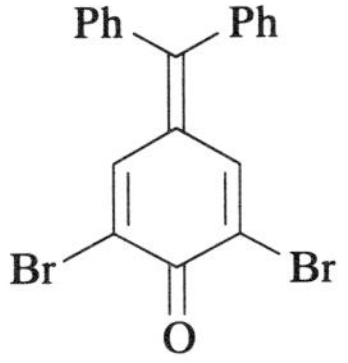 4-(Diphenylmethylene)-2,6-dibromocyclohexa-2,5-dienone (—)	152

TABLE XXI. OLEFINS FROM REACTION OF KETENES WITH CARBONYL AND THIOCARBONYL COMPOUNDS (*Continued*)

Reactant	Ketene or Ketene Source	Conditions	Product(s) and Yield(s) (%)	Refs.
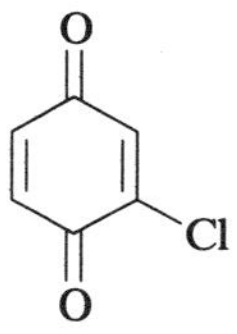 O, O, Cl	$Ph_2C=C=O$	110°	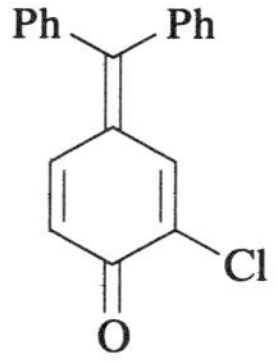 Ph, Ph, Cl, O (—)	152
	$Ph_2C=C=O$	120°	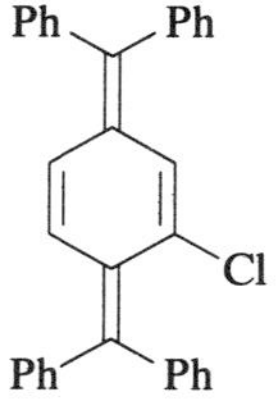 Ph, Ph, Cl, Ph, Ph (—)	152
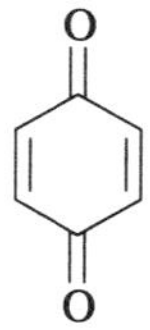 O, O	$Ph_2C=C=O$	Xylene, 140°	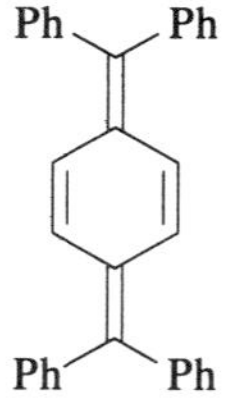 Ph, Ph, Ph, Ph (—)	718
C_7 PhCHO	BrF_2CCOCl	Zn, -10°	$PhCH=CF_2$ (11)	715
	TMS(H)C=C=O	1. $BF_3 \cdot Et_2O$, -50° 2. 150-160°	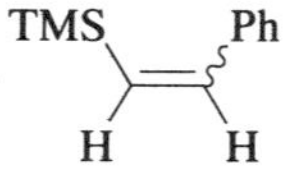 TMS(H)C=CH(Ph) (75)	167

TABLE XXI. OLEFINS FROM REACTION OF KETENES WITH CARBONYL AND THIOCARBONYL COMPOUNDS (*Continued*)

Reactant	Ketene or Ketene Source	Conditions	Product(s) and Yield(s) (%)	Refs.
XC_6H_4CHO	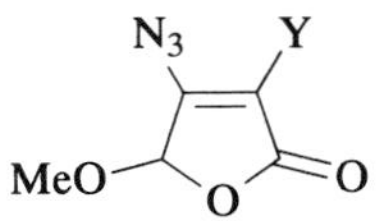 N_3, Y, MeO, O, O	C_6H_6, 80°, 5 h	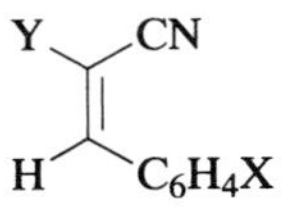 Y, CN, H, C_6H_4X	182
X	Y			
p-Cl	Cl		(32)	
p-Cl	Br		(27)	
p-O_2N	Cl		(8)	
p-O_2N	Br		(<5)	
H	Cl		(61)	
H	Br		(48)	
p-Me	Cl		(54)	
p-MeO	Cl		(73)	
p-MeO	Br		(79)	
2,4-$(MeO)_2$	Cl		(92)	
p-NMe_2	Cl		(78)	
p-MeO_2C	Cl		(61)	
p-MeO_2C	Br		(51)	
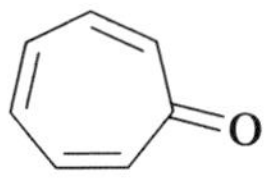 O	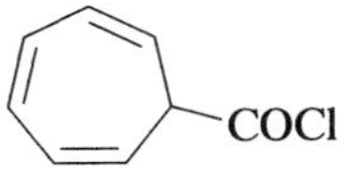 COCl	Et_3N	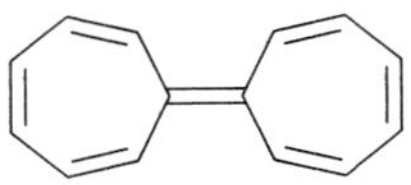 (1) 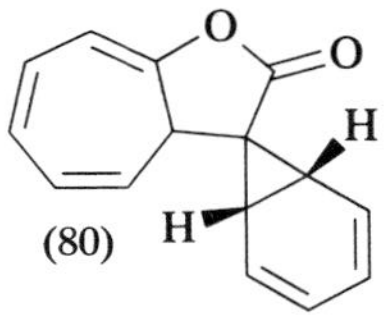 + O, O, H, H (80)	187
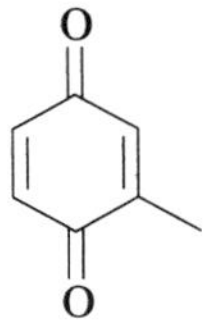 O, O	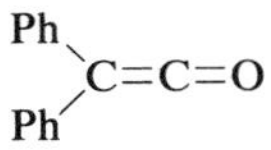 Ph, Ph, C=C=O	110°	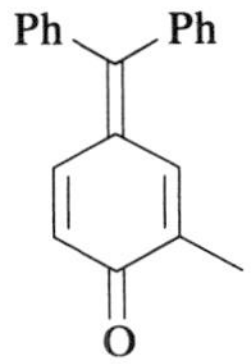 Ph, Ph, O (—)	152

TABLE XXI. OLEFINS FROM REACTION OF KETENES WITH CARBONYL AND THIOCARBONYL COMPOUNDS (*Continued*)

Reactant	Ketene or Ketene Source	Conditions	Product(s) and Yield(s) (%)	Refs.
	$Ph_2C{=}C{=}O$	—	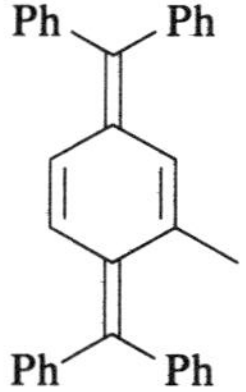 Ph, Ph; Ph, Ph (—)	152
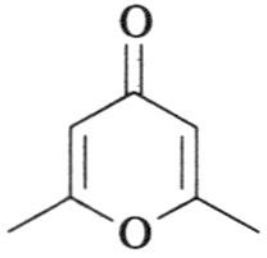 O, O	$Ph_2C{=}C{=}O$	150°	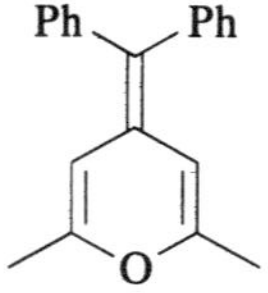 Ph, Ph; O (73)	728
C_8 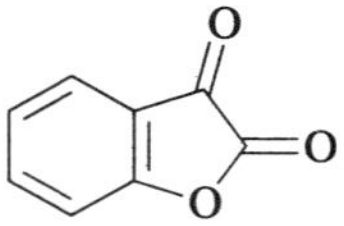 O, O, O	$Cl_2CHCOCl$	Et_3N, Et_2O	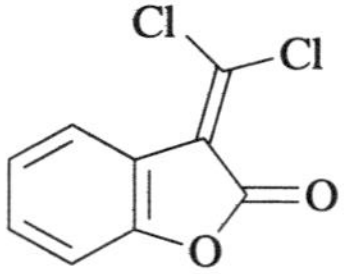 Cl, Cl; O, O (61)	324
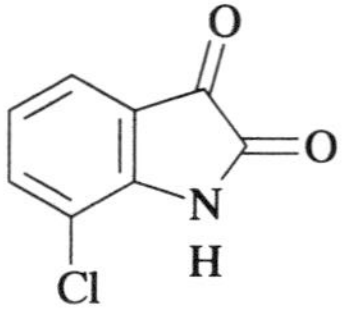 O, O, N, H, Cl	$Cl_2CHCOCl$	Et_3N, $CHCl_3$, 20°	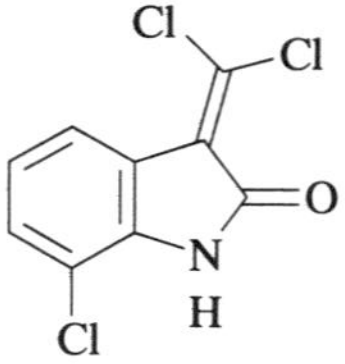 Cl, Cl; O, N, H, Cl (61)	324, 692

TABLE XXI. OLEFINS FROM REACTION OF KETENES WITH CARBONYL A ND THIOCARBONYL COMPOUNDS (*Continued*)

Reactant	Ketene or Ketene Source	Conditions	Product(s) and Yield(s) (%)	Refs.
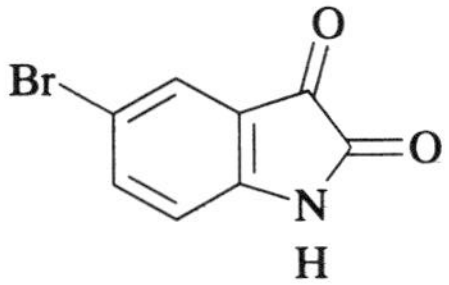 5-bromoisatin (Br, O, O, N, H)	$Cl_2CHCOCl$	Et_3N, $CHCl_3$, 20°	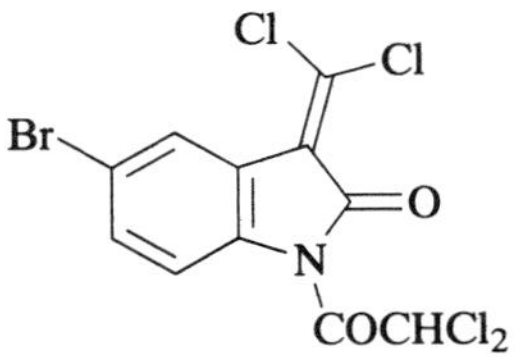 Br, Cl, Cl, O, N, $COCHCl_2$ (67)	324, 692
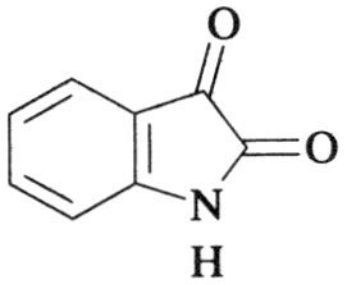 isatin (O, O, N, H)	$Cl_2CHCOCl$	Et_3N, $CHCl_3$, 20°	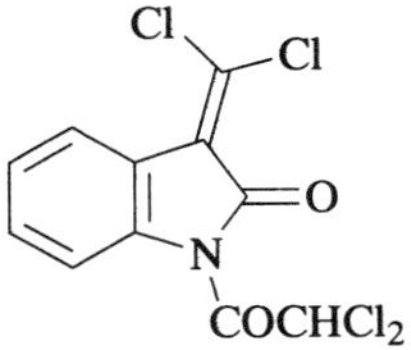 Cl, Cl, O, N, $COCHCl_2$ (80)	324, 692
	$Br_2CHCOBr$	Et_3N, $CHCl_3$, 20°	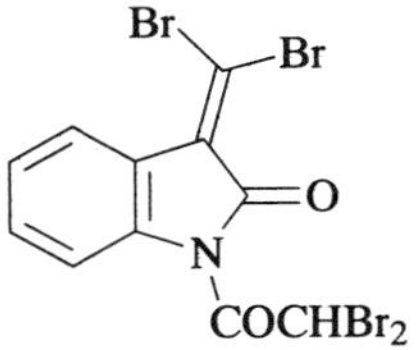 Br, Br, O, N, $COCHBr_2$ (44)	324, 692
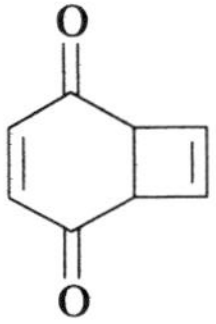 O, O	$Ph_2C{=}C{=}O$	PhMe, 111°	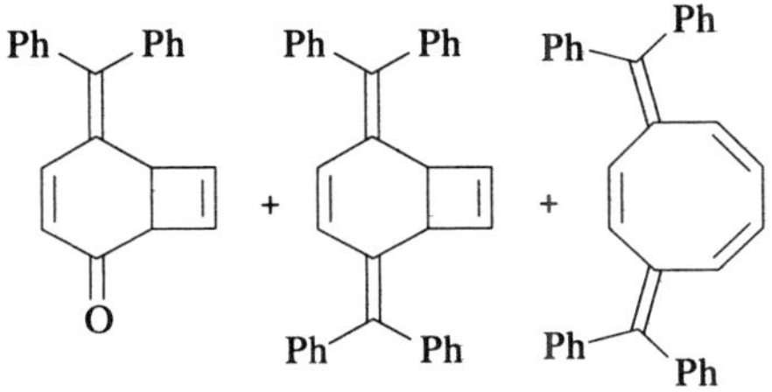 Ph, Ph, O (35) + Ph, Ph, Ph, Ph (24) + Ph, Ph, Ph, Ph (12)	731

TABLE XXI. OLEFINS FROM REACTION OF KETENES WITH CARBONYL AND THIOCARBONYL COMPOUNDS (*Continued*)

Reactant	Ketene or Ketene Source	Conditions	Product(s) and Yield(s) (%)	Refs.
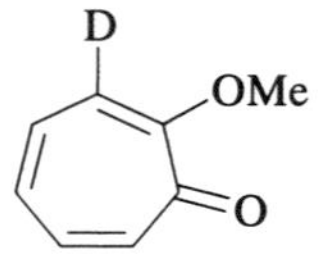 D, OMe, O (structure)	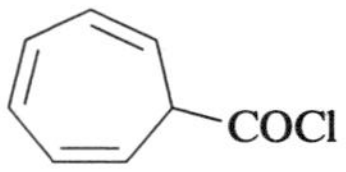 COCl (structure)	Et_3N, C_6H_{14}	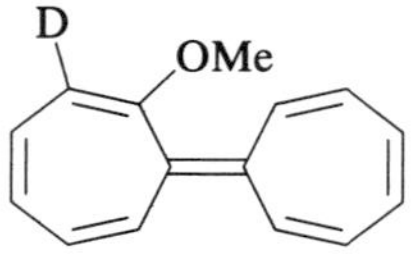 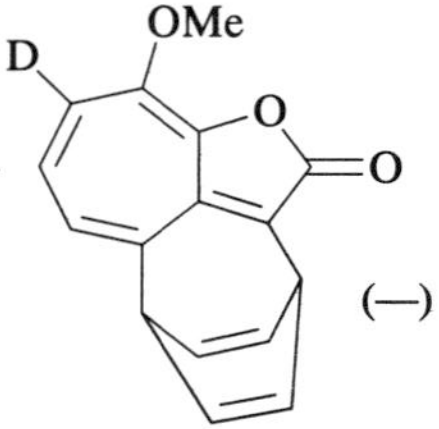 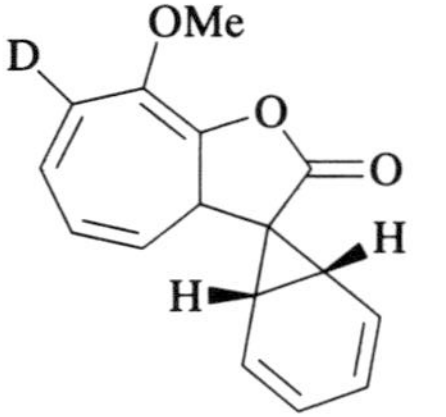 D, OMe (structure) (—) + D, OMe, O, O (structure) (—) + D, OMe, O, O, H, H (structure) (—)	732
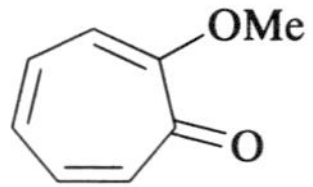 OMe, O (structure)	Ph, Ph C=C=O	C_6H_6, 80°	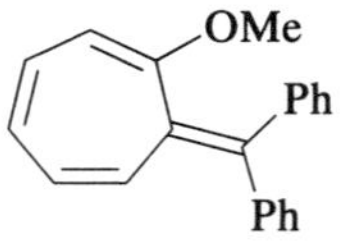 OMe, Ph, Ph (structure) (7) 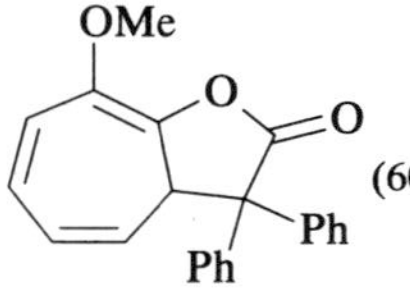 + OMe, O, O, Ph, Ph (structure) (60)	733
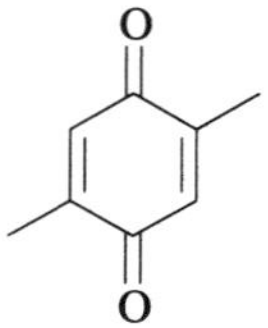 O, O (structure)	Ph, Ph C=C=O	Xylene, 140°	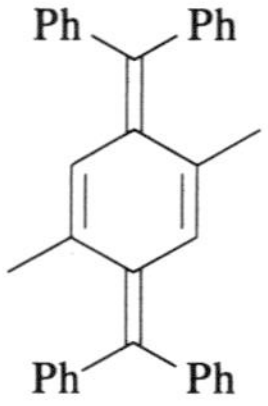 Ph, Ph, Ph, Ph (structure) (—)	718

TABLE XXI. OLEFINS FROM REACTION OF KETENES WITH CARBONYL AND THIOCARBONYL COMPOUNDS (*Continued*)

Reactant	Ketene or Ketene Source	Conditions	Product(s) and Yield(s) (%)	Refs.
CHO	$CH_2{=}C{=}O$	1. rt 2. 170-190°	(94)	180
C_9 p-ClC_6H_4 ; S S S	$Ph_2C{=}C{=}O$	Xylene, 137°	p-ClC_6H_4 ; S S Ph Ph (75)	186
O O	$Ph_2C{=}C{=}O$	150°	Ph Ph O (65)	728
Ph ; S S S	$Ph_2C{=}C{=}O$	Xylene, 137°	Ph ; S S Ph Ph (80)	186
O O N Me	$Cl_2CHCOCl$	Et_3N, $CHCl_3$, 20°	Cl Cl O N Me (69)	734

TABLE XXI. OLEFINS FROM REACTION OF KETENES WITH CARBONYL AND THIOCARBONYL COMPOUNDS (*Continued*)

Reactant	Ketene or Ketene Source	Conditions	Product(s) and Yield(s) (%)	Refs.
Ph–CH=CH–CHO	Cl_3CCOCl	Zn, Et_2O	Ph–CH=CH–CH=C(Cl)Cl (35)	698
	4-azido-3-chloro-5-methoxy-2(5H)-furanone (N_3, Cl, MeO, O)	C_6H_6, 80°	Ph–CH=CH–CH=C(Cl)CN (85)	23
	$Ph_2C{=}C{=}O$	$BF_3{\bullet}Et_2O$, Et_2O	Ph–CH=CH–CH=C(Ph)Ph (52)	698
2-OAc-tropone	cycloheptatrienyl-COCl	Et_3N	OAc-substituted heptafulvalene (9) + OAc, O, O, H, H spiro-lactone (44)	735
2-NMe_2-tropone	$Ph_2C{=}C{=}O$	C_6H_6, 80°	NMe_2, Ph, Ph heptafulvene (8)	733
methylnorbornene-CHO	$CH_2{=}C{=}O$	—	methyl-vinylnorbornene (85)	736

TABLE XXI. OLEFINS FROM REACTION OF KETENES WITH CARBONYL AND THIOCARBONYL COMPOUNDS (*Continued*)

Reactant	Ketene or Ketene Source	Conditions	Product(s) and Yield(s) (%)	Refs.
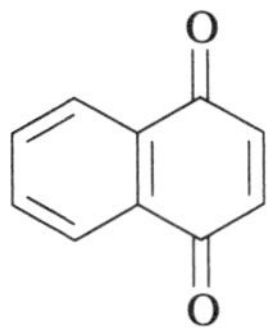 [structure: 1,4-naphthoquinone]	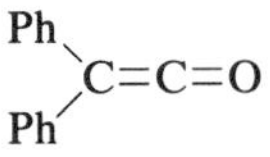 $Ph_2C{=}C{=}O$	PhMe, 100°	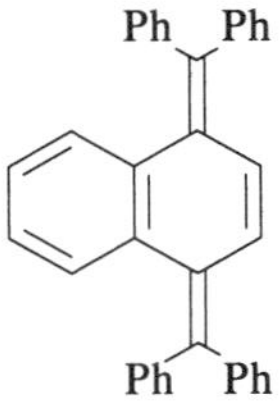 [structure: Ph, Ph / Ph, Ph] (—)	152, 718
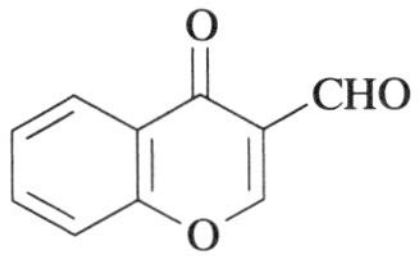 [structure: CHO, O]	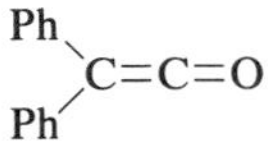 $Ph_2C{=}C{=}O$	180-200°	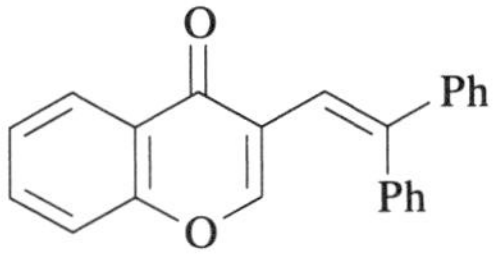 [structure: O, Ph, Ph] (30)	256
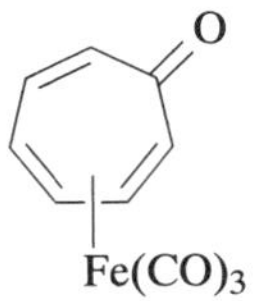 [structure: O, $Fe(CO)_3$]	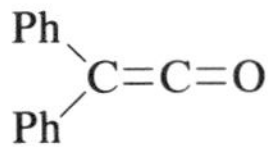 $Ph_2C{=}C{=}O$	C_6H_6, 25°	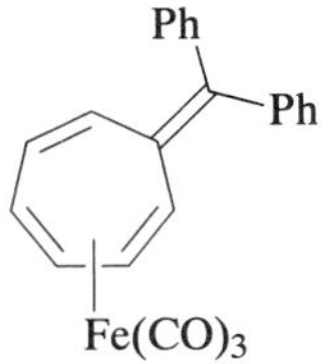 [structure: Ph, Ph, $Fe(CO)_3$] (20)	737
	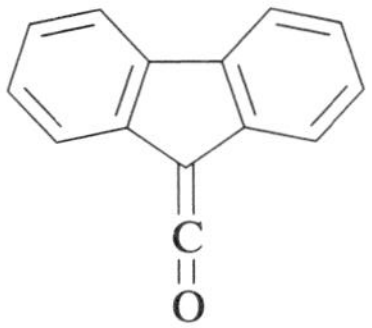 [structure: fluorenylidene ketene, C=O]	C_6H_{12}, 25°	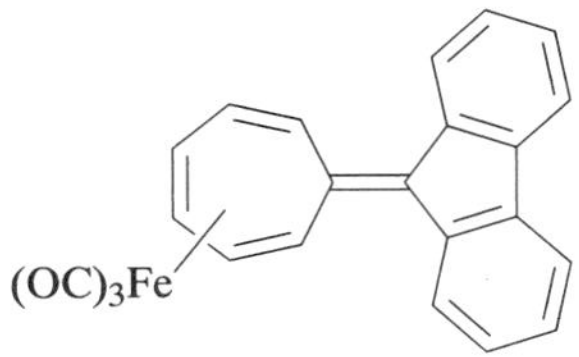 [structure: $(OC)_3Fe$] (—)	737

TABLE XXI. OLEFINS FROM REACTION OF KETENES WITH CARBONYL AND THIOCARBONYL COMPOUNDS (*Continued*)

Reactant	Ketene or Ketene Source	Conditions	Product(s) and Yield(s) (%)	Refs.
p-$MeOC_6H_4$ (dithiolethione)	$Ph_2C=C=O$	Xylene, 137°	*p*-$MeOC_6H_4$ (dithiole), =C(Ph)Ph (40)	186
Ph (methyl dithiolethione)	$Ph_2C=C=O$	Xylene, 137°	Ph (methyl dithiole), =C(Ph)Ph (60)	186
4,6-dimethylisatin (N–H)	$Cl_2CHCOCl$	Et_3N, $CHCl_3$, 20°	3-(dichloromethylene)-N-$COCHCl_2$ oxindole (38)	324, 692
2-CHO-phenoxy–CH(Me)CO_2H	—	1. $(COCl)_2$ 2. Et_3N, C_6H_6, 80°	2-methylbenzofuran (57)	185, 534
PhCH=C(Me)CHO	Cl_3CCOCl	Zn, Et_2O	PhCH=C(Me)CH=CCl_2 (32)	698
	N_3, Cl, MeO furanone	C_6H_6, 80°	PhCH=C(Me)CH=C(CN)Cl (79)	23

TABLE XXI. OLEFINS FROM REACTION OF KETENES WITH CARBONYL AND THIOCARBONYL COMPOUNDS (*Continued*)

Reactant	Ketene or Ketene Source	Conditions	Product(s) and Yield(s) (%)	Refs.
C_{11}				
6,8-dichloro thiophene-fused ketone (structure)	$R^1R^2C{=}C{=}O$	C_6H_6, 80°	Cl, Cl-substituted olefin with $=CR^1R^2$ (structure)	184
	R^1 / R^2			
	Me / CN		(34)	
	t-Bu / CN		(72)	
	Ph / CN		(42)	
	Ph / Me		(21)	
	Ph / Ph		(33)	
	Ph / 2,4,6-$Me_3C_6H_2$		(54)	
	Ph / 1-$C_{10}H_7$		(74)	
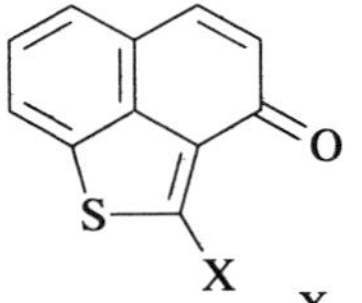 Thiophene-fused ketone with X (structure)	$R^1R^2C{=}C{=}O$	C_6H_6, 80°	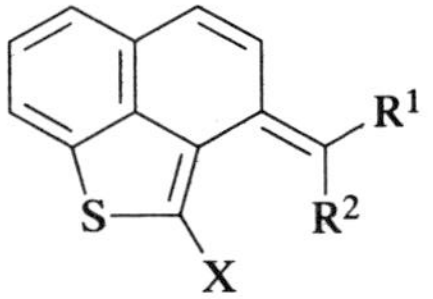 Olefin with X and $=CR^1R^2$ (structure)	325
X	R^1 / R^2			
H	Ph / Ph		(35)	
Me	Me / CN		(37)	
Me	*t*-Bu / CN		(75)	
Me	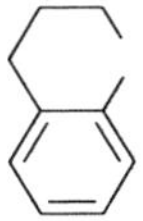 (tetralinylidene structure)		(47)	
Me	Ph / Ph		(56)	
Me	9-Fluorenylidene		(61)	

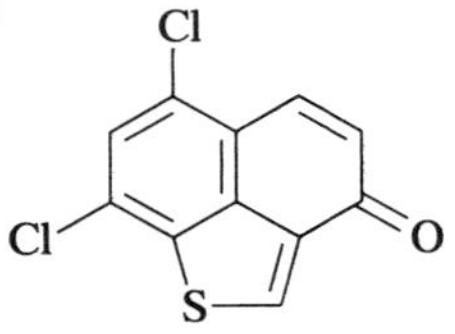

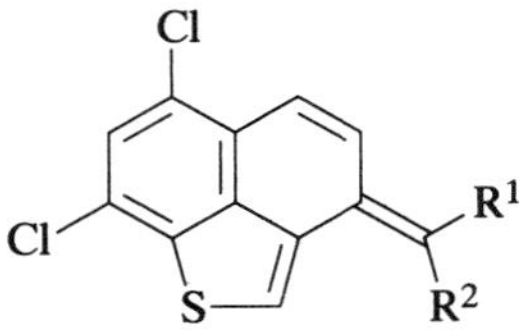

TABLE XXI. OLEFINS FROM REACTION OF KETENES WITH CARBONYL AND THIOCARBONYL COMPOUNDS (*Continued*)

Reactant	Ketene or Ketene Source	Conditions	Product(s) and Yield(s) (%)	Refs.
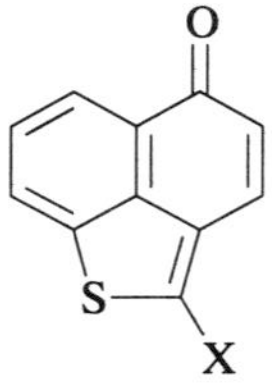 O, S, X	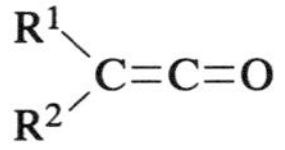 $R^1R^2C{=}C{=}O$	C_6H_6, 80°	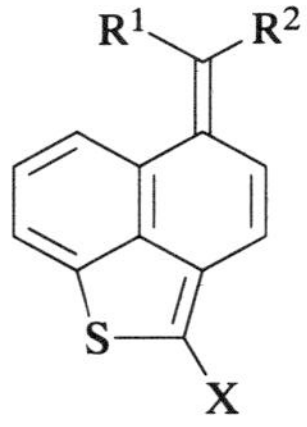 R^1, R^2, S, X	
X	R^1 R^2			
Br	t-Bu CN		(49)	325
Br	Ph Ph		(43)	325
H	Me CN		(34)	184
H	Ph CN		(42)	184
H	Ph Ph		(30)	184
H	Ph 2,4,6-$Me_3C_6H_2$		(3)	184
H	Ph 1-$C_{10}H_7$		(63)	184
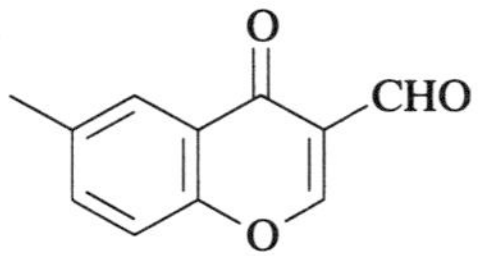 O, CHO, O	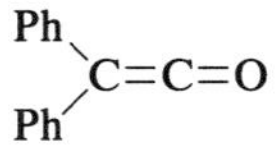 $Ph_2C{=}C{=}O$	180-200°	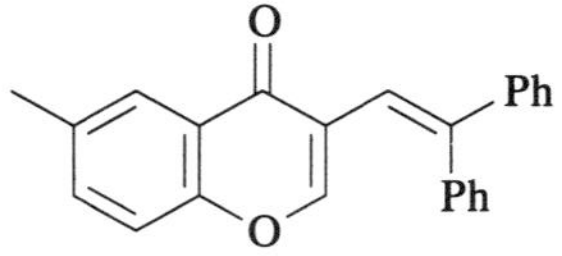 O, Ph, Ph, O (27)	256
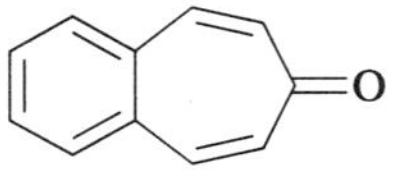 O	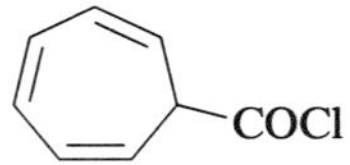 COCl	Et_3N	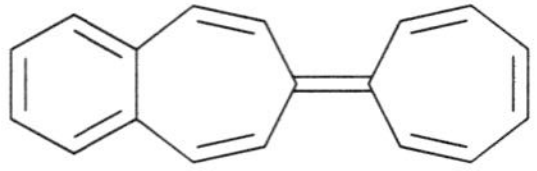 (50)	187
	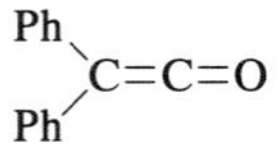 $Ph_2C{=}C{=}O$	121°	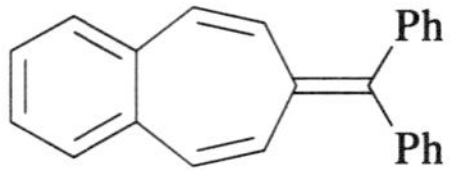 Ph, Ph (69)	539, 738

TABLE XXI. OLEFINS FROM REACTION OF KETENES WITH CARBONYL AND THIOCARBONYL COMPOUNDS (*Continued*)

Reactant	Ketene or Ketene Source	Conditions	Product(s) and Yield(s) (%)	Refs.
O, EtO_2C, O, CO_2Et	Ph, C=C=O, Ph	150°	Ph, Ph, EtO_2C, O, CO_2Et (81)	728
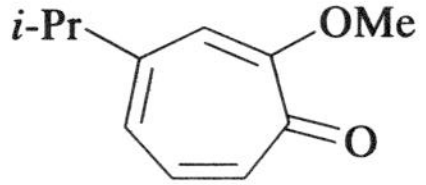 *i*-Pr, OMe, O	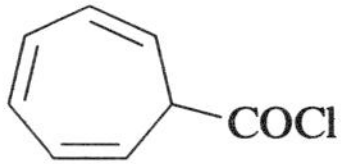 COCl	Et_3N, C_6H_{14}	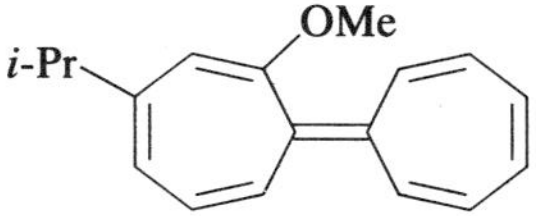 *i*-Pr, OMe (24) +	732
			i-Pr, OMe, O, O (24) +	
			OMe, O, O, H, *i*-Pr, H (15)	
C_{12} O, O, CO_2Et	Ph, C=C=O, Ph	150°	Ph, Ph, O, CO_2Et (94)	728

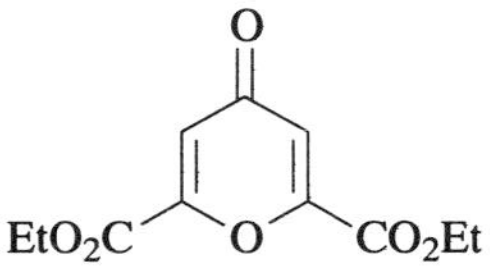

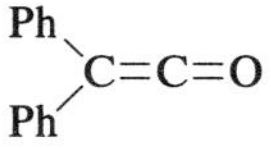

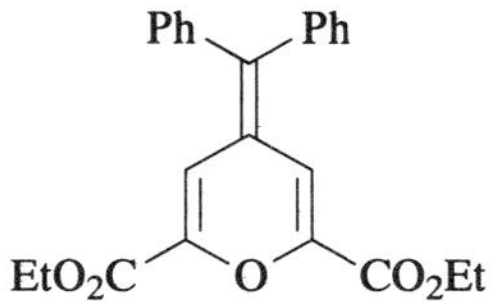

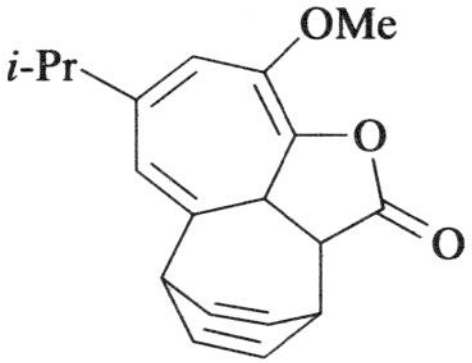

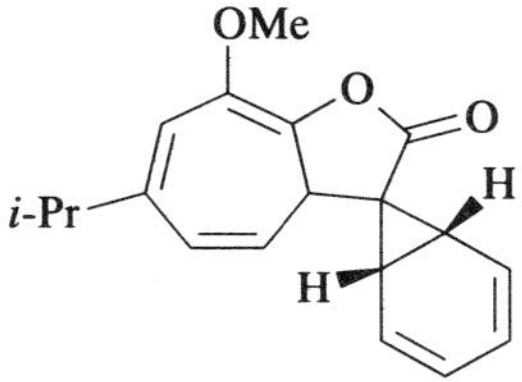

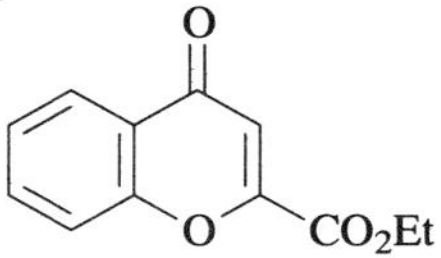

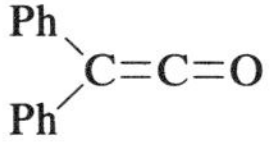

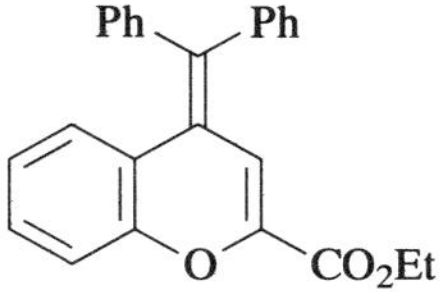

TABLE XXI. OLEFINS FROM REACTION OF KETENES WITH CARBONYL AND THIOCARBONYL COMPOUNDS (*Continued*)

Reactant	Ketene or Ketene Source	Conditions	Product(s) and Yield(s) (%)	Refs.
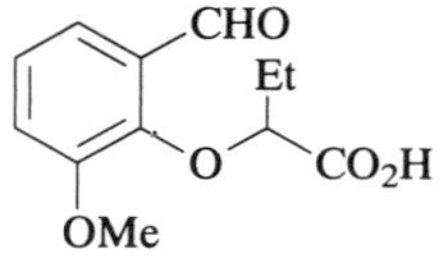 CHO, Et, O, CO_2H, OMe	—	$(COCl)_2$, C_6H_6, Et_3N, 80°	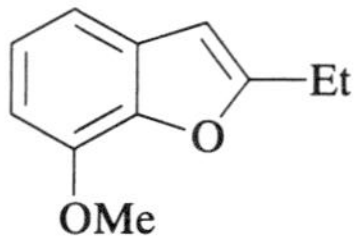 Et, O, OMe (53)	185, 534
C_{13} 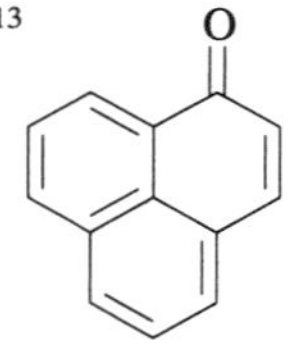 O	$Cl_2CHCOCl$	C_6H_6, Et_3N	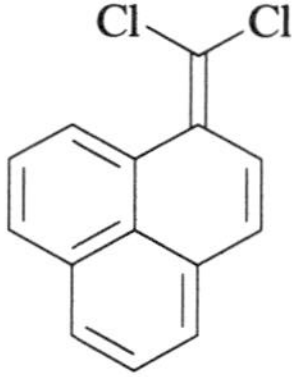 Cl, Cl (>80)	739
	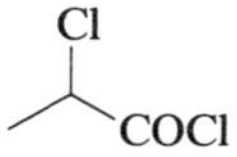 Cl, COCl	C_6H_6, Et_3N	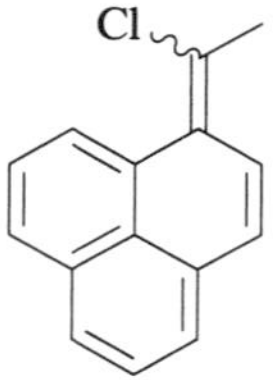 Cl (>80)	739
	i-PrCOCl	C_6H_6, Et_3N	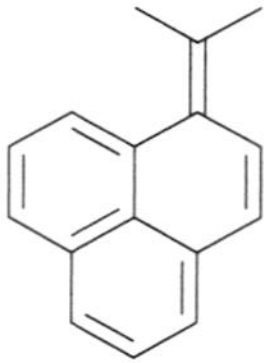 (>80)	739

TABLE XXI. OLEFINS FROM REACTION OF KETENES WITH CARBONYL AND THIOCARBONYL COMPOUNDS (*Continued*)

Reactant	Ketene or Ketene Source	Conditions	Product(s) and Yield(s) (%)	Refs.
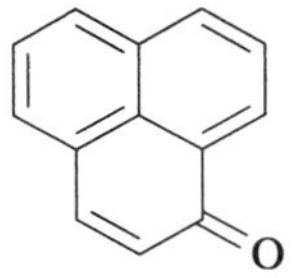 O	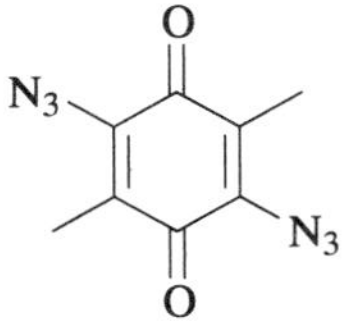 N_3, N_3, O, O	C_6H_6, 80°	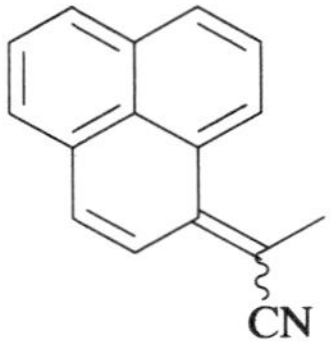 CN (69)	325
	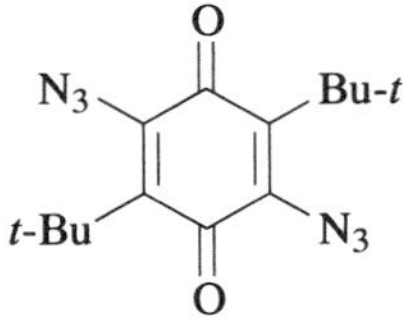 N_3, Bu-*t*, *t*-Bu, N_3, O, O	C_6H_6, 80°	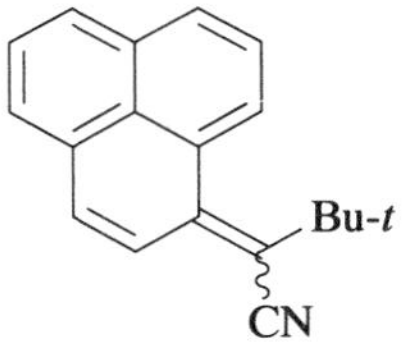 Bu-*t*, CN (75)	325
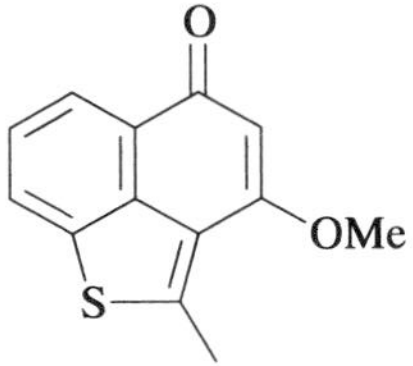 O, OMe, S	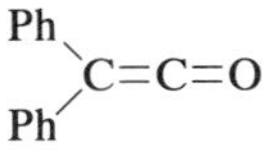 Ph, Ph, C=C=O	C_6H_6, 66°	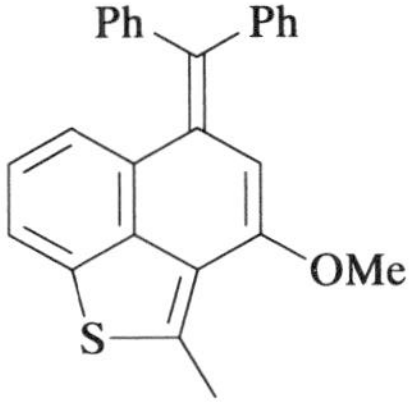 Ph, Ph, OMe, S (52)	740
	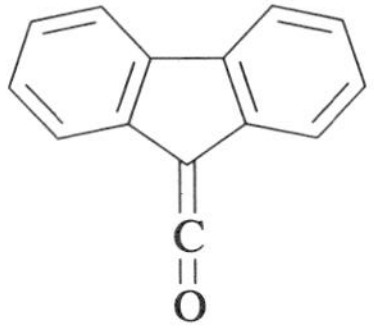 C=O	45°, 1 h	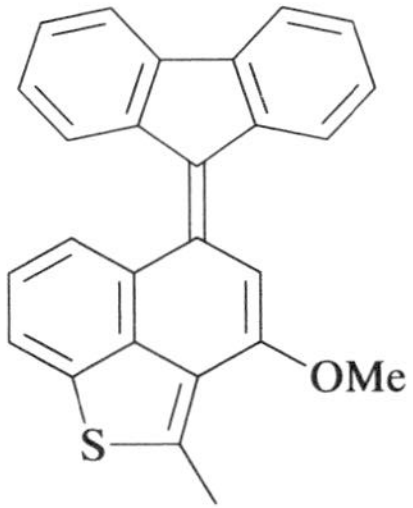 OMe, S (64)	740

TABLE XXI. OLEFINS FROM REACTION OF KETENES WITH CARBONYL AND THIOCARBONYL COMPOUNDS (*Continued*)

Reactant	Ketene or Ketene Source	Conditions	Product(s) and Yield(s) (%)	Refs.
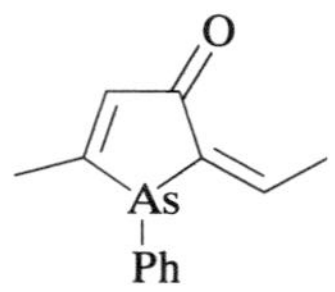 [structure: As-Ph, O]	$Ph_2C{=}C{=}O$	150-200°	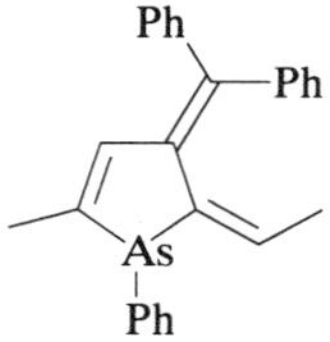 [structure: Ph, Ph, As-Ph] (34)	741
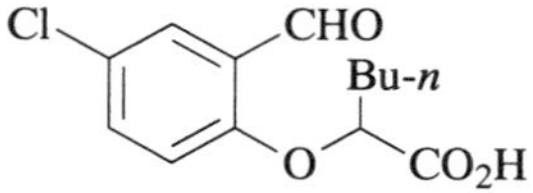 [structure: Cl, CHO, O, Bu-*n*, CO_2H]	—	TsCl, C_6H_6, Et_3N, 80°	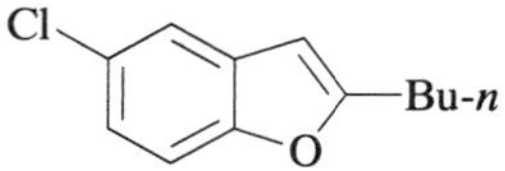 [structure: Cl, O, Bu-*n*] (72)	185
C_{14} 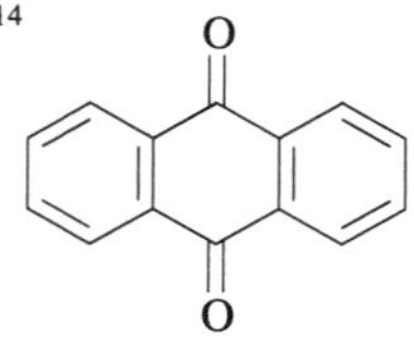 [structure: O, O]	$Ph_2C{=}C{=}O$	Xylene, 140°	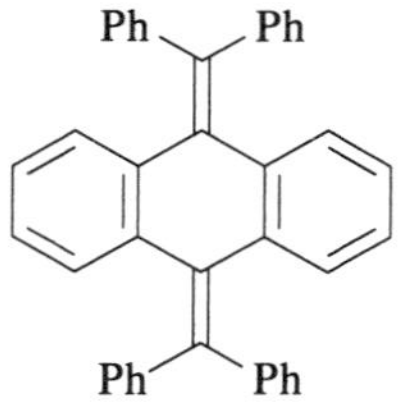 [structure: Ph, Ph, Ph, Ph] (—)	718
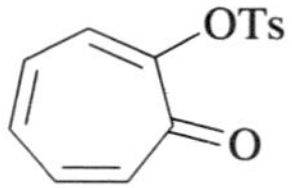 [structure: OTs, O]	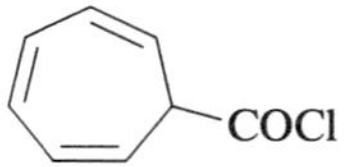 [structure: COCl]	Et_3N	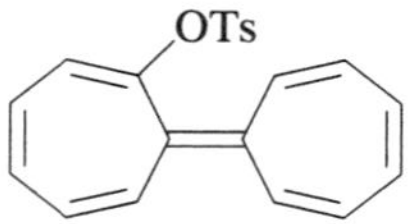 [structure: OTs] (10) 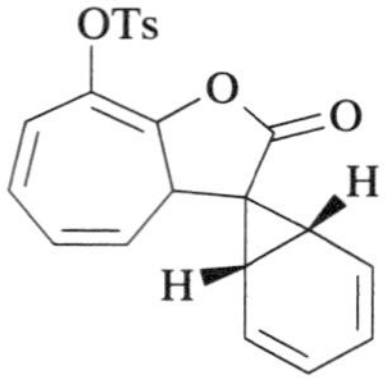 + [structure: OTs, O, O, H, H] (49)	735

TABLE XXI. OLEFINS FROM REACTION OF KETENES WITH CARBONYL AND THIOCARBONYL COMPOUNDS (*Continued*)

Reactant	Ketene or Ketene Source	Conditions	Product(s) and Yield(s) (%)	Refs.
C_{15}				
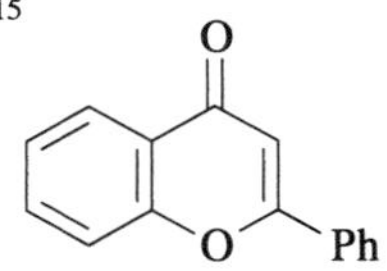 O, O, Ph	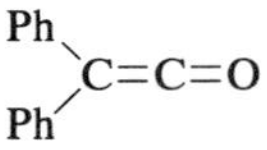 Ph, Ph, C=C=O	150°	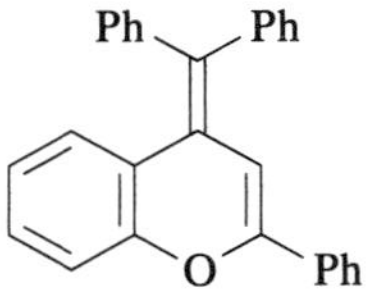 Ph, Ph, O, Ph (70)	728
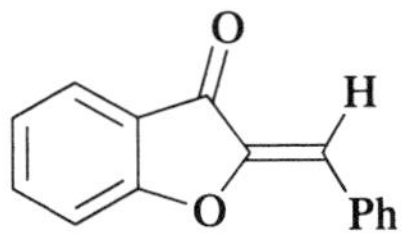 O, H, O, Ph	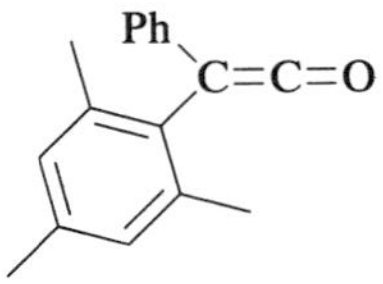 Ph, C=C=O	140°, 4 h	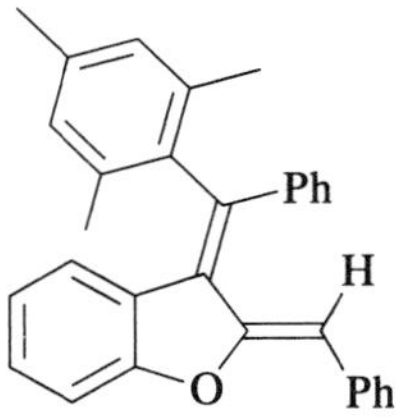 Ph, H, O, Ph (42)	742
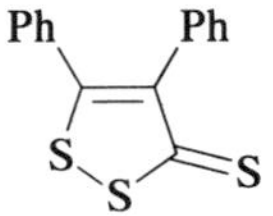 Ph, Ph, S, S, S	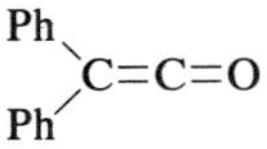 Ph, Ph, C=C=O	Xylene, 137°	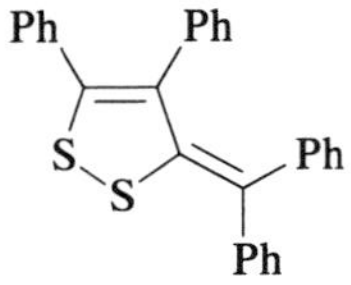 Ph, Ph, S, S, Ph, Ph (35)	186
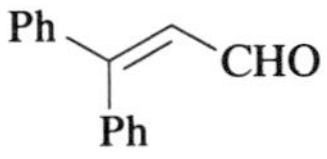 Ph, CHO, Ph	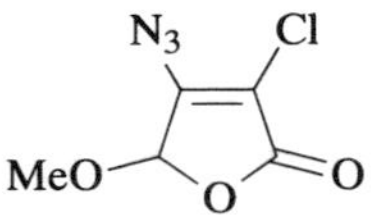 N_3, Cl, MeO, O, O	—	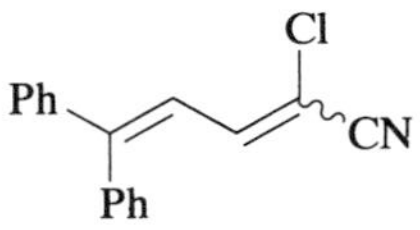 Cl, Ph, CN, Ph (82)	23

TABLE XXI. OLEFINS FROM REACTION OF KETENES WITH CARBONYL AND THIOCARBONYL COMPOUNDS (*Continued*)

Reactant	Ketene or Ketene Source	Conditions	Product(s) and Yield(s) (%)	Refs.
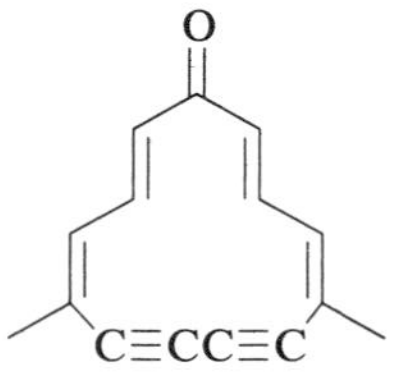 O; C≡CC≡C	$Ph_2CHCOCl$	Et_3N, Et_2O	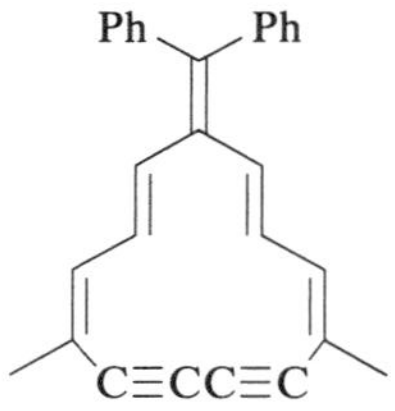 Ph, Ph; C≡CC≡C (32)	743, 744
	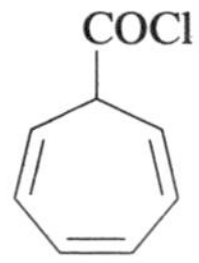 COCl	Et_3N, C_6H_6, 80°	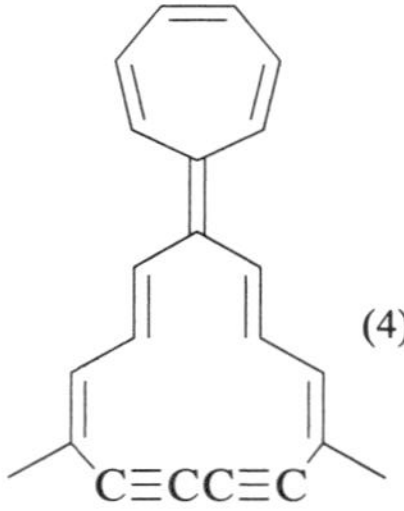 C≡CC≡C (4) 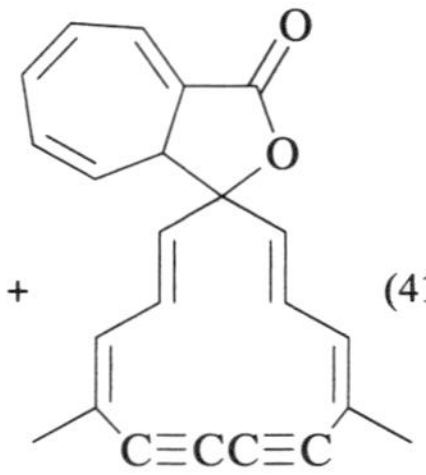 + O; O; C≡CC≡C (41)	745
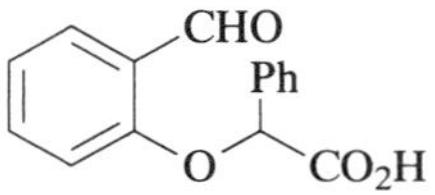 CHO; Ph; O; CO_2H	—	1. $(COCl)_2$ 2. Et_3N, C_6H_6, 80°	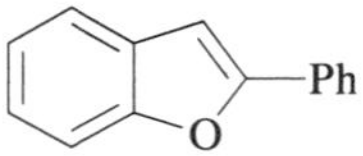 Ph; O (75)	185, 534

TABLE XXI. OLEFINS FROM REACTION OF KETENES WITH CARBONYL AND THIOCARBONYL COMPOUNDS (*Continued*)

Reactant	Ketene or Ketene Source	Conditions	Product(s) and Yield(s) (%)	Refs.
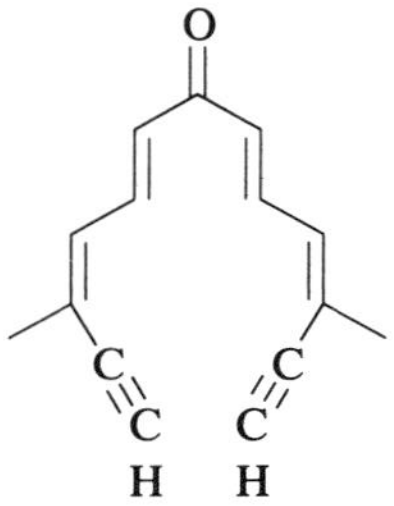 [structure: dienyne ketone with O, C≡C–H termini]	$Cl_2CHCOCl$	Et_3N, C_6H_6, 80°, 3-5 h	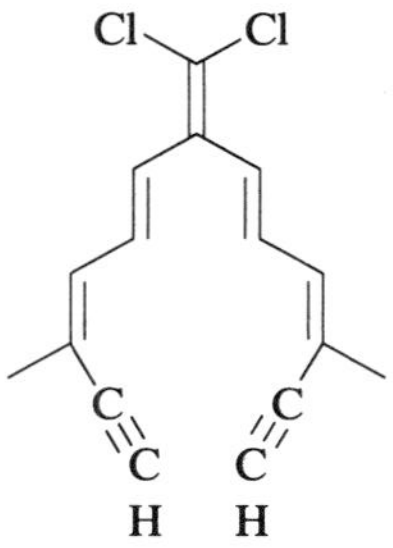 [structure: Cl, Cl-substituted olefin, C≡C–H termini] (15)	746
	$Ph_2CHCOCl$	Et_3N, Et_2O	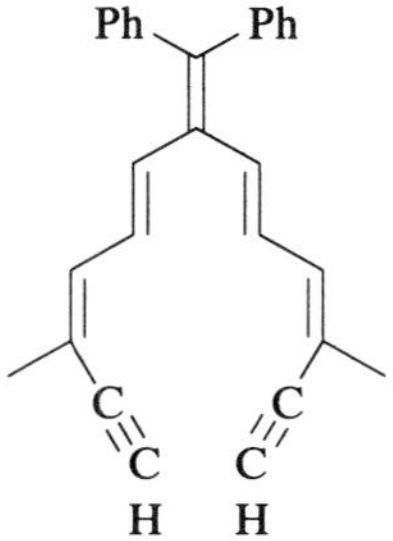 [structure: Ph, Ph-substituted olefin, C≡C–H termini] (36)	743, 744
$(p\text{-}MeC_6H_4)_2C{=}O$	$(p\text{-}MeC_6H_4)_2C{=}C{=}O$	—	$(p\text{-}MeC_6H_4)_2C{=}C(C_6H_4Me\text{-}p)_2$ (—)	747
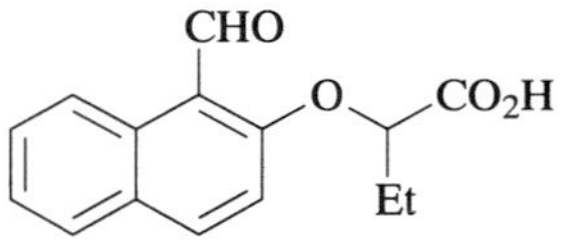 [structure: 1-CHO-naphthalen-2-yl–O–CH(Et)CO_2H]	—	TsCl, C_6H_6, Et_3N, 80°	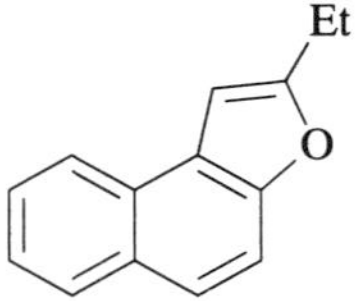 [structure: 2-Et-naphthofuran] (74)	185

TABLE XXI. OLEFINS FROM REACTION OF KETENES WITH CARBONYL AND THIOCARBONYL COMPOUNDS (*Continued*)

Reactant	Ketene or Ketene Source	Conditions	Product(s) and Yield(s) (%)	Refs.
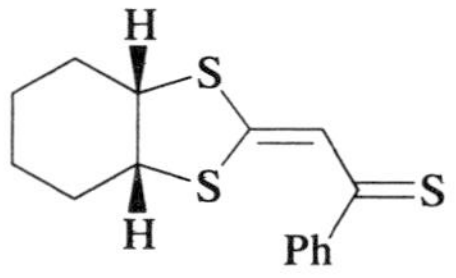 [structure]	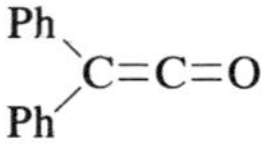 $Ph_2C=C=O$	DMF, 25°	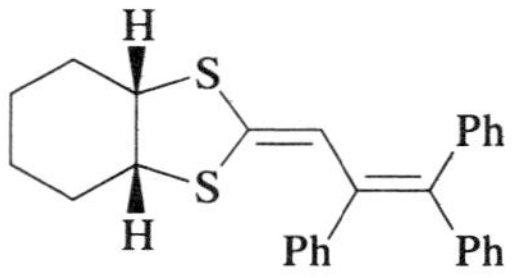 [structure] (88)	270
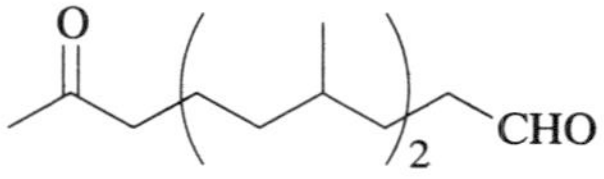 [structure]	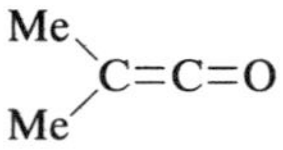 $Me_2C=C=O$	1. $BF_3•Et_2O$, *i*-PrOAc, 85° 2. 150°, 2 h	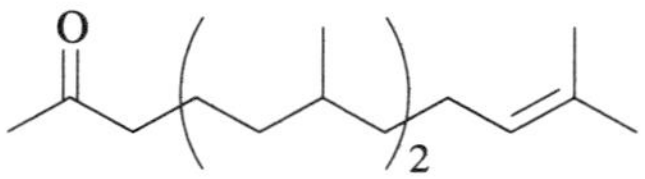 [structure] (82)	162
C_{16} 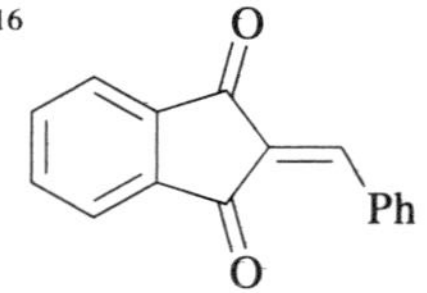 [structure]	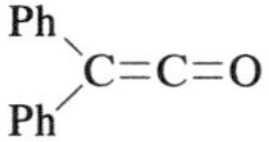 $Ph_2C=C=O$	160°	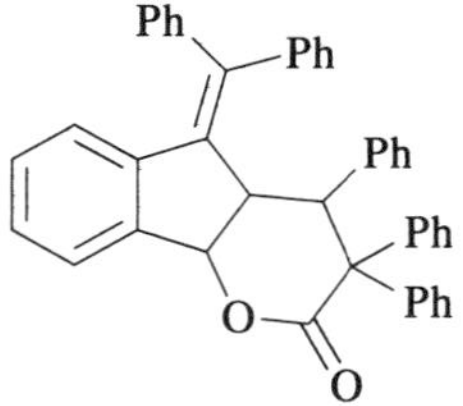 [structure] (21)	748
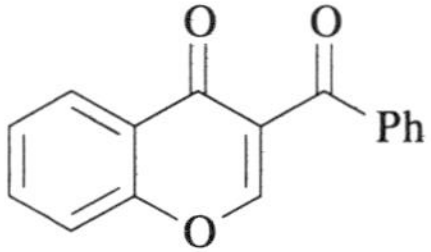 [structure]	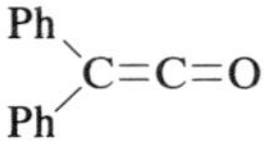 $Ph_2C=C=O$	190-250°, 4 h	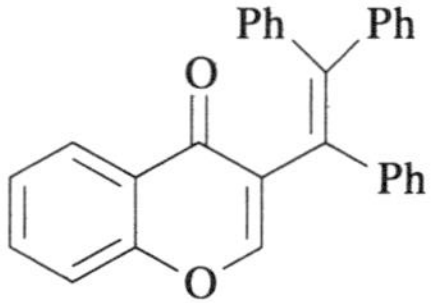 [structure] (—)	749
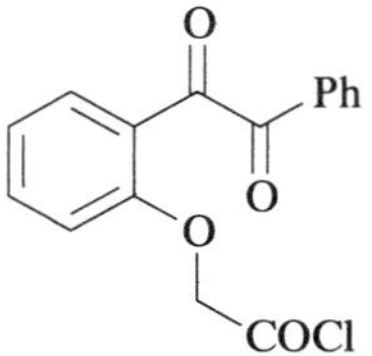 [structure]	—	Et_3N, C_6H_6, 50°	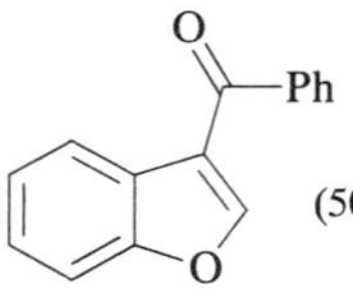 [structure] (50) 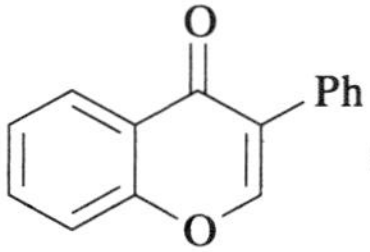 + [structure] (23)	724

TABLE XXI. OLEFINS FROM REACTION OF KETENES WITH CARBONYL AND THIOCARBONYL COMPOUNDS (*Continued*)

Reactant	Ketene or Ketene Source	Conditions	Product(s) and Yield(s) (%)	Refs.
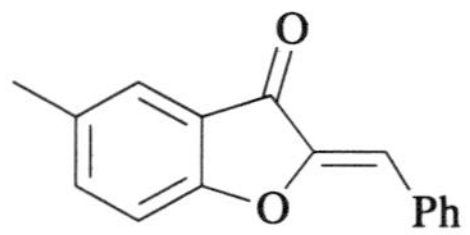 [structure: O, O, Ph]	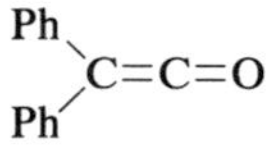 $Ph_2C{=}C{=}O$	56°	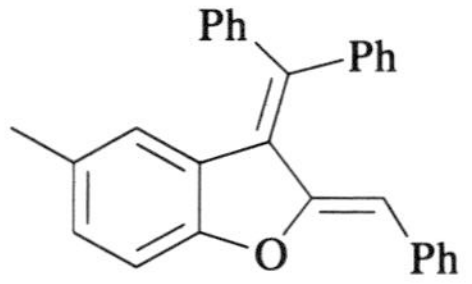 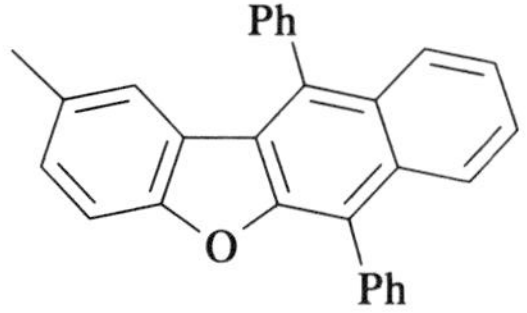 [structure: Ph, Ph, O, Ph] (—) + [structure: Ph, O, Ph] (—)	750
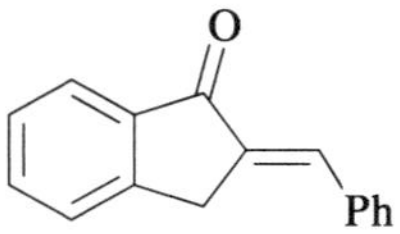 [structure: O, Ph]	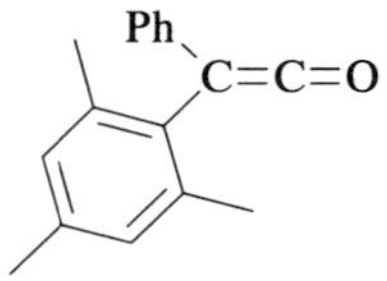 [structure: Ph, C=C=O]	140°, 4 h	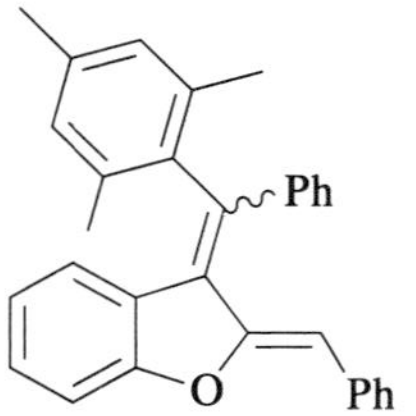 [structure: Ph, O, Ph] (low)	751
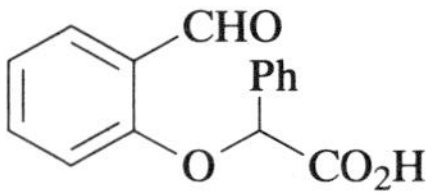 [structure: CHO, Ph, O, CO_2H]	—	1. $(COCl)_2$ 2. Et_3N, C_6H_6, 80°	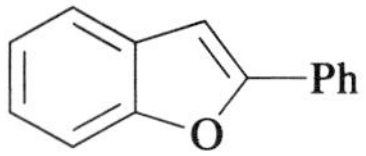 [structure: O, Ph] (78)	185, 534
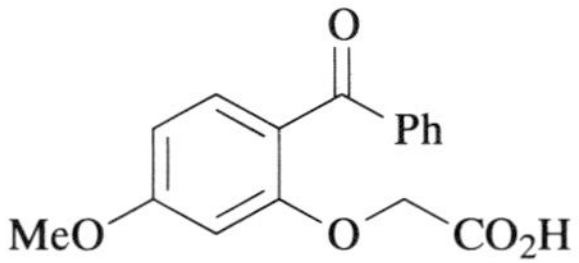 [structure: O, Ph, MeO, O, CO_2H]	—	1. $(COCl)_2$ 2. Et_3N, C_6H_6, 80°	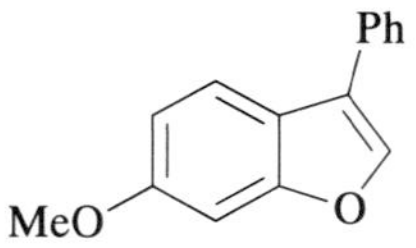 [structure: Ph, MeO, O] (74)	185, 534

TABLE XXI. OLEFINS FROM REACTION OF KETENES WITH CARBONYL AND THIOCARBONYL COMPOUNDS (*Continued*)

Reactant	Ketene or Ketene Source	Conditions	Product(s) and Yield(s) (%)	Refs.
C_{17}				
$(PhC\equiv C)_2C=O$	$Ph_2C=C=O$	130°	$(PhC\equiv C)_2C=CPh_2$ (18)	741
$(p\text{-}ClC_6H_4CH=CH)_2C=O$	$Ph_2C=C=O$	PhMe, 111°	$(p\text{-}ClC_6H_4CH=CH)_2C=CPh_2$ (—)	752
2,6-Ph₂-pyran-4-one (ring O; Ph, Ph)	$Ph_2C=C=O$	150°	4-($Ph_2C=$)-2,6-Ph₂-4H-pyran (ring O; Ph, Ph) (75)	728
2,6-Ph₂-thiopyran-4-one (ring S; Ph, Ph)	$Ph_2C=C=O$	150°	4-($Ph_2C=$)-2,6-Ph₂-4H-thiopyran (ring S; Ph, Ph) (90)	728
3-($p\text{-}MeC_6H_4$)-N-Ph-5-thioxo-pyrrolin-2-one (O, S, N–Ph)	$Ph_2C=C=O$	PhMe, 111°	3-($p\text{-}MeC_6H_4$)-N-Ph-5-($Ph_2C=$)-pyrrolin-2-one (O, N–Ph) (53)	753

TABLE XXI. OLEFINS FROM REACTION OF KETENES WITH CARBONYL AND THIOCARBONYL COMPOUNDS (*Continued*)

Reactant	Ketene or Ketene Source		Conditions	Product(s) and Yield(s) (%)	Refs.
2-($ClCOCH_2O$)$C_6H_4COCOC_6H_4OMe$-*o*	—		Et_3N, C_6H_6, 50°	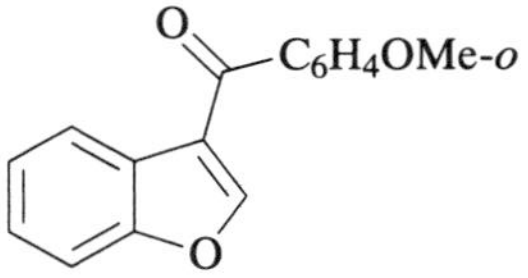 3-(o-$MeOC_6H_4CO$)benzofuran (15) + 3-(C_6H_4OMe-*o*)chromone (55)	724
2-Phenyl-X-heterocyclic ketone (X ring, C=O)	$R^1R^2C{=}C{=}O$		C_6H_6, rt to 80°	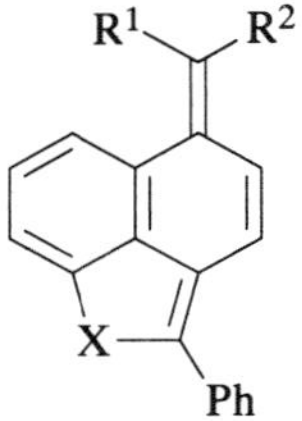 $R^1R^2C{=}$ olefin product (X, Ph)	
X	R^1	R^2			
O	CN	CN		(40)	754
O	Me	CN		(45)	754
O	*t*-Bu	CN		(51)	754
S	*t*-Bu	CN		(64)	325
O	Ph	CN		(40)	754
O	Ph	Ph		(76)	754
S	Ph	Ph		(84)	755, 756
S	9-Fluorenylidene			(38)	756
O	Ph	1-$C_{10}H_7$		(57)	754

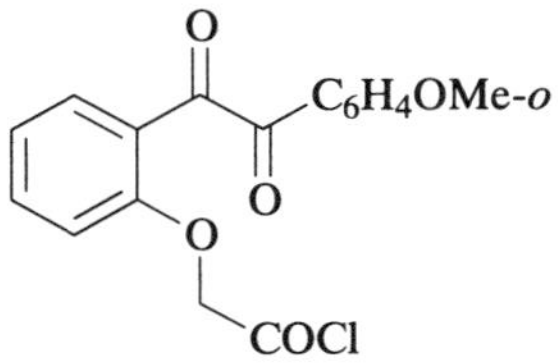

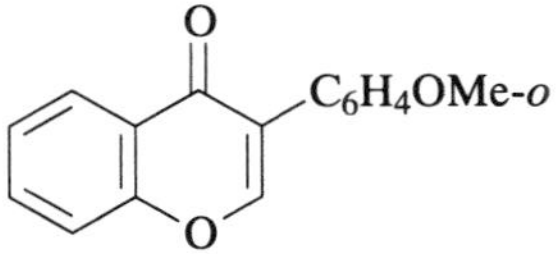

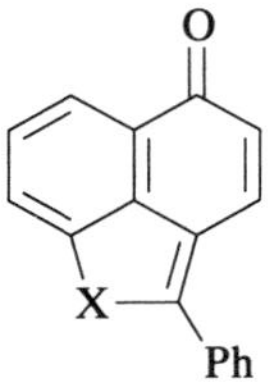

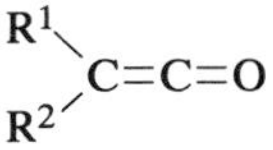

TABLE XXI. OLEFINS FROM REACTION OF KETENES WITH CARBONYL AND THIOCARBONYL COMPOUNDS (*Continued*)

Reactant	Ketene or Ketene Source	Conditions	Product(s) and Yield(s) (%)	Refs.
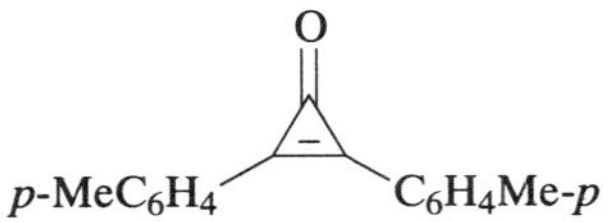 O; *p*-MeC_6H_4, C_6H_4Me-*p*	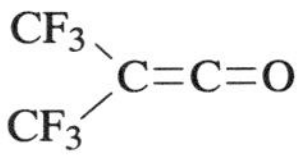 CF_3, CF_3 C=C=O	PhMe, rt, 5 d	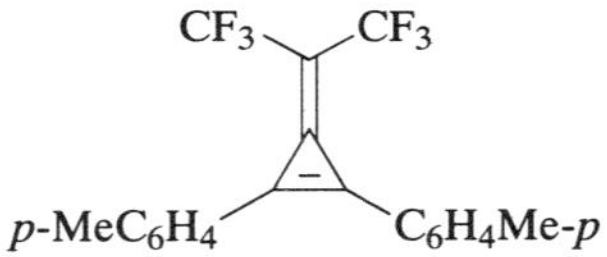 CF_3, CF_3; *p*-MeC_6H_4, C_6H_4Me-*p* (59)	757
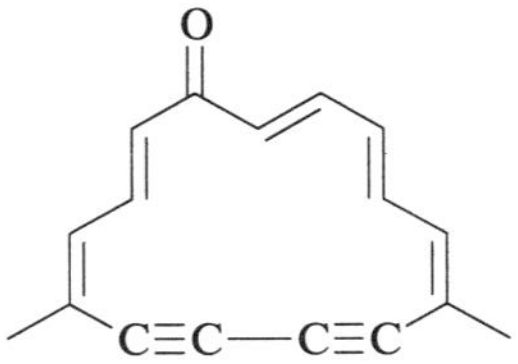 O; C≡C—C≡C	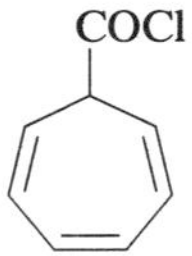 COCl	Et_3N, C_6H_6, 80°	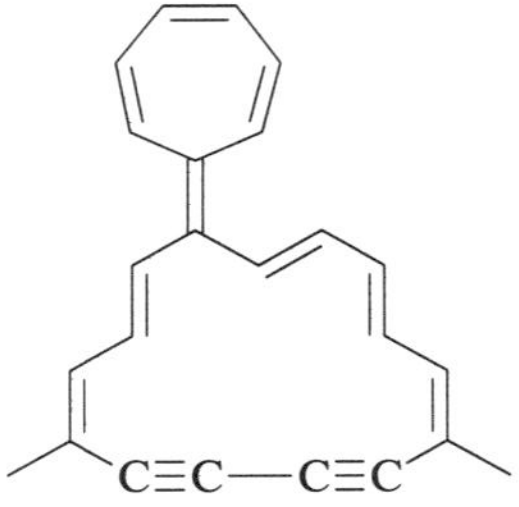 C≡C—C≡C (30)	183
	$Ph_2CHCOCl$	Et_3N, Et_2O	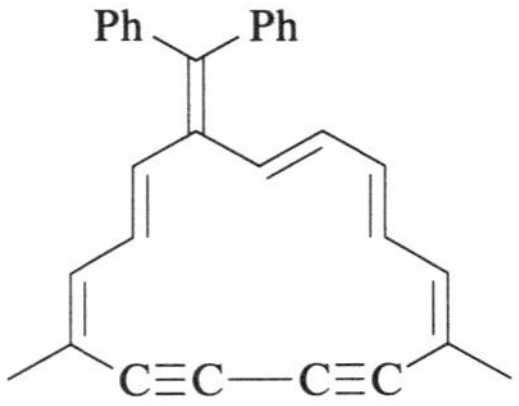 Ph, Ph; C≡C—C≡C (51)	743, 744
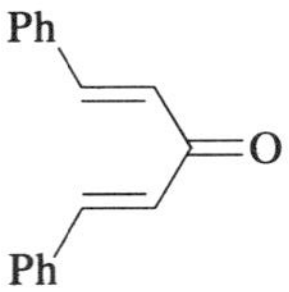 Ph, Ph; O	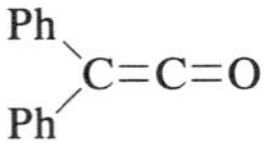 Ph, Ph C=C=O	PhMe, 111°	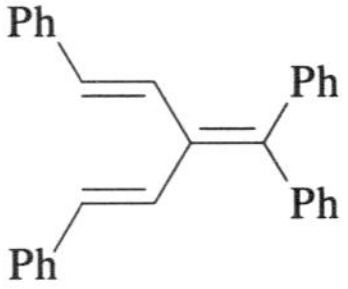 Ph, Ph, Ph, Ph (62)	752

TABLE XXI. OLEFINS FROM REACTION OF KETENES WITH CARBONYL AND THIOCARBONYL COMPOUNDS (*Continued*)

Reactant	Ketene or Ketene Source	Conditions	Product(s) and Yield(s) (%)	Refs.
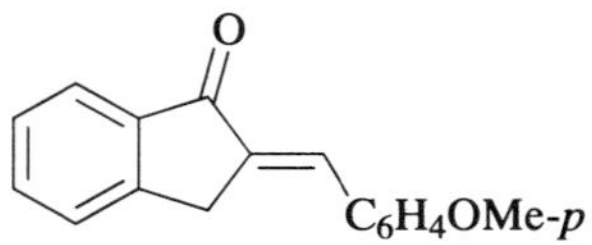 O; C_6H_4OMe-*p*	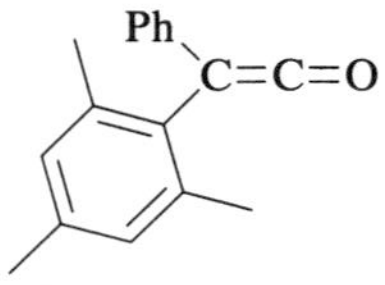 Ph, C=C=O	140°	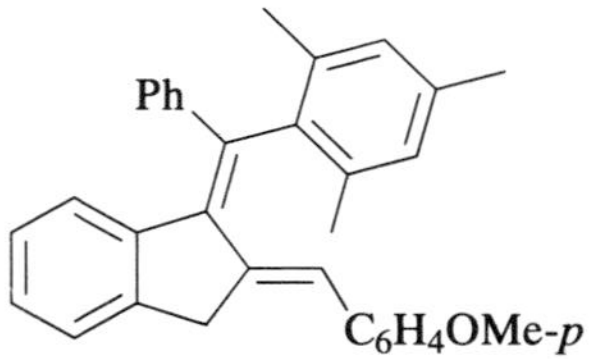 Ph; C_6H_4OMe-*p* (—)	751
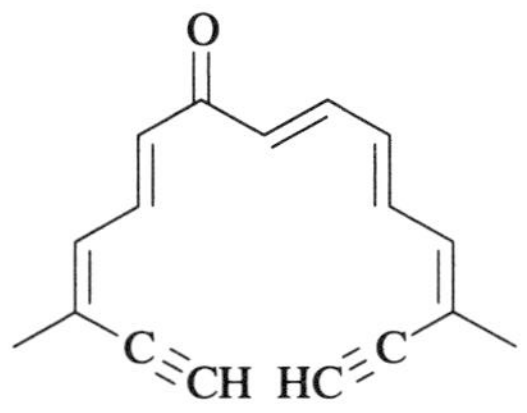 O; C≡CH HC≡C	$Cl_2CHCOCl$	Et_3N, C_6H_6, 80°, 35 h	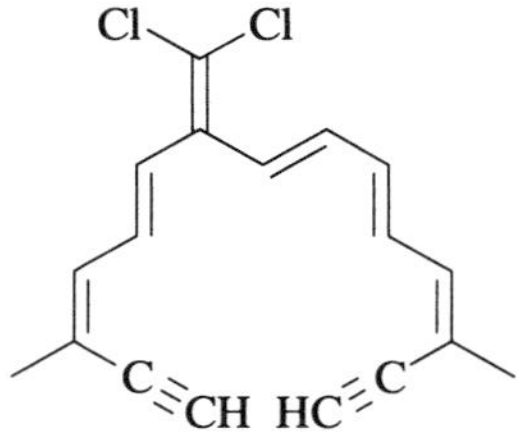 Cl, Cl; C≡CH HC≡C (12)	746
	$Ph_2CHCOCl$	Et_3N, Et_2O	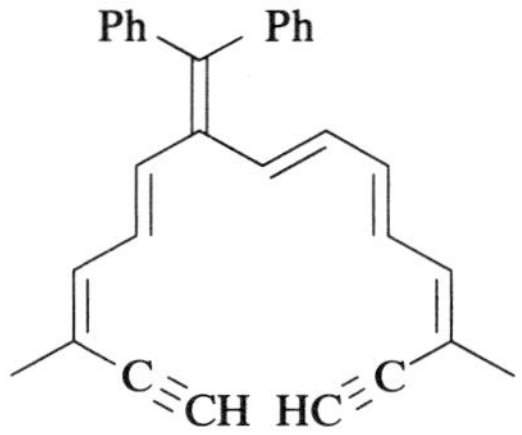 Ph, Ph; C≡CH HC≡C (35)	743, 744
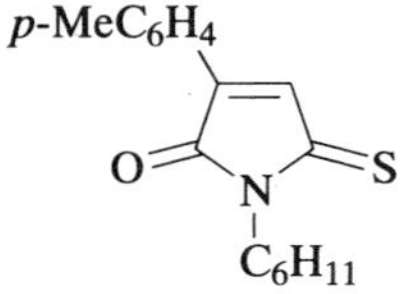 *p*-MeC_6H_4; O; N; S; C_6H_{11}	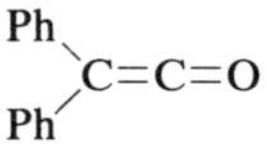 Ph, Ph, C=C=O	PhMe, 111°	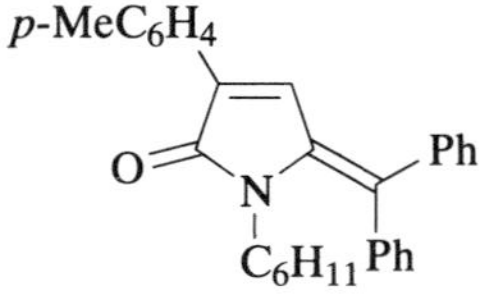 *p*-MeC_6H_4; O; N; C_6H_{11}; Ph; Ph (66)	753

TABLE XXI. OLEFINS FROM REACTION OF KETENES WITH CARBONYL AND THIOCARBONYL COMPOUNDS (*Continued*)

Reactant	Ketene or Ketene Source	Conditions	Product(s) and Yield(s) (%)	Refs.
$(p\text{-}Me_2NC_6H_4)_2C{=}S$	$Ph_2C{=}C{=}O$	Et_2O	$(p\text{-}Me_2NC_6H_4)_2C{=}CPh_2$ (—)	725
3-($C_6H_{11}NH$)-1-(C_6H_4Me-*p*)-pyrrole-2,5-dithione (S=, S=, N–C_6H_4Me-*p*)	$Ph_2C{=}C{=}O$	PhMe, 111°	3-($C_6H_{11}NH$)-1-(p-MeC_6H_4)-5-(diphenylmethylene)-2-thioxopyrrole (S=, =CPh₂) (61)	753
C_{18}				
Benzo-fused macrocyclic ketone (O=, C≡CC≡C)	cycloheptatrienyl-COCl	Et_3N, C_6H_6, 80°	Corresponding cycloheptatrienylidene olefin (C≡CC≡C) (67)	183
Methyl-substituted macrocyclic ketone (O=, C≡C—C≡C)	cycloheptatrienyl-COCl	Et_3N, C_6H_6, 80°	Corresponding cycloheptatrienylidene olefin (C≡C—C≡C) (86)	183

TABLE XXI. OLEFINS FROM REACTION OF KETENES WITH CARBONYL AND THIOCARBONYL COMPOUNDS (*Continued*)

Reactant	Ketene or Ketene Source	Conditions	Product(s) and Yield(s) (%)	Refs.
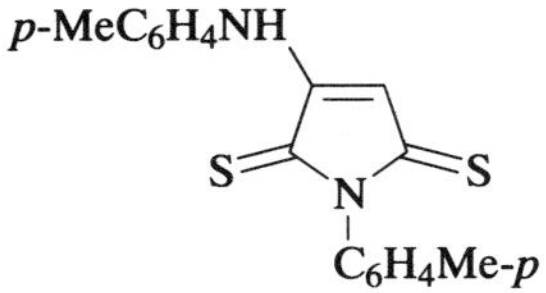 p-MeC_6H_4NH; S; S; N; C_6H_4Me-p	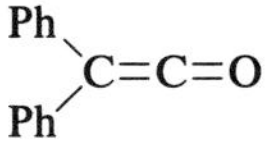 Ph, Ph, C=C=O	PhMe, 111°	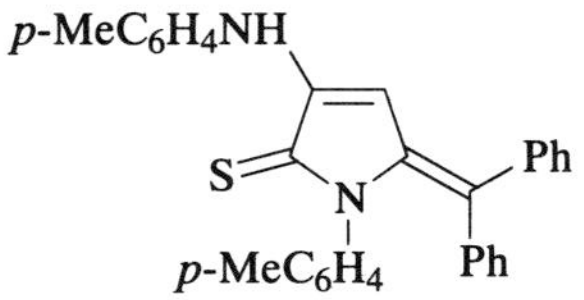 p-MeC_6H_4NH; S; N; p-MeC_6H_4; Ph; Ph (65)	753
C_{19} 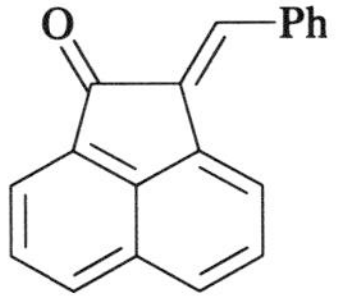 O; Ph	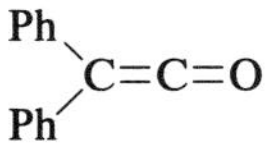 Ph, Ph, C=C=O	130°, 2 h	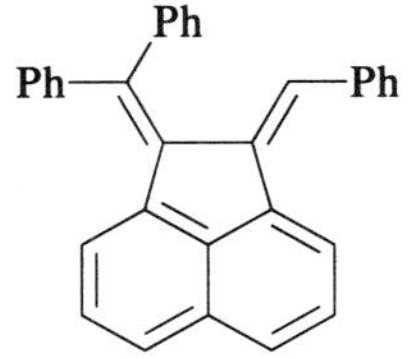 Ph; Ph; Ph (—)[a]	254
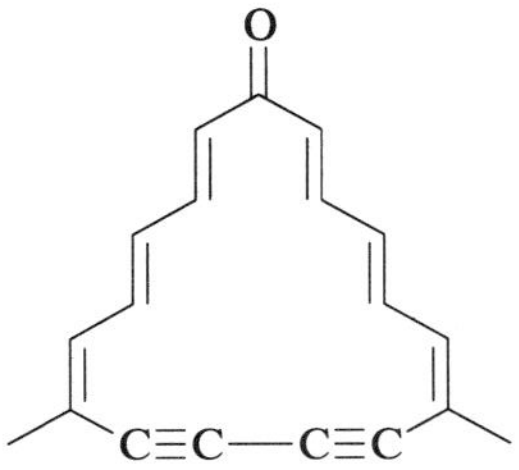 O; C≡C—C≡C	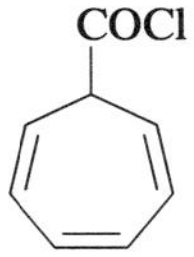 COCl	Et_3N, C_6H_6, 80°	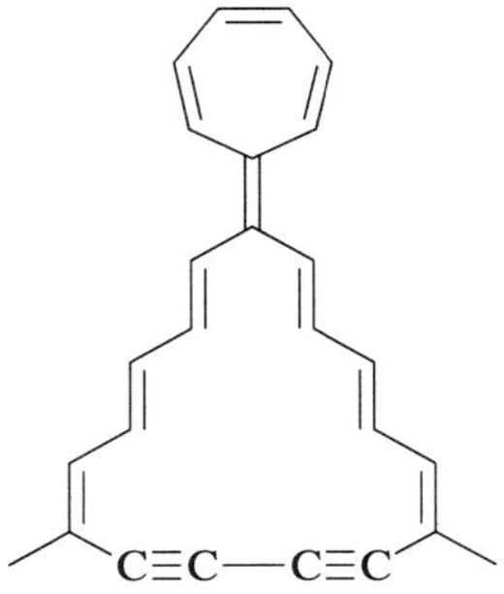 C≡C—C≡C (52)	183

TABLE XXI. OLEFINS FROM REACTION OF KETENES WITH CARBONYL AND THIOCARBONYL COMPOUNDS (*Continued*)

Reactant	Ketene or Ketene Source	Conditions	Product(s) and Yield(s) (%)	Refs.
	$Ph_2CHCOCl$	Et_3N, C_6H_6	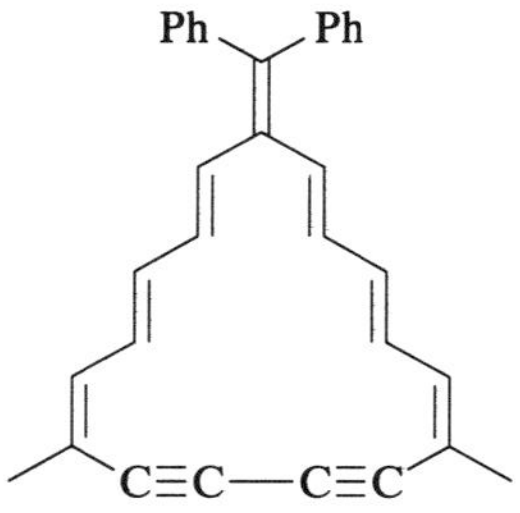 Ph, Ph; C≡C—C≡C (58)	743, 744
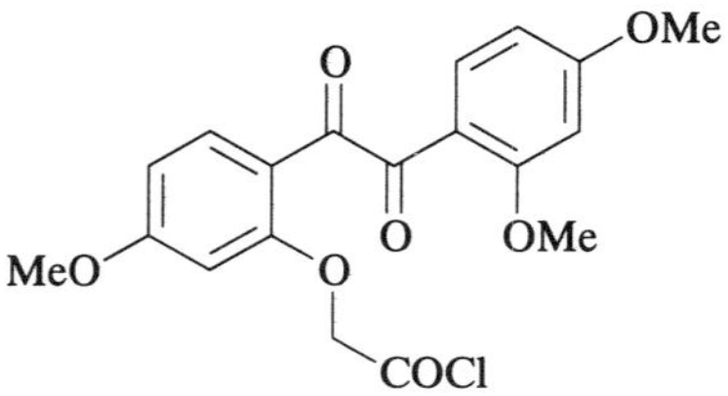 MeO, O, O, OMe, OMe, O, COCl	—	Et_3N, C_6H_6, 50°	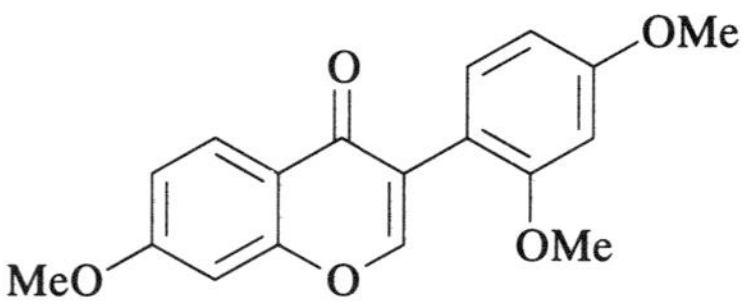 O, OMe, MeO, O, OMe (59)	724
 O; C≡CH HC≡C	$Cl_2CHCOCl$	Et_3N, C_6H_6, 80°, 3-5 h	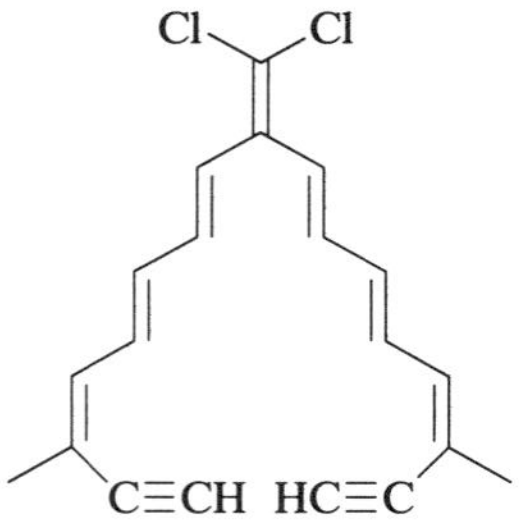 Cl, Cl; C≡CH HC≡C (11)	746

TABLE XXI. OLEFINS FROM REACTION OF KETENES WITH CARBONYL AND THIOCARBONYL COMPOUNDS (*Continued*)

Reactant	Ketene or Ketene Source	Conditions	Product(s) and Yield(s) (%)	Refs.
	$Ph_2CHCOCl$	Et_3N, Et_2O	Ph Ph C≡CH HC≡C (22)	743, 744
p-$MeOC_6H_4$ O *p*-$MeOC_6H_4$	Ph C=C=O Ph	PhMe, 111°	*p*-$MeOC_6H_4$ Ph Ph *p*-$MeOC_6H_4$ (—)	752
O *t*-Bu As Ph Bu-*t*	Ph C=C=O Ph	150-200°	Ph Ph *t*-Bu As Ph Bu-*t* (—)	741

TABLE XXI. OLEFINS FROM REACTION OF KETENES WITH CARBONYL AND THIOCARBONYL COMPOUNDS (*Continued*)

Reactant	Ketene or Ketene Source	Conditions	Product(s) and Yield(s) (%)	Refs.
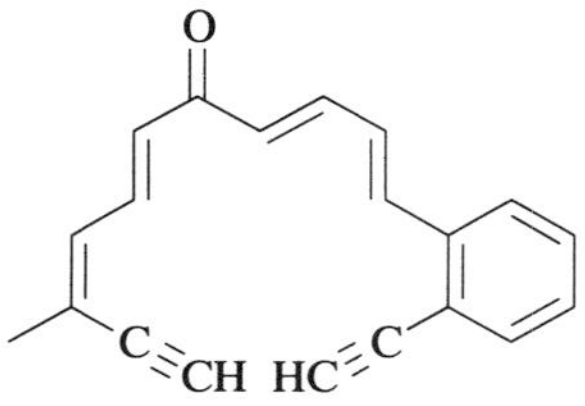 O; C≡CH HC≡C	$Ph_2CHCOCl$	Et_3N, Et_2O	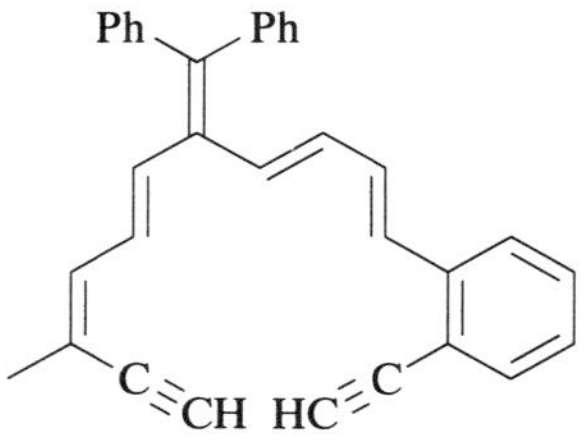 Ph Ph; C≡CH HC≡C (34)	743, 758
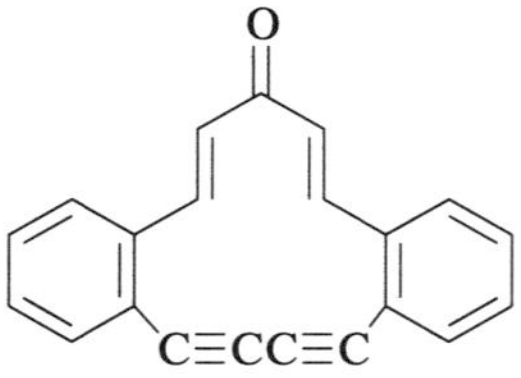 O; C≡CC≡C	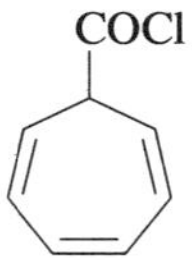 COCl	Et_3N, C_6H_6, 80°	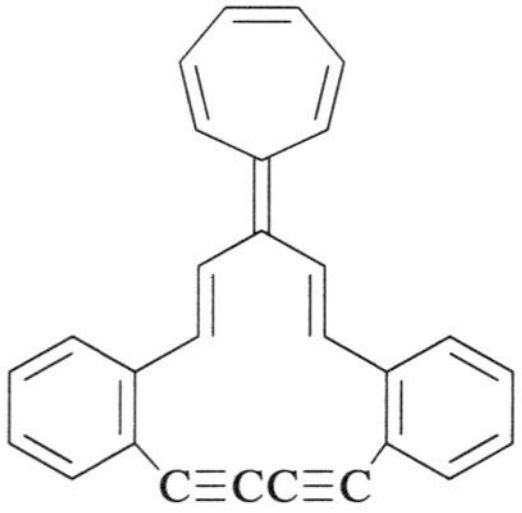 C≡CC≡C (24)	745
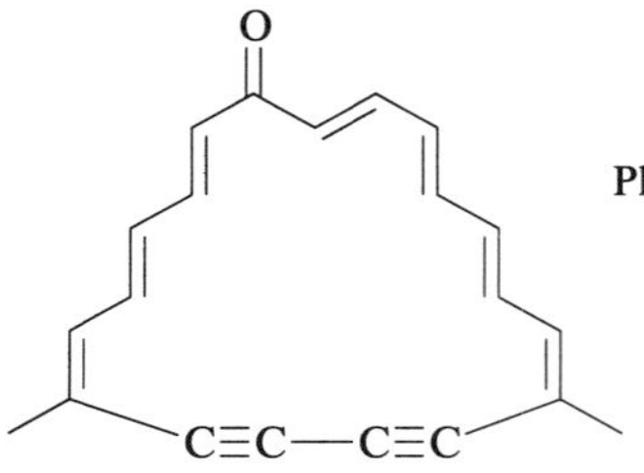 O; C≡C—C≡C	$Ph_2CHCOCl$	Et_3N, Et_2O	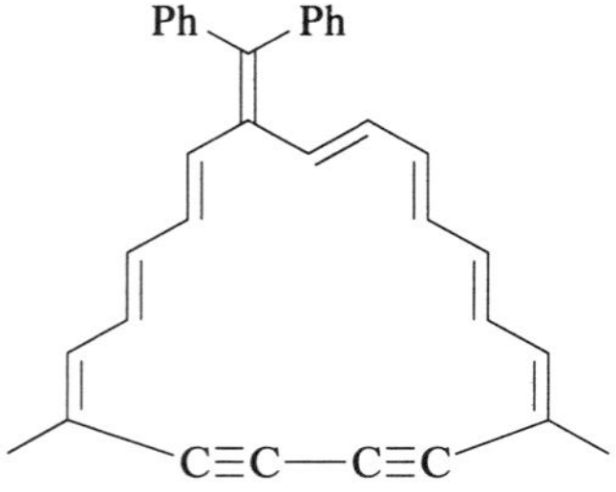 Ph Ph; C≡C—C≡C (35)	743, 744

TABLE XXI. OLEFINS FROM REACTION OF KETENES WITH CARBONYL AND THIOCARBONYL COMPOUNDS (*Continued*)

Reactant	Ketene or Ketene Source	Conditions	Product(s) and Yield(s) (%)	Refs.
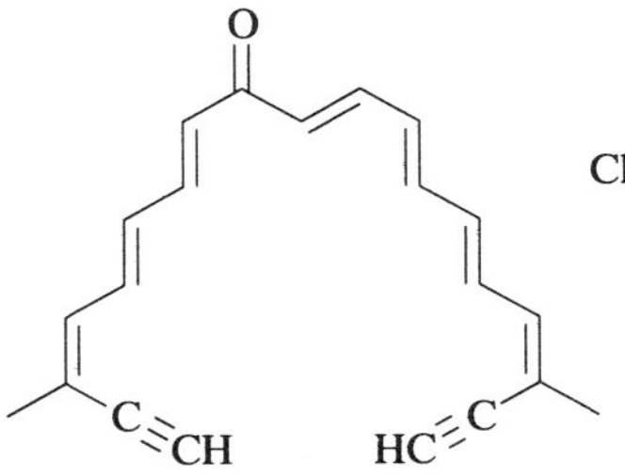 O; C≡CH; HC≡C	$Cl_2CHCOCl$	Et_3N, C_6H_6, 80°, 3-5 h	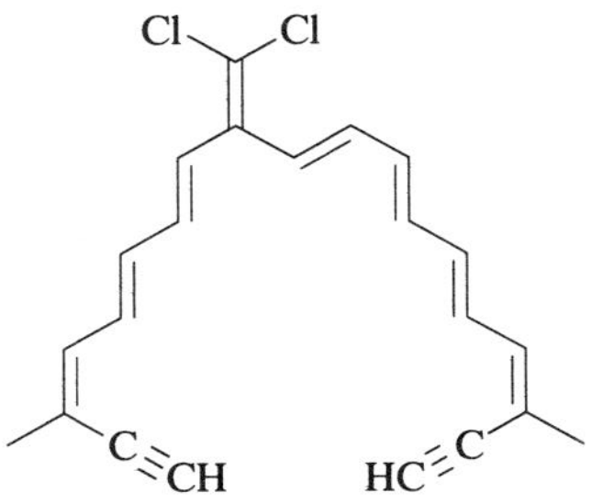 Cl, Cl; C≡CH; HC≡C (14)	746
	$Ph_2CHCOCl$	Et_3N, Et_2O	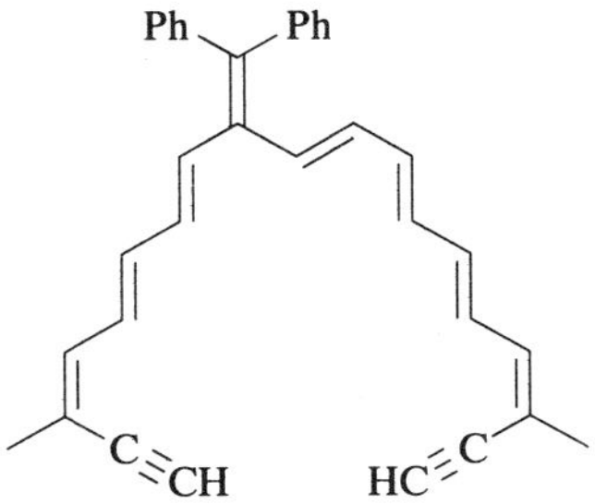 Ph, Ph; C≡CH; HC≡C (18)	743, 744
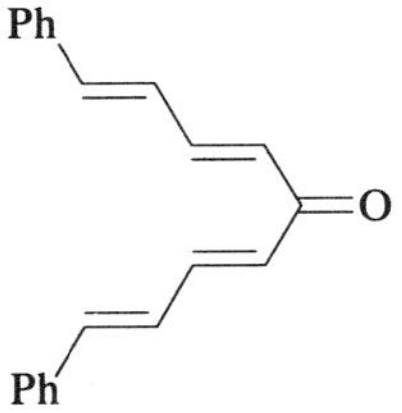 Ph; O; Ph	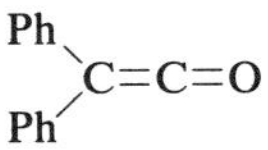 $Ph_2C{=}C{=}O$	PhMe, 111°	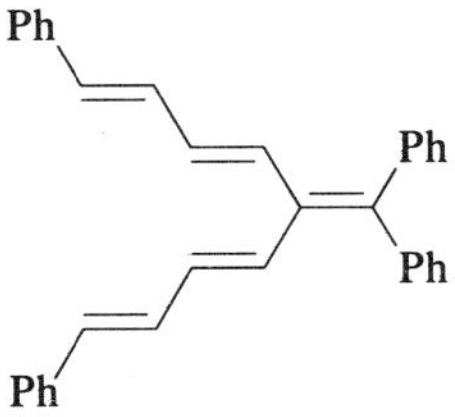 Ph; Ph; Ph; Ph (—)	752

TABLE XXI. OLEFINS FROM REACTION OF KETENES WITH CARBONYL AND THIOCARBONYL COMPOUNDS (*Continued*)

Reactant	Ketene or Ketene Source	Conditions	Product(s) and Yield(s) (%)	Refs.
C_{22}				
	$Ph_2C=C=O$	160°, 6 h	(72)	759
	—	1. $(COCl)_2$ 2. Et_3N, C_6H_6, 80°	(82)	185, 534
C_{23}				
	$Ph_2CHCOCl$	Et_3N, Et_2O	(54)	743, 758
	$Ph_2C=C=O$	150-200°	(69)	741

TABLE XXI. OLEFINS FROM REACTION OF KETENES WITH CARBONYL AND THIOCARBONYL COMPOUNDS (*Continued*)

Reactant	Ketene or Ketene Source	Conditions	Product(s) and Yield(s) (%)	Refs.
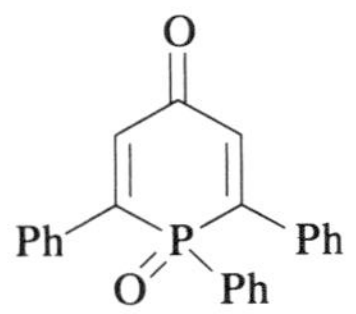 O; Ph, Ph; P; O, Ph	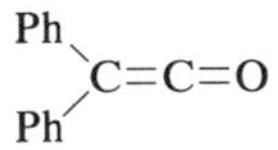 Ph, Ph, C=C=O	170°	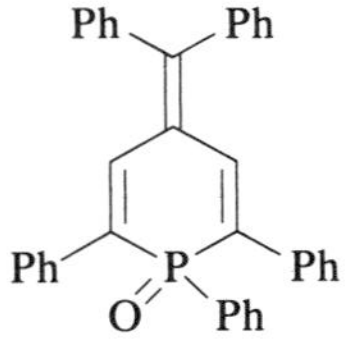 Ph, Ph; Ph, Ph; P; O, Ph (80)	760
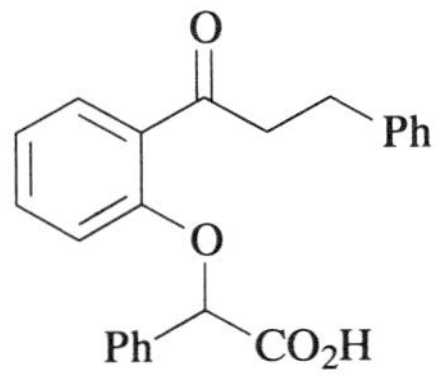 O; Ph; O; Ph, CO_2H	—	TsCl, C_6H_6, Et_3N, 80°	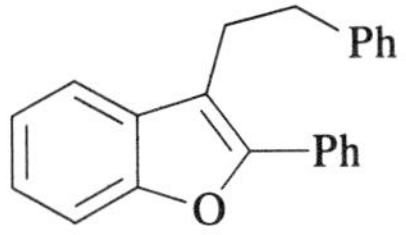 Ph; Ph; O (89)	185
C_{24} 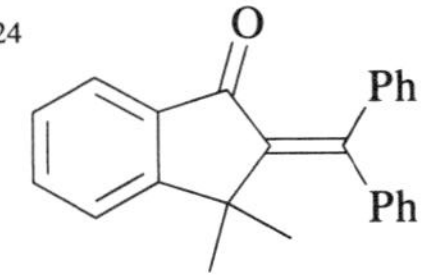 O; Ph, Ph	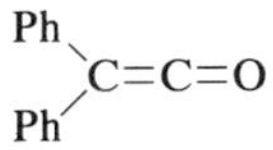 Ph, Ph, C=C=O	160°, 1 h	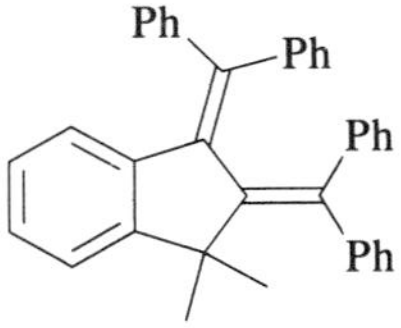 Ph, Ph; Ph, Ph (53)	759
C_{30} 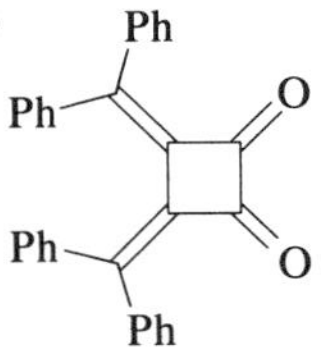 Ph, Ph, O, O, Ph, Ph	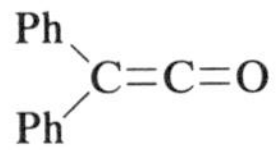 Ph, Ph, C=C=O	Xylene, 137°	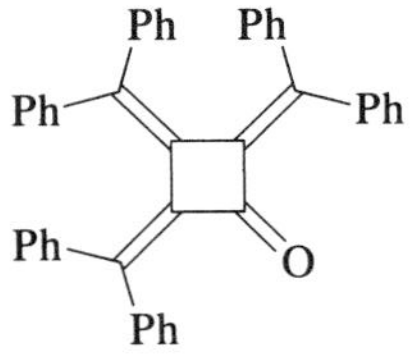 Ph, Ph, Ph, Ph, Ph, Ph, O (45)	761

[a] The product could not be isolated.

TABLE XXII. [2+2] CYCLOADDITION OF KETENES TO ISOCYANATES

Reactant	Ketene or Ketene Source	Conditions	Product(s) and Yield(s) (%)	Refs.
C_1				
FSO_2NCO	$CH_2{=}C{=}O$	$CHCl_3$, -50°	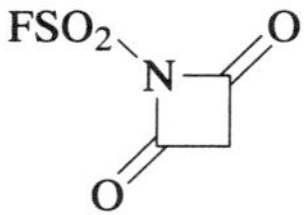 N-(FSO_2)azetidine-2,4-dione (61)	190, 189
$ClSO_2NCO$	$CH_2{=}C{=}O$	$CHCl_3$, -50°	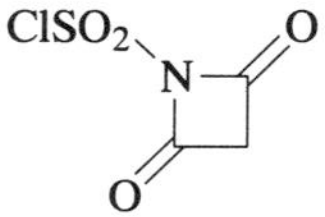 N-($ClSO_2$)azetidine-2,4-dione (54)	190, 189
C_2				
$SO_2(NCO)_2$	$CH_2{=}C{=}O$	Me_2CO, -20°	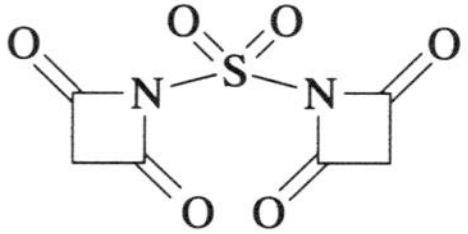 SO_2-bis(N-azetidine-2,4-dione) (74)	190, 762
MeNCO	$Ph_2C{=}C{=}O$	220°, 5 h	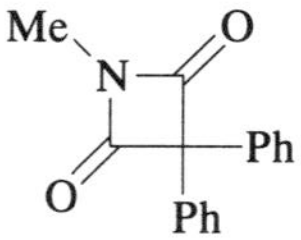 1-Me-3,3-Ph$_2$-azetidine-2,4-dione (18)	763
C_7				
p-$ClC_6H_4SO_2NCO$	$CH_2{=}C{=}O$	$CHCl_3$, -30°	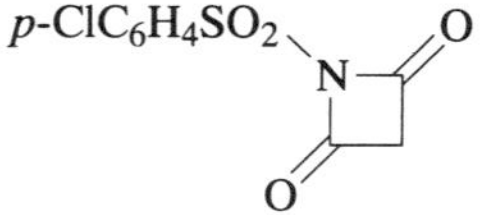 N-(p-$ClC_6H_4SO_2$)azetidine-2,4-dione (47)	189, 190
PhNCO	$Me_2C{=}C{=}O$	MeCN, 82°	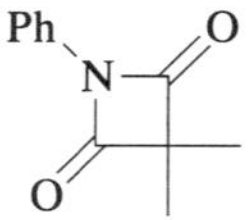 1-Ph-3,3-Me$_2$-azetidine-2,4-dione (30)	188

TABLE XXII. [2+2] CYCLOADDITION OF KETENES TO ISOCYANATES (*Continued*)

Reactant	Ketene or Ketene Source	Conditions	Product(s) and Yield(s) (%)	Refs.
	$C_6H_{11}COCl$	Et_3N, C_6H_6	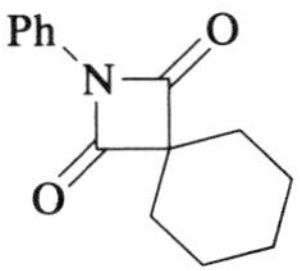 (50)	764
	n-Bu(Et)C=C=O	180°, 5 h	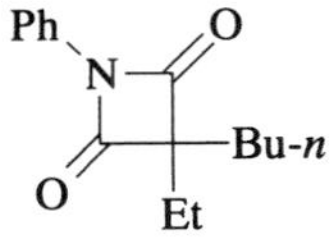 (70)	188
	$Ph_2C=C=O$	150°, 15 h	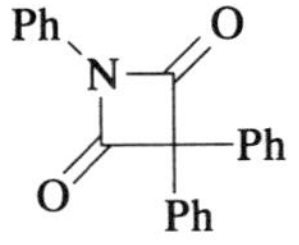 (20)	765, 766
$C_6H_{11}NCO$	$Ph_2C=C=O$	220°, 5 h	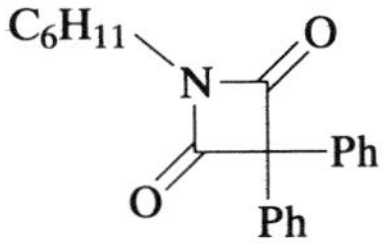 (26)	763
C_8				
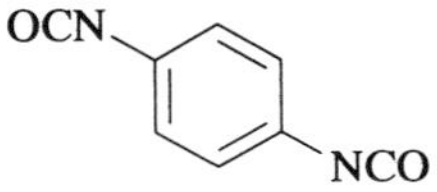 OCN-C6H4-NCO	$Et_2C=C=O$	PhMe, 180°, 5 h	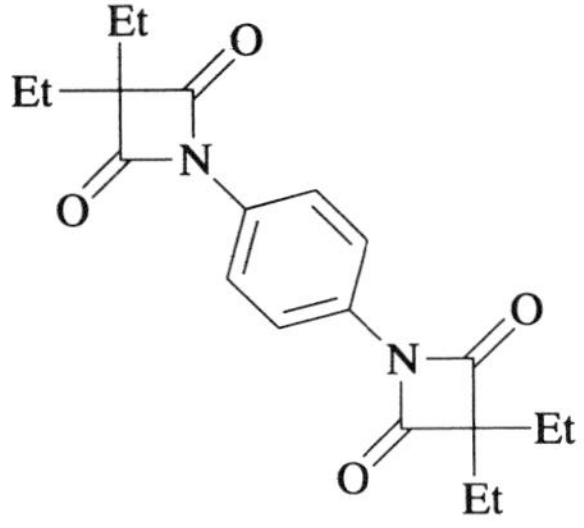 (91)	188

TABLE XXII. [2+2] CYCLOADDITION OF KETENES TO ISOCYANATES (*Continued*)

Reactant	Ketene or Ketene Source	Conditions	Product(s) and Yield(s) (%)	Refs.
	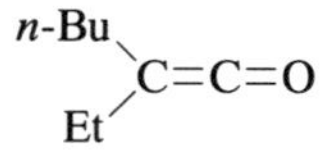 n-Bu(Et)C=C=O	PhMe, 180°, 5 h	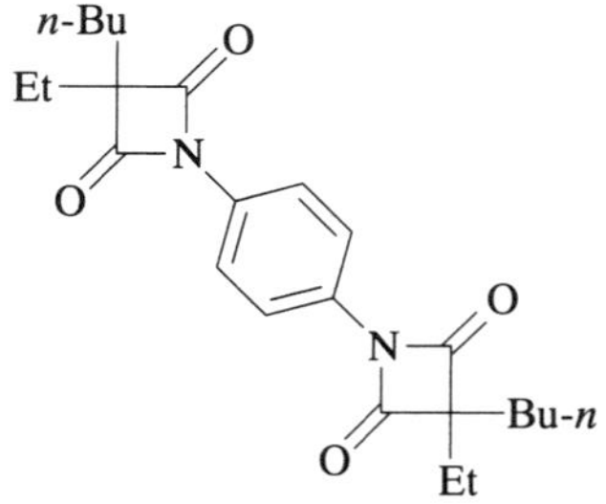 1,4-bis(3-n-Bu-3-Et-azetidine-2,4-dion-1-yl)benzene (78)	188
TsNCO	CH_2=C=O	MeCN	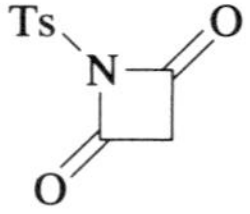 1-Ts-azetidine-2,4-dione (54)	188
	CH_2=C=O	CH_2Cl_2, -10°	" (42)	190
	CH_2=C=O	$CHCl_3$, -30°	" (42)	189, 766, 767
	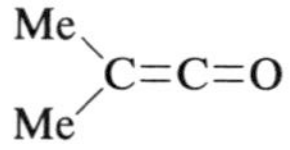 Me(Me)C=C=O	C_6H_6, 10-20°	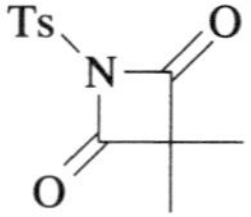 1-Ts-3,3-dimethylazetidine-2,4-dione (94)	188
$OCN(CH_2)_2NCO$	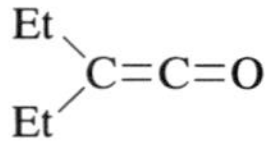 Et(Et)C=C=O	PhMe, 180°	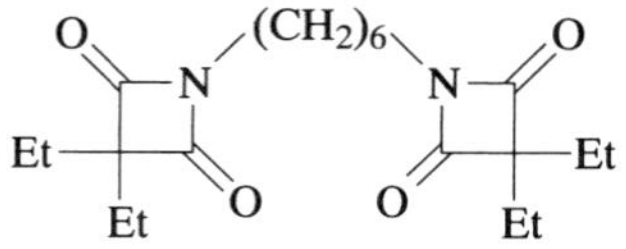 1,1'-$(CH_2)_6$-bis(3,3-diethylazetidine-2,4-dione) (30)	188

TABLE XXII. [2+2] CYCLOADDITION OF KETENES TO ISOCYANATES (*Continued*)

Reactant	Ketene or Ketene Source	Conditions	Product(s) and Yield(s) (%)		Refs.
	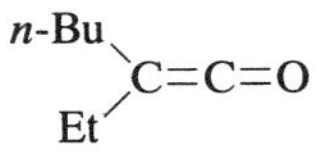 n-Bu(Et)C=C=O	PhMe, 180°	O, N, $(CH_2)_6$, N, O, Et, Bu-n, n-Bu, O, O, Et	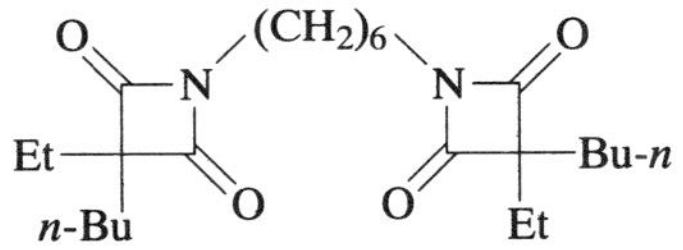 (10)	188
C_9 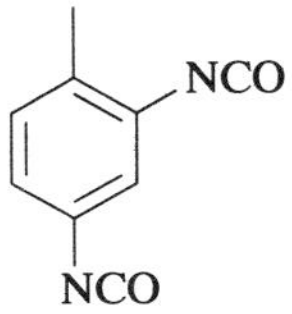 NCO, NCO	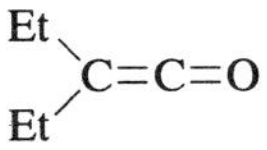 Et(Et)C=C=O	PhMe, 180°	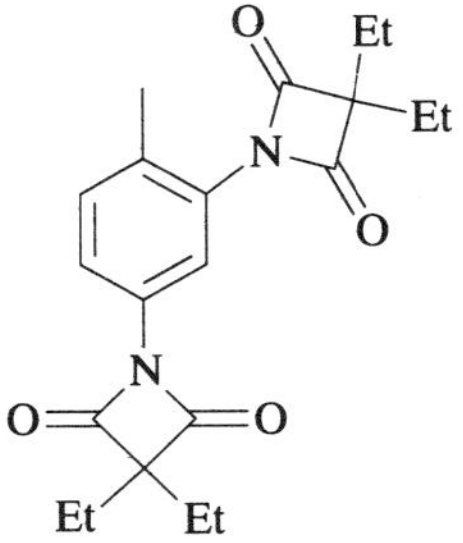 O, Et, Et, N, O, N, O, O, Et, Et	(85)	188
	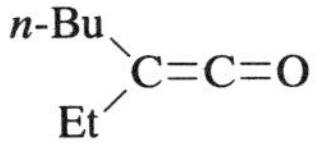 n-Bu(Et)C=C=O	PhMe, 180°	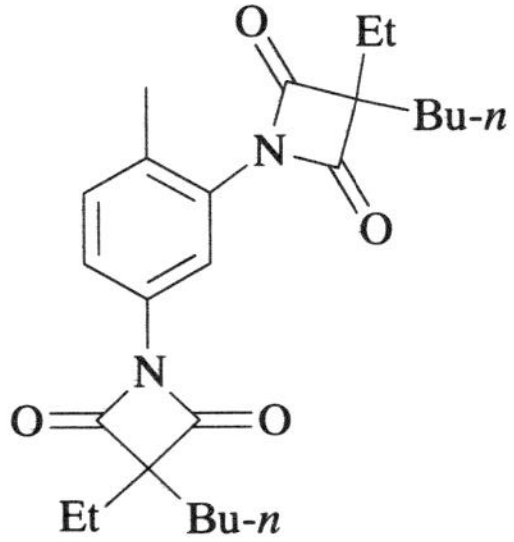 O, Et, Bu-n, N, O, N, O, O, Et, Bu-n	(80)	188

TABLE XXII. [2+2] CYCLOADDITION OF KETENES TO ISOCYANATES (*Continued*)

Reactant	Ketene or Ketene Source	Conditions	Product(s) and Yield(s) (%)	Refs.
	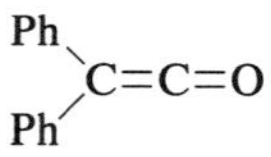 $Ph_2C=C=O$	PhMe, 180°	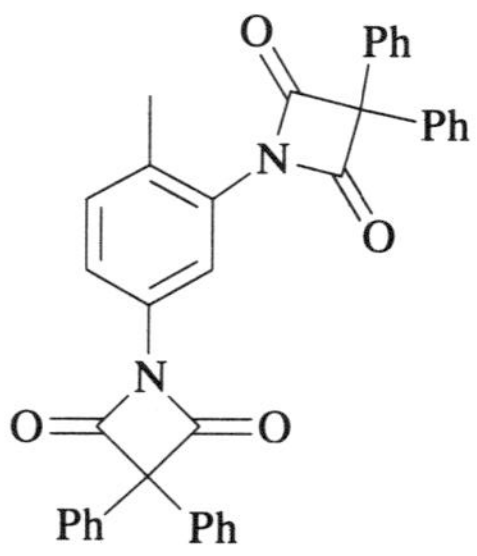 (30)	188
C_{10} 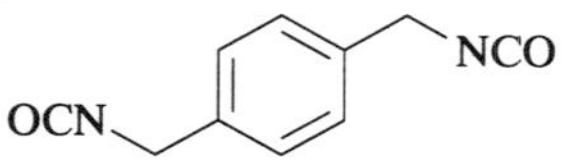 OCN–$CH_2C_6H_4CH_2$–NCO	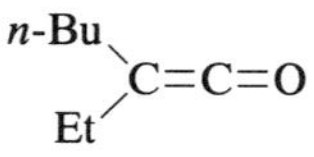 n-Bu(Et)C=C=O	PhMe, 180°	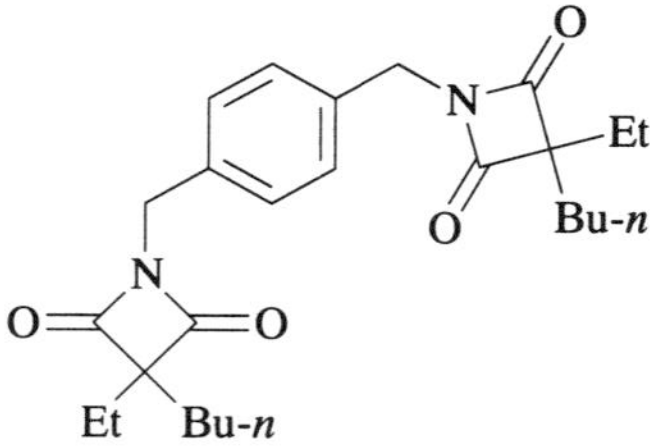 (75)	188
	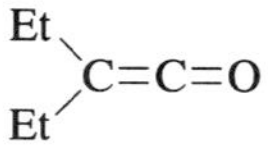 $Et_2C=C=O$	PhMe, 180°	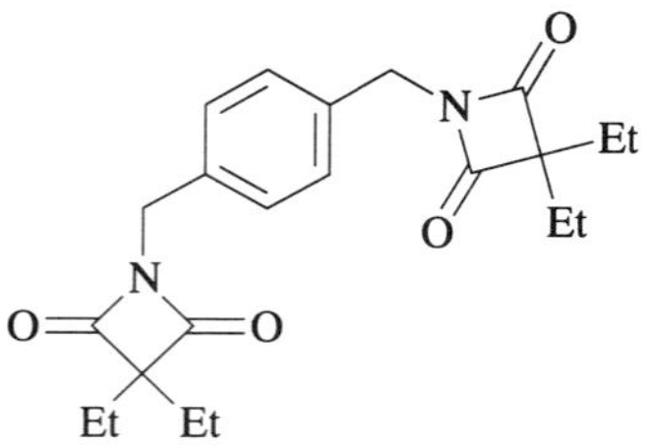 (50)	188

TABLE XXII. [2+2] CYCLOADDITION OF KETENES TO ISOCYANATES (*Continued*)

Reactant	Ketene or Ketene Source	Conditions	Product(s) and Yield(s) (%)	Refs.
C_{11}				
SO_2NCO	$CH_2{=}C{=}O$	EtOAc, 0°	O, O, S, N, O, O (50)	189
C_{15}				
OCN, NCO	Et, Et, C=C=O	PhMe, 180°	O, Et, Et, N, O, CH_2, 2 (80)	188
	n-Bu, Et, C=C=O	PhMe, 180°	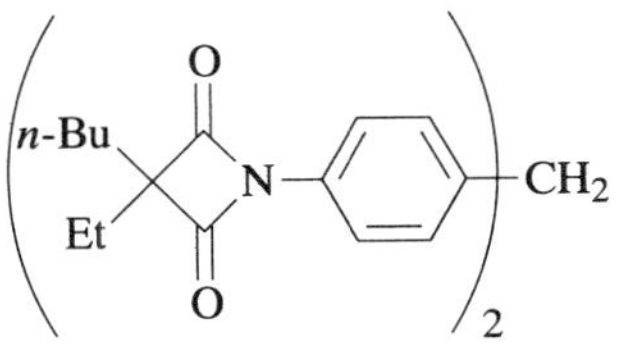 O, *n*-Bu, Et, N, O, CH_2, 2 (65)	188

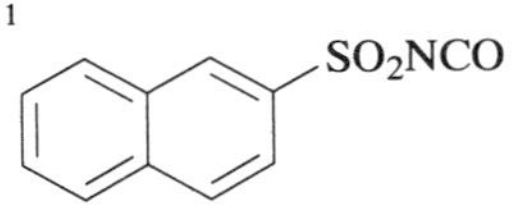

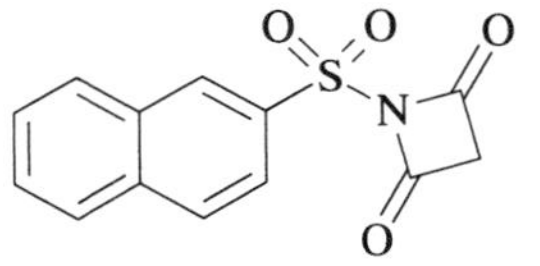

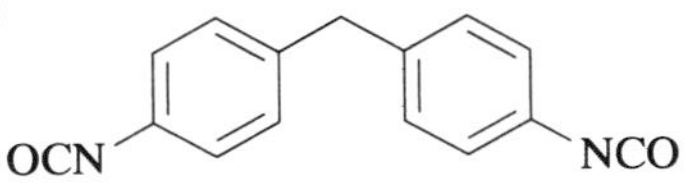

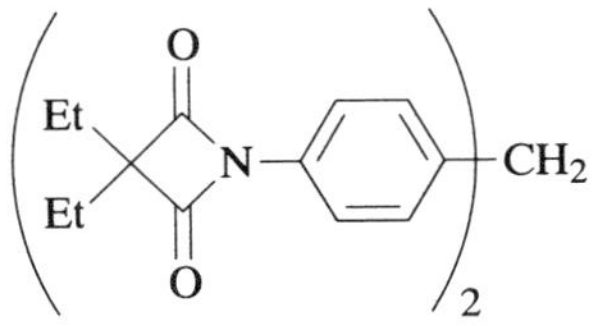

TABLE XXIII. [2+2] CYCLOADDITION OF KETENES TO CARBODIIMIDES

Reactant	Ketene or Ketene Source	Conditions	Product(s) and Yield(s) (%)	Refs.
C_5				
TMSN=C=NMe	$Cl_2CHCOCl$	Et_3N, C_6H_{14}, 50°	Me, NTMS, Cl, Cl (35)	768
C_6				
	$Cl_2CHCOCl$	Et_3N, C_6H_{14}, 50°	(60-80)	192
	$Ph(Et)C=C=O$	Et_2O, 25°, xs ketene	(91)	192
	$Ph_2C=C=O$	Et_2O, 25°	(90)	192
	$Ph_2C=C=O$	Et_2O, 25°, xs ketene	(91)	192

TABLE XXIII. [2+2] CYCLOADDITION OF KETENES TO CARBODIIMIDES (*Continued*)

Reactant	Ketene or Ketene Source	Conditions	Product(s) and Yield(s) (%)	Refs.
EtN=C=NPr-*i*	$Ph_2C=C=O$	C_6H_6, 25°	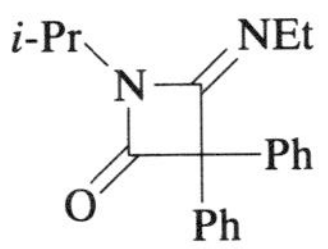 *i*-Pr, NEt, N, O, Ph, Ph (15) 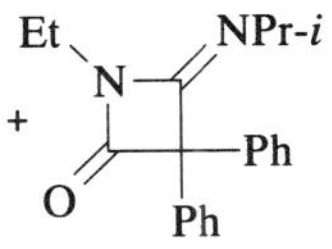 + Et, NPr-*i*, N, O, Ph, Ph (15)	769
MeN=C=NBu-*t*	$Ph_2C=C=O$	C_6H_6, 25°	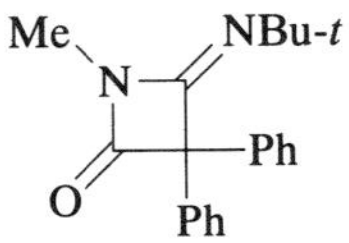 Me, NBu-*t*, N, O, Ph, Ph (70)	769
TMSN=C=NEt	$Cl_2CHCOCl$	Et_3N, C_6H_{14}, 50°	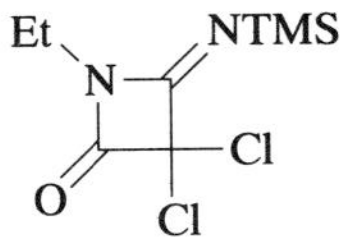 Et, NTMS, N, O, Cl, Cl (37)	768
	$Br_2CHCOCl$	Et_3N, C_6H_{14}, 50°	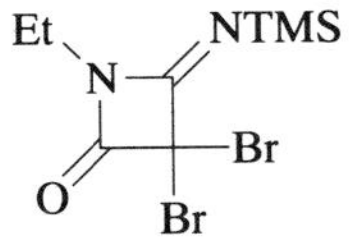 Et, NTMS, N, O, Br, Br (15)	768
C_7 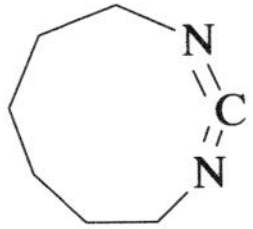 N=C=N (cyclic)	$Ph_2C=C=O$	Et_2O, 25°	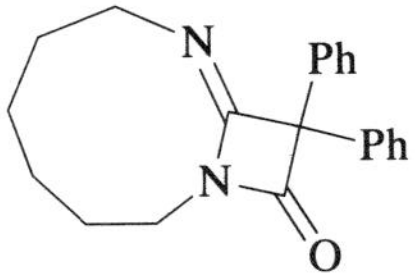 N, N, Ph, Ph, O (90)	192
EtN=C=NBu-*t*	$Ph_2C=C=O$	C_6H_6, 25°	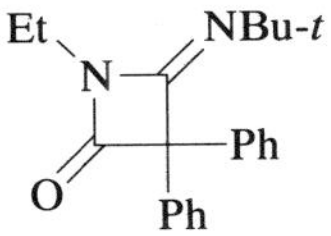 Et, NBu-*t*, N, O, Ph, Ph (71)	769

TABLE XXIII. [2+2] CYCLOADDITION OF KETENES TO CARBODIIMIDES (*Continued*)

Reactant	Ketene or Ketene Source	Conditions	Product(s) and Yield(s) (%)	Refs.
i-PrN=C=NPr-*i*	CH_2=C=O	rt	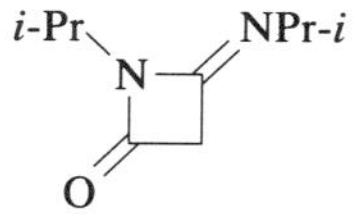 1-*i*-Pr-4-(NPr-*i*)-azetidin-2-one (5)	770
	$ClCH_2COCl$	Et_3N, C_6H_{14}, 69°	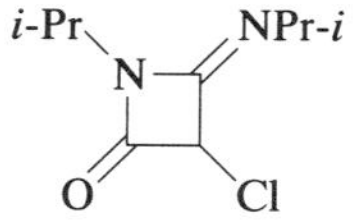 1-*i*-Pr-4-(NPr-*i*)-3-Cl-azetidin-2-one (20)	770
	FCH_2COCl	Et_3N, Et_2O, -78°	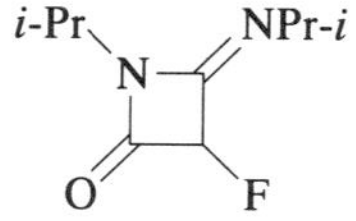 1-*i*-Pr-4-(NPr-*i*)-3-F-azetidin-2-one (40)	770
	$Cl_2CHCOCl$	Et_3N, C_6H_{14}, 50°	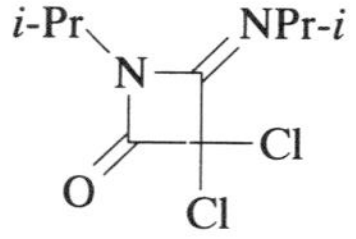 1-*i*-Pr-4-(NPr-*i*)-3,3-Cl_2-azetidin-2-one (42)	768
	$Cl_2CHCOCl$	Et_3N, C_6H_{14}, 81°	" (53)	771
	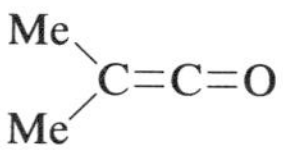 Me_2C=C=O	C_6H_{14}, 69°, 8 h	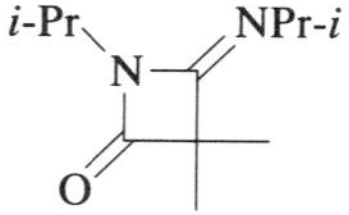 1-*i*-Pr-4-(NPr-*i*)-3,3-Me_2-azetidin-2-one (32)	770
	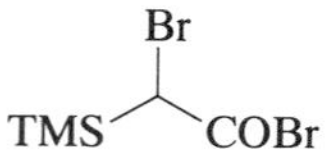 TMSCH(Br)COBr	Et_3N, C_7H_{16}, 98°	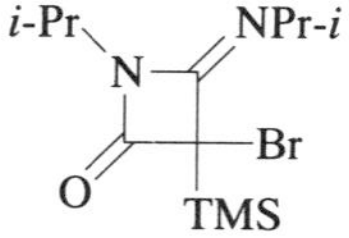 1-*i*-Pr-4-(NPr-*i*)-3-Br-3-TMS-azetidin-2-one (90)	772

TABLE XXIII. [2+2] CYCLOADDITION OF KETENES TO CARBODIIMIDES (*Continued*)

Reactant	Ketene or Ketene Source	Conditions	Product(s) and Yield(s) (%)	Refs.
	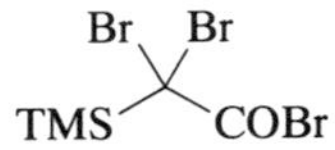 TMS(Br)₂C–COBr	Et_3N, CCl_4	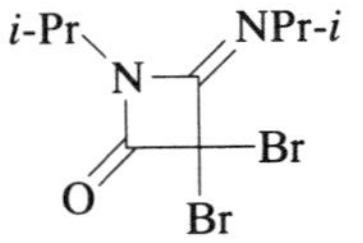 1-(i-Pr)-4-(NPr-i)-3,3-dibromoazetidin-2-one (30)	771
	$C_6H_{11}COCl$	Et_3N, C_6H_6	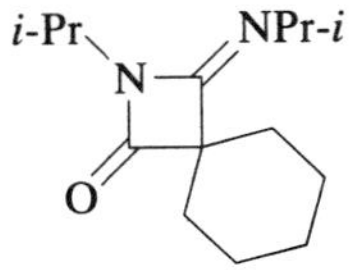 spiro[cycloheptane]-azetidinone, i-Pr–N, =NPr-i (51)	527
	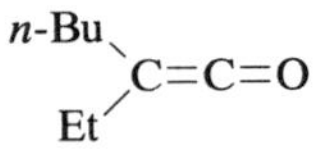 n-Bu(Et)C=C=O	C_6H_{14}, 69°	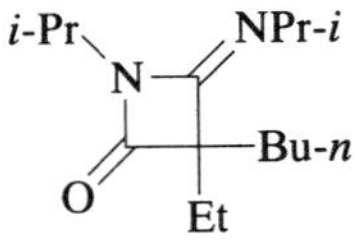 1-(i-Pr)-4-(NPr-i)-3-(Bu-n)-3-Et-azetidin-2-one (12)	770
	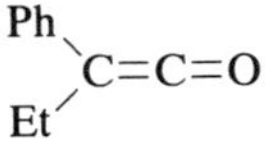 Ph(Et)C=C=O	C_6H_6, rt	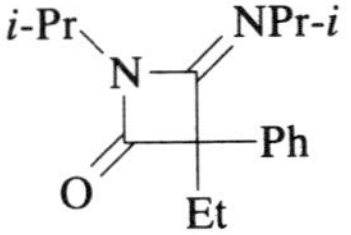 1-(i-Pr)-4-(NPr-i)-3-Ph-3-Et-azetidin-2-one (57)	770
	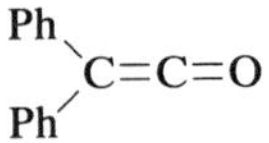 Ph(Ph)C=C=O	C_6H_6, rt	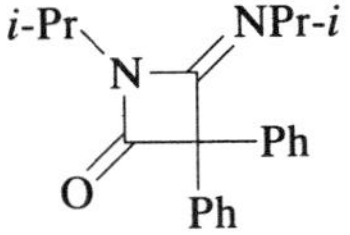 1-(i-Pr)-4-(NPr-i)-3,3-diPh-azetidin-2-one (80-88)	769, 770 773, 774
	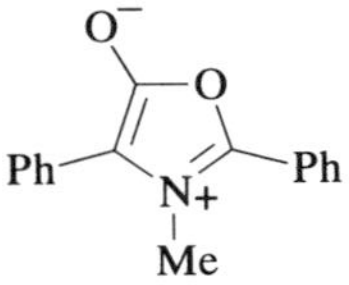 4-Ph-2-Ph-3-Me-oxazolium-5-olate (O^-, N+)	80°	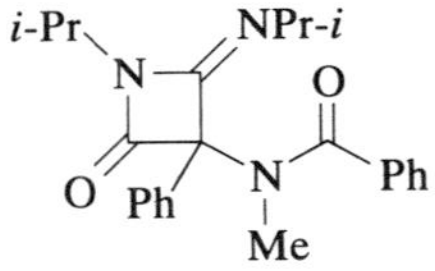 1-(i-Pr)-4-(NPr-i)-3-Ph-3-[N(Me)C(O)Ph]-azetidin-2-one (63)	775, 776

TABLE XXIII. [2+2] CYCLOADDITION OF KETENES TO CARBODIIMIDES (*Continued*)

Reactant	Ketene or Ketene Source	Conditions	Product(s) and Yield(s) (%)	Refs.
C_8				
TMSN=C=NR	XYCHCOCl	Et_3N, C_6H_{14}, 50°	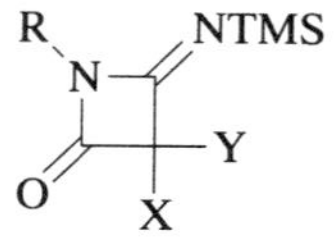 1-R-4-(NTMS)-3-X-3-Y-azetidin-2-one	768
R	X, Y			
n-Bu	Cl, Cl		(36)	
i-Bu	Cl, Cl		(40)	
t-Bu	Cl, Cl		(45)	
t-Bu	Br, Br		(28)	
t-Bu	Cl, Me		(23)	
C_9				
t-BuN=C=NBu-*t*	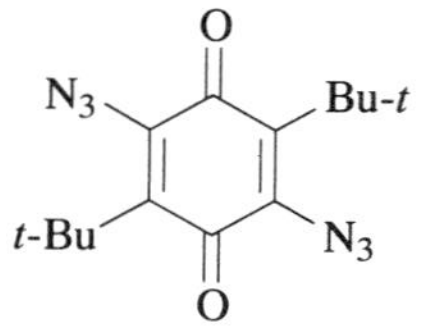 2,5-diazido-3,6-di-*t*-butyl-1,4-benzoquinone (N_3, Bu-*t*, *t*-Bu, N_3)	C_6H_6, 80°	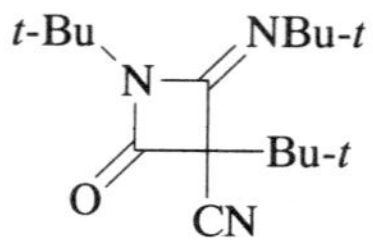 1-*t*-Bu-4-(NBu-*t*)-3-Bu-*t*-3-CN-azetidin-2-one (88)	191
	$Ph_2C=C=O$	C_6H_6, 25°	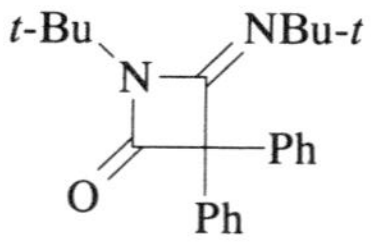 1-*t*-Bu-4-(NBu-*t*)-3,3-Ph-azetidin-2-one (75)	769
C_{10}				
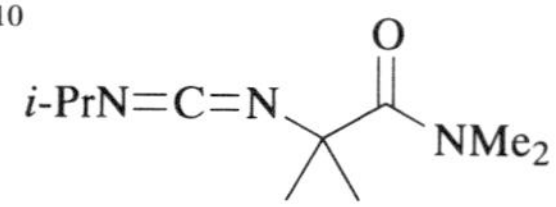 *i*-PrN=C=N–$C(Me)_2$–C(=O)NMe_2	$Ph_2C=C=O$	C_6H_6, 25°	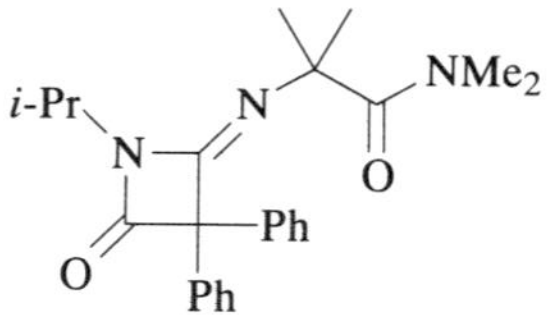 1-*i*-Pr-4-[=N–$C(Me)_2$–C(=O)NMe_2]-3,3-Ph-azetidin-2-one (59)	193

TABLE XXIII. [2+2] CYCLOADDITION OF KETENES TO CARBODIIMIDES (*Continued*)

Reactant	Ketene or Ketene Source	Conditions	Product(s) and Yield(s) (%)	Refs.
TMSN=C=NC_6H_{11}	$Cl_2CHCOCl$	Et_3N, C_6H_{14}, 50°	1-C_6H_{11}-3,3-dichloro-4-(NTMS)azetidin-2-one (28)	768
t-BuN=C=NPh	Ph_2C=C=O	C_6H_6, 25°	1-t-Bu-3,3-diphenyl-4-(NPh)azetidin-2-one (22)	769
t-BuN=C=NC_6H_4Me-p	Ph_2C=C=O	C_6H_6, 25°	1-t-Bu-3,3-diphenyl-4-(NC_6H_4Me-p)azetidin-2-one (30)	769
cyclic carbodiimide, $(CH_2)_{11}$ ring with N=C=N	$Cl_2CHCOCl$	Et_3N, C_6H_{14}	bicyclic β-lactam, $(CH_2)_{11}$ bridge linking ring N and exocyclic N, 3,3-dichloro (60-80)	192
	$CH_3CHClCOCl$	Et_3N, C_6H_{14}	bicyclic β-lactam, $(CH_2)_{11}$ bridge, 3-chloro-3-methyl (60-80)	192

TABLE XXIII. [2+2] CYCLOADDITION OF KETENES TO CARBODIIMIDES (*Continued*)

Reactant	Ketene or Ketene Source	Conditions	Product(s) and Yield(s) (%)	Refs.
	$PhCH_2COCl$	Et_3N, C_6H_{14}	(CH2)11, N, N, O, Ph (—)	192
	Ph, Ph, C=C=O	Et_2O, 25°	(CH2)11, N, N, O, Ph, Ph (90)	192
C_{12}				
i-PrN=C=N, S, N(Me)Pr-i	Ph, Ph, C=C=O	25°	i-Pr, N, N, N(Me)Pr-i, S, O, Ph, Ph (45)	193
C_{13}				
PhN=C=NPh	Ph, Ph, C=C=O	C_6H_6, 66°	Ph, N, NPh, O, Ph, Ph (60)	769
(S)-t-BuN=C=N, Ph	Bu-t, EtO_2C, COCl	Et_3N	t-Bu, N, N, Ph, O, CO_2Et, Bu-t (28.5)	194

TABLE XXIII. [2+2] CYCLOADDITION OF KETENES TO CARBODIIMIDES (*Continued*)

Reactant	Ketene or Ketene Source	Conditions	Product(s) and Yield(s) (%)	Refs.
$C_6H_{11}N{=}C{=}NC_6H_{11}$	$Cl_2CHCOCl$	Et_3N, C_6H_6, rt	C_6H_{11}, N, NC_6H_{11}, Cl, Cl, O (55)	769
	$Cl_2CHCOCl$	Et_3N, C_6H_{12}, 81°	" (72)	772
	$Br_2CHCOCl$	Et_3N, C_6H_{14}, 69°	C_6H_{11}, N, NC_6H_{11}, Br, Br, O (59)	770
	N_3, Cl, MeO, O, O	C_6H_6, 80°	C_6H_{11}, N, NC_6H_{11}, CN, Cl, O (88)	311
	Cl, COCl	Et_3N, C_6H_{14}, 69°	C_6H_{11}, N, NC_6H_{11}, Cl, O (25)	770
	O, N_3, Bu-*t*, *t*-Bu, N_3, O	C_6H_6	C_6H_{11}, N, NC_6H_{11}, Bu-*t*, CN, O (84)	353, 777

TABLE XXIII. [2+2] CYCLOADDITION OF KETENES TO CARBODIIMIDES (*Continued*)

Reactant	Ketene or Ketene Source	Conditions	Product(s) and Yield(s) (%)	Refs.
	o-$ClC_6H_4OCH_2COCl$	Et_3N, C_6H_6, 25°	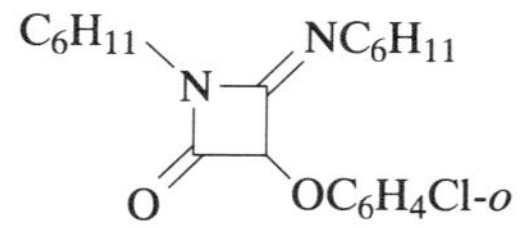 C_6H_{11}, N, NC_6H_{11}, O, OC_6H_4Cl-*o* (55)	769
	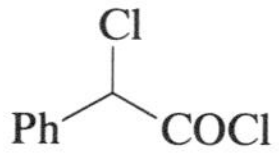 Cl, Ph, COCl	Et_3N, C_6H_{14}, 69°	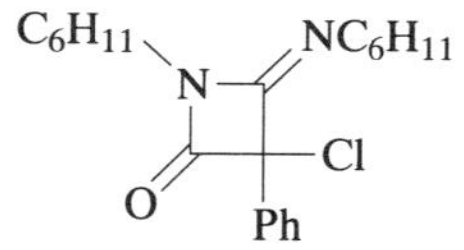 C_6H_{11}, N, NC_6H_{11}, O, Cl, Ph (65)	770
	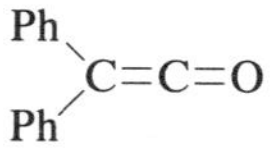 Ph, Ph, C=C=O	C_6H_{14} or C_6H_6 rt	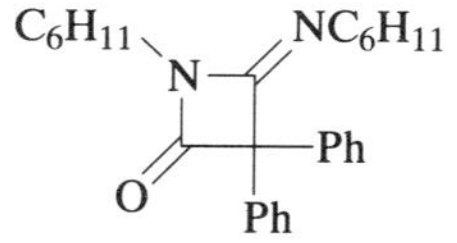 C_6H_{11}, N, NC_6H_{11}, O, Ph, Ph (88-90)	769, 770
	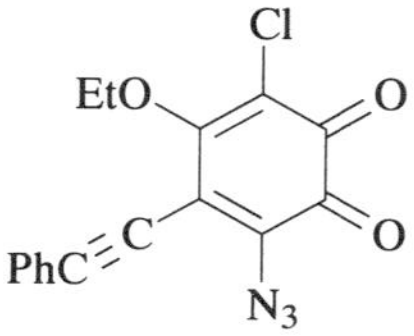 Cl, EtO, O, O, PhC≡C, N_3	C_6H_{12}, 80°	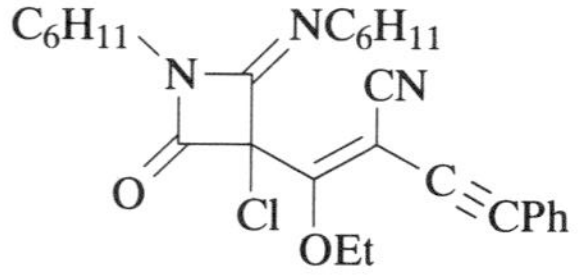 C_6H_{11}, N, NC_6H_{11}, O, Cl, CN, OEt, C≡CPh (44)	778
	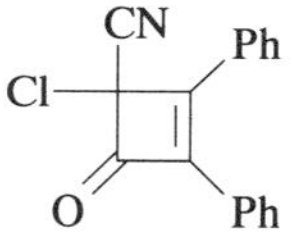 CN, Cl, Ph, O, Ph	C_6H_6, 16 h	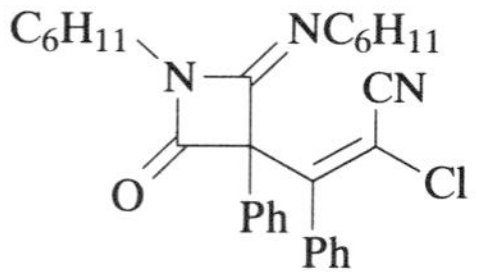 C_6H_{11}, N, NC_6H_{11}, O, CN, Ph, Ph, Cl (76)	114

TABLE XXIII. [2+2] CYCLOADDITION OF KETENES TO CARBODIIMIDES (*Continued*)

Reactant	Ketene or Ketene Source	Conditions	Product(s) and Yield(s) (%)	Refs.
(*R*,*R*)-*t*-Bu(Me)CH–N=C=N–CH(Me)Bu-*t*	Ph(CF$_3$)C=C=O	—	1-(*t*-BuCH(Me))-4-[N-CH(Me)Bu-*t*]imino-3-Ph-3-CF$_3$-azetidin-2-one (84)	194
C_{15}				
XC$_6$H$_4$CH$_2$N=C=NCH$_2$C$_6$H$_4$X	Ph$_2$C=C=O	C$_6$H$_6$, 25°	1-(XC$_6$H$_4$CH$_2$)-4-(XC$_6$H$_4$CH$_2$N=)-3,3-Ph$_2$-azetidin-2-one	769
X = H			(49)	
X = *p*-Me			(48)	
X = *p*-MeO			(62)	
C_{17}				
RN=C=NR, R = 2-methyl-1-morpholino-1-thioxopropan-2-yl [Me$_2$C–C(=S)–N(CH$_2$CH$_2$)$_2$O]	Ph$_2$C=C=O	25°	1-R-4-(RN=)-3,3-Ph$_2$-azetidin-2-one (26)	193
C_{21}				
RN=C=NR, R = 1-adamantyl	RCH=C=O	—	1-R-4-(RN=)-3-R-azetidin-2-one (20)	789

TABLE XXIII. [2+2] CYCLOADDITION OF KETENES TO CARBODIIMIDES (*Continued*)

Reactant	Ketene or Ketene Source	Conditions	Product(s) and Yield(s) (%)	Refs.
RN=C=NR R = *l*-menthyl	Ph(Me)C=C=O	—	R–N, =NR, O, Ph (37) $[\alpha]^{25}_{365}$ -439°	176
	EtO_2C–CH(Bu-*t*)–COCl	Et_3N	R–N, =NR, O, CO_2Et, Bu-*t* (64)	194
	EtO_2C–CH(CMe$_2$Ph)–COCl	Et_3N	R–N, =NR, O, CO_2Et, CMe$_2$Ph (75)	194

TABLE XXIV. [2+2] CYCLOADDITION OF KETENES TO *N*-SULFINYLAMINES

Reactant	Ketene or Ketene Source	Conditions	Product(s) and Yield(s) (%)	Refs.
C_1				
MeN=S=O	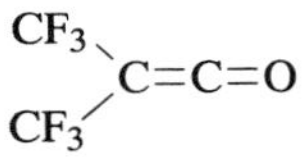 $(CF_3)_2C=C=O$	-30°	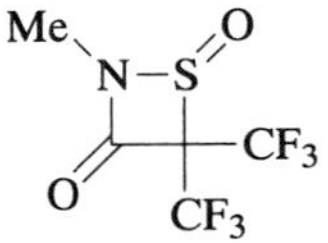 Me, N–S=O, O, CF_3, CF_3 (83)	195
C_4				
t-BuN=S=O	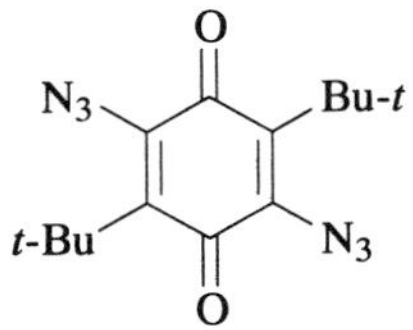 O, O, N_3, N_3, Bu-*t*, *t*-Bu	C_6H_6, 80°	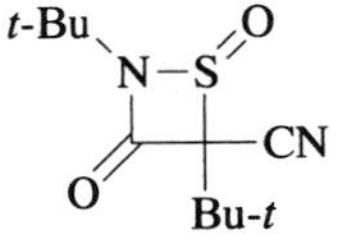 *t*-Bu, N–S=O, O, CN, Bu-*t* (88)	191
C_6				
$C_6F_5N=S=O$	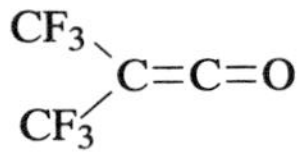 $(CF_3)_2C=C=O$	-30°	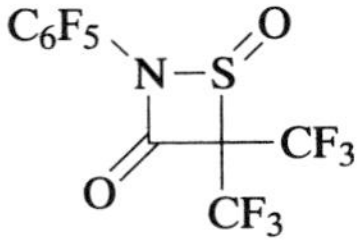 C_6F_5, N–S=O, O, CF_3, CF_3 (57)	195
$2,4\text{-}Cl_2C_6H_3N=S=O$	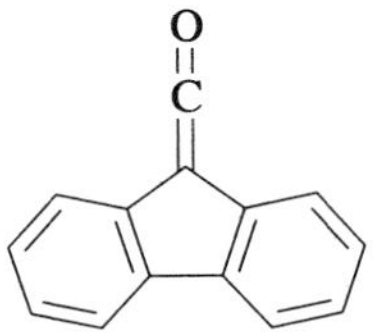 O=C= (fluorenylidene)	Pet. ether	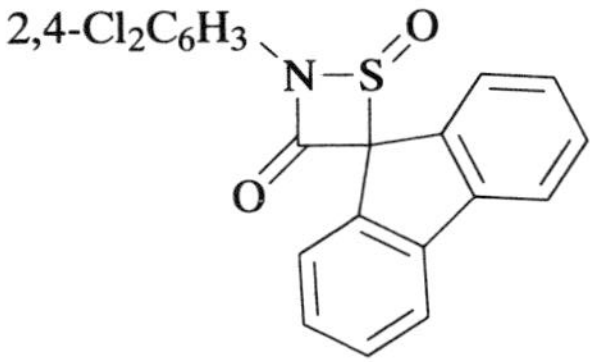 $2,4\text{-}Cl_2C_6H_3$, N–S=O, O (83)	196
m-$ClC_6H_4N=S=O$	"	60°, 72 h	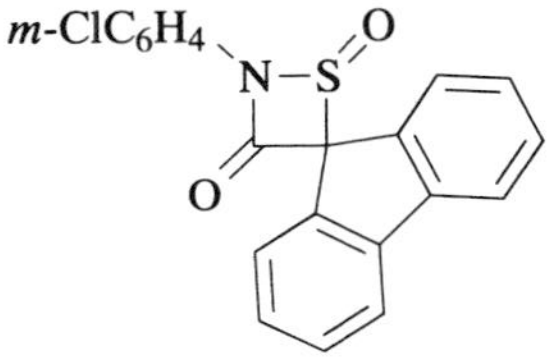 *m*-ClC_6H_4, N–S=O, O (87)	196

TABLE XXIV. [2+2] CYCLOADDITION OF KETENES TO *N*-SULFINYLAMINES (*Continued*)

Reactant	Ketene or Ketene Source	Conditions	Product(s) and Yield(s) (%)	Refs.
$XC_6H_4N{=}S{=}O$	$Ph_2C{=}C{=}O$	20°, 50 min; 60°, 72 h	2-(XC_6H_4)-3,3-diphenyl-1,2-thiazetidin-4-one 1-oxide	196
X: 2,4-Cl_2			(83)	
o-Cl			(93)	
m-Cl			(89)	
p-Cl			(99)	
p-O_2N			(86)	
H			(95)	
$PhN{=}S{=}O$	$CH_2{=}C{=}O$	Me_2CO, -78°	2-Phenyl-1,2-thiazetidin-3-one 1-oxide (100)	202, 197
	$(CF_3)_2C{=}C{=}O$	-30°	2-Phenyl-4,4-bis(CF_3)-1,2-thiazetidin-3-one 1-oxide (—)	195
	2,5-Diazido-3,6-di-*t*-butyl-1,4-benzoquinone	C_6H_6, 80°	2-Phenyl-4-cyano-4-(Bu-*t*)-1,2-thiazetidin-3-one 1-oxide (—)	191

TABLE XXIV. [2+2] CYCLOADDITION OF KETENES TO *N*-SULFINYLAMINES (*Continued*)

Reactant	Ketene or Ketene Source	Conditions	Product(s) and Yield(s) (%)	Refs.
	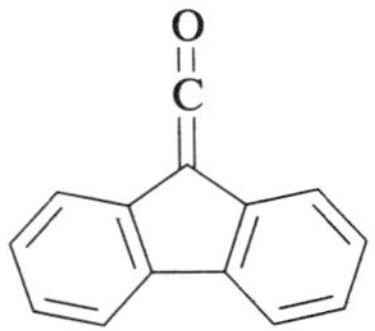 (fluorenylidene ketene, O=C=C<fluorenylidene)	Pet. ether	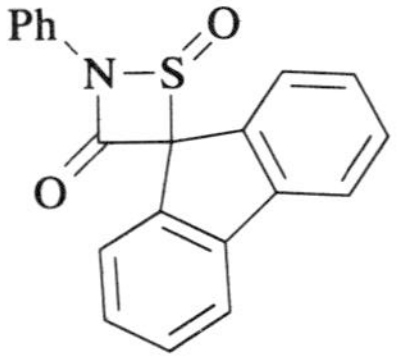 Ph–N–S(=O) spiro-β-sultam with fluorene, C=O (96)	196
	$(CF_3)_2C{=}C{=}O$	-30°	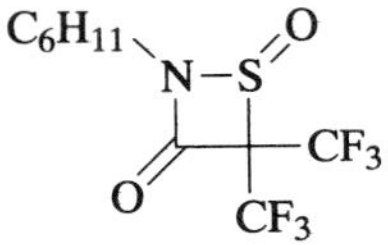 C_6H_{11}–N–S(=O)–C(CF_3)(CF_3)–C(=O) (86)	195
	$Ph_2C{=}C{=}O$	rt	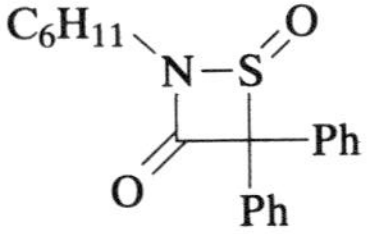 C_6H_{11}–N–S(=O)–C(Ph)(Ph)–C(=O) (99)	196
C_8				
t-BuN=S=NBu-*t*	$Cl_2CHCOCl$	Et_3N	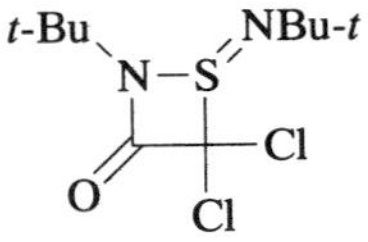 *t*-Bu–N–S(=NBu-*t*)–C(Cl)(Cl)–C(=O) (91)	198
	$Ph_2C{=}C{=}O$	Et_2O, 0°	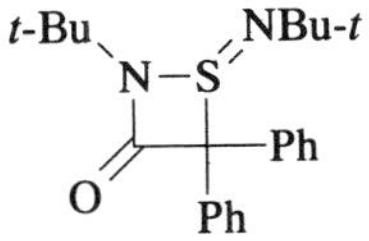 *t*-Bu–N–S(=NBu-*t*)–C(Ph)(Ph)–C(=O) (74)	780
C_{14}				
TsN=S=NTs	$Ph_2C{=}C{=}O$	$CHCl_3$, -15°	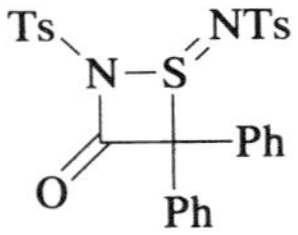 Ts–N–S(=NTs)–C(Ph)(Ph)–C(=O) (100)	781

TABLE XXV. [2+2] CYCLOADDITION OF KETENES TO NITROSO COMPOUNDS

Reactant	Ketene or Ketene Source	Conditions	Product(s) and Yield(s) (%)	Refs.
C_1				
CF_3NO	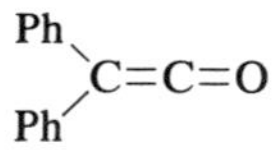 $Ph_2C=C=O$	—	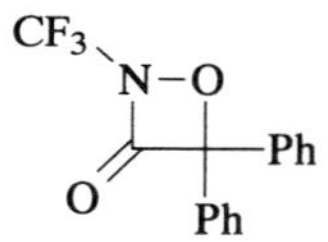 2-CF_3-4,4-diphenyl-1,2-oxazetidin-3-one (—)	782
p-ClC_6H_4NO	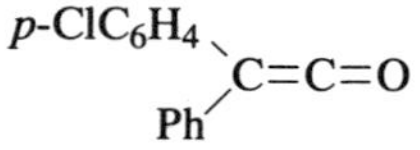 p-ClC_6H_4(Ph)C=C=O	Pet. ether	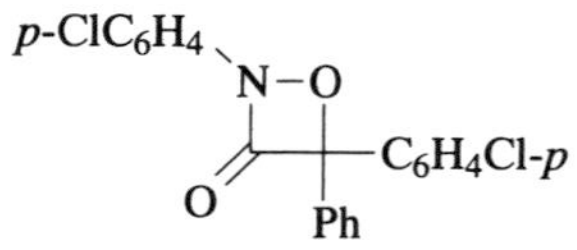 2-(p-ClC_6H_4)-4-(C_6H_4Cl-p)-4-Ph-1,2-oxazetidin-3-one (38)	200
	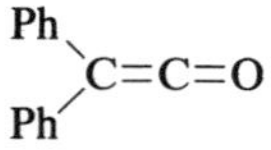 $Ph_2C=C=O$	Pet. ether	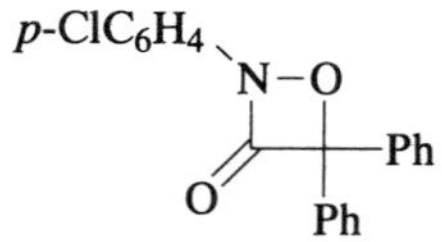 2-(p-ClC_6H_4)-4,4-diphenyl-1,2-oxazetidin-3-one (48)	200
p-BrC_6H_4NO	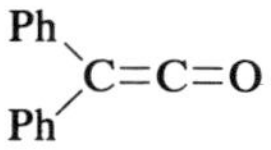 $Ph_2C=C=O$	Pet. ether	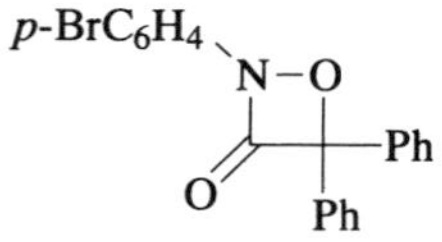 2-(p-BrC_6H_4)-4,4-diphenyl-1,2-oxazetidin-3-one (19)	200
PhNO	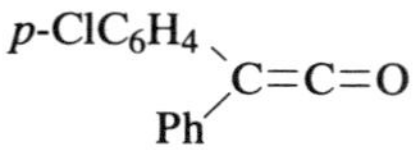 p-ClC_6H_4(Ph)C=C=O	Pet. ether	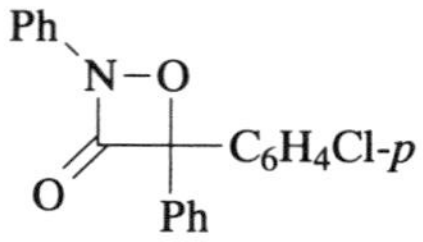 2-Ph-4-(C_6H_4Cl-p)-4-Ph-1,2-oxazetidin-3-one (52)	200
	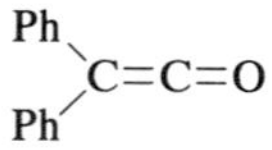 $Ph_2C=C=O$	Pet. ether	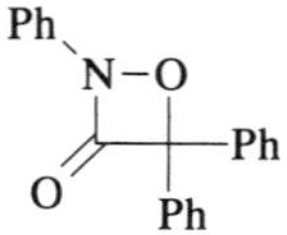 2,4,4-triphenyl-1,2-oxazetidin-3-one **I** (45-63)	200, 783

TABLE XXV. [2+2] CYCLOADDITION OF KETENES TO NITROSO COMPOUNDS (*Continued*)

Reactant	Ketene or Ketene Source	Conditions	Product(s) and Yield(s) (%)	Refs.
	$Ph_2C=C=O$	$CHCl_3$, 25°	**I** (60) + 1-Ph-3,3,4,4-tetraphenylazetidin-2-one [Ph–N, Ph, Ph, Ph, Ph, O] (13)	199
C_7				
p-MeC_6H_4NO	$Ph_2C=C=O$	Pet. ether	2-(p-MeC_6H_4)-4,4-diphenyl-1,2-oxazetidin-3-one [p-MeC_6H_4–N–O, Ph, Ph, O] (38)	200
	$Ph_2C=C=O$	$CHCl_3$, 25°	1-(p-MeC_6H_4)-3,3,4,4-tetraphenylazetidin-2-one [p-MeC_6H_4–N, Ph, Ph, Ph, Ph, O] (13)	199
p-$MeOC_6H_4NO$	$Ph_2C=C=O$	$CHCl_3$, 25°	1-(p-$MeOC_6H_4$)-3,3,4,4-tetraphenylazetidin-2-one [p-$MeOC_6H_4$–N, Ph, Ph, Ph, Ph, O] (22)	199
C_8				
p-$MeO_2CC_6H_4NO$	$Ph_2C=C=O$	$CHCl_3$, 25°	2-(p-$MeO_2CC_6H_4$)-4,4-diphenyl-1,2-oxazetidin-3-one [p-$MeO_2CC_6H_4$–N–O, Ph, Ph, O] (72) + 2-($C_6H_4CO_2Me$-p)-3,3-diphenyl-1,2-oxazetidin-4-one [O–N–$C_6H_4CO_2Me$-p, Ph, Ph, O] (28)	199

TABLE XXV. [2+2] CYCLOADDITION OF KETENES TO NITROSO COMPOUNDS (*Continued*)

Reactant	Ketene or Ketene Source	Conditions	Product(s) and Yield(s) (%)	Refs.
	$Ph_2C{=}C{=}O$	$CHCl_3$, 25°	1-(p-$Me_2NC_6H_4$)-2,2,3,3-tetraphenyl-azetidin-4-one (61-65)	199, 783

TABLE XXVI. [2+2] CYCLOADDITION OF KETENES TO AZO COMPOUNDS

Reactant	Ketene or Ketene Source	Conditions	Product(s) and Yield(s) (%)	Refs.
C_2 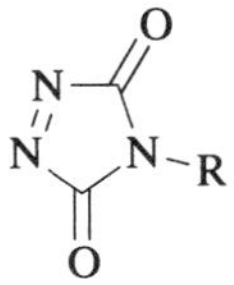 4-R-1,2,4-triazoline-3,5-dione; R = Me, Et, *n*-Pr, *n*-Bu, Ph	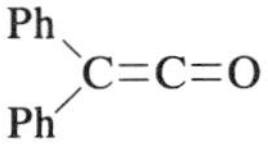 $Ph_2C{=}C{=}O$	CH_2Cl_2, 0°	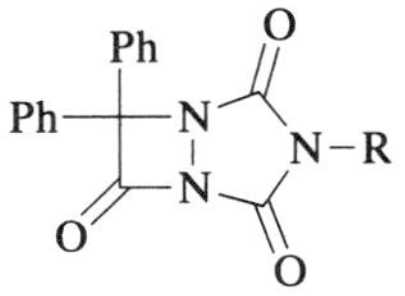 Ph, Ph, O, O, O, N–R (49-54)	783a
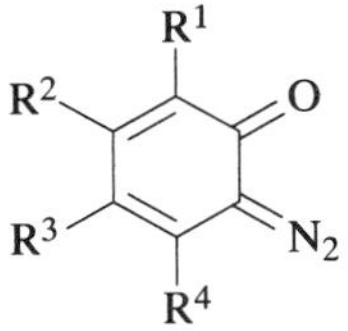 R^1, R^2, R^3, R^4, O, N_2 (diazoquinone)	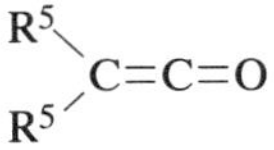 $R^5_2C{=}C{=}O$		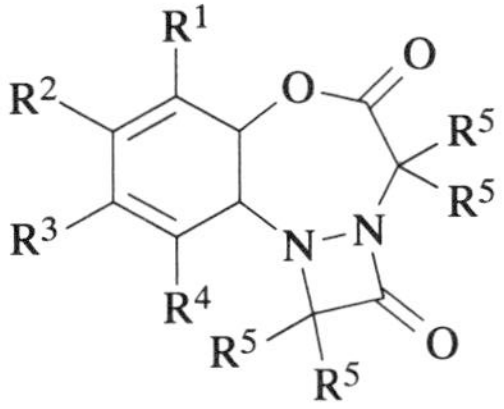 R^1, R^2, R^3, R^4, O, O, O, R^5, R^5, R^5, R^5, N–N	

R^1	R^2	R^3	R^4	R^5	Conditions	Product(s) and Yield(s) (%)	Refs.
Cl	Cl	Cl	Cl	Ph	Xylene, rt	(45)	206
H	Cl	H	Cl	Ph	Xylene, rt	(81)	206
H	Cl	H	H	Me	Et_2O, 35°, 3 h	(60)	668
H	H	Cl	H	Me	Et_2O, 35°, 3 h	(58)	668
H	Br	H	H	Me	Et_2O, 35°, 3 h	(40)	668
H	H	Br	H	Me	Et_2O, 35°, 3 h	(40)	668
H	NO_2	H	H	Ph	Xylene, rt	(51)	206
Me	Cl	Me	H	Ph	Xylene, rt	(47)	206
H	*t*-Bu	H	*t*-Bu	Ph	Xylene, rt	(85)	206
H	*t*-Bu	H	*t*-Bu	Me	Et_2O, 35°, 3 h	(52)	668

TABLE XXVI. [2+2] CYCLOADDITION OF KETENES TO AZO COMPOUNDS (*Continued*)

Reactant	Ketene or Ketene Source	Conditions	Product(s) and Yield(s) (%)	Refs.
$EtO_2CN{=}NCO_2Et$	$Ph_2C{=}C{=}O$	Pet. ether, 24 h	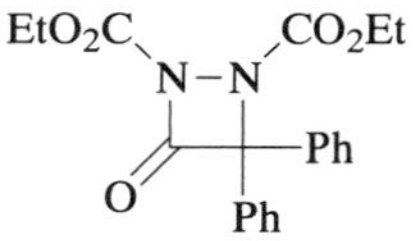 (70)	204
C_7				
NCN=NCN	$Ph_2C{=}C{=}O$	—	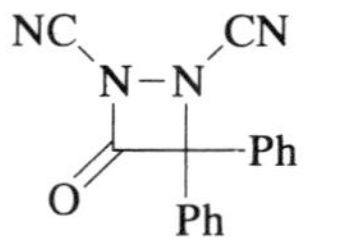 (—)	784
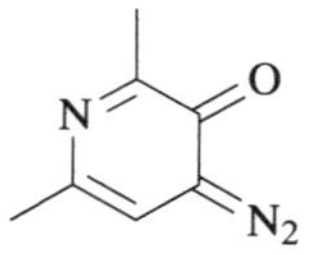	$Ph_2C{=}C{=}O$	Xylene, rt	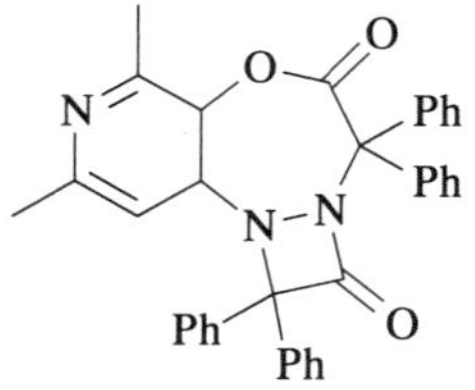 (38)	206
C_8				
PhN=NCOMe	$Ph_2C{=}C{=}O$	C_6H_6, rt	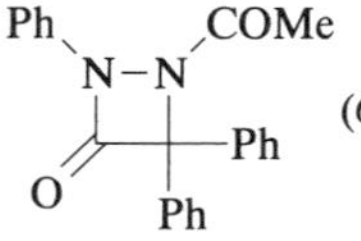 (60) + 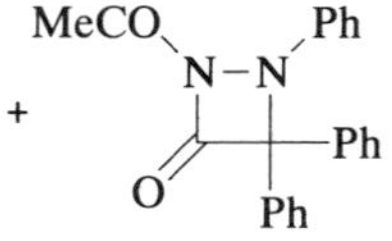 (26) + 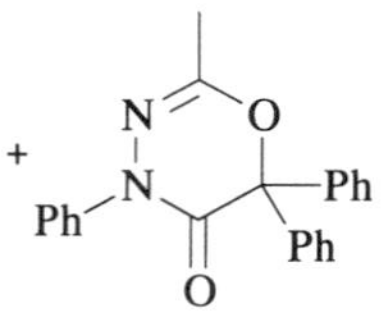 (10) + 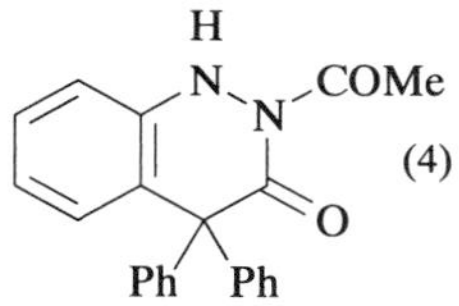 (4)	205

TABLE XXVI. [2+2] CYCLOADDITION OF KETENES TO AZO COMPOUNDS (*Continued*)

Reactant	Ketene or Ketene Source	Conditions	Product(s) and Yield(s) (%)	Refs.
$PhN{=}NNMe_2$	$Ph_2C{=}C{=}O$	$h\nu$	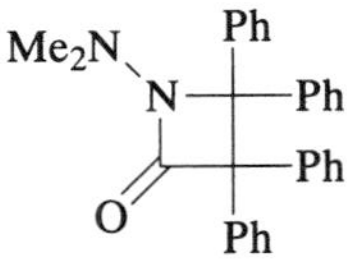 (35)	785
C_9				
$PhN{=}NPr\text{-}i$	$Ph_2C{=}C{=}O$	$h\nu$	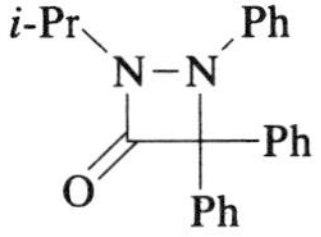 (11)	785, 786
$PhN{=}NCO_2Et$	$Ph_2C{=}C{=}O$	$h\nu$, C_6H_{12}	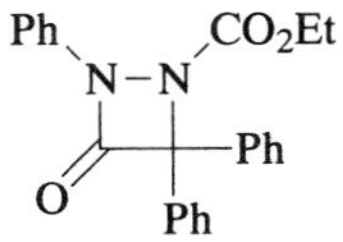 (69)	785, 786
C_{10}				
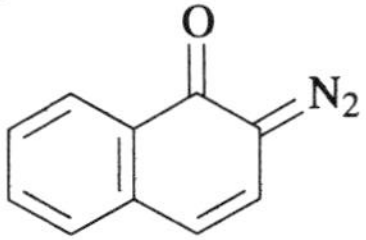	$Ph_2C{=}C{=}O$	Xylene, rt	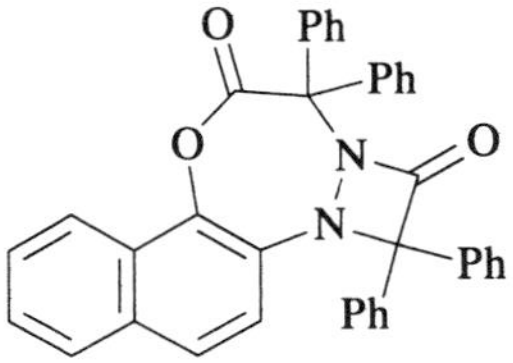 (75)	206
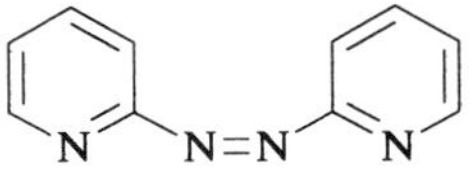	$Ph_2C{=}C{=}O$	C_6H_6	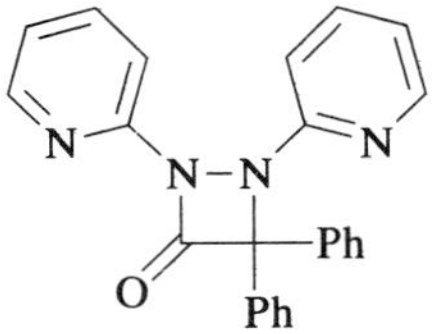 (55)	787

TABLE XXVI. [2+2] CYCLOADDITION OF KETENES TO AZO COMPOUNDS (*Continued*)

Reactant	Ketene or Ketene Source	Conditions	Product(s) and Yield(s) (%)	Refs.
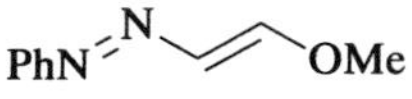 PhN=N–CH=CH–OMe	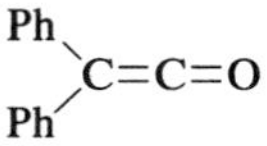 $Ph_2C=C=O$	C_6H_6	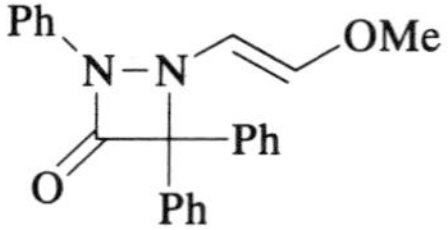 1-Ph-2-(CH=CH-OMe)-4,4-diphenyl-1,2-diazetidin-3-one (85)	268
C_{11}				
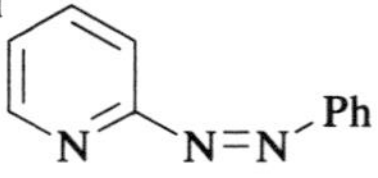 2-(PhN=N)pyridine	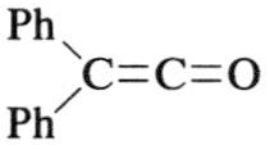 $Ph_2C=C=O$	—	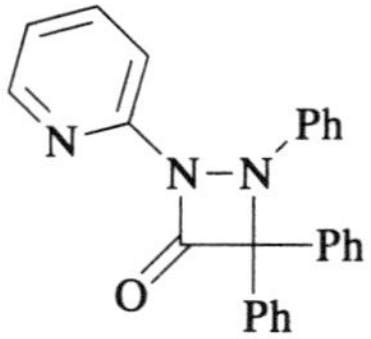 1-(2-pyridyl)-2-Ph-4,4-diphenyl-1,2-diazetidin-3-one (71)	787
C_{12}				
$PhN=NC_6H_4Cl$-*p*	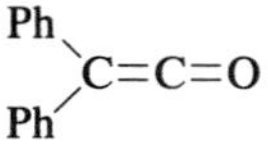 $Ph_2C=C=O$	*hν*, C_6H_6	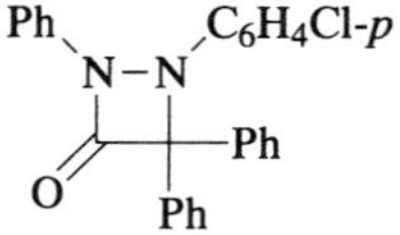 **I** 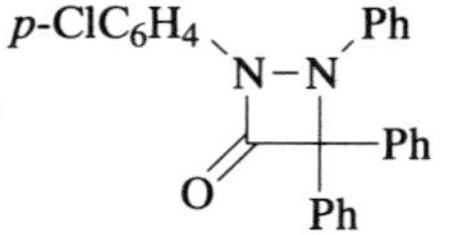 (Ph on N adjacent to C=O, C_6H_4Cl-*p* on N adjacent to CPh_2) + **II** (*p*-ClC_6H_4 on N adjacent to C=O, Ph on N adjacent to CPh_2) (78) **I:II** = 59:41	786
$PhN=NC_6H_4NO_2$-*p*	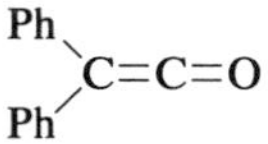 $Ph_2C=C=O$	*hν*, C_6H_6	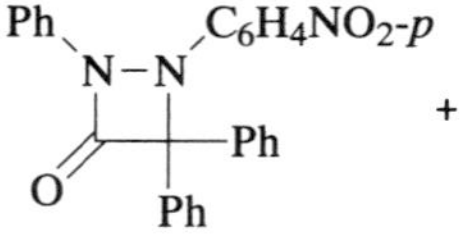 **I** 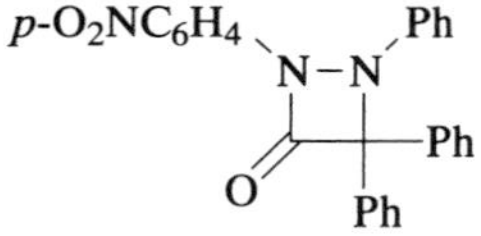 (Ph on N adjacent to C=O, $C_6H_4NO_2$-*p* on N adjacent to CPh_2) + **II** (*p*-$O_2NC_6H_4$ on N adjacent to C=O, Ph on N adjacent to CPh_2) (94) **I:II** = 61:39	786

TABLE XXVI. [2+2] CYCLOADDITION OF KETENES TO AZO COMPOUNDS (*Continued*)

Reactant	Ketene or Ketene Source	Conditions	Product(s) and Yield(s) (%)	Refs.
$XC_6H_4N{=}NC_6H_4X$	$Ph_2C{=}C{=}O$	$h\nu$, C_6H_6 or Et_2O	1,2-bis(XC_6H_4)-3,3-diphenyl-1,2-diazetidin-4-one (structure)	
X				
o-Cl			(91)	201
m-Cl			(55)	201
p-Cl			(17)	201
m-Br			(76)	201
p-Br			(14)	201
o-NO_2			(20)	201
m-NO_2			(37)	201
p-Me			(78)	788
m-Me			(15)	788
o-Me			(73)	788
o-OMe			(19)	201
m-OMe			(69)	201
p-OMe			(71)	201
p-CO_2Et			(33)	201

TABLE XXVI. [2+2] CYCLOADDITION OF KETENES TO AZO COMPOUNDS (*Continued*)

Reactant	Ketene or Ketene Source	Conditions	Product(s) and Yield(s) (%)	Refs.
PhN=NPh	$RCOCHN_2$	*hν*, solvent	1,2-diphenyl-4-R-1,2-diazetidin-3-one	
	R	Solvent		
	Ph	C_6H_6	(31)	314, 789
	1-$C_{10}H_7$	CH_2Cl_2	(30)	789
	p-$MeOC_6H_4$	CH_2Cl_2	(53)	789
	m-$MeOC_6H_4$	CH_2Cl_2	(57)	789
	p-MeC_6H_4	CH_2Cl_2	(42)	789
	p-BrC_6H_4	CH_2Cl_2	(48)	789
	o-BrC_6H_4	CH_2Cl_2	(20)	789
	p-ClC_6H_4	CH_2Cl_2	(42)	789
	m-ClC_6H_4	CH_2Cl_2	(13)	789
	o-ClC_6H_4	CH_2Cl_2	(9)	789
	p-$O_2NC_6H_4$	CH_2Cl_2	(47)	789
	m-$O_2NC_6H_4$	CH_2Cl_2	(43)	789
	o-$O_2NC_6H_4$	CH_2Cl_2	(5)	789
cis-PhN=NPh	$CH_2=C=O$	MeOH	1,2-diphenyl-1,2-diazetidin-3-one (68)	790
trans-PhN=NPh	$CH_2=C=O$	*hν*, C_6H_{14}, 15°	" (—)	791
	3-diazo-5-methylhex-4-en-2-one	*hν*	4-methyl-4-(2-methylprop-1-enyl)-1,2-diphenyl-1,2-diazetidin-3-one (—)	288

TABLE XXVI. [2+2] CYCLOADDITION OF KETENES TO AZO COMPOUNDS (*Continued*)

Reactant	Ketene or Ketene Source	Conditions	Product(s) and Yield(s) (%)	Refs.
	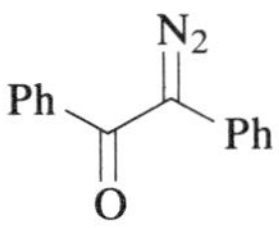 PhC(=O)C(=N_2)Ph	$h\nu$, C_6H_6	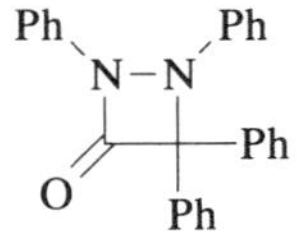 1,2,3,3-tetraphenyl-1,2-diazetidin-4-one (32)	314, 792
cis-PhN=NPh	$Me_2C=C=O$	Et_2O or MeOH	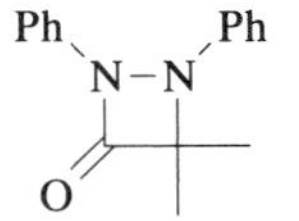 1,2-diphenyl-3,3-dimethyl-1,2-diazetidin-4-one (40)	790
trans-PhN=NPh	$Ph_2C=C=O$	$h\nu$, Et_2O	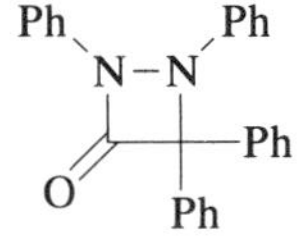 1,2,3,3-tetraphenyl-1,2-diazetidin-4-one (71)	201, 786, 788
C_{13}				
PhN=NC_6H_4CN-*p*	$Ph_2C=C=O$	$h\nu$, C_6H_6	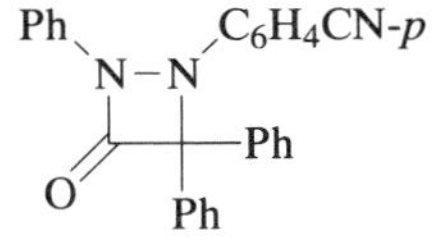 **I** 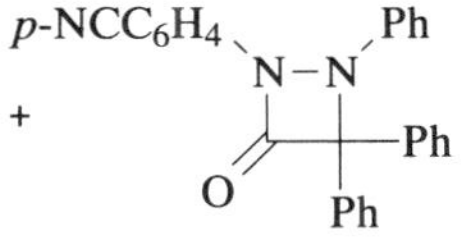 (N-Ph, N-C_6H_4CN-*p*, 3,3-Ph_2) + **II** (*p*-NCC_6H_4-N, N-Ph, 3,3-Ph_2) (—) **I**:**II** = 64:36	786
PhN=NC_6H_4Me-*o*	$CH_2=C=O$	$h\nu$	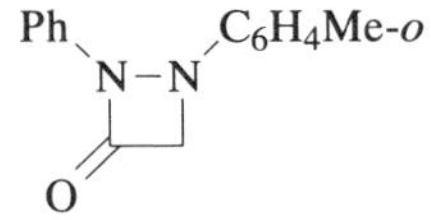 Ph-N–N-C_6H_4Me-*o* diazetidinone (81)	203
PhN=NC_6H_4Me-*m*	$CH_2=C=O$	$h\nu$	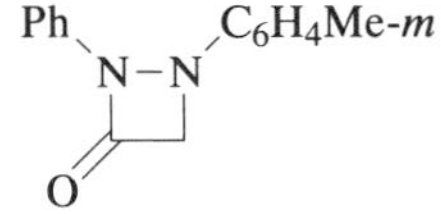 Ph-N–N-C_6H_4Me-*m* diazetidinone (29) + *m*-MeC_6H_4-N–N-Ph diazetidinone (36)	203

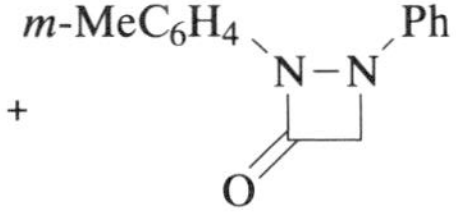

TABLE XXVI. [2+2] CYCLOADDITION OF KETENES TO AZO COMPOUNDS (*Continued*)

Reactant	Ketene or Ketene Source	Conditions	Product(s) and Yield(s) (%)	Refs.
$PhN{=}NC_6H_4Me$-*p*	$CH_2{=}C{=}O$	*hν*	1-Ph-2-(C_6H_4Me-*p*)-1,2-diazetidin-3-one (7) + 1-(*p*-MeC_6H_4)-2-Ph-1,2-diazetidin-3-one (50)	203
	$Ph_2C{=}C{=}O$	*hν*, C_6H_6	1-(*p*-MeC_6H_4)-2-Ph-4,4-Ph$_2$-1,2-diazetidin-3-one (38) + 1-Ph-2-(C_6H_4Me-*p*)-4,4-Ph$_2$-1,2-diazetidin-3-one (45)	786
$PhN{=}NC_6H_4OMe$-*p*	$Ph_2C{=}C{=}O$	*hν*, C_6H_6	1-(*p*-$MeOC_6H_4$)-2-Ph-4,4-Ph$_2$-1,2-diazetidin-3-one (25) + 1-Ph-2-(C_6H_4OMe-*p*)-4,4-Ph$_2$-1,2-diazetidin-3-one (45)	785, 786
dibenzo[c,f][1,2]diazepin-11-one (N=N)	$R_2C{=}C{=}O$	PhMe, 30-35°	fused 1,2-diazetidinone (O, N–N, R, R)	202
	R = H		(10)	
	R = Me		(30)	
	R = Ph		(80)	

TABLE XXVI. [2+2] CYCLOADDITION OF KETENES TO AZO COMPOUNDS (*Continued*)

Reactant	Ketene or Ketene Source	Conditions	Product(s) and Yield(s) (%)	Refs.
C_{14} 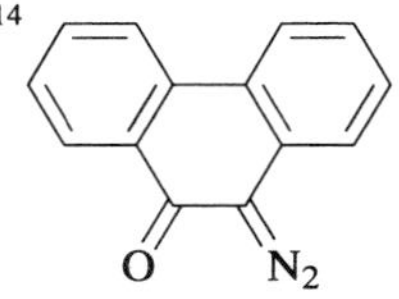	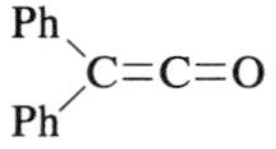		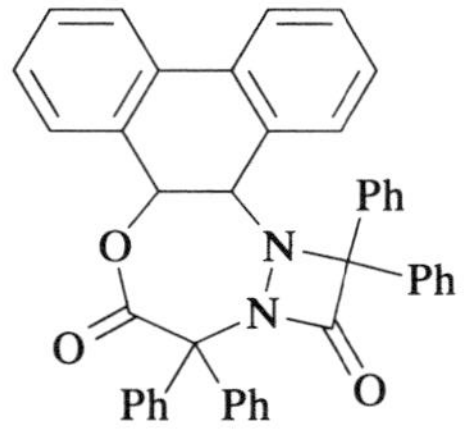	
O, N_2	Ph, Ph, C=C=O	Xylene, rt	O, O, N, N, Ph, Ph, Ph, Ph, O (70)	206
Ph, N=N, Ph	N_2, Ph, Ph, O	$h\nu$, C_6H_6	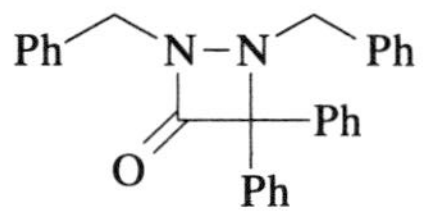 Ph, N–N, Ph, Ph, O, Ph (50)	788
C_{15}				
Ph, N=N, O, Ph	Ph, Ph, C=C=O	C_6H_6	Ph, N–N, O, Ph, Ph, O, Ph (86)	268
O, N_2	Ph, Ph, C=C=O	Xylene, rt	O, O, Ph, Ph, N–N, Ph, Ph, O (31)	206

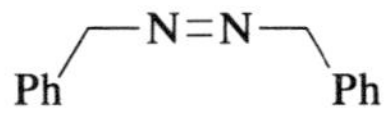

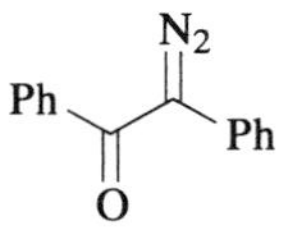

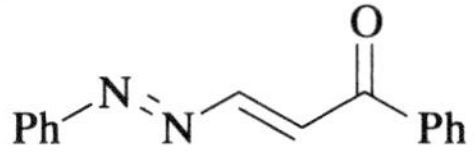

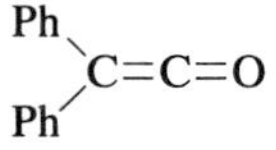

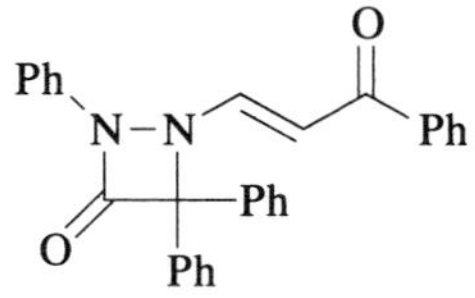

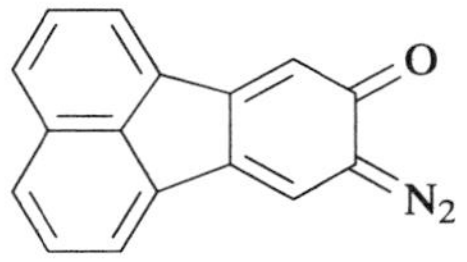

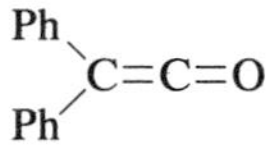

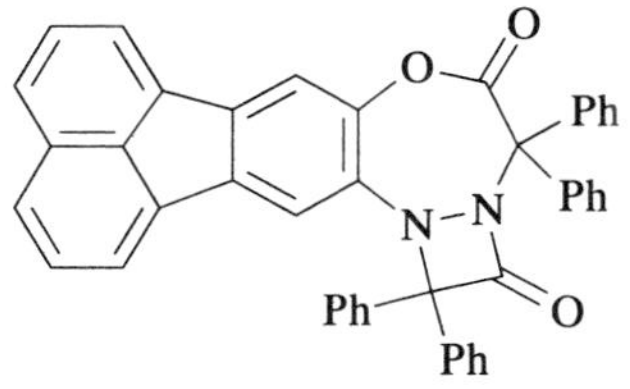

TABLE XXVI. [2+2] CYCLOADDITION OF KETENES TO AZO COMPOUNDS (*Continued*)

Reactant	Ketene or Ketene Source	Conditions	Product(s) and Yield(s) (%)	Refs.
C_{18}				
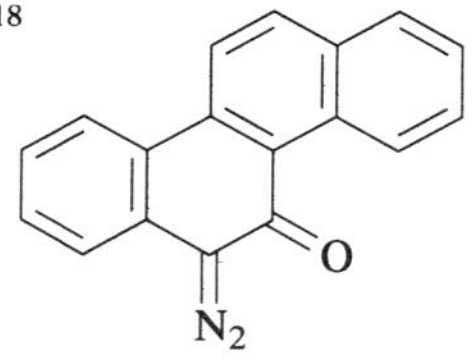 [structure: diazo ketone, N₂, O]	$Ph_2C=C=O$	Xylene, rt	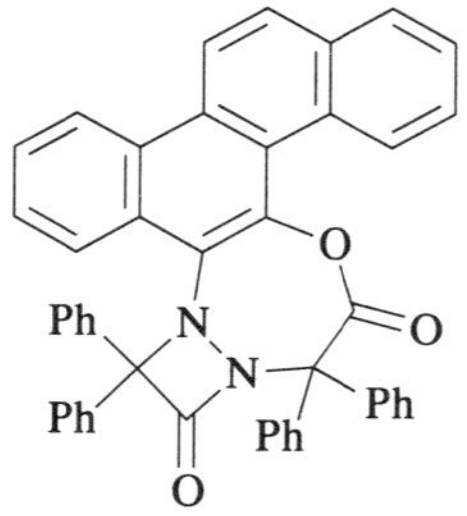 [structure: Ph, Ph, N, N, O, O, Ph, Ph] (56)	206
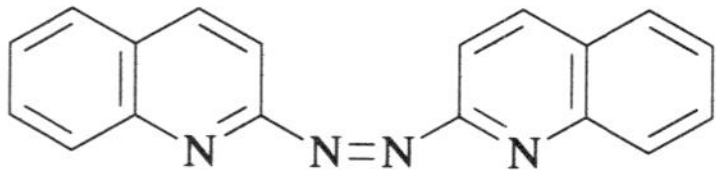 [structure: 2,2'-azoquinoline, N, N=N, N]	$Ph_2C=C=O$	Dioxane, 3 h	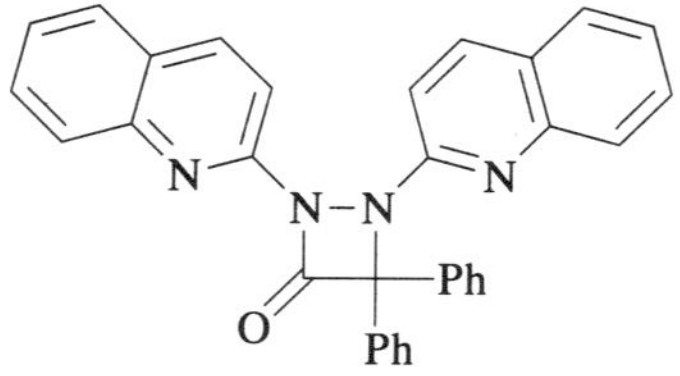 [structure: N, N–N, N, Ph, Ph, O] (31)	787
C_{20}				
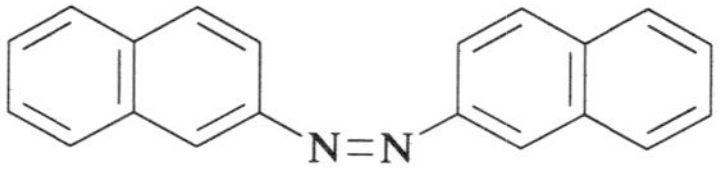 [structure: 2,2'-azonaphthalene, N=N]	$Ph_2C=C=O$	*h*ν, pet. ether	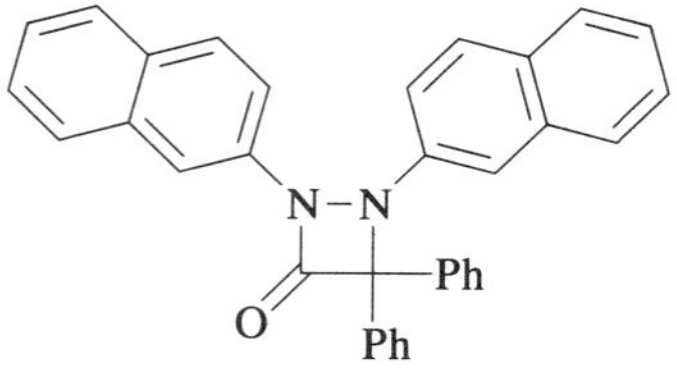 [structure: N–N, Ph, Ph, O] (9)	788

TABLE XXVII. [3+2] CYCLOADDITIONS

Reactant	Ketene or Ketene Source	Conditions	Product(s) and Yield(s) (%)	Refs.
C_6				
$RO_2C-N=N-CO_2R$	$Ph_2C=C=O$	C_6H_6	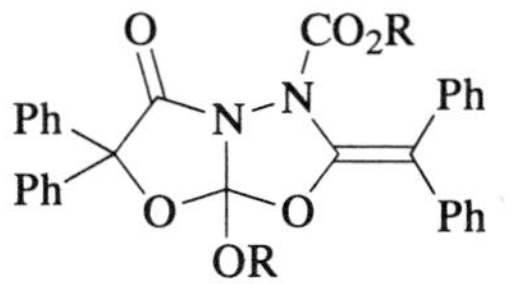 [structure: CO₂R, O, N–N, Ph, Ph, O, O, OR, Ph, Ph] R = Et (—); R = Ph (80); R = Bn (80)	793
C_7				
$RC\equiv\overset{+}{N}-O^-$	$Ph_2C=C=O$	PhMe	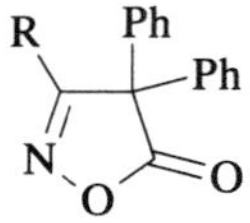 [structure: R, Ph, Ph, N, O, O]	210, 211
R: *p*-ClC_6H_4			(60)	
p-BrC_6H_4			(60)	
p-$Me_2NC_6H_4$			(75)	
o-ClC_6H_4			(65)	
m-$O_2NC_6H_4$			(70)	
t-Bu			(—)	
PhCH=CH			(80)	
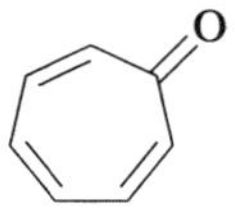 [structure: cycloheptatrienone, O]	$Ph_2C=C=O$	C_6H_6, 78°, 1 h	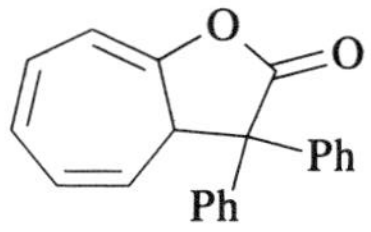 [structure: O, O, Ph, Ph] (99)	539
	$Cl_2CHCOCl$	Et_3N, Et_2O, 0°	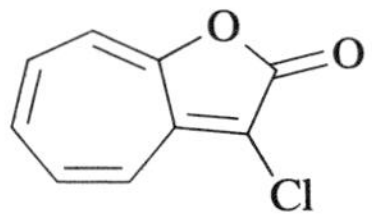 [structure: O, O, Cl] (19)	316

TABLE XXVII. [3+2] CYCLOADDITIONS (*Continued*)

Reactant	Ketene or Ketene Source	Conditions	Product(s) and Yield(s) (%)	Refs.
C_9				
R^1COCHN_2	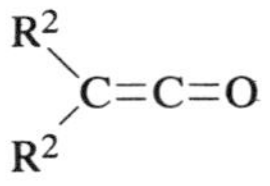 $R^2_2C{=}C{=}O$	Et_2O, 35°, 70 h	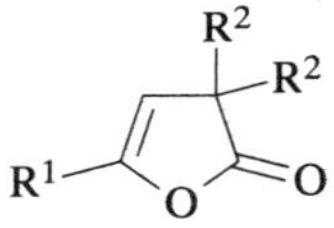 5-R^1-3,3-di-R^2-furan-2(3H)-one	222
			R^1 / R^2 / Yield	
			n-$C_{11}H_{23}$ / Me / (69)	
			n-$C_{15}H_{31}$ / Me / (55)	
			Bn / Me / (21)	
			p-$MeOC_6H_4$ / Me / (38)	
			Ph / H / (—)	
			p-$MeOC_6H_4$ / Ph / (—)	
			p-$O_2NC_6H_4$ / Ph / (—)	
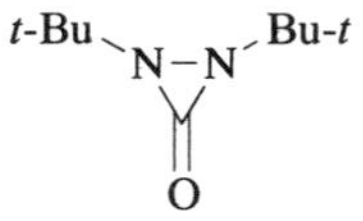 1,2-di-t-Bu-diaziridinone	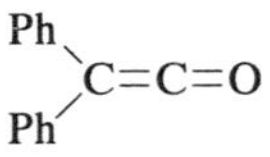 $Ph_2C{=}C{=}O$	PhMe, 110°, 10 h	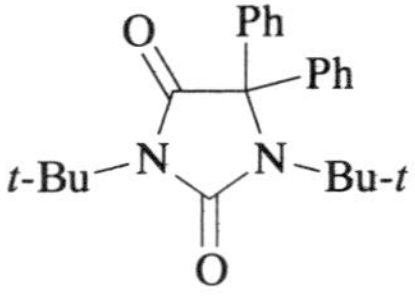 1,3-di-t-Bu-5,5-diphenylimidazolidine-2,4-dione (32) + 1,2-di-t-Bu-4,4-diphenylpyrazolidine-3,5-dione (26)	794
1-R-3-oxidopyridinium	Various haloketenes	—	3-R^1-5-R-furo[2,3-c]pyridin-2(5H)-one + 3-R^1-4-R-furo[3,2-b]pyridin-2(4H)-one	230
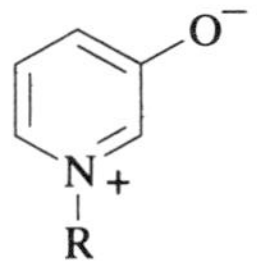			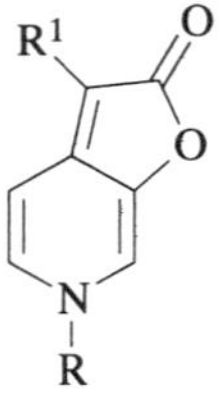 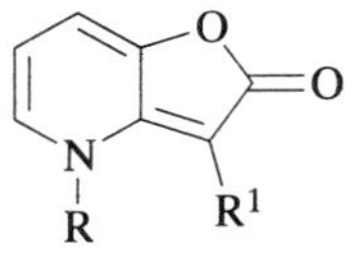 R, R^1 = various aryl and alkyl groups	

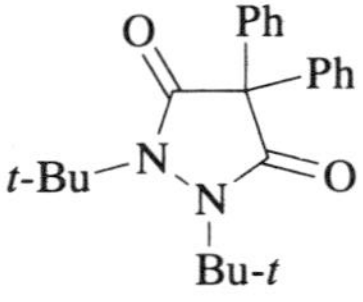

TABLE XXVII. [3+2] CYCLOADDITIONS (*Continued*)

Reactant	Ketene or Ketene Source	Conditions	Product(s) and Yield(s) (%)	Refs.
C_{10} 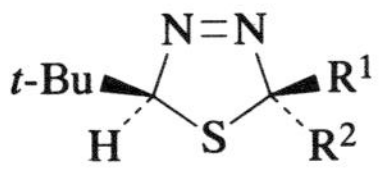 N=N, t-Bu, H, S, R^1, R^2	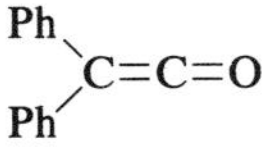 Ph(Ph)C=C=O	100°, 5 h	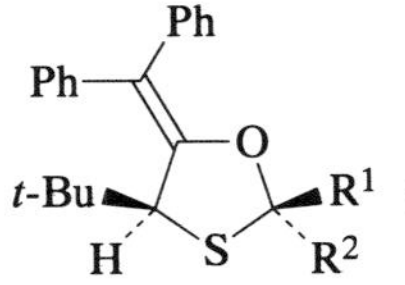 Ph, Ph, O, t-Bu, H, S, R^1, R^2 (85-94) R^1 = t-Bu, R^2 = H; R^1 = H, R^2 = t-Bu	326
C_{12} 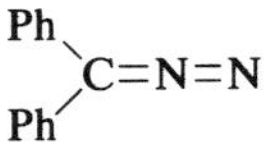 Ph(Ph)C=N=N	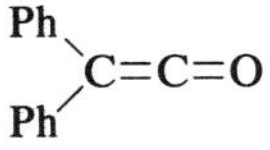 Ph(Ph)C=C=O	C_6H_6, 78° 12 h	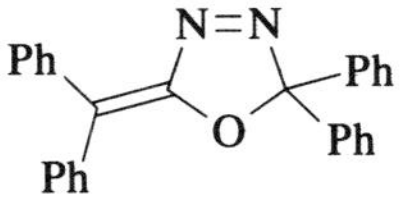 Ph, Ph, N=N, O, Ph, Ph (—)	795, 796
C_{13} 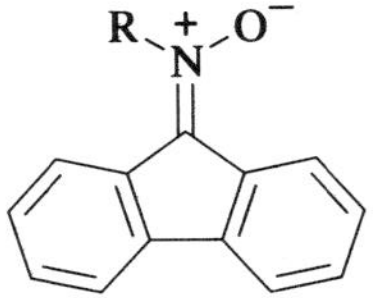 R, +N, O^-	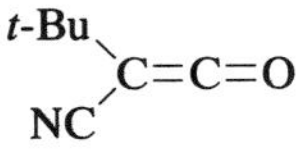 t-Bu(NC)C=C=O	110°, 12 h	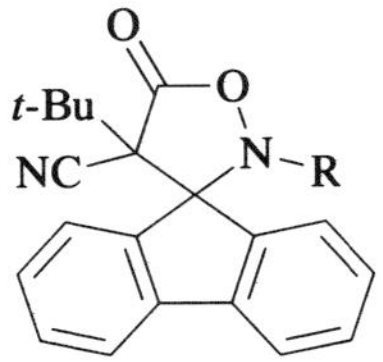 O, O, t-Bu, NC, N–R R = Me (72) R = Et (65)	209
C_{14} 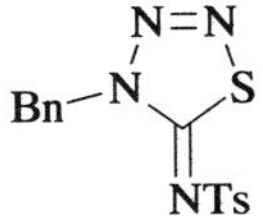 N=N, Bn–N, S, NTs	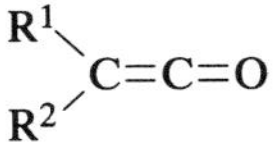 R^1(R^2)C=C=O	CCl_4	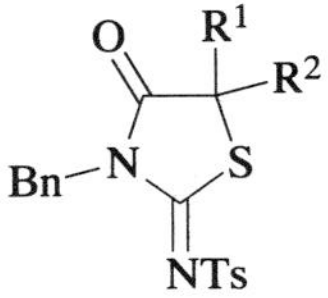 O, R^1, R^2, Bn–N, S, NTs R^1 = R^2 = Ph (54) R^1 = CN, R^2 = t-Bu (40)	797
C_{18} 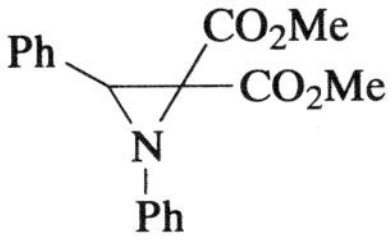 Ph, CO_2Me, CO_2Me, N, Ph	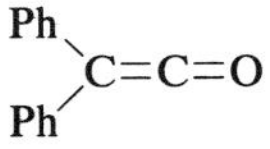 Ph(Ph)C=C=O	PhMe, 110°	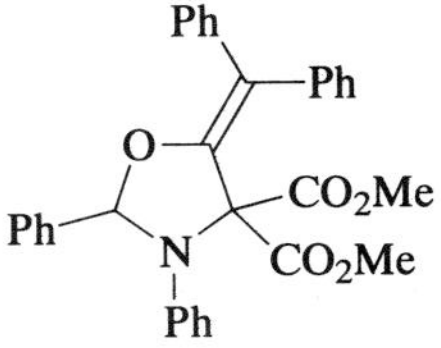 Ph, Ph, O, Ph, CO_2Me, CO_2Me, N, Ph (—)	208

TABLE XXVII. [3+2] CYCLOADDITIONS (*Continued*)

Reactant	Ketene or Ketene Source	Conditions	Product(s) and Yield(s) (%)	Refs.
	$CH_2{=}C{=}O$	PhMe, 110°	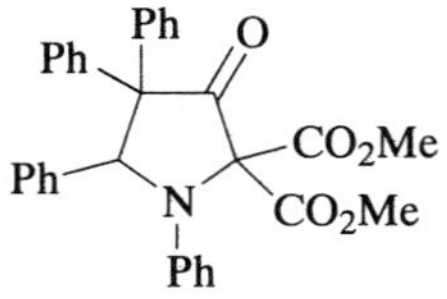 Ph, Ph, Ph, O, CO_2Me, CO_2Me, N, Ph (—)	208
C_{19} 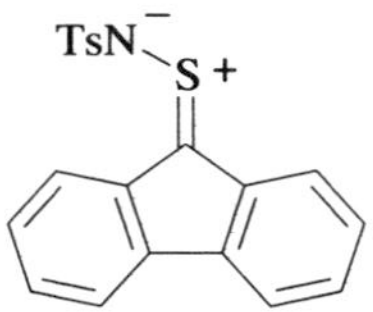 TsN$^-$ S+	Ph, Ph, $C{=}C{=}O$	$Cl(CH_2)_2Cl$ rt, 1 week	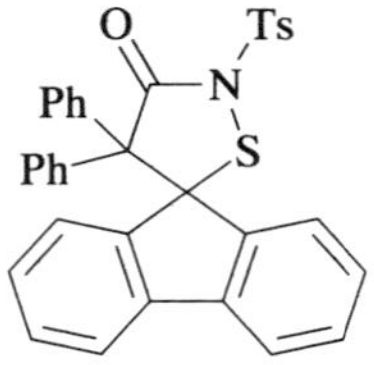 O, Ts, N, S, Ph, Ph (27) 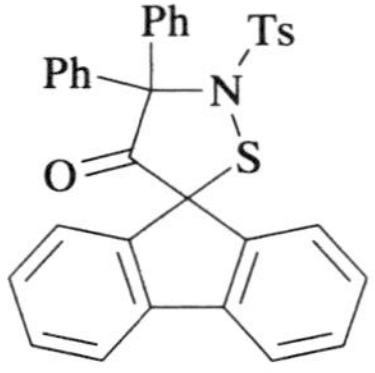 + Ph, Ts, Ph, N, S, O (9)	766

TABLE XXVIII. [4+2] CYCLOADDITION OF KETENES TO DIENES

Reactant	Ketene or Ketene Source	Conditions	Product(s) and Yield(s) (%)	Refs.
C_4				
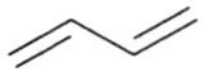	$(CF_3)_2C{=}C{=}O$	70°, 35 h	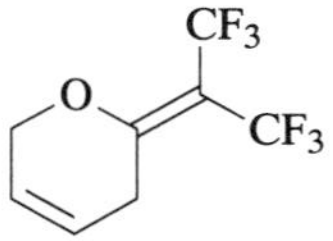 CF₃ / CF₃ (90)	231, 323
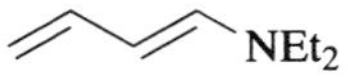 NEt₂	$CH_2{=}C{=}O$	Et_2O, 0°	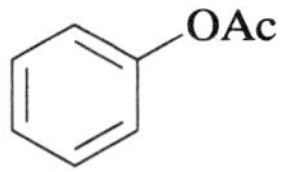 OAc (4)	240
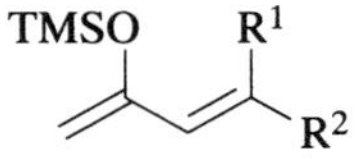 TMSO, R¹, R²	$R^3R^4CHCOCl$	Et_3N, Et_2O; MeOH, H^+	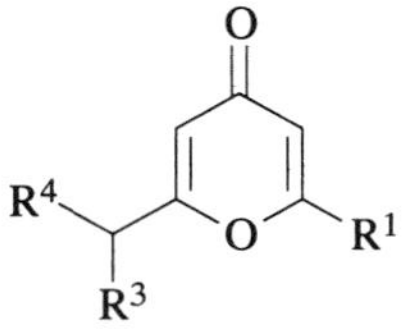 O, R⁴, R³, O, R¹	238

R^1	R^2	R^3	R^4	Yield (%)
H	OMe	Cl	Cl	(56)
H	OMe	H	Cl	(55)
H	OTMS	Ph	Ph	(55)
Me	OTMS	H	Cl	(55)
Me	OTMS	Cl	Cl	(52)
Me	OTMS	Ph	Ph	(70)

TABLE XXVIII. [4+2] CYCLOADDITION OF KETENES TO DIENES (*Continued*)

Reactant	Ketene or Ketene Source	Conditions	Product(s) and Yield(s) (%)	Refs.
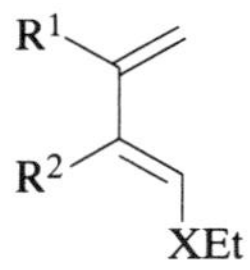 R^1, R^2 diene with XEt	$Ph_2C=C=O$	CCl_4	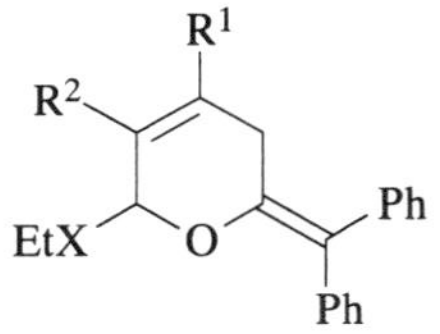 R^1, R^2, EtX, O, Ph, Ph	237
X, R^1, R^2 = O, Me, H			(—)	
S, Me, H			(—)	
O, H, Me			(—)	
O, Me, Me			(—)	
O, *i*-Pr, H			(—)	
R^1, R^2 diene with OTMS	$R^3R^4CHCOCl$	Et_3N, Et_2O, 0°	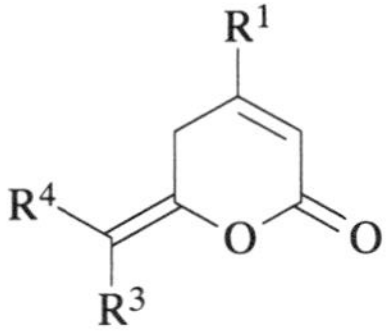 R^1, R^4, R^3, O, O	235
R^1, R^2 = OMe, OMe	R^3, R^4 = H, Cl		(23)	
OMe, OMe	Cl, Cl		(30)	
OMe, OMe	Ph, Ph		(56)	
Me, OTMS	Ph, Ph		(68)	
C_5				
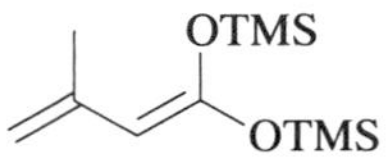 OTMS, OTMS	$Cl_2CHCOCl$	Et_3N, Et_2O, 0°	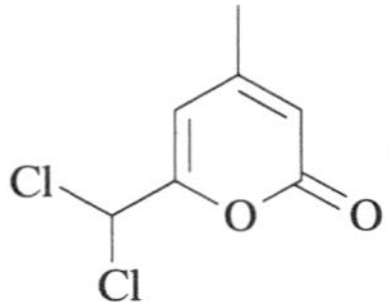 Cl, Cl, O, O (31) + Cl, Cl, O, OTMS, OTMS (55)	235

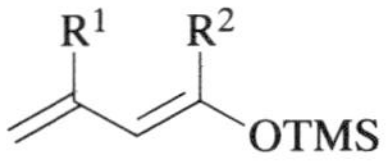

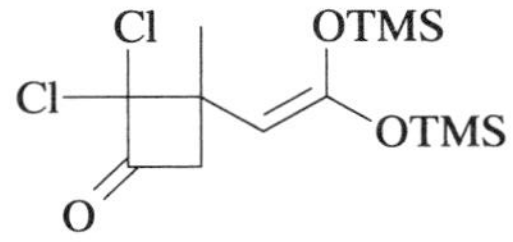

TABLE XXVIII. [4+2] CYCLOADDITION OF KETENES TO DIENES (*Continued*)

Reactant	Ketene or Ketene Source	Conditions	Product(s) and Yield(s) (%)	Refs.
C_7				
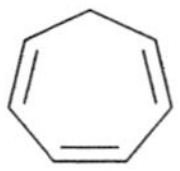 [structure]	$Ph_2C{=}C{=}O$	C_6H_6, 80°, 144 h	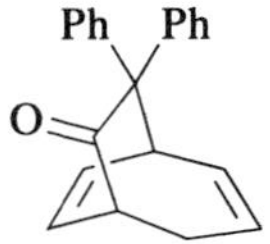 [structure: Ph, Ph, O] (11)	233
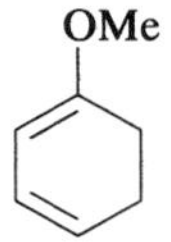 [structure: OMe]	$Ph_2C{=}C{=}O$	CCl_4	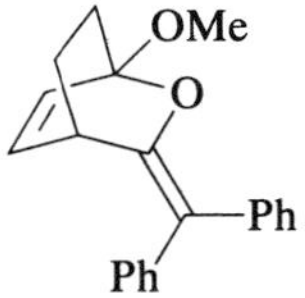 [structure: OMe, O, Ph, Ph] (—)	237
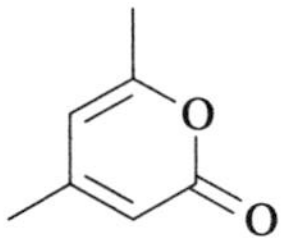 [structure: O, O]	$NCCH_2CO_2Et$	200°, 70 h	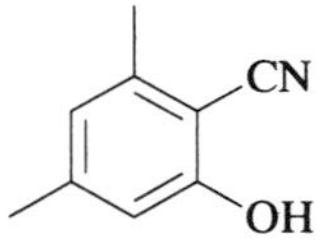 [structure: CN, OH] (10)	239
C_{10}				
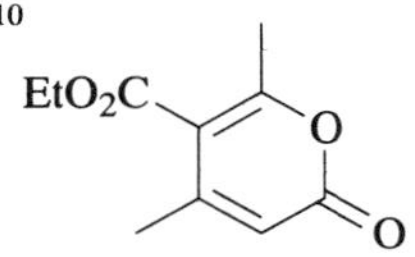 [structure: EtO_2C, O, O]	$NCCH_2CO_2Et$	195-200°, 122 h	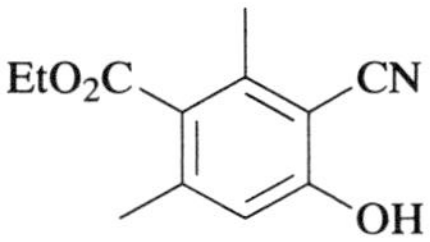 [structure: EtO_2C, CN, OH] (13)	239

TABLE XXVIII. [4+2] CYCLOADDITION OF KETENES TO DIENES (*Continued*)

Reactant	Ketene or Ketene Source	Conditions	Product(s) and Yield(s) (%)	Refs.
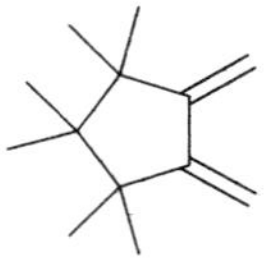	$Ph_2C{=}C{=}O$	C_6H_{12}, 55°, 15 d	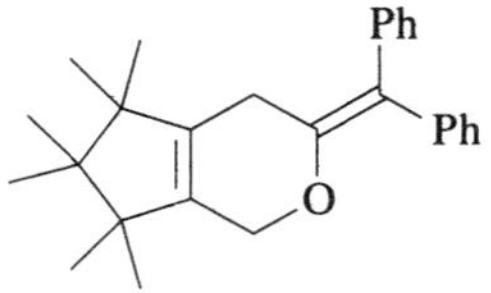 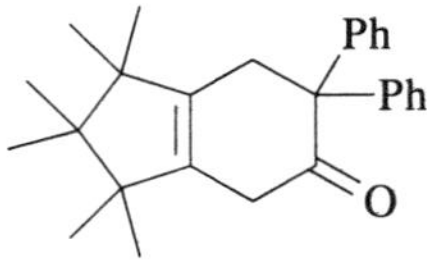 Ph, Ph, O (35) +	234
			Ph, Ph, O (35)	
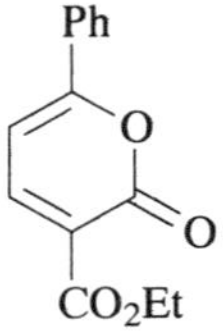 Ph, O, O, CO_2Et	$NCCH_2CO_2Et$	195-200°, 75 h	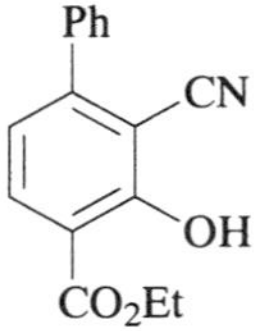 Ph, CN, OH, CO_2Et (56)	239

TABLE XXIX. [4+2] CYCLOADDITION OF KETENES TO AZADIENES

Reactant	Ketene or Ketene Source	Conditions	Product(s) and Yield(s) (%)	Refs.
C_{11}				
2-(2-(C_6H_4Cl-*p*)vinyl)thiazoline: N, S, C_6H_4Cl-*p*	Ph, Ph C=C=O	Xylene, 137°, 10 h	O, N, S, Ph, Ph, C_6H_4Cl-*p* (48)	243
N, S, $C_6H_4NO_2$-*p*	Ph, Ph C=C=O	Xylene, 137°, 10 h	O, N, S, Ph, Ph, $C_6H_4NO_2$-*p* (78)	243
N, S, Ph	Ph, Ph C=C=O	Xylene, 137°, 10 h	O, N, S, Ph, Ph, Ph (78)	243
C_{13}				
Ph, NBu-*t*	O, O, N_3, N_3, Bu-*t*, *t*-Bu	C_6H_6, 80°	NC, *t*-Bu, Ph, O, N, Bu-*t* (52) + NC, *t*-Bu, H, Ph, O, N, Bu-*t* (17) + NC, H, *t*-Bu, Ph, O, N, Bu-*t* (—)	23

TABLE XXIX. [4+2] CYCLOADDITION OF KETENES TO AZADIENES (*Continued*)

Reactant	Ketene or Ketene Source	Conditions	Product(s) and Yield(s) (%)	Refs.
C_{15}				
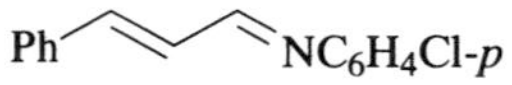 Ph‾CH=CH‾CH=NC_6H_4Cl-*p*	$BrCH_2CO_2Et$	Zn, C_6H_6	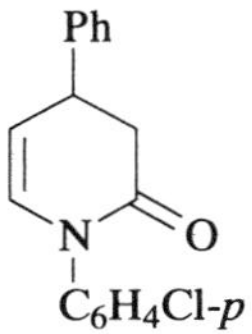 4-Ph-1-(C_6H_4Cl-*p*)-3,4-dihydropyridin-2-one (70)	241
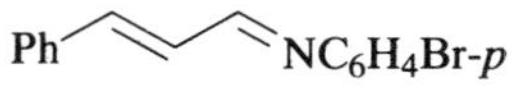 Ph‾CH=CH‾CH=NC_6H_4Br-*p*	$BrCH_2CO_2Et$	Zn, C_6H_6	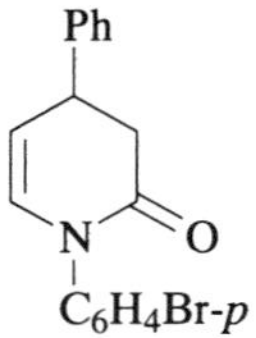 4-Ph-1-(C_6H_4Br-*p*)-3,4-dihydropyridin-2-one (90)	241
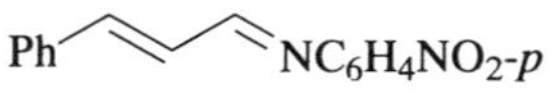 Ph‾CH=CH‾CH=$NC_6H_4NO_2$-*p*	$BrCH_2CO_2Et$	Zn, C_6H_6	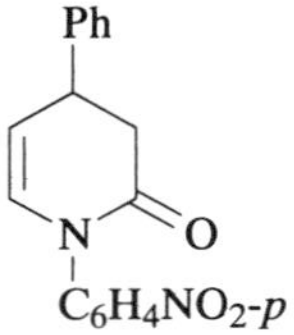 4-Ph-1-($C_6H_4NO_2$-*p*)-3,4-dihydropyridin-2-one (16)	241
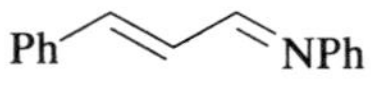 Ph‾CH=CH‾CH=NPh	$Cl_2CHCOCl$	Et_3N, C_6H_6, 20°	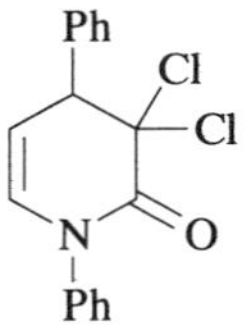 3,3-Cl₂-4-Ph-1-Ph-3,4-dihydropyridin-2-one (45)	242

TABLE XXIX. [4+2] CYCLOADDITION OF KETENES TO AZADIENES (*Continued*)

Reactant	Ketene or Ketene Source	Conditions	Product(s) and Yield(s) (%)	Refs.
	$BrCH_2CO_2Et$	Zn, C_6H_6	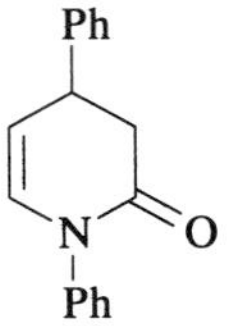 Ph, N–Ph, O (62)	241
	PhCONH, Me, C=C=O	—	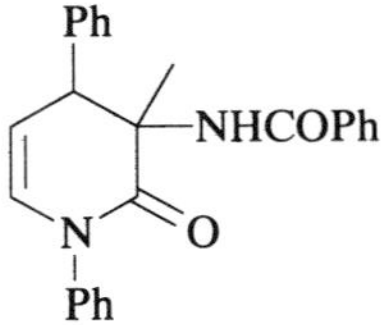 Ph, NHCOPh, N–Ph, O (—)	798
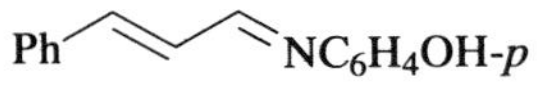 Ph, NC_6H_4OH-*p*	$BrCH_2CO_2Et$	Zn, C_6H_6	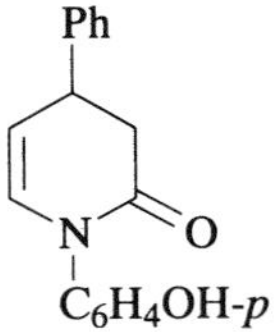 Ph, N–C_6H_4OH-*p*, O (65)	241

TABLE XXIX. [4+2] CYCLOADDITION OF KETENES TO AZADIENES (*Continued*)

Reactant	Ketene or Ketene Source	Conditions	Product(s) and Yield(s) (%)	Refs.
C_{15}				
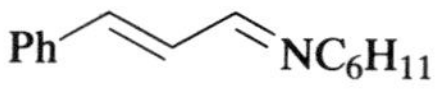 Ph–CH=CH–CH=NC_6H_{11}	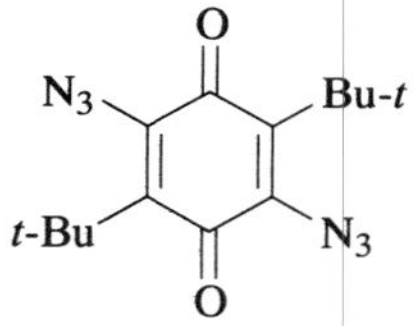 2,5-di-N_3-3,6-di-t-Bu-1,4-benzoquinone (N_3, Bu-t, t-Bu, N_3)	C_6H_6, 80°	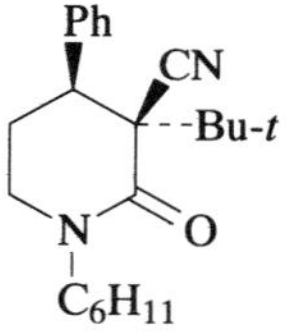 4-Ph-3-CN-3-Bu-t-1-C_6H_{11}-piperidin-2-one (23) 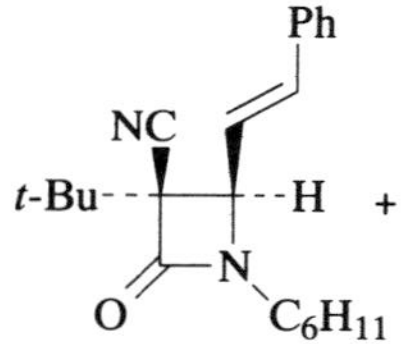 + 3-t-Bu-3-NC-4-(CH=CHPh)-4-H-1-C_6H_{11}-azetidin-2-one (64) 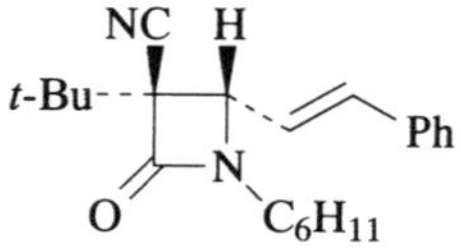 + 3-t-Bu-3-NC-4-H-4-(CH=CHPh)-1-C_6H_{11}-azetidin-2-one (10)	23
C_{16}				
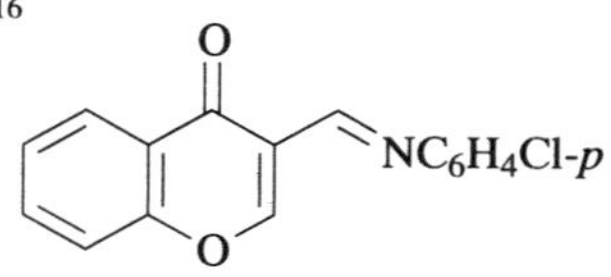 3-(CH=NC_6H_4Cl-p)-chromone	$Cl_2CHCOCl$	Et_3N, C_6H_6, 80°	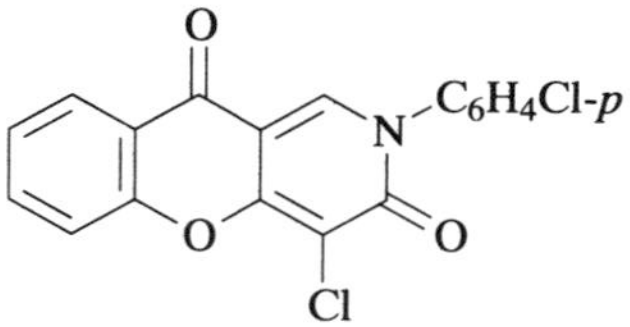 4-Cl-2-(C_6H_4Cl-p)-chromeno[2,3-c]pyridine-3,10-dione (54)	244
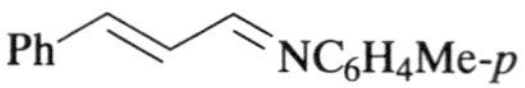 Ph–CH=CH–CH=NC_6H_4Me-p	$Cl_2CHCOCl$	Et_3N, C_6H_6, 20°	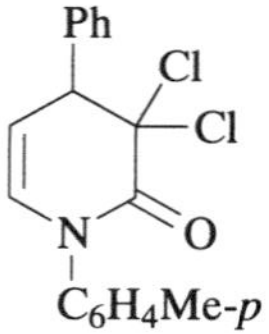 4-Ph-3,3-di-Cl-1-(C_6H_4Me-p)-3,4-dihydropyridin-2-one (67)	242

TABLE XXIX. [4+2] CYCLOADDITION OF KETENES TO AZADIENES (*Continued*)

Reactant	Ketene or Ketene Source	Conditions	Product(s) and Yield(s) (%)	Refs.
	$BrCH_2CO_2Et$	Zn, C_6H_6	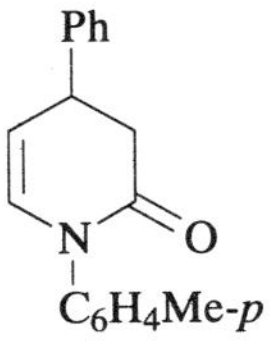 Ph; N; O; C_6H_4Me-*p* (34)	241
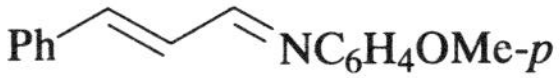 Ph NC$_6$H$_4$OMe-*p*	$BrCH_2CO_2Et$	Zn, C_6H_6	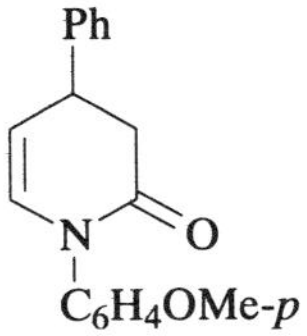 Ph; N; O; C_6H_4OMe-*p* (68)	241
	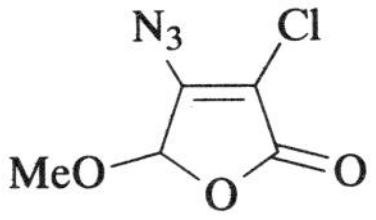 N_3; Cl; MeO; O; O	C_6H_6, 80°	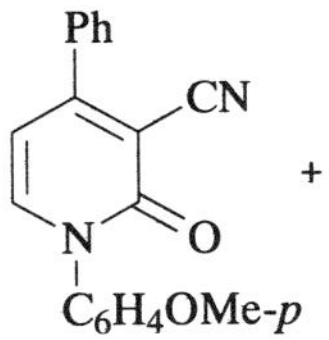 Ph; CN; N; O; C_6H_4OMe-*p* (22) 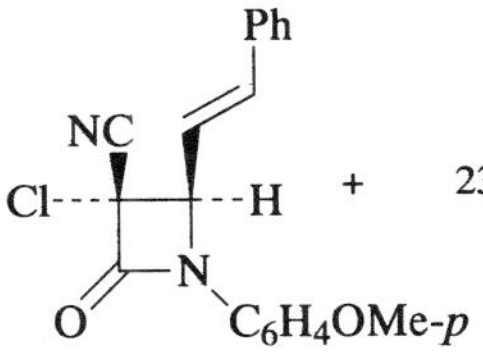 + NC; Ph; Cl; H; O; N; C_6H_4OMe-*p* (17) 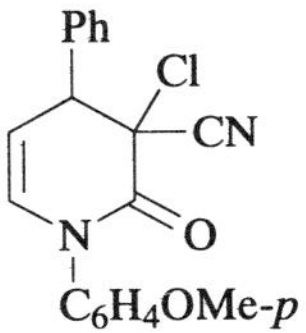 + Ph; Cl; CN; N; O; C_6H_4OMe-*p* (42)	23

TABLE XXIX. [4+2] CYCLOADDITION OF KETENES TO AZADIENES (*Continued*)

Reactant	Ketene or Ketene Source	Conditions	Product(s) and Yield(s) (%)	Refs.
$(CF_3)_2CHC(=NPh)CH=C(OEt)_2$	$(CF_3)_2C=C=O$	—	1-Ph-2-oxo-3,3-bis(CF_3)-4,4-bis(OEt)-6-[$CH(CF_3)_2$]-tetrahydropyridine (73)	799
C_{17}				
3-($CH=NC_6H_4Me$-*p*)-chromone	PhCH(Cl)COCl	Et_3N, C_6H_6, 80°	chromeno-pyridinedione, N-C_6H_4Me-*p*, Cl (21)	244
3-($CH=NC_6H_4OMe$-*p*)-chromone	$Cl_2CHCOCl$	Et_3N, C_6H_6, 80°	chromeno-pyridinedione, N-C_6H_4OMe-*p*, Cl (74)	244
$PhCH=CHCH=NC_6H_4OEt$-*p*	$BrCH_2CO_2Et$	Zn, C_6H_6	4-Ph-1-(C_6H_4OEt-*p*)-3,4-dihydro-2-pyridone (74)	241

TABLE XXIX. [4+2] CYCLOADDITION OF KETENES TO AZADIENES (*Continued*)

Reactant	Ketene or Ketene Source	Conditions	Product(s) and Yield(s) (%)	Refs.
C_{18}				
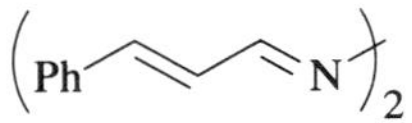 $(PhCH{=}CHCH{=}N)_2$	$Cl_2CHCOCl$	Et_3N, C_6H_6, 20°	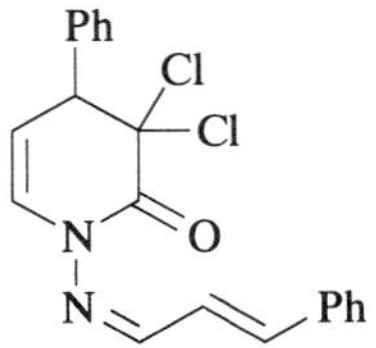 Ph, Cl, Cl, N, O, N, Ph (75)	242
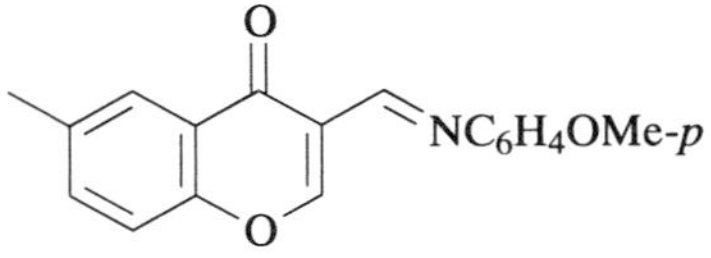 O, NC_6H_4OMe-*p*, O	$Cl_2CHCOCl$	Et_3N, C_6H_6, 80°	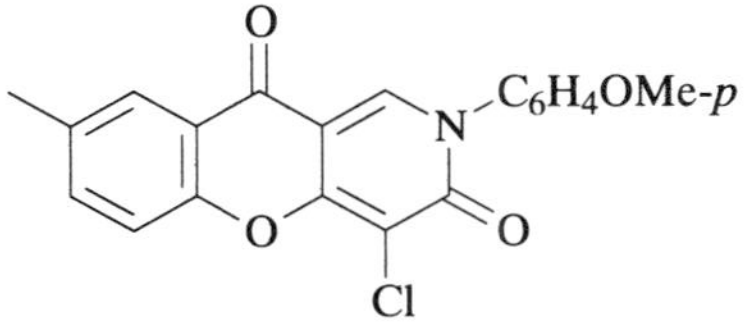 O, N, C_6H_4OMe-*p*, O, O, Cl (72)	244
C_{19}				
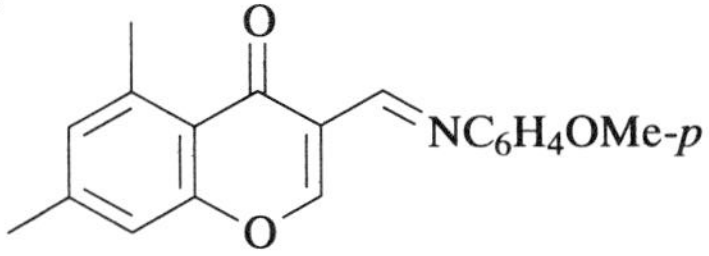 O, NC_6H_4OMe-*p*, O	$Cl_2CHCOCl$	Et_3N, C_6H_6, 80°	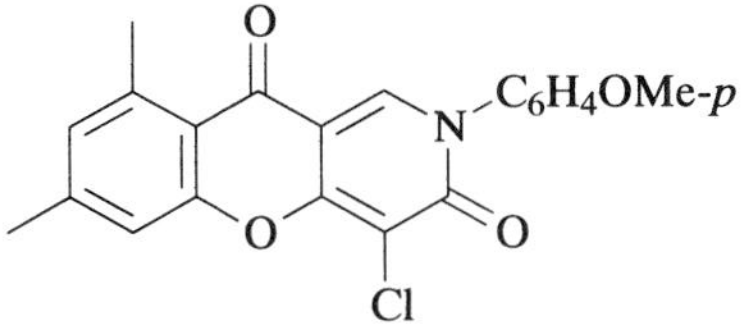 O, N, C_6H_4OMe-*p*, O, O, Cl (85)	244

TABLE XXX. [4+2] CYCLOADDITION OF KETENES TO AMIDINES

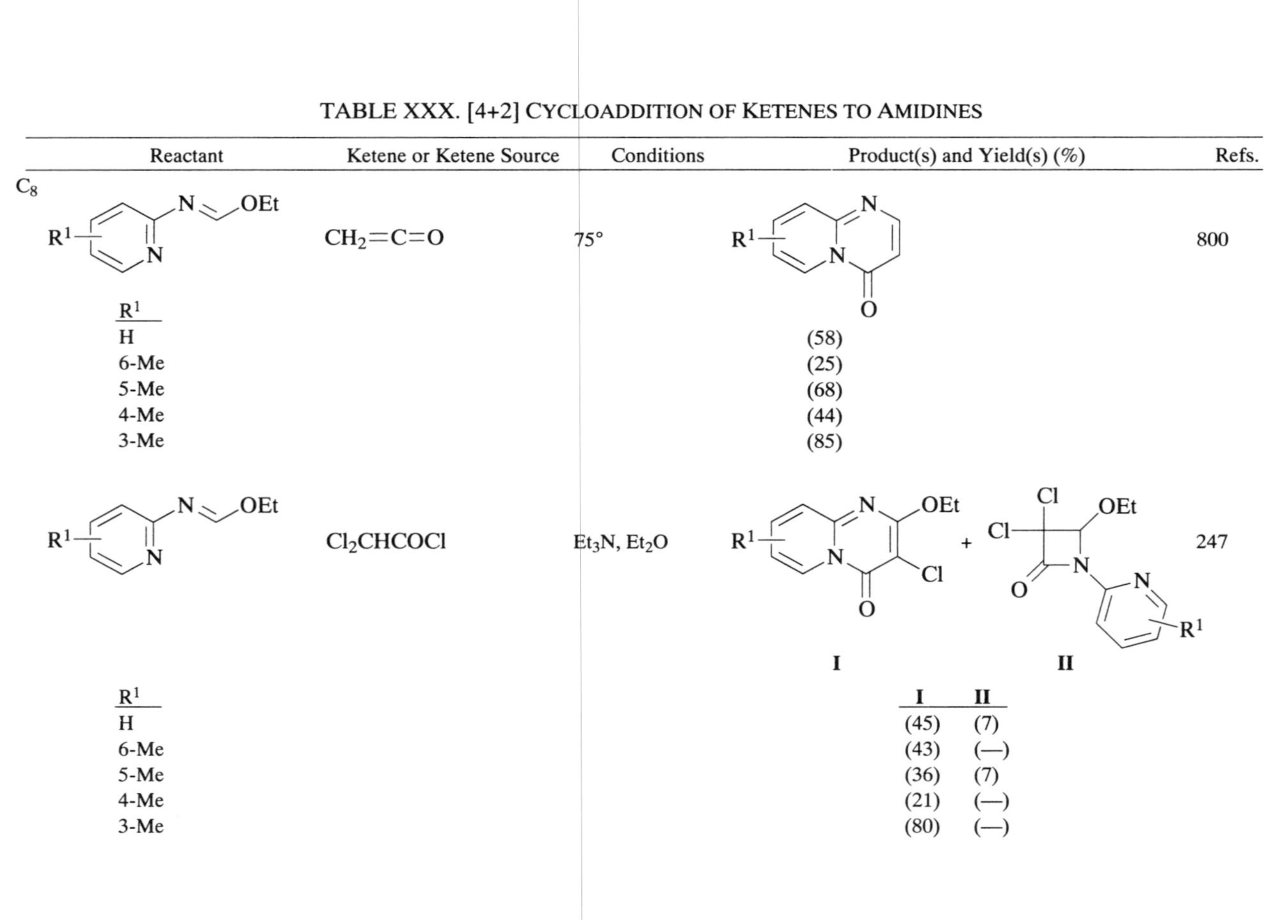

Reactant	Ketene or Ketene Source	Conditions	Product(s) and Yield(s) (%)	Refs.
C_8				
2-(N=CHOEt)-pyridine, R^1	$CH_2=C=O$	75°	pyrido[1,2-a]pyrimidin-4-one, R^1	800
R^1 = H			(58)	
6-Me			(25)	
5-Me			(68)	
4-Me			(44)	
3-Me			(85)	
2-(N=CHOEt)-pyridine, R^1	$Cl_2CHCOCl$	Et_3N, Et_2O	I + II	247
R^1			I, II	
H			(45), (7)	
6-Me			(43), (—)	
5-Me			(36), (7)	
4-Me			(21), (—)	
3-Me			(80), (—)	

TABLE XXX. [4+2] CYCLOADDITION OF KETENES TO AMIDINES (*Continued*)

Reactant	Ketene or Ketene Source	Conditions	Product(s) and Yield(s) (%)	Refs.
C_{10}				
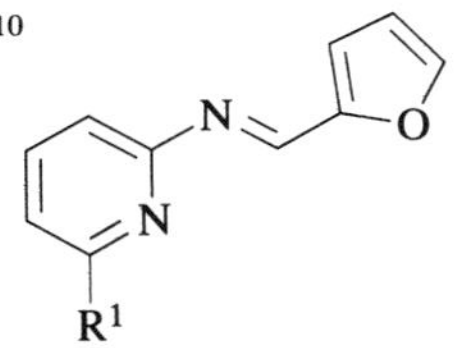 2-(furfurylideneamino)-6-R^1-pyridine	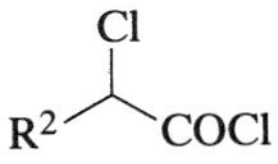 $R^2CH(Cl)COCl$	Et_3N, Et_2O	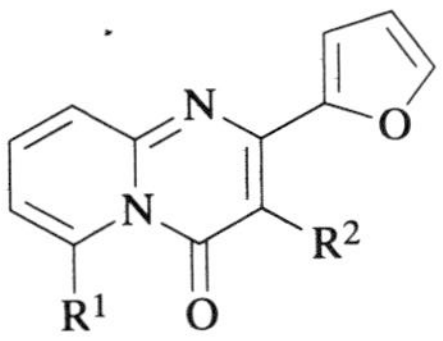 2-(2-furyl)-3-R^2-6-R^1-pyrido[1,2-a]pyrimidin-4-one	246
R^1 = H	R^2 = Cl		(78)	
R^1 = H	R^2 = Ph		(40)	
R^1 = Me	R^2 = Cl		(43)	
R^1 = Me	R^2 = H		(20)	
R^1 = Me	R^2 = Ph		(35)	
R^1-substituted N-N / X ring, $N{=}CH\text{-}C_6H_4R^2\text{-}p$	$Ph_2C{=}C{=}O$	Xylene, 137°	R^1, X, N-N fused ring, O, Ph, Ph, $C_6H_4R^2\text{-}p$	
X = S, R^1 = H, R^2 = OMe			(87)	248, 801
X = S, R^1 = Me, R^2 = NO_2			(79)	248
X = S, R^1 = Me, R^2 = OMe			(67)	248
X = S, R^1 = Me, R^2 = NMe_2			(63)	248
X = O, R^1 = Ph, R^2 = NO_2			(36)	248
X = Se, R^1 = Ph, R^2 = NO_2			(78)	248
X = O, R^1 = Ph, R^2 = OMe			(58)	248
X = Se, R^1 = Ph, R^2 = OMe			(54)	248
X = O, R^1 = Ph, R^2 = NMe_2			(33)	248
X = Se, R^1 = Ph, R^2 = NMe_2			(43)	248

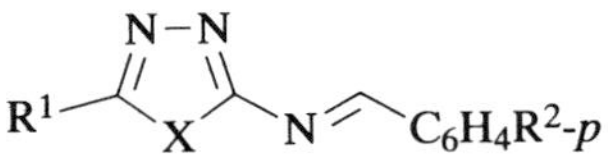

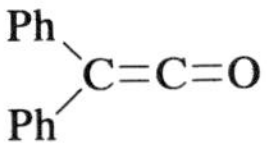

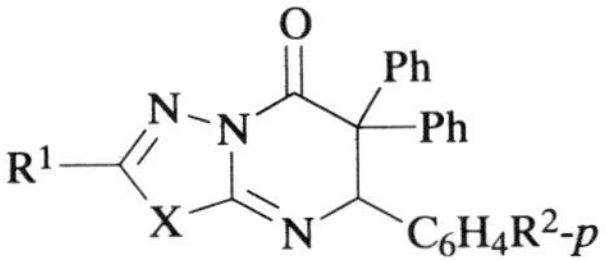

TABLE XXX. [4+2] CYCLOADDITION OF KETENES TO AMIDINES (*Continued*)

Reactant	Ketene or Ketene Source	Conditions	Product(s) and Yield(s) (%)	Refs.
C_{12} 2-($PhCH=N$)-6-R^1-pyridine	$R^2CH(Cl)COCl$	Et_3N, Et_2O	6-R^1-2-Ph-3-R^2-4H-pyrido[1,2-a]pyrimidin-4-one	246
R^1	R^2			
H	Cl		(35)	
H	Ph		(37)	
Me	H		(34)	
Me	Cl		(70)	
Me	Ph		(32)	
2-(p-$ClC_6H_4CH=N$)pyridine	$Ph_2C=C=O$	C_6H_6, 25°	2-C_6H_4Cl-p-3,3-diphenyl-2,3-dihydro-4H-pyrido[1,2-a]pyrimidin-4-one (86)	802
2-($PhCH=N$)pyridine	$Ph_2C=C=O$	C_6H_6, 25°	2,3,3-triphenyl-2,3-dihydro-4H-pyrido[1,2-a]pyrimidin-4-one (87)	802

TABLE XXX. [4+2] CYCLOADDITION OF KETENES TO AMIDINES (*Continued*)

Reactant	Ketene or Ketene Source	Conditions	Product(s) and Yield(s) (%)	Refs.
C_{14} 2-[N=CH–$C_6H_4R^1$-p]-benzazole (X in ring)	$Ph_2C=C=O$	Xylene, 137°	Benzazolo-fused pyrimidinone: 4-oxo, 3,3-diphenyl, 2-($C_6H_4R^1$-p) (X in ring)	
X = O, R^1 = NO_2			X = O (86)	248
X = S, R^1 = NO_2			X = S (70)	248
X = NH, R^1 = NO_2			X = $NCOCHPh_2$ (75)	248
X = O, R^1 = OMe			X = O (74)	248, 801
X = S, R^1 = OMe			X = S (87)	248, 801
X = NH, R^1 = OMe			X = $NCOCHPh_2$ (81)	248, 801
X = O, R^1 = NMe_2			X = O (50)	248
X = S, R^1 = NMe_2			X = S (91)	248
X = NH, R^1 = NMe_2			X = $NCOCHPh_2$ (84)	248
C_{16} $PhC(=NC_6H_4R^1$-$p)N=CHNMe_2$	$PhCH_2COCl$	Et_3N, C_6H_6	1-($C_6H_4R^1$-p)-2,5-diphenylpyrimidin-6(1H)-one	803
R^1 = Cl			(86)	
R^1 = Br			(84)	
R^1 = H			(86)	
R^1 = Me			(92)	

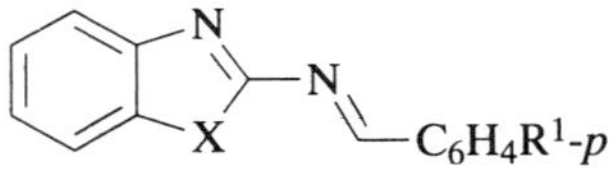

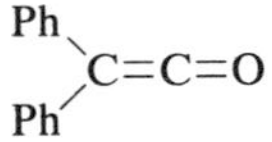

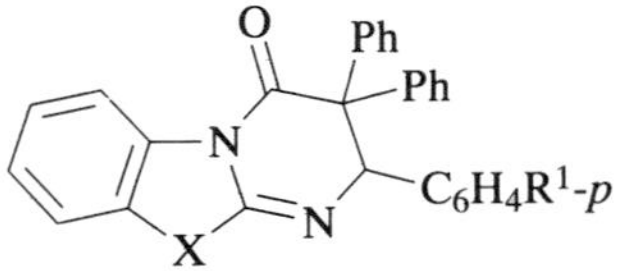

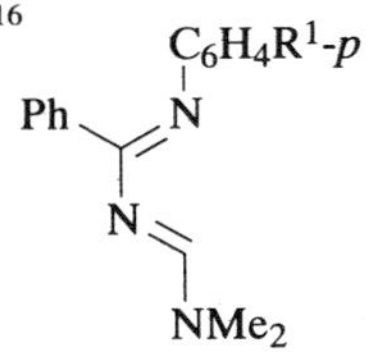

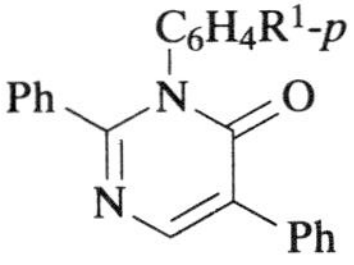

TABLE XXX. [4+2] CYCLOADDITION OF KETENES TO AMIDINES (*Continued*)

Reactant	Ketene or Ketene Source	Conditions	Product(s) and Yield(s) (%)	Refs.
$PhC(=NC_6H_4R^1\text{-}p)N=CHNMe_2$	$Ph_2CHCOCl$	Et_3N, C_6H_6	1-($C_6H_4R^1$-p)-2-Ph-4-NMe_2-5,5-Ph_2-pyrimidin-6-one	803
R^1 = Cl			(85)	
R^1 = Br			(96)	
R^1 = H			(91)	
R^1 = Me			(95)	
C_{17}				
$t\text{-}BuC(=NC_6H_4OMe\text{-}p)N=CHBu\text{-}t$	$Ph_2C=C=O$	—	1-(C_6H_4OMe-p)-2-t-Bu-4-Bu-t-5,5-Ph_2-pyrimidin-6-one (52)	245
$2,6\text{-}Me_2C_6H_3N=C(C_6H_4Cl\text{-}p)N=C=NC_6H_{11}$	$Ph_2C=C=O$	CH_2Cl_2	1-(2,6-$Me_2C_6H_3$)-2-(p-ClC_6H_4)-4-(=NC_6H_{11})-5,5-Ph_2-pyrimidin-6-one (33)	249
$2,6\text{-}Me_2C_6H_3N=C(C_6H_4NO_2\text{-}p)N=C=NC_6H_{11}$	$Ph_2C=C=O$	CH_2Cl_2	1-(2,6-$Me_2C_6H_3$)-2-(p-$O_2NC_6H_4$)-4-(=NC_6H_{11})-5,5-Ph_2-pyrimidin-6-one (90)	249

TABLE XXXI. [4+2] CYCLOADDITION OF KETENES TO *O*-QUINONES AND QUINONIMINES

Reactant	Ketene or Ketene Source	Conditions	Product(s) and Yield(s) (%)	Refs.
C_6 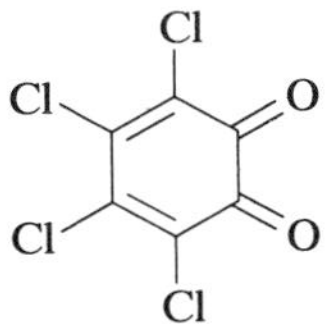 Cl, Cl, Cl, Cl, O, O	$CH_2{=}C{=}O$	-70°, 2 d	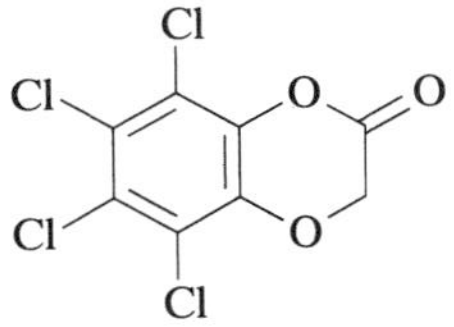 Cl, Cl, Cl, Cl, O, O, O (8)	804
	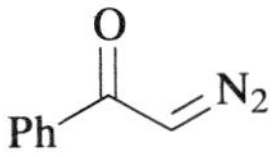 Ph, O, N_2	PhMe, 111°	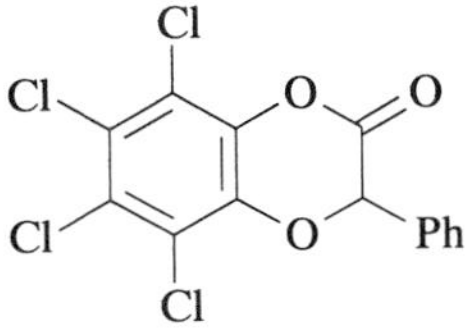 Cl, Cl, Cl, Cl, O, O, O, Ph (27)	804, 805
	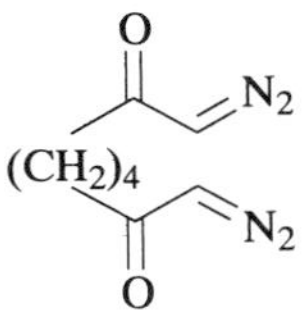 $(CH_2)_4$, O, N_2, O, N_2	PhMe, 111°	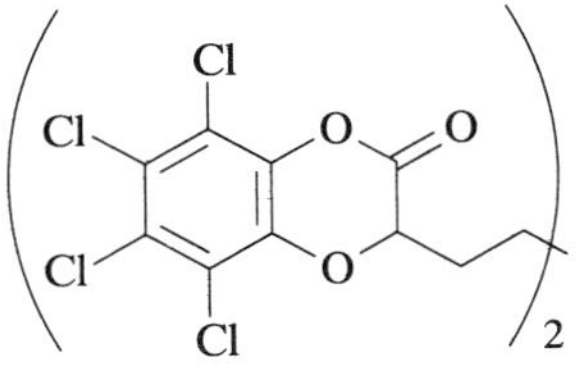 Cl, Cl, Cl, Cl, O, O, O, 2 (78)	250
	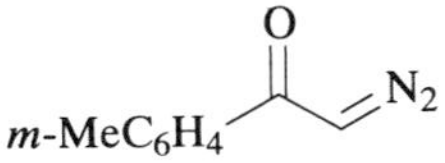 m-MeC_6H_4, O, N_2	PhMe, 111°	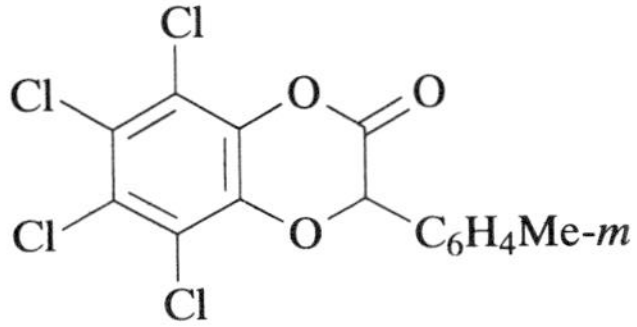 Cl, Cl, Cl, Cl, O, O, O, C_6H_4Me-m (62)	250

TABLE XXXI. [4+2] CYCLOADDITION OF KETENES TO *O*-QUINONES AND QUINONIMINES (*Continued*)

Reactant	Ketene or Ketene Source	Conditions	Product(s) and Yield(s) (%)	Refs.
	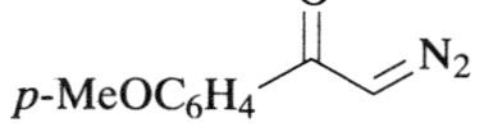 *p*-$MeOC_6H_4$, O, N_2	PhMe, 111°	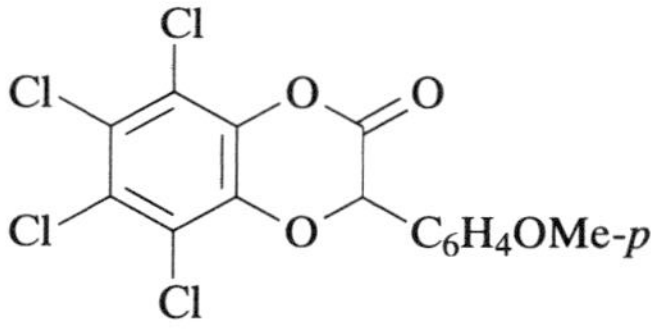 Cl, Cl, Cl, Cl, O, O, O, C_6H_4OMe-*p* (64)	250
	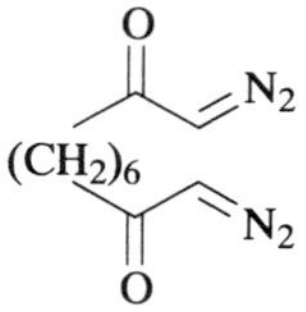 O, N_2, $(CH_2)_6$, N_2, O	PhMe, 111°	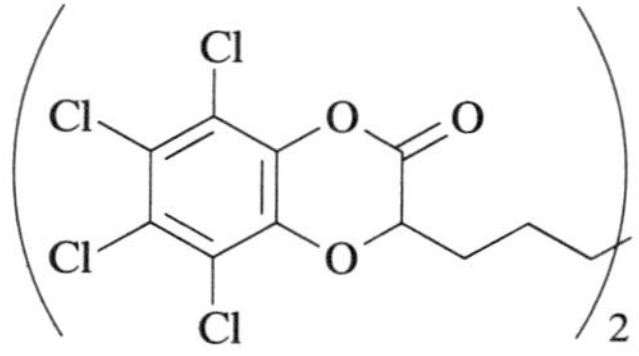 Cl, Cl, Cl, Cl, O, O, O, 2 (32)	250
	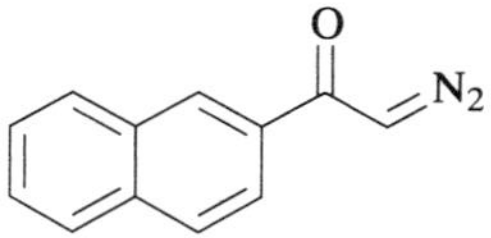 O, N_2	PhMe, 111°	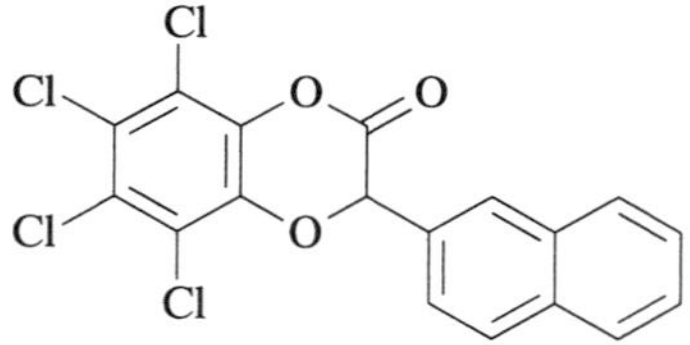 Cl, Cl, Cl, Cl, O, O, O (59)	250
	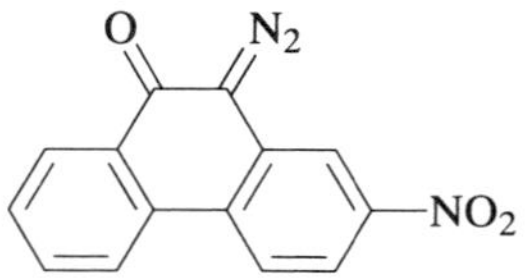 O, N_2, NO_2	—	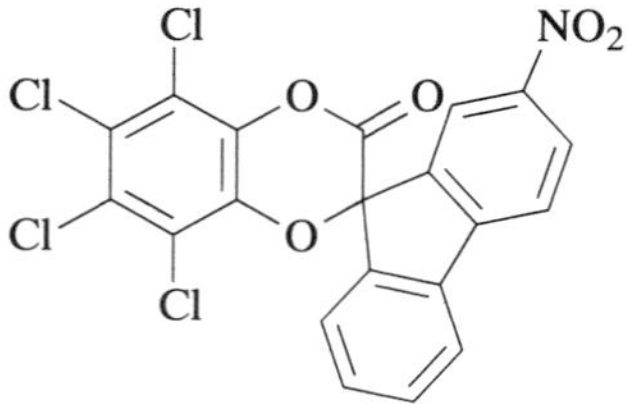 Cl, Cl, Cl, Cl, O, O, O, NO_2 (—)	805

TABLE XXXI. [4+2] CYCLOADDITION OF KETENES TO *O*-QUINONES AND QUINONIMINES (*Continued*)

Reactant	Ketene or Ketene Source	Conditions	Product(s) and Yield(s) (%)	Refs.
	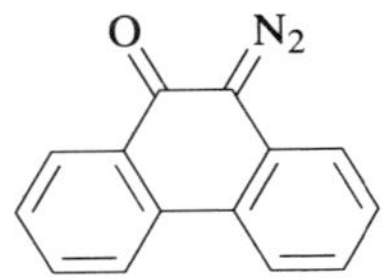 O, N_2	PhMe, 111°	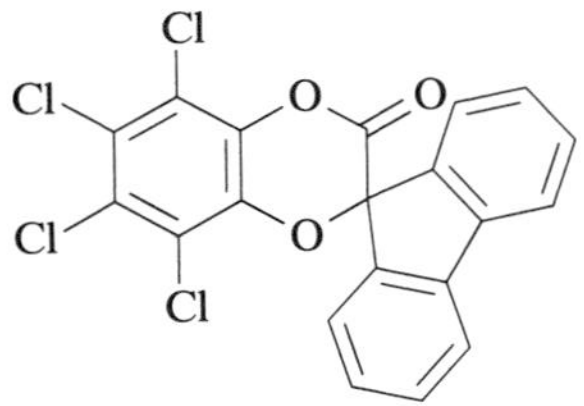 Cl, Cl, Cl, Cl, O, O, O (30)	804, 805
	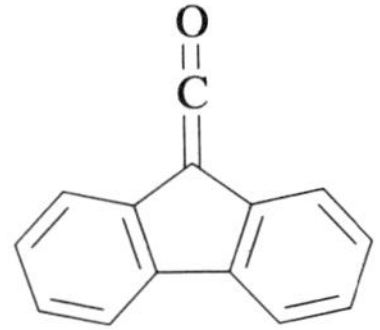 O=C=	Et_2O	" (39)	804
	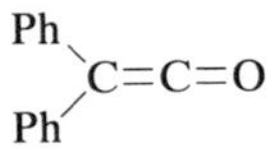 Ph(Ph)C=C=O	Et_2O	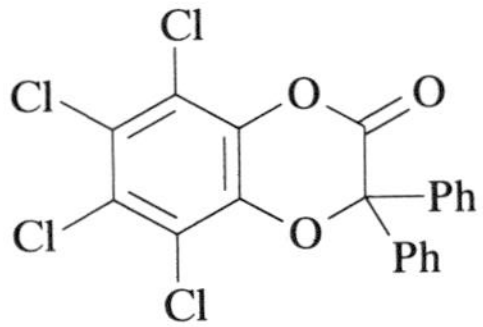 Cl, Cl, Cl, Cl, O, O, O, Ph, Ph (80)	314
	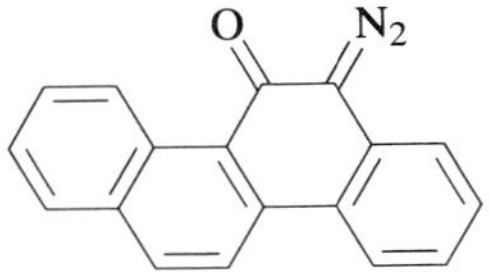 O, N_2	PhMe, 111°	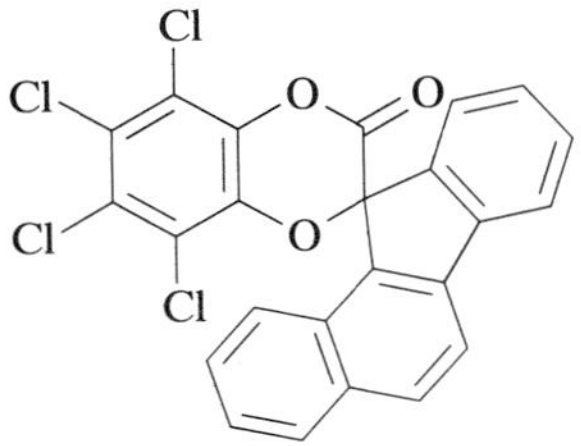 Cl, Cl, Cl, Cl, O, O, O (40)	804, 805

TABLE XXXI. [4+2] CYCLOADDITION OF KETENES TO *O*-QUINONES AND QUINONIMINES (*Continued*)

Reactant	Ketene or Ketene Source	Conditions	Product(s) and Yield(s) (%)	Refs.
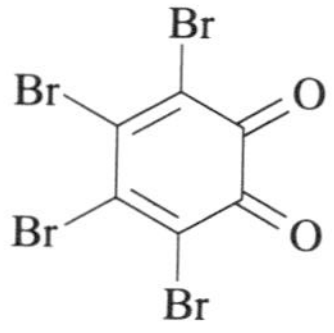	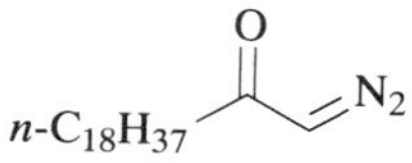 n-$C_{18}H_{37}COCH=N_2$	PhMe, 111°	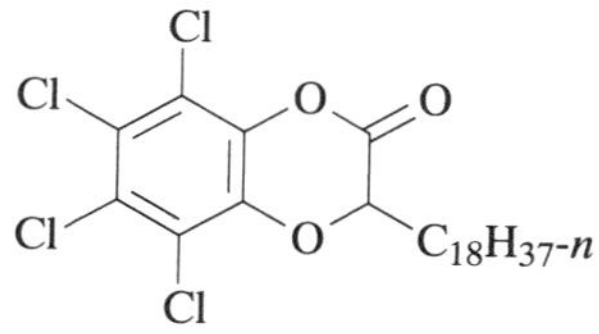 Cl, Cl, Cl, Cl; O, O; $C_{18}H_{37}$-n (77)	250
Br, Br, Br, Br; O, O	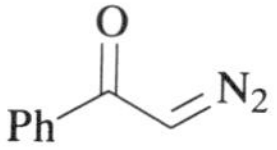 $PhCOCH=N_2$	PhMe, 111°	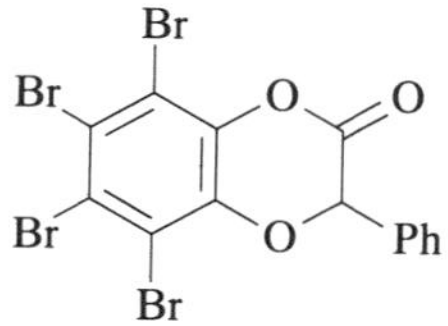 Br, Br, Br, Br; O, O; Ph (58)	250
	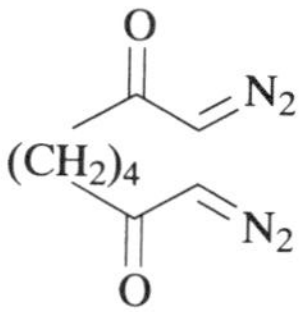 $(CH_2)_4(COCH=N_2)_2$	PhMe, 111°	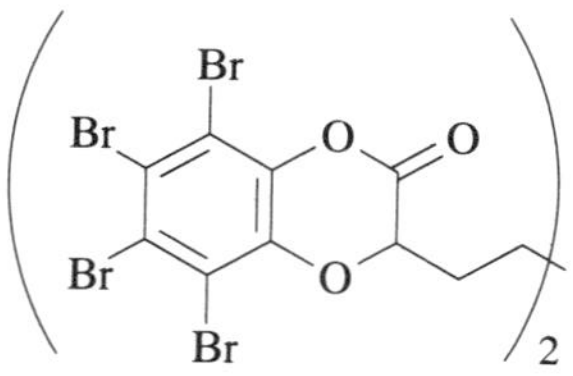 (Br, Br, Br, Br; O, O)$_2$ (71)	250
	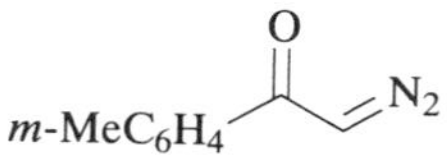 m-$MeC_6H_4COCH=N_2$	PhMe, 111°	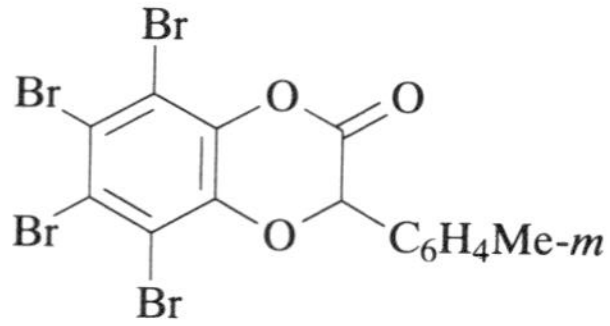 Br, Br, Br, Br; O, O; C_6H_4Me-m (45)	250

TABLE XXXI. [4+2] CYCLOADDITION OF KETENES TO *O*-QUINONES AND QUINONIMINES (*Continued*)

Reactant	Ketene or Ketene Source	Conditions	Product(s) and Yield(s) (%)	Refs.
	p-MeOC$_6$H$_4$C(O)CH=N$_2$	PhMe, 111°	Br, Br, Br, Br; O, O, O; C$_6$H$_4$OMe-p (73)	250
	(CH$_2$)$_6$; O, N$_2$; O, N$_2$	PhMe, 111°	Br, Br, Br, Br; O, O, O; 2 (15)	250
	O, N$_2$	PhMe, 111°	Br, Br, Br, Br; O, O, O (38)	250
	Ph$_2$C=C=O	Et_2O	Br, Br, Br, Br; O, O, O; Ph, Ph (48)	314

TABLE XXXI. [4+2] CYCLOADDITION OF KETENES TO *O*-QUINONES AND QUINONIMINES (*Continued*)

Reactant	Ketene or Ketene Source	Conditions	Product(s) and Yield(s) (%)	Refs.
	[structure]	Et_2O	[structure] (34)	804
	n-$C_{18}H_{37}COCH=N_2$	PhMe, 111°	[structure] $C_{18}H_{37}$-n (50)	250
[structure]	$Ph_2C=C=O$	C_6H_6	[structure] (82)	806
C_7 [structure]	$Ph_2C=C=O$	PhMe, 10 min	[structure] (65)	804

TABLE XXXI. [4+2] CYCLOADDITION OF KETENES TO *O*-QUINONES AND QUINONIMINES (*Continued*)

Reactant	Ketene or Ketene Source	Conditions	Product(s) and Yield(s) (%)	Refs.
C_{14}				
O, O	Ph, Ph, C=C=O	$h\nu$, C_6H_6, 36 d	O, Ph, Ph, O, O (—)	154, 807
Bu-*t*, O, O, *t*-Bu	O, C	Et_2O	Bu-*t*, O, O, O, *t*-Bu (43)	804
	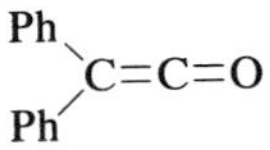 Ph, Ph, C=C=O	PhMe	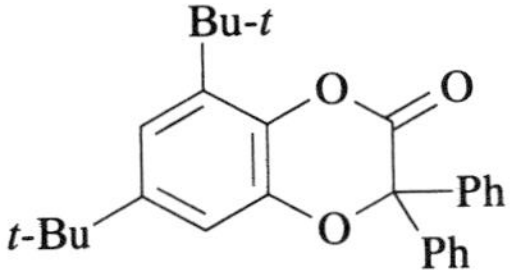 Bu-*t*, O, O, Ph, O, Ph, *t*-Bu (96)	804
C_{18}				
NSO_2Ph, NSO_2Ph, Cl	Ph, Ph, C=C=O	C_6H_6	SO_2Ph, N, O, Ph, Ph, N, SO_2Ph, Cl (32) + SO_2Ph, N, Ph, Ph, O, N, SO_2Ph, Cl (32)	251

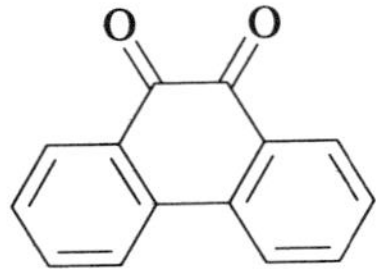

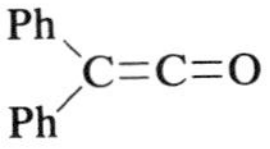

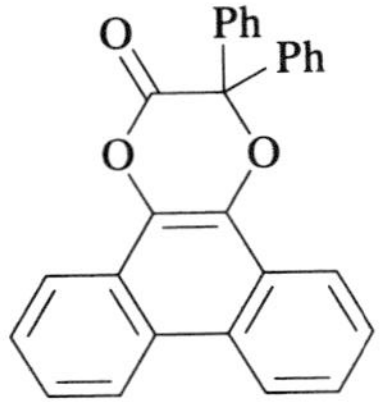

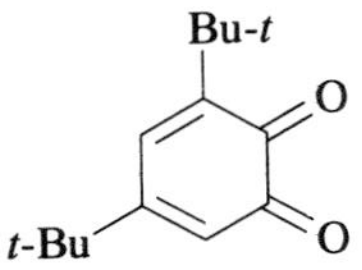

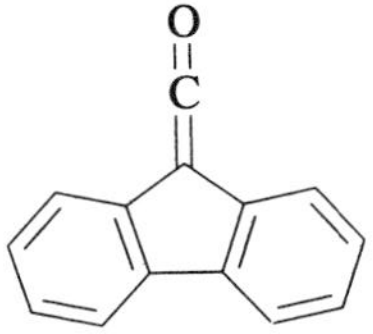

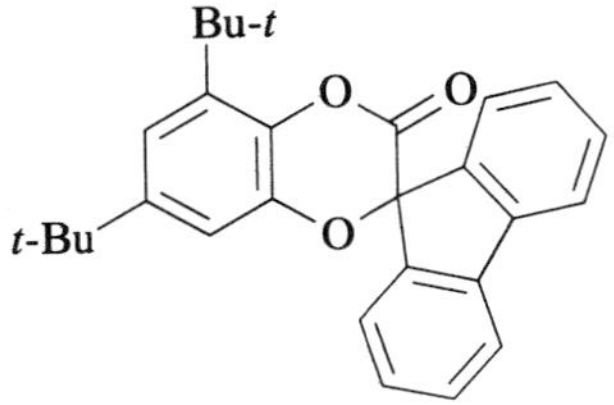

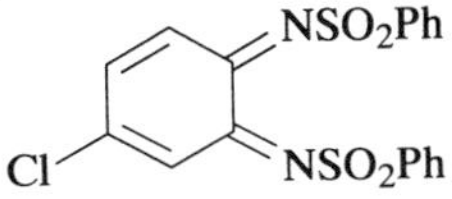

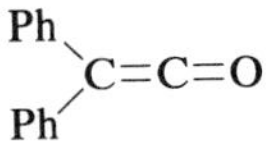

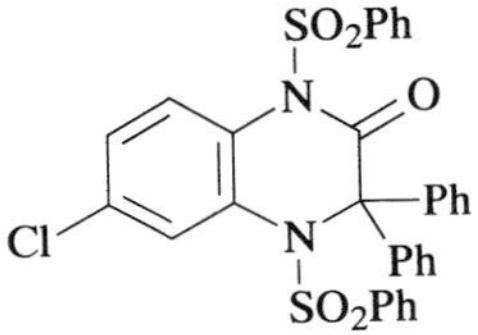

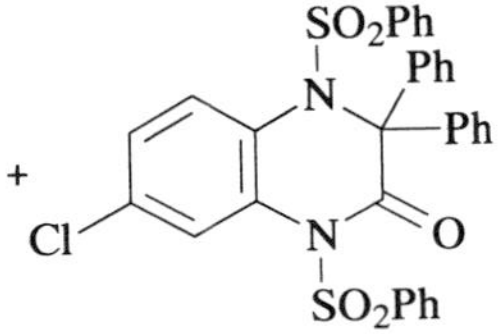

TABLE XXXI. [4+2] CYCLOADDITION OF KETENES TO *O*-QUINONES AND QUINONIMINES (*Continued*)

Reactant	Ketene or Ketene Source	Conditions	Product(s) and Yield(s) (%)	Refs.
C_{20}				
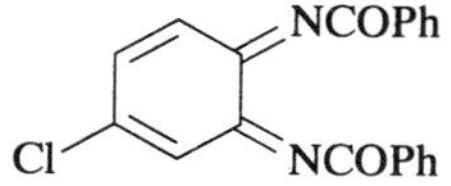 Cl; NCOPh; NCOPh	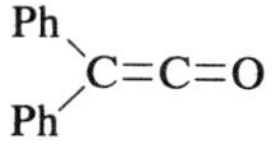 Ph, Ph C=C=O	C_6H_6	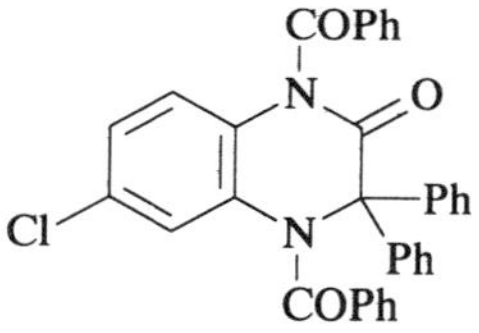 COPh, N, O, Cl, Ph, Ph, N, COPh (37) 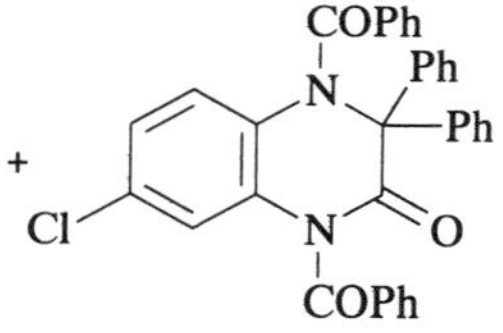 + COPh, N, Ph, Ph, Cl, N, O, COPh (44)	251
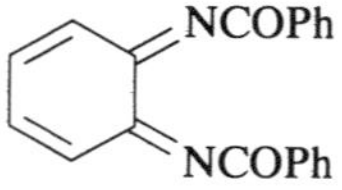 NCOPh; NCOPh	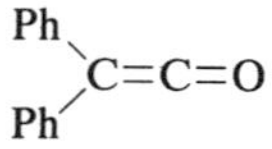 Ph, Ph C=C=O	C_6H_6	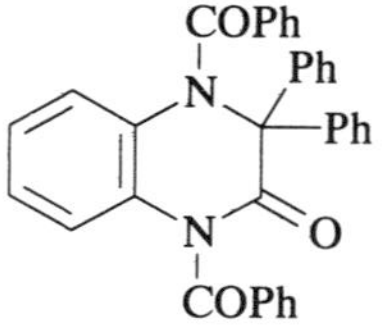 COPh, N, Ph, Ph, N, O, COPh (76)	251
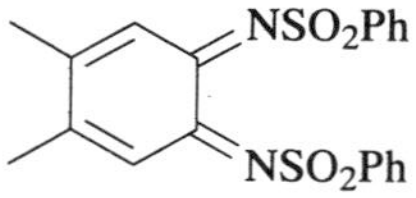 NSO_2Ph; NSO_2Ph	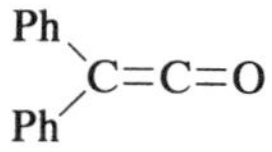 Ph, Ph C=C=O	C_6H_6	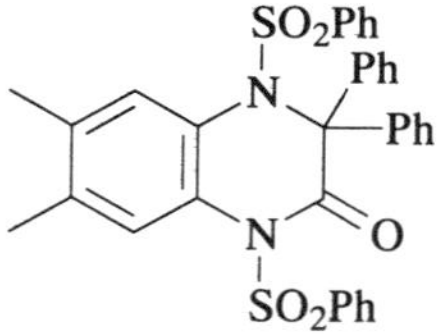 SO_2Ph, N, Ph, Ph, N, O, SO_2Ph (90)	251
	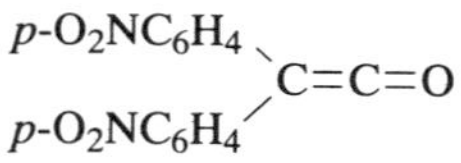 p-$O_2NC_6H_4$, p-$O_2NC_6H_4$ C=C=O	C_6H_6	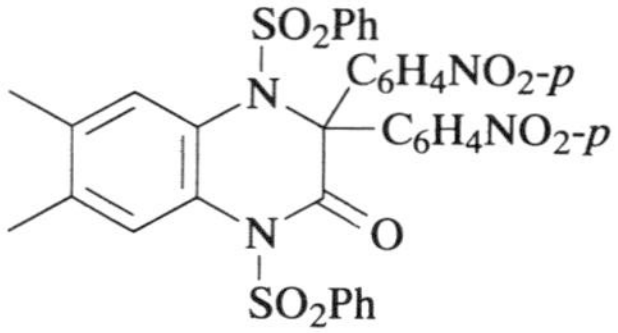 SO_2Ph, N, $C_6H_4NO_2$-p, $C_6H_4NO_2$-p, N, O, SO_2Ph (76)	251

TABLE XXXI. [4+2] CYCLOADDITION OF KETENES TO *O*-QUINONES AND QUINONIMINES (*Continued*)

Reactant	Ketene or Ketene Source	Conditions	Product(s) and Yield(s) (%)	Refs.
C_{22}				
4,5-dimethyl-1,2-bis(NCOPh)cyclohexa-3,5-diene	$Ph_2C{=}C{=}O$	C_6H_6	1,4-bis(COPh)-3,3-diphenyl-6,7-dimethyl-3,4-dihydroquinoxalin-2(1H)-one (8)	251

TABLE XXXII. [4+2] CYCLOADDITION OF KETENES TO α,β-UNSATURATED CARBONYL COMPOUNDS

Reactant	Ketene or Ketene Source	Conditions	Product(s) and Yield(s) (%)	Refs.
C_4				
(E)-4-methoxybut-3-en-2-one (MeCOCH=CHOMe)	$Cl_2CHCOCl$	Et_3N, C_6H_{14}	3,3-dichloro-4-methoxy-6-methyl-3,4-dihydro-2H-pyran-2-one (—)	327
	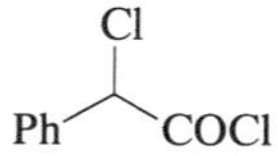 PhCH(Cl)COCl	Et_3N, C_6H_{14}	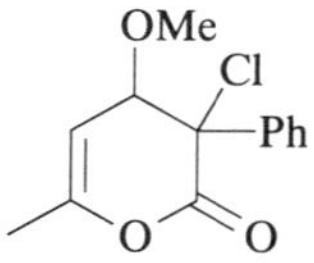 3-chloro-4-methoxy-6-methyl-3-phenyl-3,4-dihydro-2H-pyran-2-one (—)	327
	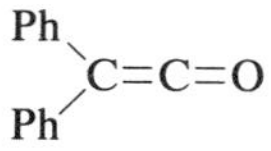 $Ph_2C{=}C{=}O$	82°	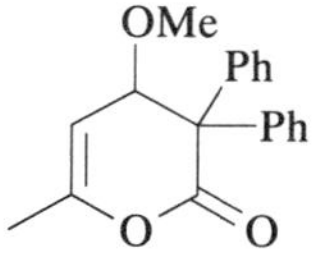 4-methoxy-6-methyl-3,3-diphenyl-3,4-dihydro-2H-pyran-2-one (78)	327
C_6				
2-(R^1R^2N-methylene)cyclopentanone	$Cl_2CHCOCl$	Et_3N, C_6H_6	4-(R^1R^2N)-3,3-dichloro-cyclopenta[b]pyran-2-one	808
R^1 = *i*-Pr, R^2 = *i*-Pr			(54)	
R^1 = Me, R^2 = Ph			(48)	
R^1 = Ph, R^2 = Ph			(42)	

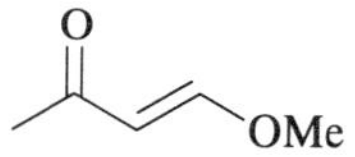

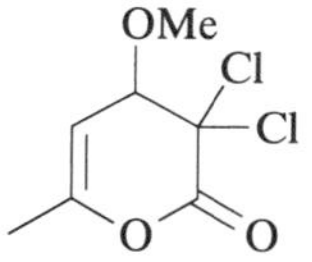

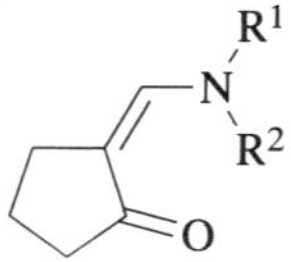

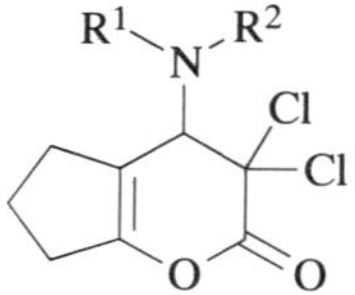

TABLE XXXII. [4+2] CYCLOADDITION OF KETENES TO α,β-UNSATURATED CARBONYL COMPOUNDS (*Continued*)

Reactant	Ketene or Ketene Source	Conditions	Product(s) and Yield(s) (%)	Refs.
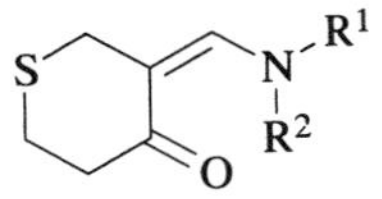 [structure] R^1 / R^2: *i*-Pr / *i*-Pr; Me / Ph; Ph / Ph	$Cl_2CHCOCl$	Et_3N, Et_2O	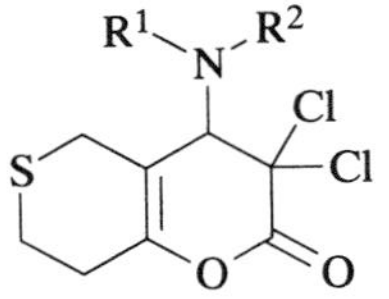 [structure] (40), (36), (42)	258
C_7 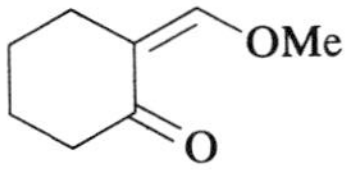 [structure]	$Cl_2CHCOCl$	Et_3N, C_6H_{14}	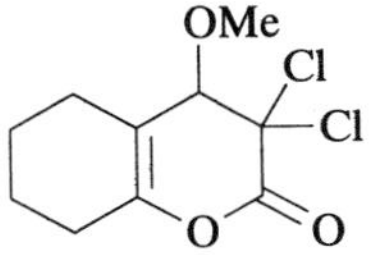 [structure] (—)	327
	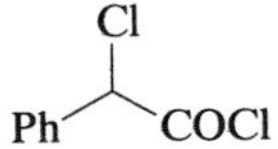 [structure]	Et_3N, C_6H_{14}	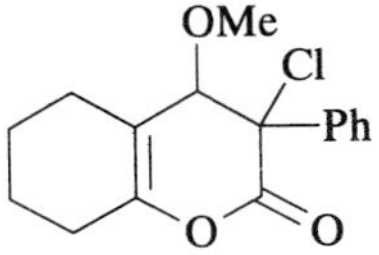 [structure] (68)	327
	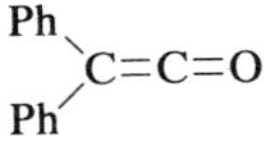 $Ph_2C=C=O$	82°	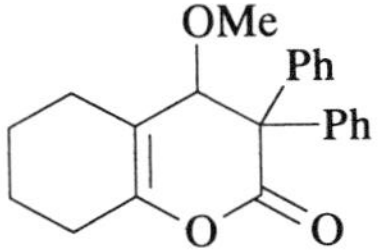 [structure] (90)	327

TABLE XXXII. [4+2] CYCLOADDITION OF KETENES TO α,β-UNSATURATED CARBONYL COMPOUNDS (*Continued*)

Reactant		Ketene or Ketene Source		Conditions	Product(s) and Yield(s) (%)	Refs.
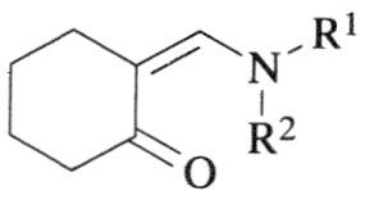 2-[(R^1R^2N)methylene]cyclohexanone		$R^3R^4CHCOCl$		Et_3N, C_6H_6	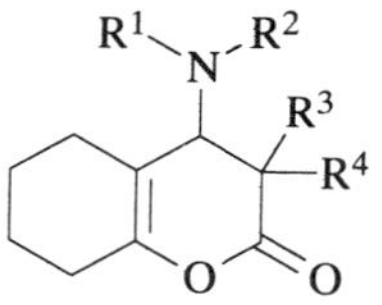 4-(R^1R^2N)-3-R^3-3-R^4-3,4,5,6,7,8-hexahydrocoumarin	
R^1	R^2	R^3	R^4			
Me	Me	SEt	SEt		(81)	809
Et	Et	S(CH$_2$)$_3$S			(63)	810
i-Pr	*i*-Pr	Cl	Cl		(61)	808
Morpholino		S(CH$_2$)$_3$S			(30)	810
Morpholino		SEt	SEt		(40)	809
Piperidino		S(CH$_2$)$_3$S			(45)	810
Piperidino		SEt	SEt		(38)	809
Me	Ph	Cl	Cl		(51)	808
Me	Ph	S(CH$_2$)$_3$S			(75)	810
Me	Ph	SEt	SEt		(35)	809
Ph	Ph	Cl	Cl		(56)	808
Ph	Ph	SEt	SEt		(15)	809

TABLE XXXII. [4+2] CYCLOADDITION OF KETENES TO α,β-UNSATURATED CARBONYL COMPOUNDS (*Continued*)

Reactant	Ketene or Ketene Source	Conditions	Product(s) and Yield(s) (%)	Refs.
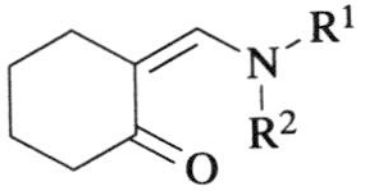 2-(R^1R^2N-methylene)cyclohexanone	$Ph_2C{=}C{=}O$	C_6H_6, 30 min	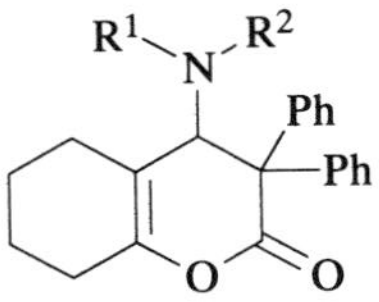 4-(R^1R^2N)-3,3-diphenyl-hexahydrochromen-2-one	811
R^1 = Me, R^2 = Me			(49)	
Morpholino			(53)	
R^1 = Et, R^2 = Et			(40)	
Piperidino			(37)	
R^1 = *i*-Pr, R^2 = *i*-Pr			(68)	
R^1 = Me, R^2 = Ph			(58)	
R^1 = Ph, R^2 = Ph			(36)	
C_8				
2-(R^1R^2N-methylene)cycloheptanone	$R^3R^4CHCOCl$	Et_3N, C_6H_6	4-(R^1R^2N)-3-R^3-3-R^4-cyclohepta[b]pyran-2-one	
R^1 = Me, R^2 = Me	R^3 = Cl, R^4 = Cl		(59)	808
R^1 = Me, R^2 = Me	R^3 = SEt, R^4 = SEt		(42)	809
Piperidino	$S(CH_2)_3S$		(70)	810
R^1 = *i*-Pr, R^2 = *i*-Pr	R^3 = Cl, R^4 = Cl		(50)	808
R^1 = Me, R^2 = Ph	R^3 = Cl, R^4 = Cl		(73)	808
R^1 = Me, R^2 = Ph	R^3 = SEt, R^4 = SEt		(33)	809
R^1 = Ph, R^2 = Ph	R^3 = Cl, R^4 = Cl		(83)	808

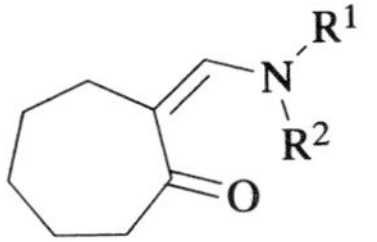

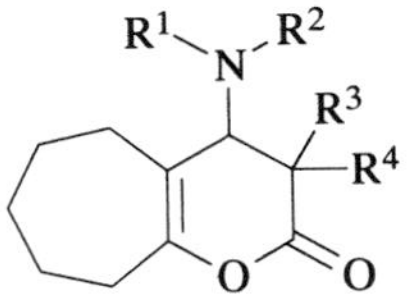

TABLE XXXII. [4+2] CYCLOADDITION OF KETENES TO α,β-UNSATURATED CARBONYL COMPOUNDS (*Continued*)

Reactant	Ketene or Ketene Source	Conditions	Product(s) and Yield(s) (%)	Refs.
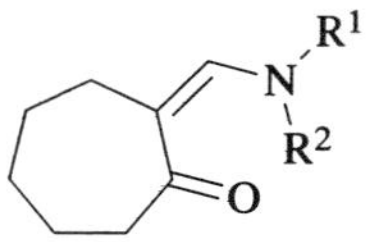 R[1] R[2]	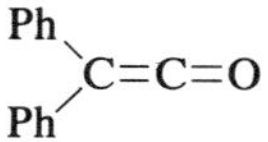 $Ph_2C=C=O$	C_6H_6, 30 min	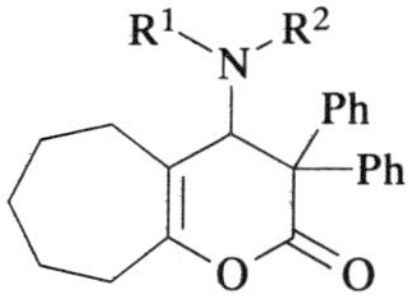	811
Me Me			(59)	
Morpholino			(49)	
Et Et			(15)	
Piperidino			(43)	
i-Pr *i*-Pr			(20)	
Me Ph			(64)	
Ph Ph			(49)	
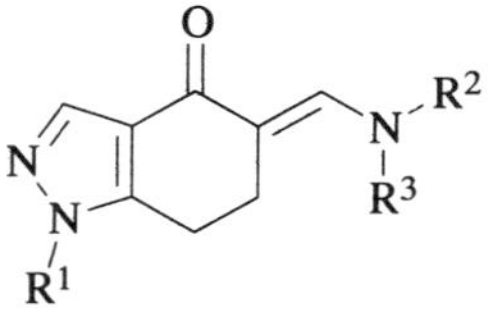 R[1] R[2] R[3]	$Cl_2CHCOCl$	Et_3N, CH_2Cl_2	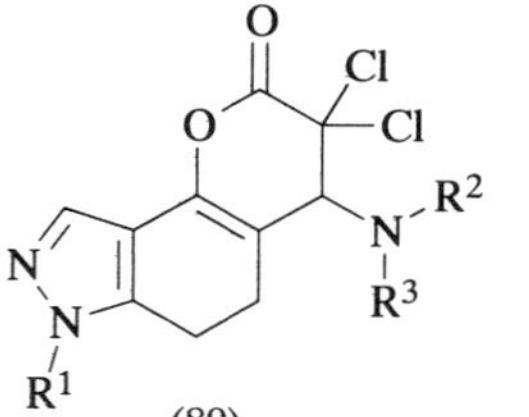	812
Me Me Me			(89)	
Me Piperidino			(49)	
Me Me Ph			(78)	
Me Ph Ph			(83)	
Ph Me Me			(67)	
Ph Piperidino			(51)	
Ph Me Ph			(90)	
Ph Ph Ph			(86)	

TABLE XXXII. [4+2] CYCLOADDITION OF KETENES TO α,β-UNSATURATED CARBONYL COMPOUNDS (*Continued*)

Reactant		Ketene or Ketene Source		Conditions	Product(s) and Yield(s) (%)	Refs.
C_9 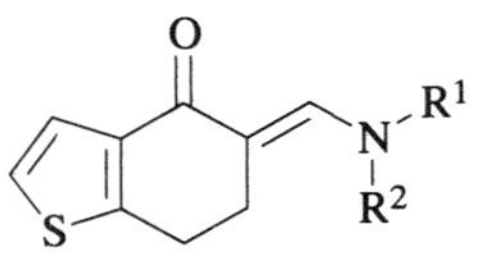 [structure]		$Cl_2CHCOCl$		Et_3N, Et_2O	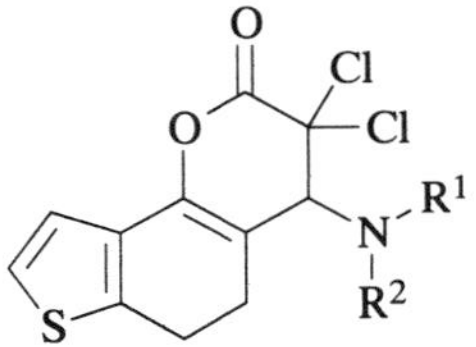 [structure]	813
R^1	R^2					
$(CH_2)_2OMe$	$(CH_2)_2OMe$				(93)	
Me	Ph				(84)	
Ph	Ph				(98)	
[structure]		$R^3R^4CHCOCl$		Et_3N, C_6H_6	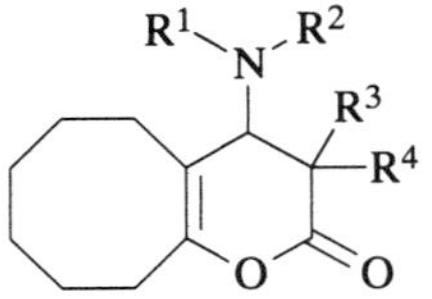 [structure]	
R^1	R^2	R^3	R^4			
Me	Me	Cl	Cl		(46)	808
Me	Me	SEt	SEt		(23)	809
Morpholino		$S(CH_2)_3S$			(27)	810
Piperidino		$S(CH_2)_3S$			(40)	810
Et	Et	$S(CH_2)_3S$			(52)	810
i-Pr	*i*-Pr	Cl	Cl		(43)	808
Me	Ph	Cl	Cl		(54)	808
Me	Ph	$S(CH_2)_3S$			(37)	810
Me	Ph	SEt	SEt		(43)	809
Ph	Ph	Cl	Cl		(36)	808

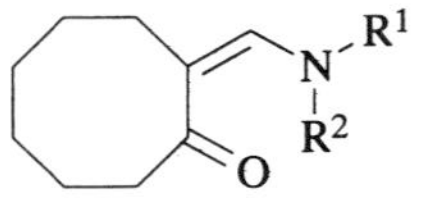

TABLE XXXII. [4+2] CYCLOADDITION OF KETENES TO α,β-UNSATURATED CARBONYL COMPOUNDS (*Continued*)

Reactant	Ketene or Ketene Source	Conditions	Product(s) and Yield(s) (%)	Refs.
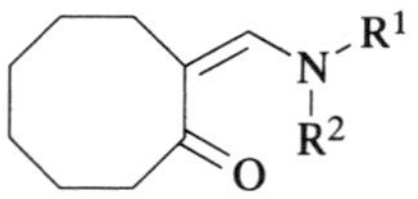 2-[(R^1R^2N)methylene]cyclooctanone	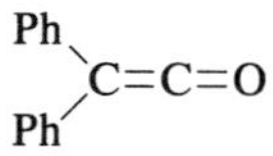 $Ph_2C=C=O$	C_6H_6, 30 min	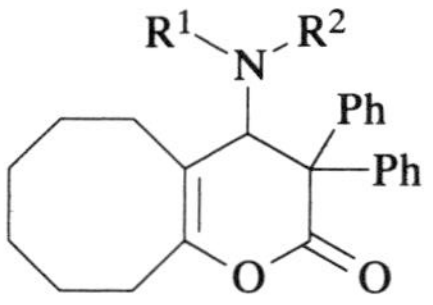 4-(R^1R^2N)-3,3-diphenyl-cycloocta-fused pyran-2-one	811
R^1 / R^2				
Me / Me			(29)	
Morpholino			(63)	
Piperidino			(68)	
Et / Et			(59)	
i-Pr / *i*-Pr			(55)	
Me / Ph			(86)	
Ph / Ph			(50)	
C_{10}				
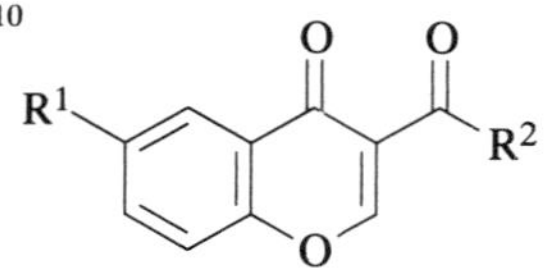 6-R^1-3-(R^2CO)chromone	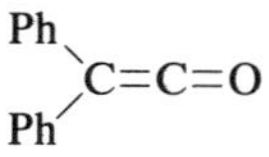 $Ph_2C=C=O$	130°	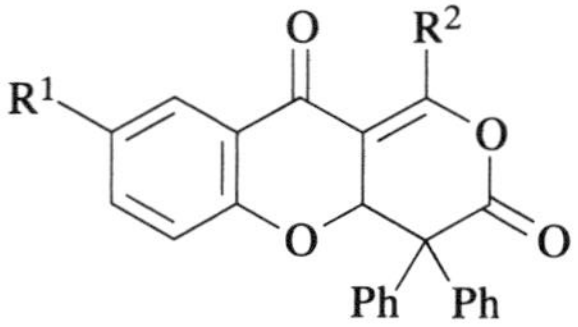 pyrano[4,3-b]chromene-dione, R^1, R^2, 4,4-Ph,Ph	
R^1 / R^2				
H / H			(60)	256
Me / H			(55)	256
H / Ph			(—)	749, 797

TABLE XXXII. [4+2] CYCLOADDITION OF KETENES TO α,β-UNSATURATED CARBONYL COMPOUNDS (*Continued*)

Reactant	Ketene or Ketene Source	Conditions	Product(s) and Yield(s) (%)	Refs.
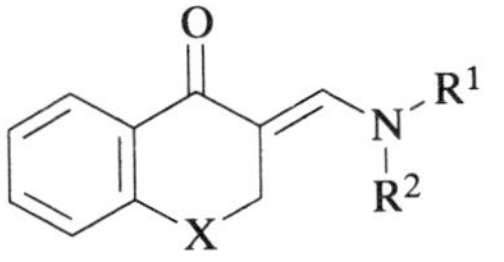 [structure: 3-(aminomethylene)-4-chromanone / thiochromanone, X, NR^1R^2]	$Cl_2CHCOCl$	Et_3N, C_6H_6	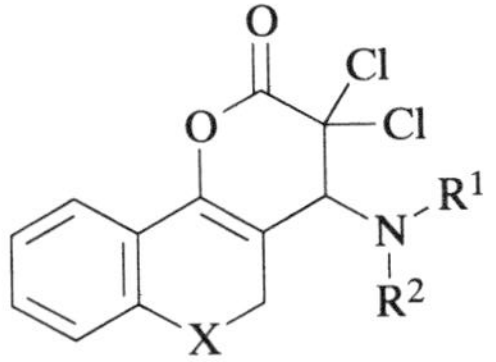 [structure: 3,3-dichloro-4-(NR^1R^2) pyranone fused product, X]	
X = O, R^1R^2 = Piperidino			(23)	814
X = O, R^1 = Me, R^2 = Ph			(79)	814
X = O, R^1 = Ph, R^2 = Ph			(90)	814
X = S, R^1 = *i*-Pr, R^2 = *i*-Pr			(81)	258
X = S, R^1 = Me, R^2 = Ph			(96)	258
C_{11}				
[structure: 2-(R^1-methylene)-1-tetralone]	$R^2R^3CHCOCl$	Et_3N	[structure: 3-R^2-3-R^3-4-R^1 dihydronaphthopyranone]	
R^1 = OMe	R^2 = Cl, R^3 = Cl		(—)	327
R^1 = OMe	R^2 = Me, R^3 = Cl		(—)	327
R^1 = OMe	R^2 = Ph, R^3 = Cl		(—)	327
R^1 = OMe	R^2 = Ph, R^3 = Et		(65)	327
R^1 = OMe	R^2 = Ph, R^3 = Ph		(82)	327
R^1 = N(Pr-*i*)$_2$	R^2 = Cl, R^3 = Cl		(65)	815
R^1 = N(Me)Ph	R^2 = Cl, R^3 = Cl		(49)	815
R^1 = NPh_2	R^2 = Cl, R^3 = Cl		(80)	815

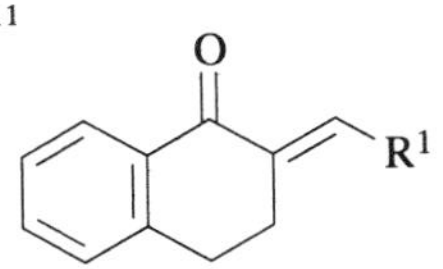

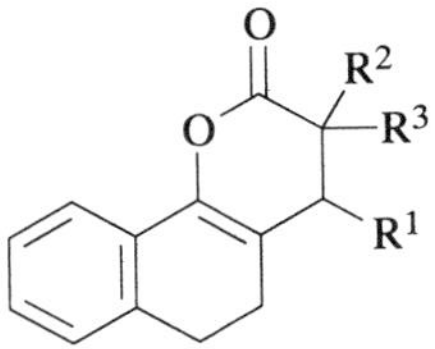

TABLE XXXII. [4+2] CYCLOADDITION OF KETENES TO α,β-UNSATURATED CARBONYL COMPOUNDS (*Continued*)

Reactant	Ketene or Ketene Source	Conditions	Product(s) and Yield(s) (%)	Refs.
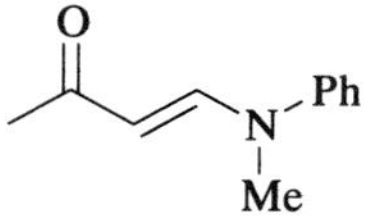 (structure: Me, Ph on N)	$Cl_2CHCOCl$	Et_3N, C_6H_6	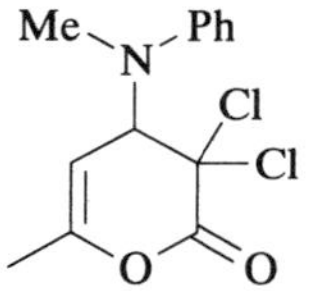 (structure: Me, Ph on N; Cl, Cl) (53)	816
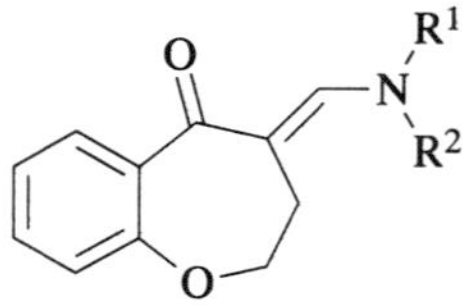 (structure: N-R[1], R[2])	$Cl_2CHCOCl$	Et_3N, C_6H_6	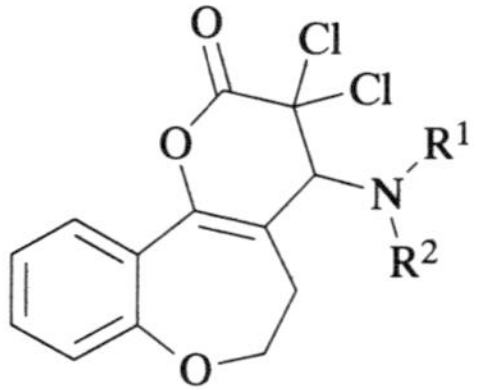 (structure: Cl, Cl; N-R[1], R[2])	817
R[1] = R[2] = Morpholino			(44)	
R[1] = i-Pr, R[2] = i-Pr			(86)	
R[1] = Me, R[2] = Ph			(70)	
R[1] = Ph, R[2] = Ph			(82)	
C_{12}				
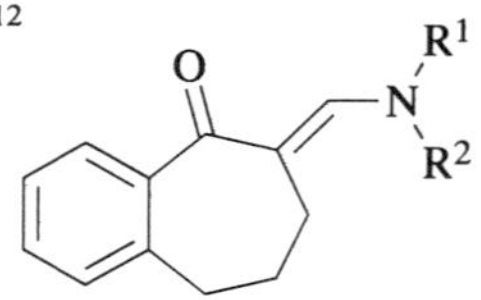 (structure: N-R[1], R[2])	$Cl_2CHCOCl$	Et_3N, C_6H_6	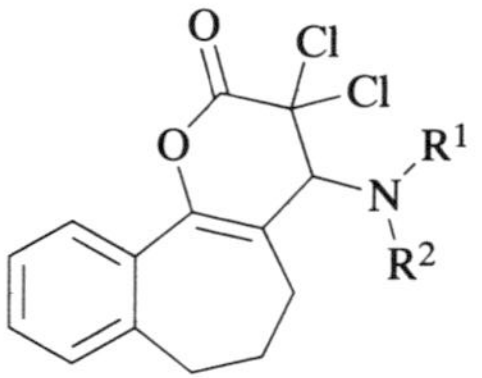 (structure: Cl, Cl; N-R[1], R[2])	818
R[1] = i-Pr, R[2] = i-Pr			(95)	
R[1] = Me, R[2] = Ph			(62)	
R[1] = Ph, R[2] = Ph			(78)	

TABLE XXXII. [4+2] CYCLOADDITION OF KETENES TO α,β-UNSATURATED CARBONYL COMPOUNDS (*Continued*)

Reactant	Ketene or Ketene Source	Conditions	Product(s) and Yield(s) (%)	Refs.
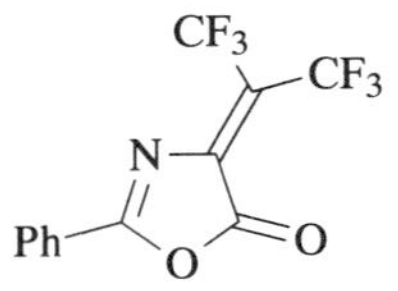 CF3, CF3, N, Ph, O, O	$CH_2{=}C{=}O$	Et_3N, Py, -70°	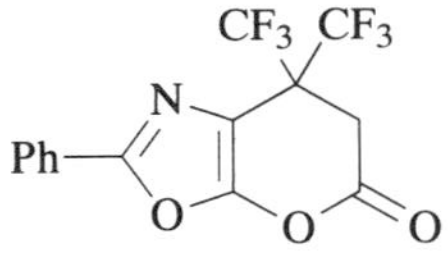 CF3, CF3, N, Ph, O, O, O (66)	257
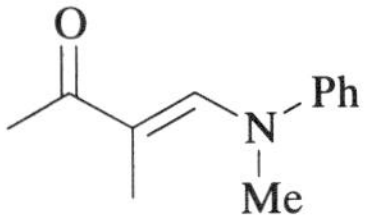 O, Ph, N, Me	$Cl_2CHCOCl$	Et_3N, C_6H_6	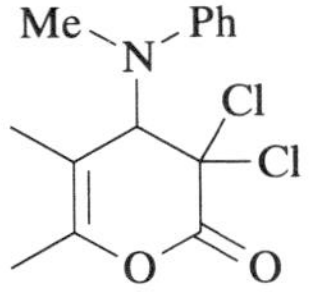 Me, N, Ph, Cl, Cl, O, O (56)	816
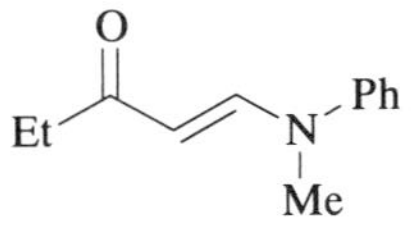 O, Et, Ph, N, Me	$Cl_2CHCOCl$	Et_3N, C_6H_6	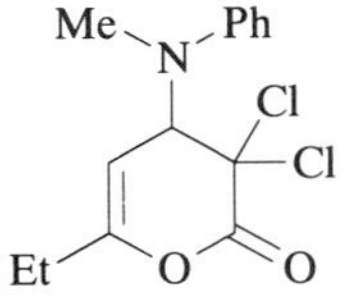 Me, N, Ph, Cl, Cl, Et, O, O (77)	816
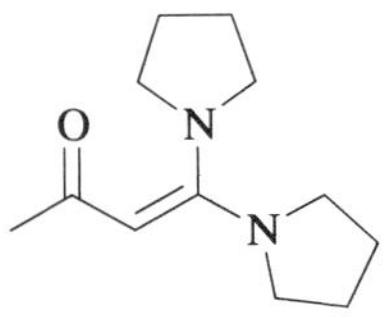 O, N, N	$CH_2{=}C{=}O$	Et_2O, 0°	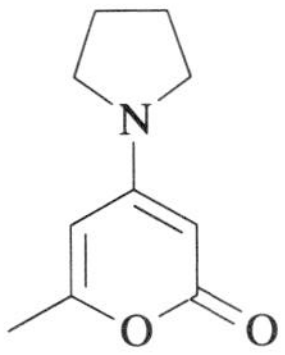 N, O, O (58)	259
C_{13}				
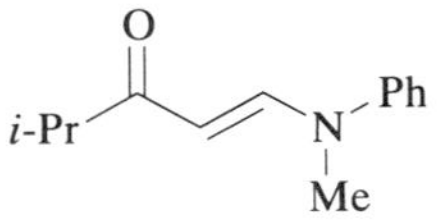 O, *i*-Pr, Ph, N, Me	$Cl_2CHCOCl$	Et_3N, C_6H_6	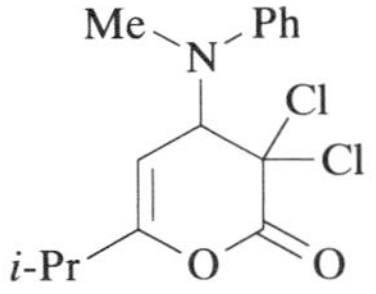 Me, N, Ph, Cl, Cl, *i*-Pr, O, O (77)	819

TABLE XXXII. [4+2] CYCLOADDITION OF KETENES TO α,β-UNSATURATED CARBONYL COMPOUNDS (*Continued*)

Reactant	Ketene or Ketene Source	Conditions	Product(s) and Yield(s) (%)	Refs.
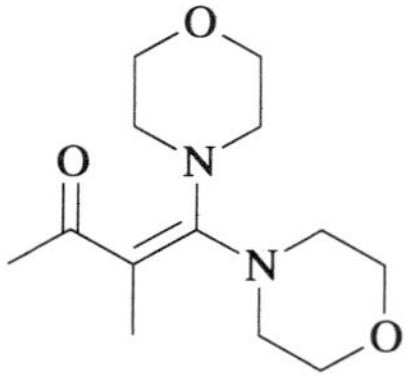 O, N, N, O	$CH_2{=}C{=}O$	Et_2O, 0°	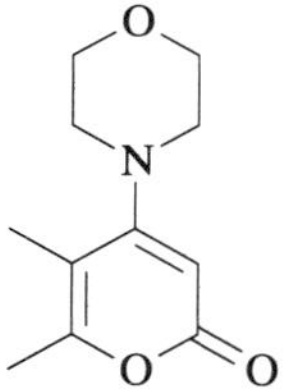 O, N, O, O (51)	259
C_{14}				
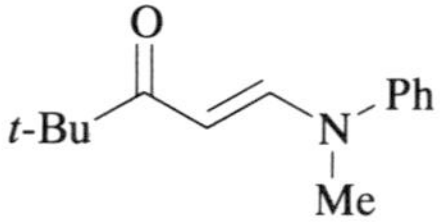 O, t-Bu, N, Ph, Me	$Cl_2CHCOCl$	Et_3N, C_6H_6	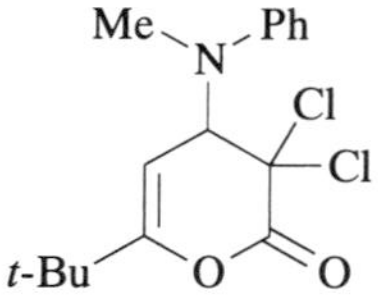 Me, N, Ph, Cl, Cl, t-Bu, O, O (29)	819
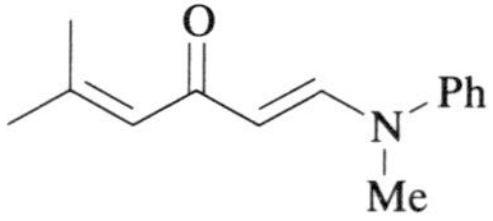 O, N, Ph, Me	$Cl_2CHCOCl$	Et_3N, Et_2O	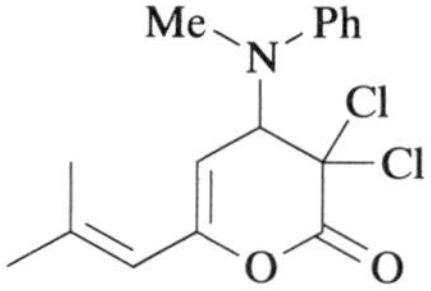 Me, N, Ph, Cl, Cl, O, O (45)	820
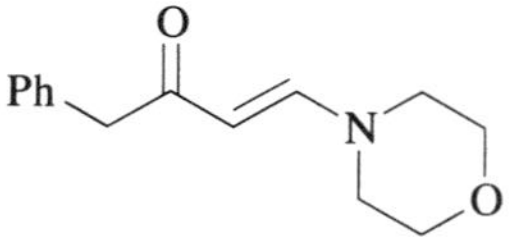 O, Ph, N, O	$Cl_2CHCOCl$	Et_3N, C_6H_6	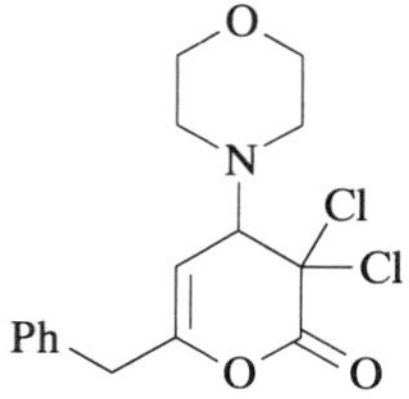 O, N, Cl, Cl, Ph, O, O (10)	821

TABLE XXXII. [4+2] CYCLOADDITION OF KETENES TO α,β-UNSATURATED CARBONYL COMPOUNDS (*Continued*)

Reactant	Ketene or Ketene Source	Conditions	Product(s) and Yield(s) (%)	Refs.
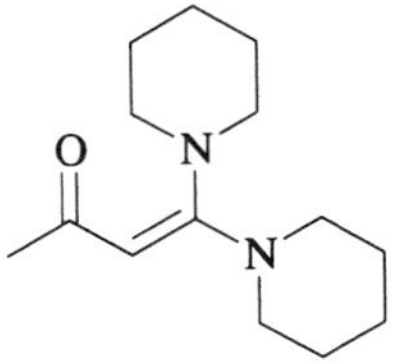 N	$CH_2{=}C{=}O$	Et_2O, 0°	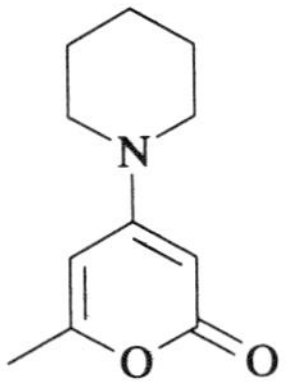 N (54)	259
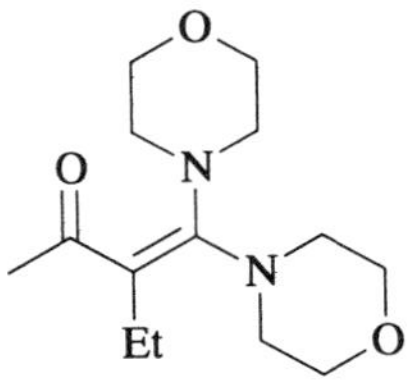 O N N Et	$CH_2{=}C{=}O$	Et_2O, 0°	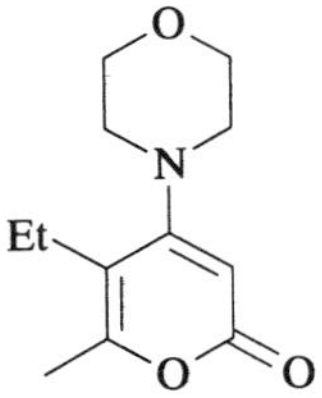 O N Et (51)	259
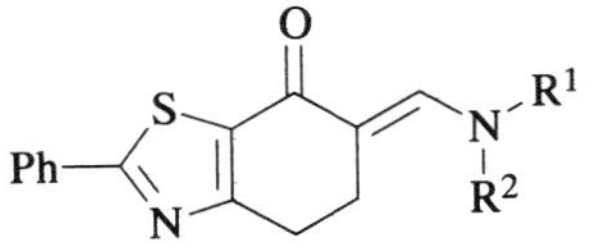 Ph S N O N R[1] R[2]	$Cl_2CHCOCl$	Et_3N, C_6H_6	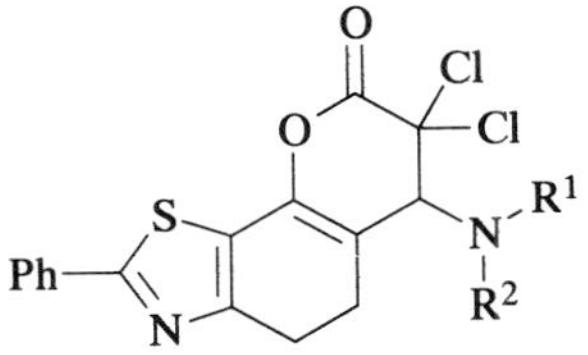 Ph S N O O Cl Cl N R[1] R[2]	822
R[1] / R[2]: Me / Me			(87)	
Morpholino			(84)	
Et / Et			(61)	
Piperidino			(67)	
i-Pr / *i*-Pr			(90)	
Me / Ph			(90)	
Ph / Ph			(92)	

TABLE XXXII. [4+2] CYCLOADDITION OF KETENES TO α,β-UNSATURATED CARBONYL COMPOUNDS (*Continued*)

Reactant	Ketene or Ketene Source	Conditions	Product(s) and Yield(s) (%)	Refs.
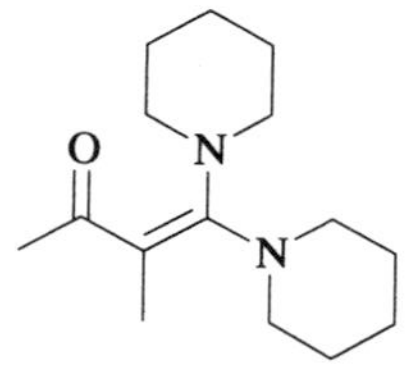 [structure]	$CH_2{=}C{=}O$	Et_2O, 0°	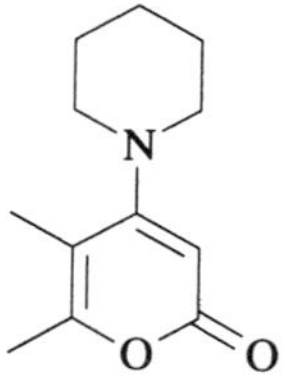 [structure] (55)	259
C_{15}				
[structure]	$Ph_2C{=}C{=}O$	130-140°, 4 h	[structure] (19)	252
C_{16}				
[structure]	$Ph_2C{=}C{=}O$	140-145°, 3.5 h	[structure] (—) + [structure] (—) + [structure] (—)	253

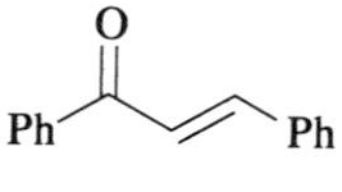

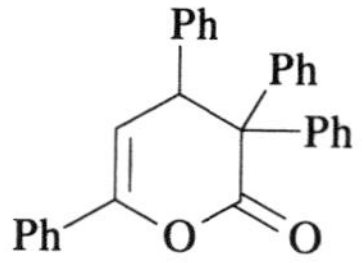

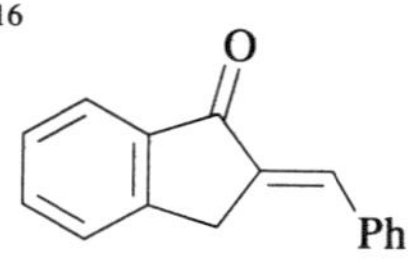

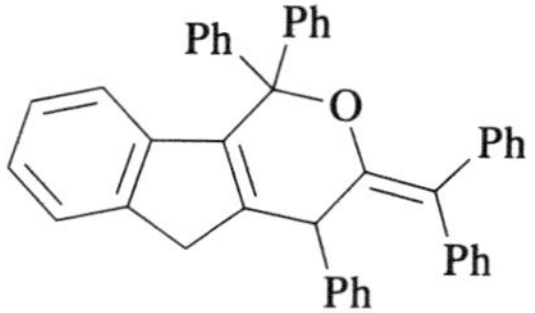

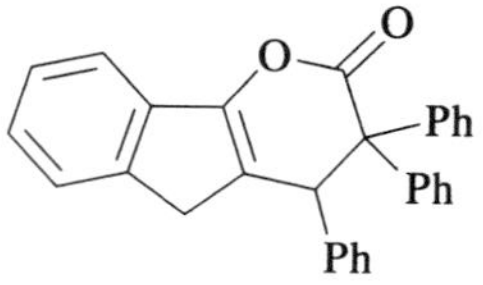

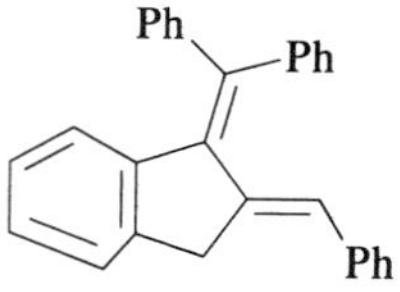

TABLE XXXII. [4+2] CYCLOADDITION OF KETENES TO α,β-UNSATURATED CARBONYL COMPOUNDS (*Continued*)

Reactant	Ketene or Ketene Source	Conditions	Product(s) and Yield(s) (%)	Refs.
PhCOCH=CHN(Me)Ph	$Cl_2CHCOCl$	Et_3N, C_6H_6	4-[N(Me)Ph]-3,3-dichloro-6-phenyl-3,4-dihydro-2-pyranone (76)	823
PhCOCH=CHC$_6$H$_4$OMe-*p*	$Ph_2C=C=O$	130-140°	4-(*p*-MeOC$_6$H$_4$)-3,3,6-triphenyl-3,4-dihydro-2-pyranone (36)	252
$CH_3COC(Et)=C(NC_5H_{10})_2$	$CH_2=C=O$	Et_2O, 0°	5-Et-6-methyl-4-piperidino-2-pyranone (53)	259
C_{17} 1-[(N-methyl-N-phenylamino)methylene]-2-indanone	$Cl_2CHCOCl$	Et_3N, C_6H_6	indeno-fused 3,3-dichloro-4-[N(Me)Ph]-dihydropyranone (58)	824

TABLE XXXII. [4+2] CYCLOADDITION OF KETENES TO α,β-UNSATURATED CARBONYL COMPOUNDS (*Continued*)

Reactant	Ketene or Ketene Source	Conditions	Product(s) and Yield(s) (%)	Refs.
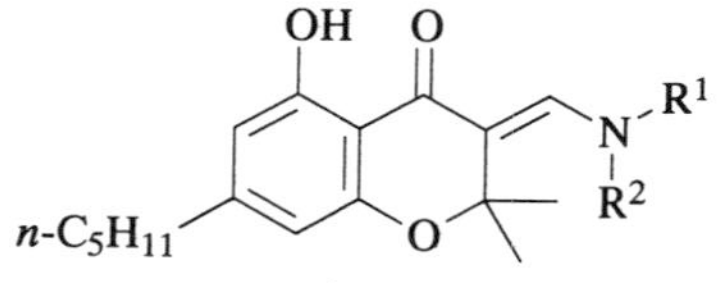 OH O, N-R¹, R², n-C_5H_{11} R¹ / R²: Morpholino; Et, Et; Piperidino; Me, Ph	$Cl_2CHCOCl$	Et_3N, Et_2O	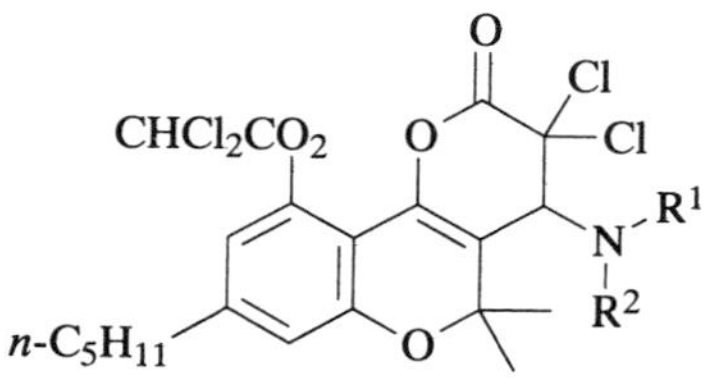 $CHCl_2CO_2$, O, Cl, Cl, N-R¹, R², n-C_5H_{11} (68) (71) (76) (80)	825
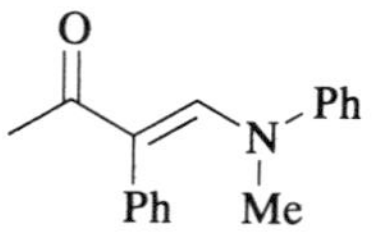 O, Ph, N-Ph, Me	$Cl_2CHCOCl$	Et_3N, C_6H_6	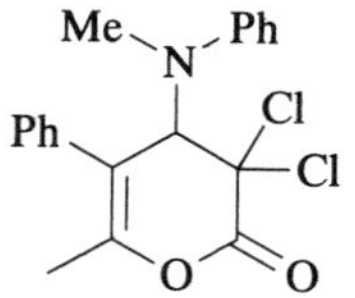 Me-N-Ph, Ph, Cl, Cl, O, O (25)	821
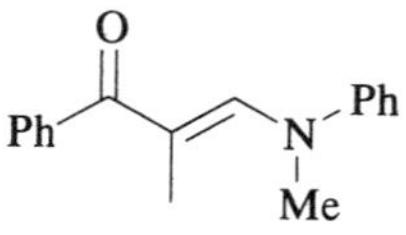 O, Ph, N-Ph, Me	$Cl_2CHCOCl$	Et_3N, C_6H_6	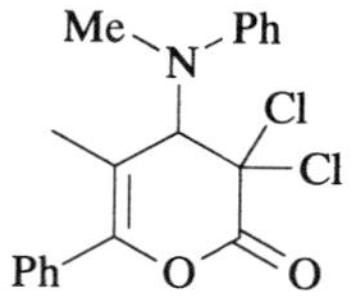 Me-N-Ph, Cl, Cl, Ph, O, O (47)	826
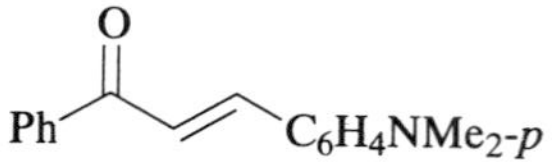 O, Ph, $C_6H_4NMe_2$-p	$Ph_2C=C=O$	130-140°	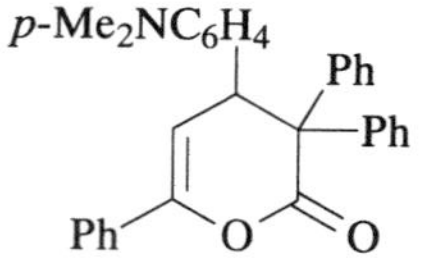 p-$Me_2NC_6H_4$, Ph, Ph, Ph, O, O (46)	252

TABLE XXXII. [4+2] CYCLOADDITION OF KETENES TO α,β-UNSATURATED CARBONYL COMPOUNDS (*Continued*)

Reactant	Ketene or Ketene Source	Conditions	Product(s) and Yield(s) (%)	Refs.
C_{18}				
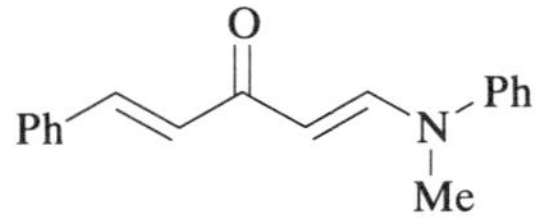	$Cl_2CHCOCl$	Et_3N, C_6H_6	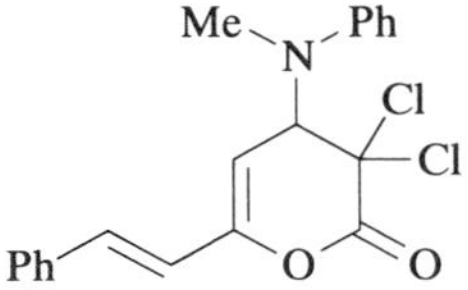 (82)	820
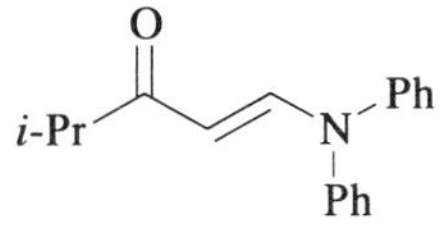	$Cl_2CHCOCl$	Et_3N, C_6H_6	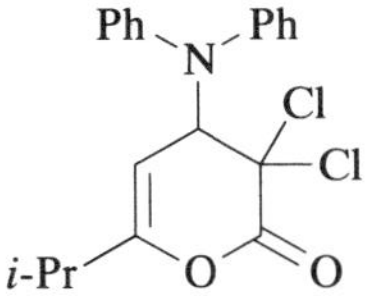 (60)	819
C_{19}				
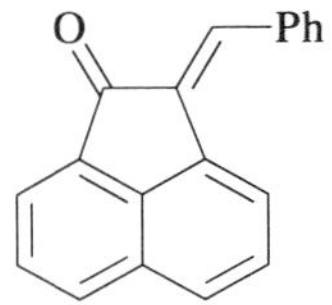	$Ph_2C{=}C{=}O$	130°, 2 h	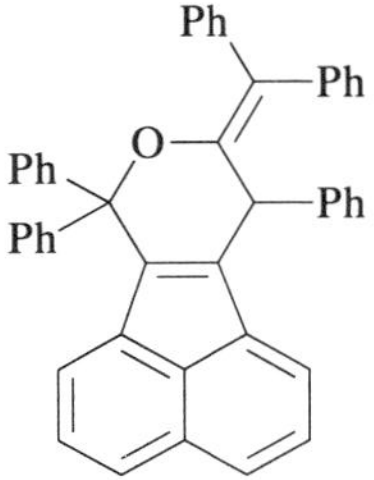 (—)	254
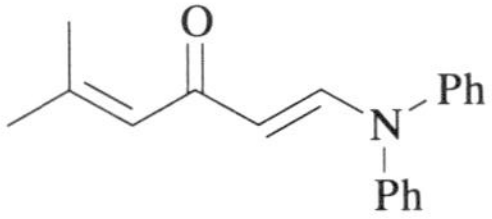	$Cl_2CHCOCl$	Et_3N, Et_2O	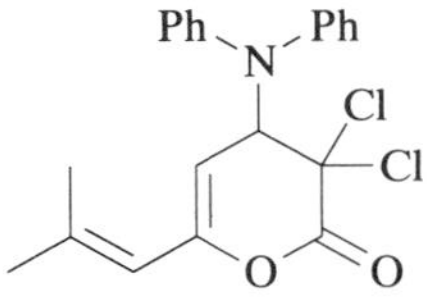 (35)	820

TABLE XXXII. [4+2] CYCLOADDITION OF KETENES TO α,β-UNSATURATED CARBONYL COMPOUNDS (*Continued*)

Reactant	Ketene or Ketene Source	Conditions	Product(s) and Yield(s) (%)	Refs.
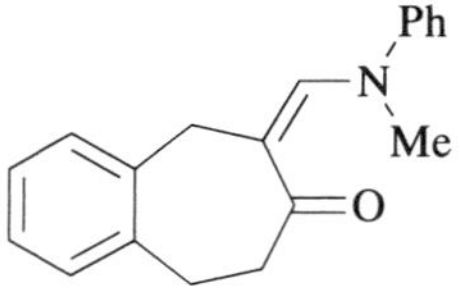 Ph, N, Me, O	$Cl_2CHCOCl$	Et_3N, CH_2Cl_2	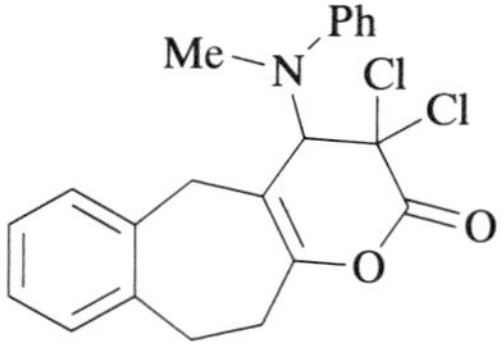 Me—N, Ph, Cl, Cl, O, O (92)	255
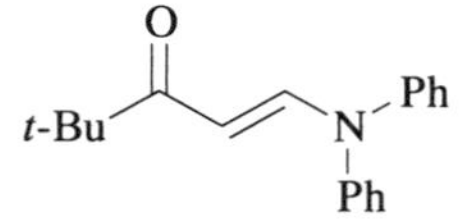 O, *t*-Bu, N, Ph, Ph	$Cl_2CHCOCl$	Et_3N, C_6H_6	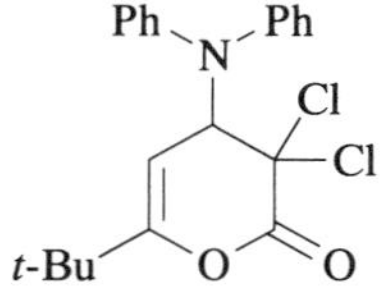 Ph, N, Ph, Cl, Cl, *t*-Bu, O, O (54)	819
C_{21}				
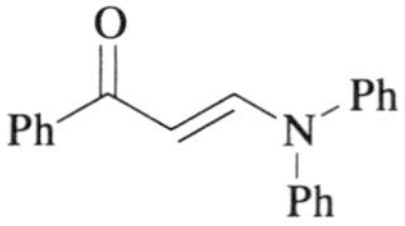 O, Ph, N, Ph, Ph	$Cl_2CHCOCl$	Et_3N, C_6H_6	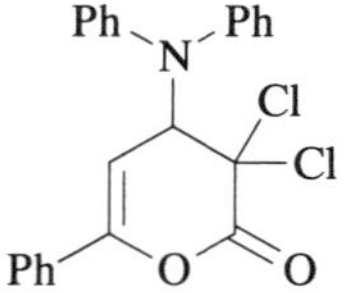 Ph, N, Ph, Cl, Cl, Ph, O, O (45)	823
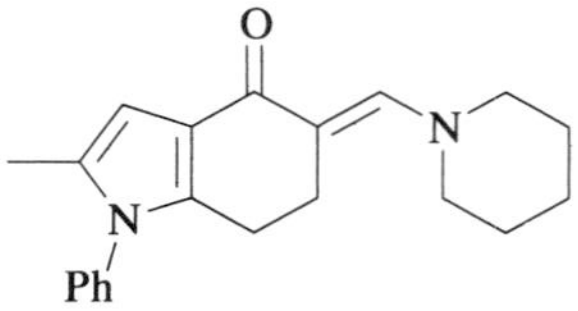 O, N, N, Ph	$Cl_2CHCOCl$	Et_3N, C_6H_6	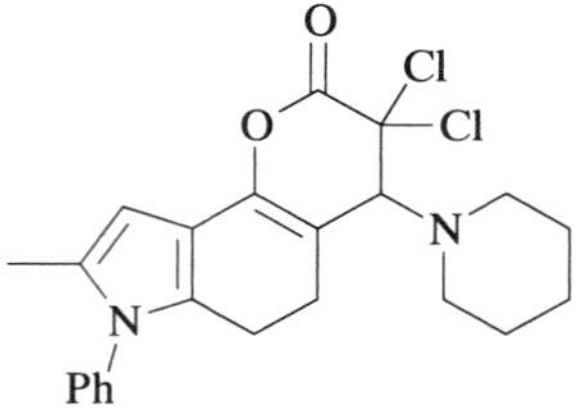 O, Cl, Cl, O, N, N, Ph (64)	827

TABLE XXXII. [4+2] CYCLOADDITION OF KETENES TO α,β-UNSATURATED CARBONYL COMPOUNDS (*Continued*)

Reactant	Ketene or Ketene Source	Conditions	Product(s) and Yield(s) (%)	Refs.
C_{22} 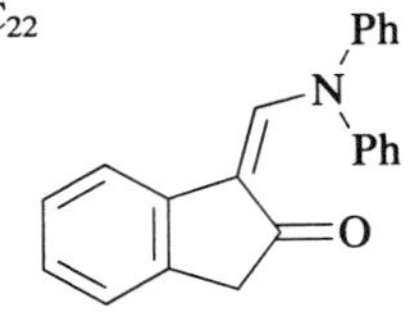	$Cl_2CHCOCl$	Et_3N, C_6H_6	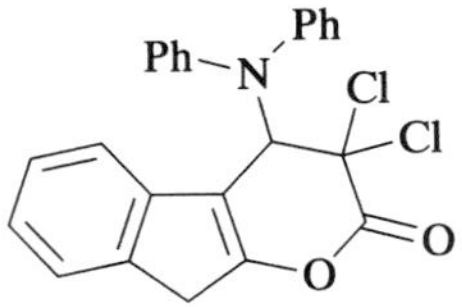 (55)	824
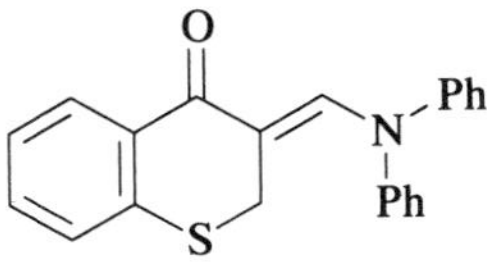	$Cl_2CHCOCl$	Et_3N, Et_2O	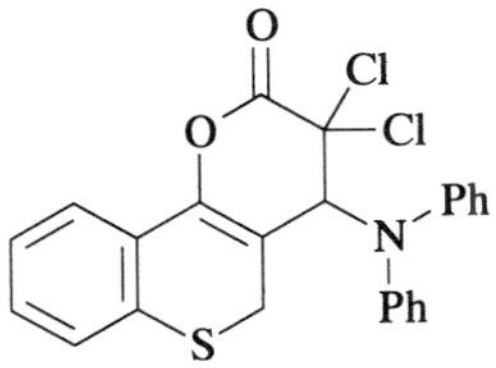 (74)	258
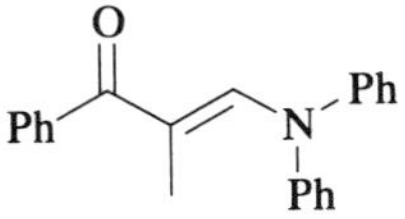	$Cl_2CHCOCl$	Et_3N, C_6H_6	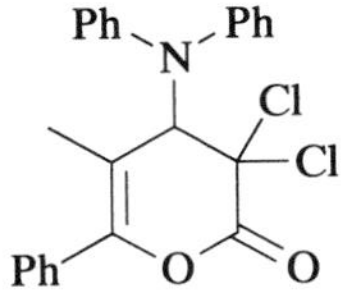 (39)	826
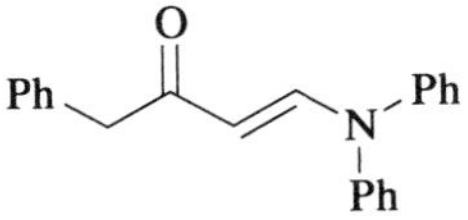	$Cl_2CHCOCl$	Et_3N, C_6H_6	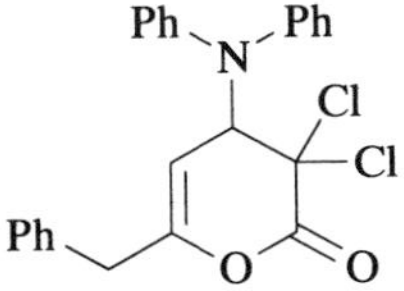 (30)	821

TABLE XXXII. [4+2] CYCLOADDITION OF KETENES TO α,β-UNSATURATED CARBONYL COMPOUNDS (*Continued*)

Reactant	Ketene or Ketene Source	Conditions	Product(s) and Yield(s) (%)	Refs.
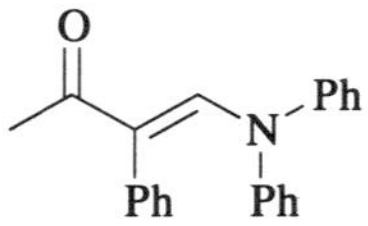	$Cl_2CHCOCl$	Et_3N, C_6H_6	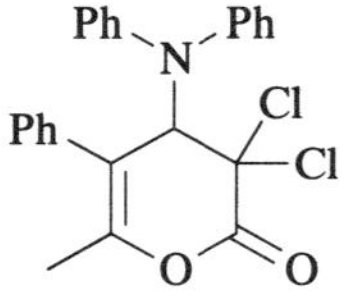 (31)	821
C_{23} 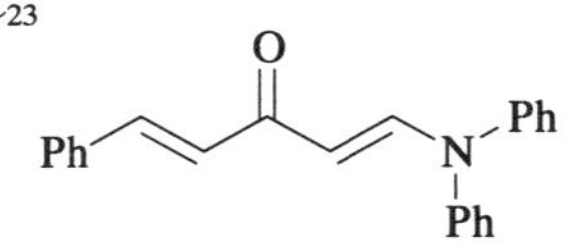	$Cl_2CHCOCl$	Et_3N, Et_2O	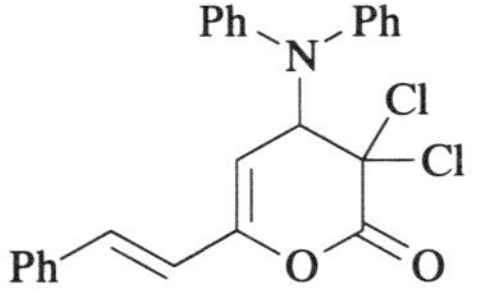 (80)	820
C_{24} 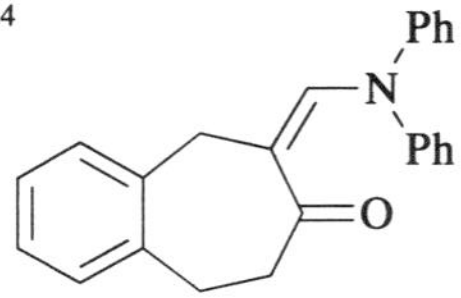	$Cl_2CHCOCl$	Et_3N, Et_2O	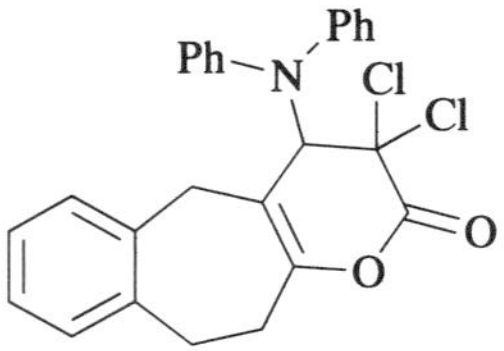 (50)	255

TABLE XXXIII. [4+2] CYCLOADDITION OF KETENES TO α,β-UNSATURATED THIOCARBONYL COMPOUNDS

Reactant	Ketene or Ketene Source	Conditions	Product(s) and Yield(s) (%)	Refs.
C_9				
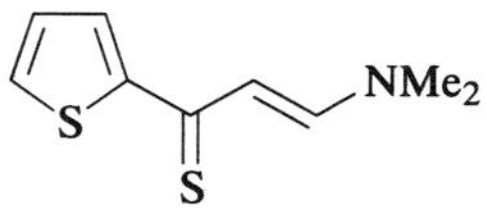 NMe$_2$ S S	$CH_2{=}C{=}O$	Me_2CO, 1 h	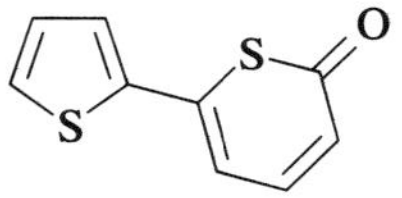 S S O (59)	260
	$PhCH_2COCl$	Et_3N, C_6H_6, 1 h	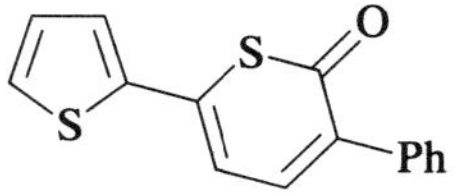 S S O Ph (53)	260
C_{11}				
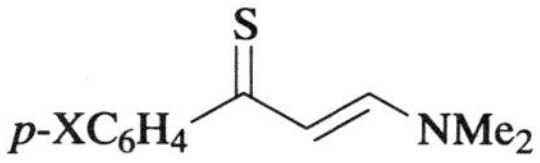 p-XC_6H_4 S NMe$_2$	$CH_2{=}C{=}O$	Me_2CO, 1 h	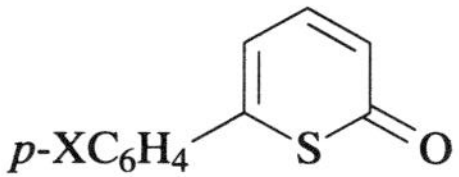 p-XC_6H_4 S O	260, 261
X				
Cl			(78)	
Br			(63)	
H			(29)	
Me			(62)	
OMe			(45)	

TABLE XXXIII. [4+2] CYCLOADDITION OF KETENES TO α,β-UNSATURATED THIOCARBONYL COMPOUNDS (*Continued*)

Reactant	Ketene or Ketene Source	Conditions	Product(s) and Yield(s) (%)	Refs.
p-$XC_6H_4C(=S)CH=CHNMe_2$	$R^1R^2CHCOCl$	Et_3N, C_6H_6, 1 h	6-(p-XC_6H_4)-4-R^3-3-R^1-2H-thiopyran-2-one	
X	R^1 R^2		R^3	
Cl	Cl Cl		NMe_2 (41)	260
Cl	Ph H		H (38)	260
Br	Cl Cl		NMe_2 (24)	260
Br	Ph H		H (34)	260
H	Cl Cl		NMe_2 (18)	260
H	Ph H		H (53)	260, 828
Me	Cl Cl		NMe_2 (16)	260
Me	Ph H		H (21)	260, 828
OMe	Cl Cl		NMe_2 (20)	260
OMe	Ph H		H (23)	260, 828
C_{13} [structure: fused bicyclic 1,3-dithiolane (H, H) ylidene cyclohexadienethione]	$Ph_2C=C=O$	C_6H_6, 40°	**I** + **II** (73) [structures: spiro benzoxathiin–dithiolane with $=CPh_2$ (**I**); spiro benzothiopyranone–dithiolane, 3,3-diphenyl (**II**)]	829

TABLE XXXIII. [4+2] CYCLOADDITION OF KETENES TO α,β-UNSATURATED THIOCARBONYL COMPOUNDS (*Continued*)

Reactant	Ketene or Ketene Source	Conditions	Product(s) and Yield(s) (%)	Refs.
C_{14}				
p-ClC_6H_4, S, N	$Cl_2CHCOCl$	Et_3N, C_6H_6	N, Cl, p-ClC_6H_4, S, O (41)	830
p-BrC_6H_4, S, N	$Cl_2CHCOCl$	Et_3N, C_6H_6	N, Cl, p-BrC_6H_4, S, O (24)	830
MeS, S, Ph, N	$CH_2{=}C{=}O$	C_6H_6	Ph, MeS, S, O (73)	831
Ph, S, N	$Cl_2CHCOCl$	Et_3N, C_6H_6	N, Cl, Ph, S, O (18)	830

TABLE XXXIII. [4+2] CYCLOADDITION OF KETENES TO α,β-UNSATURATED THIOCARBONYL COMPOUNDS (*Continued*)

Reactant	Ketene or Ketene Source	Conditions	Product(s) and Yield(s) (%)		Refs.
p-$MeOC_6H_4$–C(=S)–CH=CH–N(morpholino)	$Cl_2CHCOCl$	Et_3N, C_6H_6	p-$MeOC_6H_4$, S, O, Cl, N, O	(20)	830
C_{15} p-$MeOC_6H_4$–C(=S)–CH=CH–N(piperidino)	$CD_2{=}C{=}O$	Me_2CO	p-$MeOC_6H_4$, S, O, D	(—)	260
	$Cl_2CHCOCl$	Et_3N, C_6H_6	p-$MeOC_6H_4$, S, O, Cl, N	(16)	830
C_{16} p-$MeOC_6H_4$–C(=S)–CH=CH–Ph	$Ph_2C{=}C{=}O$	C_6H_6, 80°	p-$MeOC_6H_4$, S, O, Ph, Ph, Ph	(11)	832

TABLE XXXIII. [4+2] CYCLOADDITION OF KETENES TO α,β-UNSATURATED THIOCARBONYL COMPOUNDS (*Continued*)

Reactant	Ketene or Ketene Source	Conditions	Product(s) and Yield(s) (%)	Refs.
C_{17}				
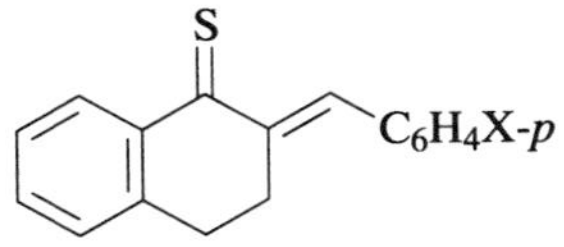 S, C_6H_4X-*p*	Ph, Ph, C=C=O	C_6H_6, 80°	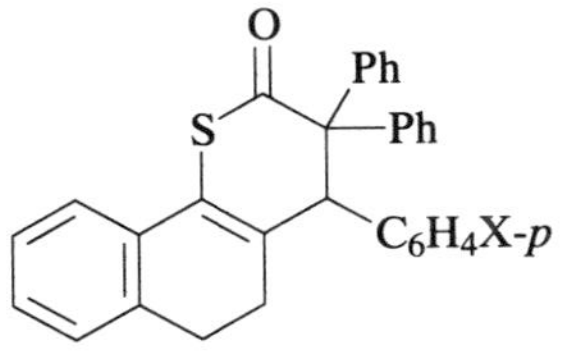 O, S, Ph, Ph, C_6H_4X-*p*	832
X: Cl			(17)	
X: H			(51)	
X: OMe			(35)	
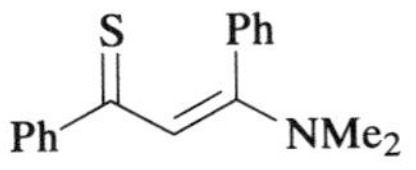 S, Ph, Ph, NMe_2	CH_2=C=O	—	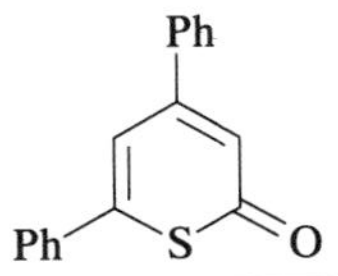 Ph, Ph, S, O (36)	261

TABLE XXXIV. [4+2] CYCLOADDITION OF KETENES TO ISOCYANATES AND ISOTHIOCYANATES

Reactant	Ketene or Ketene Source	Conditions	Product(s) and Yield(s) (%)	Refs.
C_3				
$CCl_3C(O)NCO$	$Me_2C{=}C{=}O$	C_6H_6	2-CCl_3-5,5-dimethyl-1,3-oxazine-4,6-dione (77)	188
$CH_2{=}CHNCO$	$Ph_3P{=}C{=}C{=}O$	C_6H_6	HN pyridine-2,4-dione, 3-($=PPh_3$) (98) + N-(vinylcarbamoyl) analogue, 3-($=PPh_3$) (Tr)	266
$CH_2{=}CHNCS$	$Ph_3P{=}C{=}C{=}O$	C_6H_6	HN 2-thioxo-pyridin-4-one, 3-($=PPh_3$) (94)	266
C_4				
$EtSC(S)NCS$	$Ph_2C{=}C{=}O$	Et_2O, 48 h	2-EtS-5,5-diphenyl-4-thioxo-1,3-thiazin-6-one (37)	264

TABLE XXXIV. [4+2] CYCLOADDITION OF KETENES TO ISOCYANATES AND ISOTHIOCYANATES (*Continued*)

Reactant	Ketene or Ketene Source	Conditions	Product(s) and Yield(s) (%)	Refs.
C_8				
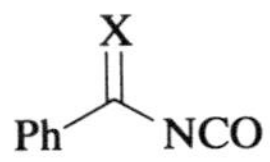 PhC(=X)NCO	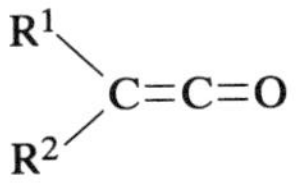 $R^1R^2C=C=O$	Et_2O or PhMe	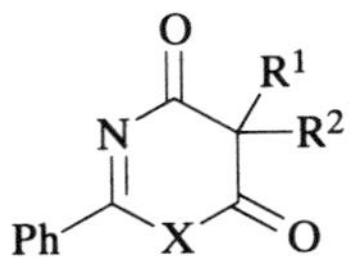 [structure]	
X	R^1, R^2			
O	H, H		(51)	188
O	Me, Me		(25)	188
S	[fluorenylidene structure]		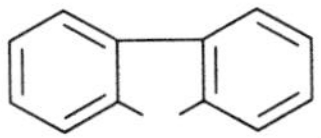 (42)	263
S	Ph, Ph		(78)	263
S	Ph, 2,4,6-$Me_3C_6H_2$		(82)	263
S	Ph, 1-$C_{10}H_7$		(94)	263
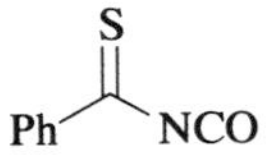 PhC(=S)NCO	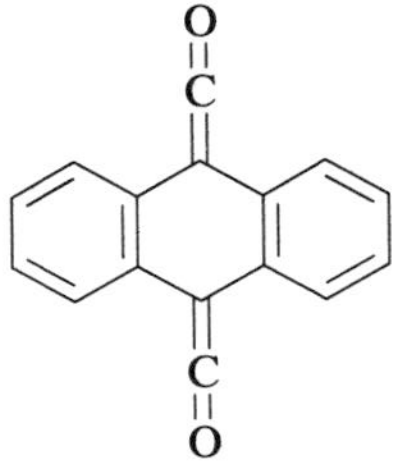 [structure]	C_6H_6	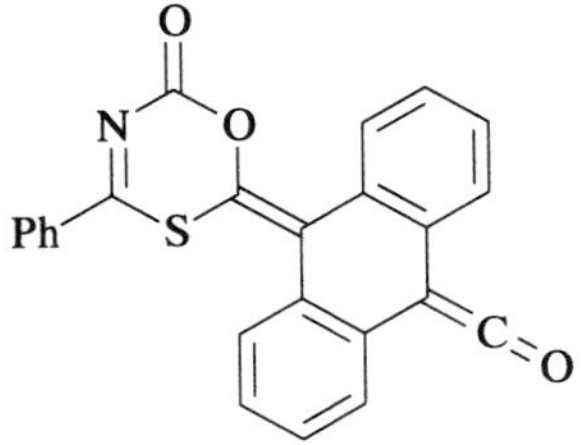 [structure] (43)	263
C_9				
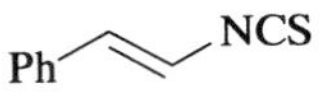 PhCH=CHNCS	$Ph_3P=C=C=O$	C_6H_6	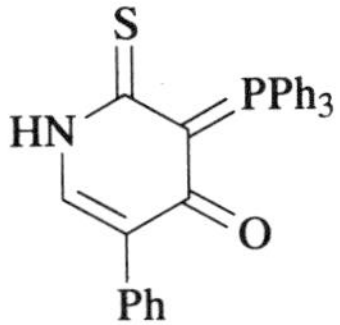 [structure] (73)	266

TABLE XXXV. [4+2] CYCLOADDITION OF KETENES TO AZO COMPOUNDS

Reactant	Ketene or Ketene Source	Conditions	Product(s) and Yield(s) (%)	Refs.
C_6				
$EtO_2C-N=N-CO_2Et$	$Ph_2C=C=O$	—	CO₂Et, O, N, N, O, Ph, Ph, O, Ph, Ph, EtO (—)	267
C_{10}				
p-$ClC_6H_4-N=N-C(CH_3)=CH-CN$	$Ph_2C=C=O$	C_6H_6	CN, Ph, Ph, N, N, O, C_6H_4Cl-p (10) + Ph, Ph, N, N, CN, O, C_6H_4Cl-p (75)	268
C_{11}				
$Ph-N=N-C(CH_3)=CH-OAc$	$Ph_2C=C=O$	C_6H_6	OAc, Ph, Ph, N, N, O, Ph (13) + Ph, Ph, N, N, OAc, O, Ph (46)	268
C_{12}				
1-cyclohexenyl-N=NPh	$Ph_2C=C=O$	C_6H_6	N, N, Ph, O, Ph, Ph (72)	268

TABLE XXXV. [4+2] CYCLOADDITION OF KETENES TO AZO COMPOUNDS (*Continued*)

Reactant	Ketene or Ketene Source	Conditions	Product(s) and Yield(s) (%)	Refs.
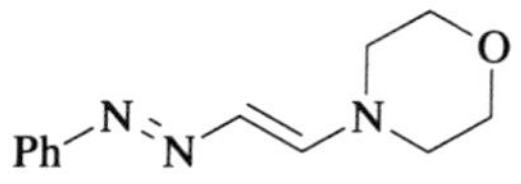 PhN=NCH=CH–N(morpholino)	$Ph_2C=C=O$	C_6H_6	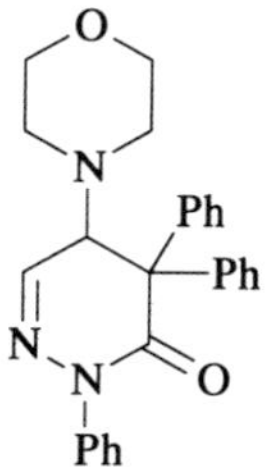 4-morpholino-2,5,5-triphenyl-2,5-dihydropyridazin-3(4H)-one [structure] (84)	268
C_{14}				
PhCON=NCOPh	$Ph_2C=C=O$	—	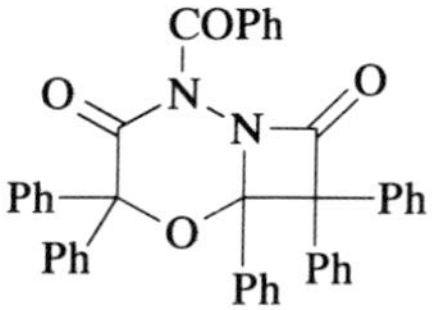 [structure: N-COPh bicyclic oxadiazine-azetidinone, Ph substituents] (—)	267

TABLE XXXVI. [4+2] CYCLOADDITION OF KETENES TO THIOACYL IMINES

Reactant	Ketene or Ketene Source	Conditions	Product(s) and Yield(s) (%)	Refs.
C_{10}				
Ph, S, N, NMe_2	$CH_2{=}C{=}O$	C_6H_6	N, Ph, S, O (40)	269
	$PhCH_2COCl$	Et_3N, C_6H_6	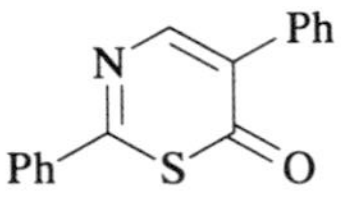 N, Ph, Ph, S, O (22)	269
C_{11}				
p-MeC_6H_4, S, N, NMe_2	$CH_2{=}C{=}O$	C_6H_6	N, p-MeC_6H_4, S, O (40)	269
	$PhCH_2COCl$	Et_3N, C_6H_6	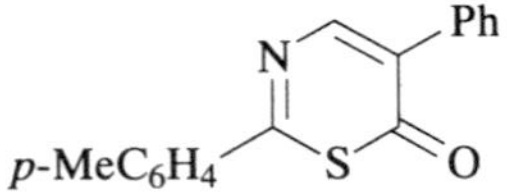 N, Ph, p-MeC_6H_4, S, O (18)	269
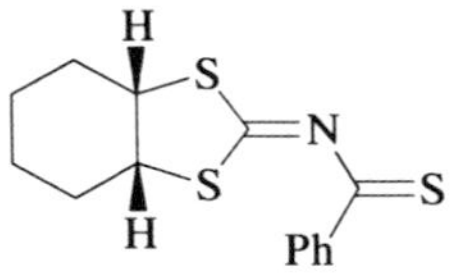 H, S, N, S, S, H, Ph	$Ph_2C{=}C{=}O$	DMF, 25°	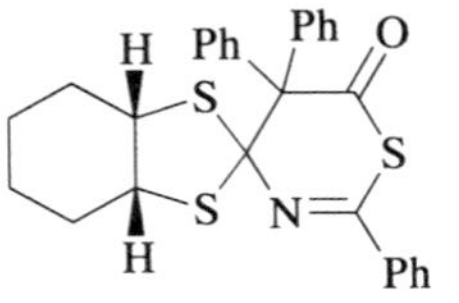 H, Ph, Ph, O, S, S, S, N, H, Ph (80)	270, 271

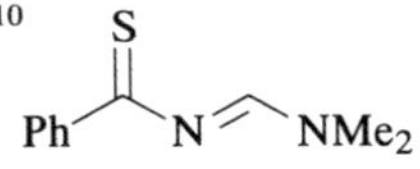

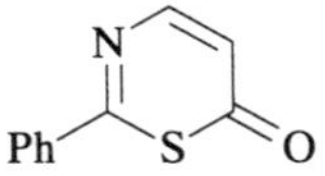

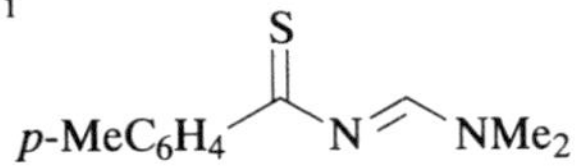

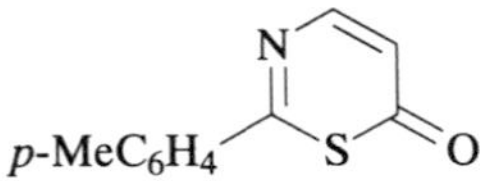

TABLE XXXVII. [4+2] CYCLOADDITION OF KETENES TO MESOIONIC COMPOUNDS

Reactant		Ketene or Ketene Source		Conditions	Product(s) and Yield(s) (%)	Refs.
[mesoionic pyrimidinium-olate: R^1, R^2, N-Ph, N⁺-Ph, O⁻, O]		$R^3R^4C{=}C{=}O$		MeCN, 20°	[bicyclic adduct: $R^3R^4C{=}$, R^1, R^2, O, N-Ph, N-Ph]	272
R^1	R^2	R^3	R^4			
Me	H	H	H		(99)	
Me	H	EtO_2C	EtO_2C		(88)	
Me	H	Ph	Ph		(74)	
Me	Me	H	H		(68)	
Me	Me	CF_3	CF_3		(87)	
H	Ph	H	H		(83)	
H	Ph	Me	Me		(93)	
H	Ph	Ph	Ph		(88)	
Me	Ph	EtO_2C	EtO_2C		(74)	

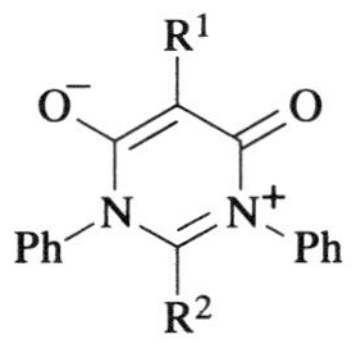

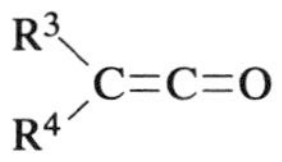

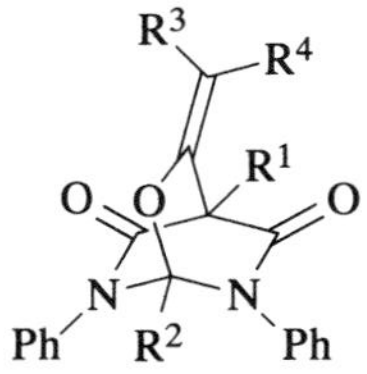

TABLE XXXVIII. [4+2] CYCLOADDITION OF ACYL AND VINYL KETENES TO OLEFINS

Reactant	Ketene or Ketene Source	Conditions	Product(s) and Yield(s) (%)	Refs.
C_4				
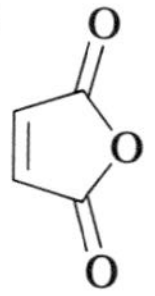 O, O, O	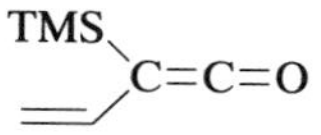 TMS, C=C=O	$CHCl_3$, 40°, 12 h	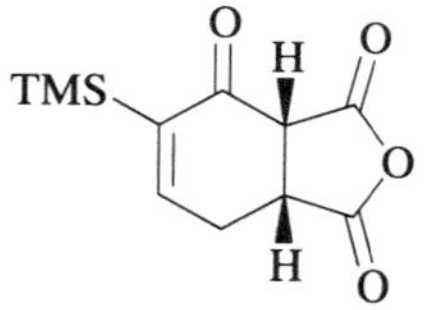 TMS, O, H, O, O, H, O (89)	273
C_6				
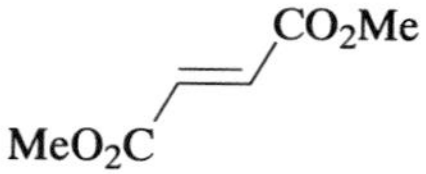 CO_2Me, MeO_2C	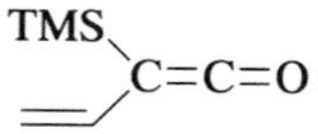 TMS, C=C=O	PhMe, 95°, 38 h	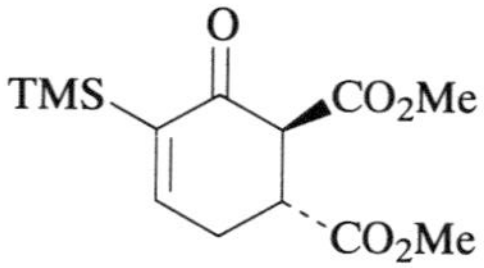 TMS, O, CO_2Me, CO_2Me (62)	273
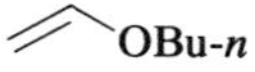 OBu-*n*	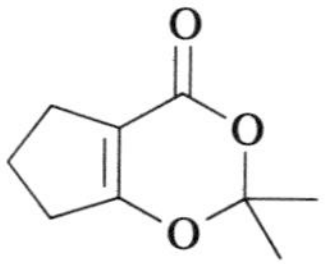 O, O, O	140-150°	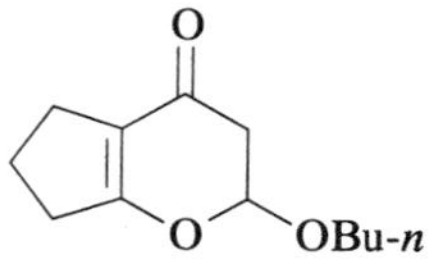 O, O, OBu-*n* (74.5)	274
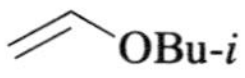 OBu-*i*	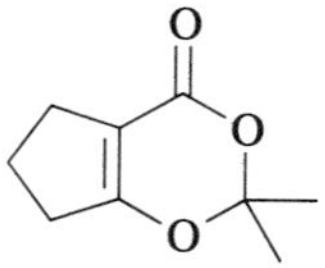 O, O, O	120-140°	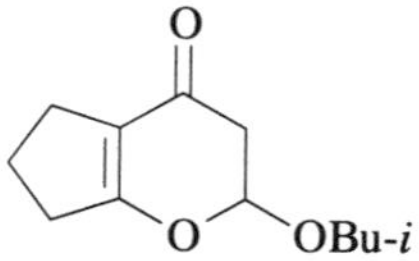 O, O, OBu-*i* (76)	274
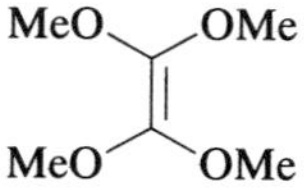 MeO, OMe, MeO, OMe	MeCOC≡COEt	Xylene, 120-140°	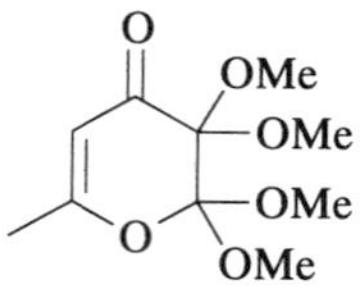 O, OMe, OMe, OMe, OMe, O (38)	282

TABLE XXXVIII. [4+2] CYCLOADDITION OF ACYL AND VINYL KETENES TO OLEFINS (*Continued*)

Reactant	Ketene or Ketene Source	Conditions	Product(s) and Yield(s) (%)	Refs.
C_8				
N	COCl	Et_3N, C_6H_{14}, 70°	O, N (28) + N, O (28)	276
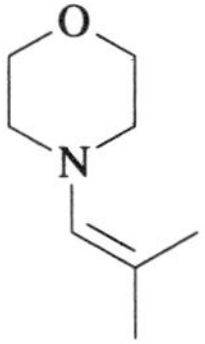 O, N	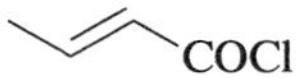 COCl	Et_3N, C_6H_{14}, 70°	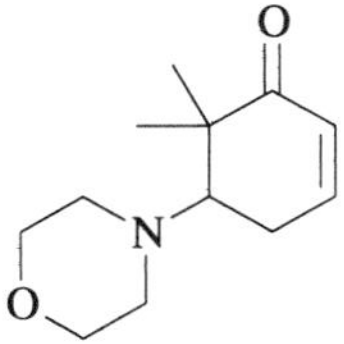 O, N, O (46)	276
	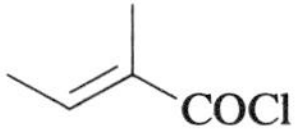 COCl	Et_3N, C_6H_{14}, 70°	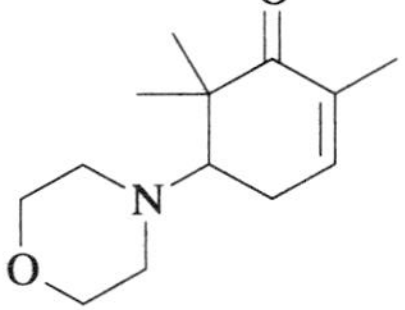 O, N, O (23) + O, N, O (3)	276
	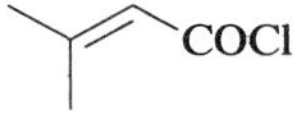 COCl	Et_3N, CH_2Cl_2, 25°	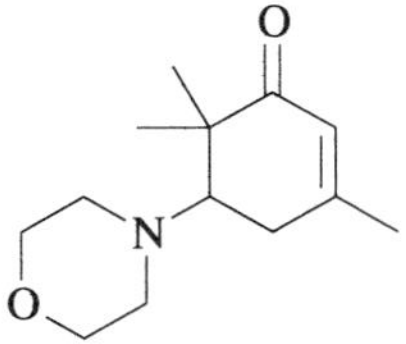 O, N, O (—)	276

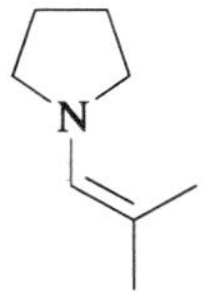

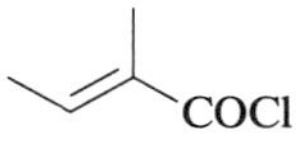

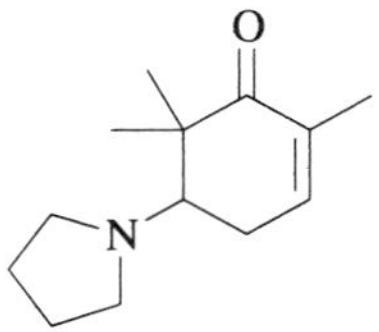

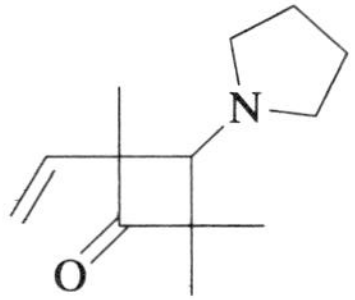

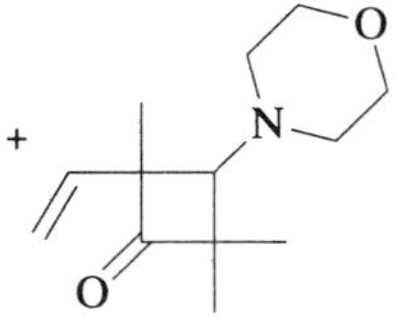

TABLE XXXVIII. [4+2] CYCLOADDITION OF ACYL AND VINYL KETENES TO OLEFINS (*Continued*)

Reactant	Ketene or Ketene Source	Conditions	Product(s) and Yield(s) (%)	Refs.
C_9 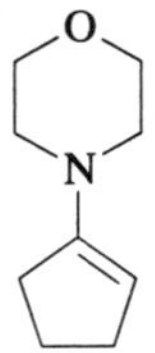	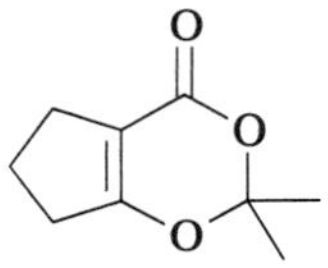	140-150°	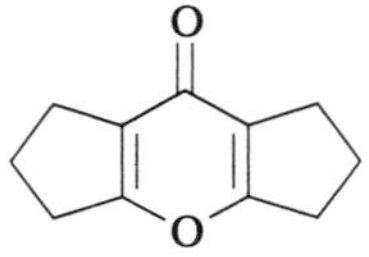 (8)	274
C_{10} 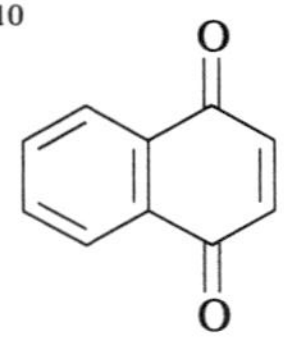	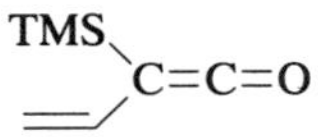 TMS C=C=O	$CHCl_3$, 60°, 41 h	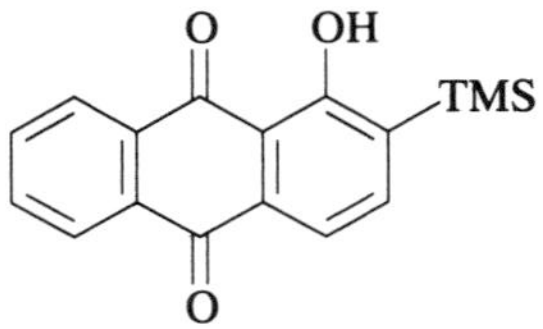 O OH TMS (28)	273
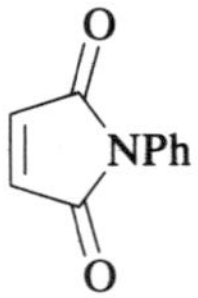 NPh	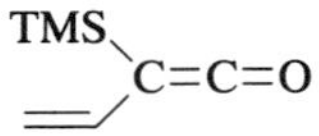 TMS C=C=O	$CHCl_3$, 40°, 24 h	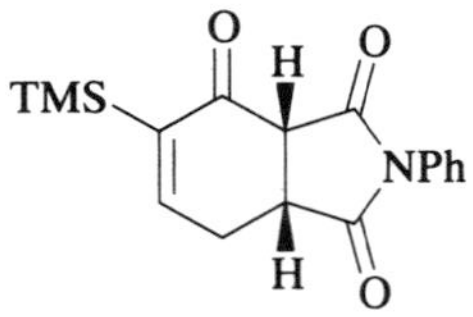 TMS H NPh H (74)	273
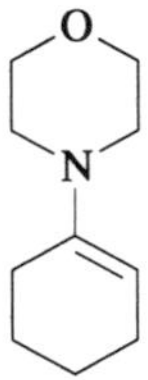	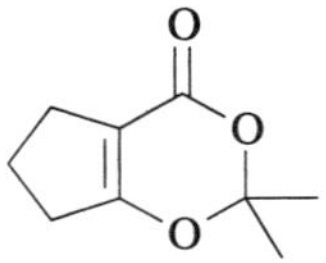	140-150°	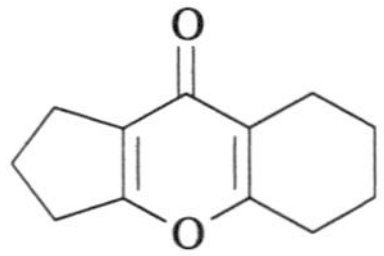 (31.5)	274

TABLE XXXIX. [4+2] CYCLOADDITION OF ACYL AND VINYL KETENES TO ACETYLENES

Reactant	Ketene or Ketene Source	Conditions	Product(s) and Yield(s) (%)	Refs.
C_4				
$HC{\equiv}CCO_2Me$	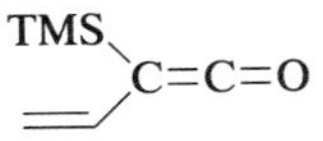 TMS, C=C=O	PhMe, 95°, 63 h	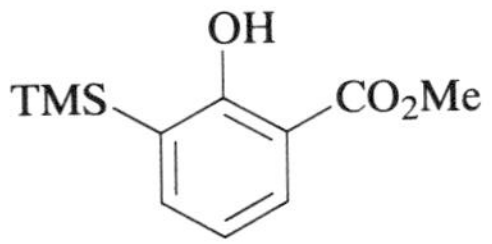 TMS, OH, CO_2Me (45)	273
C_6				
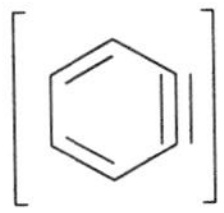 [ ]	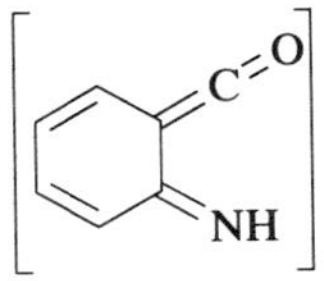 [C=O, NH]	240°	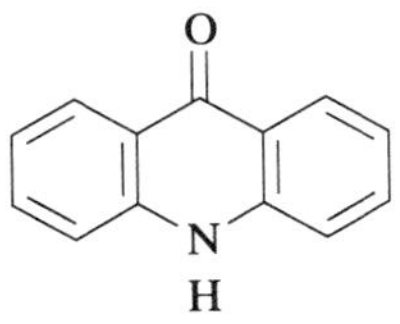 O, N, H (—)	277
$MeO_2CC{\equiv}CCO_2Me$	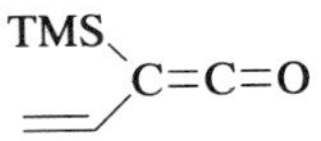 TMS, C=C=O	1. $CHCl_3$, 40° 2. CF_3CO_2H	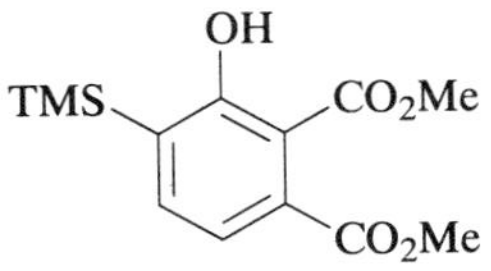 TMS, OH, CO_2Me, CO_2Me (60)	273
C_7				
$MeC{\equiv}CNEt_2$	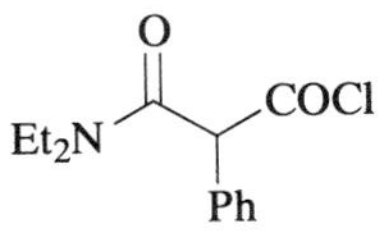 O, Et_2N, COCl, Ph	Et_3N	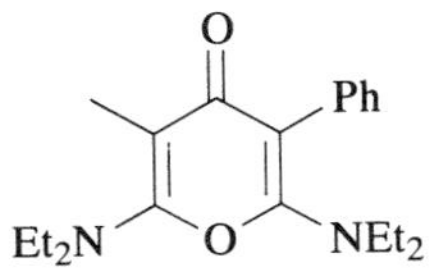 O, Ph, Et_2N, O, NEt_2 (70)	278
$PhC{\equiv}CNEt_2$	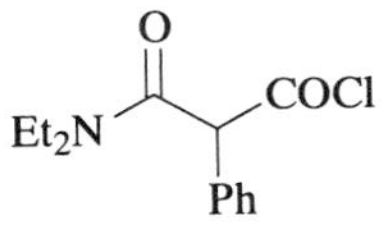 O, Et_2N, COCl, Ph	Et_3N	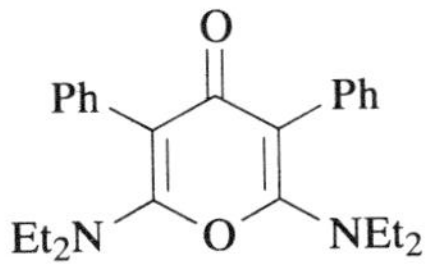 O, Ph, Ph, Et_2N, O, NEt_2 (50)	278

TABLE XXXIX. [4+2] CYCLOADDITION OF ACYL AND VINYL KETENES TO ACETYLENES (*Continued*)

Reactant	Ketene or Ketene Source	Conditions	Product(s) and Yield(s) (%)	Refs.
C_{12-32}				
$(R^1)_3XC{\equiv}CNR^2R^3$	$(EtO_2C)_2C{=}C{=}O$	Et_2O, 25°	4-pyrone: 3-$(R^1)_3X$, 5-CO_2Et, 6-OEt, 2-NR^2R^3	279
X = Sn, R^1 = Me, R^2 = Me, R^3 = Ph			(—)	
X = Sn, R^1 = Me, R^2 = Ph, R^3 = Ph			(—)	
X = Si, R^1 = Me, R^2 = Ph, R^3 = Ph			(—)	
X = Si, R^1 = Ph, R^2 = Me, R^3 = Me			(—)	
X = Si, R^1 = Ph, R^2 = Et, R^3 = Et			(—)	
X = Sn, R^1 = Ph, R^2 = Me, R^3 = Ph			(—)	
X = Ge, R^1 = Ph, R^2 = Me, R^3 = Ph			(—)	
X = Si, R^1 = Ph, R^2 = Ph, R^3 = Ph			(—)	
X = Sn, R^1 = Ph, R^2 = Ph, R^3 = Ph			(—)	

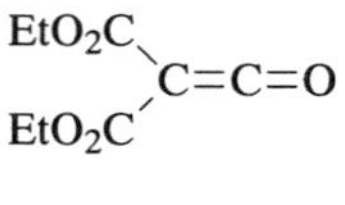

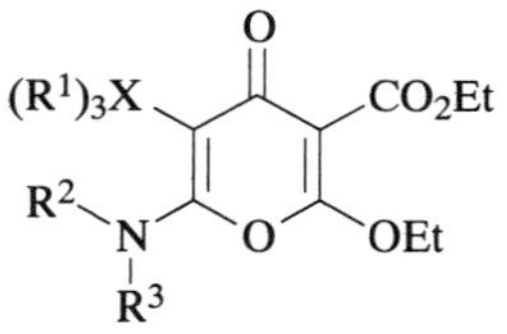

TABLE XL. [4+2] CYCLOADDITION OF ACYL AND VINYL KETENES TO ALDEHYDES

Reactant	Ketene or Ketene Source	Conditions	Product(s) and Yield(s) (%)	Refs.
RCHO	$ClCO(CH_2)_4COCl$	Et_3N, Et_2O, 35°, 30 min	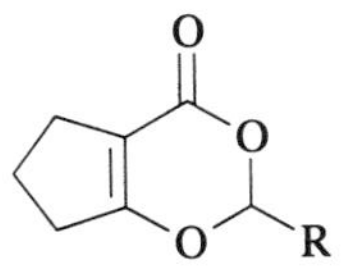 [structure: cyclopenta-fused 1,3-dioxin-4-one bearing R]	280
R				
CCl_3			(65.6)	
CBr_3			(21)	
Et			(88.5)	
2-Furyl			(54)	
4-Pyridyl			(45)	
o-ClC_6H_4			(40)	
p-ClC_6H_4			(65)	
p-$O_2NC_6H_4$			(33)	
Ph			(71)	
o-MeC_6H_4			(61.5)	
p-MeC_6H_4			(59)	
p-$MeOC_6H_4$			(71.5)	
o-CH_2=CHC_6H_4			(53.5)	
RCHO	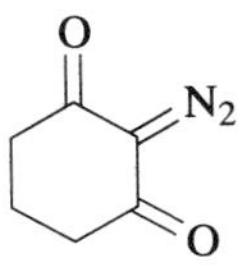 [structure: 2-diazocyclohexane-1,3-dione]	$HgCl_2$, xylene 130-140°	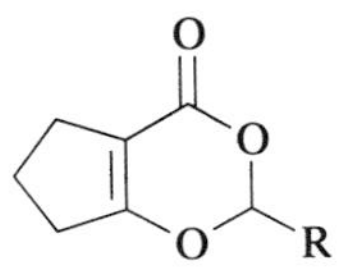 [structure: cyclopenta-fused 1,3-dioxin-4-one bearing R]	
R				
CCl_3			(66)	290
o-ClC_6H_4			(61)	290
p-ClC_6H_4p			(54)	290
p-$O_2NC_6H_4$			(74)	281, 290

TABLE XL. [4+2] CYCLOADDITION OF ACYL AND VINYL KETENES TO ALDEHYDES (*Continued*)

Reactant	Ketene or Ketene Source	Conditions	Product(s) and Yield(s) (%)	Refs.
RCHO	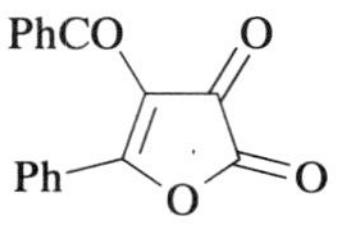 PhCO, Ph, O, O, O	Xylene, 130°, -CO	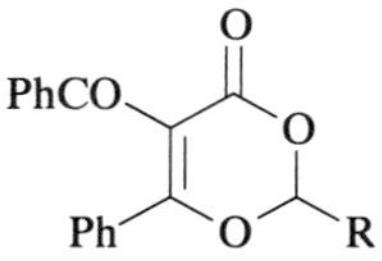 PhCO, O, O, Ph, O, R	833
R: CCl_3			(74)	
3-Pyridyl			(89)	
p-ClC_6H_4			(55)	
m-$O_2NC_6H_4$			(60)	
p-MeC_6H_4			(68)	
C_7				
PhCHO	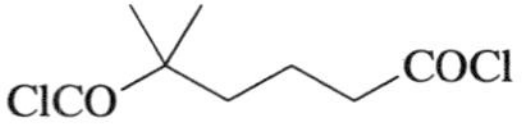 ClCO, COCl	Et_3N, Et_2O	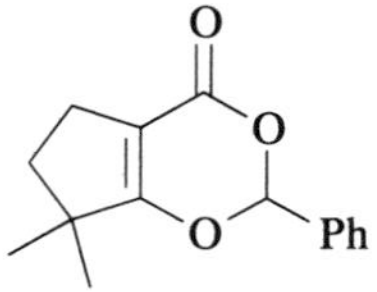 O, O, O, Ph (56)	280
	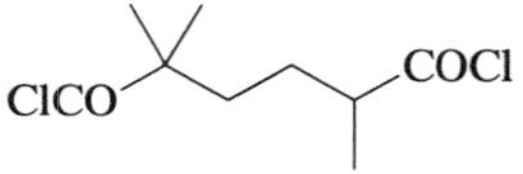 ClCO, COCl	Et_3N, Et_2O	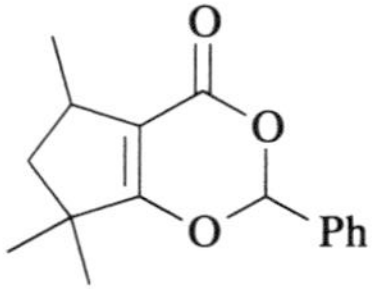 O, O, O, Ph (60)	280
C_8				
p-$MeOC_6H_4CHO$	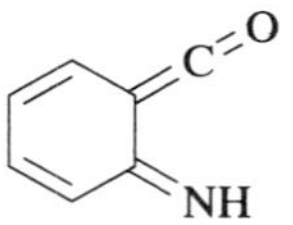 C=O, NH	248°	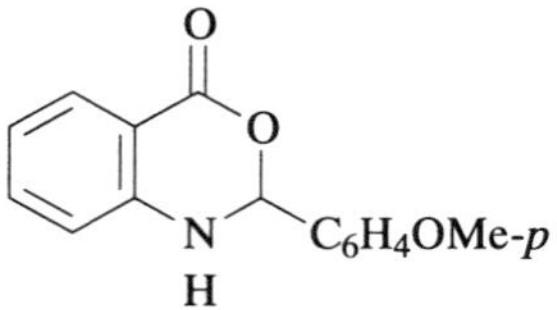 O, O, N, H, C_6H_4OMe-p (—)	277

TABLE XLI. [4+2] CYCLOADDITION OF ACYL KETENES TO KETONES

Reactant	Ketene or Ketene Source	Conditions	Product(s) and Yield(s) (%)	Refs.
C_3				
Me_2CO	$ClCO(CH_2)_4COCl$	Et_3N, Et_2O, 35°, 30 min	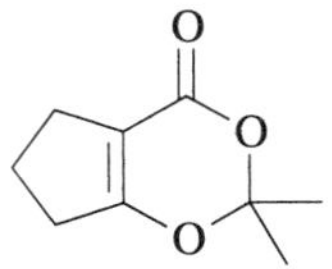 (71.5)	280
	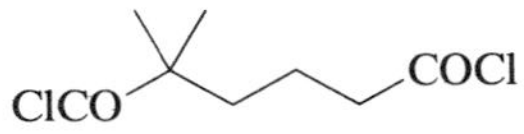 ClCO, COCl	Et_3N, Et_2O, 35°, 30 min	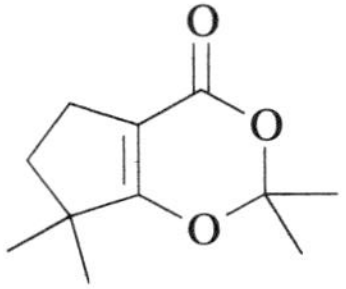 (85.5)	280
	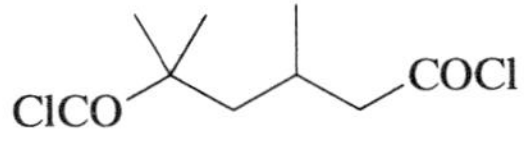 ClCO, COCl	Et_3N, Et_2O, 35°, 30 min	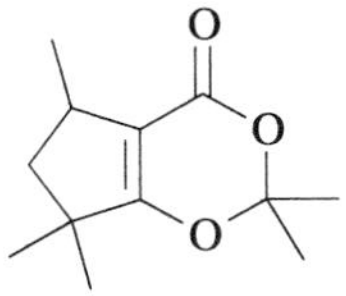 (85)	280
C_4				
MeCOEt	$ClCO(CH_2)_4COCl$	Et_3N, Et_2O, 35°, 30 min	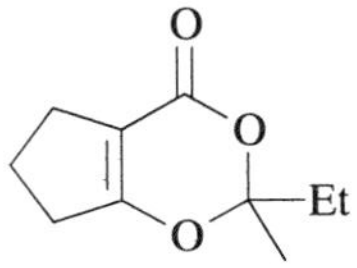 O, O, O, Et (51.5)	280
C_6				
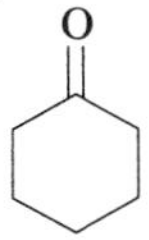 O	$ClCO(CH_2)_4COCl$	Et_3N, Et_2O, 35°, 30 min	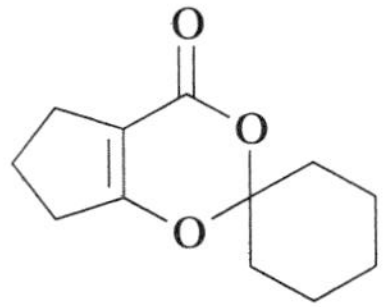 (71)	280

TABLE XLI. [4+2] CYCLOADDITION OF ACYL KETENES TO KETONES (*Continued*)

Reactant	Ketene or Ketene Source	Conditions	Product(s) and Yield(s) (%)	Refs.
	[structure: O, N₂, O]	$HgCl_2$, xylene, 130-140°	[structure: O, O, O] (72.5)	281
	MeCOC≡COEt	Xylene, 130°	[structure: O, O, O] (94)	282
	[structure: PhCO, O, Ph, O, O]	Xylene, 130°	[structure: O, PhCO, O, Ph, O] (71)	833
C_8 MeCOPh	[structure: PhCO, O, Ph, O, O]	Xylene, 130°	[structure: O, PhCO, O, Ph, O, Ph] (70)	833

TABLE XLI. [4+2] CYCLOADDITION OF ACYL KETENES TO KETONES (*Continued*)

Reactant	Ketene or Ketene Source	Conditions	Product(s) and Yield(s) (%)	Refs.
	$ClCO(CH_2)_4COCl$	Et_3N, Et_2O, 35°, 30 min	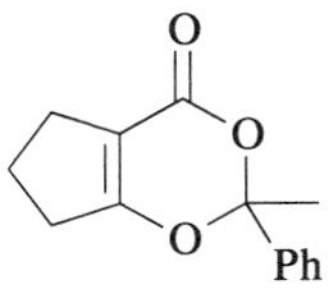 (9)	280
C_{13}				
Ph_2CO	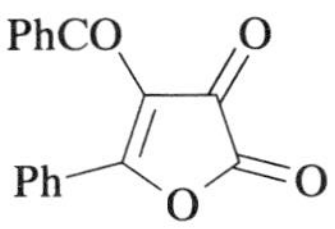 PhCO, O, Ph, O, O	Xylene, 130°	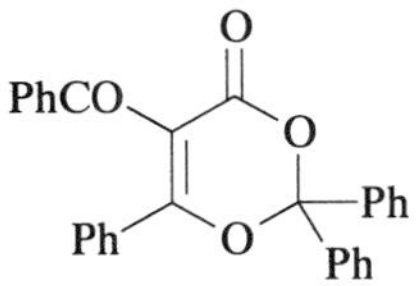 PhCO, O, O, Ph, Ph, O, Ph (55)	833

TABLE XLII. [4+2] CYCLOADDITION OF ACYL AND VINYL KETENES TO NITRILES

Reactant	Ketene or Ketene Source	Conditions	Product(s) and Yield(s) (%)	Refs.
C_3				
Me_2NCN	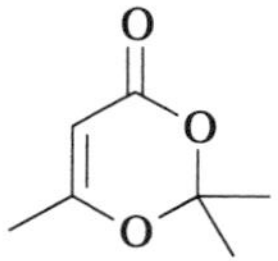	120-130°	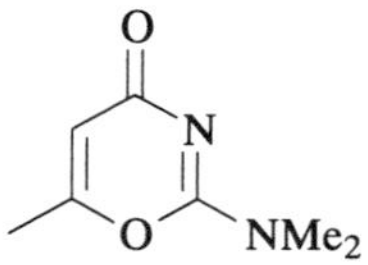 NMe_2 (85)	274
	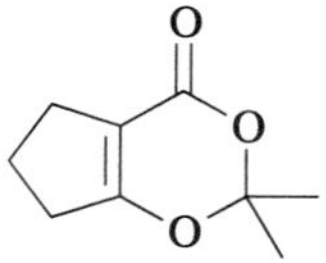	120-140°	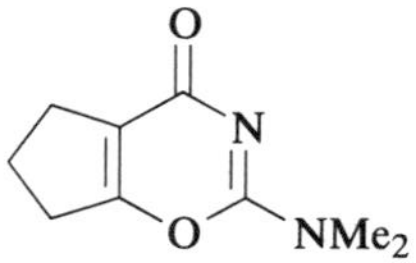 NMe_2 (84)	274
C_5				
Et_2NCN	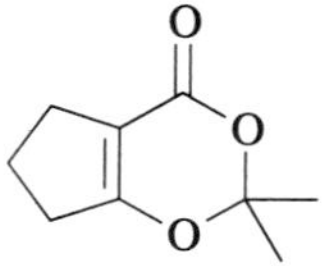	120-140°	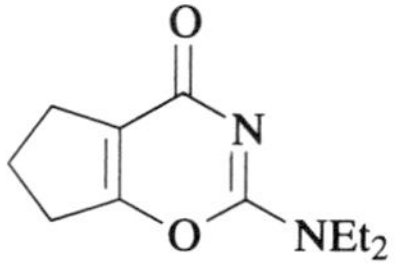 NEt_2 (93.5)	274
C_7				
PhCN	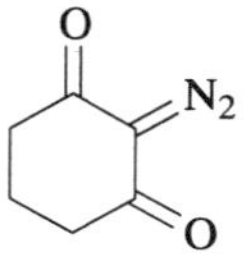 N_2	145-150°	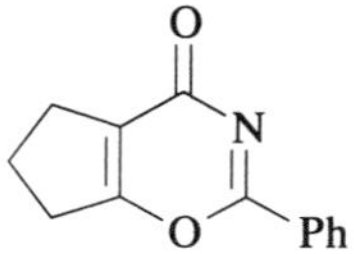 Ph (66)	281, 290
XC_6H_4CN	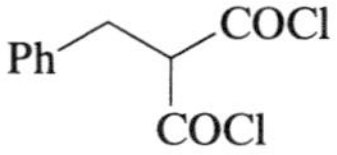 Ph, COCl, COCl	120-130° 25 min	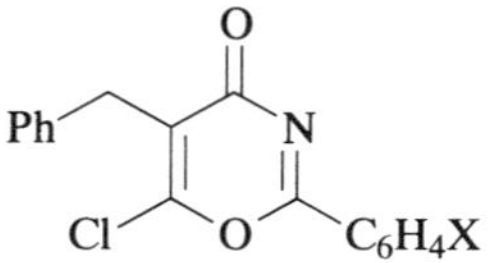 Ph, Cl, C_6H_4X	283
X				
p-Cl			(29)	
o-O_2N			(22)	
H			(40)	
p-Me			(17)	
3,4-Me_2			(33)	

TABLE XLII. [4+2] CYCLOADDITION OF ACYL AND VINYL KETENES TO NITRILES (*Continued*)

Reactant	Ketene or Ketene Source	Conditions	Product(s) and Yield(s) (%)	Refs.
XC_6H_4CN	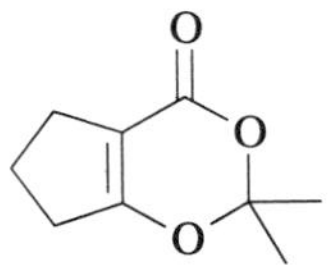 [structure]	140-150°, 15-20 min	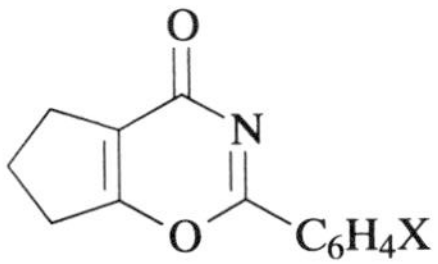 [structure] C_6H_4X	274
X				
H			(55)	
o-Cl			(1)	
p-Cl			(30)	
o-Me			(19)	
m-Me			(46)	
p-Me			(35)	
p-OMe			(56)	
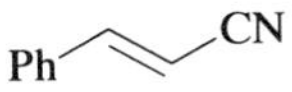 Ph-CH=CH-CN	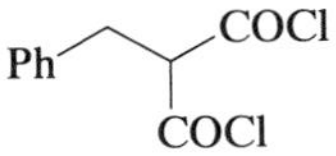 $PhCH_2CH(COCl)_2$	140°, 4 min	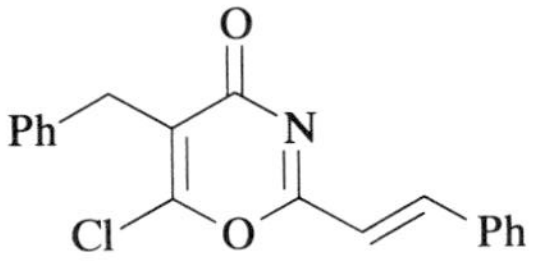 [structure] Ph, Cl, Ph (5)	283
RCN	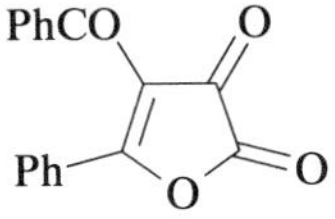 [structure] PhCO, Ph	130°	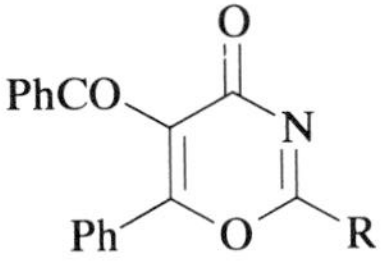 [structure] PhCO, Ph, R	834
R				
Ph			(47)	
3,4-$Me_2C_6H_3$			(44)	
1-$C_{10}H_7$			(35)	

TABLE XLIII. [4+2] CYCLOADDITION OF ACYL KETENES TO CYANATES

Reactant	Ketene or Ketene Source	Conditions	Product(s) and Yield(s) (%)	Refs.
ROCN	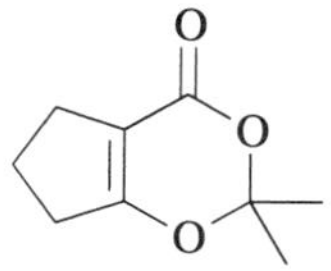 O, O, O	Xylene, 140-150°	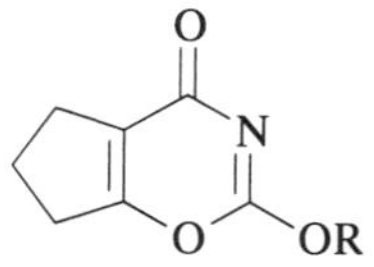 O, N, O, OR	274
R				
2,4-$Cl_2C_6H_3$			(43)	
m-ClC_6H_4			(57)	
p-ClC_6H_4			(70.5)	
p-$O_2NC_6H_4$			(44)	
Ph			(76)	
2-Me-4-ClC_6H_3			(50.5)	
p-MeC_6H_4			(47.5)	
2,4-$Me_2C_6H_3$			(19)	
p-(*t*-Bu)C_6H_4			(37.5)	

TABLE XLIV. [4+2] CYCLOADDITION OF ACYL KETENES TO ISOCYANATES AND ISOTHIOCYANATES

Reactant	Ketene or Ketene Source	Conditions	Product(s) and Yield(s) (%)	Refs.
RNCO	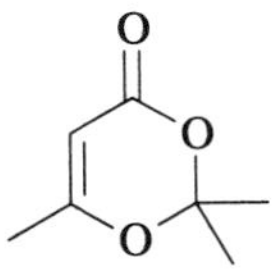 2,2,6-trimethyl-4H-1,3-dioxin-4-one (structure)	Xylene, 120-130°, 15-20 min	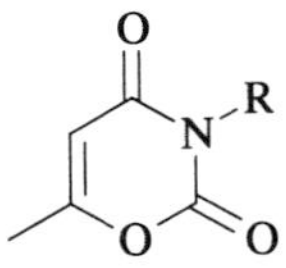 6-methyl-3-R-1,3-oxazine-2,4-dione (structure) **I**	
RNCO			**I**	
MeNCO			(63)	274
ClC_2H_4NCO			(65)	274
EtNCO			(68)	274
$MeOCH_2NCO$			(78)	274
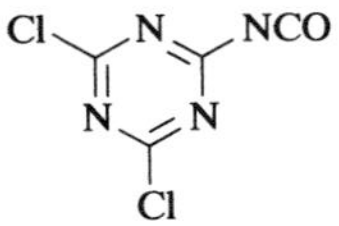 4,6-dichloro-1,3,5-triazin-2-yl NCO (structure)			(33.5)	274
i-PrNCO			(68)	274
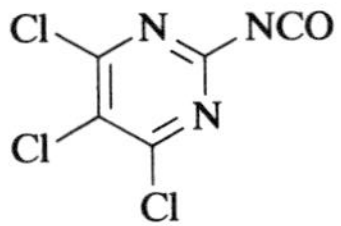 4,5,6-trichloropyrimidin-2-yl NCO (structure)			(55)	274
n-BuNCO			(65)	274
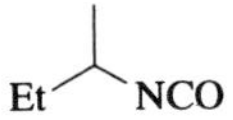 Et–CH(CH$_3$)–NCO (structure)			(63)	274
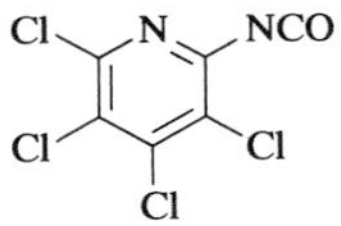 3,4,5,6-tetrachloropyridin-2-yl NCO (structure)			(57)	274

TABLE XLIV. [4+2] CYCLOADDITION OF ACYL KETENES TO ISOCYANATES AND ISOTHIOCYANATES (*Continued*)

Reactant	Ketene or Ketene Source	Conditions	Product(s) and Yield(s) (%)	Refs.
RNCO			I	
2-CCl_3-4-NCO-5,6-Cl_2-pyrimidine (structure)			(65)	274
2,4-$(CCl_3)_2$-6-NCO-1,3,5-triazine (structure)			(73.5)	274
$NCCH_2CH(Et)NCO$ (structure)			(48)	274
$EtO_2C(CH_2)_2NCO$			(70)	274
3-Cl-4-F-C_6H_3NCO (structure)			(72)	274
3,4-$Cl_2C_6H_3NCO$ (structure)			(78)	274
p-FC_6H_4NCO			(76)	274
m-ClC_6H_4NCO			(72)	274
p-ClC_6H_4NCO			(85)	274
PhNCO			(68)	274
$C_6H_{11}NCO$			(52)	274

TABLE XLIV. [4+2] CYCLOADDITION OF ACYL KETENES TO ISOCYANATES AND ISOTHIOCYANATES (*Continued*)

Reactant	Ketene or Ketene Source	Conditions	Product(s) and Yield(s) (%)	Refs.
RNCO			I	
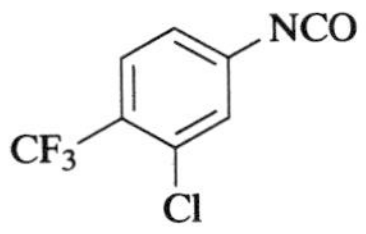 NCO; CF_3; Cl			(87.5)	274
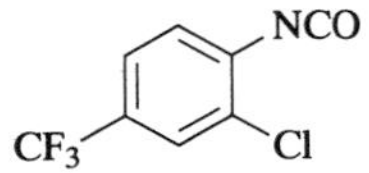 NCO; CF_3; Cl			(40)	274
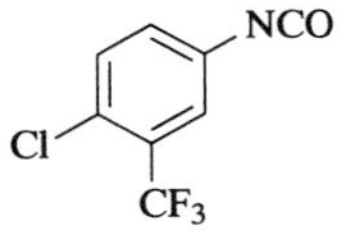 NCO; Cl; CF_3			(75)	274
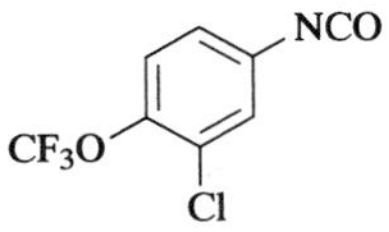 NCO; CF_3O; Cl			(78)	274
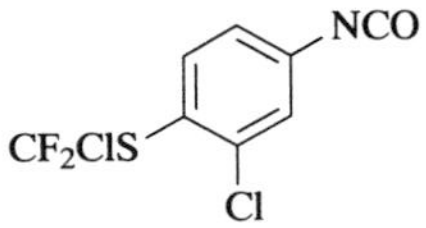 NCO; CF_2ClS; Cl			(62)	274
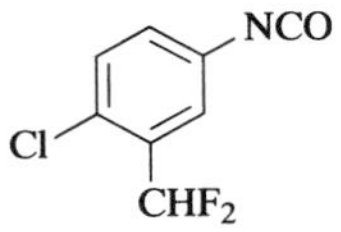 NCO; Cl; CHF_2			(67)	274

TABLE XLIV. [4+2] CYCLOADDITION OF ACYL KETENES TO ISOCYANATES AND ISOTHIOCYANATES (*Continued*)

Reactant	Ketene or Ketene Source	Conditions	Product(s) and Yield(s) (%)	Refs.
RNCO			I	
o-$CF_3C_6H_4NCO$			(42)	274
p-$CF_3OC_6H_4NCO$			(69)	274
m-$CF_3C_6H_4NCO$			(73)	274
p-$CF_3C_6H_4NCO$			(88)	274
m-$CHF_2C_6H_4NCO$			(60)	274
3-Cl-4-$MeOC_6H_3NCO$			(75)	274
m-MeC_6H_4NCO			(67)	274
$PhCH_2NCO$			(60)	274
p-$MeOC_6H_4NCO$			(86)	274
i-$BuO_2C(CH_2)_2NCO$			(63)	274
cycloheptyl-NCO			(63)	274
$C_6H_{11}CH_2NCO$			(72)	274
n-$C_7H_{15}NCO$			(52)	274
3-CN-4-Me-C_6H_3NCO			(62)	274
3-CF_3-4-MeO-C_6H_3NCO			(79)	274
3-MeO-4-CF_3-C_6H_3NCO			(52)	274

TABLE XLIV. [4+2] CYCLOADDITION OF ACYL KETENES TO ISOCYANATES AND ISOTHIOCYANATES (*Continued*)

Reactant	Ketene or Ketene Source	Conditions	Product(s) and Yield(s) (%)	Refs.
RNCO			I	
p-$EtOC_6H_4NCO$			(79)	274
$Ph(CH_2)_2NCO$			(65)	274
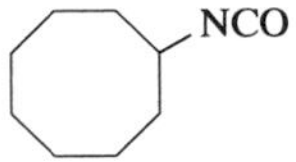 cyclooctyl-NCO			(49)	274
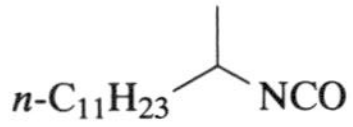 *n*-$C_{11}H_{23}CH(CH_3)NCO$			(45)	274
n-$C_{14}H_{29}NCO$			(48)	274
n-$C_{18}H_{37}NCO$			(48)	274
RNCO	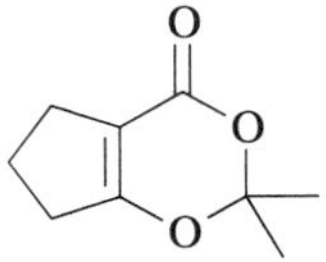 2,2-dimethyl-cyclopenta-1,3-dioxin-4-one	Xylene, 120-140°	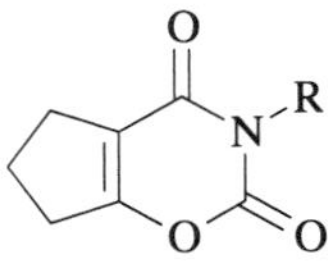 3-R-cyclopenta[e][1,3]oxazine-2,4-dione I	
RNCO			I	
Me			(70.5)	274
MeO_2CNCO			(29)	274
$Cl(CH_2)_2NCO$			(73)	274
EtNCO			(74)	274
$MeOCH_2NCO$			(54)	274
n-PrNCO			(86)	274
i-PrNCO			(78)	274
EtO_2CCH_2NCO			(68)	274
n-BuNCO			(78)	274
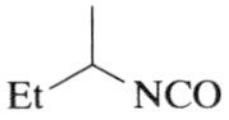 EtCH(CH_3)NCO			(75)	274

TABLE XLIV. [4+2] CYCLOADDITION OF ACYL KETENES TO ISOCYANATES AND ISOTHIOCYANATES (*Continued*)

Reactant	Ketene or Ketene Source	Conditions	Product(s) and Yield(s) (%)	Refs.
RNCO			I	
i-PrNCO			(82)	274
n-BuSCONCO			(12.5)	274
n-$BuOCH_2NCO$			(73)	274
3,4-$Cl_2C_6H_3NCO$			(75)	274
2,5-$Cl_2C_6H_3NCO$			(33.5)	274
2,4-$Cl_2C_6H_3NCO$			(52.5)	274
o-ClC_6H_4NCO			(64.5)	274
m-ClC_6H_4NCO			(64.5)	274
p-ClC_6H_4NCO			(65)	274
p-$O_2NC_6H_4NCO$			(30)	274
PhNCO			(61)	274
PhNCS			(29)	274
$C_6H_{11}NCO$			(80)	274
$Cl(CH_2)_6NCO$			(68)	274
i-Bu–CH(CH_3)–NCO			(62)	274
4-CF_3O-3-Cl-C_6H_3NCO			(52)	274
4-Cl-3-CF_3-C_6H_3NCO			(42)	274

TABLE XLIV. [4+2] CYCLOADDITION OF ACYL KETENES TO ISOCYANATES AND ISOTHIOCYANATES (*Continued*)

Reactant	Ketene or Ketene Source	Conditions	Product(s) and Yield(s) (%)	Refs.
RNCO			I	
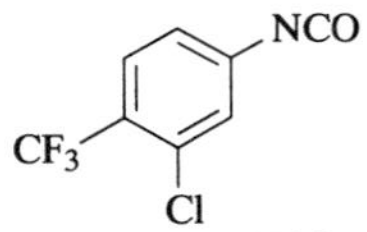 NCO, CF_3, Cl			(60)	274
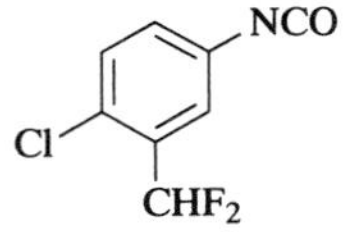 NCO, Cl, CHF_2			(40)	274
p-$CF_3C_6H_4NCO$			(78.5)	274
m-$CF_3C_6H_4NCO$			(46.5)	274
o-$CF_3C_6H_4NCO$			(46.5)	274
PhSCONCO			(57)	274
PhO_2CNCO			(52)	274
$PhCH_2NCO$			(89)	274
m-MeC_6H_4NCO			(61.5)	274
p-$MeOC_6H_4NCO$			(70)	274
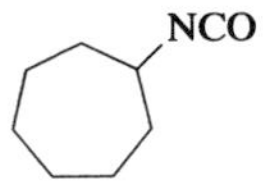 NCO			(73)	274
$C_6H_{11}CH_2NCO$			(82)	274
i-$BuO_2C(CH_2)_2NCO$			(62)	274
2,4-$(CF_3)_2C_6H_3NCO$			(52.5)	274
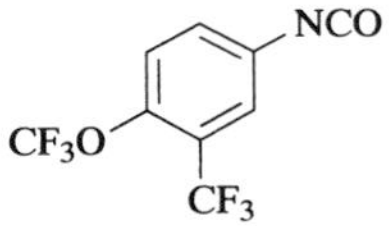 NCO, CF_3O, CF_3			(56.5)	274

TABLE XLIV. [4+2] CYCLOADDITION OF ACYL KETENES TO ISOCYANATES AND ISOTHIOCYANATES (*Continued*)

Reactant	Ketene or Ketene Source	Conditions	Product(s) and Yield(s) (%)	Refs.
RNCO			I	
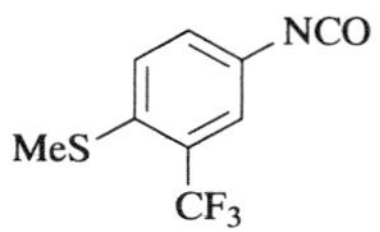 4-MeS-3-$CF_3C_6H_3NCO$ (structure: NCO, MeS, CF_3)			(36)	274
2,6-$Me_2C_6H_3NCO$			(18.5)	274
o-EtC_6H_4NCO			(34)	274
p-$EtOC_6H_4NCO$			(71.5)	274
$Ph(CH_2)_2NCO$			(54.5	274
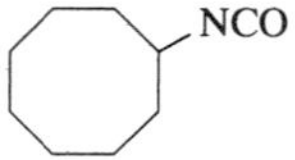 Cyclooctyl NCO (structure)			(65)	274
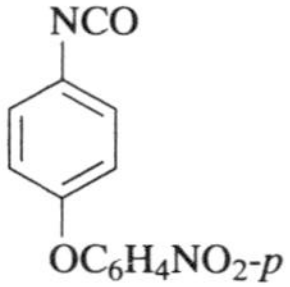 NCO / $OC_6H_4NO_2$-*p* (structure, para-substituted benzene)			(45)	274
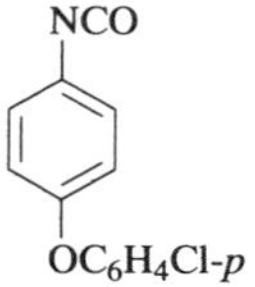 NCO / OC_6H_4Cl-*p* (structure, para-substituted benzene)			(51)	274
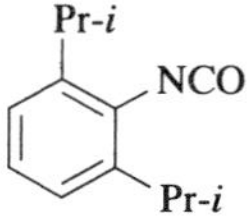 Pr-*i* / NCO / Pr-*i* (structure, 2,6-diisopropylphenyl)			(17.5)	274

TABLE XLIV. [4+2] CYCLOADDITION OF ACYL KETENES TO ISOCYANATES AND ISOTHIOCYANATES (*Continued*)

Reactant	Ketene or Ketene Source	Conditions	Product(s) and Yield(s) (%)	Refs.
RNCO			**I**	
n-$C_{12}H_{25}NCO$			(62)	274
$C_{11}H_{23}CH(CH_3)NCO$			(45)	274
C_7				
p-ClC_6H_4NCO	$PhCH_2CH(COCl)_2$	170-180°, 20 min	5-benzyl-6-chloro-3-(C_6H_4Cl-p)-1,3-oxazine-2,4-dione (88)	283
PhNCO	MeCOC≡COEt	CCl_4, 90°, 2 h	6-methyl-3-phenyl-1,3-oxazine-2,4-dione (41)	282
	2-bromocyclohexane-1,3-dione	Et_3N, C_6H_6	3-phenyl-cyclopenta-fused 1,3-oxazine-2,4-dione (25)	284
	$EtCH(COCl)_2$	160-170°, 1 h	6-chloro-5-i-Pr-3-phenyl-1,3-oxazine-2,4-dione (98)	283
	i-$PrCH(COCl)_2$	170-180°, 1 h	6-chloro-5-ethyl-3-phenyl-1,3-oxazine-2,4-dione (65)	283

TABLE XLIV. [4+2] CYCLOADDITION OF ACYL KETENES TO ISOCYANATES AND ISOTHIOCYANATES (*Continued*)

Reactant	Ketene or Ketene Source	Conditions	Product(s) and Yield(s) (%)	Refs.
	n-Bu, COCl, COCl	165°, 30 min	*n*-Bu, O, N, Ph, Cl, O, O (64)	283
	Ph, COCl, COCl	170-180°, 30 min	Ph, O, N, Ph, Cl, O, O (97)	283
PhNCS	PhCO, O, Ph, O, O	C_6H_6, 80°	PhCO, O, N, Ph, Ph, O, O (43)	835
	O, Br, O	Et_3N, C_6H_6	O, N, Ph, O, S (12)	284
	O, Br, O	Et_3N, C_6H_6	O, N, Ph, O, S (22)	284

TABLE XLIV. [4+2] CYCLOADDITION OF ACYL KETENES TO ISOCYANATES AND ISOTHIOCYANATES (*Continued*)

Reactant	Ketene or Ketene Source	Conditions	Product(s) and Yield(s) (%)	Refs.
$C_6H_{11}NCO$	Ph, COCl, COCl	175-185°, 28 min	Ph, O, N, C_6H_{11}, Cl, O, O (94)	283
C_8				
m-$CF_3C_6H_4NCO$	Ph, COCl, COCl	175-185°, 28 min	Ph, O, N, $C_6H_4CF_3$-m, Cl, O, O (89)	283
p-MeC_6H_4NCO	PhCO, O, Ph, O, O	C_6H_6, 80°	PhCO, O, N, C_6H_4Me-p, Ph, O, O (—)	835
$OCN(CH_2)_6NCO$	Et, COCl, COCl	170-180°, 10 min	Et, O, N, $(CH_2)_6$, N, O, Et, Cl, O, O, O, O, Cl (46)	283
	i-Pr, COCl, COCl	170-180°, 10 min	i-Pr, O, N, $(CH_2)_6$, N, O, Pr-i, Cl, O, O, O, O, Cl (56)	283

TABLE XLIV. [4+2] CYCLOADDITION OF ACYL KETENES TO ISOCYANATES AND ISOTHIOCYANATES (*Continued*)

Reactant	Ketene or Ketene Source	Conditions	Product(s) and Yield(s) (%)	Refs.
C_{11} 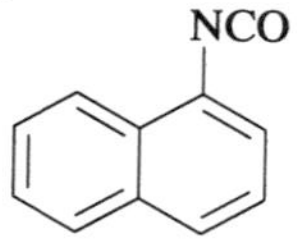 NCO (1-naphthyl isocyanate)	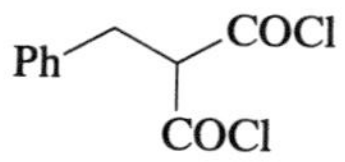 Ph, COCl, COCl	170-180°, 10 min	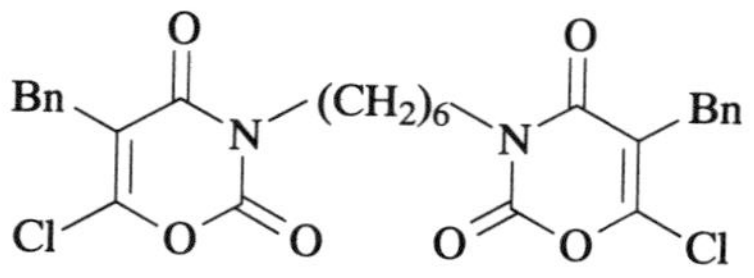 Bn, O, N, $(CH_2)_6$, N, O, Bn, Cl, O, O, O, O, Cl (88)	283
	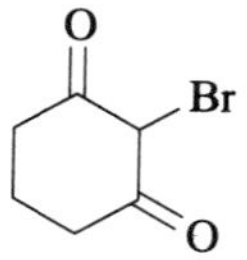 O, Br, O	Et_3N, C_6H_6	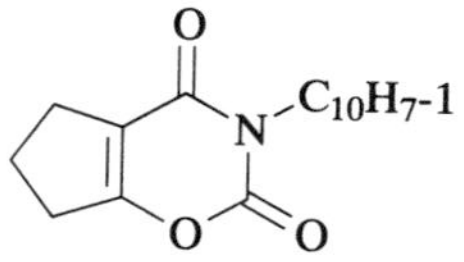 O, N, $C_{10}H_7$-1, O, O (5)	284
	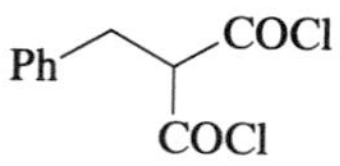 Ph, COCl, COCl	180-190°	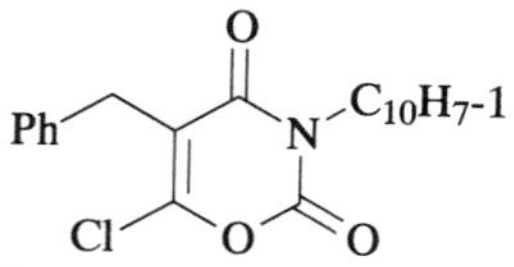 Ph, O, N, $C_{10}H_7$-1, Cl, O, O (80)	283

TABLE XLV. [4+2] CYCLOADDITION OF ACYL KETENES TO *N*-SULFINYLAMINES

Reactant	Ketene or Ketene Source	Conditions	Product(s) and Yield(s) (%)	Refs.
C_3				
n-PrNSO	Ph, Ph, O, O, N_2	PhMe, 111°	Ph, Ph, O, N, Pr-*n*, S, O (20)	286
C_6				
p-ClC_6H_4NSO	Ph, O, O, O	C_6H_6, 80°	Ph, O, N, C_6H_4Cl-*p*, S, O (—)	287
	O, O, O	C_6H_6, 80°	O, S, N, C_6H_4Cl-*p*, O (—)	287
PhNSO	Ph, O, O, O	C_6H_6, 80°	Ph, O, N, Ph, S, O (49)	287
	O, O, O	C_6H_6, 80°	O, S, N, Ph, O (74)	287

TABLE XLV. [4+2] CYCLOADDITION OF ACYL KETENES TO *N*-SULFINYLAMINES (*Continued*)

Reactant	Ketene or Ketene Source	Conditions	Product(s) and Yield(s) (%)	Refs.
	Ph, O, O, Ph, N_2	PhMe, 111°	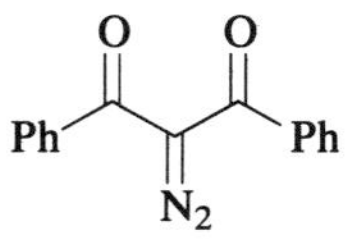 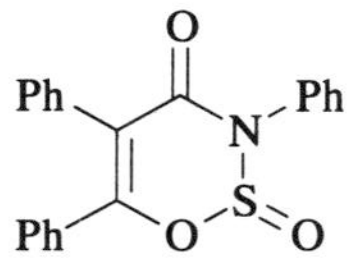 Ph, O, N, Ph, S, O, Ph, O (38)	286
$C_6H_{11}NSO$	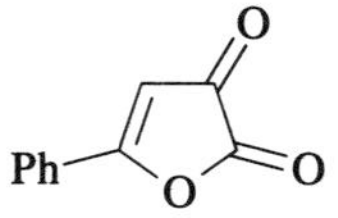 O, Ph, O, O	C_6H_6, 80° 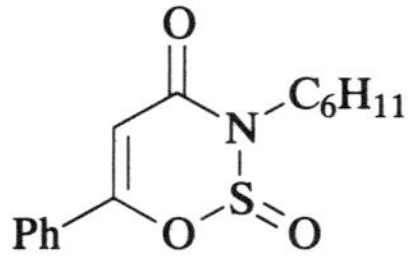	O, N, C_6H_{11}, S, Ph, O, O (—)	287
	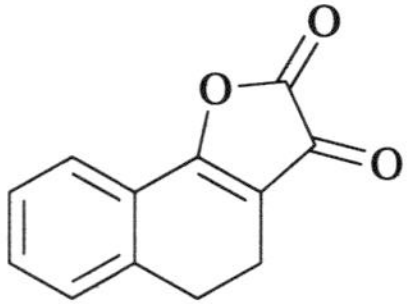 O, O, O	C_6H_6, 80° 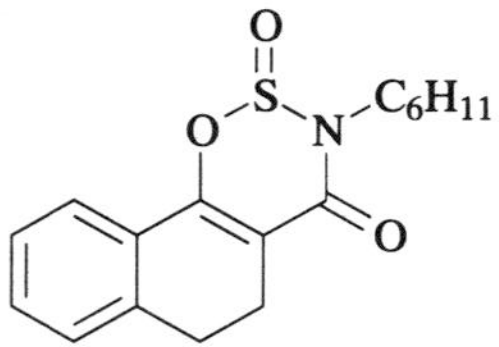	O, S, O, N, C_6H_{11}, O (—)	287
	Ph, O, O, Ph, N_2	PhMe, 111°	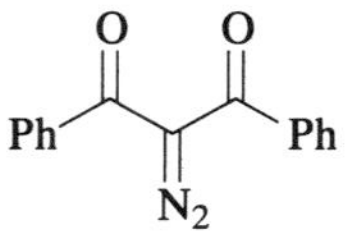 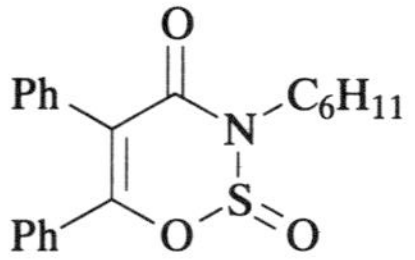 Ph, O, N, C_6H_{11}, S, O, Ph, O (34)	286
C_7				
p-MeC_6H_4NSO	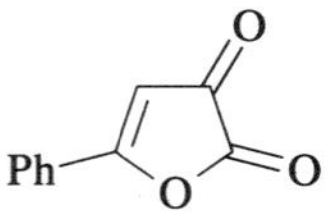 O, Ph, O, O	C_6H_6, 80° 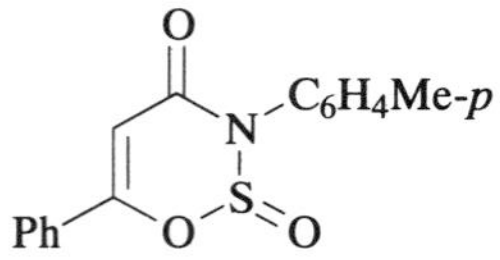	O, N, C_6H_4Me-*p*, S, Ph, O, O (—)	287

TABLE XLV. [4+2] CYCLOADDITION OF ACYL KETENES TO *N*-SULFINYLAMINES (*Continued*)

Reactant	Ketene or Ketene Source	Conditions	Product(s) and Yield(s) (%)	Refs.
	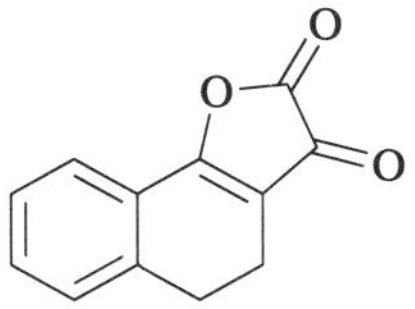 [structure]	C_6H_6, 80°	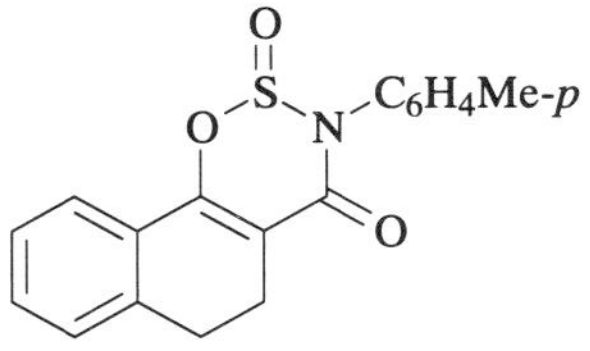 [structure] C_6H_4Me-*p* (—)	287
	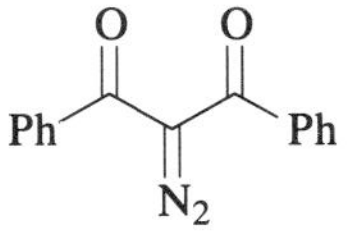 PhC(O)C(=N_2)C(O)Ph	PhMe, 111°	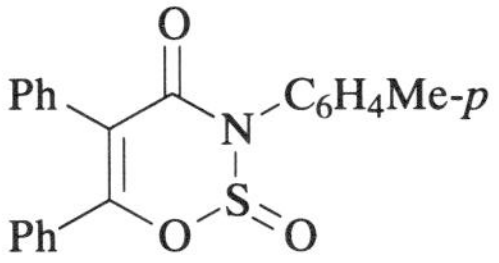 [structure] Ph, Ph, C_6H_4Me-*p* (30)	286
p-$MeOC_6H_4NSO$	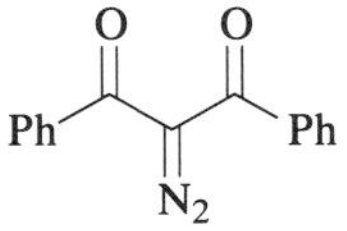 PhC(O)C(=N_2)C(O)Ph	PhMe, 111°	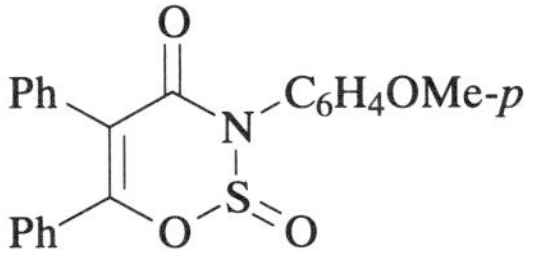 [structure] Ph, Ph, C_6H_4OMe-*p* (29)	286
	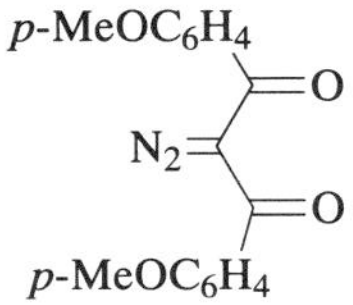 *p*-$MeOC_6H_4$C(O)C(=N_2)C(O)C_6H_4OMe-*p*	PhMe, 111°	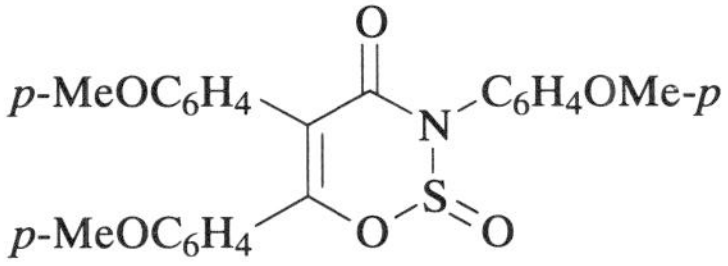 [structure] *p*-$MeOC_6H_4$, *p*-$MeOC_6H_4$, C_6H_4OMe-*p* (16)	286

TABLE XLV. [4+2] CYCLOADDITION OF ACYL KETENES TO *N*-SULFINYLAMINES (*Continued*)

Reactant	Ketene or Ketene Source	Conditions	Product(s) and Yield(s) (%)	Refs.
TsNSO	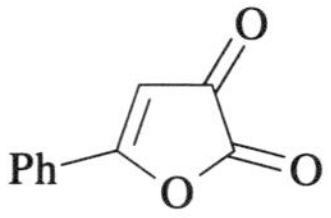 Ph, O, O, O	C_6H_6, 80°	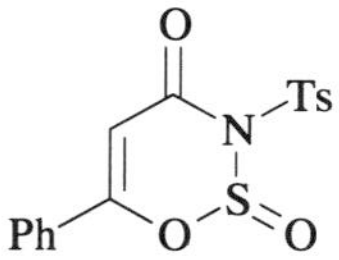 O, N, Ts, Ph, O, S, O (—)	287
C_8 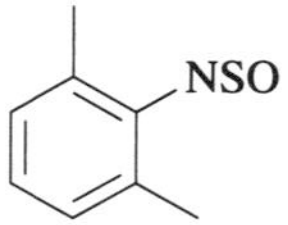 NSO	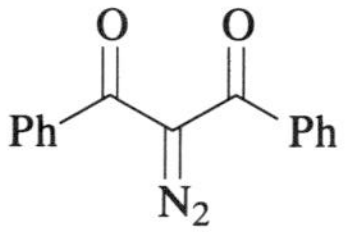 O, O, Ph, Ph, N_2	PhMe, 111°	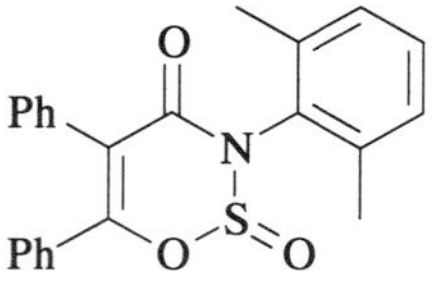 O, Ph, N, Ph, O, S, O (18)	286

TABLE XLVI. [4+2] CYCLOADDITION OF ACYL KETENES TO CARBODIIMIDES

Reactant	Ketene or Ketene Source	Conditions	Product(s) and Yield(s) (%)	Refs.
C_7				
i-PrN=C=NPr-i	Ph(C=S)C(CN)=C=O	—	5-NC-6-Ph-3-Pr-i-2-(NPr-i)-1,3-thiazin-4-one (—)	285
	Ph(C=S)C(CO_2Et)=C=O	—	5-EtO_2C-6-Ph-3-Pr-i-2-(NPr-i)-1,3-thiazin-4-one (—)	285
	2,2-dimethyl-cyclopenta[b]pyran-4-one	Xylene, 120-140°	cyclopenta-fused 3-Pr-i-2-(NPr-i)-1,3-oxazin-4-one (76)	274
C_{13}				
p-ClC_6H_4N=C=NC_6H_4Cl-p	"	"	cyclopenta-fused 3-(C_6H_4Cl-p)-2-(NC_6H_4Cl-p)-1,3-oxazin-4-one (80)	274
PhN=C=NPh	"	"	cyclopenta-fused 3-Ph-2-(NPh)-1,3-oxazin-4-one (93)	274
C_{15}				
p-MeC_6H_4N=C=NC_6H_4Me-p	"	"	cyclopenta-fused 3-(C_6H_4Me-p)-2-(NC_6H_4Me-p)-1,3-oxazin-4-one (72)	274

TABLE XLVII. [4+2] CYCLOADDITION OF ACYL AND VINYL KETENES TO IMINES AND AZO COMPOUNDS

Reactant	Ketene or Ketene Source	Conditions	Product(s) and Yield(s) (%)	Refs.
C_7				
COMe, N, N	N_2, COMe	*hν*	MeOC, N, N, O (46)	288
C_{10}				
H, N, O, N_2, N	O, O, Ph, Ph, N_2	Xylene, 120°	H, N, O, N_2, N, O, O, Ph, Ph (72)	291
PhCH=NPr-*i*	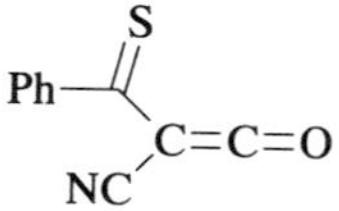 S, Ph, C=C=O, NC	—	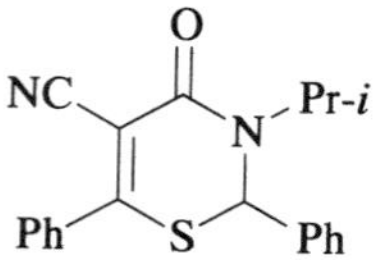 O, NC, N, Pr-*i*, Ph, S, Ph (—)	285
	S, Ph, C=C=O, EtO_2C	—	O, EtO_2C, N, Pr-*i*, Ph, S, Ph (—)	285

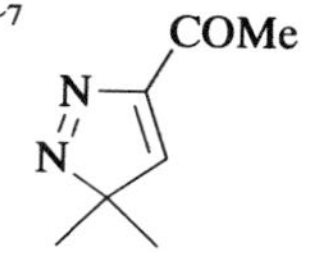

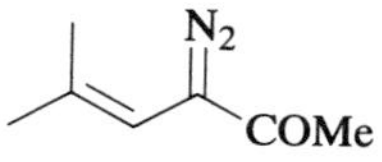

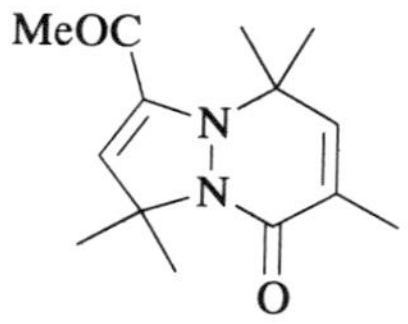

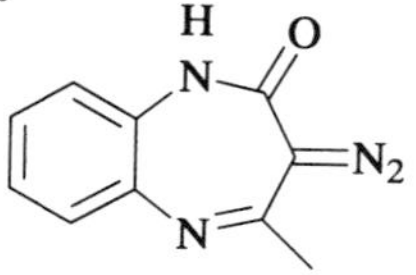

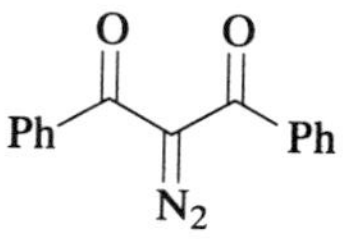

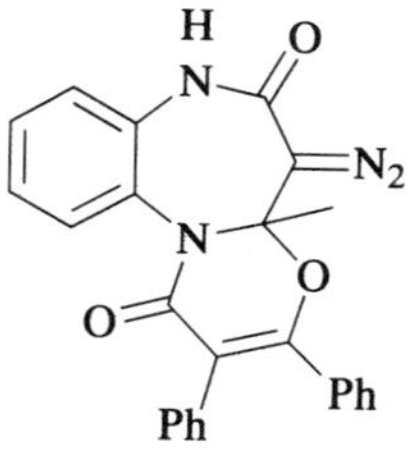

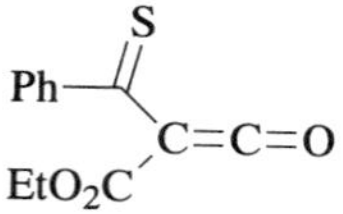

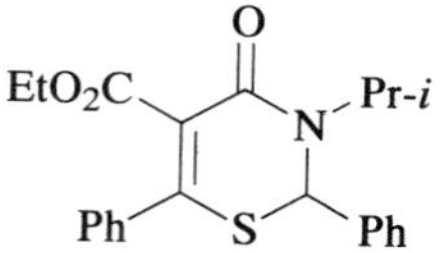

TABLE XLVII. [4+2] CYCLOADDITION OF ACYL AND VINYL KETENES TO IMINES AND AZO COMPOUNDS (*Continued*)

Reactant	Ketene or Ketene Source	Conditions	Product(s) and Yield(s) (%)	Refs.
C_{11} 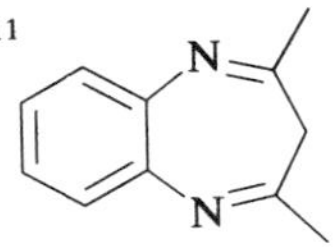 N, N, Ph, O, O, N_2, Ph	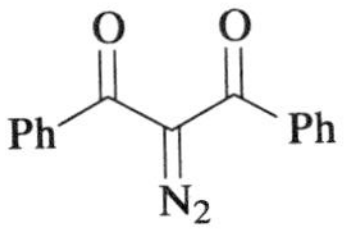	Xylene, 120°	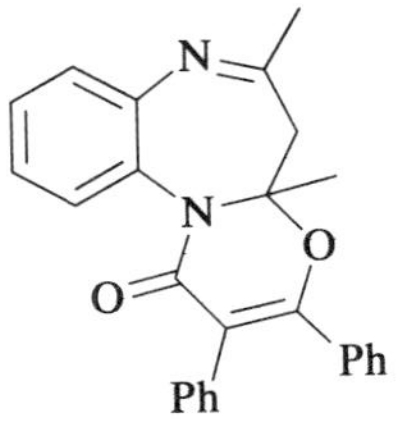 N, N, O, O, Ph, Ph (44)	291
	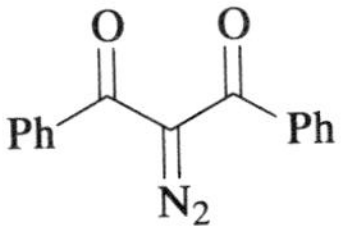 Ph, O, O, N_2, Ph	Excess ketene, xylene, 120°	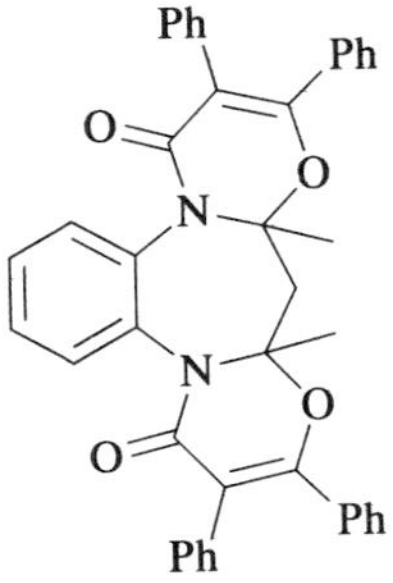 Ph, Ph, O, N, O, N, O, O, Ph, Ph (66)	291
C_{13} 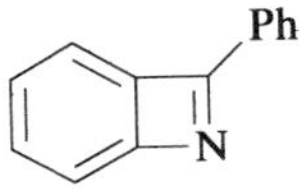 Ph, N	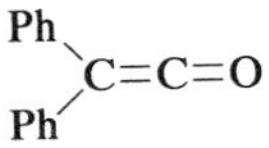 $Ph_2C{=}C{=}O$	CH_2Cl_2	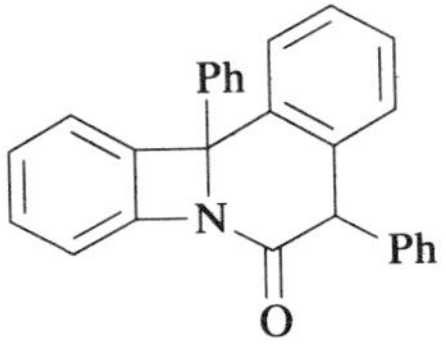 Ph, N, O, Ph (30)	289

TABLE XLVII. [4+2] CYCLOADDITION OF ACYL AND VINYL KETENES TO IMINES AND AZO COMPOUNDS (*Continued*)

Reactant	Ketene or Ketene Source	Conditions	Product(s) and Yield(s) (%)	Refs.
PhCH=NPh	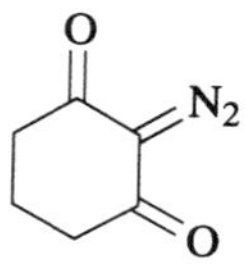 2-diazocyclohexane-1,3-dione (N_2, O, O)	Xylene, 140-145°	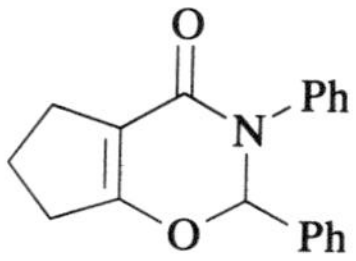 (25) 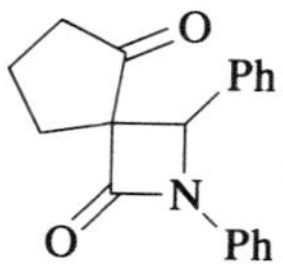 + (17)	281, 290
	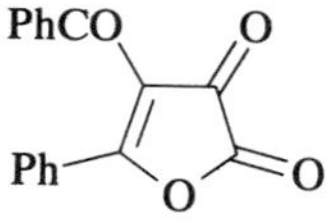 PhCO, Ph, O, O, O	Xylene, 130°	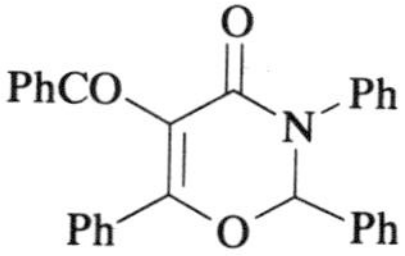 PhCO, Ph, N–Ph, O, Ph (27)	836
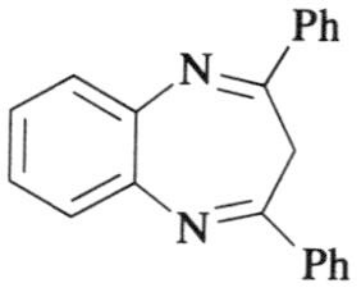 2,4-diphenyl-3H-1,5-benzodiazepine (N, N, Ph, Ph)	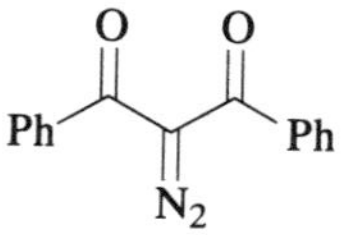 PhCO–C(N_2)–COPh	Xylene, 120°	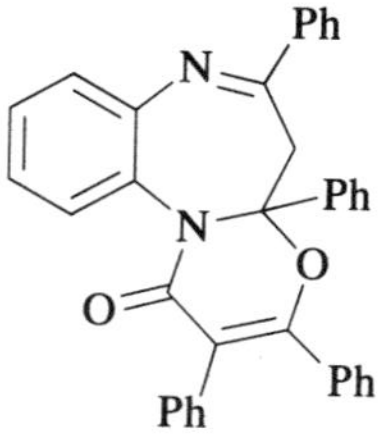 (54)	291

REFERENCES

[1] Chick, F.; Wilsmore, N. T. M. *J. Chem. Soc.* **1908**, 946.
[2] Staudinger, H.; Klever, H. W. *Ber.* **1908**, *41*, 594.
[3] Staudinger, H.; Klever, H. W. *Ber.* **1908**, *41*, 1516, cited in *C.A.* **1908**, *2*, 2224.
[4] Staudinger, H. *Die Ketene*, Verlag von Ferdinand Enke, Stuttgart, 1912.
[5] Ghosez, L.; Montaigne, R.; Mollet, P. *Tetrahedron Lett.* **1966**, 135.
[6] Stevens, H. C.; Reich, D. A.; Brandt, D. R.; Fountain, K. R.; Gaughan, E. J. *J. Am. Chem. Soc.* **1965**, *87*, 5257.
[7] Brady, W. T.; Liddell, H. G.; Vaughn, W. L. *J. Org. Chem.* **1966**, *31*, 626.
[8] Woodward, R. B.; Hoffmann, R. *The Conservation of Orbital Symmetry*, Academic, New York, 1971.
[9] Luknitskii, F. I.; Vovsi, B. A. *Russ. Chem. Rev. (Engl. Transl.)* **1969**, *38*, 487.
[10] Seikaly, H. R.; Tidwell, T. T. *Tetrahedron* **1986**, *42*, 2587.
[11] Scheeren, J. W. *Recl. Trav. Chim. Pays-Bas* **1986**, *105*, 71.
[12] Holder, R. W. *J. Chem. Educ.* **1976**, *53*, 81.
[13] Hasek, R. H. in *Kirk-Othmer Encyclopedia of Chemical Technology*, 3rd ed., Grayson, M., Eckroth, D., Eds., Vol. 13, Wiley, New York, 1981, p. 874.
[14] Roberts, J. D.; Sharts, C. M. *Org. React.* **1962**, *12*, 1.
[15] Miller, R.; Claudio, A.; Adel, S. in *Ullmann's Encyclopedia of Industrial Chemistry*, Elvers, B., Hawkins, S., Schulz, G., Eds., Vol. A15, VCH Verlagsgesellschaft, Weinheim, Germany, 1990, p. 63.
[16] Ziegler, E. *Chimia* **1970**, *24*, 62.
[17] Quadbeck, G. *Angew. Chem.* **1956**, *68*, 361.
[18] Ulrich, H. *Cycloaddition Reactions of Heterocumulenes*, Academic, New York, 1967.
[19] Brady, W. T. *Tetrahedron* **1981**, *37*, 2949.
[20] Brady, W. T. *Synthesis* **1971**, 415.
[21] Snyder, E. I. *J. Org. Chem.* **1970**, *35*, 4287.
[22] Cheburkov, Y. A.; Knunyants, I. L. *Fluorine Chem. Rev.* **1967**, *1*, 107.
[23] Moore, H. W.; Gheorghiu, M. D. *Chem. Soc. Rev.* **1981**, 289.
[24] Reisig, H. *Nachr. Chem. Tech. Lab.* **1986**, *34*, 880.
[25] Snider, B. B. *Chem. Rev.* **1988**, *88*, 793.
[26] Moore, H. W.; Decker, O. H. *Chem. Rev.* **1986**, *86*, 821.
[27] Manhas, M. S.; Bose, A. K. *Synthesis of Penicillin, Cephalosporin C. and Analogs* Marcel Dekker, New York, 1969.
[28] Morin, R. B.; Gorman, M. *Chemistry and Biology of beta-Lactam Antibiotics*, Vols. 1 and 2, Academic, New York, 1982.
[29] Schaumann, E.; Kettcham, R. *Angew. Chem., Int. Ed. Engl.* **1982**, *21*, 225.
[30] Ranganathan, S.; Ranganathan, D.; Mehrotra, A. K. *Synthesis* **1977**, 289.
[31] Houk, K. N. *Acc. Chem. Res.* **1975**, *8*, 361.
[32] Gompper, R. *Angew. Chem., Int. Ed. Engl.* **1969**, *8*, 312.
[33] Belluš, D.; Ernst, B. *Angew. Chem., Int. Ed. Engl.* **1988**, *27*, 797.
[34] Wong, H. N. C.; Lau, K.; Tam, K. in *Topics in Current Chemistry*, Vol. 133, Springer-Verlag, Berlin, 1986, p. 83.
[35] Trost, B. M. in *Topics in Current Chemistry*, Vol. 133, Springer-Verlag, Berlin, 1986, p. 3.
[36] Huisgen, R.; Otto, P. *Tetrahedron Lett.* **1968**, 4491.
[37] Isaacs, N. S.; Stanbury, P. F. *J. Chem. Soc., Chem. Commun.* **1970**, 1061.
[38] Rey, M.; Roberts, S.; Dieffenbacher, A.; Dreiding, A. S. *Helv. Chim. Acta* **1970**, *53*, 417.
[39] Taylor, E. C.; McKillop, A.; Hawks, H. H. *Org. Synth. Coll. Vol. VI*, Wiley, New York, 1988, p. 549.
[40] Al-Husaini, A. H.; Moore, H. W. *J. Org. Chem.* **1985**, *50*, 2595.
[41] Moore, H. W.; Wilbur, D. S. *J. Am. Chem. Soc.* **1978**, *100*, 6523.
[42] Huisgen, R.; Mayr, H. *Tetrahedron Lett.* **1975**, 2969.
[43] Moore, H. W.; Wilbur, D. S. *J. Org. Chem.* **1980**, *45*, 4483.

[44] Wagner, H. U.; Gompper, R. *Tetrahedron Lett.* **1970**, 2819.
[45] Huisgen, R.; Mayr, H. *Tetrahedron Lett.* **1975**, 2965.
[46] Rey, M.; Roberts, S. M.; Dreiding, A. S.; Roussel, A.; Vanlierde, H.; Toppet, S.; Ghosez, L. *Helv. Chim. Acta* **1982**, *65,* 703.
[47] Huisgen, R.; Feiler, L. A.; Binsch, G. *Chem. Ber.* **1969**, *102,* 3460.
[48] Huisgen, R.; Feiler, L. A.; Otto, P. *Tetrahedron Lett.* **1968**, 4485.
[49] Huisgen, R.; Otto, P. *J. Am. Chem. Soc.* **1968**, *90,* 5342.
[50] Baldwin, J. E.; Kapecki, J. A. *J. Am. Chem. Soc.* **1969**, *91,* 3106.
[51] Baldwin, J. E.; Kapecki, J. A. *J. Am. Chem. Soc.* **1970**, *92,* 4874.
[52] Holder, R. W.; Graf, N. A.; Duesler, E.; Moss, J. C. *J. Am. Chem. Soc.* **1983**, *105,* 2929.
[53] Katz, T. J.; Dessau, R. *J. Am. Chem. Soc.* **1963**, *85,* 2172.
[54] Collins, C. J.; Benjamin, B. M.; Kabalka, G. W. *J. Am. Chem. Soc.* **1978**, *100,* 2570.
[55] Becker, D.; Brodsky, N. C. *J. Chem. Soc., Chem. Commun.* **1978**, 237.
[56] Hanford, W. E.; Sauer, J. C. *Org. React.* **1946**, *3,* 108.
[57] Montaigne, R.; Ghosez, L. *Angew. Chem., Int. Ed. Engl.* **1968**, *7,* 221.
[58] Kende, A. S. *Tetrahedron Lett.* **1967**, 2661.
[59] Tenud, L.; Weilenmann, M.; Dallwigk, E. *Helv. Chim. Acta* **1977**, *60,* 975.
[60] Krebs, A.; Kimling, H. *Justus Liebigs, Ann. Chem.* **1974**, 2074.
[61] Das, H.; Kooyman, E. C. *Recl. Trav. Chim. Pays-Bas* **1965**, *84,* 965.
[62] Pregaglia, G. F.; Mazzanti, G.; Binaghi, M. *Makromol. Chem.* **1966**, *48,* 234.
[63] Ashworth, G.; Berry, D.; Smith, D. C. C. *J. Chem. Soc., Perkin Trans. 1* **1979**, 2995.
[64] Berchtold, G. A.; Harvey, G. R.; Wilson, G. E. *J. Org. Chem.* **1961**, *26,* 4776.
[65] England, D. C.; Krespan, C. G. *J. Am. Chem. Soc.* **1966**, *88,* 5582.
[66] Cheburkov, Y. A.; Aronov, Y. E.; Mirzabekyants, N. S.; Knunyants, I. L. *Bull. Acad. Sci USSR, Div. Chem. Sci.* **1966**, 745 [*C.A.* **1966**, *65,* 8755e].
[67] Cheburkov, Y. A.; Bargamova, M. D.; Knunyants, I. L. *Bull. Acad. Sci USSR, Div. Chem. Sci.* **1967**, 2052 [*C.A.* **1966**, *64,* 17411b].
[68] Moore, H. W.; Duncan, W. G. *J. Org. Chem.* **1973**, *38,* 156.
[69] Newman, M. S.; Zuech, E. A. *J. Org. Chem.* **1962**, *27,* 1436.
[70] Ruden, R. A. *J. Org. Chem.* **1974**, *39,* 3607.
[71] Farnum, D. G.; Johnson, J. R.; Hess, R. E.; Marshall, T. B.; Webster, B. *J. Am. Chem. Soc.* **1965**, *87,* 5191.
[72] Wasserman, H. H.; Hearn, M. J.; Cochoy, R. E. *J. Org. Chem.* **1980**, *45,* 2874.
[73] Brady, W. T.; Ting, P. L. *J. Org. Chem.* **1975**, *40,* 3417.
[74] Andreades, S.; Carlson, H. D. *Org. Synth. Coll. Vol. V,* **1973**, 679.
[75] McMurry, J. E.; Miller, D. D. *Tetrahedron Lett.* **1983**, *24,* 1885.
[76] Erden, I.; de Meijere, A. *Tetrahedron Lett.* **1983**, *24,* 3811.
[77] Becker, K. B.; Labhart, M. P. *Helv. Chim. Acta* **1983**, *66,* 1090.
[78] Donskaya, N. A.; Bessmertnykh, A. G.; Drobysh, V. A.; Shabarov, Y. S. *Zh. Org. Khim.* **1987**, *23,* 745 [*C.A.* **1988**, *108,* 74843x].
[79] Aue, D. H.; Helwig, G. S. *J. Chem. Soc., Chem. Commun.* **1975**, 604.
[80] Baldwin, J. E.; Johnson, D. S. *J. Org. Chem.* **1975**, *38,* 2147.
[81] Huisgen, R.; Koppitz, P. *Chem.-Ztg.* **1974**, *98,* 461 [*C.A.* **1975**, *82,* 43079t].
[82] Deprés, J. P.; Coelho, F.; Greene, A. E. *J. Org. Chem.* **1985**, *50,* 1972.
[83] Greene, A. E.; Deprés, J. P. *J. Am. Chem. Soc.* **1979**, *101,* 4003.
[84] England, D. C.; Krespan, C. G. *J. Org. Chem.* **1970**, *35,* 3300.
[85] Cheburkov, Y. A.; Mukhamadaliev, N.; Knunyants, I. L. *Bull. Acad. Sci USSR, Div. Chem. Sci.* **1967**, 361 [*C.A.* **1968**, *68,* 59149n].
[86] Clennan, E. L.; Lewis, K. K. *J. Am. Chem. Soc.* **1987**, *109,* 2475.
[87] Vogel, E.; Müller, K. *Justus Liebigs Ann. Chem.* **1958**, *615,* 29.
[88] Corey, E. J.; Ravindranathan, T. *Tetrahedron Lett.* **1971**, 4753.
[89] Corey, E. J.; Arnold, Z.; Hutton, J. *Tetrahedron Lett.* **1970**, 307.
[90] Corey, E. J.; Nicolaou, K. C.; Beames, D. J. *Tetrahedron Lett.* **1974**, 2439.
[91] Corey, E. J.; Mann, J. *J. Am. Chem. Soc.* **1973**, *95,* 6832.

[92] Stadler, H.; Rey, M.; Dreiding, A. S. *Helv. Chim. Acta* **1984**, *67,* 1854.
[93] Bertrand, M.; Maurin, R.; Gras, J. L.; Gil, G. *Tetrahedron* **1975**, *31,* 849.
[94] Duncan, W. G.; Weyler, W.; Moore, H. W. *Tetrahedron Lett.* **1973**, 4391.
[95] Bampfield, H. A.; Brook, P. R.; McDonald, W. S. *J. Chem. Soc., Chem. Commun.* **1975**, 132.
[96] Hünig, S.; Buysch, H.; Hoch, H.; Lendle, W. *Chem. Ber.* **1967**, *100,* 3996.
[97] Berge, J. M.; Rey, M.; Dreiding, A. S. *Helv. Chim. Acta* **1982**, *65,* 2230.
[98] Berchtold, G. A.; Harvey, G. R.; Wilson, G. E. *J. Org. Chem.* **1965**, *30,* 2642.
[99] Feiler, L. A.; Huisgen, R. *Chem. Ber.* **1969**, *102,* 3428.
[100] Martin, J. C. U.S. Patent 3,328,392, 1967.
[101] Aben, R. W.; Scheeren, H. W. *J. Chem. Soc., Perkin Trans. 1* **1980**, 3132.
[102] Raynolds, P. W.; DeLoach, J. A. *J. Am. Chem. Soc.* **1984**, *106,* 4566.
[103] Huisgen, R.; Feiler, L. A.; Otto, P. *Chem. Ber.* **1969**, *102,* 3405.
[104] Brady, W. T.; Lloyd, R. M. *J. Org. Chem.* **1979**, *44,* 2560.
[105] Scarpati, R.; Sica, D. *Gazz. Chim. Ital.* **1967**, *97,* 1073 [*C.A.* **1967**, *67,* 63990b].
[106] Hoffmann, R. W.; Bressel, U.; Gehlhaus, J.; Häuser, H. *Chem. Ber.* **1971**, *104,* 873.
[107] Snider, B. B.; Allentoff, A. J.; Kulkarni, Y. S. *J. Org. Chem.* **1988**, *53,* 5320.
[108] Lee, S. Y.; Kulkarni, Y. S.; Burbaum, B. W.; Johnston, M. I.; Snider, B. B. *J. Org. Chem.* **1988**, *53,* 1848.
[109] Kulkarni, Y. S.; Niwa, M.; Ron, E.; Snider, B. B. *J. Org. Chem.* **1987**, *52,* 1568.
[110] Snider, B. B.; Ron, E.; Burbaum, B. W. *J. Org. Chem.* **1987**, *52,* 5413.
[111] Hassner, A.; Dillon, J. *Synthesis* **1979**, 689.
[112] Danheiser, R. L.; Savariar, S. *Tetrahedron Lett.* **1987**, *28,* 3299.
[113] Hassner, A.; Dilon, J. L. *J. Org. Chem.* **1983**, *48,* 3382.
[114] Fishbein, P. L.; Moore, H. W. *J. Org. Chem.* **1985**, *50,* 3226.
[115] Hasek, R. H.; Gott, P. G.; Martin, J. C. *J. Org. Chem.* **1964**, *29,* 2510.
[116] Ammann, A.; Rey, M.; Dreiding, A. S. *Helv. Chim. Acta* **1987**, *70,* 321.
[117] Chow, K.; Moore, H. W. *Tetrahedron Lett.* **1987**, *28,* 5013.
[118] Smith, L. I.; Hoehn, H. H. *J. Am. Chem. Soc.* **1939**, *61,* 2619.
[119] Smith, L. I.; Hoehn, H. H. *J. Am. Chem. Soc.* **1941**, *63,* 1180.
[120] Danheiser, R. L.; Sard, H. *Tetrahedron Lett.* **1983**, *24,* 23.
[121] Zaitseva, G. S.; Livantsova, L. I.; Baukov, Y. I.; Lutsenko, I. F. *Zh. Obshch. Khim.* **1984**, *54,* 1323 [*C.A.* **1984**, *101,* 171389s].
[122] Hong, P.; Sonogashira, K.; Hagihara, N. *J. Organomet. Chem.* **1981**, *219,* 363.
[123] Martin, J. C. British UK Patent GB 1,049,326 [*C.A.* **1967**, *66,* 29592t].
[124] Martin, J. C. U.S. Patent 3,288,854 [*C.A.* **1967**, *66,* 37502d].
[125] Hasek, R. H.; Martin, J. C. *J. Org. Chem.* **1962**, *27,* 3743.
[126] Wasserman, H. H.; Dehmlow, E. V. *Tetrahedron Lett.* **1962**, 1031.
[127] Wasserman, H. H.; Piper, J. U.; Dehmlow, E. V. *J. Org. Chem.* **1973**, *38,* 1451.
[128] McCarney, C. C.; Ward, R. S.; Roberts, D. W. *Tetrahedron* **1976**, *32,* 1189.
[129] Nieuwenhuis, J.; Arens, J. F. *Recl. Trav. Chim. Pays-Bas* **1958**, *77,* 761.
[130] Pericas, M.; Serratosa, F. *Tetrahedron Lett.* **1977**, 4437.
[131] Danheiser, R. L.; Gee, S. K. *J. Org. Chem.* **1984**, *49,* 1672.
[132] Nieuwenhuis, J.; Arens, J. F. *Recl. Trav. Chim. Pays-Bas* **1958**, *77,* 1153.
[133] England, D. C.; Krespan, C. G. *J. Org. Chem.* **1970**, *35,* 3312.
[134] Teufel, H.; Jenny, E. F. *Tetrahedron Lett.* **1971**, 1769.
[135] Barton, D. H.; Gardner, J. N.; Petterson, R. C.; Stamm, O. A. *Proc. Chem. Soc.* **1962**, 21 [*C.A.* **1962**, *57,* 685d].
[136] Jenny, E. F.; Schenker, K.; Woodward, R. B. *Angew. Chem.* **1961**, *73,* 756.
[137] Barton, D. H.; Gardner, J. N.; Petterson, R. C.; Stamm, O. A. *J. Chem. Soc.* **1962**, 2708.
[138] Druey, J.; Jenny, E. F.; Schenker, K.; Woodward, R. B. *Helv. Chim. Acta* **1962**, *45,* 600.
[139] Delaunois, M.; Ghosez, L. *Angew. Chem., Int. Ed. Engl.* **1969**, *8,* 72.
[140] Himbert, G. *Justus Liebigs Ann. Chem.* **1979**, 829.
[141] Himbert, G. *Justus Liebigs Ann. Chem.* **1979**, 1828.
[142] Himbert, G.; Henn, L. *Z. Naturforsch.* **1981**, *36b,* 218.

[143] Dotz, K. H.; Muhlemeier, J.; Trenkle, B. *J. Organomet. Chem.* **1985**, *289,* 257.

[144] Dotz, K. H.; Trenkle, B.; Schubert, U. *Angew. Chem., Int. Ed. Engl.* **1981**, *20,* 287.

[145] Grzegorzewska, U.; Leplawy, M.; Redlinski, A. *Rocz. Chem.* **1975**, 1859 [*C.A.* **1976**, *84,* 135059f].

[146] Knunyants, I. L.; Cheburkov, Y. A. *Izv. Akad. Nauk SSSR, Ser. Khim* **1960**, 678 [*C.A.* **1960**, *54,* 22349c].

[147] Zaitseva, G. S.; Vasil'eva, L. I.; Vinokurova, N. G.; Safronova, O. A.; Baukov, Y. I. *Zh. Obshch. Khim.* **1978**, *48,* 1363 [*C.A.* **1978**, *89,* 146998r].

[148] England, D. C.; Krespan, C. G. *J. Am. Chem. Soc.* **1965**, *87,* 4019.

[149] Nations, R. G.; Hasek, R. H. U.S. Patent 3,221,028 [*C.A.* **1966**, *64,* 4950e].

[150] Brady, W. T.; Smith, L. *Tetrahedron Lett.* **1970**, 2963.

[151] Brady, W. T.; Patel, A. D. *J. Heterocycl. Chem.* **1971**, *8,* 739.

[152] Staudinger, H.; Bereza, S. *Justus Liebigs Ann. Chem.* **1911**, *380,* 243.

[153] Ogino, K.; Matsumoto, T.; Kozuka, S. *J. Chem. Soc., Chem. Commun.* **1979**, 643.

[154] Burpitt, R. D.; Brannock, K. C.; Nations, R. G.; Martin, J. C. *J. Org. Chem.* **1971**, *36,* 2222.

[155] Arjona, O.; de la Pradilla, R.; Perez, S.; Plumet, J.; Carrupt, P.; Vogel, P. *Tetrahedron Lett.* **1986**, *27,* 5505.

[156] Metzger, C.; Borrmann, D.; Wegler, R. *Chem. Ber.* **1967**, *100,* 1817.

[157] Turro, N. J.; Edelson, S.; Gagosian, R. *J. Org. Chem.* **1970**, *35,* 2058.

[158] Noels, A. F.; Herman, J. J.; Tyssie, P. *J. Org. Chem.* **1976**, *41,* 2527.

[159] Sheehan, J. J.; Buhle, E. L.; Corey, E. J.; Laubach, G. D.; Ryan, J. J. *J. Am. Chem. Soc.* **1950**, *72,* 3828.

[160] Williams, P. H.; Payne, G. B.; Sullivan, W. J.; Van Ess, P. R. *J. Am. Chem. Soc.* **1960**, *82,* 4883.

[161] Krabbenhoft, H. O. *J. Org. Chem.* **1978**, *43,* 1305.

[162] Pond, D. A. U.S. Patent 3,862,988, 1975 [*C.A.* **1976**, *84,* 43304d].

[163] Jowitt, H. Brit. Patent UK 885,217, 1961 [*C.A.* **1962**, *57,* 4550a].

[164] Probst, O.; Fernholz, H.; Mundlos, E.; Oheme, H.; Orth, A. U.S. Patent 3,113,149, 1963.

[165] Chitwood, J. L.; Gott, P. G.; Krutak, J. J.; Martin, J. C. *J. Org. Chem.* **1971**, *36,* 2216.

[166] Brady, W. T.; Smith, L. *J. Org. Chem.* **1971**, *36,* 1637.

[167] Zaitseva, G. S.; Vinokurova, N. G.; Baukov, Y. I. *Zh. Obshch. Khim.* **1975**, *45,* 1398 [*C.A.* **1975**, *83,* 114548d].

[168] Borrmann, D.; Wegler, R. *Chem. Ber.* **1966**, *99,* 1245.

[169] Borrmann, D.; Wegler, R. *Chem. Ber.* **1967**, *100,* 1575.

[170] Wynberg, H.; Staring, E. J. *J. Am. Chem. Soc.* **1982**, *104,* 166.

[171] Wynberg, H.; Staring, E. J. *J. Org. Chem.* **1985**, *50,* 1977.

[172] Staring, E. J.; Moorlag, H.; Wynberg, H. *Recl. Trav. Chim. Pays-Bas* **1986**, *105,* 374.

[173] Ketelaar, P. F.; Staring, E. J.; Wynberg, H. *Tetrahedron Lett.* **1985**, *26,* 4665.

[174] Wynberg, H.; Staring, E. G. J. PCT Int. Appl WO 84 01,577, 1984 [*C.A.* **1984**, *101,* 230848r].

[175] Lavallee, C.; Lemay, G.; Leborgne, A.; Spassky, N.; Prudhomme, R. *Macromolecules* **1984**, *17,* 2457.

[176] Voyer, R.; Prudhomme, R. *J. Poly. Sci.: Part A: Polym. Chem. Ed.* **1986**, *24,* 2773.

[177] Coyle, J. D.; Rapley, P. A.; Kamphuis, J.; Bos, H. J. *J. Chem. Soc., Perkin Trans. 1* **1985**, 1957.

[178] Kohn, H.; Charumilind, P.; Gopichand, Y. *J. Org. Chem.* **1978**, *43,* 4961.

[179] Drozd, V. N.; Popova, O. A. *Zh. Obshch. Khim.* **1979**, *15,* 2602 [*C.A.* **1980**, *92,* 163777x].

[180] Takasu, I.; Higuchi, M.; Suzuki, R. Jpn. Kokai JP 73 12,736.

[181] Vanderkooi, N.; Huthwaite, H. J. U.S. Patent 3,894,097, 1975 [*C.A.* **1975**, *83,* 148015t].

[182] Moore, H. W.; Mercer, F.; Kunert, D.; Albaugh, P. *J. Am. Chem. Soc.* **1979**, *101,* 5435.

[183] Morita, N.; Asao, T.; Ojima, J.; Wada, K. *Chem. Lett.* **1981**, 57.

[184] Neidlein, R.; Humburg, G. *Justus Liebigs Ann. Chem.* **1978**, 1974.

[185] Brady, W. T.; Giang, Y. F., Marchand, A. P.; Wu, A. *J. Org. Chem.* **1987**, *52,* 3457.

[186] Rioult, P.; Vialle, J. *Bull. Soc. Chim. Fr.* **1967**, 2883.

[187] Asao, T. *Synth. Commun.* **1972**, *2,* 353.

[188] Martin, J. C.; Burpitt, R. D.; Gott, P. G.; Harris, M.; Meen, R. H. *J. Org. Chem.* **1971**, *36,* 2205.

[189] Farbwerke Hoechst, British UK Pat. GB 939,844, 1963.
[190] Mundlos, E.; Graf, R. *Justus Liebigs Ann. Chem.* **1964**, *677*, 108.
[191] Goldish, D. M.; Axon, B. W.; Moore, H. W. *Heterocycles* **1983**, *20*, 187.
[192] Brady, W. T.; Schieh, C. H. *J. Heterocycl. Chem.* **1985**, *22*, 357.
[193] Schaumann, E.; Behr, H.; Adiwidjaja, G. *Justus Liebigs Ann. Chem.* **1979**, 1322.
[194] Belzecki, C.; Krawczyk, Z. *J. Chem. Soc., Chem. Commun.* **1977**, 302.
[195] Jäger, U.; Schwab, M.; Sundermeyer, W. *Chem. Ber.* **1986**, *119*, 1127.
[196] Beecken, H.; Korte, F. *Tetrahedron* **1962**, *18*, 1527.
[197] Semple, J. E.; Joullie, M. M. *J. Org. Chem.* **1978**, *43*, 3066.
[198] Neidlein, R.; Lehr, W. *Heterocycles* **1981**, *16*, 1187.
[199] Kerber, R. C.; Cann, M. C. *J. Org. Chem.* **1974**, *39*, 2552.
[200] Kresze, G.; Trede, A. *Tetrahedron* **1963**, *19*, 133.
[201] Hall, J. H.; Kellogg, R. M. *J. Org. Chem.* **31**, *1966*, 1079.
[202] Ried, W.; Piesch, S. *Chem. Ber.* **1966**, *99*, 233.
[203] Fahr, E.; Fischer, W.; Jung, A.; Sauer, L. *Tetrahedron Lett.* **1967**, 161.
[204] Ingold, C. K.; Weaver, S. D. *J. Chem. Soc.* **1925**, *127*, 378.
[205] Sommer, S. *Angew. Chem.* **1976**, *88*, 449.
[206] Ried, W.; Dietrich, R. *Justus Liebigs Ann. Chem.* **1963**, *666*, 113.
[207] Staudinger, H. *Ber.* **1911**, *44*, 1619.
[208] Texier, F.; Carrié, R.; Jaz, J. *J. Chem. Soc., Chem. Commun.* **1972**, 199.
[209] Abou-Gharbia, M. A.; Joullie, M. M. *Heterocycles* **1978**, *9*, 457.
[210] Singal, K. K. *Chem. Ind. (London)* **1985**, 761.
[211] Scarpati, R.; Sorrentino, P. *Gazz. Chim. Ital.* **1959**, *89*, 1525 [*C.A.* **1960**, *54*, 22570d].
[212] Evans, A. R.; Taylor, G. A. *J. Chem. Soc., Perkin Trans. 1* **1983**, 979.
[213] Alsop, S. H.; Barr, J. J.; Storr, R. C.; Cameron, A. F.; Freer, A. A. *J. Chem. Soc., Chem. Commun.* **1976**, 888.
[214] van Tilborg, W. J. M.; Steinberg, H.; de Boer, T. J. *Synth. Commun.* **1973**, *3*, 189.
[215] Rothgery, E. F.; Holt, R. J.; McGee, H. A. *J. Am. Chem. Soc.* **1975**, *97*, 4971.
[216] Turro, N. J.; Hammond, W. B. *J. Am. Chem. Soc.* **1966**, *88*, 3672.
[217] Schaafsma, S. E.; Steinberg, H.; De Boer, T. J. *Recl. Trav. Chim. Pays-Bas* **1966**, *85*, 1170.
[218] Hammond, W. B.; Turro, N. J. *J. Am. Chem. Soc.* **1966**, *88*, 2880.
[219] Saitseva, G. S.; Bogdanova, G. S.; Baukov, Y. I.; Lutsenko, I. F. *Zh. Obshch. Khim.* **1978**, *48*, 131 [*C.A.* **1978**, *88*, 170250r].
[220] Zaitseva, G. S.; Bogdanova, G. S.; Baukov, Y. I.; Lutsenko, I. F. *J. Organomet. Chem.* **1976**, *121*, C21.
[221] Ter-Gabrielyan, E. G.; Avetisyan, E. A.; Gambaryan, N. P. *Izv. Akad. Nauk SSSR, Ser. Khim.* **1973**, 2562 [*C.A.* **1974**, *80*, 59886y].
[222] Ried, W.; Kraemer, R. *Justus Liebigs Ann. Chem.* **1973**, 1952.
[223] Yates, P.; Clark, T. J. *Tetrahedron Lett.* **1961**, 435.
[224] Toppet, S.; L'abbé, G.; Smets, G. *Chem. Ind. (London)* **1973**, 1110.
[225] Ikeda, K.; L'abbé, G.; Smets, G. *Chem. Ind. (London)* **1973**, 327.
[226] Tsuge, O.; Urano, S.; Oe, K. *J. Org. Chem.* **1980**, *45*, 5130.
[227] Moriarty, R. M.; White, K. B.; Chin, A. *J. Am. Chem. Soc.* **1978**, *100*, 5582.
[228] Hatakeyama, S.; Honda, S.; Akimoto, H. *Bull. Chem. Soc. Jpn.* **1985**, *58*, 2411.
[229] Wheland, R.; Bartlett, P. D. *J. Am. Chem. Soc.* **1970**, *92*, 6057.
[230] Dennis, N.; Katritzky, A. R.; Sabounji, G. J. *Tetrahedron Lett.* **1976**, 2959.
[231] Cheburkov, Y. A.; Mukhamadaliev, N.; Knunyants, I. L. *Tetrahedron* **1968**, *24*, 1341.
[232] Cheburkov, Y. A.; Mukhamadaliev, N.; Knunyants, I. L. *Izv. Akad. Nauk SSSR, Ser. Khim.* **1966**, 383 [*C.A.* **1966**, *64*, 17411d].
[233] Falshaw, C.; Lakoues, A.; Taylor, G. *J. Chem. Res. (S)* **1985**, 106.
[234] Mayr, H.; Heigl, U. W. *J. Chem. Soc., Chem. Commun.* **1987**, 1804.
[235] Brady, W. T.; Agho, M. O. *J. Heterocycl. Chem.* **1983**, *20*, 501.
[236] Katritzky, A. R.; Cutler, A. T.; Dennis, N.; Sabongi, G. J.; Rahimi-Rastgoo, S.; Fischer, G. W.; Fletcher, I. J. *J. Chem. Soc., Perkin Trans. 1* **1980**, 1176.

[237] Gouesnard, J. P. *Tetrahedron* **1974**, *30*, 3117.
[238] Brady, W. T.; Agho, M. O. *Synthesis* **1982**, 500.
[239] Jaworski, T.; Kwiatkowski, S. *Roczniki Chemii* **1975**, *49*, 63 [*C.A.* **1975**, *83*, 58593a].
[240] Tanimoto, S.; Matsumura, Y.; Sugimoto, T.; Okano, M. *Tetrahedron Lett.* **1977**, 2899.
[241] Krishan, K.; Singh, A.; Singh, B.; Kumar, S. *Synth. Commun.* **1984**, *14*, 219.
[242] Duran, F.; Ghosez, L. *Tetrahedron Lett.* **1970**, 245.
[243] Sakamoto, M.; Miyazawa, K.; Kuwabara, K.; Tomimatsu, Y. *Heterocycles* **1979**, *12*, 231.
[244] Fitton, A. O.; Frost, J. R.; Houghton, P. G.; Suschitzky, H. *J. Chem. Soc., Perkin Trans. 1* **1977**, 1450.
[245] Luthardt, P.; Würthwein, E. *Tetrahedron Lett.* **1988**, *29*, 921.
[246] Katagiri, N.; Kato, T.; Niwa, R. *J. Heterocycl. Chem.* **1984**, *21*, 407.
[247] Satsumabayashi, S.; Motoki, S.; Nakano, H. *J. Org. Chem.* **1976**, *41*, 156.
[248] Sakamoto, M.; Miyazawa, K.; Tomimatsu, Y. *Chem. Pharm. Bull.* **1976**, *24*, 2532.
[249] Goerdeler, J.; Lohmann, H. *Chem. Ber.* **1977**, *110*, 2996.
[250] Ried, W.; Radt, W. *Justus Liebigs Ann. Chem.* **1965**, *688*, 170.
[251] Friedrichsen, W.; Oeser, H. *Chem. Ber.* **1975**, *108*, 31.
[252] Staudinger, H.; Endle, R. *Justus Liebigs Ann. Chem.* **1913**, *401*, 263.
[253] Campbell, N.; Davison, P. S.; Heller, H. G. *J. Chem. Soc.* **1963**, 993.
[254] Campbell, N.; Davison, P. S. *Chem. Ind. (London)* **1962**, 2012.
[255] Mosti, L.; Menozzi, G.; Bignardi, G.; Schenone, P. *Il Farmaco (Ed. Sci.)* **1977**, *32*, 794 [*C.A.* **1978**, *88*, 62262n].
[256] Eiden, F.; Breugst, I. *Chem. Ber.* **1979**, *112*, 1791.
[257] Rokhlin, E. M.; Gambaryan, N. P.; Knunyants, I. L. *Izv. Akad. Nauk SSSR, Ser. Khim.* **1963**, 1952 [*C.A.* **1964**, *60*, 7454a].
[258] Mosti, L.; Schenone, P.; Menozzi, G. *J. Heterocycl. Chem.* **1978**, *15*, 181.
[259] Opitz, G.; Zimmerman, F. *Chem. Ber.* **1964**, *97*, 1266.
[260] Meslin, J. C.; N'Guessan, Y. T.; Quiniou, H.; Tonnard, F. *Tetrahedron* **1975**, *31*, 2679.
[261] Meslin, J. C.; Quiniou, H. *Bull. Soc. Chim. Fr.* **1972**, 2517.
[262] Bachi, M. D.; Rothfield, M. *J. Chem. Soc., Perkin Trans. 1* **1972**, 2326.
[263] Goerdeler, J.; Schimpf, R.; Tiedt, M. *Chem. Ber.* **1972**, *105*, 3322.
[264] Goerdeler, J.; Hohage, H. *Chem. Ber.* **1973**, *106*, 1487.
[265] Katrizky, A. R.; Cutler, A. T.; Dennis, N.; Rahimi-Rastgoo, S.; Sabongi, G. J. *Z. Chem.* **1979**, *19*, 20.
[266] Kniezo, L.; Kristian, P.; Imrich, J.; Ugozzoli, F.; Andreetti, G. *Tetrahedron* **1988**, *44*, 543.
[267] Fahr, E. *Angew. Chem.* **1964**, *76*, 505.
[268] Sommer, S. *Angew. Chem.* **1977**, *89*, 59.
[269] Meslin, J. C.; Quiniou, H. *Synthesis* **1974**, 298.
[270] Ooka, M.; Kitamura, A.; Okazaki, R.; Inamoto, N. *Bull. Chem. Soc. Jpn.* **1978**, 51, 301.
[271] Okazaki, R.; Ooka, M.; Inamoto, N. *J. Chem. Soc., Chem. Commun.* **1976**, 562.
[272] Gotthardt, H.; Schenk, K. *Angew. Chem.* **1985**, *97*, 604.
[273] Danheiser, R. L.; Sard, H. *J. Org. Chem.* **1980**, *45*, 4810.
[274] Jäger, G.; Wenzelburger, J. *Justus Liebigs Ann. Chem.* **1976**, 1689.
[275] Hünig, S.; Benzing, E.; Hubner, K. *Chem. Ber.* **1961**, *94*, 486.
[276] Berge, J. M.; Rey, M.; Dreiding, A. S. *Helv. Chim. Acta* **1982**, *65*, 2230.
[277] Crabtree, H. E.; Smalley, R. K.; Suschitzky, H. *J. Chem. Soc. (C)* **1968**, 2730.
[278] Ficini, J.; Pouliquen, J. *Tetrahedron Lett.* **1972**, 1135.
[279] Himbert, G.; Henn, L. *Justus Liebigs Ann. Chem.* **1987**, 381.
[280] Jäger, G. *Chem. Ber.* **1972**, *105*, 137.
[281] Stetter, H.; Kiehs, K. *Chem. Ber.* **1965**, *98*, 2099.
[282] Hyatt, J. A.; Feldman, P. L.; Clemens, R. J. *J. Org. Chem.* **1984**, *49*, 5105.
[283] Ziegler, E.; Kleineberg, G.; Meindl, H. *Monatsh. Chem.* **1966**, *97*, 10.
[284] Selvarajan, R.; Narasimhan, K.; Swaminathan, S. *Tetrahedron Lett.* **1967**, 2089.
[285] Goerdeler, J.; Köhler, H. *Tetrahedron Lett.* **1976**, 2961.
[286] Capuano, L.; Urhahn, G.; Willmes, A. *Chem. Ber.* **1979**, *112*, 1012.

[287] Minami, T.; Yamauchi, Y.; Ohshiro, Y.; Murai, S.; Sonoda, N. *J. Chem. Soc., Perkin Trans. 1* **1977**, 904.
[288] Day, A. C.; McDonald, A. N.; Anderson, B. F.; Bartczak, T. J.; Hodder, O. J. *J. Chem. Soc., Chem. Commun.* **1973**, 247.
[289] Rees, C. W.; Somanathan, R.; Storr, R. C.; Woolhouse, A. D. *J. Chem. Soc., Chem. Commun.* **1976**, 125.
[290] Stetter, H.; Kiehs, K. *Chem. Ber.* **1965**, *98*, 2099.
[291] Capuano, L.; Gärtner, K. *J. Heterocycl. Chem.* **1981**, *18*, 1341.
[292] Turro, N. J.; Chow, M.-F.; Ito, Y. *J. Am. Chem. Soc.* **1978**, *100*, 5580.
[293] Bestian, H.; Günther, D. *Angew. Chem., Int. Ed. Engl.* **1963**, *2*, 608.
[294] Frey, H. M.; Isaacs, N. S. *J. Chem. Soc. (B)* **1970**, 830.
[295] Bartlett, P. D.; McCluney, R. E. *J. Org. Chem.* **1983**, *48*, 4165.
[296] Williams, J. W.; Hurd, C. D. *J. Org. Chem.* **1940**, *5*, 122.
[297] Smith, C. W.; Norton, D. G. *Org. Synth. Coll. Vol. IV*, 348, 1963, Wiley, New York, 1963.
[298] Krepski, L. R.; Hassner, A. *J. Org. Chem.* **1978**, *43*, 2879.
[299] Johnston, B. D.; Czyzewska, E.; Oehlschlager, A. C. *J. Org. Chem.* **1987**, *52*, 3693.
[300] Weyler, W.; Duncan, W. G.; Liewen, M. B.; Moore, H. W. *Org. Synth.* **1976**, *55*, 32.
[301] Weyler, W.; Duncan, W. G.; Liewen, M. B.; Moore, H. W. *Org. Synth. Coll. Vol. VII*, 210, Wiley, New York, 1990.
[302] Arya, F.; Bouquant, J.; Chuche, J. *Tetrahedron Lett.* **1986**, *27*, 1913.
[303] Huber, U. A.; Dreiding, A. S. *Helv. Chim. Acta* **1970**, *53*, 495.
[304] Baxter, G. J.; Brown, R. F. C.; Eastwood, F. W.; Harrington, K. J. *Tetrahedron Lett.* **1975**, 4283.
[305] Danheiser, R. L.; Gee, S. K.; Sard, H. *J. Am. Chem. Soc.* **1982**, *104*, 7670.
[306] Hassner, A.; Naidorf, S. *Tetrahedron Lett.* **1986**, *27*, 6389.
[307] Wright, B. B. *J. Am. Chem. Soc.* **1988**, *110*, 4456.
[308] Ainsworth, C.; Chen, F.; Kuo, Y. *J. Organomet. Chem.* **1972**, *46*, 59.
[309] Ripoll, J. L. *Tetrahedron* **1977**, *33*, 389.
[310] Kunert, D. M.; Chambers, R.; Mercer, F.; Hernandez, L.; Moore, H. W. *Tetrahedron Lett.* **1978**, 929.
[311] Moore, H. W.; Hernandez, L.; Sing, A. *J. Am. Chem. Soc.* **1976**, *98*, 3728.
[312] Fishbein, P. L.; Moore, H. W. *J. Org. Chem.* **1984**, *49*, 2190.
[313] Yates, P.; Crawford, R. J. *J. Am. Chem. Soc.* **1966**, *88*, 1562.
[314] Horner, L.; Spietschka, E.; Gross, A. *Justus Liebigs Ann. Chem.* **1951**, *573*, 17.
[315] Becker, D.; Birnbaum, D. *J. Org. Chem.* **1980**, *45*, 578.
[316] Ciabattoni, J.; Anderson, H. *Tetrahedron Lett.* **1967**, 3377.
[317] Greene, A. E.; Lansard, J. P.; Luche, J. L.; Petrier, C. *J. Org. Chem.* **1984**, *49*, 931.
[318] Grieco, P. A. *J. Org. Chem.* **1972**, *37*, 2363.
[319] Nee, M.; Roberts, J. D. *J. Org. Chem.* **1981**, *46*, 67.
[320] Hasek, R. H.; Elam, E. U. Canadian Patent CA 618,772, 1962.
[321] Hasek, R. H.; Gott, P. G.; Martin, J. C. *J. Org. Chem.* **1964**, *29*, 1239.
[322] Brady, W. T.; Marchand, A. P.; Giang, Y. F.; Wu, A. *Synthesis* **1987**, 395.
[323] Hyatt, J. A. U.S. Patent 4,621,146 [*C.A.* **1987**, *106*, 119677y].
[324] Borrmann, D.; Wegler, R. *Chem. Ber.* **1969**, *102*, 64.
[325] Neidlein, R.; Cepera, K. *Chem. Ber.* **1978**, *111*, 1824.
[326] Kellogg, R. M. *J. Org. Chem.* **1973**, *38*, 844.
[327] Brady, W. T.; Agho, M. O. *J. Org. Chem.* **1983**, *48*, 5337.
[328] Brady, W. T.; Hoff, E. F., Jr. *J. Am. Chem. Soc.* **1968**, *90*, 6256.
[329] Grimme, W.; Schneider, E. *Angew. Chem.* **1977**, *89*, 754.
[330] Krügerke, T.; Buschmann, J.; Kleemann, G.; Luger, P.; Seppelt, K. *Angew. Chem., Int. Ed. Engl.* **1987**, *26*, 799.
[331] Sauer, J. C. *J. Am. Chem. Soc.* **1947**, *69*, 2444.
[332] Reid, E. B.; Groszos, S. J. *J. Am. Chem. Soc.* **1953**, *75*, 1655.
[333] Staudinger, H. *Ber.* **1911**, *44*, 533.

[334] McCarney, C. C.; Ward, R. S. *J. Chem. Soc., Perkin Trans. 1* **1975**, 1600.
[335] Enk, E.; Spes, H. *Angew. Chem.* **1961**, *73,* 334.
[336] England, D. C.; Krespan, C. G. *J. Org. Chem.* **1970**, *35,* 3322.
[337] Trahanovsky, W. S.; Serber, B. W.; Wilkes, M. C.; Preckel, M. M. *J. Am. Chem. Soc.* **1982**, *104,* 6779.
[338] Wear, R. L. *J. Am. Chem. Soc.* **1951**, *73,* 2390.
[339] Wedekind, E.; Weisswange, W. *Ber.* **1906**, *39,* 1631.
[340] Staudinger, H. *Justus Liebigs Ann. Chem.* **1907**, *356,* 51.
[341] Staudinger, H.; Schneider, H.; Schotz, P.; Strong, P. M. *Helv. Chim. Acta* **1923**, *6,* 291.
[342] Elam, E. U. *J. Org. Chem.* **1967**, *32,* 215.
[343] Payne, G. B. *J. Org. Chem.* **1966**, *31,* 718.
[344] Berkowitz, W. F.; Ozorio, A. A. *J. Org. Chem.* **1975**, *40,* 527.
[345] Erickson, J. L. E.; Collins, F. E.; Owen, B. L. *J. Org. Chem.* **1966**, *31,* 480.
[346] England, D. C. *J. Org. Chem.* **1981**, *46,* 147.
[347] Brady, W. T.; Ting, P. L. *J. Org. Chem.* **1976**, *41,* 2336.
[348] Staudinger, H.; Maier, J. *Justus Liebigs Ann. Chem.* **1913**, *401,* 292.
[349] Baldwin, J. E. *J. Org. Chem.* **1963**, *28,* 3112.
[350] Berkowitz, W. F.; Ozorio, A. A. *J. Org. Chem.* **1971**, *36,* 3787.
[351] Hoffmann, H. M. R.; Wulff, J. M.; Kütz, A.; Wartchow, R. *Angew. Chem. Suppl.* **1982**, 17.
[352] Brady, W. T.; Cheng, T. C. *J. Org. Chem.* **1977**, *42,* 732.
[353] Moore, H. W.; Weyler, W. *J. Am. Chem. Soc.* **1970**, *92,* 4132.
[354] Walborsky, H. M.; Buchman, E. R. *J. Am. Chem. Soc.* **1953**, *75,* 6339.
[355] Hill, C. M.; Senter, G. W. *J. Am. Chem. Soc.* **1949**, *71,* 364.
[356] Farnum, D. G.; Webster, B. *J. Am. Chem. Soc.* **1963**, *85,* 3502.
[357] Baldwin, J. E.; Roberts, J. D. *J. Am. Chem. Soc.* **1963**, *84,* 2444.
[358] Franke, W. K. R.; Ahne, H. *Angew. Makromol. Chem.* **1972**, *21,* 195.
[359] Mormann, W.; Hoffmann, S.; Hoffmann, W. *Chem. Ber.* **1987**, *120,* 285.
[360] Buchta, E.; Geibel, K. *Justus Liebigs Ann. Chem.* **1964**, *678,* 53.
[361] Baldwin, J. E. *J. Org. Chem.* **1964**, *29,* 1882.
[362] Buchta, E.; Fischer, M. *Ber.* **1966**, *99,* 1509.
[363] Dehmlow, E. V.; Slopianka, M.; Pickardt, J. *Justus Liebigs Ann. Chem.* **1979**, 572.
[364] Farnum, D. G.; Webster, B.; Wolf, A. D. *Tetrahedron Lett.* **1968**, 5003.
[365] Buchta, E.; Theuer, W. *Justus Liebigs Ann. Chem.* **1963**, *666,* 81.
[366] Bamberger, E.; Baum, M.; Schlein, L. *J. Prakt. Chem.* **1921**, *105,* 266.
[367] Karmarkar, S. S.; Sharma, M. M. *Indian J. Chem., Sect. A* **1970**, *8,* 413 [*C.A.* **1970**, *73,* 35148p].
[368] Buchta, E.; Merk, W. *Justus Liebigs Ann. Chem.* **1966**, *694,* 1.
[369] Corbellini, A.; Maggioni, P.; Pietta, P. G. *Gazz. Chim. Ital.* **1967**, *97,* 1126 [*C.A.* **1968**, *68,* 50047v].
[370] Strating, J.; Scharp, J.; Wynberg, H. *Recl. Trav. Chim. Pays-Bas* **1970**, *89,* 23.
[371] Baldwin, J. E. *J. Org. Chem.* **1964**, *29,* 1880.
[372] Staudinger, H.; Suter, E. *Ber.* **1912**, *53,* 1092.
[373] Erickson, J. L. E.; Kitchens, G. C. *J. Org. Chem.* **1962**, *27,* 460.
[374] Kimbrough, R. D., Jr. *J. Org. Chem.* **1964**, *29,* 1246.
[375] Anet, R. *Chem. Ind. (London)* **1961**, 1313.
[376] Dehmlow, S. S.; Dehmlow, E. V. *Justus Liebigs Ann. Chem.* **1975**, 209.
[377] Wentrup, C.; Winter, H.; Gross, G.; Netsch, K.; Kollenz, G.; Ott, W.; Biedermann, A. *Angew. Chem., Int. Ed. Engl.* **1984**, *23,* 800.
[378] Leuchs, H.; Wutke, J.; Gieseler, E. *Ber.* **1913**, *46,* 2200.
[379] Rothman, E. S. *J. Org. Chem.* **1967**, *32,* 1683.
[380] Chickos, J. S.; Sherwood, D. E.; Jug, K. *J. Org. Chem.* **1978**, *43,* 1146.
[381] Birum, G. H.; Matthews, C. N. *J. Am. Chem. Soc.* **1968**, *90,* 3842.
[382] Roberts, J. D.; Armstrong, R.; Trimble, R. F.; Burg, M. *J. Am. Chem. Soc.* **1949**, *71,* 843.
[383] Brady, W. T.; Ting, P. L. *Tetrahedron Lett.* **1974**, 2619.

[384] Livantsova, L. I.; Zaitseva, G. S.; Bekker, R. A.; Baukov, Y. A.; Lutsenko, I. F. *Zh. Obshch. Khim.* **1980**, *50,* 475. [*C.A.* **1980**, *92,* 215485x].
[385] Stetter, H.; Kiehs, K. *Tetrahedron Lett.* **1964**, 3531.
[386] Huisgen, R.; Feiler, L. A. *Chem. Ber.* **1969**, *102,* 3391.
[387] Cheburkov, Y. A.; Mukhamadaliev, N.; Knunyants, I. L. *Bull. Acad. Sci USSR, Div. Chem. Sci.* **1966**, 362 [*C.A.* **1966**, *64,* 17438a].
[388] Martin, P.; Greuter, H.; Bellůš, D. *J. Am. Chem. Soc.* **1979**, *101,* 5853.
[389] Bellůš, D. *Pure Appl. Chem.* **1985**, *57,* 1827.
[390] Huisgen, R.; Mayr, H. *Tetrahedron Lett.* **1975**, 2969.
[391] DoMinh, T.; Strausz, O. P. *J. Am. Chem. Soc.* **1970**, *92,* 1766.
[392] Brook, P. R.; Hunt, K. *J. Chem. Soc., Chem. Commun.* **1974**, 989.
[393] Brady, W. T.; Hieble, J. P. *J. Org. Chem.* **1971**, *36,* 2033.
[394] Neidlein, R.; Bernhard, E. *Angew. Chem., Int. Ed. Engl.* **1978**, *17,* 369.
[395] Binsch, G.; Feiler, L. A.; Huisgen, R. *Tetrahedron Lett.* **1968**, 4497.
[396] Farooq, M. O.; Vahidy, T. A.; Husain, S. M. *Bull. Soc. Chim. Fr.* **1958**, 830.
[397] Erden, I.; Sorenson, E. M. *Tetrahedron Lett.* **1983**, *24,* 2731.
[398] Brook, P. R.; Griffiths, J. G. *J. Chem. Soc., Chem. Commun.* **1970**, 1344.
[399] Brady, W. T.; Patel, A. D. *J. Org. Chem.* **1973**, *38,* 4106.
[400] Bak, D. A.; Brady, W. T. *J. Org. Chem.* **1979**, *44,* 107.
[401] Brook, P. R.; Eldeeb, A. M.; Hunt, K.; McDonald, W. S. *J. Chem. Soc., Chem. Commun.* **1978**, 10.
[402] Nguyen, N. V.; Chow, K.; Moore, H. W. *J. Org. Chem.* **1987**, *52,* 1315.
[403] Brady, W. T.; Waters, O. H. *J. Org. Chem.* **1967**, *32,* 3703.
[404] Malherbe, R.; Rist, G.; Belluš, D. *J. Org. Chem.* **1983**, *48,* 860.
[405] Hasek, R. H.; Gott, P. G.; Martin, J. C. *J. Org. Chem.* **1966**, *31,* 1931.
[406] de Meijere, A.; Erden, I.; Weber, W.; Kaufmann, D. *J. Org. Chem.* **1988**, *53,* 152.
[407] Weber, W.; Erden, I.; de Meijere, A. *Angew. Chem.* **1980**, *92,* 387.
[408] Maurin, R.; Bertrand, M. *Bull. Soc. Chim. Fr.* **1970**, 998.
[409] Jackson, D. A.; Rey, M.; Dreiding, A. S. *Helv. Chim. Acta* **1983**, *66,* 2330.
[410] van der Bij, J. R.; Kooyman, E. C. *Recl. Trav. Chim. Pays-Bas* **1952**, *71,* 837.
[411] Danheiser, R. L.; Martinez-Davila, C.; Sard, H. *Tetrahedron* **1981**, *37,* 3943.
[412] Farmer, E. H.; Farooq, M. O. *Chem. Ind. (London)* **1937**, 1079.
[413] Fitjer, L.; Kanschik, A.; Majewski, M. *Tetrahedron Lett.* **1988**, *29,* 5525.
[414] Dunkelblum, E. *Tetrahedron* **1976**, *32,* 975.
[415] Wiseman, J. R.; Chan, H. F. *J. Am. Chem. Soc.* **1970**, *92,* 4749.
[416] Hassner, A.; Pinnick, H. W.; Ansell, J. M. *J. Org. Chem.* **1978**, *43,* 1774.
[417] Aljancic-Solaja, I.; Rey, M.; Dreiding, A. S. *Helv. Chim. Acta* **1987**, *70,* 1302.
[418] Becherer, J.; Hauel, N.; Hoffmann, R. W. *Justus Liebigs Ann. Chem.* **1978**, 312.
[419] Donskaya, N. A.; Bessmertnykh, A. G.; Drobysh, V. A.; Shabarov, Y. S. *Zh. Org. Khim.* **1987**, *23,* 745 [*C.A.* **1988**, *108,* 74844y].
[420] Fitjer, L.; Kanschik, A.; Majewski, M. *Tetrahedron Lett.* **1985**, *26,* 5277.
[421] Fitjer, L.; Quabeck, U. *Angew. Chem., Int. Ed. Engl.* **1987**, *26,* 1023.
[422] Wiel, J.; Rouessac, F. *Bull. Soc. Chim. Fr.* **1979**, II-273.
[423] Senft, E.; Maurin, R. *C. R. Acad. Sci. Paris, Série C* **1972**, *275,* 1113.
[424] Mehta, G.; Rao, H. S. P. *Synth. Commun.* **1985**, *15,* 991.
[425] Trah, S.; Weidmann, K.; Fritz, H.; Prinzbach, H. *Tetrahedron Lett.* **1987**, *28,* 4399.
[426] Sasaki, T.; Hayakawa, K.; Manabe, T.; Nishida, S. *J. Am. Chem. Soc.* **1981**, *103,* 565.
[427] Yang, C. H.; Lin H. C.; Whang, M. J. *J. Chin. Chem. Soc. (Taipei)* **1986**, *33,* 139.
[428] Terlinden, R.; Boland, W.; Jaenicke, L. *Helv. Chim. Acta* **1983**, *66,* 466.
[429] Aue, D. H.; Shellhamer, D. F.; Helwig, G. S. *J. Chem. Soc., Chem. Commun.* **1975**, 603.
[430] Harding, K. E.; Trotter, J. W.; May, L. *Synth. Commun.* **1972**, *2,* 231.
[431] Ghosez, L.; Montaigne, R.; Roussel, A.; Vanlierde, H.; Mollet, P. *Tetrahedron* **1971**, *27,* 615.
[432] Ali, S. M.; Roberts, S. M. *J. Chem. Soc., Chem. Commun.* **1975**, 887.
[433] Hassner, A.; Dillon, J.; Onan, K. D. *J. Org. Chem.* **1986**, *51,* 3315.
[434] Smith, L. R.; Gream, G. E.; Meinwald, J. *J. Org. Chem.* **1977**, *42,* 927.

[435] Gream, G. E.; Smith L. R.; Meinwald, J. *J. Org. Chem.* **1974**, *39*, 3461.
[436] Wiberg, K. B.; Jason, M. E. *J. Am. Chem. Soc.* **1976**, *98*, 3393.
[437] Aue, D. H.; Helwig, G. S. *J. Chem. Soc., Chem. Commun.* **1974**, 925.
[438] Witzeman, J. S.; Ph.D. Dissertation, University of California, Santa Barbara, 1984.
[439] Jeffs, P. W.; Molina, G.; Cass, M. W.; Cortese, N. A. *J. Org. Chem.* **1982**, *47*, 3871.
[440] Jeffs, P. W.; Molina, G. *J. Chem. Soc., Chem. Commun.* **1973**, 3.
[441] Ghosez, L.; Montaigne, R.; Vanlierde, H.; Dumay, F. *Angew. Chem., Int. Ed. Engl.* **1968**, *7*, 643.
[442] Brady, W. T.; Roe, R. *J. Am. Chem. Soc.* **1971**, *93*, 1662.
[443] Dehmlow, E. V.; Birkhahn, M. *Justus Liebigs Ann. Chem.* **1987**, 701.
[444] Staudinger, H.; Suter, E. *Ber.* **1912**, *53*, 1092.
[445] Smith, L. I.; Agre, C. L.; Leekley, R. M.; Prichard, W. W. *J. Am. Chem. Soc.* **1939**, *61*, 7.
[446] DeSelms R. C.; Delay, F. *J. Org. Chem.* **1972**, *37*, 2908.
[447] Gheorghiu, M. D.; Filip, P.; Drăghici, C.; Pârvulescu, L. *J. Chem. Soc., Chem. Commun.* **1975**, 635.
[448] Feiler, L. A.; Huisgen, R.; Koppitz, P. *J. Am. Chem. Soc.* **1974**, *96*, 2270.
[449] Lange, G.; Savard, M. E.; Viswanatha, T.; Dmitrienko, G. I. *Tetrahedron Lett.* **1985**, *26*, 1791.
[450] Hassner, A.; Cory, R. M.; Sartoris, N. *J. Am. Chem. Soc.* **1976**, *98*, 7698.
[451] Cory, R. M.; Hassner, A. *Tetrahedron Lett.* **1972**, 1245.
[452] Greene, A. E. *Tetrahedron Lett.* **1980**, *21*, 3059.
[453] Mehta, G.; Rao, K. S. *Tetrahedron Lett.* **1984**, *25*, 1839.
[454] Jendralla, H. *Tetrahedron* **1983**, *39*, 1359.
[455] Erden, I. *Tetrahedron Lett.* **1984**, *25*, 1535.
[456] Mehta, G.; Padma, S.; Rao, K. S. *Synth. Commun.* **1985**, *15*, 1137.
[457] Krapcho, A. P.; Lesser, J. H. *J. Org. Chem.* **1966**, *31*, 2030.
[458] Ziegler, K.; Sauer, H.; Bruns, L.; Froitzheim-Kühlhorn, H.; Schneider, J. *Justus Liebigs Ann. Chem.* **1954**, *589*, 122.
[459] Skattebøl, L.; Stenstrøm, Y. *Acta Chem. Scand., Ser. B* **1985**, *39*, 291.
[460] Hoch, H.; Hünig, S. *Chem. Ber.* **1972**, *105*, 2660.
[461] Greene, A. E.; Coelho, F.; Deprés, J. P.; Brocksom, T. J. *Tetrahedron Lett.* **1988**, *29*, 5661.
[462] Johnston, B. D.; Slessor, K. N.; Oehlschlager, A. C. *J. Org. Chem.* **1985**, *50*, 114.
[463] Turner, R. W.; Seden, T. *J. Chem. Soc., Chem. Commun.* **1966**, 399.
[464] Asao, T.; Morita, N.; Kitahara, Y. *J. Am. Chem. Soc.* **1972**, *94*, 3655.
[465] Campbell, N.; Heller, H. G. *J. Chem. Soc.* **1965**, 5473.
[466] Becker, K. B.; Hohermuth, M. K.; Rihs, G. *Helv. Chim. Acta* **1982**, *65*, 235.
[467] Wiel, J. B.; Rouessac, F. *J. Chem. Soc., Chem. Commun.* **1975**, 180.
[468] Stevens, H. C. US Patent 3,440,284 (1969) [*C. A.* **1969**, *71*, 60866n].
[469] Hassner, A.; Fletcher V. R.; Hamon, D. P. G. *J. Am. Chem. Soc.* **1971**, *93*, 264.
[470] Tsunetsugu, J.; Asai, M.; Hiruma, S.; Kurata, Y.; Mori, A.; Ono, K.; Uchiyama, H.; Sato, M.; Ebine, S. *J. Chem. Soc., Perkin Trans. 1* **1983**, 285.
[471] Oda, M.; Morita, N.; Asao, T. *Chem. Lett.* **1981**, 1165.
[472] Scott, L. T.; Erden, I.; Brunsvold, W. R.; Schultz, T. H.; Houk, K. N.; Paddon-Row, M. N. *J. Am. Chem. Soc.* **1982**, *104*, 3659.
[473] Mehta, G.; Nair, M. S. *J. Chem. Soc., Chem. Commun.* **1983**, 439.
[474] Mehta, G.; Nair, M. S. *J. Am. Chem. Soc.* **1985**, *107*, 7519.
[475] Ried, W.; Bellinger, O. *Synthesis* **1982**, 729.
[476] Rosini, G.; Ballini, R.; Zanotti, V. *Tetrahedron* **1983**, *39*, 1085.
[477] Inamoto, Y.; Fujikura, Y.; Ikeda, H.; Takaishi, N. Jpn. Kokai JP 77 10,252 [*C. A.* **1977**, *87*, 52880c].
[478] Inamoto, Y.; Fujikura, Y.; Ikeda, H.; Takaishi, N. Jpn. Kokai JP 77 10,249 [*C. A.* **1977**, *87*, 52881d].
[479] Greene, A. E.; Luche, M. J.; Deprés, J. P. *J. Am. Chem. Soc.* **1983**, *105*, 2435.
[480] Tsunetsugu, J.; Kanda, M.; Takahashi, M.; Yoshida, K.; Koyama, H.; Shiraishi, K.; Takano, Y.; Sato, M.; Ebine, S. *J. Chem. Soc., Perkin Trans. 1* **1984**, 1465.
[481] Tsunetsugu, J.; Sato, M.; Kanda, M.; Takahashi, M.; Ebine, S. *Chem. Lett.* **1977**, 885.

[482] Gleiter, R.; Dobler, W. *Chem. Ber.* **1985**, *118*, 4725.
[483] Kelly, R. C.; VanRheenen, V.; Schletter, I.; Pillai, M. D. *J. Am. Chem. Soc.* **1973**, *95*, 2746.
[484] Sugihara, Y.; Fujita, H.; Murata, I. *J. Chem. Soc. Chem. Commun.* **1986**, 1130.
[485] Jeffs, P. W.; Cortese, N. A.; Wolfram, J. *J. Org. Chem.* **1982**, *47*, 3881.
[486] Mehta, G.; Raja Reddy, K. *Tetrahedron Lett.* **1988**, *29*, 3607.
[487] Taylor, G. A.; Yildirir, S. *J. Chem. Soc., Perkin Trans. 1* **1981**, 3129.
[488] Wakamatsu, T.; Miyachi, N.; Ozaki, F.; Shibasaki, M.; Ban, Y. *Tetrahedron Lett.* **1988**, *29*, 3829.
[489] Cragg, G. M. L. *J. Chem. Soc. (C)* **1970**, 1829.
[490] Fletcher, V. R.; Hassner, A. *Tetrahedron Lett.* **1970**, 1071.
[491] Brady, W. T.; O'Neal, H. R. *J. Org. Chem.* **1967**, *32*, 2704.
[492] Chickos, J. S.; Frey, H. M. *J. Chem. Soc., Perkin Trans. 2* **1987**, 365.
[493] Martin, J. C.; Gott, P. G.; Goodlett, V. W.; Hasek, R. H. *J. Org. Chem.* **1965**, *30*, 4175.
[494] Huisgen, R.; Otto, P. *Chem. Ber.* **1969**, *102*, 3475.
[495] Gouesnard, J. P. *Tetrahedron* **1974**, *30*, 3113.
[496] Gouesnard, J. P. *C. R. Acad. Sci. Paris, Série C* **1973**, *277*, 883.
[497] Grimme, W.; Köser, H. G. *Angew. Chem.* **1980**, *92*, 307.
[498] Coates, R. M.; Robinson, W. H. *J. Am. Chem. Soc.* **1972**, *94*, 5920.
[499] Brady, W. T.; Lloyd, R. M. *J. Org. Chem.* **1981**, *46*, 1322.
[500] Kakiuchi, K.; Hiramatsu, Y.; Tobe, Y.; Odaira, Y. *Bull. Chem. Soc. Jpn.* **1980**, *53*, 1779.
[501] Sasaki, T.; Manabe, T.; Hayakawa, K. *Tetrahedron Lett.* **1981**, *22*, 2579.
[502] Harmon, R. E.; Barta, W. D.; Gupta, S. K.; Slomp, G. *J. Chem. Soc. (C)* **1971**, 3645.
[503] Harmon, R. E.; Barta, W. D.; Gupta, S. K.; Slomp, G. *J. Chem. Soc., Chem. Commun.* **1970**, 935.
[504] Brady, W. T. *J. Org. Chem.* **1966**, *31*, 2676.
[505] Stevens, H. C.; Rinehart, J. K.; Lavanish, J. M.; Trenta, G. M. *J. Org. Chem.* **1971**, *36*, 2780.
[506] Rey, M.; Huber, U. A.; Dreiding, A. S. *Tetrahedron Lett.* **1968**, 3583.
[507] Rehn, K.; Dehmer, K.; Roechling, H. Ger. Offen. DE 2,428,410, 1976 [*C. A.* **1975**, *85*, 46093r].
[508] Deprés, J. P.; Greene, A. E.; Crabbé, P. *Tetrahedron* **1981**, *37*, 621.
[509] Brady, W. T.; Hoff, E. F., Jr. *J. Org. Chem.* **1970**, *35*, 3733.
[510] Brook, P. R.; Harrison, J. M.; Duke, A. J. *J. Chem. Soc., Chem. Commun.* **1970**, 589.
[511] Brook, P. R.; Duke, A. J.; Harrison, J. M.; Hunt, K. *J. Chem. Soc., Perkin Trans. 1* **1974**, 927.
[512] Brook, P. R.; Duke, A. J.; Duke, J. R. C. *J. Chem. Soc., Chem. Commun.* **1970**, 574.
[513] Berson, J. A.; Patton, J. W. *J. Am. Chem. Soc.* **1962**, *84*, 3406.
[514] Blomquist, A. T.; Kwiatek, J. *J. Am. Chem. Soc.* **1951**, *73*, 2098.
[515] Brooks, B. T.; Wilbert, G. *J. Am. Chem. Soc.* **1941**, *63*, 870.
[516] Brady, W. T.; Roe, R. *J. Am. Chem. Soc.* **1970**, *92*, 4618.
[517] Brady, W. T.; Roe, R.; Hoff, E. F.; Parry, F. H. *J. Am. Chem. Soc.* **1970**, *92*, 146.
[518] Brady, W. T.; Holifield, B. M. *Tetrahedron* **1967**, *23*, 4251.
[519] Fitjer, L.; Kanschik, A.; Majewski, M. *Tetrahedron Lett.* **1985**, *26*, 5277.
[520] Russell, G. A.; Schmitt, K. D.; Mattox, J. *J. Am. Chem. Soc.* **1975**, *97*, 1882.
[521] Michel, P.; O'Donnell, M.; Binamé R.; Hesbain-Frisque, A. M.; Ghosez, L.; Declercq, J. P.; Germain, G.; Arte, E.; Van Meerssche, M. *Tetrahedron Lett.* **1980**, *21*, 2577.
[522] Owen, M. D. *J. Indian Chem. Soc.* **1943**, *20*, 343 [*C. A.* **1944**, *38*, 3634[7]].
[523] Staudinger, H.; Meyer, P. J. *Helv. Chim. Acta* **1924**, *7*, 19.
[524] Dawson, T. L.; Ramage, G. R. *J. Chem. Soc.* **1950**, 3523.
[525] Cossement, E.; Binamé, R.; Ghosez, L. *Tetrahedron Lett.* **1974**, 997.
[526] Rellensmann, W.; Hafner, K. *Chem. Ber.* **1962**, *95*, 2579.
[527] Brady, W. T.; Ting, P. L. *J. Org. Chem.* **1974**, *39*, 763.
[528] Brady, W. T.; Hieble, J. P. *J. Am. Chem. Soc.* **1972**, *94*, 4278.
[529] Brady, W. T.; Parry, F. H.; Roe, R.; Hoff, E. F. *Tetrahedron Lett.* **1970**, 819.
[530] Ohshiro, Y.; Ishida, M.; Shibata, J.; Minami, T.; Agawa, T. *Chem. Lett.* **1982**, 587.
[531] Staudinger, H. *Ber.* **1907**, *40*, 1145.
[532] Lewis, J. R.; Ramage, G. R.; Wainwright, J. L.; Simonsen W. G. *J. Chem. Soc.* **1937**, 1837.
[533] Dao, L. H.; Hopkinson, A. C.; Lee-Ruff, E. *Tetrahedron Lett.* **1978**, 1413.

534 Brady, W. T.; Giang, Y. F. *J. Org. Chem.* **1986**, *51*, 2145.
535 Carpino, L. A.; Gund, P.; Springer, J. P.; Gund, T. *Tetrahedron Lett.* **1981**, *22*, 371.
536 Brady, W. T.; Ting, P. L. *J. Chem. Soc., Perkin Trans. 1* **1975**, 456.
537 Lee-Ruff, E.; Ablenas, F. J. *Can. J. Chem.* **1987**, *65*, 1663.
538 Chung, Y. S.; Kruk, H.; Barizo, O. M.; Katz, M.; Lee-Ruff, E. *J. Org. Chem.* **1987**, *52*, 1284.
539 Jutz, C.; Rommel, I.; Lengyel, I.; Feeney, J. *Tetrahedron* **1966**, *22*, 1809.
540 Asao, T.; Machiguchi, T.; Kitamura, T.; Kitahara, Y. *J. Chem. Soc., Chem. Commun.* **1970**, 89.
541 Paquette, L. A.; Andrews, D. R.; Springer, J. P. *J. Org. Chem.* **1983**, *48*, 1147.
542 Gordon, E. M.; Plušč̌ec, J.; Ondetti, M. A. *Tetrahedron Lett.* **1981**, *22*, 1871.
543 Tanaka, K.; Yoshikoshi, A. *Tetrahedron* **1971**, *27*, 4889.
544 Au-Yeung, B.; Fleming, I. *J. Chem. Soc., Chem. Commun.* **1977**, 79.
545 Fleming, I.; Au-Yeung, B. *Tetrahedron* **1981**, *37*, Supplement No. 1, 13.
546 Goldschmidt, Z.; Antebi, S. *Tetrahedron Lett.* **1978**, 271.
547 Kuroda, S.; Asao, T. *Tetrahedron Lett.* **1977**, 285.
548 Grieco, P. A.; Hiroi, K. *Tetrahedron Lett.* **1974**, 3467.
549 Fleming, I.; Williams, R. V. *J. Chem. Soc., Perkin Trans. 1* **1981**, 684.
550 Greenlee, M. L. *J. Am. Chem. Soc.* **1981**, *103*, 2425.
551 Friedrichsen, W.; Debaerdemaeker, T.; Böttcher, A.; Hahnemann, S.; Schmidt, R. *Z. Naturforsch., Teil B* **1983**, *38*, 504.
552 Harding, K. E.; Strickland, J. B.; Pommerville, J. *J. Org. Chem.* **1988**, *53*, 4877.
553 Kato, M.; Kido, F.; Wu, M. D.; Yoshikoshi, A. *Bull. Chem. Soc. Jpn.* **1974**, *47*, 1516.
554 Grieco, P. A.; Oguri, T.; Gilman, S.; DeTitta, G. T. *J. Am. Chem. Soc.* **1978**, *100*, 1616.
555 Grieco, P. A.; Oguri, T.; Wang, C. L. J.; Williams, E. *J. Org. Chem.* **1977**, *42*, 4113.
556 Baldwin, J. E.; Kapecki, J. A. *J. Am. Chem. Soc.* **1970**, *92*, 4868.
557 Isaacs, N. S.; Hatcher, B. G. *J. Chem. Soc., Chem. Commun.* **1974**, 593.
558 Gheorghiu, M. D.; Pârvulescu, L.; Drâghici, C.; Elian, M. *Tetrahedron,* **1981**, *37*, Supplement No. 1, 143.
559 Farooq, M. O.; Abraham, N. A. *Bull. Soc. Chim. Fr.* **1958**, 832.
560 Greene, A. E.; Lansard, J. P.; Luche, J. L.; Petrier, C. *J. Org. Chem.* **1983**, *48*, 4763.
561 Leriverend, M. L.; Vazeux, M. *J. Chem. Soc., Chem. Commun.* **1982**, 866.
562 Baldwin, J. E.; Breckinridge, M. F.; Johnson, D. S. *Tetrahedron Lett.* **1972**, 1635.
563 Gott, P. G.; French Patent 1,414,456, 1965 [*C. A.* **1966**, *64*, 6523f].
564 Brook, P. R.; Harrison, J. M.; Hunt, K. *J. Chem. Soc., Chem. Commun.* **1973**, 733.
565 Bertrand, M.; Gras, J. L.; Goré, J. *Tetrahedron Lett.* **1972**, 2499.
566 Bertrand, M.; Gras, J. L.; Goré, J. *Tetrahedron* **1975**, *31*, 857.
567 Bertrand, M.; Gras, J. L.; Gil, G. *Tetrahedron Lett.* **1974**, 37.
568 Bampfield, H. A.; Brook, P. R. *J. Chem. Soc., Chem. Commun.* **1974**, 171.
569 Bertrand, M.; Maurin R.; Gras, J. L. *C. R. Acad. Sci. Paris, Série C* **1968**, *267*, 417.
570 Brady, W. T.; Stockton, J. D.; Patel, A. D. *J. Org. Chem.* **1974**, *39*, 236.
571 Bampfield, H. A.; Brook, P. R. *J. Chem. Soc., Chem. Commun.* **1974**, 172.
572 Gras, J. L.; Maurin, R.; Bertrand, M. *Tetrahedron Lett.* **1969**, 3533.
573 Weyler, W.; Byrd, L. R.; Caserio, M. C.; Moore, H. W. *J. Am. Chem. Soc.* **1972**, *94*, 1027.
574 Boan, C.; Skattebøl, L. *J. Chem. Soc., Perkin Trans. 1* **1978**, 1568.
575 Bertrand, M; Gras, J. L. *Tetrahedron* **1974**, *30*, 793.
576 Opitz, G.; Zimmermann, F. *Justus Liebigs Ann. Chem.* **1963**, *662*, 178.
577 Hasek, R. H.; Martin, J. C. *J. Org. Chem.* **1961**, *26*, 4775.
578 Hasek, R. H.; Martin, J. C. *J. Org. Chem.* **1963**, *28*, 1468.
579 Martin, J. C. U.S. Patent 3,297,705, 1967 [*C. A.* **1967**, *67*, 99998z].
580 Otto, P.; Feiler, L. A.; Huisgen, R. *Angew. Chem., Int. Ed. Engl.* **1968**, *7*, 737.
581 Huisgen, R.; Otto, P. *J. Am. Chem. Soc.* **1969**, *91*, 5922.
582 Opitz, G.; Kleemann, M.; Zimmermann, F. *Angew. Chem.* **1962**, *74*, 32.
583 Martin, J. C.; Gott, P. G.; Hostettler, H. U. *J. Org. Chem.* **1967**, *32*, 1654.
584 Martin, J. C. *J. Org. Chem.* **1965**, *30*, 4311.
585 Opitz, G.; Adolph, H.; Kleemann, M.; Zimmermann, F. *Angew. Chem.* **1961**, *73*, 654.

586 Hünig, S.; Hoch, H. *Chem. Ber.* **1972**, *105,* 2216.
587 Buysch, H. J.; Hünig, S. *Angew. Chem., Int. Ed. Engl.* **1966**, *5,* 128.
588 Hünig, S.; Hoch, H. *Tetrahedron Lett.* **1966**, 5215.
589 Hassner, A.; Haddadin, M. J.; Levy, A. B. *Tetrahedron Lett.* **1973**, 1015.
590 Sieja, J. B. *J. Am. Chem. Soc.* **1971**, *93,* 130.
591 Hurd, C. D.; Kimbrough, R. D. *J. Am. Chem. Soc.* **1960**, *82,* 1373.
592 Lee-Ruff, E.; Chung, Y. S. *J. Heterocycl. Chem.* **1986**, *23,* 1551.
593 Effenberger, F.; Fischer, P.; Prossel, G.; Kiefer, G. *Chem. Ber.* **1971**, *104,* 1987.
594 Bisceglia, R. H.; Cheer, C. J. *J. Chem. Soc., Chem. Commun.* **1973**, 165.
595 Potman, R. P.; Janssen, N. J. M. L.; Scheeren, J. W.; Nivard, R. J. F. *J. Org. Chem.* **1984**, *49,* 3628.
596 Huisgen, R.; Feiler, L.; Binsch, G. *Angew. Chem., Int. Ed. Engl.* **1964**, *3,* 753.
597 Huisgen, R.; Feiler, L.; Binsch, G. *Angew. Chem.* **1964**, *76,* 892.
598 Martin, J. C.; Goodlett, V. W.; Burpitt, R. L. *J. Org. Chem.* **1965**, *30,* 4309.
599 Fráter, G.; Müller, U.; Günther, W. *Helv. Chim. Acta* **1986**, *69,* 1858.
600 Krepski, L. R.; Hassner, A. *J. Org. Chem.* **1978**, *43,* 3173.
601 Vedejs, E.; Larsen, S. D. *J. Am. Chem. Soc.* **1984**, *106,* 3030.
602 Pirrung, M. C.; DeAmicis, C. V. *Tetrahedron Lett.* **1988**, *29,* 159.
603 Shipov, A. G.; Savost'yanova, I. A.; Zaitseva, G. S.; Baukova, Y. I.; Lutsenko, I. F. *Zh. Obshch. Khim.* **1977**, *47,* 1198 [*C. A.* **1977**, *87,* 85078a].
604 Cocuzza, A. J.; Boswell, G. A. *Tetrahedron Lett.* **1985**, *26,* 5363.
605 Annis, G. D.; Paquette, L. A. *J. Am. Chem. Soc.* **1982**, *104,* 4504.
606 Marino, J. P.; Neisser, M. *J. Am. Chem. Soc.* **1981**, *103,* 7687.
607 Greene, A. E.; Charbonnier, F. *Tetrahedron Lett.* **1985**, *26,* 5525.
608 Martin, J. C. U.S. Patent 3,410,892, 1968 [*C. A.* **1969**, *70,* 28475c].
609 Gompper, R.; Studeneer, A.; Elser, W. *Tetrahedron Lett.* **1968**, 1019.
610 Scarpati, R.; Sica, D.; Santacroce, C. *Tetrahedron* **1964**, *20,* 2735.
611 Brady, W. T.; Watts, R. D. *J. Org. Chem.* **1981**, *46,* 4047.
612 Zaitseva, G. S.; Bsukov, Y. I.; Mal'tsev, V. V.; Lutsenko, I. P. *Zh. Obshch. Khim.* **1974**, *44,* 1415 [*C. A.* **1974**, *81,* 105619j].
613 Bellůš, D.; Fischer, H.; Greuter, H; Martin, P. *Helv. Chim. Acta* **1978**, *61,* 1784.
614 Bellůš, D. *J. Am. Chem. Soc.* **1978**, *100,* 8026.
615 Brady, W. T.; Saidi, K. *J. Org. Chem.* **1980**, *45,* 727.
616 Scarpati, R.; Sica, D. *Rend. Acad. Sci. Fis. Mat., Naples* **1961**, *28,* 70 [*C. A.* **1965**, *62,* 6425a].
617 Habibi, M. H.; Saidi, K.; Sams, L. C. *J. Fluorine Chem.* **1987**, *37,* 177.
618 Brady, W. T.; Giang, Y. F. *J. Org. Chem.* **1985**, *50,* 5177.
619 Snider, B. B.; Hui, R. A. H. F.; Kulkarni, Y. S. *J. Am. Chem. Soc.* **1985**, *107,* 2194.
620 Snider, B. B.; Hui, R. A. H. F. *J. Org. Chem.* **1985**, *50,* 5167.
621 Maujean, A.; Marcy, G.; Chuche, J. *J. Chem. Soc., Chem. Commun.* **1980**, 92.
622 Leyendecker, F.; Bloch, R.; Conia, J. M. *Tetrahedron Lett.* **1972**, 3703.
623 Leyendecker, F. *Tetrahedron* **1976**, *32,* 349.
624 Aryal-Kaloustian, S.; Wolff, S.; Agosta, W. C. *J. Org. Chem.* **1978**, *43,* 3314.
625 Kulkarni, Y. S.; Burbaum, B. W.; Snider, B. B. *Tetrahedron Lett.* **1985**, *26,* 5619.
626 Markó, I.; Ronsmans, B.; Hesbain-Frisque, A. M.; Dumas, S.; Ghosez, L.; Ernst, B.; Greuter, H. *J. Am. Chem. Soc.* **1985**, *107,* 2192.
627 Snider, B. B.; Kulkarni, Y. S. *J. Org. Chem.* **1987**, *52,* 307.
628 Baldwin, S. W.; Page, E. H. *J. Chem. Soc., Chem. Commun.* **1972**, 1337.
629 Veenstra, S. J.; De Mesmaeker, A.; Ernst, B. *Tetrahedron Lett.* **1988**, *29,* 2303.
630 Yadav, J. S.; Joshi, B. V.; Gadgil, V. R. *Indian J. Chem., Sect. A* **1987**, *26B,* 399 [*C. A.* **1988**, *108,* 131097h].
631 Snider, B. B.; Niwa, M. *Tetrahedron Lett.* **1988**, *29,* 3175.
632 Goldschmidt, Z.; Gutman, U.; Bakal, Y.; Worchel, A. *Tetrahedron Lett.* **1973**, 3759.
633 Mori, K.; Miyake, M. *Tetrahedron* **1987**, *43,* 2229.
634 Kulkarni, Y. S.; Snider, B. B. *J. Org. Chem.* **1985**, *50,* 2809.

635 Beereboom, J. J. *J. Org. Chem.* **1965**, *30,* 4230.
636 Beereboom, J. J. *J. Am. Chem. Soc.* **1963**, *85,* 3525.
637 Lee, S. Y.; Niwa, M.; Snider, B. B. *J. Org. Chem.* **1988**, *53,* 2356.
638 Corey, E. J.; Desai, M. C. *Tetrahedron Lett.* **1985**, *26,* 3535.
639 Sauers, R. R.; Kelly, K. W. *J. Org. Chem.* **1970**, *35,* 3286.
640 De Mesmaeker, A.; Veenstra, S. J.; Ernst, B. *Tetrahedron Lett.* **1988**, *29,* 459.
641 Becker, D.; Nagler, M.; Birnbaum, D. *J. Am. Chem. Soc.* **1972**, *94,* 4771.
642 Murray, R. K.; Goff, D. L.; Ford, T. M. *J. Org. Chem.* **1977**, *42,* 3870.
643 Moon, S.; Kolesar, T. F. *J. Org. Chem.* **1974**, *39,* 995.
644 Quinkert, G.; Kleiner, E.; Freitag, B. J.; Glenneberg, J; Billhardt, U. M.; Cech, F.; Schmieder, K. R.; Schudok, C.; Steinmetzer, H. C.; Bats, J. W.; Zimmerman, G.; Dürner, G.; Rehm, D.; Paulus, E. F. *Helv. Chim. Acta* **1986**, *69,* 469.
645 Gleiter, R.; Kissler, B.; Ganter, C. *Angew. Chem., Int. Ed. Engl.* **1987**, *26,* 1252.
646 Sasaki, T.; Eguchi, S.; Hirako, Y. *J. Org. Chem.* **1977**, *42,* 2981.
647 Snider, B. B.; Beal, R. B. *J. Org. Chem.* **1988**, *53,* 4508.
648 Yates, P.; Fallis, A. G. *Tetrahedron Lett.* **1968**, 2493.
649 Becker, D.; Harel, Z.; Birnbaum, D. *J. Chem. Soc., Chem. Commun.* **1975**, 377.
650 Funk, R. L.; Novak, P. M.; Abelman, M. M. *Tetrahedron Lett.* **1988**, *29,* 1493.
651 Kuwajima, I.; Higuchi, Y.; Iwasawa, H.; Sato, T. *Chem. Lett.* **1976**, 1271.
652 Ireland, R. E.; Dow, W. C.; Godfrey, J. D.; Thaisrivongs, S. *J. Org. Chem.* **1984**, *49,* 1001.
653 Smit, A.; Kok, J. G. J.; Geluk, H. W. *J. Chem. Soc., Chem. Commun.* **1975**, 513.
654 Oppolzer, W.; Nakao, A. *Tetrahedron Lett.* **1986**, *27,* 5471.
655 Hart, H.; Love, G. M. *J. Am. Chem. Soc.* **1971**, *93,* 6266.
656 Masamune, S.; Fukumoto, K. *Tetrahedron Lett.* **1965**, 4647.
657 Corey, E. J.; Desai, M. C.; Engler, T. A. *J. Am. Chem. Soc.* **1985**, *107,* 4339.
658 Ireland, R. E.; Godfrey, J. D.; Thaisrivongs, S. *J. Am. Chem. Soc.* **1981**, *103,* 2446.
659 Ireland, R. E.; Aristoff, P. A. *J. Org. Chem.* **1979**, *44,* 4323.
660 Knoche, H. *Justus Liebigs Ann. Chem.* **1969**, *722,* 232.
661 England, D. C.; Krespan, C. *J. Org. Chem.* **1970**, *35,* 3308.
662 Martin, J. C. U.S. Patent 3,408,398, 1968 [*C. A.* **1969**, *70,* 19658n].
663 Gheorghiu, M. D.; Drăghici, C.; Stanescu, L.; Avram, M. *Tetrahedron Lett.* **1973**, 9.
664 Smith, L. I.; Hoehn, H. H. *J. Am. Chem. Soc.* **1941**, *63,* 1175.
665 Kinugasa, K.; Agawa, T. *Organomet. Chem. Synth.* **1972**, *1,* 427.
666 Wong, H. N.; Sondheimer, F.; Goodin, R.; Breslow, R. *Tetrahedron Lett.* **1976**, 2715.
667 Springer, J. P.; Clardy, J.; Cole, R. I.; Kirkney, J. W.; Hill, R. K.; Carlson, R. M.; Isidor, J. L. *J. Am. Chem. Soc.* **1974**, *96,* 2268.
668 Ried, W.; Kraemer, R. *Justus Liebigs Ann. Chem.* **1965**, *681,* 52.
669 Johns, R. B.; Kreigler, A. B. *Aust. J. Chem.* **1964**, *17,* 765.
670 Canonica, L.; Corbella, A.; Jommi, G.; Pelizzoni, F.; Scholastico, C. *Tetrahedron Lett.* **1966**, 3031.
671 Dorsey, D. A.; King, S. M.; Moore, H. W. *J. Org. Chem.* **1986**, *51,* 2814.
672 Rosebeek, B.; Arens, J. F. *Recl. Trav. Chim. Pays-Bas* **1962**, *81,* 549.
673 Pericas, M.; Serratosa, F.; Valenti, E. *Synthesis* **1985**, 1118.
674 Daalen, J. J.; Kraak, A.; Arens, J. F. *Recl. Trav. Chim. Pays-Bas* **1961**, *80,* 810.
675 Danheiser, R. L.; Gee, S. K. *J. Org. Chem.* **1984**, *49,* 1672.
676 Danheiser, R. L.; Gee, S. K.; Perez, J. J. *J. Am. Chem. Soc.* **1986**, *108,* 806.
677 Truce, W. E.; Bavry, R. H.; Bailey, P. S. *Tetrahedron Lett.* **1968**, 5651.
678 Kuehne, M. E.; Sheehan, P. J. *J. Org. Chem.* **1968**, *33,* 4406.
679 Henn, L.; Himbert, G.; Diehl, K.; Kaftory, M. *Chem. Ber.* **1986**, *119,* 1953.
680 Himbert, G. *J. Chem. Res. (S)* **1978**, 104.
681 Himbert, G.; Schwickerath, W. *Justus Liebigs Ann. Chem.* **1984**, 85.
682 Himbert, G. *J. Chem. Res. (S)* **1978**, 442.
683 Henn, L.; Himbert, G. *Chem. Ber.* **1981**, *114,* 1015.
684 Wheeler, E. N.; Fisher, G. J. U.S. Patent 3,029,253, 1962 [*C. A.* **1962**, *57,* 7970c].

[685] Nagato, S.; Kato, R. *Yuki Gosei Kagaku Kyokai Shi* **1970**, *28,* 326 [*C. A.* **1970**, *73,* 44855u].
[686] Nagato, S.; Omori, Y.; Kato, R. *Yuki Gosei Kagaku Kyokai Shi* **1970**, *28,* 243 [*C. A.* **1970**, *72,* 110735n].
[687] Schilling, B.; Eberle, H. Ger. Offen. DE 3227108 A1, 1984 [*C. A.* **1984**, *100,* 156482c].
[688] Newman, M. S.; Leegwater, A. *J. Org. Chem.* **1968**, *33,* 2144.
[689] Oshe, H.; Palm, R.; Cherdron, H. *Monatsh. Chem.* **1967**, *98,* 2138.
[690] Schimmelschmidt, Mundlos, E. Ger. DE 1,136,323, 1960 [*C. A.* **1963**, *58,* 3321d].
[691] Del'tsova, D. P.; Koshtoyan, S. O.; Zeifman, Y. V. *Izv. Akad. Nauk SSSR, Ser. Khim.* **1970**, 2140 [*C. A.* **1971**, 76378c].
[692] Bormann, D.; Wegler, R. *Chem. Ber.* **1969**, *102,* 64.
[693] Reich, D. A.; Stevens, H. C. U.S. Patent 3,719,689, 1973 [*C. A.* **1973**, *78,* 136991s].
[694] Nakajima, K.; Yamashita, A. Jpn. Kokai JP 76 56445, 1976 [*C. A.* **1976**, *85,* 176859f].
[695] Nakajima, K.; Yamashita, A.; Okamoto, S. Jpn. Kokai JP 76 59850, 1976 [*C. A.* **1976**, *85,* 93868g].
[696] White, E. in Smith, C. W. *Acrolein,* Wiley, New York, 1962.
[697] McCain, J. H.; Marcus, E. *J. Org. Chem.* **1970**, *35,* 2414.
[698] Brady, W. T.; Saidi, K. *J. Org. Chem.* **1979**, *44,* 733.
[699] Fomina, T. B.; Artem'eva, V. N.; Sazanov, Y. N. *Zh. Org. Khim.* **1971**, *7,* 2295 [*C. A.* **1972**, *76,* 59154a].
[700] Fomina, T. B.; Artem'eva, V. N.; Sozanov, Y. N.; Koton, M. M. *Dokl. Akad. Nauk SSSR* **1970**, *193,* 838 [*C. A.* **1975**, *73,* 99247c].
[701] Higuchi, H.; Arimoto, K. Jpn. Kokai Tokkyo Koho, 73 00462, 1973.
[702] Ishida, M.; Minami, T.; Agawa, T. *J. Org. Chem.* **1979**, *14,* 2067.
[703] Takasu, I.; Higuchi, M.; Suzuki, R. Jpn. Kokai JP 75 05191, (1975) [*C. A.* **1975**, *83,* 131238v].
[704] Kondo, K.; Ryu, Y. Jpn. Kokai JP 77 06977, 1977 [*C. A.* **1973**, *78,* 29239f].
[705] Kondo, K.; Ryu, Y.; Dobashi, S. Jpn. Kokai JP Tokkyo Koho 74 61153, 1974.
[706] Kondo, K.; Ryu, Y. Jpn. Kokai Tokkyo Koho JP 73 61420, 1973.
[707] Kondo, K.; Ryu, Y. Jpn. Kokai Tokkyo Koho JP 77 020969, 1977 [*C. A.* **1974**, *80,* 59469q].
[708] Achmatowicz, O.; Leplawy, M. *Bull. Akad. Polon. Sci., Ser. Sci., Chem. Geol. Geograph.* **1958**, *6,* 417 [*C. A.* **1959**, *53,* 3183g].
[709] Zubovics, Z.; Ishikawa, N. *J. Fluorine Chem.* **1976**, *8,* 43.
[710] Chse, H.; Pilgrim, K. *Tetrahedron Lett.* **1968**, 1949.
[711] Pilgrim, K.; Ohse, H. *J. Org. Chem.* **1969**, *34,* 1586.
[712] Knunyants, I. L.; Cheburkov, Y. A. *Bull. Acad. Sci. USSR. Div. Chem. Sci.* **1961**, 747 [*C. A.* **1960**, *54,* 22349c].
[713] Bergman, E.; Cohen, S.; Hoffman, E.; Rand-Meir, Z. *J. Chem. Soc.* **1961**, 3452.
[714] Cornforth, R. H. *J. Chem. Soc.* **1959**, 4052.
[715] England, D. C.; Krespan, C. G. *J. Org. Chem.* **1968**, *33,* 816.
[716] Wiley, R. H. U.S. Patent 3,356,721, 1967 [*C. A.* **1968**, *68,* 49085z].
[717] Knunyants, I. L.; Gambaryan, N.; Tyuleneva, V. *Izv. Akad. Nauk SSSR, Ser. Khim.* **1967**, 2662 [*C. A.* **1968**, *69,* 67299x].
[718] Staudinger, H. *Ber.* **1908**, *41,* 1355.
[719] Sazonov, Y. N.; Glukhov, N. A.; Koton, M. M. *Dokl. Akad. Nauk SSSR* **1967**, *177,* 363 [*C. A.* **1968**, *69,* 2559n].
[720] Rosowski, A.; Tarbell, D. S. *J. Org. Chem.* **1961**, *26,* 2255.
[721] Cornforth, J. W.; Cornforth, R. H.; Pelter, A.; Horning, M. G.; Popjak, G. *Tetrahedron* **1959**, *5,* 311.
[722] Reid, W.; Bellinger, O. *Justus Liebigs Ann. Chem.* **1984**, 1778.
[723] Agranat, I.; Cohen, S.; Aharon-Shalom, E.; Bergmann, E. *Tetrahedron* **1975**, *31,* 1163.
[724] Brady, W. T.; Gu, Y. *J. Org. Chem.* **1988**, *53,* 1353.
[725] Staudinger, H. *Helv. Chim. Acta* **1918**, *3,* 862.
[726] Brady, W. T.; Patel, A. D. *J. Chem. Soc., Chem. Commun.* **1971**, 1642.
[727] Van Der Puy, M.; Anello, L. C.; Sukornick, B.; Sweeney, R.; Wiles, R. A. U.S. Patent 4,244,891, 1981 [*C. A.* **1981**, *94,* 208337p].

[728] Eiden, F.; Peglow, M. *Arch. Pharm. (Weinheim, Ger.)* **1970**, *303,* 61 [*C. A.* **1970**, *72,* 78797w].
[729] Gompper, R.; Studeneer, A.; Elser, W. *Tetrahedron Lett.* **1968**, 1019.
[730] Nazarov, I. N.; Kuznetsov, N. V. *J. Gen. Chem. USSR (Eng. Transl.)* **1959**, *29,* 754.
[731] Kitahara, Y.; Oda, M.; Kayama, Y. *Angew. Chem.* **1976**, *88,* 536.
[732] Morita, N.; Asao, T. *Chem. Lett.* **1973**, 67.
[733] Asao, T.; Morita, N.; Iwagame, N. *Bull. Chem. Soc. Jpn.* **1974**, *47,* 773.
[734] Wittig, G.; Hesse, A. *Justus Liebigs Ann. Chem.* **1976**, 500.
[735] Morita, N.; Asao, T.; Kitahara, Y. *Tetrahedron Lett.* **1974**, 2083.
[736] Higuchi, M.; Suzuki, A. Jpn. Kokai, 74 25,942.
[737] Goldschmidt, Z.; Antebi, S. *Tetrahedron Lett.* **1978**, 1225.
[738] Staudinger, H.; Kon, N. *Justus Liebigs Ann. Chem.* **1912**, *384,* 38.
[739] Morita, N.; Asao, T. *Chem. Lett.* **1975**, 71.
[740] Neidlein, R.; Salzl, M. *Arch. Pharm. (Weinheim, Ger.)* **1977**, *310,* 685 [*C. A.* **1978**, *88,* 6637b].
[741] Markl, G.; Hauptmann, H. *Tetrahedron* **1978**, *32,* 2131.
[742] Hastings, J. S.; Heller, H. G.; Salisbury, K. *J. Chem. Soc., Perkin Trans. 1* **1975**, 1995.
[743] Kuroda, S.; Ojima, J.; Kitatani, K.; Kirita, M.; Nakada, T. *Chem. Soc., Perkin Trans. 1* **1983**, 2987.
[744] Kuroda, S.; Kitatani, K.; Ojima, J. *Tetrahedron Lett.* **1982**, *23,* 2657.
[745] Asao, T.; Morita, N.; Ojima, J.; Fujiyoshi, M. *Tetrahedron Lett.* **1978**, 2795.
[746] Ojima, J.; Itagawa, K.; Hamai, S.; Nakada, T.; Kuroda, S. *J. Chem. Soc., Perkin Trans. 1* **1983**, 2997.
[747] Gilman, H.; Adams, C. E. *Recl. Trav. Chim. Pays-Bas* **1929**, *48,* 464.
[748] Heller, H. G.; Auld, D.; Salisbury, K. *J. Chem. Soc. (C)* **1967**, 1552.
[749] Eiden, F.; Peglow, M. *Arch. Pharm. (Weinheim, Ger.)* **1970**, *303,* 825 [*C. A.* **1971**, *74,* 3460p].
[750] Hastings, J. S.; Heller, H. G. *J. Chem. Soc., Perkin Trans. 2* **1972**, 1839.
[751] Hastings, J. S.; Heller, H. G.; Tucker, H.; Smith, K. *J. Chem. Soc., Perkin Trans. 1* **1975**, 1545.
[752] Staudinger, H. *Ber.* **1908**, *41,* 1493.
[753] Augustin, M.; Kohler, M. *Z. Chem.* **1983**, *23,* 402.
[754] Neidlein, R.; Bernhard, E. *Justus Liebigs Ann. Chem.* **1979**, 959.
[755] Neidlein, R.; Kraemer, A. D. *Angew. Chem.* **1977**, *89,* 48.
[756] Neidlein, R.; Kraemer, A. D. *J. Heterocycl. Chem.* **1977**, *14,* 1369.
[757] Agranat, I.; Pick, M. *Tetrahedron Lett.* **1973**, 4079.
[758] Ojima, J.; Kuroda, S.; Kirita, M. *Chem. Lett.* **1982**, 1371.
[759] Heller, H. G.; Auld, D.; Salisbury, K. *J. Chem. Soc. (C)* **1967**, 682.
[760] Markl, G.; Olbrich, H. *Angew. Chem.* **1966**, *78,* 598.
[761] Tanaka, K.; Toda, F. *Tetrahedron Lett.* **1980**, *21,* 2713.
[762] Fuchs, R. Ger. Offen. DE 2,634,540, 1977 [*C. A.* **1978**, *88,* 89136m].
[763] Ebnöther, A.; Jucker, E.; Rissi, E.; Rutschmann, J.; Schreier, E.; Steiner, R.; Süess, R.; Vogel, A. *Helv. Chim. Acta* **1959**, *42,* 918.
[764] Poshkus, A. C.; Herweh, J. E. *J. Org. Chem.* **1965**, *30,* 2466.
[765] Staudinger, H.; Göhring, O.; Scholler, M. *Chem. Ber.* **1914**, *47,* 40.
[766] Saito, T.; Oikawa, I.; Motoki, S. *Bull. Chem. Soc. Jpn.* **1980**, *53,* 2582.
[767] Boosen, K. J.; Azer, N. Swiss Patent CH 602,554, 1978 [*C. A.* **1978**, *89,* 214910h].
[768] Birkofer, L.; Lukenhaus, W. *Justus Liebigs Ann. Chem.* **1984**, 1193.
[769] Metzger, C.; Kurz, J. *Chem. Ber.* **1971**, *104,* 50.
[770] Brady, W. T.; Dorsey, E. D.; Parry, F. H. *J. Org. Chem.* **1969**, *34,* 2846.
[771] Hull, R. *J. Chem. Soc. (C)* **1967**, 1154.
[772] Brady, W. T.; Owens, R. A. *Tetrahedron Lett.* **1976**, 1553.
[773] Brady, W. T.; Dorsey, E. D. *J. Chem. Soc., Chem. Commun.* **1968**, 1638.
[774] Brady, W. T.; Dorsey, E. D. *J. Org. Chem.* **1970**, *35,* 2732.
[775] Huisgen, R.; Funke, E.; Schaefer, F.; Knorr, R. *Angew. Chem.* **1967**, *79,* 321.
[776] Funke, E.; Huisgen, R. *Chem. Ber.* **1971**, *104,* 3222.
[777] Weyler, W.; Duncan, W. G.; Moore, H. W. *J. Am. Chem. Soc.* **1975**, *97,* 6187.

778 Moore, H. W.; Chow., K.; Nguyen, N. *J. Org. Chem.* **1987**, *52,* 2530.
779 Talaty, E. R.; Dupuy, A. E.; Untermoehlen, C. M.; Sterkoll, L. H. *J. Chem. Soc., Chem. Commun.* **1973**, 48.
780 Minami, T.; Yamakata, K.; Ohshiro, Y.; Agawa, T.; Yasuoka, N.; Kasai, N. *J. Org. Chem.* **1972**, *37,* 3810.
781 Grill, H.; Kresze, G. *Tetrahedron Lett.* **1970**, 1427.
782 Makarov, S. P.; Shpanskii, V. A.; Ginsberg, V. A.; Shchekotikhin, A. I.; Filatov, A. S.; Martynova, L. L.; Pavlovskaya, I. V.; Golovaneva A. F.; Yakubovich, A. Y. *Dokl. Akad. Nauk SSSR* **1961**, *142,* 596 [*C. A.* **1962**, *57,* 4228a].
783 Staudinger, H.; Jelagin, S. *Ber.* **1911**, *44,* 365.
783a Hall, J.; Krishnan, G. *J. Org. Chem.* **1984**, *49,* 2498.
784 Bird, C. W. *Chem. Ind. (London)* **1963**, 1556.
785 Kerber, R. C.; Ryan, T. J. *Tetrahedron Lett.* **1970**, 703.
786 Kerber, R. C.; Ryan, T. J.; Hsu, S. D. *J. Org. Chem.* **1974**, *39,* 1215.
787 Colonna, M.; Risalti, A. *Gazz. Chim. Ital.* **1960**, *90,* 1165 [*C. A.* **1958**, *52,* 377c].
788 Cook, A. H.; Jones, D. G. *J. Chem. Soc.* **1941**, 184.
789 Fischer, W.; Fahr, E. *Tetrahedron Lett.* **1966**, 5245.
790 Brooks, G.; Shaw, M. A.; Taylor, G. A. *J. Chem. Soc., Perkin Trans. 1* **1973**, 1297.
791 Yamamoto, A.; Abe, I.; Nozawa, M.; Kotani, M.; Motoyoshiya, J.; Gotoh, H.; Matsuzaki, K. *J. Chem. Soc., Perkin Trans. 1* **1983**, 2297.
792 Horner, L.; Spietschka, E. *Chem. Ber.* **1956**, *89,* 2765.
793 Fahr, E.; Büttner, E.; Keil, K. H.; Markert, J.; Schenkenbach, F.; Thiedemann, R.; Fontaine, J. *Justus Liebigs Ann. Chem.* **1981**, 1433.
794 Oshiro, Y.; Komatsu, M.; Yamamoto, Y.; Takari, K.; Agawa, T. *Chem. Lett.* **1974**, 383.
795 Kirmse, W. *Chem. Ber.* **1960**, *93,* 2357.
796 Staudinger, H.; Anthes, E.; Pfenninger, F. *Ber.* **1916**, *49,* 1928.
797 L'abbé, G.; Verhelst, G.; Yu, C.; Toppet, S. *J. Org. Chem.* **1975**, *40,* 1728.
798 Mohan, S.; Kumar, B.; Sandhu, J. *Chem. Ind. (London)* **1971**, 671.
799 Del'tsova, D. P.; Gambaryan, N. P. *Izv. Akad. Nauk SSSR, Ser. Khim.* **1976**, 858 [*C. A.* **1976**, *85,* 123732c].
800 Katagiri, N.; Niwa, R.; Kato, T. *Heterocycles* **1983**, *20,* 597.
801 Sakamoto, M.; Miyazawa, K.; Yamamoto, K.; Tomimatsu, Y. *Chem. Pharm. Bull.* **1974**, *22,* 2201.
802 Bödeker, J.; Courault, K. *Tetrahedron* **1978**, *34,* 101.
803 Mazumdar, S. N.; Ibnusaud, I.; Mahajan, M. P. *Tetrahedron Lett.* **1986**, *27,* 5875.
804 Ried, W.; Radt, W. *Justus Liebigs Ann. Chem.* **1964**, *676,* 110.
805 Ried, W.; Radt, W. *Angew. Chem., Int. Ed. Engl.* **1963**, *2,* 397.
806 Erickson, J. L. E.; Dechary, J. M. *J. Am. Chem. Soc.* **1952**, *74,* 2644.
807 Schönberg, A.; Mustafa, A. *J. Chem. Soc.* **1947**, 997.
808 Bignardi, G.; Evangelisti, F.; Schenone, P.; Bargagna, A. *J. Heterocycl. Chem.* **1972**, *9,* 1071.
809 Bargagna, A.; Cafaggi, S.; Schenone, P. *J. Heterocycl. Chem.* **1977**, *14,* 249.
810 Schenone, P.; Bargagna, A.; Bignardi, G.; Evangelisti, F. *J. Heterocycl. Chem.* **1976**, *13,* 1105.
811 Bignardi, G.; Schenone, P.; Evangelisti, F. *Ann. Chim. (Rome)* **1971**, *61,* 326 [*C. A.* **1971**, *75,* 129608g].
812 Mosti, L.; Menozzi, G.; Schenone, P. *J. Heterocycl. Chem.* **1984**, *21,* 361.
813 Mosti, L.; Schenone, P.; Menozzi, G.; Romussi, G.; Baccichetti, F.; Carlassare, F.; Bordin, F. *Il Farmaco (Ed. Sci.)* **1984**, *39,* 81 [*C. A.* **1983**, *99,* 88080d].
814 Mosti, L.; Schenone, P.; Menozzi, G. *J. Heterocycl. Chem.* **1980**, *17,* 61.
815 Schenone, P.; Evangilisti, F.; Bignardi, G.; Bargagna, A. *Ann. Chim. (Rome)* **1974**, *64,* 613 [*C. A.* **1976**, *84,* 121596s].
816 Bargagna, A.; Schenone, P.; Bondavalli, F.; Longobardi, M. *J. Heterocycl. Chem.* **1980**, *17,* 33.
817 Menozzi, G.; Mosti, L.; Schenone, P. *J. Heterocycl. Chem.* **1983**, *20,* 539.
818 Bignardi, G.; Mosti, L.; Schenone, P.; Menozzi, G. *J. Heterocycl. Chem.* **1977**, *14,* 1023.
819 Bargagna, A.; Cafaggi, S.; Schenone, P. *J. Heterocycl. Chem.* **1980**, *17,* 507.

[820] Bargagna, A.; Schenone, P.; Bondavalli, F.; Longobardi, M. *J. Heterocycl. Chem.* **1982**, *19,* 257.
[821] Bargagna, A.; Evangilisti, F.; Schenone, P. *J. Heterocycl. Chem.* **1981**, *18,* 111.
[822] Mosti, L.; Menozzi, G.; Schenone, P. *J. Heterocycl. Chem.* **1981**, *18,* 1263.
[823] Bargagna, A.; Evangilisti, F.; Schenone, P. *J. Heterocycl. Chem.* **1979**, *16,* 93.
[824] Mosti, L.; Bignardi, G.; Evangelisti, F.; Schenone, P. *J. Heterocycl. Chem.* **1976**, *13,* 1201.
[825] Cafaggi, S.; Romussi, G.; Ciarallo, C.; Bignardi, G. *Il Farmaco (Ed. Sci.)* **1983**, *38,* 775 [*C. A.* **1984**, *100,* 68199k].
[826] Bargagna, A.; Schenone, P.; Bondavalli, F.; Longobardi, M. *J. Heterocycl. Chem.* **1980**, *17,* 1201.
[827] Mosti, L.; Schenone, P.; Menozzi, G. *J. Heterocycl. Chem.* **1979**, *16,* 913.
[828] Meslin, J. C.; Quiniou, H. *C. R. Acad. Sci. Paris, Série C* **1971**, *273,* 148.
[829] Okazaki, R.; Sunagawa, K.; Kang, K.; Inamoto, N. *Bull. Chem. Soc. Jpn.* **1979**, *52,* 496.
[830] Meslin, J. C. *C. R. Acad. Sci. Paris, Série C* **1973**, *277,* 1391.
[831] Meslin, J. C.; Pradere, J. P.; Quiniou, H. *Bull. Soc. Chim. Fr.* **1976**, 1195.
[832] Karakasa, T.; Yamaguchi, H.; Motoki, S. *J. Org. Chem.* **1980**, *45,* 927.
[833] Kollenz, G.; Ziegler, E.; Ott, W.; Kriwetz, G. *Z. Naturforsch., Teil B* **1977**, *32,* 701.
[834] Kollenz, G.; Ziegler, E.; Ott, W. *Org. Prep. Proc. Int.* **1973**, *5,* 261.
[835] Kollenz, G.; Igel, H.; Ziegler, E. *Monatsh. Chem.* **1972**, *103,* 450.
[836] Ziegler, E.; Ott, W. *Synthesis* **1973**, 679.

CUMULATIVE CHAPTER TITLES BY VOLUME

Volume 5 (1949)

1. **The Synthesis of Acetylenes**: Thomas L. Jacobs
2. **Cyanoethylation**: Herman A. Bruson
3. **The Diels-Alder Reaction: Quinones and Other Cyclenones**: Lewis W. Butz and Anton W. Rytina
4. **Preparation of Aromatic Fluorine Compounds from Diazonium Fluoborates: The Schiemann Reaction**: Arthur Roe
5. **The Friedel and Crafts Reaction with Aliphatic Dibasic Acid Anhydrides**: Ernst Berliner
6. **The Gattermann-Koch Reaction**: Nathan N. Crounse
7. **The Leuckart Reaction**: Maurice L. Moore
8. **Selenium Dioxide Oxidation**: Norman Rabjohn
9. **The Hoesch Synthesis**: Paul E. Spoerri and Adrien S. DuBois
10. **The Darzens Glycidic Ester Condensation**: Melvin S. Newman and Barney J. Magerlein

Volume 6 (1951)

1. **The Stobbe Condensation**: William S. Johnson and Guido H. Daub
2. **The Preparation of 3,4-Dihydroisoquinolines and Related Compounds by the Bischler-Napieralski Reaction**: Wilson M. Whaley and Tuticorin R. Govindachari
3. **The Pictet-Spengler Synthesis of Tetrahydroisoquinolines and Related Compounds**: Wilson M. Whaley and Tuticorin R. Govindachari
4. **The Synthesis of Isoquinolines by the Pomeranz-Fritsch Reaction**: Walter J. Gensler
5. **The Oppenauer Oxidation**: Carl Djerassi
6. **The Synthesis of Phosphonic and Phosphinic Acids**: Gennady M. Kosolapoff
7. **The Halogen-Metal Interconversion Reaction with Organolithium Compounds**: Reuben G. Jones and Henry Gilman
8. **The Preparation of Thiazoles**: Richard H. Wiley, D. C. England, and Lyell C. Behr
9. **The Preparation of Thiophenes and Tetrahydrothiophenes**: Donald E. Wolf and Karl Folkers
10. **Reductions by Lithium Aluminum Hydride**: Weldon G. Brown

Volume 7 (1953)

1. **The Pechmann Reaction**: Suresh Sethna and Ragini Phadke
2. **The Skraup Synthesis of Quinolines**: R. H. F. Manske and Marshall Kulka
3. **Carbon-Carbon Alkylations with Amines and Ammonium Salts**: James H. Brewster and Ernest L. Eliel
4. **The von Braun Cyanogen Bromide Reaction**: Howard A. Hageman
5. **Hydrogenolysis of Benzyl Groups Attached to Oxygen, Nitrogen, or Sulfur**: Walter H. Hartung and Robert Simonoff
6. **The Nitrosation of Aliphatic Carbon Atoms**: Oscar Touster
7. **Epoxidation and Hydroxylation of Ethylenic Compounds with Organic Peracids**: Daniel Swern

Volume 8 (1954)

1. **Catalytic Hydrogenation of Esters to Alcohols**: Homer Adkins
2. **The Synthesis of Ketones from Acid Halides and Organometallic Compounds of Magnesium, Zinc, and Cadmium**: David A. Shirley
3. **The Acylation of Ketones to Form β-Diketones or β-Keto Aldehydes**: Charles R. Hauser, Frederic W. Swamer, and Joe T. Adams
4. **The Sommelet Reaction**: S. J. Angyal
5. **The Synthesis of Aldehydes from Carboxylic Acids**: Erich Mosettig
6. **The Metalation Reaction with Organolithium Compounds**: Henry Gilman and John W. Morton, Jr.
7. **β-Lactones**: Harold E. Zaugg
8. **The Reaction of Diazomethane and Its Derivatives with Aldehydes and Ketones**: C. David Gutsche

Volume 9 (1957)

1. **The Cleavage of Non-enolizable Ketones with Sodium Amide**: K. E. Hamlin and Arthur W. Weston
2. **The Gattermann Synthesis of Aldehydes**: William E. Truce
3. **The Baeyer-Villiger Oxidation of Aldehydes and Ketones**: C. H. Hassall
4. **The Alkylation of Esters and Nitriles**: Arthur C. Cope, H. L. Holmes, and Herbert O. House

5. **The Reaction of Halogens with Silver Salts of Carboxylic Acids**: C. V. Wilson

6. **The Synthesis of β-Lactams**: John C. Sheehan and Elias J. Corey

7. **The Pschorr Synthesis and Related Diazonium Ring Closure Reactions**: DeLos F. DeTar

Volume 10 (1959)

1. **The Coupling of Diazonium Salts with Aliphatic Carbon Atoms**: Stanley M. Parmerter

2. **The Japp-Klingemann Reaction**: Robert R. Phillips

3. **The Michael Reaction**: Ernst D. Bergmann, David Ginsburg, and Raphael Pappo

Volume 11 (1960)

1. **The Beckmann Rearrangement**: L. Guy Donaruma and Walter Z. Heldt

2. **The Demjanov and Tiffeneau-Demjanov Ring Expansions**: Peter A. S. Smith and Donald R. Baer

3. **Arylation of Unsaturated Compounds by Diazonium Salts**: Christian S. Rondestvedt, Jr.

4. **The Favorskii Rearrangement of Haloketones**: Andrew S. Kende

5. **Olefins from Amines: The Hofmann Elimination Reaction and Amine Oxide Pyrolysis**: Arthur C. Cope and Elmer R. Trumbull

Volume 12 (1962)

1. **Cyclobutane Derivatives from Thermal Cycloaddition Reactions**: John D. Roberts and Clay M. Sharts

2. **The Preparation of Olefins by the Pyrolysis of Xanthates. The Chugaev Reaction**: Harold R. Nace

3. **The Synthesis of Aliphatic and Alicyclic Nitro Compounds**: Nathan Kornblum

4. **Synthesis of Peptides with Mixed Anhydrides**: Noel F. Albertson

5. **Desulfurization with Raney Nickel**: George R. Pettit and Eugene E. van Tamelen

Volume 13 (1963)

1. **Hydration of Olefins, Dienes, and Acetylenes via Hydroboration**: George Zweifel and Herbert C. Brown

2. **Halocyclopropanes from Halocarbenes**: William E. Parham and Edward E. Schweizer

3. **Free Radical Addition to Olefins to Form Carbon-Carbon Bonds**: Cheves Walling and Earl S. Huyser

4. **Formation of Carbon-Heteroatom Bonds by Free Radical Chain Additions to Carbon-Carbon Multiple Bonds**: F. W. Stacey and J. F. Harris, Jr.

Volume 14 (1965)

1. **The Chapman Rearrangement**: J. W. Schulenberg and S. Archer

2. **α-Amidoalkylations at Carbon**: Harold E. Zaugg and William B. Martin

3. **The Wittig Reaction**: Adalbert Maercker

Volume 15 (1967)

1. **The Dieckmann Condensation**: John P. Schaefer and Jordan J. Bloomfield

2. **The Knoevenagel Condensation**: G. Jones

Volume 16 (1968)

1. **The Aldol Condensation**: Arnold T. Nielsen and William J. Houlihan

Volume 17 (1969)

1. **The Synthesis of Substituted Ferrocenes and Other π-Cyclopentadienyl-Transition Metal Compounds**: Donald E. Bublitz and Kenneth L. Rinehart, Jr.

2. **The γ-Alkylation and γ-Arylation of Dianions of β-Dicarbonyl Compounds**: Thomas M. Harris and Constance M. Harris

3. **The Ritter Reaction**: L. I. Krimen and Donald J. Cota

Volume 18 (1970)

1. **Preparation of Ketones from the Reaction of Organolithium Reagents with Carboxylic Acids**: Margaret J. Jorgenson

2. **The Smiles and Related Rearrangements of Aromatic Systems**: W. E. Truce, Eunice M. Kreider, and William W. Brand

3. **The Reactions of Diazoacetic Esters with Alkenes, Alkynes, Heterocyclic, and Aromatic Compounds**: Vinod David and E. W. Warnhoff

4. **The Base-Promoted Rearrangements of Quaternary Ammonium Salts**: Stanley H. Pine

Volume 19 (1972)

1. **Conjugate Addition Reactions of Organocopper Reagents**: Gary H. Posner

2. **Formation of Carbon-Carbon Bonds via π-Allylnickel Compounds**: Martin F. Semmelhack

3. **The Thiele-Winter Acetoxylation of Quinones**: J. F. W. McOmie and J. M. Blatchly

4. **Oxidative Decarboxylation of Acids by Lead Tetraacetate**: Roger A. Sheldon and Jay K. Kochi

Volume 20 (1973)

1. **Cyclopropanes from Unsaturated Compounds, Methylene Iodide, and Zinc-Copper Couple**: H. E. Simmons, T. L. Cairns, Susan A. Vladuchick, and Connie M. Hoiness

2. **Sensitized Photooxygenation of Olefins**: R. W. Denny and A. Nickon

3. **The Synthesis of 5-Hydroxyindoles by the Nenitzescu Reaction**: George R. Allen, Jr.

4. **The Zinin Reduction of Nitroarenes**: H. K. Porter

Volume 21 (1974)

1. **Fluorination with Sulfur Tetrafluoride**: G. A. Boswell, Jr., W. C. Ripka, R. M. Scribner, and C. W. Tullock

2. **Modern Methods to Prepare Monofluoroaliphatic Compounds**: William A. Sheppard

Volume 22 (1975)

1. **The Claisen and Cope Rearrangements**: Sara Jane Rhoads and N. Rebecca Raulins

2. **Substitution Reactions Using Organocopper Reagents**: Gary H. Posner

3. **Clemmensen Reduction of Ketones in Anhydrous Organic Solvents**: E. Vedejs

4. **The Reformatsky Reaction**: Michael W. Rathke

Volume 23 (1976)

1. **Reduction and Related Reactions of α,β-Unsaturated Compounds with Metals in Liquid Ammonia**: Drury Caine

2. **The Acyloin Condensation**: Jordan J. Bloomfield, Dennis C. Owsley, and Janice M. Nelke

3. **Alkenes from Tosylhydrazones**: Robert H. Shapiro

Volume 24 (1976)

1. **Homogeneous Hydrogenation Catalysts in Organic Synthesis**: Arthur J. Birch and David H. Williamson
2. **Ester Cleavages via S_N2-Type Dealkylation**: John E. McMurry
3. **Arylation of Unsaturated Compounds by Diazonium Salts (The Meerwein Arylation Reaction)**: Christian S. Rondestvedt, Jr.
4. **Selenium Dioxide Oxidation**: Norman Rabjohn

Volume 25 (1977)

1. **The Ramberg-Bäcklund Rearrangement**: Leo A. Paquette
2. **Synthetic Applications of Phosphoryl-Stabilized Anions**: William S. Wadsworth, Jr.
3. **Hydrocyanation of Conjugated Carbonyl Compounds**: Wataru Nagata and Mitsuru Yoshioka

Volume 26 (1979)

1. **Heteroatom-Facilitated Lithiations**: Heinz W. Gschwend and Herman R. Rodriguez
2. **Intramolecular Reactions of Diazocarbonyl Compounds**: Steven D. Burke and Paul A. Grieco

Volume 27 (1982)

1. **Allylic and Benzylic Carbanions Substituted by Heteroatoms**: Jean-François Biellmann and Jean-Bernard Ducep
2. **Palladium-Catalyzed Vinylation of Organic Halides**: Richard F. Heck

Volume 28 (1982)

1. **The Reimer-Tiemann Reaction**: Hans Wynberg and Egbert W. Meijer
2. **The Friedländer Synthesis of Quinolines**: Chia-Chung Cheng and Shou-Jen Yan
3. **The Directed Aldol Reaction**: Teruaki Mukaiyama

Volume 29 (1983)

1. **Replacement of Alcoholic Hydroxy Groups by Halogens and Other Nucleophiles via Oxyphosphonium Intermediates**: Bertrand R. Castro

2. **Reductive Dehalogenation of Polyhalo Ketones with Low-Valent Metals and Related Reducing Agents**: Ryoji Noyori and Yoshihiro Hayakawa

3. **Base-Promoted Isomerizations of Epoxides**: Jack K. Crandall and Marcel Apparu

Volume 30 (1984)

1. **Photocyclization of Stilbenes and Related Molecules**: Frank B. Mallory and Clelia W. Mallory

2. **Olefin Synthesis via Deoxygenation of Vicinal Diols**: Eric Block

Volume 31 (1984)

1. **Addition and Substitution Reactions of Nitrile-Stabilized Carbanions**: Simeon Arseniyadis, Keith S. Kyler, and David S. Watt

Volume 32 (1984)

1. **The Intramolecular Diels-Alder Reaction**: Engelbert Ciganek

2. **Synthesis Using Alkyne-Derived Alkenyl- and Alkynylaluminum Compounds**: George Zweifel and Joseph A. Miller

Volume 33 (1985)

1. **Formation of Carbon-Carbon and Carbon-Heteroatom Bonds via Organoboranes and Organoborates**: Ei-Ichi Negishi and Michael J. Idacavage

2. **The Vinylcyclopropane-Cyclopentene Rearrangement**: Tomáš Hudlický, Toni M. Kutchan, and Saiyid M. Naqvi

Volume 34 (1985)

1. **Reductions by Metal Alkoxyaluminum Hydrides**: Jaroslav Málek

2. **Fluorination by Sulfur Tetrafluoride**: Chia-Lin J. Wang

Volume 35 (1988)

1. **The Beckmann Reactions: Rearrangements, Elimination-Additions, Fragmentations, and Rearrangement-Cyclizations**: Robert E. Gawley

2. **The Persulfate Oxidation of Phenols and Arylamines (The Elbs and the Boyland-Sims Oxidations)**: E. J. Behrman

3. **Fluorination with Diethylaminosulfur Trifluoride and Related Aminofluorosulfuranes**: Miloš Hudlický

Volume 36 (1988)

1. **The [3 + 2] Nitrone-Olefin Cycloaddition Reaction**: Pat N. Confalone and Edward M. Huie
2. **Phosphorus Addition at sp^2 Carbon**: Robert Engel
3. **Reduction by Metal Alkoxyaluminum Hydrides. Part II. Carboxylic Acids and Derivatives, Nitrogen Compounds, and Sulfur Compounds**: Jaroslav Málek

Volume 37 (1989)

1. **Chiral Synthons by Ester Hydrolysis Catalyzed by Pig Liver Esterase**: Masaji Ohno and Masami Otsuka
2. **The Electrophilic Substitution of Allylsilanes and Vinylsilanes**: Ian Fleming, Jacques Dunoguès, and Roger Smithers

Volume 38 (1990)

1. **The Peterson Olefination Reaction**: David J. Ager
2. **Tandem Vicinal Difunctionalization: β-Addition to α,β-Unsaturated Carbonyl Substrates Followed by α-Functionalization**: Marc J. Chapdelaine and Martin Hulce
3. **The Nef Reaction**: Harold W. Pinnick

Volume 39 (1990)

1. **Lithioalkenes from Arenesulfonylhydrazones**: A. Richard Chamberlin and Steven H. Bloom
2. **The Polonovski Reaction**: David Grierson
3. **Oxidation of Alcohols to Carbonyl Compounds via Alkoxysulfonium Ylides: The Moffatt, Swern, and Related Oxidations**: Thomas T. Tidwell

Volume 40 (1991)

1. **The Pauson-Khand Cycloaddition Reaction for Synthesis of Cyclopentenones**: Neil E. Schore
2. **Reduction with Diimide**: Daniel J. Pasto and Richard T. Taylor
3. **The Pummerer Reaction of Sulfinyl Compounds**: Ottorino DeLucchi, Umberto Miotti, and Giorgio Modena
4. **The Catalyzed Nucleophilic Addition of Aldehydes to Electrophilic Double Bonds**: Hermann Stetter and Heinrich Kuhlmann

Volume 41 (1992)

1. **Divinylcyclopropane–Cycloheptadiene Rearrangement**: Thomáš Hudlicky, Rulin Fan, Josephine W. Reed, and Kumar G. Gadamasetti

2. **Organocopper Reagents: Substitution, Conjugate Addition, Carbo/Metallocupration, and Other Reactions**: Bruce H. Lipshutz and Saumitra Sengupta

Volume 42 (1992)

1. **The Birch Reduction of Aromatic Compounds**: Peter W. Rabideau and Zbigniew Marcinow

2. **The Mitsunobu Reaction**: David L. Hughes

Volume 43 (1993)

1. **Carbonyl Methylenation and Alkylidenation Using Titanium-Based Reagents**: Stanley H. Pine

2. **Anion-Assisted Sigmatropic Rearrangements**: Stephen R. Wilson

3. **The Baeyer–Villiger Oxidation of Ketones and Aldehydes**: Grant R. Krow

Volume 44 (1993)

1. **Preparation of α,β-Unsaturated Carbonyl Compounds and Nitriles by Selenoxide Elimination**: Hans J. Reich and Susan Wollowitz

2. **Enone Olefin [2 + 2] Photochemical Cycloadditions**: Michael T. Crimmins and Tracy L. Reinhold

AUTHOR INDEX, VOLUMES 1–45

Volume number only is designated in this index.

CHAPTER AND TOPIC INDEX, VOLUMES 1–45

Many chapters contain brief discussions of reactions and comparisons of alternative synthetic methods related to the reaction that is the subject of the chapter. These related reactions and alternative methods are not usually listed in this index. In this index, the volume number is in **boldface**, the chapter number is in ordinary type.